国外优秀数学著作
原版系列

A Course in Analysis

—Vol. II, Differentiation and Integration of Functions of Several Variables, Vector Calculus

分析学教程

——第2卷，多元函数的微分和积分，向量微积分

[英] 尼尔斯·雅各布 （Niels Jacob）

[英] 克里斯蒂安·P. 埃文斯 （Kristian P. Evans）

著

（英文）

HITP

哈尔滨工业大学出版社

HARBIN INSTITUTE OF TECHNOLOGY PRESS

黑版贸审字 08－2020－188 号

图书在版编目(CIP)数据

分析学教程. 第 2 卷,多元函数的微分和积分,向量微积分 = A Course in Analysis: Vol. II, Differentiation and Integration of Functions of Several Variables, Vector Calculus:英文/(英)尼尔斯·雅各布(Niels Jacob),(英)克里斯蒂安·P. 埃文斯(Kristian P. Evans)著. —哈尔滨:哈尔滨工业大学出版社,2023.3
　　ISBN 978－7－5767－0612－3

　　I.①分… Ⅱ.①尼… ②克… Ⅲ.①数学分析－英文 Ⅳ.①O17

中国国家版本馆 CIP 数据核字(2023)第 030377 号

FENXIXUE JIAOCHENG: DI-ER JUAN, DUOYUAN HANSHU DE WEIFEN HE JIFEN, XIANGLIANG WEIJIFEN

W **World Scientific**

策划编辑	刘培杰　杜莹雪
责任编辑	刘家琳
封面设计	孙茵艾
出版发行	哈尔滨工业大学出版社
社　　址	哈尔滨市南岗区复华四道街 10 号　邮编 150006
传　　真	0451－86414749
网　　址	http://hitpress.hit.edu.cn
印　　刷	哈尔滨博奇印刷有限公司
开　　本	720 mm×1 000 mm　1/16　印张 51.25　字数 890 千字
版　　次	2023 年 3 月第 1 版　2023 年 3 月第 1 次印刷
书　　号	ISBN 978－7－5767－0612－3
定　　价	118.00 元

(如因印装质量问题影响阅读,我社负责调换)

Preface

A detailed description of the content of Volume II of our Course in Analysis will be provided in the introduction. Here we would like to take the opportunity to thank those who have supported us in writing this volume. We owe a debt of gratitude to James Harris who has typewritten a major part of the manuscript. Thanks for typewriting further parts are expressed to Huw Fry and Yelena Liskevich who also provided a lot of additional support. Lewis Bray, James Harris and Elian Rhind did a lot of proofreading for which we are grateful. We also want to thank the Department of Mathematics, the College of Science and Swansea University for providing us with funding for typewriting.

Finally we want to thank our publisher, in particular Tan Rok Ting and Ng Qi Wen, for a pleasant collaboration.

Niels Jacob
Kristian P. Evans
Swansea, January 2016

Introduction

This is the second volume of our Course in Analysis and it is designed for second year students or above. In the first volume, in particular in Part 1, the transition from school to university determined the style and approach to the material, for example by introducing abstract concepts slowly or by giving very detailed (elementary) calculations. Now we use an approach that is suitable for students who are more used to the university style of doing mathematics. As we go through the volumes our intention is to guide and develop students to nurture a more professional and rigorous approach to mathematics. We will still take care with motivations (some lengthy) when introducing new concepts, exploring new notions by examples and their limitations by counter examples. However some routine calculations are taken for granted as is the willingness to "fight" through more abstract concepts.

In addition we start to change the way we use references and students should pick up on this. Calculus and analysis in one dimension is so widely taught that it is difficult to trace back in textbooks the origin of how we present and prove results nowadays. Some comments about this were made in Volume I. The more advanced the material becomes, the more appropriate it becomes to point out in more detail the existing literature and our own sources. Still we are in a territory where a lot of material is customary and covered by "standard approaches". However in some cases authors may claim more originality and students should know about the existing literature and give it fair credit - as authors of books are obliged to do. We hope that the more experienced reader will consider our referencing as being fair, please see further details below.

The goal of this volume is to extend analysis for real-valued functions of one real variable to mappings from domains in $\mathbb{R}^m$ to $\mathbb{R}^n$, i.e to vector-valued mappings of several variables. At a first glance we need to address three wider fields: convergence and continuity; linear approximation and differentiability; and integration. As it turns out, to follow this programme, we need to learn in addition much more about geometry. Some of the geometry is related to topological notions, e.g. does a set have several components? Does a set have "holes"? What is the boundary of a set? Other geometric notions will be related to the vector space structure of $\mathbb{R}^n$, e.g. quadratic forms, orthogonality, convexity, some types of symmetries, say rotationally invariant functions. But we also need a proper understanding of elementary differential geometric concepts such as parametric curves and surfaces (and later on manifolds and sub-manifolds). Therefore we have included a fair amount of geometry in our treatise starting with this volume.

The situation where we introduce integrals is more difficult. The problem is to define a volume or an area for certain subsets in $G \subset \mathbb{R}^n$. Once this is done for a reasonably large class of subsets a construction of the integral along the lines of the one-dimensional case is possible. It turns out that Lebesgue's theory of measure and integration is much better suited to this and we will develop this theory in the next volume. Our approach to volume (and surface) integrals following Riemann's ideas (as transformed by Darboux) is only a first, incomplete attempt to solve the integration problem. However it is essentially sufficient to solve concrete problems in analysis, geometry as well as in mathematical physics or mechanics.

Let us now discuss the content of this volume in more detail. In the first four chapters we cover convergence and continuity. Although our main interest is in handling mappings $f : G \to \mathbb{R}^n$, $G \subset \mathbb{R}^m$, in order to be prepared for dealing with convergence of sequences of functions, continuity of linear operators, etc., we discuss convergence and continuity in metric spaces as we introduce the basic concepts of point set topology. However we also spend time on normed spaces. We then turn to continuous mappings and study their basic properties in the context of metric and topological spaces. Eventually we consider mappings $f : G \to \mathbb{R}^n$ and for this we investigate some topological properties of subsets of $\mathbb{R}^n$. In particular we discuss the notion of compactness and its consequences. The main theoretical concepts are developed along the lines of N. Bourbaki, i.e. J. Dieudonné [10], however when working in the more concrete Euclidean context we used several different sources, in particular for dealing with connectivity we preferred the approach of M. Heins [24]. A general reference is also [9]

Differentiability is the topic in Chapter 5 and Chapter 6. First we discuss partial derivatives of functions $f : G \to \mathbb{R}$, $G \subset \mathbb{R}^n$, and then the differential of mappings $f : G \to \mathbb{R}^n$, $G \subset \mathbb{R}^m$. These chapters must be viewed as "standard" and our approach does not differ from any of the approaches known to us. Once differentiability is established we turn to applications and here geometry is needed. An appropriate name for Chapter 7 and Chapter 8 would be G. Monge's classical one: "Applications of Analysis to Geometry". We deal with parametric curves in $\mathbb{R}^n$, with some more details in the case $n = 3$, and we have a first look at parametric surfaces in $\mathbb{R}^3$. In addition to having interesting and important applications of differential calculus we prepare our discussion of the integral theorems of vector calculus where we have to consider boundaries of sets either as parametric curves or as parametric surfaces. These two chapters benefit greatly from M. DoCarmo's textbook [11].

In Chapter 9 to Chapter 11 we extend the differential calculus for functions of several variables as we give more applications, many of them going back to the times of J. d'Alembert, L. Euler and the Bernoulli family. Key phrases are Taylor formula, local extreme values under constraints (Lagrange multipliers) or envelopes. However note that Taylor series or more generally power series for functions of several variables are much more difficult to handle due to the structure of convergence domains and we postpone this until we discuss complex-valued functions of complex variables (in Volume III). Of a more theoretical nature, but with a lot of applications, is the implicit function theorem and its consequence, the inverse mapping theorem. In general in these chapters we follow different sources and merge them together. In particular, since some of the classical books include nice applications, however in their theoretical parts they are now outdated, some effort is needed to obtain a coherent view. The book of O. Forster [19] was quite helpful in treating the implicit function theorem. In Chapter 12 curvilinear coordinates are addressed - a topic which is all too often neglected nowadays. However, when dealing with problems that have symmetry, for example in mathematical physics, curvilinear coordinates are essential. In our understanding, they also form a part of classical differential geometry, as can be already learned from G. Lamé's classical treatise.

Almost every book about differential calculus in several variables discusses convexity, e.g. convex sets which are useful when dealing with the mean value theorem, or convex functions when handling local extreme values. We have decided to treat convex sets and functions in much more detail than what other authors do by including more on the geometry of convex sets (for example separating hyperplanes) where S. Hildebrandt's treatise [26] was quite helpful. We further do this by discussing extreme points (Minkowski's theorem) and its applications to extreme values of convex functions on compact convex sets. We also look at a characterisation of differentiable convex functions by variational inequalities as do we discuss the Legendre transform (conjugate functions) of convex functions and metric projections onto convex sets. All of this can be done in $\mathbb{R}^n$ at this stage of our Course and it will be helpful in other parts such as the calculus of variations, functional analysis and differential geometry. Most of all, these are beautiful results.

After introducing continuity or differentiability (or integrability) we can consider vector spaces of functions having these (and some additional) properties. For example we may look at the space $C(K)$ of all continuous functions defined on a compact set $K \subset \mathbb{R}^n$, which is equipped with the norm $\|g\|_\infty := \sup_{x \in K} |g(x)|$ a Banach space. Already in classical analysis the question whether an arbitrary continuous function can be approximated by simpler functions such as polynomials or trigonometrical functions was discussed. It was also discussed whether a uniformly bounded sequence of continuous functions always has a uniformly convergent subsequence. We can now interpret the first question as the problem to find a "nice" dense subset in the Banach space $C(K)$ whereas the second problem can be seen as to find or characterise (pre-) compact sets in $C(K)$. We deliberately put these problems into the context of Banach spaces, i.e. we treat the problems as problems in functional analysis. We prove the general Stone-Weierstrass theorem, partly by a detour, by first proving Korovkin's approximatiuon theorem, and we prove the Arzela-Ascoli theorem. We strongly believe at this stage of the Course that students should start to understand the benefit of reformulating concrete classical problems as problems of functional analysis.

The final chapter of Part 3 deals with line integrals. We locate line integrals in Part 3 and not Part 4 since eventually they are reduced to integrals of functions defined on an interval and not on a domain in $\mathbb{R}^n$, $n > 1$. We discuss the basic definition, the problem of rectifying curves and we start to examine the integrability conditions.

As already indicated, defining an integral in the sense of Riemann for a bounded function $f : G \to \mathbb{R}$, $G \subset \mathbb{R}$ compact, is not as straightforward as it seems. In Chapter 16 we give more details about the problems and we indicate our strategy to overcome these difficulties. A first step is to look at iterated integrals for functions defined on a hyper-rectangle (which we assume to be axes parallel) and this is done in the natural frame of parameter dependent integrals. In the following chapter we introduce and investigate Riemann integrals (volume integrals) for functions defined on hyper-rectangles. This can be done rather closely along the lines we followed in the one-dimensional case. Identifying volume integrals with iterated integrals allows us to reduce the actual integration problem to one-dimensional integrals.

Integrating functions on sets G other than hyper-rectangles is much more difficult. The main point is that we do not know what the volume of a set in $\mathbb{R}^n$ is, hence Riemann sums are difficult to introduce. Even the definition of an integral for step functions causes a problem. It turns out that the boundary ∂G determines whether we can define, say for a bounded continuous function $f : G \to \mathbb{R}$ an integral. This leads to a rather detailed study of boundaries and their "content" or "measure". Basically it is the intertwining of the topological notion "boundary" with the (hidden) measure theoretical notion "set of measure zero" which causes difficulties. We devote Chapter 19 to these problems and once we end up with the concept of (bounded) Jordan measurable sets, we can construct integrals for bounded (continuous) functions defined on bounded Jordan measurable sets. In our presentation of this part we combine parts of the approaches of [20], [25] and [26].

In order to evaluate volume integrals we need further tools, in particular the transformation theorem. Within the Riemann context this theorem is notoriously difficult and lengthy to prove which is essentially due to the problems mentioned above, i.e. the mixture of topological and measure theoretical notions. In the context of Lebesgue's theory of integration the transformation theorem admits a much more transparent proof, we also refer to our remarks in Chapter 21. For this reason we do not provide a proof here but we clearly state the result and give many applications. Eventually we return to improper and parameter dependent integrals, but now in the context of volume integrals. Many of these considerations will become of central importance when investigating partial differential equations.

The final part of this volume is devoted to vector calculus in $\mathbb{R}^2$, but most of all in $\mathbb{R}^3$. A pure mathematician's point of view could be to first introduce manifolds including E. Cartan's exterior calculus, then to introduce integrals for k-forms over m-dimensional sub-manifolds of n-dimensional manifolds, and then to eventually prove the general Stokes' theorem. By specialising we can now derive the classical theorems of Gauss, Green and Stokes. This programme neither takes the historical development into account nor is it suitable for second year students. Thus we decided on a more classical approach. Chapter 23 gives in some sense a separate introduction to Part 5, hence we can be more brief here.

In Chapter 24 we discuss the problem of how to define the area of a parametric surface and then we turn to surface integrals for scalar-valued functions as well as for vector fields. With line and surface integrals at our disposal we can prove Gauss' theorem (in $\mathbb{R}^3$ and later on in $\mathbb{R}^n$), Stokes' theorem in $\mathbb{R}^3$ and Green's theorem in the plane. One part of our investigations is devoted to the question of in what type of domain can we prove these theorems? Another part deals with applications. Our aim is to give students who are interested in applied mathematics, mathematical physics or mechanics the tools (and the ideas of the mathematical background) needed to solve such problems. Only in Volume VI will we provide a rigorous proof of the general Stokes' theorem.

As in Volume I we have provided solutions to all problems (ca. 275) and since we depend on a lot of results from linear algebra we have collected these results in an appendix. Since many of our considerations are geometry related, the text contains a substantial number of figures (ca. 150). All of these figures were done by the second named author using LaTex. Finally a remark about referring to Volume I. When referring to a theorem, a lemma, a definition, etc. in Volume I we write, for example, Theorem I.25.9 etc., and when referring to a formula we write, for example, (I.25.10) etc.

As in Volume I, problems marked with a * are more challenging.

Contents

CONTENTS

List of Symbols

In general, symbols already introduced in Volume I are not listed here and we refer to the List of Symbols in Volume I.

$\mathbb{N}_0^n$ the set of all multi-indices

$\alpha! = \alpha_1! \cdot \ldots \cdot \alpha_n!$ for $\alpha = (\alpha_1, \ldots, \alpha_n)$

$\binom{\alpha}{\beta} := \binom{\alpha_1}{\beta_1} \cdot \ldots \cdot \binom{\alpha_n}{\beta_n}$ for $\alpha = (\alpha_1, \ldots, \alpha_n)$

$\alpha \leq \beta$ $\alpha_j \leq \beta_j, \alpha, \beta \in \mathbb{N}_0^n$

$\alpha + \beta = (\alpha_1 + \beta_1, \ldots, \alpha_n + \beta_n), \alpha, \beta \in \mathbb{N}_0^n$

$x^\alpha = x_1^{\alpha_1} \cdot \ldots \cdot x_n^{\alpha_n}$ for $\alpha = (\alpha_1, \ldots, \alpha_n)$ and $x \in \mathbb{R}^n$

$\mathcal{P}(X)$ power set of X

f_j the jth component of f

pr_j projection on the jth factor or component

$f = (f_1, \ldots, f_n)$ vector-valued f with components

$R(f) = \operatorname{ran}(f)$ range of f

$\Gamma(f)$ graph of f

$f|_K$ restriction of f to K

$(f \vee g)(x) := \max(f(x), g(x))$

$(f \wedge g)(x) := \min(f(x), g(x))$

$f * g$ convolution of f and g

$\operatorname{supp} f$ support of f

A^t transpose of the matrix A

$\beta_A(x, y)$ bilinear form associated with the matrix A

$\beta_A(x) = \beta_A(x, x)$ quadratic form associated with A

$\det(A)$ determinant of the matrix A

$\otimes_a$ algebraic tensor product

$x \perp y$ x and y are orthogonal

$\sphericalangle(a, b)$ angle between a and b

$\langle x, y \rangle, = x \cdot y$ scalar product of x and y

$x \times y$ cross product of x and y

$M(m, n; \mathbb{R})$ vector space of all real $m \times n$ matrices

$M(n; \mathbb{R}) = M(n, n; \mathbb{R})$

$GL(n; \mathbb{R})$ general linear group

$O(n)$ orthogonal group in $\mathbb{R}^n$

$SO(2)$ special orthogonal group

(X, d) metric space

$d(x, y)$ distance between x and y (metric)

$B_r(y)$ open ball with centre y and radius $r > 0$

$\text{diam}(A)$ diameter of A

$C(x)$ connectivity component of x

$\overset{\circ}{Y}$ interior of Y

$\overline{Y}$ closure of Y

∂Y boundary of Y

$\text{dist}(A, B)$ distance between two sets A and B

$\text{dist}(x, A)$ distance between a point x and a set A

$\text{dist}_\infty(x, H) := \inf\{\|x - y\|_\infty | y \in H\}$

$(V, \|.\|)$ normed space

$\|.\|$ norm

$\|x\|_p = \left(\sum_{j=1}^n |x_j|^p\right)^{\frac{1}{p}}, \quad x = (x_1, \dots, x_n) \in \mathbb{R}^n$

$\|x\|_\infty := \max_{1 \leq j \leq n}\{|x_j| \, | x = (x_1, \dots, x_n)\}$

$d_p(x, y) = \|x - y\|_p$

$\|u\|_\infty = \sup_{x \in G} |u(x)|$

$\|u\|_{k,\infty} := \sum_{n=0}^k \|u^{(n)}\|_\infty$

$\|\|u\|\|_{k,\infty} := \max_{0 \leq n \leq k} \|u^{(n)}\|_\infty$

$\|h\|_{\infty,X} := \sup_{x \in X} \|h(x)\|$

$p_{\alpha,\beta}(u) := \sup_{x \in \mathbb{R}^n} |x^\alpha \partial^\alpha u(x)|$

$\text{dis}(h)$ point(s) of discontinuity of h

S^{n-1} unit sphere in $\mathbb{R}^n$

S_+^{n-1} upper unit sphere in $\mathbb{R}^n$

$\frac{\partial}{\partial x_j}$ partial derivative with respect to x_j

$\frac{\partial^2}{\partial x_j \partial x_k}$ second order partial derivative first with respect to x_k and then with respect to x_j

$\partial^\alpha, D^\alpha, \frac{\partial^{|\alpha|}}{\partial x_1^{\alpha_1} \cdot \dots \cdot \partial x_n^{\alpha_n}}$ higher order partial derivatives

D_ν derivative in the ν direction

$\frac{\partial}{\partial n}$ normal derivative (with respect to outer normal)

$J_f(x)$ Jacobi matrix of f at x

$(\mathrm{Hess}\,f)(x)$ Hesse matrix of f at x

$\mathrm{grad}\varphi(x) = \nabla\varphi(x)$ gradient of $\varphi(x)$

$\mathrm{div}\,A$ divergence of A

$\mathrm{curl}A$ curl or rotation of A

$\Delta_n u$ n-dimensional Laplace operator applied to u

$\mathrm{tr}(\gamma)$ trace of γ

$\dot{\gamma}(t) = \frac{d\gamma}{dt}(t)$

$\vec{t}$ tangent vector

$\vec{n}$ normal vector

l_γ length of a curve

$\gamma_1 \oplus \gamma_2$ sum of two curves

$\mathrm{epi}(f)$ epi-graph of f

$\mathrm{conv}(A)$ convex hull of A

$\mathrm{ext}(K)$ set of extreme points of K

$\mathrm{vol}_n(G)$ n dimensional volume of G

$J^{(n)}(G)$ Jordan content of G

$\mathrm{mesh}(Z)$ mesh size or width of the partition Z

$\int_*$ lower integral

$\int^*$ upper integral

$\int_\gamma \psi \cdot dp$ line integral of a function

$\int_\gamma X_p \cdot dp$ line integral of a vector field

$\int_S \psi \cdot dp$ surface integral of a function

$\int_S A \cdot dp$ surface integral of a vector field

$V_Z(\gamma)$ Z-variation of a curve

$V(\gamma)$ total variation of a curve

$\mathcal{M}(X;\mathbb{R})$ vector space of all bounded functions $f : X \to \mathbb{R}$

$C([a,b])$ continuous functions on $[a,b]$

$C^k([a,b])$ k times continuously differentiable functions on $[a,b]$

$C(X)$ continuous functions on X

$C_b(X)$ space of bounded functions on the metric space X

$C_0(X)$ space of all continuous functions with compact support

$C^k(G)$ k-times continuously differentiable functions on G

$C^\infty(G) = \cap_{k=1}^\infty C^k(G)$

$C_0^\infty(G) = C^\infty(G) \cap C_0(G)$

$C_\infty(\mathbb{R}^n)$ space of all continuous functions vanishing at infinity

C_{per} space of all continuous 2π-periodic functions

Part 3: Differentiation of Functions of Several Variables

1 Metric Spaces

In the first volume of our treatise we discussed sequences and series of real numbers or of functions from subsets of the real numbers to the real line. In every case our investigations depended heavily on the concept of a limit: limits of sequences and series; limits of functions; continuity; and differentiability and integrability. In fact we have already studied limits of sequences of functions. Even a proper understanding of the real numbers needs the notion of a limit, namely the limits of sequences of rational numbers. All these definitions of limits make use of the absolute value of a real number. A more careful analysis shows that when dealing with limits we use a function of two variables derived from the absolute value. We always look at a term $|x - y|$ for (real) numbers $x, y \in \mathbb{R}$, and we interpret $|x - y|$ as the distance between x and y. When working with limits we use the following three properties:

(a) $|x - y| \geq 0$ and $|x - y| = 0$ if and only if $x = y$;

(b) $|x - y| = |y - x|$;

(c) $|x - y| \leq |x - z| + |z - y|$.

These properties have simple and natural interpretations:

(a') the distance is non-negative and two distinct points have a strict positive distance while the distance from x to itself is 0;

(b') the distance from x to y is equal to the distance from y to x, i.e. distance is symmetric;

(c') the triangle inequality holds, meaning that the distance of "going from x to y" should be shorter than the distance of "going first from x to z" and then "going from z to y".

The idea of a limit, say of a function $f : [a, b] \to \mathbb{R}$ at a point $x_0 \in [a, b]$, was that given an error bound $\epsilon > 0$ we can find a $\delta > 0$ such that if the distance from $x \in [a, b]$ to x_0 is less than δ then the distance from $f(x)$ to $f(x_0)$ is less than ϵ:
$$0 < |x - x_0| < \delta \text{ implies } |f(x) - f(x_0)| < \epsilon.$$

Suppose that for each pair (x, y) of points x, y belonging to a set X we can define a distance $d(x, y)$ such that d satisfies (a') - (c'). It is natural to ask the following question: can we transfer our theory of limits established in $\mathbb{R}$ to X using the distance d? The answer to this question is yes and it leads to the theory of metric spaces.

Definition 1.1. *Let $\emptyset \neq X$ be a set. A **metric** (or distance or distance function) on X is a mapping $d : X \times X \to \mathbb{R}$ such that the following properties*

(i) $d(x, y) \geq 0$ and $d(x, y) = 0$ if and only if $x = y$;

(ii) $d(x, y) = d(y, x)$;

(iii) $d(x, y) \leq d(x, z) + d(z, y)$,

*hold for all $x, y, z \in X$. The pair (X, d) is called a **metric space**.*

Remark 1.2. A. It is possible to replace in Definition 1.1 condition (i) by the second part of (i) : $d(x, y) = 0$ if and only if $x = y$. Indeed, for $x, y \in X$ we find

$$0 = d(x, x) \leq d(x, y) + d(y, x) = 2d(x, y) \tag{1.1}$$

implying $d(x, y) \geq 0$.
B. Of course (ii) is a symmetry condition and we will again call (iii) the **triangle inequality**.

First of all we want to look at some examples of metric spaces.

Example 1.3. A. Of course we can associate on $\mathbb{R}$ a metric with the absolute value by defining for $x, y \in \mathbb{R}$ the metric $d(x, y) := |x - y|$.
B. Let (X, d) be a metric space and $\emptyset \neq Y \subset X$ be a subset. Define

$$d_Y : Y \times Y \to \mathbb{R}$$
$$(x, y) \mapsto d_Y(x, y) := d(x, y).$$

Then (Y, d_Y) is again a metric space. The proof is obvious, the interpretation is that if we can define the distance for every pair of points in X, we can of course restrict this distance to subsets and the restriction defines a distance on this subset.
C. On every $\emptyset \neq X$ we can define a metric by

$$d(x, y) := \begin{cases} 0, & x = y \\ 1, & x \neq y \end{cases}$$

and (X, d) is a metric space. Indeed, (i) is trivial by the definition as is (ii). In order to see (iii), i.e. the triangle inequality, note that for $x \neq y$ we find

$$d(x, y) = 1 \leq d(x, z) + d(z, y) \, (\leq 2).$$

Thus on every non-empty set we can introduce at least one metric, which however is not very interesting. But it is worth to note a corollary of this fact: on a given set we may have several metrics.

D. We want to return to $X = \mathbb{R}$ and add a class of examples which later on will turn out to be quite useful. Let $\varphi : \mathbb{R} \to \mathbb{R}_+$ be an even function such that $\varphi(x) = 0$ if and only if $x = 0$. Further we assume that φ is **sub-additive**, i.e. for all $x, y \in \mathbb{R}$ we have

$$\varphi(x + y) \leq \varphi(x) + \varphi(y). \tag{1.2}$$

If we define

$$d_\varphi(x, y) := \varphi(x - y) \tag{1.3}$$

then d_φ is a metric on $\mathbb{R}$. From our assumptions it follows immediately that $d_\varphi(x, y) \geq 0$ and $d_\varphi(x, y) = 0$ if and only if $x = y$, as well as

$$d_\varphi(x, y) = \varphi(x - y) = \varphi(-(y - x)) = \varphi(y - x) = d_\varphi(y, x).$$

Using sub-additivity we find also the triangle inequality

$$d_\varphi(x, z) = \varphi(x - z) = \varphi(x - y + y - z)$$
$$\leq \varphi(x - y) + \varphi(y - z) = d_\varphi(x, y) + d_\varphi(y, z).$$

Now let $\psi : \mathbb{R}_+ \to \mathbb{R}_+$ be a monotone increasing function such that $\psi(x) = 0$ if and only if $x = 0$. Moreover assume that ψ has a continuous derivative on $\mathbb{R}_+$ ($\psi'(0)$ is considered as one-sided derivative) which is monotone decreasing. For $0 \leq x < y$ we get

$$\psi(x + y) - \psi(x) = \int_x^{x+y} \psi'(t) dt = \int_0^y \psi'(s + x) ds \leq \int_0^y \psi'(s) ds = \psi(y),$$

or

$$\psi(x + y) \leq \psi(x) + \psi(y).$$

Now it follows that

$$d_\psi(x, y) := \psi(|x - y|) \tag{1.4}$$

is again a metric on $\mathbb{R}$. Clearly $d_\psi(x, y) \geq 0$ and $d_\psi(x, y) = 0$ if and only if $x = y$, as well as $d_\psi(x, y) = d_\psi(y, x)$. The triangle inequality follows now as above:

$$\begin{aligned} d_\psi(x, z) = \psi(|x - z|) &= \psi(|x - y + y - z|) \\ &\leq \psi(|x - y| + |y - z|) \leq \psi(|x - y|) + \psi(|y - z|) \\ &= d_\psi(x, y) + d_\psi(y, z). \end{aligned}$$

More concrete examples are $\psi_1(t) = \arctan t$ or $\psi_2(s) = \ln(1 + s)$. We only have to note that $\psi_1'(t) = \frac{1}{1+t^2}$ and $\psi_2'(s) = \frac{1}{1+s}$. In Problem 1 we will prove that for $0 < \alpha < 1$ a metric on $\mathbb{R}$ is given by $d_{(\alpha)}(x, y) := |x - y|^\alpha$.

Lemma 1.4. *For a metric d on X we have for all $x, y, z \in X$ that*

$$|d(x, z) - d(z, y)| \leq d(x, y). \tag{1.5}$$

Proof. The triangle inequality yields together with the symmetry of d

$$d(x, z) \leq d(x, y) + d(y, z)$$

or

$$d(x, z) - d(z, y) \leq d(x, y) \tag{1.6}$$

as well as

$$d(z, y) \leq d(z, x) + d(x, y)$$

or

$$-(d(x, z) - d(z, y)) \leq d(x, y) \tag{1.7}$$

which together with (1.6) gives (1.5)). $\qquad\square$

We already know the notion of a norm on a vector space (over $\mathbb{R}$), see Definition I.23.12. We recall the definition and prove that every norm induces a metric.

Definition 1.5. *Let V be a vector space over $\mathbb{R}$ or $\mathbb{C}$. A **norm** $\|.\|$ on V is a mapping*

$$\begin{aligned} \|.\| : V &\to \mathbb{R} \\ x &\mapsto \|x\| \end{aligned} \tag{1.8}$$

with the properties

6

(i) $\|x\| = 0$ *if and only if* $x = 0$;

(ii) $\|\lambda x\| = |\lambda|\,\|x\|$ *for all* $x \in V$ *and* $\lambda \in \mathbb{R}$ *(or* $\mathbb{C}$*)*;

(iii) $\|x + y\| \le \|x\| + \|y\|$ *for all* $x, y \in V$.

If $\|.\|$ *is a norm on* V *we call* $(V, \|.\|)$ *a **normed** (vector) **space**.*

Remark 1.6. A. Since $0 = \|x - x\| \le \|x\| + \|x\| = 2\,\|x\|$ it follows that $\|x\| \ge 0$ for all $x \in V$.
B. A norm is defined on a vector space V. Since a subset Y of a vector space need not be a vector space we can restrict a norm to Y, but the restriction will not in general be a norm on Y.
C. Condition (ii) is called the **homogeneity** of the norm and (iii) is also referred to as the **triangle inequality**.

Proposition 1.7. *Let* $(V, \|.\|)$ *be a normed space. Then by*

$$d(x, y) := \|x - y\| \tag{1.9}$$

a metric is defined on V.

Proof. Obviously we have $\|x - y\| = d(x, y) \ge 0$ and $0 = d(x, y) = \|x - y\|$ if and only if $x = y$. Moreover we find

$$d(x, y) = \|x - y\| = \|-(y - x)\| = |-1|\,\|y - x\| = d(y, x).$$

Furthermore, for $x, y, z \in V$ we have

$$d(x, y) = \|x - y\| = \|x - z + z - y\|$$
$$\le \|x - z\| + \|z - y\| = d(x, z) + d(z, y).$$

$\square$

Example 1.8. A. By Corollary I.23.18 we know that for $1 \le p < \infty$ a norm is given on $\mathbb{R}^n$ by

$$\|x\|_p = \left(\sum_{j=1}^{n} |x_j|^p\right)^{\frac{1}{p}}, \qquad x = (x_1, \ldots, x_n) \in \mathbb{R}^n. \tag{1.10}$$

B. On $\mathbb{R}^n$ we define

$$\|x\|_\infty := \max_{1 \le j \le n} \left\{|x_j|\,\big|\,x = (x_1, \ldots, x_n)\right\} \tag{1.11}$$

which is a further norm on $\mathbb{R}^n$. Indeed $\|x\|_\infty = 0$ holds if and only if $x_j = 0$ for $j = 1, \ldots, n$, i.e. $x = 0$. Moreover we have

$$
\begin{aligned}
\|\lambda x\| &= \max_{1 \leq j \leq n} \left\{ |\lambda x_j| \, \big| \, x = (x_1, \ldots, x_n) \right\} \\
&= \max_{1 \leq j \leq n} \left\{ |\lambda||x_j| \, \big| \, x = (x_1, \ldots, x_n) \right\} \\
&= |\lambda| \max_{1 \leq j \leq n} \left\{ |x_j| \, \big| \, x = (x_1, \ldots, x_n) \right\} \\
&= |\lambda| \, \|x\|_\infty \, .
\end{aligned}
$$

Finally we observe that

$$
\begin{aligned}
\|x + y\|_\infty &= \max_{1 \leq j \leq n} \left\{ |x_j + y_j| \, \big| \, x = (x_1, \ldots, x_n), y = (y_1, \ldots, y_n) \right\} \\
&\leq \max_{1 \leq j \leq n} \left\{ |x_j| + |y_j| \big| x = (x_1, \ldots, x_n), y = (y_1, \ldots, y_n) \right\} \\
&\leq \max_{1 \leq j \leq n} \left\{ |x_j| \big| x = (x_1, \ldots, x_n) \right\} + \max_{1 \leq j \leq n} \left\{ |y_j| \big| y = (y_1, \ldots, y_n) \right\} \\
&= \|x\|_\infty + \|y\|_\infty \, .
\end{aligned}
$$

Note that

$$
\|x\|_\infty \leq \|x\|_2 = \left(\sum_{j=1}^{n} |x_j|^2 \right)^{\frac{1}{2}} \leq \sqrt{n} \, \|x\|_\infty \, , \tag{1.12}
$$

and by the Cauchy-Schwarz inequality we find

$$
\|x\|_2 \leq \|x\|_1 = \sum_{j=1}^{n} |x_j| \leq \left(\sum_{j=1}^{n} 1 \right)^{\frac{1}{2}} \left(\sum_{j=1}^{n} |x_j|^2 \right)^{\frac{1}{2}} \tag{1.13}
$$
$$
\leq \sqrt{n} \, \|x\|_2 \, ,
$$

where the first inequality follows from

$$
\left(\sum_{j=1}^{n} |x_j|^2 \right) \leq \left(\sum_{j=1}^{n} |x_j| \right)^2 = \sum_{l,j=1}^{n} |x_l| \, |x_j|.
$$

In addition we have

$$
\|x\|_\infty \leq \|x\|_1 \leq n \, \|x\|_\infty \, , \tag{1.14}
$$

since

$$
\sum_{j=1}^{n} |x_j| \leq n \max_{1 \leq j \leq n} \left\{ |x_j| \, \big| \, x = (x_1, \ldots, x_n) \right\} .
$$

In Problem 6 we will prove that for $p, q \geq 1$ there exists constants $c_{p,q} > 0$ and $C_{p,q} > 0$ such that

$$c_{p,q} \|x\|_q \leq \|x\|_p \leq C_{p,q} \|x\|_q \tag{1.15}$$

holds for all $x \in \mathbb{R}^n$. Combining (1.15) with (1.12) or (1.14) we see that (1.15) holds for all $1 \leq p, q \leq \infty$.

Example 1.9. (Compare with Lemma I.24.5) Let $X \neq \emptyset$ be a set and $\mathcal{M}_b(X; \mathbb{R})$ the vector space of all bounded real-valued functions on X, i.e. $f \in \mathcal{M}_b(X; \mathbb{R})$ if $f : X \to \mathbb{R}$ and $|f(x)| \leq M_f < \infty$ for all $x \in X$. On $\mathcal{M}_b(X; \mathbb{R})$ we have the norm

$$\|f\|_\infty := \sup_{x \in X} |f(x)|. \tag{1.16}$$

In Problem 9 we will see that on every real vector space a positive definite symmetric bilinear form will induce a norm. We want to combine the metrics considered in Example 1.3.D with norms. Let $(V, \|.\|)$ be a normed space and $\psi : \mathbb{R}_+ \to \mathbb{R}_+$ be as in Example 1.3.D, i.e. ψ is monotone increasing such that $\psi(t) = 0$ if and only if $t = 0$ and ψ has on $[0, \infty)$ a continuous monotone decreasing derivative. On V we can define the metric

$$d_\psi(x, y) := \psi(\|x - y\|). \tag{1.17}$$

It follows as in Example 1.3.D that d_ψ is indeed a metric. Clearly, $d_\psi(x, 0)$ need not be a norm. For example on $\mathbb{R}^n$ with the Euclidean norm $\|.\|_2$ we find that

$$d_{\arctan}(x, y) := \arctan(\|x - y\|_2)$$

is a metric but $x \mapsto d_{\arctan}(x, 0)$ cannot define a norm since the homogeneity condition fails to hold: In general

$$\arctan \|\lambda x\|_2 \neq |\lambda| \arctan \|x\|_2$$

which follows from the fact that for $x \neq 0$ the right hand side is unbounded with respect to $\lambda \in \mathbb{R}$ while the left hand side is of course bounded with respect to λ.

Thus on $\mathbb{R}^n$, or more generally, on every normed space we have a lot of different metrics, certain ones are derived from the norms given by (1.10), but others are not necessarily derived from norms.

There is an important difference between a norm and a metric with regard to subsets. We know by Example 1.3.B that we can restrict every metric to any subset and the restriction is again a metric. This does not apply to a norm. A norm is always defined on a vector space and in general a subset of a vector space is not a vector space. However we can always restrict the metric induced by a norm to any subset of a normed space and we will get a metric space.

Example 1.10. A. Let (X, d_X) and (Y, d_Y) be two metric spaces. On $X \times Y$ a metric $d_{X \times Y}$ is defined by

$$d_{X \times Y} \left((x_1, y_1), (x_2, y_2) \right) := d_X(x_1, x_2) + d_Y(y_1, y_2). \qquad (1.18)$$

Clearly $d_{X \times Y}((x_1, y_1), (x_2, y_2)) \geq 0$ and equality implies that $d_X(x_1, x_2) = 0$ and $d_Y(y_1, y_2) = 0$, or $x_1 = x_2$ and $y_1 = y_2$, i.e. $(x_1, y_1) = (x_2, y_2)$. Moreover, the symmetry of $d_{X \times Y}$ follows immediately from that of d_X and d_Y, respectively. Finally, since

$$\begin{aligned} d_{X \times Y}((x_1, y_1), (x_3, y_3)) &= d_X(x_1, x_3) + d_Y(y_1, y_3) \\ &\leq d_X(x_1, x_2) + d_X(x_2, x_3) + d_Y(y_1, y_2) + d_Y(y_2, y_3) \\ &= d_{X \times Y}((x_1, y_1), (x_2, y_2)) + d_{X \times Y}((x_2, y_2), (x_3, y_3)), \end{aligned}$$

the triangle inequality holds too.

B. For two normed spaces $(V, \|.\|_V)$ and $(W, \|.\|_W)$ we define on the product $V \times W$

$$\|(v, w)\|_{V \times W} := \|v\|_V + \|w\|_W \qquad (1.19)$$

and claim that $(V \times W, \|.\|_{V \times W})$ is a normed space too. The proof goes along the lines of part A, we refer also to Problem 5.

As pointed out in the introduction to this chapter open intervals, i.e. sets of the form $\{y \in \mathbb{R} \,|\, |x - y| < \eta\}$ are playing a central role when introducing limits for sequences of real numbers or real-valued functions defined on a subset of $\mathbb{R}$. We now introduce a substitute for open intervals in general metric spaces.

Definition 1.11. *Let (X, d) be a metric space and $y \in X$.*
***A.** We call*

$$B_r(y) := \left\{ x \in X \,\big|\, d(y, x) < r \right\} \qquad (1.20)$$

*the **open ball** with centre y and radius $r > 0$.*
***B.** A set $U \subset X$ is called a **neighbourhood** of $y \in X$ if there exists $\epsilon > 0$ such that $B_\epsilon(y) \subset U$. In particular $y \in U$ and $B_\epsilon(y)$ is a neighbourhood of y.*

In the case where d is generated by a norm, i.e. X is a vector space and $d(x, y) = \|x - y\|$ for a norm on X, we find

$$B_r(y) = \left\{ x \in X \mid \|y - x\| < r \right\}, \qquad (1.21)$$

and for $X = \mathbb{R}$ and $d(x, y) = |x - y|$ we have

$$B_r(y) = \left\{ x \in X \mid |x - y| < r \right\} = (y - r, y + r), \qquad (1.22)$$

i.e. we recover open intervals.

Example 1.12. For every $1 \le p \le \infty$ we can consider on $\mathbb{R}^n$ the norm $\|.\|_p$ and the associated open balls $B_r^{(p)}(y)$. We refer to Chapter I.23 where for $p = 1,\ 2,\ \infty$ and centre $y = 0 \in \mathbb{R}^2$, the balls $B_1^{(p)}(0) \subset \mathbb{R}^2$ were discussed and sketched. Note that (1.15) implies that for every pair $1 \le p, q \le \infty$ we can find $r_1, r_2, r_3 > 0$ such that

$$B_{r_1}^{(p)}(0) \subset B_{r_2}^{(q)}(0) \subset B_{r_3}^{(p)}(0). \qquad (1.23)$$

Let us add a further observation. If $(X, \|.\|)$ is a normed vector space then for $y \in X$ and $r > 0$ we have

$$B_r(y) = y + B_r(0), \qquad (1.24)$$

where as usual $a + A = \{ b \in X \mid b = a + a' \text{ and } a' \in A \}$ for A a subset in a vector space Y and $a, a', b \in Y$. Thus we obtain all open balls in a normed space $(X, \|.\|)$ by translating the open balls with centre 0.

Theorem 1.13. *For every two points* $x, y \in X$, $x \neq y$, *in a metric space* (X, d) *there exist neighbourhoods* $U = U(x)$ *of* x *and* $V = V(y)$ *of* y *such that* $U \cap V = \emptyset$, *i.e we may separate* x *and* y *by neighbourhoods.*

Proof. Let $\epsilon := \frac{1}{2} d(x, y) > 0$ and define $U = B_\epsilon(x)$ and $V = B_\epsilon(y)$, see Figure 1.1. Clearly U is a neighbourhood of x and V is a neighbourhood of y. We claim $U \cap V = \emptyset$. For $z \in U \cap V$ we would have $2\epsilon = d(x, y) \le d(x, z) + d(z, y) < 2\epsilon$ which is a contradiction and the theorem is proved.

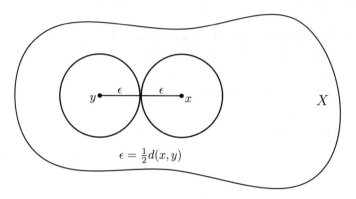

$$\epsilon = \tfrac{1}{2}d(x,y)$$

Figure 1.1

□

Definition 1.14. *A subset U of a metric space (X,d) is called **open** if U is a neighbourhood of all its points, i.e. for $x \in U$ there exists some $\epsilon > 0$ such that $B_\epsilon(x) \subset U$.*

Example 1.15. A. The open interval $(a,b) \subset \mathbb{R}$, $a < b$, is an open set in $\mathbb{R}$ equipped with the metric $d(x,y) = |x - y|$, see Lemma I.19.2.
B. If (X,d) is a metric space and $B_r(z)$ is an open ball in X, then $B_r(z)$ is an open set, i.e. open balls are open. To see this take $x \in B_r(z)$. For $\epsilon := r - d(x,z) > 0$ it follows for $y \in B_\epsilon(x)$ that $d(x,y) < r - d(x,z)$, or (see Figure 1.2) $d(y,z) \leq d(y,x) + d(x,z) = d(x,y) + d(x,z) < r$ i.e. $B_{r-d(x,z)}(x) \subset B_r(z)$. □

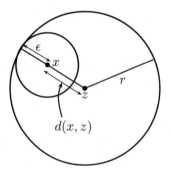

Figure 1.2

Our next result prepares the introduction of a fundamental notion of mathematics, namely that of a topology.

Theorem 1.16. *Denote by $\mathcal{O}$ the system of all open sets of a metric space (X, d). Then we have*

(i) $\emptyset$ and X are open, i.e. $\emptyset, X \in \mathcal{O}$;

(ii) if U and V are open, then $U \cap V$ is open, i.e. $U, V \in \mathcal{O}$ implies $U \cap V \in \mathcal{O}$;

(iii) if $(U_j)_{j \in I}$ is an arbitrary family of open sets, then $\bigcup_{j \in I} U_j$ is open, i.e. $U_j \in \mathcal{O}$ for $j \in I$ yields $\bigcup_{j \in I} U_j \subset \mathcal{O}$.

Proof. (i) The set X is open since for every $x \in X$ we clearly have that X is a neighbourhood of x. Further, $\emptyset$ is open since there is no point $x \in \emptyset$ for which a neighbourhood $U \subset \emptyset$ has to exist.

(ii) Let $x \in U \cap V$, $U, V \subset X$ open. Then there exist $\epsilon_1 > 0$ and $\epsilon_2 > 0$ such that $B_{\epsilon_1}(x) \subset U$ and $B_{\epsilon_2}(x) \subset V$, see Figure 1.3.

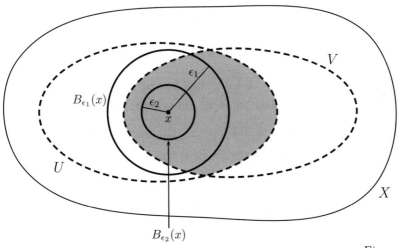

Figure 1.3

For $\epsilon := \min(\epsilon_1, \epsilon_2) > 0$ we have $B_\epsilon(x) \subset U \cap V$, i.e. $U \cap V$ is open.

(iii) If $x \in \bigcup_{j \in I} U_j$, $U_j \in \mathcal{O}$, then there exists $j_0 \in I$ such that $x \in U_{j_0}$. Since U_{j_0} is open there exists $\epsilon > 0$ such that

$$B_\epsilon(x) \subset U_{j_0} \subset \bigcup_{j \in I} U_j,$$

implying that $\bigcup_{j \in I} U_j$ is open. $\qquad\square$

Remark 1.17. From (ii) we deduce immediately that the finite intersections of open sets are open, i.e. if $U_1, \ldots, U_k$ are open then $U_1 \cap \ldots \cap U_k$ is open. However the infinite intersections of open sets need not be open:

$$[0, 1] = \bigcap_{n=1} \left(-\frac{1}{n}, 1 + \frac{1}{n} \right).$$

Definition 1.18. *Given an arbitrary set X and a family $\mathcal{O}$ of subsets of X satisfying (i) - (iii) from Theorem 1.16. Then $\mathcal{O}$ is called a **topology** on X and $(X, \mathcal{O})$ is called a **topological space**.*

Example 1.19. A. For every set X the power set $\mathcal{P}(X)$ and the family $\mathcal{O}_{\text{trivial}} = \{\emptyset, X\}$ are topologies on X, the proof of which is trivial.
B. If (X, d) is a metric space then the family of open sets form a topology in X. In particular in normed spaces the norm induces a topology. Note: on a vector space we may have different norms on a set X, as we may have different metrics. In general they do not necessarily induce different topologies.

Extending Definition I.19.5 with Theorem I.19.6 in mind we get

Definition 1.20. *A subset $A \subset X$ of a topological space $(X, \mathcal{O})$ is called **closed** if $A^{\complement}$ is open. In particular we have defined the notion of closed sets for metric and normed spaces.*

Taking complements we find in the topological space $(X, \mathcal{O})$ that:

X and $\emptyset$ are closed; $\hspace{4cm}$ (1.25)

the finite unions of closed sets are closed; $\hspace{2cm}$ (1.26)

the arbitrary intersections of closed sets are closed, i.e. if C_j, $\hspace{0.5cm}$ (1.27)

$j \in I$, are closed in $(X, \mathcal{O})$ then $\bigcap_{j \in I} C_j$ is closed in $(X, \mathcal{O})$.

Example 1.21. A. Closed intervals are closed, see Lemma I.19.7.
B. Since

$$(0, 1) = \bigcup_{j=1}^{\infty} \left[\frac{1}{j}, 1 - \frac{1}{j} \right]$$

it does not follow in general that an infinite union of closed sets is closed.

We know that for $1 \leq p \leq \infty$ a norm is given on $\mathbb{R}^n$ by $\|.\|_p$ and each of these norms induces a topology $\mathcal{O}_p$. From (1.23) we know that each ball $B_{r_1}^{(p)}(0)$ is contained in a ball $B_{\rho_1}^{(2)}(0)$ and further each ball $B_{\rho_2}^{(2)}(0)$ is contained in a ball $B_{r_2}^{(p)}(0)$. Hence an open set in $\mathcal{O}_2$ is an open set in $\mathcal{O}_p$ and an open set in $\mathcal{O}_p$ is an open set in $\mathcal{O}_2$, or

$$\mathcal{O}_p = \mathcal{O}_2 \tag{1.28}$$

for all $1 \leq p \leq \infty$. We call $\mathcal{O}_2$ the **Euclidean topology** on $\mathbb{R}^n$. Thus all the norms $\|.\|_p$ induce on $\mathbb{R}^n$ the Euclidean topology, however they induce different metrics as they are different norms. From now on, whenever we consider $\mathbb{R}^n$ as a topological space (i.e. use topological notions) if not stated otherwise we mean the Euclidean topology and we will also usually work with the Euclidean norm $\|.\|_2$ and the corresponding metric.

Example 1.22. Let $A_1 \subset \mathbb{R}^k$ and $A_2 \subset \mathbb{R}^m$ be closed sets. Then $A_1 \times A_2 \subset \mathbb{R}^{k+m}$ is closed too. We prove that $(A_1 \times A_2)^{\complement}$ is open. For $(x,y) \notin A_1 \times A_2$ we have $x \notin A_1$ or $y \notin A_2$. Suppose $x \notin A_1$. Since A_1 is closed there exists $\epsilon > 0$ such that $B_\epsilon(x) \subset A_1^{\complement} = \mathbb{R}^k \setminus A_1$. This implies however

$$B_\epsilon((x,y)) \subset \mathbb{R}^{k+m} \setminus (A_1 \times A_2),$$

i.e. $A_1 \times A_2$ is closed. The case $y \notin A_2$ goes analogously. In particular all closed n-dimensional cuboids or rectangles

$$Q := \{x = (x_1, \ldots, x_n) \in \mathbb{R}^n \mid a_j \leq x_j \leq b_j\} = \bigtimes_{j=1}^{n} [a_j, b_j]$$

are closed.

Example 1.23. The half-open interval $[a,b) \subset \mathbb{R}$ is neither closed nor open, whereas $\mathbb{R}^n$ and $\emptyset$ are both open and closed. Thus subsets in a metric (topological) space are not necessarily open nor closed, and further there are sets which are both, closed and open.

We have seen that if (X,d) is a metric space and $Y \subset X$ is a subset then (Y, d_Y) is again a metric space. A similar situation occurs in the case of topological spaces.

Proposition 1.24. *Let $(X, \mathcal{O})$ be a topological space and $Y \subset X$ a set. Then*

$$\mathcal{O}_Y := \{U \cap Y \mid U \in \mathcal{O}\} \tag{1.29}$$

*defines on Y a topology, called the **relative topology** or the **trace topology** on Y induced by $\mathcal{O}$.*

Proof. Since $Y \cap \emptyset = \emptyset$ and $Y \cap X = Y$ we find $\emptyset, Y \in \mathcal{O}_Y$. Suppose $U_Y, V_Y \in \mathcal{O}_Y$. Then there exists $U, V \in \mathcal{O}$ such that $U_Y = U \cap Y$ and $V_Y = V \cap Y$, and therefore $U_Y \cap V_Y = U \cap Y \cap V \cap Y = (U \cap V) \cap Y$. But $U \cap V \in \mathcal{O}$, so $U_Y \cap V_Y \in \mathcal{O}_Y$. Finally, let $U_{Y,j}, j \in I$, be a collection of sets in $\mathcal{O}_Y$. For each $j \in I$ there exists $U_j \in \mathcal{O}, j \in I$, such that $U_{Y,j} = U_j \cap Y$ implying

$$\bigcup_{j \in I} U_{Y,j} = \bigcup_{j \in I} (U_j \cap Y) = \left(\bigcup_{j \in I} U_j \right) \cap Y,$$

again we observe that $\bigcup_{j \in I} U_j \in \mathcal{O}$ and conclude that $\bigcup_{j \in I} U_{Y,j} \in \mathcal{O}_Y$. $\square$

Remark 1.25. If $(X, \mathcal{O})$ is a topological space and $Y \subset X, Y \notin \mathcal{O}$, i.e Y is not open, then $\mathcal{O}_Y$ contains elements which are not elements in $\mathcal{O}$ since $Y \in \mathcal{O}_Y$ but by assumption $Y \notin \mathcal{O}$. Thus open sets in the relative topology induced in Y by $\mathcal{O}$ are not necessarily open in X equipped with the topology $\mathcal{O}$.

Lemma 1.26. *Let $(X, \mathcal{O})$ be a topological space and $Y \subset X$ an open set, i.e. $Y \in \mathcal{O}$. Then $\mathcal{O}_Y$ consists of all open sets $Z \in \mathcal{O}$ such that $Z \subset Y$, i.e.*

$$\mathcal{O}_Y = \{Z \subset Y \mid Z \in \mathcal{O}\}.$$

Proof. If $A \in \mathcal{O}_Y$ then $A = U_A \cap Y$ with $U_A \in \mathcal{O}$, hence $U_A \cap Y \in \mathcal{O}$ implying $A \in \mathcal{O}$, i.e. $\mathcal{O} \subset \{Z \subset Y \mid Z \in \mathcal{O}\}$. The converse is however trivial. If $A \in \{Z \cap Y \mid Z \in \mathcal{O}\}$ then $A = Z_A \cap Y$ for some $Z_A \in \mathcal{O}$, i.e. $A \in \mathcal{O}_Y$. $\square$

Finally we want to introduce the boundary of a subset of a metric space.

Definition 1.27. *Let (X, d) be a metric space and $Y \subset X$. A point $x \in X$ is called a **boundary point** of Y if every neighbourhood of x contains points of Y and $Y^{\complement} (= X \setminus Y)$. The set of all boundary points of Y is called the **boundary** of Y and is denoted by ∂Y.*

Example 1.28. For any interval $I \subset \mathbb{R}$ with end points $a < b$, $a, b \in \mathbb{R}$, the boundary ∂I is the set $\{a, b\}$. Consider for $0 < \epsilon < \frac{b-a}{2}$ the open intervals (open balls) $(a-\epsilon, a+\epsilon)$ and $(b-\epsilon, b+\epsilon)$. In each possible case we have $a-\frac{\epsilon}{2} \in (a-\epsilon, a+\epsilon)$ and $a-\frac{\epsilon}{2} \notin I$ while $a+\frac{\epsilon}{2} \in I$, and $b+\frac{\epsilon}{2} \in (b-\epsilon, b+\epsilon)$ and $b+\frac{\epsilon}{2} \notin I$ while $b-\frac{\epsilon}{2} \in I$. Since every neighbourhood of a and b will contain an interval of the type $(a-\epsilon, a+\epsilon)$ and $(b-\epsilon, b+\epsilon)$, respectively, every neighbourhood of a will contain points of I and $I^{\complement}$ as will every neighbourhood of b contain points of I and $I^{\complement}$. Thus $a, b \in \partial I$. Since $(a, b) \subset I$ it is clear that every $x \in I$, $a < x < b$, is not a boundary point of I: the interval $(x - \eta, x + \eta)$, $\eta = \frac{1}{2} \min(x-a, b-x)$ is disjoint to $I^{\complement}$. On the other hand if $y \in [a, b]^{\complement}$, then there exists $\epsilon > 0$ such that $(y - \epsilon, y + \epsilon) \subset [a, b]^{\complement}$, note that $[a, b]^{\complement}$ is open. Thus $[a, b]^{\complement}$ is disjoint to I. Therefore we have proved $\partial I = \{a, b\}$.

Example 1.29. A. Let (X, d) be a metric space then $\partial X = \emptyset$ since $X^{\complement} = \emptyset$. Thus the interval $(a, b) \subset \mathbb{R}$ as subset in $\mathbb{R}$ has a boundary $\partial(a, b) = \{a, b\}$ by Example 1.28, but the metric space $((a, b), d)$, where $d(x, y) = |x - y|$ is the restriction of the Euclidean metric to (a, b), has no boundary i.e. the boundary is empty.
B. In $\mathbb{R}^n$ the closed unit ball $\overline{B_1(0)}$ is the set

$$\overline{B_1(0)} := \{x \in \mathbb{R}^n \mid \|x\|_2 \leq 1\}$$

which has the boundary

$$\partial\overline{B_1(0)} = S^{n-1} := \{x \in \mathbb{R}^n \mid \|x\|_2 = 1\}. \tag{1.30}$$

Often $S^{n-1} \subset \mathbb{R}^n$ is called the **(n − 1)-sphere**.
C. The boundary of $\mathbb{Q} \subset \mathbb{R}$ is $\mathbb{R}$, i.e. $\partial\mathbb{Q} = \mathbb{R}$. Hence the boundary of a set can be a much larger set. The proof is trivial: we know that in every open interval $I \subset \mathbb{R}$ there are rational as well as irrational points, see Theorem I.18.30.

Theorem 1.30. *Let (X, d) be a metric space and $Y \subset X$. Then the following hold:*

$$\partial Y = \partial(Y^{\complement}); \tag{1.31}$$

$$Y \setminus \partial Y \text{ is open;} \tag{1.32}$$

$$Y \cup \partial Y \text{ is closed;} \tag{1.33}$$

$$\partial Y \text{ is closed.} \tag{1.34}$$

Proof. The first assertion is trivial since $\left(Y^{\complement}\right)^{\complement} = Y$. In order to see (1.32) let $x \in Y \setminus \partial Y$. Then there exists $\epsilon > 0$ such that $B_{\epsilon}(x) \cap (X \setminus Y) = \emptyset$, otherwise x would be a boundary point. Thus for this $\epsilon > 0$ we have $B_{\epsilon}(x) \cap \partial Y = \emptyset$, otherwise, if $y \in B_{\epsilon}(x) \cap \partial Y$, then some point of $X \setminus Y$ would belong to $B_{\epsilon}(x)$ since $B_{\epsilon}(x)$ is a neighbourhood of y. Thus $B_{\epsilon}(x) \subset Y \setminus \partial Y$, i.e. $Y \setminus \partial Y$ is open. Using (1.31) we find with (1.32) that $Y^{\complement} \setminus (\partial Y^{\complement})$ is open, and therefore

$$X \setminus \left(Y^{\complement} \setminus \partial(Y^{\complement})\right) = \left(X \setminus Y^{\complement}\right) \cup \partial(Y^{\complement}) = Y \cup \partial Y,$$

i.e. $Y \cup \partial Y$ is closed proving (1.33). Finally we observe the identity $\partial Y = (Y \cup \partial Y) \setminus (Y \setminus \partial Y)$, or

$$X \setminus \partial Y = X \setminus (Y \cup \partial Y) \cup (Y \setminus \partial Y).$$

But we know that $X \setminus (Y \cup \partial Y)$ and $Y \setminus \partial Y$ are open, which implies that $X \setminus \partial Y$ is open, i.e. ∂Y is closed. $\qquad\square$

Remark 1.31. The definition of ∂Y only needs topological notions as does Theorem 1.30 and its proof. Thus we can define the boundary for a set in a topological space using neighbourhoods and the assertions of Theorem 1.30 still hold.

We introduce two further notations. If (X, d) is a metric space (or $(X, \mathcal{O})$ a topological space) and $Y \subset X$ then we call

$$\mathring{Y} := Y \setminus \partial Y \tag{1.35}$$

the **interior** of Y and

$$\bar{Y} := Y \cup \partial Y \tag{1.36}$$

the **closure** of Y.

Problems

1. Let $f : [0, \infty] \to \mathbb{R}$ be a non-negative monotone increasing continuous function such that $f(t) = 0$ if and only if $t = 0$. Assume further that f is continuously differentiable on $(0, \infty)$ with decreasing, non-negative derivative f'. Prove that $f(s+t) \leq f(s) + f(t)$ for $x, y \geq 0$ and deduce that $d_f : \mathbb{R}^n \times \mathbb{R}^n \to \mathbb{R}$ defined by $d_f(x, y) := f(\|x - y\|)$ is for every norm $\|.\|$ on $\mathbb{R}^n$ a metric.

2. Let $\|.\|$ be a norm on $\mathbb{R}^n$. For which $\alpha > 0$ is a metric defined on $\mathbb{R}^n$ by $d^{(\alpha)}(x, y) := \|x - y\|^\alpha$? Further, define for $\alpha > 0$

$$\|x\|_{(\alpha)} := \left(\sum_{j=1}^n |x_j|^\alpha \right)^{\frac{1}{\alpha}}$$

and decide for which α this determines a norm on $\mathbb{R}^n$.

3. Let $H := \{(a_k)_{k \in \mathbb{N}} \,|\, 0 \leq a_k \leq 1\}$. For $(a_k)_{k \in \mathbb{N}}, (b_k)_{k \in \mathbb{N}} \in H$ define

$$d\left((a_k)_{k \in \mathbb{N}}, (b_k)_{k \in \mathbb{N}}\right) := \sum_{k=1}^n 2^{-k} |a_k - b_k|$$

and prove that (H, d) is a metric space.

4. Let $\gamma : [a, b] \to \mathbb{R}$ be a strictly positive continuous function and define on $C([a, b]) \times C([a, b])$

$$d_\gamma(f, g) := \|\gamma f - \gamma g\|_\infty = \sup_{x \in [a,b]} \gamma(x)|f(x) - g(x)|.$$

Prove that d_γ is a metric on $C([a, b])$.

5. Let $\|.\|$ be a norm on $\mathbb{R}^n$ and $(V_j, \|.\|_j)$, $1 \leq j \leq n$, be normed spaces over $\mathbb{R}$. Consider the vector space $V := V_1 \times \ldots \times V_n$ and define for $v := (v_1, \ldots, v_n) \in V$, $v_j \in V_j$,

$$\|v\|_V := \sum_{j=1}^n \|v_j\|_j.$$

Prove that $\|.\|_V$ is a norm on V.

6. For $1 \leq p$ denote by $\|.\|_p$ the norm $\|x\|_p := \left(\sum_{j=1}^n |x_j|^p \right)^{\frac{1}{p}}$ on $\mathbb{R}^n$. Prove that for all $p, q \geq 1$ there exists constants $c_{p,q} > 0$ and $C_{p,q}$ such that

$$c_{p,q} \|x\|_q \leq \|x\|_p \leq C_{p,q} \|x\|_q$$

holds for all $x \in \mathbb{R}^n$.

19

7. Denote by $M(n, \mathbb{R})$ the vector space of all real $n \times n$ matrices

$$A = \begin{pmatrix} a_{11} & \cdots & a_{1n} \\ a_{21} & \cdots & a_{2n} \\ . & & . \\ . & & . \\ . & & . \\ a_{n1} & \cdots & a_{nn} \end{pmatrix}.$$

Prove that by

$$\|A\| := \left(\sum_{k,l=1}^{n} a_{kl}^2 \right)^{\frac{1}{2}}$$

a norm is given on $M(n, \mathbb{R})$ and by using the Cauchy-Schwarz inequality show for the Euclidean norm $\|.\|_2$ in $\mathbb{R}^n$

$$\|Ax\|_2 \leq \|A\| \, \|x\|_2 \, .$$

8. a) Consider the vector space $P := \{p : \mathbb{R} \to \mathbb{R} \,|\, p \text{ is a polynomial}\}$. For $p \in P$, $p(x) = \sum_{k=0}^{m} a_k x^k$, $a_k \in \mathbb{R}$, define

$$\|p\| := \sum_{k=0}^{m} |a_k|.$$

Prove that $(P, \|.\|)$ is a normed space.

b) Let $[a, b] \subset \mathbb{R}$, $a < b$, be a compact interval and let

$$C([a, b]) := \{f : [a, b] \to \mathbb{R} \,|\, f \text{ is continuous}\} \, .$$

Denote by $\|.\|_1$ the mapping $\|.\|_1 : C([a, b]) \to \mathbb{R}$,

$$\|f\|_1 := \int_a^b |f(t)|dt.$$

Prove that $(C([a, b]), \|.\|_1)$ is a normed space.
Hint: to prove that $\|f\|_1 = 0$ implies $f = 0$, i.e. $f(t) = 0$ for all $t \in [a, b]$, suppose that $\int_a^b |f(t)|dt \neq 0$. Deduce that $f(t_0) \neq 0$ for some $t_0 \in (a, b)$ and now give reasons why for some $\eta > 0$ we have $|f(t)| > 0$ for all $t \in (t_0 - \eta, t_0 + \eta) \subset (a, b)$ and consequently

$$0 < \int_{t_0 - \frac{\eta}{2}}^{t_0 + \frac{\eta}{2}} |f(t)|dt \leq \int_a^b |f(t)|dt.$$

9. Let H be a real vector space and $B : H \times H \to \mathbb{R}$ a symmetric positive definite bilinear form on H. Prove that for B the Cauchy-Schwarz inequality $|B(u,v)| \le B^{\frac{1}{2}}(u,u)B^{\frac{1}{2}}(v,v)$ holds and that $\|u\|_B := B^{\frac{1}{2}}(u,u)$ is a norm on H.

10. For $A := [0,1) \cup ([2,3] \cap \mathbb{Q}) \cup \{5\} \subset \mathbb{R}$ find $\bar{A}$, $\mathring{A}$ and ∂A.

11. Let (X,d) be a metric space and $A \subset X$ a subset. Prove that A is open if and only if for every $x \in A$ there exists an open ball $B_\epsilon(x) \subset A$. Further show that if N_1 and N_2 are neighbourhoods of $y \in X$ then $N_1 \cap N_2$ is a further neighbourhood of y and every set N such that $N_1 \subset N$ is a neighbourhood of y.

12. Let (X,d) be a metric space and $\emptyset \ne A \subset X$ a subset. Prove that if $B \subset X$ is open and $B \subset A$ then B is open in (A, d_A).

13. Consider on $A = (a,b] \subset \mathbb{R}$, $a < b$, the metric $d|_A$ where d is the usual metric on $\mathbb{R}$, i.e. $d(x,y) = |x - y|$. Prove that in the metric space (A, d_A) the sets $(c,b]$, $c > a$, are open.

14. Let (X,d) be a metric space and $A \subset X$ be a non-empty open set. Prove that if $x_0 \in A$ then x_0 is a boundary point of $A \setminus \{x_0\}$. Give an example of a metric space (X,d), a closed set $B \subset X$, and a point $y_0 \in B$ such that y_0 is not a boundary point of $B \setminus \{y_0\}$.

15. Prove that for every set $X \ne \emptyset$ the power set $\mathcal{P}(X)$ is a topology. Now let $\mathcal{A} \subset \mathcal{P}(X)$ and prove that on X there exists a "smallest" topology $\mathcal{O}_\mathcal{A}$ containing $\mathcal{A}$.
    Hint: prove that the arbitrary intersection of topologies $\mathcal{O}_j$, $j \in I$, in a set X is a topology in X and then investigate

$$\mathcal{O}_\mathcal{A} := \bigcap \{\mathcal{O} \subset \mathcal{P}(X) \mid \mathcal{A} \subset \mathcal{O} \text{ and } \mathcal{O} \text{ is a topology on } X\}.$$

2 Convergence and Continuity in Metric Spaces

We have introduced metrics and metric spaces in order to extend the notions of convergence and continuity. We start our investigations with convergence in metric spaces.

Definition 2.1. A. *A sequence $(x_k)_{k \in \mathbb{N}}$, $x_k \in X$, in a metric space (X, d)* **converges** *to $x \in X$ if for every $\epsilon > 0$ there exists $N = N(\epsilon) \in \mathbb{N}$ such that $k \geq N(\epsilon)$ implies $d(x_k, x) < \epsilon$.*
B. *We call $y \in X$ an* **accumulation point** *of the sequence $(x_k)_{k \in \mathbb{N}}$, $x_k \in X$, if a subsequence $(x_{k_l})_{l \in \mathbb{N}}$ of $(x_k)_{k \in \mathbb{N}}$ converges to y.*

If $(x_k)_{k \in \mathbb{N}}$ converges to x then we write

$$\lim_{k \to \infty} x_k = x \tag{2.1}$$

which is equivalent to

$$\lim_{k \to \infty} d(x_k, x) = 0, \tag{2.2}$$

note that (2.2) is a limit in $\mathbb{R}$.
The definition of convergence extends in a straightforward way to sequences $(x_k)_{k \geq k_0}$, $x_k \in X$ and $k_0 \in \mathbb{Z}$. Further, the convergence of the sequence $(x_k)_{k \in \mathbb{N}}$ to $x \in X$ implies that for every neighbourhood U of x exists $\tilde{N} = \tilde{N}(U) \in \mathbb{N}$ such that $x_k \in U$ for $k \geq \tilde{N}(U)$, see Problem 6.

Lemma 2.2. *The limit of a sequence in a metric space is uniquely determined.*

Proof. Suppose that $(x_k)_{k \in \mathbb{N}}$ has the two limits $x, y \in X$, $x \neq y$. For $\epsilon := \frac{1}{2} d(x, y) > 0$ we can find $N \in \mathbb{N}$ such that for $k \geq N$ it follows that $d(x_k, x) < \epsilon$ and $d(x_k, y) < \epsilon$ implying for $k \geq N$

$$d(x, y) \leq d(x, x_k) + d(x_k, y) < 2\epsilon = d(x, y)$$

which is a contradiction. $\qquad \square$

In the case where X is a vector space over $\mathbb{R}$ and d is induced by a norm on X, i.e. $d(x, y) = \|x - y\|$, then we have

Corollary 2.3. *In a normed space $(X, \|.\|)$ a sequence $(x_k)_{k \in \mathbb{N}}$, $x_k \in X$, converges to $x \in X$ if and only if for every $\epsilon > 0$ there exists $N = N(\epsilon) \in \mathbb{N}$ such that $k \geq N(\epsilon)$ implies $\|x_k - x\| < \epsilon$.*

Proof. This statement is obvious when we recall that for the metric d induced by $\|.\|$ we have $d(x, y) = \|x - y\|$. □

If not stated otherwise, when working in $\mathbb{R}^n$ we choose the Euclidean norm $\|.\|_2$ which later on we often will just denote by $\|.\|$.

Theorem 2.4. *Let $(x_k)_{k \in \mathbb{N}}$ be a sequence in $\mathbb{R}^n$, $x_k = \left(x_k^{(1)}, \ldots, x_k^{(n)}\right)$. The sequence $(x_k)_{k \in \mathbb{N}}$ converges to $x = \left(x^{(1)}, \ldots, x^{(n)}\right) \in \mathbb{R}^n$ if and only if for all $1 \leq j \leq n$ the sequences $\left(x_k^{(j)}\right)_{k \in \mathbb{N}}$ converge in $\mathbb{R}$ to $x^{(j)}$, i.e. for $j = 1, \ldots, n$ we have*

$$\lim_{k \to \infty} x_k^{(j)} = x^{(j)}. \tag{2.3}$$

Proof. Suppose $\lim_{k \to \infty} x_k = x$. Then for $\epsilon > 0$ there exists $N(\epsilon) \in \mathbb{N}$ such that $k \geq N(\epsilon)$ implies

$$\left|x_k^{(j)} - x^{(j)}\right| \leq \left(\sum_{l=1}^{n} \left|x_k^{(l)} - x^{(l)}\right|^2\right)^{\frac{1}{2}} = \|x_k - x\|_2 < \epsilon,$$

i.e. $\lim_{k \to \infty} x_k^{(j)} = x^{(j)}$. Conversely, suppose that for all $1 \leq j \leq n$ the sequence $\left(x_k^{(j)}\right)_{k \in \mathbb{N}}$ converges to $x^{(j)}$. Then, given $\epsilon > 0$ there exists $N_j(\epsilon) \in \mathbb{N}$ such that $k \geq N_j(\epsilon)$ implies

$$\left|x_k^{(j)} - x^{(j)}\right| < \frac{\epsilon}{\sqrt{n}}.$$

For $k \geq N := \max \{N_1(\epsilon), \ldots, N_n(\epsilon)\}$ it follows that

$$\|x_k - x\|_2 = \left(\sum_{j=1}^{n} \left|x_k^{(j)} - x^{(j)}\right|^2\right)^{\frac{1}{2}} < \left(\sum_{j=1}^{n} \frac{\epsilon^2}{n}\right)^{\frac{1}{2}} = \epsilon,$$

implying the convergence of $(x_k)_{k \in \mathbb{N}}$ to x in $\mathbb{R}^n$. □

Example 2.5. A. The sequence $\left(\frac{k}{k+1}, \frac{\sin k}{\sqrt{k}}, e^{-k}\right)_{k \in \mathbb{N}}$ converges in $\mathbb{R}^3$ to $(1, 0, 0)$. Indeed, we know that

$$\lim_{k \to \infty} \frac{k}{k+1} = 1, \ \lim_{k \to \infty} \frac{\sin k}{\sqrt{k}} = 0, \ \lim_{k \to \infty} e^{-k} = 0.$$

B. Consider in $(C([0, 2\pi]), \|.\|_\infty)$ the sequence $\left(f_k^{(\alpha)}\right)_{k \in \mathbb{N}}$, $f_k^{(\alpha)} : [0, 2\pi] \to \mathbb{R}$, $f_k^{(\alpha)}(t) = \frac{\cos kt}{k^\alpha}$, $\alpha > 0$. Since

$$\left\|f_k^{(\alpha)}\right\|_\infty = \sup_{t \in [0, 2\pi]} \left|\frac{\cos kt}{k^\alpha}\right| = \frac{1}{k^\alpha}$$

it follows that $\lim_{k \to \infty} f_k^{(\alpha)} = 0$, i.e. $\left(\frac{\cos k \cdot}{k^\alpha}\right)_{k \in \mathbb{N}}$ converges in $(C([0, 2\pi], \|.\|_\infty)$ to the zero function.

C. Let $g_l : [0, 1] \to \mathbb{R}$, $g_l(t) = t^l$, $l \in \mathbb{N}$. Since

$$\|g_l\|_1 = \int_0^1 |g_l(t)| \, dt = \int_0^1 t^l \, dt = \frac{1}{l+1}$$

it follows that in $(C([0, 1]); \|.\|_1)$ the sequence $(g_l)_{l \in \mathbb{N}}$ converges to $g_\infty :$ $[0, 1] \to \mathbb{R}$, $g_\infty(t) = 0$ for all $t \in [0, 1]$. However, since $g_l(1) = 1$ for all $l \in \mathbb{N}$ we find

$$\|g_l\|_\infty := \sup_{t \in [0, 1]} |g_l(t)| = 1$$

which implies that $(g_l)_{l \in \mathbb{N}}$ does not converge to g_∞ in $(C([0, 1]); \|.\|_\infty)$. In fact the sequence has no limit in $(C([0, 1]); \|.\|_\infty)$ since convergence with respect to $\|.\|_\infty$ is uniform convergence, hence a limit must be continuous and coincide with the pointwise limit, but the pointwise limit is

$$\lim_{l \to \infty} g_l(t) = \lim_{l \to \infty} t^l = \begin{cases} 0, & t \in [0, 1) \\ 1, & t = 1 \end{cases}$$

which is not continuous. Thus it may happen that on a vector space we can find a norm with respect to which a given sequence converges, but with respect to a further norm the same sequence need not converge.

We are now in a position to characterise closed sets in a metric space.

Theorem 2.6. *A set $A \subset X$ in a metric space (X, d) is closed if and only if for every sequence $(x_k)_{k \in \mathbb{N}}$ in A, i.e. $x_k \in A$, which converges in X to some $x \in X$ the limit x belongs to A, i.e. a set A is closed if and only if it contains all limit points of sequences in A converging in X.*

Proof. Suppose that A is closed and $(x_k)_{k \in \mathbb{N}}$, $x_k \in A$, converges in X to x, i.e. $\lim_{k \to \infty} x_k = x \in X$. We have to prove that $x \in A$. If $x \notin A$ we note that $X \setminus A = A^{\complement}$ is open and therefore is a neighbourhood of x. Hence there exists $N \in \mathbb{N}$ such that $x_k \in X \setminus A$ for all $k \geq N$ which is a contradiction, thus $x \in A$. Conversely, suppose that every sequence in A converging in X has a limit in A. We want to prove that A is closed which is equivalent to the fact that $X \setminus A = A^{\complement}$ is open. Let $x \in X \setminus A$ and assume that for every $\epsilon > 0$ we have $B_\epsilon(x) \cap A \neq \emptyset$. Then we may find for every $k \in \mathbb{N}$ an element $x_k \in A$ such that $d(x_k, x) < \frac{1}{k}$, implying $\lim_{k \to \infty} x_k \in A$, which is again a contradiction. Hence, there is at least one $\epsilon > 0$ such that $B_\epsilon(x) \cap A = \emptyset$, or $B_\epsilon(x) \subset A^{\complement}$, implying that $A^{\complement}$ is open and therefore A is closed. $\square$

Example 2.7. Consider the metric space $((0, 1], |.|)$, where $|.|$ denotes the absolute value, and the sequence $(x_k)_{k \in \mathbb{N}}$, $x_k = \frac{1}{k} \in (0, 1]$. In the metric space $((0, 1], |.|)$ the set $(0, 1]$ is closed and the sequence $(x_k)_{k \in \mathbb{N}}$ does not converge in this metric space. Indeed, $((0, 1], |.|)$ is obtained from $(\mathbb{R}, |.|)$ by restricting the metric induced by $|.|$ on $\mathbb{R}$ to $(0, 1]$. In $(\mathbb{R}, |.|)$ the sequence $(x_k)_{k \in \mathbb{N}}$ converges to $0 \notin (0, 1]$, but the limit is uniquely determined, see Lemma 2.2. Hence $(x_k)_{k \in \mathbb{N}}$ does not converge in $((0, 1], |.|)$. Note that $(x_k)_{k \in \mathbb{N}}$ is a Cauchy sequence in $\mathbb{R}$, hence in $(0, 1]$.

To proceed further we need

Definition 2.8. *A sequence $(x_k)_{k \in \mathbb{N}}$, $x_k \in X$, in a metric space (X, d) is called a **Cauchy sequence** if for every $\epsilon > 0$ there exists $N = N(\epsilon) \in \mathbb{N}$ such that $k \geq m \geq N$ implies $d(x_k, x_m) < \epsilon$.*

Corollary 2.9. *In a metric space every convergent sequence is a Cauchy sequence.*

Proof. If $\lim_{k \to \infty} x_k = x$ then for $\epsilon > 0$ there exists $N(\epsilon) \in \mathbb{N}$ such that $k \geq N(\epsilon)$ implies $d(x_k, x) < \frac{\epsilon}{2}$. Therefore, for $k, m \geq N$ we have

$$d(x_k, x_m) \leq d(x_k, x) + d(x, x_m) < \frac{\epsilon}{2} + \frac{\epsilon}{2} = \epsilon.$$

$\square$

Definition 2.10. A. *A metric space (X, d) is called **complete** if every Cauchy sequence converges.* **B.** *A complete normed space, i.e. a normed space $(V, \|.\|)$ which is complete with respect to the metric induced by the norm, is called a **Banach space**.* **C.** *A topological space in which we can introduce a metric generating the topology such that the corresponding metric space is complete is called a **Fréchet space**. In general a topological space is called **metrizable** if we can find a metric in this space generating the corresponding topology.*

Theorem 2.11. *The space $(\mathbb{R}^n, \|.\|_2)$ is complete.*

Proof. For $x_k = \left(x_k^{(1)}, \ldots, x_k^{(n)}\right) \in \mathbb{R}^n$ let $(x_k)_{k \in \mathbb{N}}$ be a Cauchy sequence in $\mathbb{R}^n$. Since $\left|x_k^{(j)} - x_m^{(j)}\right| \leq \|x_k - x_m\|_2$ for every $1 \leq j \leq n$, it follows that each of the sequences $\left(x_k^{(j)}\right)_{k \in \mathbb{N}}$, $1 \leq j \leq n$, is a Cauchy sequence in $\mathbb{R}$, hence convergent to some $x^{(j)} \in \mathbb{R}$. According to Theorem 2.4 the sequence $(x_k)_{k \in \mathbb{N}}$ must converge to $x = (x^{(1)}, \ldots, x^{(n)}) \in \mathbb{R}^n$. $\square$

We can now interpret Theorem 2.6 and Example 2.7 in the following way: if (X, d) is a complete metric space and $A \subset X$, then (A, d_A) is not necessarily a complete metric space. The next result clarifies the situation.

Theorem 2.12. *Let (X, d) be a complete metric space and $A \subset X$. The metric space (A, d_A) is complete if and only if A is closed in X.*

Proof. Let $(x_k)_{k \in \mathbb{N}}$ be a sequence in A which is also a Cauchy sequence in A, hence in X. Since (X, d) is complete $(x_k)_{k \in \mathbb{N}}$ must have a limit $x \in X$. If A is closed in X then $x \in A$, thus for a closed set $A \subset X$ a Cauchy sequence in (A, d_A) converges, i.e. (A, d_A) is complete. Conversely suppose that (A, d_A) is complete and let $(x_k)_{k \in \mathbb{N}}$ be a sequence in A converging to $x \in X$. Since a convergent sequence is a Cauchy sequence, $(x_k)_{k \in \mathbb{N}}$ is a Cauchy sequence in A. The completeness of (A, d_A) implies that x belongs to A, i.e. A is closed in X. $\square$

Definition 2.13. *Let $A \subset X$ be a subset of a metric space (X, d). The **diameter** of A is defined by*

$$\text{diam}(A) := \sup \{d(x, y) \mid x, y \in A\}. \tag{2.4}$$

*We call $A \subset X$ **bounded** if $\text{diam}(A) < \infty$.*

For every ball $B_r(x)$ we have

$$\operatorname{diam}(B_r(x)) \leq 2r. \tag{2.5}$$

Obviously we have

Corollary 2.14. A. *A subset $A \subset X$ of a metric space (X, d) is bounded if and only if for some ball $B_r(x)$ we have $A \subset B_r(x)$. In a normed space $(V, \|.\|)$ we can choose $x = 0$.*
B. *A convergent sequence $(x_k)_{k \in \mathbb{N}}$, $x_k \in X$, in a metric space is bounded.*

Proof. **A.** Only the last statement still requires a proof. But if $A \subset B_r(x)$ then $A \subset B_{r+\|x\|}(0)$.
B. Let $x \in X$ be the limit of $(x_k)_{k \in \mathbb{N}}$. For $\epsilon = 1$ there exists $N \in \mathbb{N}$ such that $x_k \in B_1(x)$ for $k \geq N$. Now if $R := \max\{1, d(x_1, x), \ldots, d(x_{N-1}, x)\}$ then $x_k \in B_{2R}(x)$ for all $k \in \mathbb{N}$. $\square$

Note that a metric space (X, d) itself can be bounded, i.e.

$$\operatorname{diam}(X) < \infty.$$

An example is $B_1(0) \subset \mathbb{R}^n$ with Euclidean metric, i.e. we consider the metric space $(B_1(0), d_{B_1(0)})$ where $d_{B_1(0)}$ is the metric obtained by restricting the Euclidean metric to $B_1(0)$. It follows that $\operatorname{diam}(B_1(0)) = 2$, hence the metric space $(B_1(0), d_{B_1(0)})$ is bounded.
However a normed space $(V, \|.\|)$ cannot be bounded. Recall $(V, \|.\|)$ is a metric space with metric $d(x, y) = \|x - y\|$. Suppose $V \subset B_R(0)$ for some $R > 0$. For $x \in V$, $x \neq 0$, this implies $\|x\| < R$ and therefore with $\lambda = \frac{2R}{\|x\|}$ we find $\|\lambda x\| = \frac{2R}{\|x\|} \|x\| = 2R < R$ which is a contradiction.

The next result is due to G. Cantor and generalises the **Principle of Nested Intervals**, Theorem I.17.15.

Theorem 2.15 (G. Cantor). *Let (X, d) be a complete metric space and $(A_k)_{k \in \mathbb{N}_0}$ be a sequence of non-empty, closed subsets $\emptyset \neq A_k \subset X$ such that $A_j \supset A_{j+1}$ for $j \in \mathbb{N}_0$ and $\lim_{k \to \infty} \operatorname{diam}(A_k) = 0$. Then there exists exactly one point $x \in X$ which belongs to all sets A_k, $k \in \mathbb{N}_0$, i.e.*

$$\{x\} = \bigcap_{k \in \mathbb{N}_0} A_k. \tag{2.6}$$

28

Proof. First we prove the uniqueness. Suppose $x_1, x_2 \in \bigcap_{k \in \mathbb{N}_0} A_k$ and $x_1 \neq x_2$. It follows that

$$\text{diam}(A_k) \geq \text{diam}\left(\bigcap_{j \in \mathbb{N}_0} A_j\right) \geq d(x_1, x_2) > 0$$

which contradicts the assumption $\lim_{k \to \infty} \text{diam}(A_k) = 0$.

In order to prove the existence of $x \in \bigcap_{k \in \mathbb{N}_0} A_k$ we choose for every $n \in \mathbb{N}_0$ an element $x_n \in A_n$. Since for $n, m \geq N$

$$d(x_n, x_m) \leq \text{diam}(A_N)$$

it follows that $(x_n)_{n \in \mathbb{N}}$ is a Cauchy sequence, hence it converges in X to some $x \in X$. Furthermore, since $x_n \in A_k$ for $n \geq k$, using Theorem 2.6, we deduce $x \in A_k$ for all $k \in \mathbb{N}_0$, i.e. $x \in \bigcap_{k \in \mathbb{N}_0} A_k$ and the theorem is proved. $\square$

Let (X, d_1) and (Y, d_2) be two metric spaces and $f : X \to Y$ a mapping. For $x \in X$ and a sequence $(x_k)_{k \in \mathbb{N}}$, $x_k \in X$, converging to x, i.e. $\lim_{k \to \infty} d_1(x_k, x) = 0$, we may investigate the sequence $(f(x_k))_{k \in \mathbb{N}}$ in the metric space (Y, d_2). For example $(f(x_k))_{k \in \mathbb{N}}$ may converge to some value $y \in Y$, or certain subsequences of $(f(x_k))_{k \in \mathbb{N}}$ may converge but to different limits, etc.

Definition 2.16. *Let (X, d_1) and (Y, d_2) be two metric spaces and $f : X \to Y$ be a mapping. We say that f has a **limit** $y_0 \in Y$ at $x_0 \in X$ if for every sequence $(x_k)_{k \in \mathbb{N}}$, $x_k \in X$, converging in (X, d_1) to x_0 the sequence $(f(x_k))_{k \in \mathbb{N}}$ converges in (Y, d_2) to y_0. For this we write*

$$\lim_{x \to x_0} f(x) = y_0. \tag{2.7}$$

A few remarks are now required. First, as in the case of real-valued functions of a real variable we need convergence of $(f(x_k))_{k \in \mathbb{N}}$ to y_0 for **all** sequences $(x_k)_{k \in \mathbb{N}}$ converging to x_0. Secondly, in the case of a real-valued function of a real variable, say $g : D \to \mathbb{R}$, $D \subset \mathbb{R}$, we have taken in D and $\mathbb{R}$, i.e. in the domain and co-domain, the absolute value to define convergence, we have now in the domain X of f and its co-domain Y in general two different metrics d_1 and d_2, respectively. Moreover, we must be aware that even in the case where $X = Y$ still $d_1 \neq d_2$ is possible and convergence in (X, d_1) does not necessarily imply convergence in (X, d_2) and vice versa. However we do not need to worry about domains in the sense that if (Z, d_1) is a metric space, $X \subset Z$ and $f : X \to Y$, then we can consider $(X, d_{1,X})$, $d_{1,X} := d_1|_{X \times X}$, as a new metric space and investigate f as a mapping defined on $(X, d_{1,X})$. Continuity is now easily defined:

Definition 2.17. Let (X, d_1) and (Y, d_2) be two metric spaces and $f : X \to Y$ a mapping. We call f **continuous at** x_0, $x_0 \in X$, if

$$\lim_{x \to x_0} f(x) = f(x_0). \tag{2.8}$$

If f is continuous for all $x_0 \in X$ we call f **continuous on X**.

An easy consequence of this definition is

Theorem 2.18. Let (X_j, d_j), $j = 1, 2, 3$, be three metric spaces and $f : X_1 \to X_2$, $g : X_2 \to X_3$ be two mappings. If f is continuous at $x_0 \in X_1$ and if g is continuous at $y_0 = f(x_0)$ then $g \circ f : X_1 \to X_3$ is continuous at x_0.

Proof. Let $(x_k)_{k \in \mathbb{N}}$, $x_k \in X_1$, be a sequence converging in (X_1, d_1) to x_0. By the continuity of f it follows that $(f(x_k))_{k \in \mathbb{N}}$ converges in (X_2, d_2) to $y_0 = f(x_0)$. The continuity of g at y_0 now implies the convergence of $(g(f(x_k)))_{k \in \mathbb{N}}$ to $g(y_0) = g(f(x_0))$, and therefore $g \circ f$ is continuous at x_0. $\square$

We want to consider mappings $f : X \to \mathbb{R}^n$ where X is temporarily a non-empty set. For $x \in X$ it follows that $f(x) \in \mathbb{R}^n$, hence $f(x) = (f(x)_1, \ldots, f(x)_n)$. We define

$$f_j(x) := (f(x))_j \tag{2.9}$$

and with $f_j : X \to \mathbb{R}$, $f_j(x) = (f(x))_j$, we have

$$f(x) = (f_1(x), \ldots, f_n(x)). \tag{2.10}$$

We call f_j the jth **component** of f.

Theorem 2.19. Let (X, d) be a metric space and as usual consider on $\mathbb{R}^n$ the Euclidean metric induced by the norm $\|.\|_2$. A mapping $f = (f_1, \ldots, f_n) : X \to \mathbb{R}^n$ is continuous (at $x_0 \in X$) if and only if each of its components $f_j : X \to \mathbb{R}$, $1 \leq j \leq n$, is continuous (at $x_0 \in X$).

Proof. We just have to apply Theorem 2.4. $\square$

Theorem 2.20. *The following mappings are continuous:*

$$\text{add} : \mathbb{R}^2 \to \mathbb{R}, \qquad \text{mult} : \mathbb{R}^2 \to \mathbb{R}, \qquad \text{quot} : \mathbb{R} \times (\mathbb{R} \setminus \{0\}) \to \mathbb{R}.$$
$$(x, y) \mapsto x + y \qquad (x, y) \mapsto x \cdot y \qquad (x, y) \mapsto \frac{x}{y}$$

Proof. Let $((x_k, y_k))_{k \in \mathbb{N}}$, $(x_k, y_k) \in \mathbb{R}^2$, be a sequence in $\mathbb{R}^2$ such that $\lim_{k \to \infty}(x_k, y_k) = (x, y) \in \mathbb{R}^2$. It follows that $\lim_{k \to \infty} x_k = x$ and $\lim_{k \to \infty} y_k = y$ implying

$$\lim_{k \to \infty} \text{add}(x_k, y_k) = \lim_{k \to \infty} (x_k + y_k) = \lim_{k \to \infty} x_k + \lim_{k \to \infty} y_k = x + y = \text{add}(x, y)$$

and

$$\lim_{k \to \infty} \text{mult}(x_k, y_k) = \lim_{k \to \infty} x_k \cdot y_k = \left(\lim_{k \to \infty} x_k \right) \cdot \left(\lim_{k \to \infty} y_k \right) = x \cdot y = \text{mult}(x, y).$$

If in addition $y_k, y \in \mathbb{R} \setminus \{0\}$ then we find

$$\lim_{k \to \infty} \text{quot}(x_k, y_k) = \lim_{k \to \infty} \frac{x_k}{y_k} = \frac{\lim_{k \to \infty} x_k}{\lim_{k \to \infty} y_k} = \frac{x}{y} = \text{quot}(x, y).$$

$\square$

Corollary 2.21. *Let (X, d) be a metric space and let $f, g : X \to \mathbb{R}$ be continuous functions. The functions $f + g : X \to \mathbb{R}$, $x \mapsto f(x) + g(x)$, $f \cdot g : X \to \mathbb{R}$, $x \mapsto f(x) \cdot g(x)$, are continuous and if in addition $g(x) \neq 0$ for all $x \in X$ then $\frac{f}{g} : X \to \mathbb{R}$, $x \mapsto \frac{f(x)}{g(x)}$, is also continuous.*

Proof. By Theorem 2.19 the mapping $(f, g) : X \to \mathbb{R}^2$, $x \mapsto (f(x), g(x))$ is continuous. The result now follows from Theorem 2.20 when observing that $f + g = \text{add}(f, g)$, $f \cdot g = \text{mult}(f, g)$ and $\frac{f}{g} = \text{quot}(f, g)$. $\square$

Example 2.22. For $\alpha \in \mathbb{N}_0^n$, $\alpha = (\alpha_1, \ldots, \alpha_n)$ and $\alpha_j \in \mathbb{N}_0$, we define on $\mathbb{R}^n$ the function $x \mapsto x^\alpha$ by

$$x^\alpha := x_1^{\alpha_1} \cdot x_2^{\alpha_2} \cdot \ldots \cdot x_n^{\alpha_n}, \ x = (x_1, \ldots, x_n). \tag{2.11}$$

Clearly, $x \mapsto x^\alpha$ is continuous from $\mathbb{R}^n$ to $\mathbb{R}$. If we set for $\alpha \in \mathbb{N}_0^n$

$$|\alpha| := \alpha_1 + \ldots + \alpha_n, \tag{2.12}$$

then for every choice of $c_\alpha \in \mathbb{R}$, $|\alpha| \leq k$, the function $p : \mathbb{R}^n \to \mathbb{R}$,

$$p(x) = \sum_{|\alpha| \leq k} c_\alpha x^\alpha \tag{2.13}$$

is continuous. We call $\mathbb{N}_0^n$ the set of all **multi-indices** (with n components) and a function of the type (2.13) is called a **polynomial** in n variables.

In $\mathbb{R}^n$ we have introduced a whole family of norms $\|.\|_p$, $1 \leq p \leq \infty$, each inducing a metric which we denote for the moment by d_p,

$$d_p(x, y) = \|x - y\|_p . \tag{2.14}$$

From (1.15) we deduce that for all $x, y \in \mathbb{R}^n$ the following holds for all $1 \leq p, q \leq \infty$ with $0 < c_{p,q} \leq C_{p,q}$

$$c_{p,q} d_q(x, y) \leq d_p(x, y) \leq C_{p,q} d_q(x, y), \tag{2.15}$$

and in particular for a sequence $(x_k)_{k \in \mathbb{N}}$ we have

$$c_{p,q} d_q(x_k, x_l) \leq d_p(x_k, x_l) \leq C_{p,q} d_q(x_k, x_l), \tag{2.16}$$

i.e. if $(x_k)_{k \in \mathbb{N}}$, $x_k \in \mathbb{R}^n$, is a Cauchy sequence with respect to d_p it is a Cauchy sequence with respect to d_q for all q, $1 \leq q \leq \infty$. Moreover, if $(x_k)_{k \in \mathbb{N}}$ converges with respect to d_p for one p, $1 \leq p \leq \infty$, to $x \in \mathbb{R}^n$, then it converges to x with respect to d_q for all q, $1 \leq q \leq \infty$, since

$$c_{p,q} d_q(x_k, x) \leq d_p(x_k, x) \leq C_{p,q} d_q(x_k, x). \tag{2.17}$$

Consequently $(\mathbb{R}^n, d_q)$ is for every q, $1 \leq q \leq \infty$, complete. Thus convergence of a sequence in $\mathbb{R}^n$ is independent of the choice of d_p (or $\|.\|_p$) and therefore our convention to always choose the Euclidean norm is not a serious restriction. Let $D \subset \mathbb{R}^m$ and consider the metric space $(D, d_{p,D})$ where $d_{p,D}$ is the restriction of d_p to $D \times D$. From our previous consideration it follows that if $f : D \to \mathbb{R}^n$ is continuous as a mapping between the metric spaces $(D, d_{p_0,D})$ and $(\mathbb{R}^n, d_{q_0})$ then it is continuous between $(D, d_{p,D})$ and $(\mathbb{R}^n, d_q)$ for every choice of $1 \leq p, q \leq \infty$. Thus we find that the continuity of $f : D \to \mathbb{R}^n$, $D \subset \mathbb{R}^m$, also does not depend on the choice of the pair of metrics d_p and d_q, and this again justifies choosing in general the Euclidean metric for both. As in the one-dimensional case we can give an $\epsilon-\delta$-**criterion for continuity** of mappings between metric spaces.

Theorem 2.23. *Let (X, d_1) and (Y, d_2) be two metric spaces and $f : X \to Y$ a mapping. The mapping f is continuous at $x_0 \in X$ if and only if for every $\epsilon > 0$ there exists $\delta > 0$ such that for $x \in X$ satisfying $0 < d_1(x, x_0) < \delta$ it follows that $d_2(f(x), f(x_0)) < \epsilon$.*

Proof. (Compare with Theorem I.20.2) Suppose first that $\lim_{x \to x_0} f(x) = f(x_0)$ but that the $\epsilon - \delta$-criterion does not hold. Then there exists $\epsilon > 0$ such that for every $\delta > 0$ there exists $x \in X$ such that

$$0 < d_1(x, x_0) < \delta \ \text{ but } \ d_2(f(x), f(x_0)) \geq \epsilon.$$

Thus for $\delta = \frac{1}{k}$, $k \in \mathbb{N}$, we find $x_k \in X$ such that $d_1(x_k, x_0) < \frac{1}{k}$, i.e. $\lim_{k \to \infty} x_k = x_0$, but $d_2(f(x_k), f(x_0)) \geq \epsilon$ implying that $(f(x_k))_{k \in \mathbb{N}}$ does not converge with respect to d_2 to $f(x_0)$ which is a contradiction. Suppose now that the $\epsilon - \delta$-criterion holds and let $(x_k)_{k \in \mathbb{N}}$, $x_k \in X$, be a sequence converging in (X, d_1) to x_0, i.e. $\lim_{k \to \infty} x_k = x_0$. We have to prove that $\lim_{k \to \infty} f(x_k) = f(x_0)$. Given $\epsilon > 0$ there exists $\delta > 0$ such that $d_1(x, x_0) < \delta$ implies that $d_2(f(x), f(x_0)) < \epsilon$. Since $\lim_{k \to \infty} x_k = x_0$ there exists $N \in \mathbb{N}$ such that $d_1(x, x_0) < \delta$ for $k \geq N$ and it follows for this $N = N(\delta(\epsilon))$ that $d_2(f(x_k), f(x_0)) < \epsilon$ for $k \geq N$, i.e. $\lim_{k \to \infty} f(x_k) = f(x_0)$. $\square$

From Lemma 1.4 we now deduce

Example 2.24. In a metric space (X, d) the function $f_{x_0} : X \to \mathbb{R}$, $f_{x_0}(x) := d(x, x_0)$, is for every $x_0 \in X$ continuous, where as usual on $\mathbb{R}$ the metric is the one induced by the absolute value. Indeed from (1.5) it follows that

$$|f_{x_0}(x) - f_{x_0}(y)| = |d(x, x_0) - d(y, x_0)| \leq d(x, y),$$

i.e. $|f_{x_0}(x) - f_{x_0}(y)| < \epsilon$ provided that $d(x, y) < \delta = \epsilon$.

Example 2.25. The projections $\mathrm{pr}_j : \mathbb{R}^n \to \mathbb{R}$, $\mathrm{pr}_j(x) = x_j$, where $x = (x_1, \ldots, x_n) \in \mathbb{R}^n$, are continuous. This follows from

$$|\mathrm{pr}_j(x) - \mathrm{pr}_j(y)| = |x_j - y_j| \leq \|x - y\|_2$$

for all $x, y \in \mathbb{R}^n$.

Note that in the last two examples we are dealing with a generalisation of Lipschitz continuity.

Definition 2.26. *Let $f : X \to Y$ be a mapping between two metric spaces (X, d_1) and (Y, d_2). We call f **Lipschitz continuous** if there exists a constant $\kappa > 0$ such that*

$$d_2(f(x), f(y)) \leq \kappa d_1(x, y) \tag{2.18}$$

for all $x, y \in X$.

Clearly, a Lipschitz continuous mapping is continuous: For $\epsilon > 0$ choose $\delta = \frac{\epsilon}{\kappa}$ to find that for $d_1(x, y) < \delta$ it follows that

$$d_2(f(x), f(y)) \leq \kappa d_1(x, y) < \kappa \frac{\epsilon}{\kappa} = \epsilon.$$

It turns out that linear continuous mappings between normed spaces are always Lipschitz continuous.

Theorem 2.27. *Let $(V, \|.\|_V)$ and $(W, \|.\|_W)$ be two normed spaces (over $\mathbb{R}$) and $A : V \to W$ be a linear mapping. The mapping A is continuous if and only if there exists $\kappa \geq 0$ such that*

$$\|Ax\|_W \leq \kappa \|x\|_V \tag{2.19}$$

holds for all $x \in V$.

Remark 2.28. With $d_V(x, y) = \|x - y\|_V$ and $d_W(a, b) = \|a - b\|_W$ we find using the linearity of A that (2.19) implies

$$d_W(Ax, Ay) = \|Ax - Ay\|_W = \|A(x - y)\|_W \leq \kappa \|x - y\|_V = \kappa d_V(x, y),$$

i.e. the Lipschitz continuity of A.

Proof of Theorem 2.27. By Remark 2.28 we already know that (2.19) implies the continuity of A. Conversely, if A is continuous it is continuous at $0 \in V$. Therefore, for $\epsilon = 1$ there exists $\delta > 0$ such that $\|z\|_V < \delta$, $z \in V$, implies $\|Az\|_W < 1$. With $\kappa := \frac{2}{\delta}$ we find for $x \in V \setminus \{0\}$ and with $\lambda := (\kappa \|x\|_V)^{-1}$ that $z := \lambda x$ satisfies $\|z\|_V = |\lambda| \|x\|_V = \frac{\delta}{2} < \delta$ which yields

$$Az = A(\lambda x) = \lambda Ax = \frac{1}{\kappa \|x\|_V} Ax,$$

i.e.

$$1 > \|Az\|_W = \frac{1}{\kappa \|x\|_V} \|Ax\|_W,$$

or $\|Ax\|_W \leq \kappa \|x\|_V$. The case $x = 0$ is trivial. $\square$

Example 2.29. A. On $(C([a,b]), \|.\|_\infty)$ we consider the linear mapping $I :$ $C([a,b]) \to \mathbb{R}$, $I(f) = \int_a^b f(t)\,\mathrm{d}t$. Since

$$|I(f)| = \left| \int_a^b f(t)\,\mathrm{d}t \right| \le \int_a^b |f(t)|\,\mathrm{d}t \le \|f\|_\infty \int_a^b 1\,\mathrm{d}t = (b-a)\,\|f\|_\infty$$

it follows that I is continuous.

B. Let $C^1([0,1]) \subset C([0,1])$ be the vector (sub-)space of all continuous functions $f : [0,1] \to \mathbb{R}$ which are continuously differentiable on $(0,1)$ with the derivative f' having one-sided finite limits at 0 and 1. Consider on $C([0,1])$ the norm $\|.\|_\infty$ which we also consider on $C^1([0,1])$. We claim that the linear mapping (operator)

$$\frac{\mathrm{d}}{\mathrm{d}x} : \left(C^1([0,1]), \|.\|_\infty\right) \to \left(C([0,1]), \|.\|_\infty\right)$$

$$f \mapsto \frac{\mathrm{d}}{\mathrm{d}x} f := f'$$

is **not** continuous. For this take $f_n(x) = x^n$. Clearly $f_n \in C^1([0,1])$. Moreover we have $\|f_n\|_\infty = 1$ for all $n \in \mathbb{N}$ but

$$\left\| \frac{\mathrm{d}}{\mathrm{d}x} f_n \right\|_\infty = \sup_{x \in [0,1]} \left| nx^{n-1} \right| = n.$$

Since $\frac{\mathrm{d}}{\mathrm{d}x}$ is linear its continuity would imply

$$n = \left\| \frac{\mathrm{d}}{\mathrm{d}x} f_n \right\|_\infty \le c\,\|f_n\| = c$$

for a fixed constant $c \ge 0$ and all $n \in \mathbb{N}$ which is of course impossible.

In finite dimensions the situation is quite convenient: all linear mappings between finite dimensional normed vector spaces are continuous. We prove here a first variant of this result.

Theorem 2.30. *A linear mapping $A : \mathbb{R}^m \to \mathbb{R}^n$ is continuous.*

Proof. A linear mapping $A : \mathbb{R}^m \to \mathbb{R}^n$ has a matrix representation with respect to the canonical basis in $\mathbb{R}^m$ and $\mathbb{R}^n$, respectively, given by the matrix

$$A = \begin{pmatrix} a_{11} & \cdots & a_{1m} \\ . & & . \\ . & & . \\ . & & . \\ a_{n1} & \cdots & a_{nm} \end{pmatrix}$$

with $a_{kl} \in \mathbb{R}$. Further, with the canonical basis $\{e_1, \ldots, e_n\}$ in $\mathbb{R}^n$ we find for $x \in \mathbb{R}^m$

$$Ax = \sum_{l=1}^{n} \left(\sum_{j=1}^{m} a_{lj} x_j \right) e_l$$

and the Cauchy-Schwarz inequality yields

$$\|Ax\|_2 = \left(\sum_{l=1}^{n} \left(\sum_{j=1}^{m} a_{lj} x_j \right)^2 \right)^{\frac{1}{2}}$$

$$\leq \left(\sum_{l=1}^{n} \left(\left(\sum_{j=1}^{m} a_{lj}^2 \right)^{\frac{1}{2}} \left(\sum_{j=1}^{n} x_j^2 \right)^{\frac{1}{2}} \right)^2 \right)^{\frac{1}{2}}$$

$$= \left(\sum_{l=1}^{n} \sum_{j=1}^{m} a_{lj}^2 \right)^{\frac{1}{2}} \|x\|_2 = C_A \|x\|_2.$$

$\square$

From our previous considerations it now follows that every linear mapping $A : (\mathbb{R}^m, \|.\|_p) \to (\mathbb{R}^n, \|.\|_q)$ is for every pair p, q, $1 \leq p, q \leq \infty$, continuous.

Finally in this section we want to study sequences of continuous functions. We know from our one-dimensional studies that pointwise limits of continuous functions are not necessarily continuous (see Example I.24.7), however the uniform limit is. The first step is to extend the notion of uniform convergence to metric spaces.

Definition 2.31. *Let (X, d_1) and (Y, d_2) be two metric spaces and $f_n, f :$ $X \to Y$, $n \in \mathbb{N}$, be mappings.* **A.** *We say that the sequence $(f_n)_{n \in \mathbb{N}}$* ***converges pointwisely*** *to f if for every $x \in X$ the sequence $(f_n(x))_{n \in \mathbb{N}}$ converges to $f(x)$.*
B. *If for every $\epsilon > 0$ there exists $N \in \mathbb{N}$ such that $n \geq N$ implies for all $x \in X$ the estimate $d_2(f_n(x), f(x)) < \epsilon$, then $(f_n)_{n \in \mathbb{N}}$ is called* ***uniformly convergent*** *to f.*

Thus $(f_n)_{n \in \mathbb{N}}$ converges uniformly to f if for every $\epsilon > 0$ there exists $N \in \mathbb{N}$ such that $n \geq N$ implies

$$\sup_{x \in X} d_2(f_n(x), f(x)) < \epsilon. \tag{2.20}$$

Theorem 2.32. *Let (X, d_1) and (Y, d_2) be two metric spaces and $(f_n)_{n \in \mathbb{N}}$, $f_n : X \to Y$, $n \in \mathbb{N}$, a sequence of continuous mappings which converges uniformly to $f : X \to Y$. Then f is continuous.*

Proof. (Compare with Theorem I.24.6) We have to prove that f is continuous for every $x_0 \in X$. Let $\epsilon > 0$ be given. By uniform convergence there exists $N \in \mathbb{N}$ such that $d_2(f_N(x), f(x)) < \frac{\epsilon}{3}$ for all $x \in X$. Since f_N is continuous at x_0 there exists $\delta > 0$ such that $d_2(f_N(x), f_N(x_0)) < \frac{\epsilon}{3}$ for $x \in X$ provided that $d_1(x, x_0) < \delta$. Now for $x \in X$ such that $d_1(x, x_0) < \delta$ we find

$$d_2(f(x), f(x_0)) \leq d_2(f(x), f_N(x)) + d_2(f_N(x), f_N(x_0)) + d_2(f_N(x_0), f(x_0))$$
$$< \frac{\epsilon}{3} + \frac{\epsilon}{3} + \frac{\epsilon}{3} = \epsilon$$

proving the theorem. $\qquad\square$

Problems

1. We call two metrics d_1 and d_2 equivalent on X if there exists constants $\kappa_1 > 0$ and κ_2 such that for all $x, y \in X$

   $$\kappa_1 d_1(x, y) \leq d_2(x, y) \leq \kappa_2 d_1(x, y)$$

   holds. Prove that a sequence converges in (X, d_1) if and only if it converges in (X, d_2).

2. Let $\psi : [0, \infty) \to \mathbb{R}_+$ be a monotone increasing and continuous function with $\psi(x) = 0$ if and only if $x = 0$ and suppose that on $(0, \infty)$ the first derivative ψ' exists and is monotone decreasing. On $\mathbb{R}^n$ consider the metric d_ψ, $d_{\psi(x,y)} = \psi(||x - y||_2)$ and prove that a sequence converges in $(\mathbb{R}^n, d_\psi)$ if and only if it converges in $(\mathbb{R}^n, || \cdot ||_2)$.

3. Let (X, d) be a metric space. Show that by δ, where $\delta(x, y) := \min(1, d(x, y))$, a further metric is given on X and a sequence converges in (X, d) if and only if it converges in (X, δ).

4. Denote by S the set of all sequences $(a_k)_{k \in \mathbb{N}}$ of real numbers. For $a, b \in S$, $a = (a_k)_{k \in \mathbb{N}}$ and $b = (b_k)_{k \in \mathbb{N}}$, define

   $$d_S(a, b) := \sum_{k=1}^{\infty} \frac{1}{2^k} \frac{|a_k - b_k|}{1 + |a_k - b_k|}$$

and prove that d_S is a metric and a sequence $(a^{(l)})_{l \in \mathbb{N}}$ converges to $a \in S$ if and only if for all $k \in \mathbb{N}$ the sequences $(a_k^{(l)})_{l \in \mathbb{N}}$ converge to a_k, $a = (a_k)_{k \in \mathbb{N}}$.

5. Consider the metric space $(C([-\pi, \pi]), || \cdot ||_1)$, compare with Problem 8b) in Chapter 1, and the sequence $g_k : [-\pi, \pi] \to \mathbb{R}$, $g_k(x) = \left(\frac{x}{\pi}\right)^k$. Prove that g_k converges in $(C[-\pi, \pi], || \cdot ||_1)$ to the zero-function g_0, $g_0(x) = 0$ for all $x \in [-\pi, \pi]$, however $g_k(x)$ is not convergent for all $x \in [-\pi, \pi]$, and hence it is not uniformly convergent to g_0.

6. Let (X, d) be a metric space. Prove that a sequence $(x_k)_{k \in \mathbb{N}}$ converges to x if and only if for every neighbourhood U of x there exists $N = N(U)$ such that $k \geq N$ implies $x_k \in U$.

7. We call two metric spaces (X, d_X) and (Y, d_Y) isometric if there exists a mapping $j : X \to Y$ such that $d_Y(j(x_1), j(x_2)) = d_X(x_1, x_2)$ holds for all $x_1, x_2 \in X$. Prove that j must be injective and if $(x_k)_{k \in \mathbb{N}}$ converges to x, $x_k, x \in X$, then $(j(x_k))_{k \in \mathbb{N}}$ converges to $j(x)$. Does this imply that the completion of (Y, d_Y) leads to the completeness of (X, d_X)?

8. Prove that $(C([0, 1]), ||\cdot||_1)$ is not a complete space. **Hint**: approximate a discontinuous function by a sequence of continuous functions in the norm $|| \cdot ||_1$.

9. For $a < b$ define

$$C\mathcal{L}^1((a, b)) := \left\{ u \in C((a, b)) \,\middle|\, \left| \int_a^b |u(t)| dt < \infty \right. \right\}.$$

Prove that $C\mathcal{L}^1((a, b))$ is a vector space and $||u||_1$, given by $||u||_1 = \int_a^b |u(t)| dt$ is a norm on $C\mathcal{L}^1((a, b))$ (the result of Problem 8b) in Chapter 1 may be used). Give an example of a bounded set in $(C\mathcal{L}^1((a, b)), ||\cdot||_1)$ that does not necessarily consist of bounded functions.

10. Prove that on $C^k([a, b])$ norms are given by

$$||u||_{k,\infty} := \sum_{n=0}^{k} ||u^{(n)}||_\infty$$

and

$$|||u|||_{k,\infty} := \max_{0 \leq n \leq k} ||u^{(n)}||_\infty.$$

For which $n \in \{0, 1, \ldots, k\}$ is $||u^{(n)}||_\infty$ a norm on $C^k([a,b])$? (Note that $u \in C^k([a,b])$ if $u \in C^k((a,b))$ and for all $0 \leq n \leq k$ the functions $u^{(n)}$ have a continuous extension to $[a,b]$.)

11. Let (X, d_X) and (Y, d_Y) be two metric spaces and $f : X \to Y$ be a mapping. We call f **bounded** if $f(X)$ is bounded in Y, i.e.

$$\operatorname{diam}\{y \in Y | y = f(x), x \in X\} < \infty.$$

Now consider the function $f : \mathbb{R} \to \mathbb{R}$, $f(x) = x$, and the metric spaces $(\mathbb{R}, |\cdot|)$ and $(\mathbb{R}, \arctan|\cdot|)$. Decide which of the following mappings is bounded and prove that they are all continuous:

   i) $f : (\mathbb{R}, |\cdot|) \to (\mathbb{R}, |\cdot|)$;

   ii) $f : (\mathbb{R}, |\cdot|) \to (\mathbb{R}, \arctan|\cdot|)$;

   iii) $f : (\mathbb{R}, \arctan|\cdot|) \to (\mathbb{R}, |\cdot|)$.

12. Define on $\mathbb{R}^n$ the mapping $||\cdot||_\alpha : \mathbb{R}^n \to \mathbb{R}$, $0 < \alpha < 1$, by $||x||_\alpha := \left(\sum_{j=1}^{n} |x_j|^\alpha\right)^{\frac{1}{\alpha}}$. Prove that $||\cdot||_\alpha$ is not a norm, however with some $c_\alpha > 1$ the following holds

$$||x + y||_\alpha \leq c_\alpha(||x||_\alpha + ||y||_\alpha).$$

13. Let $(V, ||\cdot||)$ be a normed space and $x, y \in V, x \neq y$. The line segment joining x and y is given by

$$L(x,y) := \{z = \lambda x + (1 - \lambda)y | 0 \leq \lambda \leq 1\}.$$

For $z \in L(x,y)$ prove

$$||x - y|| = ||x - z|| + ||z - y||,$$

i.e. we have equality in the triangle inequality.
For the metric $||x - y||^\alpha, 0 < \alpha < 1, n \geq 2$, prove that there is no point $z \in L(x,y), x, y \in \mathbb{R}^n, x \neq y$, such that

$$||x - y||^\alpha = ||x - z||^\alpha + ||z - y||^\alpha.$$

14. Let (X, d) be a metric space and $f, g : X \to \mathbb{R}$ continuous functions. Prove that $f \vee g : X \to \mathbb{R}$ and $f \wedge g : X \to \mathbb{R}$ are continuous too, where

$$(f \vee g)(x) := \max(f(x), g(x)) \quad \text{and} \quad (f \wedge g)(x) := \min(f(x), g(x)).$$

Deduce that $f^+ = f \vee 0$, $f^- = (-f)^+$ and $|f| = f^+ + f^-$ are continuous.

15. Let $x = (x_1, \ldots, x_m)$, $x_j \in [a, b]$, and define

$$\text{pr}_x : (C([a, b]), || \cdot ||_\infty) \to (\mathbb{R}^m, || \cdot ||_2)$$
$$\text{pr}_x(u) := (u(x_1), \ldots, u(x_m)).$$

Prove that pr_x is continuous.

16. By inspection prove that on $(C([a, b]), \mathbb{R}^n)$, the space of all continuous functions $u : [a, b] \to \mathbb{R}^n$, a norm is given by $||u||_\infty := \max_{1 \le k \le n} ||u_j||_\infty$, where $u = (u_1, \ldots, u_n)$. Now fix $A \in \mathbb{R}^n$ and define

$$h : (C([a, b], \mathbb{R}^n), || \cdot ||_\infty) \to (C([a, b], \mathbb{R}^n), || \cdot ||_\infty) \tag{2.21}$$

$$h(u) := \langle A, u \rangle = \sum_{j=1}^n A_j u_j. \tag{2.22}$$

Use the $\epsilon - \delta$-criterion to prove the continuity of h.

17. With $|| \cdot ||_{\infty,1}$ defined as in Problem 10, prove that

$$\frac{d}{dx} : (C^1([a, b]), || \cdot ||_{\infty,1}) \to (C([a, b]), || \cdot ||_\infty)$$

is continuous. Is the mapping

$$\frac{d}{dx} : (C^1([a, b]), || \cdot ||_\infty) \to (C([a, b]), || \cdot ||_\infty)$$

continuous?

18. Justify that for a continuous function $u : [0, 1] \to \mathbb{R}$ the function $x \mapsto \int_0^1 (x - y)^2 u(y) dy$ is continuous too. Now prove that $T : (C([0, 1]), || \cdot ||_\infty) \to (C([0, 1]), || \cdot ||_\infty)$ is Lipschitz continuous.

19. Let $f : X \to Y$ be a Lipschitz continuous mapping between the metric spaces (X, d_X) and (Y, d_Y). Prove that f maps bounded sets onto bounded sets.

20. Let $(V, || \cdot ||)$ be a normed space and $(x_k)_{k \in \mathbb{N}_0}, x_k \in V$, a sequence. We say that the series $\sum_{k=0}^{\infty} x_k$ converges to $S \in V$ if the sequence of partial sums $S_N = \sum_{k=0}^{N} x_k$ converges to S in $(V, || \cdot ||)$. Formulate a Cauchy criterion for convergence of series in a normed space.

21. Suppose that $(X, || \cdot ||)$ is a Banach space and that $||x_k|| \le ||x_0||^k$ holds for the sequence $(x_k)_{k \in \mathbb{N}_0}, x_k \in X$. Prove that if $||x_0|| \le R$ and the power series of real numbers $\sum_{k=0}^{\infty} a_k t^k$ converges for all t, $|t| < \rho$, then if $R < \rho$ the series $\sum_{k=0}^{\infty} a_k x_k$ converges in $(X, || \cdot ||)$.

3   More on Metric Spaces and Continuous Functions

There are still some properties of $\mathbb{R}$ or of continuous real-valued functions defined on a subset of $\mathbb{R}$ which we want to extend to metric spaces. We may ask for an analogue of the statement that the rational numbers are dense in any interval I in the sense that every element in I can be approximated by a sequence of rational numbers belonging to I. Further questions refer to a substitute for the Heine-Borel or the Bolzano-Weierstrass theorem, and we may ask for a generalisation of compactness and whether continuous functions on compact sets are "nicely" behaved as real-valued functions on compact subsets of $\mathbb{R}$ are. For example they are bounded, attain infimum and supremum, and they are uniformly continuous. This chapter deals with all these questions and more. It contains much more material than needed for most of the next few chapters to follow. Therefore, at the end of this chapter we will point to those results eventually needed in the following chapters when dealing with mappings $f : G \to \mathbb{R}^n$, $G \subset \mathbb{R}^n$.

Definition 3.1. *Let (X, d) be a metric space. We call $x \in X$ a **limit point** or an **accumulation point** of the subset $Y \subset X$ if every neighbourhood of x contains a point $y \in Y$, $y \neq x$.*

Corollary 3.2. *Let $x \in X$ be a limit point of Y and $x \notin Y$. Then for every neighbourhood U of x the intersection $U \cap Y$ contains at least countably many points.*

Proof. Suppose that $U \cap Y$ contains only finitely many points, i.e. $U \cap Y = \{y_1, \ldots, y_N\}$. Since $x \notin Y$ it follows for $k = 1, \ldots, N$ that $r_k := d(x, y_k) > 0$ and with $r < \min\{r_1, \ldots, r_N\}$ we have $B_r(x) \cap Y = \emptyset$. But $B_r(x)$ is a neighbourhood of x and by assumption x is a limit point. Hence we have constructed a contradiction. $\square$

Proposition 3.3. *In a metric space (X, d) the closure $\bar{Y}$ of a subset $Y \subset X$ is the union of Y with all its limit points.*

Proof. We have to prove that if $y \in \bar{Y} \setminus Y$ then y is a limit point of Y. Now $y \in \bar{Y} \setminus Y$ means $y \in \partial Y$ and hence every neighbourhood of y must contain points of Y implying that y is a limit point of Y. $\square$

Definition 3.4. A. *A set $Y \subset X$ of a metric space (X, d) is called **dense in X** if $\bar{Y} = X$, i.e. the closure of Y is equal to X. **B.** For $Z \subset X$ we call $Y \subset X$ **dense with respect to Z** if $Z \subset \bar{Y}$. **C.** A metric space (X, d) is called **separable** if there exists a countable dense subset $Y \subset X$, i.e. Y is countable and $X = \bar{Y}$.*

Remark 3.5. A. Sometimes a set Y dense in X is called **everywhere dense** in X. **B.** From Theorem I.18.33 it follows that for every subset $D \subset \mathbb{R}$ the set $D \cap \mathbb{Q}$ is dense.

Since $\mathbb{Q}$ is countable and dense in $\mathbb{R}$ we immediately have

Corollary 3.6. *The real line $\mathbb{R}$ with the metric induced by the absolute value is separable.*

Definition 3.7. *A family $(U_j)_{j \in I}$ of non-empty open sets of a metric space (X, d) is called a **base** of the open sets of X, i.e. of the topology of X, if every non-empty open set $U \subset X$ is the union of a subfamily of $(U_j)_{j \in I}$.*

Lemma 3.8. *A family $(U_j)_{j \in I}$, $U_j \subset X$, of non-empty open sets is a base of the topology of X if and only if for every $x \in X$ and every neighbourhood V of x there exists $j_0 \in I$ such that $x \in U_{j_0} \subset V$.*

Proof. If V is a neighbourhood of x there exists an open set $U \subset X$ such that $x \in U \subset V$. Thus for some $I_1 \subset I$ we have $x \in U = \cup_{j \in I_1} U_j \subset V$ implying the existence of some $j_0 \in I$ such that $U_{j_0} \subset V$. Conversely, if $U \subset X$ is an open set then there exists for every $x \in U$ some index $j(x) \in I$ such that $x \in U_{j(x)} \subset U$ and therefore $U \subset \cup_{x \in U} U_{j(x)} \subset U$. $\square$

Theorem 3.9. *A metric space (X, d) is separable if and only if its topology has a denumerable base.*

Proof. First let $(U_j)_{j \in \mathbb{N}}$ be a base of X and $a_j \in U_j$. For an open set $U \subset X$ there exists a subset $I \subset \mathbb{N}$ such that $U = \cup_{j \in I} U_j$ and therefore $U \cap \{a_j \mid j \in \mathbb{N}\} \neq \emptyset$ proving that $\{a_j \mid j \in \mathbb{N}\}$ is dense in X, i.e. each $x \in X$ is a limit point of $\{a_j \mid j \in \mathbb{N}\}$. Indeed, let $x \in X$ and U be any neighbourhood of x, then we find elements of $\{a_j \mid j \in \mathbb{N}\}$ in U, i.e. a subsequence of $(a_j)_{j \in \mathbb{N}}$ converges to x. Suppose now that $\{a_j \mid j \in \mathbb{N}\}$ is dense in X. The set $\left\{ B_{\frac{1}{m}}(a_n) \mid n, m \in \mathbb{N} \right\}$ is countable since $B_{\frac{1}{m}}(a_n) \mapsto (m, n)$ provides a bijective mapping to $\mathbb{N} \times \mathbb{N}$. We claim that $\left\{ B_{\frac{1}{m}}(a_n) \mid n, m \in \mathbb{N} \right\}$ is a base of the topology of X. By density, for $x \in X$ and every $r > 0$ there exist $m, n \in \mathbb{N}$ such that $\frac{1}{m} < \frac{r}{2}$ and $a_n \in B_{\frac{1}{m}}(x)$, or $x \in B_{\frac{1}{m}}(a_n)$. For $y \in B_{\frac{1}{m}}(a_n)$ the triangle inequality yields $d(x, y) \leq d(x, a_n) + d(a_n, y) < \frac{2}{m} < r$, or $B_{\frac{1}{m}}(a_n) \subset B_r(x)$ and the result follows from Lemma 3.8.   □

Corollary 3.10. *If $\emptyset \neq Y \subset X$ and if (X, d) is separable then (Y, d_Y) is separable too.*

Proof. If $(U_j)_{j \in \mathbb{N}}$ is a base for X then $(Y \cap U_j)_{j \in \mathbb{N}}$ is a base for (Y, d_Y).   □

Example 3.11. For $\mathbb{R}$ with the topology induced by the absolute value the family $\left(B_{\frac{1}{m}}(q) \right)_{m \in \mathbb{N}, q \in \mathbb{Q}}$ forms a countable base for the topology.

Following the ideas of Chapter I.20, we extend the notion of compactness to a general metric space.

Definition 3.12. A. *Let $Y \subset X$ be a subset of the metric space (X, d). A family $(U_j)_{j \in I}$ of open sets $U_j \subset X$ is called an **open covering** of Y if*

$$Y \subset \bigcup_{j \in I} U_j.$$

B. *A subset $K \subset X$ of a metric space is called **compact** if every open covering $(U_j)_{j \in I}$ of K contains a finite subcovering of K, i.e. if $K \subset \cup_{j \in I} U_j$ then there exists $j_1, \ldots, j_k \in I$, $k \in \mathbb{N}$, such that $K \subset \cup_{l=1}^{k} U_{j_l}$. If X itself is compact we call (X, d) a **compact metric space**.*

Suppose that $Y \subset X$ is compact, then (Y, d_Y) is a compact metric space. Here is a first example of a compact set.

Lemma 3.13. *If the sequence $(x_k)_{k \in \mathbb{N}}$, $x_k \in X$, converges in the metric space (X, d) to $x \in X$ then the set $K := \{ x_k \in X \mid k \in \mathbb{N} \} \cup \{ x \}$ is compact.*

Proof. Let $(U_j)_{j \in I}$ be any open covering of K. Since $x \in K$ there exists $j_0 \in I$ such that $x \in U_{j_0}$. Further, since U_{j_0} is open it is a neighbourhood of x. The convergence of $(x_k)_{k \in \mathbb{N}}$ to x now implies the existence of $N \in \mathbb{N}$ such that $k \geq N$ yields $x_k \in U_{j_0}$. Therefore we find

$$K \subset U_{j_1} \cup \ldots \cup U_{j_{k-1}} \cup U_{j_0}$$

where for $k < N$ we have $x_k \in U_{j_k}$. Thus $(U_j)_{j=0,\ldots,N-1}$ is a finite subcovering of $(U_j)_{j \in I}$ which covers K.   □

Example 3.14. The set $\left\{\frac{1}{n} \,\middle|\, n \in \mathbb{N}\right\} \subset \mathbb{R}$ is not compact in $(\mathbb{R}, |.|)$.

Proof. We may argue that this set is not closed in $(\mathbb{R}, |.|)$ and hence according to Proposition I.20.24 it cannot be compact. However we prefer to give here a further proof using coverings. Consider the open set $U_1 := \left(\frac{1}{2}, 2\right)$ and $U_n := \left(\frac{1}{n+1}, \frac{1}{n-1}\right)$, $n \geq 2$. Since $\frac{1}{n} \in U_n$, $(U_n)_{n \in \mathbb{N}}$ is an open covering of $\left\{\frac{1}{n} \,\middle|\, n \in \mathbb{N}\right\}$. But any of the sets U_n contains only one point of $\left\{\frac{1}{n} \,\middle|\, n \in \mathbb{N}\right\}$, namely $\frac{1}{n}$. Thus we can never cover $\left\{\frac{1}{n} \,\middle|\, n \in \mathbb{N}\right\}$ by finitely many of these sets. $\square$

Note that the closure of $\left\{\frac{1}{n} \,\middle|\, n \in \mathbb{N}\right\}$ is the set $\left\{\frac{1}{n} \,\middle|\, n \in \mathbb{N}\right\} \cup \{0\}$ which by Lemma 3.13 is compact. This leads to

Definition 3.15. A. *A set $K \subset X$ in a metric space (X, d) is called **relatively compact** if its closure $\bar{K}$ is compact.*
B. *We call $Y \subset X$ **totally bounded** if for every $\epsilon > 0$ there exists a finite covering $(U_j)_{j=1,\ldots,m}$, $m \in \mathbb{N}$, of Y with open sets $U_j \subset X$ such that $\operatorname{diam} U_j < \epsilon$.*

Remark 3.16. A. Many authors also call totally bounded sets **pre-compact**.
B. If X is totally bounded, we call the metric space (X, d) totally bounded.
C. If $Y \subset X$ is totally bounded, for $\epsilon > 0$ we can find a finite subset $F \subset Y$ such that $d(x, y) < \epsilon$ for all $y \in F$ and $x \in Y$, i.e. $\operatorname{dist}(x, F) := \inf \left\{ d(x, y) \,\middle|\, y \in F \right\} < \epsilon$.

Lemma 3.17. *Let (X, d) be a metric space and $K \subset X$.*
A. *If K is totally bounded it is bounded.*
B. *If K is compact it is bounded and closed.*
C. *A closed subset $\tilde{K}$ of a compact set K is compact.*
D. *A subset of a relatively compact set is relatively compact.*
E. *A subset of a totally bounded set is totally bounded.*

Proof. **A.** Since K is totally bounded for $\epsilon = 1$ we can find a finite covering $(U_j)_{j=1,\ldots,m}$ of K with $\operatorname{diam} U_j < 1$, implying by the triangle inequality that $\operatorname{diam} K \leq m$, compare also with Problem 4 and Problem 6.
B. For the set K an open covering is given by $(B_1(x))_{x \in K}$ and the compactness of K implies the existence of $m \in \mathbb{N}$ and points $x_1, \ldots, x_m \in K$ such that

$$K \subset \bigcup_{j=1}^{m} B_1(x) \subset B_r(x_1)$$

for $r > \max_{2 \le j \le m}(d(x_j, x_1) + 1)$, i.e. K is bounded. For $x \in K^{\complement}$ we find that

$$U_n := \left\{ y \in X \,\middle|\, d(x, y) > \frac{1}{n} \right\} = \left(\overline{B_{\frac{1}{n}}(x)} \right)^{\complement}$$

is open and we have

$$K \subset X \setminus \{x\} = \bigcup_{n=1}^{\infty} U_n,$$

i.e. $(U_n)_{n \in \mathbb{N}}$ is an open covering of K. Hence we can find a finite subcovering $(U_{n_j})_{j=1,\ldots,m}$ such that $K \subset \bigcup_{j=1}^{m} U_{n_j}$ and for $k := \max\{n_1, \ldots, n_m\}$ it follows that $B_{\frac{1}{k}}(x) \subset K^{\complement}$, i.e. K^c is open and K is therefore closed.

C. Let $(U_j)_{j \in I}$ be an open covering of $\tilde{K}$. Since $\tilde{K}$ is closed $\tilde{K}^{\complement}$ is open and

$$\tilde{K}^{\complement} \cup \bigcup_{j \in I} U_j = X \supset K.$$

The compactness of K yields the existence of $j_1, \ldots, j_k \in I$, $k \in \mathbb{N}$, such that

$$\tilde{K}^{\complement} \cup U_{j_1} \cup \ldots \cup U_{j_k} \supset K,$$

or, using the fact that $\tilde{K} \subset K$ we have

$$\tilde{K} \subset U_{j_1} \cup \ldots \cup U_{j_k},$$

implying the compactness of $\tilde{K}$.

D. If K_1 is a subset of a relatively compact set K then $\bar{K}_1$ is a closed subset of the compact set $\bar{K}$, hence it is compact and therefore K_1 is relatively compact.

E. This is a trivial consequence of the definition. $\square$

A proof of the following result will be provided in Appendix II.

Theorem 3.18. *For a metric space (X, d) the following are equivalent:*

i) *X is compact;*

ii) *every infinite sequence in X has at least one accumulation point;*

iii) *X is complete and totally bounded.*

Proposition 3.19. *A totally bounded metric space is separable.*

Proof. If X is totally bounded we can find for every $n \in \mathbb{N}$ a finite set $F_n \subset X$ such that for all $x \in X$ we have $d(x, F_n) = \inf \{d(x, y) \,|\, y \in F_n\} < \frac{1}{n}$. The set $F := \bigcup_{n \in \mathbb{N}} F_n$ is countable and for every $x \in X$ and every $n \in \mathbb{N}$ it follows that $d(x, F) \leq d(x, F_n) < \frac{1}{n}$ implying $d(x, F) = 0$ for all $x \in X$ i.e. $X = \bar{F}$. Note, since $d(x, F) = 0$ we can approximate every $x \in X$ by a sequence of elements belonging to F, see Problem 3. $\qquad\square$

We want to characterise the compact sets in $\mathbb{R}^n$ by generalising the Heine-Borel theorem, compare Theorem I.20.26. We start with

Proposition 3.20. *Let $n \in \mathbb{N}$ and $a_\nu \leq b_\nu$, $\nu = 1, \ldots, n$, be $2n$ real numbers. Then the closed hyper-rectangle $Q = \{x \in \mathbb{R}^n \,|\, a_\nu \leq x \leq b_\nu\}$ is compact.*

Proof. Let $(U_j)_{j \in I}$ be an open covering of Q. We aim to construct a contradiction to the assumption that Q cannot be covered by a finite number of sets U_j. By induction we construct a sequence of closed hyper-rectangles $Q_0 \supset Q_1 \supset \ldots$ having the following properties

i) Q_m cannot be covered by finitely many U_{j_k},

ii) $\operatorname{diam}(Q_m) = 2^{-m} \operatorname{diam}(Q)$.

We set $Q_0 = Q$. Suppose that Q_m is already constructed

$$Q_m = I_1 \times \ldots \times I_n,$$

where $I_\nu \subset \mathbb{R}$, $1 \leq \nu \leq n$, is a closed interval. We divide I_ν into two closed intervals of half the length

$$I_\nu = I_\nu^{(1)} \cup I_\nu^{(2)}, \quad I_\nu^{(1)} \cap I_\nu^{(2)} \text{ consists of one point,}$$

and for $s_\nu = 1, 2$ we set

$$Q_m^{(s_1, \ldots, s_m)} = I_1^{(s_1)} \times I_2^{(s_2)} \times \ldots I_n^{(s_n)}.$$

Thus we have 2^n hyper-rectangles such that

$$\bigcup_{s_1, \ldots, s_n} Q_m^{(s_1, \ldots, s_n)} = Q_m.$$

Since we cannot cover Q_m by finitely many U_{j_k}, at least one of the hyper-rectangles $Q_m^{(s_1,\ldots,s_n)}$ could also not be covered by finitely many of the sets U_{j_k}. This hyper-rectangle will be Q_{m+1}. Thus we have

$$\operatorname{diam}(Q_{m+1}) = \frac{1}{2}\operatorname{diam}(Q_m) = 2^{-m-1}\operatorname{diam}(Q),$$

i.e. Q_{m+1} has the properties i) - ii). According to Theorem 2.15 there exists a point $a \in \bigcap_{m\in\mathbb{N}_0} Q_m$. Since $(U_j)_{j\in I}$ is a covering of Q, there exists $j_0 \in I$ such that $a \in U_{j_0}$. But U_{j_0} is open, hence there is $\epsilon > 0$ such that $B_\epsilon(a) \subset U_{j_0}$. Now for m large such that $\operatorname{diam}(Q_m) < \epsilon$ it follows $Q_m \subset B_\epsilon(a) \subset U_{j_0}$ which is a contradiction to i). $\qquad\square$

Now we can prove

Theorem 3.21 (Heine-Borel). *A subset $K \subset \mathbb{R}^n$ is compact if and only if it is bounded and closed.*

Proof. It remains to prove that bounded and closed subsets of $\mathbb{R}^n$ are compact. But every bounded set is contained in a (large) closed hyper-rectangle and now the result follows from Proposition 3.20 and Lemma 3.17.C. $\qquad\square$

This theorem implies immediately that the product of a compact set $K_1 \subset \mathbb{R}^n$ with a compact set $K_2 \subset \mathbb{R}^m$ is a compact set $K_1 \times K_2 \subset \mathbb{R}^{n+m}$. The following result is already known to us

Lemma 3.22. *Let $K \subset \mathbb{R}$ be compact. Then we have*

$$\sup K := \sup\{x \in K\} \in K \quad \text{and} \quad \inf K := \inf\{x \in K\} \in K.$$

Proof. Since K is compact it is bounded. Therefore $\sup K$ and $\inf K$ are finite. Further there are sequences $(x_k)_{k\in\mathbb{N}}$ and $(y_k)_{k\in\mathbb{N}}$, $x_k, y_k \in K$, such that

$$\lim_{k\to\infty} x_k = \sup K \quad \text{and} \quad \lim_{k\to\infty} y_k = \inf K.$$

Since K is closed these limits must belong to K. $\qquad\square$

As stated at the beginning of this chapter, we already know that real-valued continuous functions defined on a compact set $K \subset \mathbb{R}$ have a lot of "nice" properties. In order to extend these results for continuous mappings between metric spaces we need a further characterisation of continuous functions, see Problem 2 in Chapter I.20.

Theorem 3.23. *Let (X, d_X) and (Y, d_Y) be two metric spaces and $f : X \to Y$ be a mapping.*
A. *The mapping f is continuous at $x \in X$ if and only if for every neighbourhood V of $f(x)$ there exists a neighbourhood U of x such that $f(U) \subset V$.*
B. *The mapping f is continuous on X if and only if the pre-image $f^{-1}(V)$ of every open set $V \subset Y$ is open in X.*
C. *The mapping f is continuous on X if and only if the pre-image $f^{-1}(B)$ of every closed set $B \subset Y$ is closed in X.*

Proof. **A.** Part A is just a reformulation of Theorem 2.23.
B. Suppose that f is continuous and $V \subset Y$ is open. We have to prove that $f^{-1}(V)$ is open (in X). Let $a \in f^{-1}(V)$. Since V is a neighbourhood of $f(a) \in V$, there is a neighbourhood U of a such that $f(U) \subset V$ implying that $U \subset f^{-1}(V)$. Thus $f^{-1}(V)$ is a neighbourhood of a and we have proved that $f^{-1}(V)$ is an open set.
Suppose conversely that every pre-image of an open set is open and take $a \in X$. For a neighbourhood V of $f(a)$ we have: there exists an open set V_1 such that $f(a) \in V_1 \subset V$. Therefore $U := f^{-1}(V_1)$ is open. But $a \in U$, i.e. U is a neighbourhood of a and $f(U) \subset V$ holds, i.e. f is continuous at a and the assertion is proved.
C. Since for every set $D \subset Y$ we have

$$\left(f^{-1}(D)\right)^{\complement} = f^{-1}(D^{\complement}),$$

part B implies C, recall B is closed if and only if $B^{\complement}$ is open. □

For later purposes we note

Definition 3.24. **A.** *Let $(X, \mathcal{O}_X)$ and $(Y, \mathcal{O}_Y)$ be two topological spaces. We call $f : X \to Y$ **continuous** if every pre-image of an open set (in Y) is open (in X), i.e. $f^{-1}(U) \in \mathcal{O}_X$ for $U \in \mathcal{O}_Y$.*
B. *A bijective continuous mapping $f : X \to Y$ between two topological spaces with continuous inverse is called a **homeomorphism** and X and Y are called **homeomorphic**.*

Remark 3.25. Two metric spaces (X, d_X) and (Y, d_Y) are homeomorphic if and only if there exists a bijective continuous mapping $f : X \to Y$ with continuous inverse, i.e. homeomorphy is a topological and not a metric property.

As a consequence of Theorem 3.23 we have

Theorem 3.26. *Let* (X, d_X) *and* (Y, d_Y) *be two metric spaces and* $f : X \to Y$ *a continuous mapping. If* $K \subset X$ *is compact, then* $f(K) \subset Y$ *is compact, i.e. continuous images of compact sets are compact.*

Proof. Let $(V_j)_{j \in I}$ be an open covering of $f(K)$. By Theorem 3.23 the sets $U_j := f^{-1}(V_j)$ are open in X and $K \subset \bigcup_{j \in I} U_j$ holds. Since K is compact there exists finitely many $j_1, \ldots, j_k \in I$ such that $K \subset \bigcup_{l=1}^{k} U_{j_l}$ which yields $f(K) \subset \bigcup_{l=1}^{k} V_{j_l}$, i.e. $f(K)$ is compact. $\qquad\square$

For real-valued functions $f : X \to \mathbb{R}$ we find further

Theorem 3.27. *If* (X, d) *is a compact metric space and* $f : X \to \mathbb{R}$ *is continuous then* f *is bounded, i.e.* $f(X) \subset \mathbb{R}$ *is a bounded set, and* f *attains its supremum and its infimum, i.e. there are points* $p, q \in X$ *such that*

$$f(p) = \sup \left\{ f(x) \,\middle|\, x \in X \right\} \quad and \quad f(q) = \inf \left\{ f(x) \,\middle|\, x \in X \right\}.$$

Proof. Since $A := f(X) \subset \mathbb{R}$ is compact, the result follows by the boundedness of compact sets in $\mathbb{R}$, Theorem 3.23 and Lemma 3.22. $\qquad\square$

Example 3.28. Let (X, d) be a metric space and $A \subset X$ be a subset as well as $x \in X$. The **distance** of x to A is defined by

$$\text{dist}(x, A) := \inf \left\{ d(x, y) \,\middle|\, y \in A \right\}. \tag{3.1}$$

The function $x \mapsto \text{dist}(x, A)$ is continuous on X since by the triangle inequality we have

$$\text{dist}(x', A) \leq d(x', x'') + \text{dist}(x'', A)$$

or

$$|\text{dist}(x', A) - \text{dist}(x'', A)| < \epsilon \quad \text{if} \quad d(x', x'') < \delta = \epsilon.$$

If $K \subset X$ is a second subset of X we define

$$\text{dist}(K, A) := \inf \left\{ \text{dist}(x, A) \,\middle|\, x \in K \right\} = \inf \left\{ d(x, y) \,\middle|\, x \in K, y \in A \right\}, \tag{3.2}$$

and we claim:
If A is closed and K compact, then $A \cap K = \emptyset$ implies $\text{dist}(K, A) > 0$.

Proof. Since K is compact and $x \mapsto \operatorname{dist}(x, A)$ is continuous, there exists $q \in K$ such that

$$\operatorname{dist}(q, A) = \operatorname{dist}(K, A)$$

Further, since A is closed and $q \notin A$, there exists $\epsilon > 0$ such that $B_\epsilon(q) \subset A^{\complement}$ implying $\operatorname{dist}(q, A) \geq \epsilon$. $\qquad\square$

Remark 3.29. For two bounded but not closed sets the assertion of Example 3.28 is in general not correct. For example take $B_1(0) \subset \mathbb{R}^2$ and the point $(1,0) \in \mathbb{R}^2$, recall $B_1(0) = \{x \in \mathbb{R}^2 \,|\, |x| < 1\}$. Obviously $B_1(0) \cap \{(1,0)\} = \emptyset$, but $\operatorname{dist}(B_1(0), \{(1,0)\}) = 0$.

Further, if A_1 and A_2 are disjoint and closed, but not bounded, then again we cannot expect the result of Example 3.28 to hold : In $\mathbb{R}^2$ the sets

$$A_1 := \{(x,y) \in \mathbb{R}^2 \,|\, x \cdot y = 0\}$$

and

$$A_2 := \{(x,y) \in \mathbb{R}^2 \,|\, x \cdot y = 1\}$$

are closed and $A_1 \cap A_2 = \emptyset$, but $\operatorname{dist}(A_1, A_2) = 0$.

Let us recall the intermediate value theorem, Theorem I.20.17 which states that if $f : [a, b] \to \mathbb{R}$ is continuous and $\eta_0 \in [\min\{f(a), f(b)\}, \max\{f(a), f(b)\}]$ then there exists $\xi_0 \in [a, b]$ such that $f(\xi_0) = \eta_0$. We can modify the result in the following way: since $[a, b]$ is compact and f is continuous there exists $\xi_1, \xi_2 \in [a, b]$ such that $f(\xi_1) = \min\{f(\xi) \,|\, \xi \in [a, b]\}$ and $f(\xi_2) = \max\{f(\xi) \,|\, \xi \in [a, b]\}$. Denote by I_{ξ_1, ξ_2} the closed interval with end points ξ_1 and ξ_2, i.e. $I_{\xi_1, \xi_2} = [\xi_1, \xi_2]$ or $I_{\xi_1, \xi_2} = [\xi_2, \xi_1]$. By the intermediate value theorem for $\eta \in [f(\xi_1), f(\xi_2)]$ there exists $\xi \in I_{\xi_1, \xi_2} \subset [a, b]$ such that $f(\xi) = \eta$, i.e. the image of $[a, b]$ under f is the interval $[f(\xi_1), f(\xi_2)]$ and hence $f([a, b])$ is connected.

This is a result we can extend to metric spaces. We start our considerations by introducing connected sets in a metric space, see Definition I.19.14, and compare with M. Heins [24]

Definition 3.30. *Let (X, d) be a metric space.*
A. *We call a pair $\{U_1, U_2\}$ of non-empty open subsets $U_j \subset X$ a **splitting** of X if $U_1 \cup U_2 = X$ and $U_1 \cap U_2 = \emptyset$.*
B. *A metric space is called **connected** if a splitting of X does not exist.*

C. *A subset* $Y \subset X$ *is called connected if the metric space* (Y, d_Y) *is connected.*

D. *A* **region** *in* X *is a non-empty open and connected subset of* X.

Remark 3.31. A metric space (X, d) is connected if and only if X and $\emptyset$ are the only subsets which are both open and closed. Indeed, suppose $Y \subset X$, $Y \neq \emptyset$, is open and closed. Then $\{Y, Y^{\complement}\}$ is a splitting of X and if $Y \neq X$ then X is not connected.

Theorem 3.32. *Let* (X, d_X) *and* (Y, d_Y) *be two metric spaces and* $f : X \to Y$ *a continuous mapping. If* $A \subset X$ *is connected then* $f(A) \subset Y$ *is connected too, i.e. the image of a connected set under a continuous mapping is connected.*

Proof. The theorem is proved once we have shown that if X is a connected metric space and $f : X \to Y$ is continuous with $f(X) = Y$, then Y is connected. This follows from the fact that we can alway consider $f|_A$ and look at A as a metric space (A, d_A). Now suppose that $f(X) = Y$ and assume that Y is not connected. Then there exists a splitting $\{U_1, U_2\}$ of Y. By continuity we find that $f^{-1}(U_1)$ and $f^{-1}(U_2)$ are open in X and hence $\{f^{-1}(U_1), f^{-1}(U_2)\}$ is a splitting of X which is a contradiction. $\qquad\square$

The following corollary is a more obvious extension of the intermediate value theorem.

Corollary 3.33. *Let* (X, d) *be a connected metric space and* $f : X \to \mathbb{R}$ *a continuous mapping. If* $a, b \in f(X)$, $a < b$, *and* $\eta \in [a, b]$ *then there exists* $\xi \in X$ *such that* $f(\xi) = \eta$.

Proof. Since $f(X) \subset \mathbb{R}$ is connected it must be an interval implying the assertion. $\qquad\square$

Let (X, d) be a metric space and $(Y_j)_{j \in I}$ be a family of connected subsets of X such that $Y := \bigcap_{j \in I} Y_j \neq \emptyset$. Let $\{U_1, U_2\}$ be a splitting of X such that for some $y \in \bigcap_{j \in I} Y_j$ we have $y \in U_1$. We claim that $U_2 \cap Y_j = \emptyset$ for all $j \in I$. Indeed, since $y \in \bigcap_{j \in I} Y_j$ it follows that $y \in Y_j$ for all $j \in I$, and hence $U_1 \cap Y_j \neq \emptyset$ for all $j \in I$. Since Y_j is connected we must have $U_2 \cap Y_j = \emptyset$ for all $j \in I$, otherwise we would have for some j_0 that $(U_1 \cap Y_{j_0}) \cup (U_2 \cap Y_{j_0}) = Y_{j_0}$ and $(U_1 \cap Y_{j_0}) \cap (U_2 \cap Y_{j_0}) \neq \emptyset$ which is a contradiction. Thus $U_2 \cap \bigcup_{j \in I} Y_j = \emptyset$. Therefore, if we replace X by $\bigcup_{j \in I} Y_j$ we have proved

Proposition 3.34. *If a family of non-empty connected sets in a metric space has a non-empty intersection then its union is connected.*

A simple induction shows

Corollary 3.35. *Let $(Y_j)_{j=1,\ldots,N}$ be a finite family of connected sets in the metric space (X,d) such that $Y_j \cap Y_{j+1} \neq \emptyset$ for $1 \leq j \leq N-1$. Then $\bigcup_{j=1}^{N} Y_j$ is connected.*

Corollary 3.36. *Let $x \in X$ and denote by $C(x)$ the union of all connected sets containing x. Then $C(x)$ is connected.*

The set $C(x)$ is obviously the largest connected set containing x and it is called the **component** or **connectivity component** of x. If $y \in C(x)$ then $C(y) = C(x)$, and if $y \neq C(x)$ then $C(x) \cap C(y) = \emptyset$. Thus the components $C(x)$, $x \in X$, give a partition of X.

Proposition 3.37. *Let (X,d) be a metric space and $Y \subset X$ a connected set. Every set $Z \subset X$ such that $Y \subset Z \subset \bar{Y}$ is connected too, in particular $\bar{Y}$ is connected.*

Proof. Suppose that Z is not connected and $U_1, U_2 \subset X$ are open sets such that $(Z \cap U_1) \cup (Z \cap U_2) = Z$ and $(Z \cap U_1) \cap (Z \cap U_2) = \emptyset$ as well as $Z \cap U_j \neq \emptyset$. It follows that $(Y \cap U_1) \cup (Y \cap U_2) = Y$ and $(Y \cap U_1) \cap (Y \cap U_2) = \emptyset$ since Y is connected. Let $z_j \in Z \cap U_j$ which yields $z_j \in \bar{Y}$ and therefore $U \cap Y \neq \emptyset$ for every open set $U \subset X$ such that $z_j \in U$. In particular it follows that $U_j \cap Y \neq \emptyset$ which is a contradiction. $\square$

Let (X, d_X) and (Y, d_Y) be two metric spaces. Suppose that $g : X \to Y$ is an injective mapping, i.e. $g : X \to g(X)$ is bijective. Suppose in addition that g is continuous. We are interested to know whether the inverse mapping $g^{-1} : g(X) \to X$ is continuous. Choosing on $g(X)$ the relative topology we may reduce the problem to the question when a bijective continuous function $f : X \to Y$ has a continuous inverse. One way to paraphrase continuity is that neighbourhood relations are not disturbed under continuous mappings: if $U \subset Y$ is a neighbourhood of $y = f(x)$ then there exists a neighbourhood $V \subset X$ of x such that $f(V) \subset U$.

Now look at the half-open interval $[0, 2\pi)$ which we can map bijectively onto the circle $S^1 \subset \mathbb{R}^2$ by $\varphi : [0, 2\pi) \to S^1$, $\varphi(t) = (\cos t, \sin t)$.

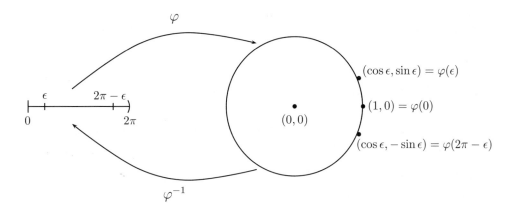

Figure 3.1

Clearly φ is continuous and bijective. However $\varphi^{-1} : S^1 \to [0, 2\pi)$ cannot be continuous. Consider the points $(\cos \epsilon, \sin \epsilon)$ and $(\cos(2\pi - \epsilon), \sin(2\pi - \epsilon)) = (\cos \epsilon, -\sin \epsilon)$ for $0 < \epsilon < \pi$ and their distance to $(1, 0) = (\cos 0, \sin 0)$. We find

$$\|(\cos \epsilon, \sin \epsilon) - (1, 0)\|^2 = \|(\cos \epsilon - 1, \sin \epsilon)\|^2 = (\cos \epsilon - 1)^2 + \sin^2 \epsilon$$

and

$$\|(\cos(2\pi - \epsilon), \sin(2\pi - \epsilon)) - (1, 0)\|^2 = \|(\cos \epsilon - 1, -\sin \epsilon)\|^2 = (\cos \epsilon - 1)^2 + \sin^2 \epsilon.$$

Thus for $\delta > 0$ we can find $\epsilon > 0$ such that

$$\|(\cos(2\pi - \epsilon), \sin(2\pi - \epsilon)) - (1, 0)\|^2 = \|(\cos \epsilon, \sin \epsilon) - (1, 0)\| < \delta.$$

However the distance from $2\pi - \epsilon$ to 0 is $2\pi - \epsilon > \pi$, thus φ^{-1} destroys neighbourhood relations and it is not continuous.

In the case of compact metric spaces (X, d_X) and (Y, d_Y) the situation is however different. (Note that $[0, 2\pi)$ is not compact but S^1 is compact.)

Theorem 3.38. *A bijective continuous mapping $f : X \to Y$ between two compact metric spaces (X, d_X) and (Y, d_Y) has a continuous inverse.*

Proof. Since f^{-1} is continuous if and only if the pre-image under f^{-1} of an open set in X is open in Y we need to prove that an open set $V \subset X$ is

55

mapped onto an open set $f(V)$ in X. Since $V^\complement$ is closed and X is compact, $V^\complement$ is compact in X, hence $f(V^\complement)$ is compact in Y, thus $f(V^\complement)$ is closed in Y since compact sets are closed. The bijectivity of f implies further $f(V) = f(V^\complement)^\complement$, i.e. $f(V)$ is open. $\qquad\square$

Definition 3.39. *A mapping $f : X \to Y$ between two metric spaces (X, d_X) and (Y, d_Y) is called **open** or an **open mapping** if it maps open sets in X to open sets in Y.*

In general continuous mappings are not open. Just consider the mapping $h : \mathbb{R} \to \mathbb{R}$, $h(x) = 1$ for all $x \in \mathbb{R}$. However the proof of Theorem 3.38 yields

Corollary 3.40. *A continuous, open and bijective mapping between two metric spaces has a continuous inverse.*

Finally we study uniformly continuous mappings.

Definition 3.41. *Let (X, d_X) and (Y, d_Y) be two metric spaces. We call $f : X \to Y$ **uniformly continuous** if for every $\epsilon > 0$ there exists $\delta > 0$ such that*

$$d_X(x, x') < \delta \quad \text{implies} \quad d_Y(f(x), f(x')) < \epsilon.$$

Theorem 3.42. *Let (X, d_X) and (Y, d_Y) be two metric spaces and $f : X \to Y$ a continuous mapping. If X is compact, then f is uniformly continuous.*

Proof. Given $\epsilon > 0$. Then for every $\xi \in X$ there is $r(\xi) > 0$ such that

$$d_Y(f(\eta), f(\xi)) < \frac{\epsilon}{2} \quad \text{for all} \quad \eta \in B_{r(\xi)}(\xi) \subset X.$$

Since $\bigcup_{\xi \in X} B_{\frac{1}{2}r(\xi)}(\xi) = X$ and X is compact, there are points $\xi_1, \ldots, \xi_k \in X$ such that $\bigcup_{j=1}^{k} B_{\frac{1}{2}r(\xi_j)}(\xi_j) = X$. Let $\delta := \frac{1}{2}\min\{r(\xi_1), \ldots, r(\xi_k)\}$. For $x, x' \in X$ and $d_X(x, x') < \delta$ we find $j \in \{1, \ldots, k\}$ such that $x' \in B_{\frac{1}{2}r(\xi_j)}(\xi_j)$ and $x \in B_{r(\xi_j)}(\xi_j)$. Thus it follows that

$$d_Y(f(x), f(\xi_j)) < \frac{\epsilon}{2} \quad \text{and} \quad d_Y(f(x'), f(\xi_j)) < \frac{\epsilon}{2}$$

implying

$$d_Y(f(x), f(x')) < \epsilon \quad \text{for} \quad d(x, x') < \delta$$

and the theorem is proved. $\qquad\square$

Remark 3.43. In order to study the following chapters the reader will need 3.1 - 3.4, 3.12 - 3.14 and 3.20 - 3.32. However, we recommend that the student works through the entire chapter in a second reading.

Problems

1. Let (X, d) be a metric space and $(x_n)_{n \in \mathbb{N}}$, $(y_n)_{n \in \mathbb{N}}$ be two sequences. Define the new sequence $(z_n)_{n \in \mathbb{N}}$ by $z_{2n-1} := x_n$ and $z_{2n} := y_n$. Prove that $(z_n)_{n \in \mathbb{N}}$ converges to $z \in X$ if and only if $(x_n)_{n \in \mathbb{N}}$ and $(y_n)_{n \in \mathbb{N}}$ converges to z.

2. Let $(x_n)_{n \in \mathbb{N}}$ be a Cauchy sequence in the metric space (X, d). Suppose that a subsequence $(x_{n_k})_{k \in \mathbb{N}}$ of $(x_n)_{n \in \mathbb{N}}$ converges to x. Prove that $(x_n)_{n \in \mathbb{N}}$ converges to x.

3. Prove that for a set $Y \subset X$, (X, d) being a metric space, the closure $\bar{Y}$ is given by $\bar{Y} = \{x \in X | \operatorname{dist}(x, y) = 0\}$.

4. Let $K_j, j = 1, \ldots, N$ be compact sets in the metric space (X, d). Prove that $\bigcup_{j=1}^{N} K_j$ is compact too, while a countable union of compact sets is not necessarily compact.

5. Let (X, d) be a metric space and $Y \subset X$ a dense subset. If $U \subset X$ is open then $U \subset \overline{Y \cap U}$.

6. Prove that in general a bounded set in a metric space is not totally bounded.

7. Give an example of a continuous mapping $f : X \to Y$ between two metric spaces (X, d_X) and (Y, d_Y) with $f(Y)$ being compact but X being not compact.

8. Let $K_1 \subset \mathbb{R}^n$ and $K_2 \subset \mathbb{R}^m$ be two compact sets and $f : K_1 \times K_2 \to \mathbb{R}$ be continuous. Define $g : K_1 \to \mathbb{R}$ by $g(x) := \sup_{y \in K_2} f(x, y)$. Prove that g is continuous.

9. Let (X, d_X) and (Y, d_Y) be two metric spaces and $f : X \to Y$ a mapping. Show that f is continuous if and only if for every compact set $K \subset X$ the mapping $f|_K$ is continuous.

10. For $1 \leq j \leq n$ let $\operatorname{pr}_j : \mathbb{R}^n \to \mathbb{R}$, $\operatorname{pr}_j(x) = x_j$, be the jth coordinate projection. Prove that pr_j is open.

11. Let $C \subset \mathbb{R}^n$ be a set with the property that $x, y \in C$ implies that the line segment connecting x and y also belongs to C, i.e. $x, y \in C$ implies

$\lambda x + (1 - \lambda)y \in C$ for $\lambda \in [0, 1]$. Such a set is called convex. Prove that C is pathwise connected, hence connected.

12.   a) We call a set $S \subset \mathbb{R}^n$ a **chain of line segments** if for some $N \in \mathbb{N}$ with $a_j \in \mathbb{R}^n, j \in \mathbb{N}_0$, we have

$$S = S_{a_0,a_1} \cup S_{a_1,a_2} \cup \cdots \cup S_{a_{N-1},a_N},$$

where $S_{a_j,a_{j+1}}$ denotes the line segment connecting a_j with a_{j+1}. Prove that a chain of line segments is connected.

   b) We call $S \subset \mathbb{R}^n$ star-shaped if for some $x_0 \in S$ we have $S = \bigcup_{x \in S} S_{x,x_0}$. Prove that a star-shaped set is connected.

13. Prove that a metric space (X, d) is connected if and only if every non-empty open set $U \neq X$ has a non-empty boundary ∂U.

14. Give a proof that in a metric space (X, d) the connectivity component of $x \in X$ is closed.

15. Prove **Dini's Theorem**: let $(f_n)_{n\in\mathbb{N}}$ be a sequence of real-valued functions $f_n : K \to \mathbb{R}$ where (K, d) is a compact metric space. Suppose that $f_n(x) \leq f_{n+1}(x)$ for all $x \in K$ and $n \in \mathbb{N}$ and that $f(x) := \lim_{n\to\infty} f_n(x)$ exists and defines a real-valued continuous function on K. Prove that $(f_n)_{n\in\mathbb{N}}$ converges uniformly on K to f.

16. Consider the sequence $p_n : [0, 1] \to \mathbb{R}$ of polynomials defined by $p_0(x) = 0$, $p_{n+1}(x) := p_n(x) + \frac{1}{2}(x - p_n(x)^2)$. Prove that for all $x \in [0, 1]$ we have $p_n(x) \leq p_{n+1}(x) \leq \sqrt{x} \leq 1$ and that $\lim_{n\to\infty} p_n(x) = \sqrt{x}, x \in [0, 1]$, holds as a pointwise limit. Deduce that $(p_n)_{n\in\mathbb{N}}$ is a sequence of polynomials converging uniformly on $[0, 1]$ to $x \mapsto \sqrt{x}$.

4 Continuous Mappings Between Subsets of Euclidean Spaces

In this chapter we want to study continuous mappings $f : G \to \mathbb{R}^m$, where $G \subset \mathbb{R}^n$ is some subset, in more detail. We will always use in the domain and co-domain of f the Euclidean metric and $\|.\|$ will denote the Euclidean norm in $\mathbb{R}^k$, $k \geq 2$, i.e. we do not introduce a notational distinction between the norm in $\mathbb{R}^m$ and $\mathbb{R}^n$, $m \neq n$. However in $\mathbb{R}$, i.e. $k = 1$, we use $|.|$, the absolute value, instead of $\|.\|$. A similar notational simplification will be applied to the canonical basis in $\mathbb{R}^k$ which we denote by $\{e_1, \ldots, e_k\}$. Only in cases where we have a clear need to make a distinction between the canonical basis in the domain and in the co-domain, we use $\{e_1, \ldots, e_n\}$ and $\{e'_1, \ldots, e'_m\}$, respectively. A vector in $\mathbb{R}^k$ has components $x_j \in \mathbb{R}$, $1 \leq j \leq k$, and we write $x = (x_1, \ldots, x_k)$. However, for $k = 2$ or $k = 3$ it is sometimes more convenient to write $(x, y) \in \mathbb{R}^2$ or $(x, y, z) \in \mathbb{R}^3$.

Given $f : G \to \mathbb{R}^m$, $G \subset \mathbb{R}^n$, then $f(x) \in \mathbb{R}^m$, hence components $f_j : G \to \mathbb{R}$, $1 \leq j \leq m$, are defined. The following holds

$$f(x) = (f_1(x), \ldots, f_n(x)) = \sum_{j=1}^{m} f_j(x)e_j, \tag{4.1}$$

and with the projection $\mathrm{pr}_j : \mathbb{R}^m \to \mathbb{R}$, $x = (x_1, \ldots, x_m) \mapsto \mathrm{pr}_j(x) = x_j$, we find

$$f_j = \mathrm{pr}_j \circ f. \tag{4.2}$$

Note that we will often write $f(x_1, \ldots, x_n)$ for $f(x) = f((x_1, \ldots, x_n))$.

As long as we do not use matrix notation we do not distinguish between row and column vectors, i.e. we write $(x_1, \ldots, x_k)$ as well as $\begin{pmatrix} x_1 \\ \vdots \\ x_k \end{pmatrix}$ for $x \in \mathbb{R}^k$.

Later on in the chapters on vector calculus and differential geometry we will be more careful.

On $G \subset \mathbb{R}^n$ we use the relative topology of the Euclidean topology in $\mathbb{R}^n$. This has the consequence that sometimes we must make a clear and careful distinction between G as a closed and open set in the metric space (G, d), where $d(x, y) = \|x - y\|$, and $G \subset \mathbb{R}^n$ which is not necessarily open nor closed.

From Theorem 2.19 we know that f is continuous if and only if its components are continuous. A different problem is whether the continuity of the functions $z_j \mapsto f_l(x_1, \ldots, x_{j-1}, z_j, x_{j+1}, \ldots, x_n)$ will imply the continuity of $z = (z_1, \ldots, z_n) \mapsto (f_1(z), \ldots, f_m(z))$. Since $f : G \to \mathbb{R}^m$, $G \subset \mathbb{R}^n$, is determined by its components $f_j : G \to \mathbb{R}$, $j = 1, \ldots, m$, we start our investigations with the study of functions $f : G \to \mathbb{R}$, $G \subset \mathbb{R}^n$. If $G \subset \mathbb{R}^n$ is an open set in $\mathbb{R}^n$ and $x \in G$ we can find $\rho > 0$ such that $B_\rho(x) \subset G$. Furthermore, since

$$|y_j - x_j| \leq \left(\sum_{l=1}^{n} |y_l - x_l|^2 \right)^{\frac{1}{2}} \tag{4.3}$$

the sets (see Figure 4.1)

$$I_{j,x} := \left\{ y \in \mathbb{R}^n \,\middle|\, y = (x_1, \ldots, x_{j-1}, y_j, x_{j+1}, \ldots, x_n), \; |y_j - x_j| < \rho \right\} \tag{4.4}$$

are for $1 \leq j \leq n$ subsets of $B_\rho(x)$, hence $I_{j,x} \subset G$, and the functions

$$J_j : (x_j - \rho, x_j + \rho) \to I_{j,x}, \; z \mapsto (x_1, \ldots, x_{j-1}, z, x_{j+1}, \ldots, x_n) \tag{4.5}$$

are bijective mappings which induce the real-valued functions

$$g_{j,x} : f \circ J_j : (x_j - \rho, x_j + \rho) \to \mathbb{R}. \tag{4.6}$$

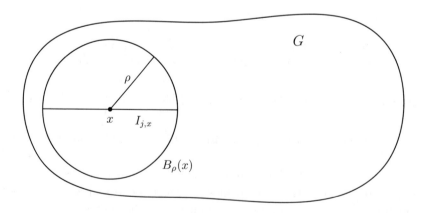

Figure 4.1

60

If f is continuous at $x \in G$ then for $1 \leq j \leq n$ the functions $g_{j,x}$ are continuous in x_j which follows from (4.3): Given $\epsilon > 0$ we can find $0 < \delta < \rho$ such that $\|y - x\| < \delta$ implies $|f(y) - f(x)| < \epsilon$, hence $z \in (x_j - \delta, x_j + \delta)$ implies $|g_{j,x}(z) - g_{j,x_j}(x_j)| < \epsilon$, since $g_{j,x}(z) = f(x_1, \ldots, x_{j-1}, z, x_{j+1}, \ldots x_n)$ and $g_j(x_j) = f(x)$.

A more informal way for saying this is that $z \mapsto f(x_1, \ldots, x_{j-1}, z, x_{j+1}, \ldots, x_n)$ is continuous at x_j. Some authors call a mapping $h : G \to \mathbb{R}$ **partially** (or in direction e_j) **continuous** at $x \in G$ if $z \mapsto h(x_1, \ldots, x_{j-1}, z, x_{j+1}, \ldots, x_n)$ is continuous at x_j.

It is of utmost importance to note that if for all $j = 1, \ldots, n$ the mappings $g_{j,x}$ are continuous at $x_j \in I_{j,x}$, i.e. for all $j = 1, \ldots, n$ the functions $z_j \mapsto f(x_1, \ldots, x_{j-1}, z_j, x_{j+1}, \ldots, x_n)$ are continuous at x_j, then $f : G \to \mathbb{R}$ is **not** necessarily continuous at $x \in G$.

Example 4.1. Consider $f : \mathbb{R}^2 \to \mathbb{R}$ given by

$$f(x, y) = \begin{cases} \frac{xy}{x^2+y^2}, & (x,y) \neq (0,0) \\ 0, & (x,y) = (0,0). \end{cases} \tag{4.7}$$

First we prove that f is not continuous at $(0,0)$. The sequence $\left(\left(\frac{1}{k}, \frac{1}{k}\right)\right)_{k \in \mathbb{N}}$ converges in $\mathbb{R}^2$ to $(0,0)$ however

$$f\left(\frac{1}{k}, \frac{1}{k}\right) = \frac{\frac{1}{k} \cdot \frac{1}{k}}{\frac{1}{k^2} + \frac{1}{k^2}} = \frac{1}{2},$$

and consequently $\lim_{k \to \infty} f\left(\frac{1}{k}, \frac{1}{k}\right) = \lim_{k \to \infty} \frac{1}{2} = \frac{1}{2} \neq 0$. Thus by Definition 2.17 the function f is not continuous at $(0,0)$. On the other hand we find

$$g_{1,(0,0)}(z) = \begin{cases} \frac{z \cdot 0}{z^2 + 0^2} = 0, & \text{for } z \neq 0 \\ 0, & \text{for } z = 0, \end{cases}$$

i.e. $g_{1,(0,0)}(z) = 0$ for all $z \in \mathbb{R}$ and hence it is continuous at $(0,0)$. Analogously we find

$$g_{2,(0,0)}(z) = \begin{cases} \frac{0 \cdot z}{0^2 + z^2} = 0, & \text{for } z \neq 0 \\ 0, & \text{for } z = 0, \end{cases}$$

i.e. $g_{1,(0,0)}$ and $g_{2,(0,0)}$ are both continuous on $\mathbb{R}$, in particuluar at $z = 0$, but f is not continuous at $(0,0)$.

In order to understand this example better we introduce continuity of a function $f : G \to \mathbb{R}$ with respect to or in a given direction. Recall that for a unit vector $v \in \mathbb{R}^n$, $\|v\| = 1$, the set

$$\gamma_{v,x,\rho} := \{y \in \mathbb{R}^n \mid y = tv + x, \ |t| < \rho\} \tag{4.8}$$

is a line segment passing through x in the direction of v and length 2ρ:

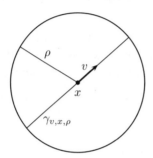

Figure 4.2

For $G \subset \mathbb{R}^n$ open, $x \in G$ and $B_\rho(x) \subset G$ every straight line segment $\gamma_{v,x,\rho}$ belongs to G since $\gamma_{v,x,\rho} \subset B_\rho(x) \subset G$. For $y \in \gamma_{v,x,\rho}$ we have a representation $y = t_0 v + x$ for some t_0, $|t_0| < \rho$. Thus we can define for $f : G \to \mathbb{R}$ the function

$$h_{\gamma_{v,x,\rho}} : (-\rho, \rho) \to \mathbb{R}, \ h_{\gamma_{v,x,\rho}}(t) = f(tv + x), \tag{4.9}$$

note that $h_{\gamma_{v,x,\rho}}(0) = f(x)$.

Definition 4.2. *We call $f : G \to \mathbb{R}$, $G \subset \mathbb{R}^n$ open, **continuous** at $x \in G$ in the direction of v, $\|v\| = 1$, if $h_{\gamma_{v,x,\rho}}$ is continuous at 0.*

Remark 4.3. Note that the continuity of $h_{\gamma_{v,x,\rho}}$ at 0 does not depend on ρ.

Corollary 4.4. *If $f : G \to \mathbb{R}$, $G \subset \mathbb{R}^n$ open, is continuous at $x \in G$ then for every direction $v \in \mathbb{R}^n$, $\|v\| = 1$, f is continuous at x in the direction v.*

Proof. Let $x \in G$, $B_\rho(x) \subset G$ and $v \in \mathbb{R}^n$, $\|v\| = 1$. For $y \in \gamma_{v,x,\rho}$, $y = tv + x$, $|t| < \rho$, it follows that $|t| = \|x - y\|$. Hence, given $\epsilon > 0$ we can find $0 < \delta < \rho$ such that for $y \in G$ it follows that $\|y - x\| < \delta$ implies $|f(y) - f(x)| < \epsilon$, and in particular, for $y \in \gamma_{v,x,\rho}$ and $|t| < \delta$ it follows $|h_{\gamma_{v,x,\rho}}(t) - h_{\gamma_{v,x,\rho}}(0)| < \epsilon$. $\square$

It is at first glance maybe surprising that even if $f : G \to \mathbb{R}$ is continuous at $x \in G$ for all directions $v \in \mathbb{R}^n$, $\|v\| = 1$, then f is still not necessarily continuous at x.

Example 4.5. Consider in $\mathbb{R}^2$ the function

$$f(x, y) = \begin{cases} 0, & x \neq y^2 \\ 0, & x = y = 0 \\ 1, & x = y^2,\ x > 0. \end{cases} \tag{4.10}$$

In order to understand the behaviour of f at $(0, 0)$ we look at Figure 4.3

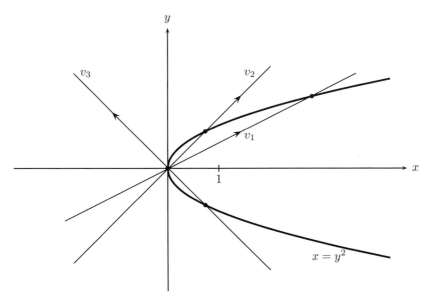

Figure 4.3

Every straight line passing through $(0, 0)$ intersects the parabola $\{(x, y) \mid x = y^2\}$ at exactly one point. Therefore f restricted to such a line is zero everywhere except at the intersection point with the parabola where the value is 1. The intersection of $\gamma_{v,(0,0),\rho}$ with the parabola $\{(x, y) \mid x = y^2\}$ has a strict positive distance to $(0, 0)$: if $v = (v_1, v_2)$ then the intersection point is $\left(\frac{v_1^2}{v_2^2}, \frac{v_1}{v_2}\right)$ and $\left\| (0, 0) - \left(\frac{v_1^2}{v_2^2}, \frac{v_1}{v_2}\right) \right\| = \frac{|v_1|}{|v_2|} \sqrt{1 + \frac{v_1^2}{v_2^2}} =: \rho_v > 0$, where we used that $v_1 \neq 0$. Thus $f|_{\gamma_{v,(0,0)},\rho_v} = 0$ and therefore for every v, $v_1 \neq 0$, the function f is continuous at $(0,0)$ in the direction of v, and the continuity of f at $(0,0)$ in the direction of $\pm e_2$ is trivial. However for the sequence $\left(\frac{1}{k^2}, \frac{1}{k}\right)_{k\in\mathbb{N}}$ we have $\lim_{k\to\infty} \left(\frac{1}{k^2}, \frac{1}{k}\right) = (0,0)$ but $f\left(\left(\frac{1}{k^2}, \frac{1}{k}\right)\right) = 1$, hence $\lim_{k\to\infty} f\left(\left(\frac{1}{k^2}, \frac{1}{k}\right)\right) = 1 \neq f(0,0)$ and f is not continuous at $(0,0)$.

We want to explore this situation in more detail. Let $\{b^{(1)}, \ldots, b^{(n)}\}$ be a basis in $\mathbb{R}^n$ such that for the canonical basis we have

$$e_j = \sum_{l=1}^{n} a_{jl} b^{(l)}$$

with invertible matrix $A = (a_{jl})_{j,l=1,\ldots,n}$. For $x \in \mathbb{R}^n$ we find

$$x = \sum_{j=1}^{n} x^{(j)} e_j = \sum_{j,l=1}^{n} a_{jl} x^{(j)} b^{(l)} = \sum_{l=1}^{n} \left(\sum_{j=1}^{n} a_{jl} x^{(j)}\right) b^{(l)}$$

$$= \sum_{l=1}^{n} \beta^l(x) b^{(l)}$$

and it follows that $x \mapsto \beta^l(x) = \sum_{j=1}^{n} a_{jl} x^{(j)}$ is continuous. Consequently, with Definition 2.17 in mind, we find

Lemma 4.6. *A sequence $(x_k)_{k\in\mathbb{N}}$ in $\mathbb{R}^n$ converges to $x \in \mathbb{R}^n$ if for every basis $\{b^{(1)}, \ldots, b^{(n)}\}$ of $\mathbb{R}^n$ the sequences $\left(\beta^{(l)}(x_k)\right)_{k\in\mathbb{N}}$, $l = 1, \ldots, n$, converges to $\beta^{(l)}(x)$.*

Let $f : G \to \mathbb{R}$, $G \subset \mathbb{R}^n$ open, be continuous at $x \in G$. On $G - \{x\} := \{y \in \mathbb{R}^n \mid y = z - x, z \in G\}$ we define $h(z) := f(z + x)$ implying that h is continuous at $0 \in G - \{x\}$ if and only if f is continuous at $x \in G$. Thus, instead of investigating the continuity of f at x we may investigate the continuity of h at 0.

Proposition 4.7. *Let $h : G \to \mathbb{R}$, $0 \in G$, G open in $\mathbb{R}^n$. Suppose that $\{v^{(1)}, \ldots, v^{(n)}\}$ is an independent family of directions in $\mathbb{R}^n$, i.e. $\{v^{(1)}, \ldots, v^{(n)}\}$ is a basis in $\mathbb{R}^n$ with all basis vectors of length 1. Let $\rho > 0$ be such that $B_\rho(0) \subset G$ and suppose that $h|_{\gamma_{v^{(j)},0,\rho}}$, $j = 1, \ldots, n$, is Lipschitz continuous in the sense that for all $y, z \in \gamma_{v^{(j)},0,\rho}$ the following holds*

$$\left| h|_{\gamma_{v^{(j)},0,\rho}}(y) - h|_{\gamma_{v^{(j)},0,\rho}}(z) \right| \leq L \|y - z\|.$$

Then h is continuous at 0.

Proof. We need to estimate $|h(y) - h(0)|$ for any $y \in \mathbb{R}^n$, $\|y\|$ sufficiently small. Since $\{v^{(1)}, \ldots, v^{(n)}\}$ is a basis we have

$$y = \sum_{j=0}^{n} \lambda_j(y) v^{(j)}$$

and by Lemma 4.6 we know that $\lim_{y \to \infty} \lambda_j(y) = 0$. Since

$$h(y) - h(0) = h \left(\sum_{l=1}^{n} \lambda_l v^{(l)} \right) - h(0)$$

$$= \sum_{j=1}^{n} \left(h \left(\sum_{l=j}^{n} \lambda_l(y) v^{(l)} \right) - h \left(\sum_{l=j+1}^{n} \lambda_l(y) v^{(l)} \right) \right)$$

with the convention that $\sum_{l=n+1}^{n} \lambda_l(y) v^{(l)} = 0$, we find

$$|h(y) - h(0)| \leq \sum_{j=1}^{n} \left| h \left(\sum_{l=j}^{n} \lambda_l(y) v^{(l)} \right) - h \left(\sum_{l=j+1}^{n} \lambda_l(y) v^{(l)} \right) \right|$$

$$\leq \sum_{j=1}^{n} L |\lambda_j(y)|.$$

Thus, with our previous remark in mind, given $\epsilon > 0$ we can find $\delta > 0$ such that $\|y\| < \delta$ implies $|\lambda_j(y)| < \frac{\epsilon}{nL}$ implying the result. $\square$

Corollary 4.8. *If* $f : G \to \mathbb{R}$, $G \subset \mathbb{R}^n$ *open, has the property that for* $1 \leq j \leq n$ *the functions* $z_j \mapsto f(x_1, \ldots, x_{j-1}, z_j, x_{j+1}, \ldots, x_n)$ *are Lipschitz continuous on some interval* $(x_j - \rho_j, x_j + \rho_j)$, $\rho_j > 0$, *then* f *is continuous at* $x \in G$.

Consider the function $g : \mathbb{R}^2 \to \mathbb{R}$ defined by

$$g(x, y) := \begin{cases} \frac{x^2 - y^2}{x^2 + y^2}, & (x, y) \neq (0, 0) \\ 0, & (x, y) = (0, 0). \end{cases} \tag{4.11}$$

For $x \in \mathbb{R}$, $x \neq 0$, fixed we find

$$\lim_{\substack{y \to 0 \\ y \neq 0}} g(x, y) = \frac{x^2}{x^2} = 1$$

and therefore

$$\lim_{x \to 0} \left(\lim_{y \to 0} g(x, y) \right) = \lim_{x \to 0} 1 = 1.$$

On the other hand we have for $y \in \mathbb{R}$, $y \neq 0$, fixed

$$\lim_{\substack{x \to 0 \\ x \neq 0}} g(x, y) = \frac{-y^2}{y^2} = -1$$

implying

$$\lim_{y \to 0} \left(\lim_{x \to 0} g(x, y) \right) = \lim_{y \to 0} (-1) = -1.$$

Hence in general we have

$$\lim_{x \to 0} \left(\lim_{y \to 0} g(x, y) \right) \neq \lim_{y \to 0} \left(\lim_{x \to 0} g(x, y) \right), \qquad (4.12)$$

and consequently g cannot be continuous at $(0, 0)$.

Let $f : G \to \mathbb{R}$, $G \subset \mathbb{R}^2$, be a function and $(x_0, y_0) \in G$. We call

$$\lim_{y \to y_0} \left(\lim_{x \to x_0} f(x, y) \right) \text{ and } \lim_{x \to x_0} \left(\lim_{y \to y_0} f(x, y) \right) \qquad (4.13)$$

the **iterated limits** of f at (x_0, y_0). Our previous consideration shows that both iterated limits may exist but they need not coincide. Of course we have

Corollary 4.9. *If $f : G \to \mathbb{R}$, $G \subset \mathbb{R}^2$ is continuous at $(x_0, y_0) \in G$ then both iterated limits at (x_0, y_0) exist and coincide.*

Note that even in the case where both iterated limits exist at (x_0, y_0) and coincide, by Example 4.5 we know that f is not necessarily continuous at (x_0, y_0). It is obvious how to extend the notion of iterated limits to more than two variables and that Corollary 4.9 holds in higher dimensions.

We want to formulate some results from Chapter 3 for functions $f : G \to \mathbb{R}$, $G \subset \mathbb{R}^n$.

Proposition 4.10. A. *Let $f : G \to \mathbb{R}$ be a continuous function on the compact set $G \subset \mathbb{R}^n$. Then f attains its infimum as well as its supremum, i.e. there are points $x_{\min}, x_{\max} \in G$ such that*

$$f(x_{\min}) = \inf \{ f(x) \, | \, x \in G \} \text{ and } f(x_{\max}) = \sup \{ f(x) \, | \, x \in G \}. \qquad (4.14)$$

Furthermore $\text{ran}(f)$ *is a compact subset of* $[f(x_{\min}), f(x_{\max})]$ *and* f *is uniformly continuous.* **B.** *If* $G \subset \mathbb{R}^n$ *is a connected set and* $f : G \to \mathbb{R}$ *is continuous then* $f(G) \subset \mathbb{R}$ *is connected, i.e. an interval which in the case of a compact set* G *is* $[f(x_{\min}), f(x_{\max})]$.

Proof. For part A compare with Theorem 3.26 and Theorem 3.27, as well as Theorem 3.42, while part B follows from Theorem 3.32 (and part A). □

We now turn to mappings $f : I \to \mathbb{R}^n$ where I is an interval with end points $a < b$. Thus f is given by its components $f_j : I \to \mathbb{R}$, $j = 1, \ldots, n$, and we may write $f(t) = (f_1(t), \ldots, f_n(t))$. Clearly, f is continuous (at t_0) if and only if all components f_j are continuous (at t_0), see Theorem 2.19, and since $I \subset \mathbb{R}$ is connected, if f is continuous then $f(I) \subset \mathbb{R}^n$ is connected too. Moreover, if I is compact f is uniformly continuous as well as all its components f_j and each component is a bounded function as is f. Note that boundedness of f means that for some $R > 0$ we have

$$\{f(t) \,|\, t \in I\} \subset \overline{B_R(0)} \tag{4.15}$$

or equivalently

$$f_1^2(t) + \cdots + f_n(t)^2 \leq R^2, \tag{4.16}$$

for all $t \in I$. Using (1.15) we can replace (4.15) or (4.16) by

$$\|f(t)\|_p \leq R_p \tag{4.17}$$

for some $R_p > 0$.
From the geometrical (or topological) point of view mappings $\gamma : I \to \mathbb{R}^n$, $I \subset \mathbb{R}$ being an interval, lead to some new tools and observations.

Definition 4.11. *A continuous mapping* $\gamma : [a, b] \to G$, $G \subset \mathbb{R}^n$ *and* $a < b$, *is called a* **path** *or an* **arc** *in* G *with* **initial point** $\gamma(a) \in G$, **terminal point** *or* **end point** $\gamma(b) \in G$ *and* **parameter interval** $[a, b]$. *By definition* $\text{tr}(\gamma)$, *the* **trace** *of* γ *is the range* $\text{ran}(\gamma) \subset G$. *If* $q, p \in G$ *we say that a path* $\gamma : [a, b] \to G$ **connects** q *and* p *if* $\{q, p\} = \{\gamma(a), \gamma(b)\}$, *i.e.* q *is either the initial or terminal point of* γ *and then* p *is the terminal or the initial point of* γ, *respectively.*

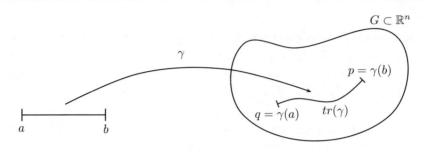

$G \subset \mathbb{R}^n$

γ

$p = \gamma(b)$

$q = \gamma(a)$ $tr(\gamma)$

a b

Figure 4.4

Remark 4.12. We will mainly use the word *path*, but in the case of $n = 2$ or in the complex plane it is more common to speak of an *arc*. Later on, when adding differentiability we will use the name *curve*.

Example 4.13. A. For $n \geq 2$ the path $s : [0, T] \to \mathbb{R}^n$, $s(t) = vt + q$, has as trace the line segment connecting $q = s(0)$ with $p = s(T) = vT + q$, see Figure 4.5. **B.** The mapping $\gamma : [0, 2\pi] \to \mathbb{R}^2$, $\gamma(t) = (\cos(t), \sin(t))$, is a path the trace of which is the unit circle. Here the initial point $\gamma(0) = (1, 0)$ is equal to the terminal point $\gamma(2\pi) = (1, 0)$, see Figure 4.6. **C.** If $\gamma : [a, b] \to G$ is a path with initial point $q = \gamma(a)$ and terminal point $\gamma(b) = p$. The path $\tilde{\gamma} : [a, b] \to G$ defined by $\tilde{\gamma}(t) = \gamma(-t + b + a)$ has the same trace as γ but initial point $p = \tilde{\gamma}(a) = \gamma(b)$ and terminal point $q = \tilde{\gamma}(b) = \gamma(a)$, see Figure 4.7.

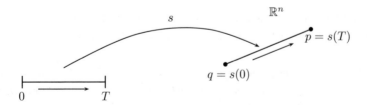

s

$\mathbb{R}^n$

$p = s(T)$

$q = s(0)$

0 T

Figure 4.5

68

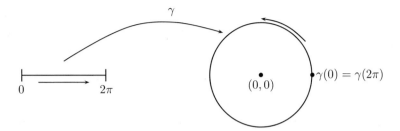

Figure 4.6

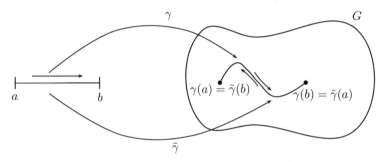

Figure 4.7

These three examples already show some possible features a path may have. Let us start with Example 4.13.C. It may happen that two paths have identical traces but different **orientation**, initial and end points are reversed. Further, Example 4.13.B leads to

Definition 4.14. *A path* $\gamma : [a, b] \to G$, $G \subset \mathbb{R}^n$, *is called* ***closed*** *if* $\gamma(a) = \gamma(b)$. *If in addition* $\gamma|_{[a,b)}$ *is injective we call* γ *a* ***simple closed*** *path.*

The circle, considered as the trace of γ as in Example 4.13.B is a simply connected path, while the circle as the trace of $\eta : [0, 4\pi] \to \mathbb{R}^2$, $\eta(t) = (\cos(t), \sin(t))$ is closed but not simply closed since $\eta|_{[0,4\pi)}$ is not injective: for $0 \le t < 2\pi$ we have $t + 2\pi \in [0, 4\pi)$ and $(\cos t, \sin t) = (\cos(t+2\pi), \sin(t+2\pi))$. Note that the path $\tilde{\gamma} : [0, 2\pi] \to \mathbb{R}^2$, $\tilde{\gamma}(t) = (-\cos t, -\sin t) = (\cos t, \sin(-t))$ is again a simply closed path the trace of which is the unit circle. Since initial and end points are identical in this situation our preliminary version of a definition of orientation does not allow us to distinguish the orientation of γ in Example 4.13.B and $\tilde{\gamma}$. Thus we will have to provide a refinement.

Let $[a, b]$ and $[\alpha, \beta]$ be two non-trivial closed intervals. The mapping $\lambda :$ $[a, b] \to [\alpha, \beta]$, $\lambda(t) = \frac{\alpha - \beta}{a - b} t + \frac{a\beta - b\alpha}{a - b}$, is bijective with $\lambda(a) = \alpha$ and $\lambda(b) = \beta$, hence for a path $\gamma : [\alpha, \beta] \to G$ we find that $\gamma \circ \lambda : [a, b] \to G$ is again a path with $\text{tr}(\gamma) = \text{tr}(\gamma \circ \lambda)$ and $\gamma(\alpha) = (\gamma \circ \lambda)(a)$ and $\gamma(\beta) = (\gamma \circ \lambda)(b)$. Therefore both paths connect the initial point $\gamma(\alpha) = (\gamma \circ \lambda)(a)$ with the terminal point $\gamma(\beta) = (\gamma \circ \lambda)(b)$ and they have an identical trace, hence in G we cannot distinguish between them. In particular, for every path $\gamma : [\alpha, \beta] \to G$ we can find a **linear parameter transformation** $\lambda : [0, 1] \to [\alpha, \beta]$ such that $\tilde{\gamma} := \gamma \circ \lambda : [0, 1] \to G$ has the same trace as γ and initial and end points are preserved, i.e. the parameter transformations preserves the orientation. The notion of a parameter transformation is not restricted to linear functions.

Definition 4.15. *Let $\gamma : [\alpha, \beta] \to G$, $G \subset \mathbb{R}^n$, be a path and $\varphi : [a, b] \to [\alpha, \beta]$ a strictly increasing continuous function with $\varphi(a) = \alpha$ and $\varphi(b) = \beta$. Then $\gamma \circ \varphi : [a, b] \to G$ is a path with $\text{tr}(\gamma) = \text{tr}(\gamma \circ \varphi)$ and the initial and end points of γ and $\gamma \circ \varphi$ coincide. We call φ a **parameter transformation** for γ.*

Given $G \subset \mathbb{R}^n$ we may ask whether it is possible to connect $q \in G$ with $p \in G$ by a path γ such that $\text{tr}\,\gamma \subset G$. Of course, in general we do not expect this. Just take $n = 1$ and $G = (-1, 0) \cup (1, 2)$. A path connecting $-\frac{1}{2} \in G$ with $\frac{3}{2} \in G$ must be a continuous mapping γ defined on some interval $[a, b]$ such that $\gamma(a) = -\frac{1}{2}$ and $\gamma(b) = \frac{3}{2}$ (or $\gamma(a) = \frac{3}{2}$ and $\gamma(b) = -\frac{1}{2}$). Thus $\text{tr}(\gamma)$ must contain the interval $[-\frac{1}{2}, \frac{3}{2}]$ since it must map the connected set $[a, b]$ onto a connected set in $\mathbb{R}$. However $[-\frac{1}{2}, \frac{3}{2}]$ is not a subset of G, i.e. $\gamma([a, b])$ is not contained in G.

Definition 4.16. *We call a subset $G \subset \mathbb{R}^n$ **pathwise** or **arcwise connected** if every pair of points $q, p \in G$ can be connected by a path, i.e. there exists a continuous mapping $\gamma : [a, b] \to G$ such that $\gamma(a) = q$ and $\gamma(b) = p$.*

In order to further study the relationship between connected and pathwise connected sets we need to introduce the sum of two paths.

Definition 4.17. *Let $\gamma_1 : [a_1, b_1] \to G$ and $\gamma_2 : [a_2, b_2] \to G$, $G \subset \mathbb{R}^n$, be two paths such that the terminal point of γ_1 is the initial point of γ_2, i.e. $\gamma_1(b_1) = \gamma_2(a_2)$. Let $\varphi_1 : [0, \frac{1}{2}] \to [a_1, b_1]$ and $\varphi_2 : [\frac{1}{2}, 1] \to [a_2, b_2]$ be a parameter transformation of γ_1 and γ_2 respectively. We define the **sum** of*

γ_1 *and* γ_2 *as the path* $\gamma := \gamma_1 \oplus \gamma_2 : [0, 1] \to G$ *where*

$$(\gamma_1 \oplus \gamma_2)(t) := \begin{cases} (\gamma_1 \circ \varphi_1)(t), & t \in [0, \tfrac{1}{2}] \\ (\gamma_2 \circ \varphi_2)(t), & t \in [\tfrac{1}{2}, 1]. \end{cases} \tag{4.18}$$

Clearly $\gamma_1 \oplus \gamma_2$ is continuous and we have $\mathrm{tr}(\gamma_1 \oplus \gamma_2) = \mathrm{tr}(\gamma_1) \cup \mathrm{tr}(\gamma_2)$. Moreover the initial point of $\gamma_1 \oplus \gamma_2$ is $\gamma_1(a_1)$ whereas its terminal point is $\gamma_2(b)$.

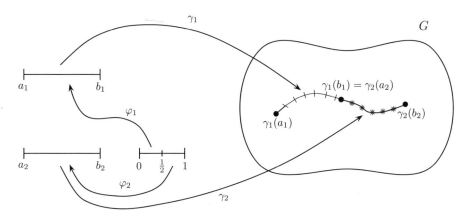

Figure 4.8

Proposition 4.18. *Every pathwise connected set* $G \subset \mathbb{R}^n$ *is connected.*

Proof. Suppose that $\{U_1, U_2\}$ is a splitting of G. Let $q \in U_1$ and $p \in U_2$ and let $\gamma : [a, b] \to G$ be a path such that $\gamma(a) = q$ and $\gamma(b) = p$. Then $\{\gamma^{-1}(U_1), \gamma^{-1}(U_2)\}$ is a splitting of $[a, b]$, but $[a, b]$ is connected. Hence we have a contradiction and deduce that a pathwise connected set is connected. $\qquad \square$

Lemma 4.19. *Let* $G \subset \mathbb{R}^n$ *be an open set and* $q_0 \in G$. *The set*

$$G_{q_0} := \bigcup \{ \mathrm{tr}(\gamma) \,|\, \gamma : [a, b] \to G \text{ is a path with interval point } \gamma(a) = q_0 \} \tag{4.19}$$

is a pathwise connected set.

71

Proof. Let $q, p \in G_{q_0}$. By the definition of G_{q_0} and Definition 4.14 we can construct $\gamma_1 : [0, 1] \to G_{q_0}$ and $\gamma_2 : [0, 1] \to G_{q_0}$ such that $\gamma_1(0) = \gamma_2(0) = q_0$, $\gamma_1(1) = q$ and $\gamma_2(1) = p$. The mapping $\gamma : [0, 1] \to G_{q_0}$ defined by

$$\gamma(t) := \begin{cases} \gamma_1(-2t + 1), & t \in [0, \frac{1}{2}] \\ \gamma_2(2t - 1), & t \in [\frac{1}{2}, 1] \end{cases}$$

is continuous, $\operatorname{tr} \gamma|_{[0, \frac{1}{2}]} = \operatorname{tr} \gamma_1$, $\operatorname{tr} \gamma|_{[\frac{1}{2}, 1]} = \operatorname{tr} \gamma_2$, and $\gamma(0) = q$ as well as $\gamma(1) = p$. $\square$

Obviously G_{q_0} is the largest subset of G which contains q_0 and is pathwise connected.

Theorem 4.20. *If $G \subset \mathbb{R}^n$ is a non-empty open set which is connected then G is pathwise connected.*

Proof. Since G is open, non-empty and connected we can pick $z \in G$ and define

$$G_1 := \{q \in G \,|\, q \text{ can be connected to } z \text{ by a path } \gamma_{zq}, \operatorname{tr} \gamma_{zq} \subset G\} \quad (4.20)$$

and

$$G_2 := \{q \in G \,|\, q \text{ cannot be connected to } z \text{ by a path } \gamma_{zq}, \operatorname{tr} \gamma_{zq} \subset G\}. \quad (4.21)$$

Thus $G = G_1 \cup G_2$ and $G_1 \cap G_2 = \emptyset$. We will prove that both sets are open. The fact that G is connected then implies that either G_1 or G_2 is empty. Since for $z \in G_1$, $\gamma_{zz} : [0, 1] \to G$, $\gamma_{zz}(t) = z$ for all $t \in [0, 1]$, is a path connecting z to itself, we find that $G = G_1$ and G is pathwise connected. First we prove that G_1 is open. For $z_1 \in G_1$ there exists $B_\epsilon(z_1) \subset G$ such that all points $y \in B_\epsilon(z_1)$ can be connected to z_1 by a path $\gamma_{z_1 y}$, just take the line segment connecting z_1 with y. Thus we can connect y to z by $\gamma := \gamma_{zz_1} \oplus \gamma_{z_1 y}$. This implies that $B_\epsilon(z_1) \subset G_1$, i.e. every point $z_1 \in G_1$ has an open neighbourhood entirely belonging to G_1, i.e. G_1 is open. Next we prove that G_2 is open too. Take $z_2 \in G_2$ and let $B_\epsilon(z_2) \subset G$. If $y \in B_\epsilon(z_2)$ can be connected to z by a path γ_{zy} then we can connect z_2 to z since there is always a path γ_{yz_2} connecting $y \in B_\epsilon(z_2)$ to z_2. Thus $\gamma_{zy} \oplus \gamma_{yz_2}$ will connect z to z_2 which is a contradiction to the definition of z_2. Consequently $B_\epsilon(z_2) \subset G_2$, i.e. G_2 is open, and the theorem is proved. $\square$

Example 4.21. Let $I \subset \mathbb{R}$ be an interval and $f : I \to \mathbb{R}$ a continuous function. Define $\Gamma_f : I \to \mathbb{R}^2$, $\Gamma_f(x) = (x, f(x))$, which is a continuous function the range of which is the graph $\Gamma(f)$ of f. Since I is connected $\Gamma(f)$ as well as $\overline{\Gamma(f)}$, compare Proposition 3.37, are connected sets in $\mathbb{R}^2$.

Example 4.22. A. The set $\left\{ (x, \sin \frac{1}{x}) \mid x \in (0, 1] \right\} \subset \mathbb{R}^2$ which can be considered as the graph of $x \mapsto \sin \frac{1}{x}$, $x \in (0, 1]$ is connected.
B. The graph of $f : \mathbb{R} \setminus \{0\} \to \mathbb{R}$, $f(x) = \frac{1}{x}$, is not connected since it admits the splitting $U_1 = \left\{ (x, \frac{1}{x}) \mid x > 0 \right\}$ and $U_2 = \left\{ (x, \frac{1}{x}) \mid x < 0 \right\}$.

Example 4.23. The set $F := \overline{\left\{ (x, \sin \frac{1}{x}) \mid x \in (0, 1] \right\}}$ is the closure of the graph of $x \mapsto \sin \frac{1}{x}$, $x \in (0, 1]$, and therefore it is connected, see Example 4.21. We claim that F is not pathwise connected. For $k \in \mathbb{N}$ the pair $\left(\frac{1}{k\pi}, \sin \left(\frac{1}{\frac{1}{k\pi}} \right) \right) = \left(\frac{1}{k\pi}, 0 \right)$ belongs to the graph of $x \mapsto \sin \frac{1}{x}$, $x \in (0, 1]$, and since $\lim_{k \to \infty} \left(\frac{1}{k\pi}, 0 \right) = (0, 0)$ it follows that $(0, 0) \in F$. The point $\left(\frac{2}{\pi}, 1 \right)$ belongs to F too. Suppose $\gamma : [0, 1] \to F$ is a path with initial point $\gamma(0) = (0, 0)$ and the terminal point $\gamma(1) = \left(\frac{2}{\pi}, 1 \right)$. For $t_k = \frac{2}{(2k+1)\pi}$ we have $\gamma(t_k) \in \{+1, -1\}$ and therefore γ cannot be continuous which is a contradiction, i.e. F is connected but not pathwise connected.

The range of a continuous function $f : I \to \mathbb{R}^n$, $I \subset \mathbb{R}$ being an interval, can be quite surprising. A result due to G. Peano shows the existence of a continuous surjective function $\gamma : [0, 1] \to [0, 1] \times [0, 1]$. A simpler construction of a continuous surjective function $\tilde{\gamma} : [0, 1] \to D$, where D is the closed unilateral triangle with sides of length $\frac{1}{2}$ is given in [4], p. 36 - 37. These results indicate that paths in our definition by no means need to look like or behave like "nice" curves in space. Mappings $\gamma : (a, b) \to \mathbb{R}^n$ which are continuously differentiable (see Chapter 7) give rise to a much better geometric theory. Indeed the behaviour of "only" continuous mappings falls into the realm of topology. We are interested in "better" mappings, i.e. differentiable mappings and to extend calculus to higher dimensions.
With this experience in mind we do not want to study continuous mappings $f : G \to \mathbb{R}^m$, $G \subset \mathbb{R}^n$ in depth. Each such mapping is given (as we already know) by components $f_j : G \to \mathbb{R}$, $j = 1, \ldots, m$, and f is continuous (at $x_0 \in G$) if and only if all f_j, $j = 1, \ldots, m$, are continuous (at $x_0 \in G$). If $G \subset \mathbb{R}^n$ is compact then f is uniformly continuous and $f(G) \subset \mathbb{R}^m$ is compact too. Hence $f(G)$ is bounded, i.e. $f(G) \subset B_R(0) \subset \mathbb{R}^m$ for some $R > 0$. Moreover, if G is connected then $f(G)$ is connected in $\mathbb{R}^m$. However, we must

expect that $f(G) \subset \mathbb{R}^m$ is quite a complicated set. The idea that for $n < m$ the set $f(G)$ is a type of n-dimensional "surface" in the m-dimensional space is misleading as long as we require f to only be continuous. However f can "transport" paths in G to paths in $f(G)$, i.e. if $\gamma : [0,1] \to G$ is a path then $f \circ \gamma : [0,1] \to f(G) \subset \mathbb{R}^m$ is a path too.

Problems

1. On $\mathbb{R}^2$ consider the function $f : \mathbb{R}^2 \to \mathbb{R}$ defined by

$$f(x,y) = \begin{cases} \frac{(x-1)(y+1)^2}{(x-1)^2+(y+1)^4}, & (x,y) \neq (1,-1) \\ 0, & (x,y) = (1,-1). \end{cases}$$

Find the following limits of $f(x,y)$ for $(x,y) \mapsto (1,-1)$:

a) along the line $x = 1$;

b) along the line $y = -1$;

c) along the line $(y+1) = a(x-1), a \neq 0$;

d) along the parabola $(y+1)^2 = (x-1)$.

2. For $g : \mathbb{R}^2 \to \mathbb{R}$ defined by

$$g(x,y) = \begin{cases} \frac{(x-2)^2(y-2)}{(x-2)^6+(y-2)^2}, & (x,y) \neq (2,2) \\ 0, & (x,y) = (2,2) \end{cases}$$

prove that for every straight line passing through $(2,2)$ the limits of $g(x,y)$ along these lines for $(x,y) \mapsto (2,2)$ exist and they are all equal to 0. However the limit of $g(x,y)$ along the parabola $(y-2) = (x-2)^3$ does not exist.

3. Let $H \subset \mathbb{R}^2$ be the open half space $H = \{(x,y) \in \mathbb{R}^2 | x > y\}$ and χ_H its characteristic function, i.e.

$$\chi_H = \begin{cases} 1, & (x,y) \in H, \\ 0, & (x,y) \notin H. \end{cases}$$

Prove that χ_H is continuous on all straight lines parallel to the line $x = y$.

4. Suppose that $f : G \to \mathbb{R}^m$, $G \subset \mathbb{R}^n$ open, satisfies a **local Hölder condition** with exponent $0 < \alpha \leq 1$, i.e. for every $x \in G$ there exists $\rho > 0$ such that $B_\rho(x) \subset G$ and for all $y \in B_\rho(x)$ the following holds with $c_x > 0$

$$(*) \qquad ||f(x) - f(y)|| \leq c_x ||x - y||^\alpha.$$

If we can chose c_x independent of $x \in G$ and $(*)$ holds for all $x, y \in G$, we call f **globally Hölder continuous** with exponent α. Prove that if f is locally Hölder continuous then f is continuous in G. Further give an example of a mapping which is locally but not globally Hölder continuous.

5. Let $f : [-2, 2] \to \mathbb{R}$ be locally Hölder continuous with exponent $0 < \alpha \leq 1$ and $f(0) = 0$. Prove that

$$\int_0^1 \frac{f(y)}{y} dy$$

exists.

6. A mapping $f : G \to \mathbb{R}^m$, $G \subset \mathbb{R}^n$ open, is called **(globally) bounded** if $f(G)$ is bounded, and f is said to be **locally bounded** if for every $x \in G$ there exists $\rho > 0$ such that $B_\rho(x) \subset G$ and $f|_{B_\rho(x)}$ is bounded. Prove that every continuous mapping $f : G \to \mathbb{R}^m$ is locally bounded and give an example of a locally but not globally bounded continuous mapping.

7.   a) Let $f : [a, b] \to \mathbb{R}$ be a continuous function. Find a continuous curve $\gamma : [a, b] \to \mathbb{R}$ such that $\gamma(a) = (b, f(b))$ and $\gamma(b) = (a, f(a))$ and $\mathrm{tr}(\gamma) = \Gamma(f)$.

   b) For $a > b > 0$ the set $\mathcal{E} := \left\{ (x, y) \in \mathbb{R}^2 \left| \frac{x^2}{a^2} + \frac{y^2}{b^2} = 1 \right. \right\}$ describes an ellipse in $\mathbb{R}^2$. Find a simply closed and continuous curve $\gamma : [0, 2\pi] \to \mathbb{R}^2$ such that $\mathrm{tr}(\gamma) = \mathcal{E}$.

   c) Give a parametrization of the unit circle in $\mathbb{R}^2$ as a simply closed continuous curve with parameter interval $[0, \frac{1}{2}]$.

8. Let $\gamma : [a, b] \to G$, $G \subset \mathbb{R}^n$ open, be a Lipschitz continuous curve, i.e. for $t, s \in [a, b]$ the following holds

$$(**) \qquad ||\gamma(t) - \gamma(s)|| \leq \kappa |t - s|$$

75

with $\kappa > 0$ independent of t and s. Let $Z = Z(t_1, \ldots, t_k)$ be a partition of $[a, b]$ (recall that $t_0 = a$ and $t_{k+1} = b$ are by definition points of this partition). Prove that

$$V(\gamma; Z) := \sum_{j=1}^{k+1} ||\gamma(t_j) - \gamma(t_{j-1})|| \leq \kappa(b - a)$$

and deduce that

$$V(\gamma) := \sup_Z V(\gamma; Z) \leq \kappa(b - a)$$

where the supremum is taken over all partitions of $[a, b]$. Is it possible to derive with the same type of argument a similar conclusion if the Lipschitz continuity $(**)$ is replaced by the Hölder condition

$$||\gamma(t) - \gamma(s)|| \leq \kappa|t - s|^\alpha, \quad 0 < \alpha < 1?$$

9. Prove that $B_1((0,0)) \cup \overline{B_1((2,0))} \subset \mathbb{R}^2$ is arcwise connected. **Hint**: show that $[0, 2] \times \{0\} \subset B_1(0,0) \cup \overline{B_1((2,0))}$.

10. Consider the general linear group $GL(n; \mathbb{R})$ as a subset of $\mathbb{R}^{n^2}$. Prove that $GL(n; \mathbb{R})$ is an open subset of $\mathbb{R}^{n^2}$ which is not connected. Find the connectivity component of the unit matrix id_n.
Hint: first give a reason why $\det : GL(n; \mathbb{R}) \to \mathbb{R}$ is continuous.

11. By $SO(2)$ we denote the special orthonormal group in $\mathbb{R}^2$, i.e. all orthonormal matrices with determinant 1. Prove that we can consider $SO(2)$ as a simply closed continuous curve in $M(2, \mathbb{R})$, in fact in $GL(2, \mathbb{R})$.
Hint: use a representation of $U \in SO(2)$ involving sin and cos.

5 Partial Derivatives

Introducing a metric instead of the absolute value made it possible to extend the notion of continuity from real-valued functions of one real variable to mappings between metric spaces, and in particular for mappings $f : G \to \mathbb{R}^m$, $G \subset \mathbb{R}^n$. However differentiability is not so straightforward to generalise. Obviously we cannot define a difference quotient of f at $x \in G \subset \mathbb{R}^n$ and pass to the limit since we cannot divide by a vector. In order to establish calculus (and analysis) for higher dimensions we proceed in two steps: we first study partial derivatives of real-valued functions of several real variables and then we introduce the differential $(\mathrm{d}f)(x)$ of $f : G \to \mathbb{R}^m$, $G \subset \mathbb{R}^n$, at a point $x \in G$ as a linear approximation of f in a neighbourhood of x. We continue to use the notion introduced in Chapter 4.

Definition 5.1. *Let $G \subset \mathbb{R}^n$ be an open set and $f : G \to \mathbb{R}$ a function. If the limit*

$$\frac{\partial f}{\partial x_j}(x_0) := \lim_{h \to 0} \frac{f(x_0 + he_j) - f(x_0)}{h} \tag{5.1}$$

*exists we call $\frac{\partial f}{\partial x_j}(x_0)$ the **partial derivative of f at $x_0 \in G$ in the direction of e_j** or with respect to the coordinate x_j.*

Remark 5.2. A. Since by assumption $G \subset \mathbb{R}^n$ is open there exists $\rho > 0$ such that $B_\rho(x_0) \subset G$ and consequently $x_0 + he_j \in G$ for $|h| < \rho$. Hence for $|h|$ sufficiently small $f(x_0 + he_j)$ is defined and $\frac{f(x_0 + he_j) - f(x_0)}{h}$ makes sense. **B.** It is convenient to introduce further notations for $\frac{\partial f}{\partial x_j}(x_0)$, namely

$$D_j f(x_0) := D_{x_j} f(x_0) := f_{x_j}(x_0) := \left(\frac{\partial}{\partial x_j} f \right)(x_0) := \frac{\partial f(x_0)}{\partial x_j} := \frac{\partial f}{\partial x_j}(x_0). \tag{5.2}$$

Definition 5.3. *Let $G \subset \mathbb{R}^n$ be an open set and $f : G \to \mathbb{R}$ a function. If for all $x \in G$ the partial derivatives $\frac{\partial f}{\partial x_j}(x)$ exist we call the function $\frac{\partial f}{\partial x_j} : G \to \mathbb{R}$, $x \mapsto \frac{\partial f}{\partial x_j}(x)$, the **partial differentiation** of f in G in the direction of e_j or with respect to the coordinate x_j.*

If $\frac{\partial f}{\partial x_j} : G \to \mathbb{R}$ exists then we can investigate whether this new real-valued functions on G has a partial derivative, but we now have n different possibilities.

Definition 5.4. *If $f : G \to \mathbb{R}$ admits the partial derivative $\frac{\partial f}{\partial x_j} : G \to \mathbb{R}$ then, for $x_0 \in G$, we call*

$$\frac{\partial^2 f}{\partial x_l \partial x_j}(x_0) := \frac{\partial}{\partial x_l}\left(\frac{\partial f}{\partial x_j}\right)(x_0) \tag{5.3}$$

*the **second (order) partial derivative of** f **at** x_0 first in the direction of e_j and then in the direction of e_l. If $\frac{\partial^2 f}{\partial x_l \partial x_j}(x)$ exists for all $x \in G$ we call $\frac{\partial^2 f}{\partial x_l \partial x_j} : G \to \mathbb{R}$ the **second (order)** partial derivative of f in G first in the direction of e_j and then in the direction of e_l.*

As we will see in Example 5.9 below, in general we have $\frac{\partial^2 f(x)}{\partial x_l \partial x_j} \neq \frac{\partial^2 f(x)}{\partial x_j \partial x_l}$, i.e. we must take care on the order in which we take partial derivatives.
Of course we may now iterate the process and introduce k^{th} order partial derivatives (at $x \in G$) by

$$\frac{\partial^k f}{\partial x_{i_k} \ldots \partial x_{i_1}}(x) := \frac{\partial}{\partial x_{i_k}}\left(\frac{\partial^{k-1} f}{\partial x_{i_{k-1}} \ldots \partial x_{i_1}}\right)(x) \tag{5.4}$$

for $1 \leq i_j \leq n$ and $1 \leq j \leq k$.

Definition 5.5. *We call $f : G \to \mathbb{R}$ k-times **continuously (partial) differentiable** in G if all partial derivatives of order less than or equal to k exist and are continuous. The set of all k-times continuously differentiable functions $f : G \to \mathbb{R}$, $G \subset \mathbb{R}^n$ open, is denoted by $C^k(G)$, and we use the convention $C(G) = C^0(G)$ as well as $C^\infty(G) = \bigcap_{k=1}^\infty C^k(G)$.*

Remark 5.6. Let $f : G \to \mathbb{R}$, $G \subset \mathbb{R}^n$ open, be a function and $x_0 \in G$. Define $I_{j,\rho}(x_0) := (x_0^{(j)} - \rho, x_0^{(j)} + \rho)$, $x_0 = (x_0^{(1)}, \ldots, x_0^{(n)})$. Since G is open and $x_0 \in G$ we can find $\rho > 0$ such that $I_{1,\rho}(x_0) \times \cdots \times I_{n,\rho}(x_0) \subset G$, and therefore we can study the n functions

$$t \mapsto f\left(x_0^{(1)}, \ldots, x_0^{(j-1)}, t, x_0^{(j+1)}, \ldots, x_0^{(n)}\right), j = 1, \ldots, n, \tag{5.5}$$

which are real-valued functions of one real variable defined on $I_{j,\rho}(x_0)$ and t in the jth position. For $x_0 \in G$ fixed we can apply our one-dimensional results to these functions, in particular we may use all our rules to study limits such as

$$\lim_{t \to t_0} f\left(x_0^{(1)}, \ldots, x_0^{(j-1)}, t, x_0^{(j+1)}, \ldots, x_0^{(n)}\right).$$

Example 5.7. A. For $f : \mathbb{R}^2 \to \mathbb{R}$, $f(x, y) = \cos xy$ we find

$$\frac{\partial f}{\partial x}(x, y) = -y \sin xy \quad \text{and} \quad \frac{\partial f}{\partial y}(x, y) = -x \sin xy.$$

B. On $G = \mathbb{R}^n \setminus \{0\}$ we consider $g(x) = g(x_1, \ldots, x_n) = \frac{1}{x_1^2 + \cdots + x_n^2} = \frac{1}{\|x\|^2}$, and with $1 \leq j \leq n$ it follows that

$$\frac{\partial}{\partial x_j} g(x) = \frac{-2x_j}{(x_1^2 + \cdots + x_n^2)^2} = \frac{-2x_j}{\|x\|^4}.$$

C. Consider $h : \mathbb{R}^3 \setminus \{0\} \to \mathbb{R}$, $h(x) = h(x_1, x_2, x_3) = \frac{1}{(x_1^2 + x_2^2 + x_3^2)^{\frac{1}{2}}} = \frac{1}{\|x\|}$. For $1 \leq j \leq 3$ we find

$$\frac{\partial h}{\partial x_j}(x) = \frac{-x_j}{(x_1^2 + x_2^2 + x_3^2)^{\frac{3}{2}}} = \frac{-x_j}{\|x\|^3}$$

and further

$$\frac{\partial^2 h}{\partial x_j^2}(x) = \frac{-1}{(x_1^2 + x_2^2 + x_3^2)^{\frac{3}{2}}} + \frac{3x_j^2}{(x_1^2 + x_2^2 + x_3^2)^{\frac{5}{2}}} = \frac{3x_j^2 - (x_1^2 + x_2^2 + x_3^2)}{(x_1^2 + x_2^2 + x_3^2)^{\frac{5}{2}}}.$$

It follows that

$$\triangle_3 h(x) := \frac{\partial^2 h}{\partial x_1^2}(x) + \frac{\partial^2 h}{\partial x_2^2}(x) + \frac{\partial^2 h}{\partial x_3^2}(x)$$
$$= \frac{3x_1^2 + 3x_2^2 + 3x_3^2 - 3(x_1^2 + x_2^2 + x_3^2)}{(x_1^2 + x_2^2 + x_3^2)^{\frac{5}{2}}} = 0,$$

i.e. $\triangle_3 h(x) = 0$, $x \neq 0$. In general we call the operator

$$\triangle_n = \frac{\partial^2}{\partial x_1^2} + \cdots + \frac{\partial^2}{\partial x_n^2} \tag{5.6}$$

the n-dimensional **Laplace operator** and we call $h : G \to \mathbb{R}$, $G \subset \mathbb{R}$ open, a **harmonic function** in G if $\triangle_n h(x) = 0$ for all $x \in G$. (When it is clear what n is from the context we will often write $\triangle$ instead of $\triangle_n$.) Thus the function $x \mapsto \frac{1}{\|x\|}$ is harmonic in $\mathbb{R}^3 \setminus \{0\}$.

Example 5.8. (Compare with Example 4.1) We want to investigate the partial derivatives of $f : \mathbb{R}^2 \to \mathbb{R}$ at $(0,0)$ where f is defined by

$$f(x,y) = \begin{cases} \frac{xy}{x^2+y^2}, & (x,y) \neq (0,0) \\ 0, & (x,y) = (0,0). \end{cases}$$

Since for $(x_0, y_0) = (0,0)$ we have with $x \neq 0$ and $y \neq 0$

$$\frac{f(x, y_0) - f(x_0, y_0)}{x - x_0} = \frac{f(x,0) - f(0,0)}{x - x_0} = 0$$

and

$$\frac{f(x_0, y) - f(x_0, y_0)}{y - y_0} = \frac{f(0, y) - f(0,0)}{y - y_0} = 0$$

we conclude that $\frac{\partial f}{\partial x}(0,0) = \frac{\partial f}{\partial y}(0,0) = 0$. Hence both partial derivatives exist, but we know from Example 4.1 that f is not continuous at $(0,0)$. If we look at $(x_0, y_0) \neq (0,0)$ we find further

$$\frac{\partial f}{\partial x}(x_0, y_0) = \frac{y_0}{x_0^2 + y_0^2} - \frac{2x_0^2 y_0}{(x_0^2 + y_0^2)^2}$$

and

$$\frac{\partial f}{\partial y}(x_0, y_0) = \frac{x_0}{x_0^2 + y_0^2} - \frac{2x_0 y_0^2}{(x_0^2 + y_0^2)^2}.$$

In particular we have

$$\frac{\partial f}{\partial x}(0, y_0) = \frac{1}{y_0} \quad \text{and} \quad \frac{\partial f}{\partial y}(x_0, 0) = \frac{1}{x_0}$$

implying that

$$\frac{\partial f}{\partial x}(0,0) \neq \lim_{y_0 \to 0} \frac{\partial f}{\partial x}(0, y_0) \quad \text{and} \quad \frac{\partial f}{\partial y}(0,0) \neq \lim_{x_0 \to 0} \frac{\partial f}{\partial y}(x_0, 0).$$

Thus for all $(x_0, y_0) \in \mathbb{R}^2$ the partial derivatives $\frac{\partial f}{\partial x}(x_0, y_0)$ and $\frac{\partial f}{\partial y}(x_0, y_0)$ exist but they are not continuous at $(0,0)$ as f is not continuous at $(0,0)$.

Example 5.9. On $\mathbb{R}^2$ we consider the function $g : \mathbb{R}^2 \to \mathbb{R}$ defined by

$$g(x,y) := \begin{cases} xy\frac{x^2-y^2}{x^2+y^2}, & (x,y) \neq (0,0) \\ 0, & (x,y) = (0,0). \end{cases}$$

At $(0,0)$ we find

$$\frac{\partial g}{\partial x}(0,0) = \lim_{h \to 0} \frac{g(h,0) - g(0,0)}{h} = 0$$

and

$$\frac{\partial g}{\partial y}(0,0) = \lim_{h \to 0} \frac{g(0,h) - g(0,0)}{h} = 0.$$

Moreover, for $(x,y) \neq (0,0)$ we have

$$\frac{\partial g}{\partial x}(x,y) = y\frac{x^4 + 4x^2y^2 - y^4}{(x^2 + y^2)^2}$$

and

$$\frac{\partial g}{\partial y}(x,y) = x\frac{x^4 - 4x^2y^2 - y^4}{(x^2 + y^2)^2}.$$

Combining these results we find

$$\frac{\partial^2 g}{\partial x \partial y}(0,0) = \lim_{h \to 0} \frac{1}{h}\left(\frac{\partial g}{\partial y}(h,0) - \frac{\partial g}{\partial y}(0,0)\right)$$

$$= \lim_{h \to 0} \frac{1}{h}\left(h\frac{h^4}{h^4}\right) = 1$$

and

$$\frac{\partial^2 g}{\partial y \partial x}(0,0) = \lim_{h \to 0} \frac{1}{h}\left(\frac{\partial g}{\partial x}(0,h) - \frac{\partial g}{\partial y}(0,0)\right)$$

$$= \lim_{h \to 0} \frac{1}{h}\left(h \cdot \frac{-h^4}{h^4}\right) = -1,$$

i.e. both second order partial derivatives exist at $(0,0)$ but

$$\frac{\partial^2 g}{\partial y \partial x}(0,0) \neq \frac{\partial^2 g}{\partial x \partial y}(0,0).$$

Example 5.10. If $p : \mathbb{R}^n \to \mathbb{R}$ is a polynomial, i.e.

$$p(x) = p(x_1, \ldots, x_n) = \sum_{\alpha_j \leq m_j} a_{\alpha_1, \ldots, \alpha_n} x_1^{\alpha_1} \cdot \ldots \cdot x_n^{\alpha_n}$$

81

with $a_{\alpha_1,\dots,\alpha_n} \in \mathbb{R}$ and $m_j, \alpha_j \in \mathbb{N}_0$, then we find for $\beta_j \leq m_j$

$$\frac{\partial^{\beta_j}}{\partial x_j^{\beta_j}} p(x_1, \dots, x_n) = \sum_{\alpha_j \leq m_j} a_{\alpha_1,\dots,\alpha_n} \alpha_j (\alpha_j - 1) \cdot \dots \cdot (\alpha_j - \beta_j + 1) x_1^{\alpha_1} \cdot \dots$$

$$\dots \cdot x_{j-1}^{\alpha_{j-1}} x_j^{\alpha_j - \beta_j} x_{j+1}^{\alpha_{j+1}} \cdot \dots \cdot x_n^{\alpha_n},$$

and for $\beta_j > m_j$ we have $\frac{\partial^{\beta_j}}{\partial x_j^{\beta_j}} p(x) = 0$.

Example 5.11. Let $G \subset \mathbb{R}^n$ be an open set and denote by $\mathrm{pr}_j : \mathbb{R}^n \to \mathbb{R}$ the jth coordinate projection. For $x \in G$ we have $\mathrm{pr}_j(x) = x_j$. Denote by

$$\hat{G}_j := \mathrm{pr}_1(G) \times \dots \times \mathrm{pr}_{j-1}(G) \times \mathrm{pr}_{j+1}(G) \times \dots \times \mathrm{pr}_n(G).$$

Let $f : G \to \mathbb{R}$ and $h : \hat{G}_j \to \mathbb{R}$ be two functions such that for all $x \in G \cap \mathrm{pr}_1(G) \times \dots \times \mathrm{pr}_{j-1}(G) \times \mathrm{pr}_j(G) \times \mathrm{pr}_{j+1}(G) \times \dots \times \mathrm{pr}_n(G)$ then we have $f(x) = h(\hat{x}_j)$, $\hat{x}_j = (x_1, \dots, x_{j-1}, x_{j+1}, \dots, x_n) \in \hat{G}_j$, i.e. f is a function defined on G but independent of the variable $x_j \in \mathrm{pr}_j(G)$. In this case it follows that $\frac{\partial f}{\partial x_j}(x) = 0$ for all $x \in G$. The converse question whether $\frac{\partial f}{\partial x_j}(x) = 0$ for all $x \in G$ implies that f is independent of x_j we will discuss later.

We can easily extend the linearity, Leibniz's rule and the chain rule to partial derivatives, we only need to take Remark 5.6 into account and apply our results from one-dimensional calculus. Thus we have

Proposition 5.12. *Let $G \subset \mathbb{R}^n$ be an open set and $f, g : G \to \mathbb{R}$ have partial derivatives $\frac{\partial f}{\partial x_j}$ and $\frac{\partial g}{\partial x_j}$ (at $x_0 \in G$), $1 \leq j \leq n$, respectively.*
***A.** For all $\lambda, \mu \in \mathbb{R}$ the functions $\lambda f + \mu g$ and $f \cdot g$ have partial derivatives $\frac{\partial}{\partial x_j}(\lambda f + \mu g)$ and $\frac{\partial}{\partial x_j}(f \cdot g)$ (at $x_0 \in G$) and we have*

$$\frac{\partial}{\partial x_j}(\lambda f + \mu g)(x) = \lambda \frac{\partial f}{\partial x_j}(x) + \mu \frac{\partial g}{\partial x_j}(x) \tag{5.7}$$

as well as

$$\frac{\partial}{\partial x_j}(f \cdot g)(x) = \left(\frac{\partial f}{\partial x_j}(x) \right) g(x) + f(x) \frac{\partial g}{\partial x_j}(x). \tag{5.8}$$

***B.** If $f : G \to \mathbb{R}$ has the partial derivative $\frac{\partial f}{\partial x_j}$ (at $x_0 \in G$) and if $h : I \to \mathbb{R}$ is a differentiable function such that $f(G) \subset I$ then $h \circ f : G \to \mathbb{R}$ has the* *partial derivative* $\frac{\partial(h\circ f)}{\partial x_j}$ *(at $x_0 \in G$) and the following holds for all $x \in G$ (or at $x_0 \in G$)*

$$\frac{\partial(h \circ f)}{\partial x_j}(x) = h'(f(x))\frac{\partial f}{\partial x_j}(x). \tag{5.9}$$

In particular we have under appropriate conditions

$$\frac{\partial}{\partial x_j}e^{h(x)} = \frac{\partial h(x)}{\partial x_j}e^{h(x)} \tag{5.10}$$

and

$$\frac{\partial}{\partial x_j}\ln h(x) = \frac{1}{h(x)}\frac{\partial h}{\partial x_j}(x). \tag{5.11}$$

Remark 5.13. Clearly these results extend to higher order partial derivatives, however we wait until we have a more suitable notation at our disposal to give explicit formulae.

Corollary 5.14. *The set $C^k(G)$, $G \subset \mathbb{R}^n$ open, with its natural algebraic operations is an algebra over $\mathbb{R}$.*

Example 5.15. We call a function $f : \mathbb{R}^n \to \mathbb{R}$ (or $f : \mathbb{R}^n \setminus \{0\} \to \mathbb{R}$) **radially symmetric** or **rotationally invariant** if for some function $h : [0,\infty) \to \mathbb{R}$ (or $h : (0,\infty) \to \mathbb{R}$) with $r = r(x) = (x_1^2 + \cdots + x_n^2)^{\frac{1}{2}}$ the following holds

$$f(x) = h(r) = h\left((x_1^2 + \cdots + x_n^2)^{\frac{1}{2}}\right). \tag{5.12}$$

Suppose that f has the partial derivative $\frac{\partial f}{\partial x_j}$ and that h is differentiable. Then it follows for $x \neq 0$, i.e. $r \neq 0$, that

$$\frac{\partial}{\partial x_j}f(x) = h'(r)\frac{\partial r}{\partial x_j} = h'(r)\frac{x_j}{r}. \tag{5.13}$$

Example 5.16. If $f, g : \mathbb{R} \to \mathbb{R}$ are two twice differentiable functions of one real variable then

$$u(x, t) := f(x + ct) + g(x - ct), \quad c > 0, \tag{5.14}$$

has second order partial derivatives and u solves the **one-dimensional wave equation**

$$\frac{1}{c^2}\frac{\partial^2 u}{\partial t^2} - \frac{\partial^2 u}{\partial x^2} = 0. \tag{5.15}$$

Indeed we have

$$\frac{\partial}{\partial t}(f(x+ct)+g(x-ct)) = \left(\frac{\partial}{\partial t}(x+ct)\right)f'(x+ct) + \left(\frac{\partial}{\partial t}(x-ct)\right)g'(x-ct)$$
$$= cf'(x+ct) - cg'(x-ct);$$

$$\frac{\partial^2}{\partial t^2}(f(x+ct)+g(x-ct)) = c^2 f''(x+ct) + c^2 g''(x-ct);$$

$$\frac{\partial}{\partial x}(f(x+ct)+g(x-ct)) = \left(\frac{\partial}{\partial x}(x+ct)\right)f'(x+ct) + \left(\frac{\partial}{\partial x}(x-ct)\right)g'(x-ct)$$
$$= f'(x+ct) + g'(x-ct);$$

$$\frac{\partial^2}{\partial x^2}(f(x+ct)+g(x-ct)) = f''(x+ct) + g''(x-ct),$$

implying that

$$\frac{1}{c^2}\frac{\partial^2}{\partial t^2}u(x,t) - \frac{\partial^2}{\partial x^2}u(x,t) = 0.$$

Example 5.17. We may interpret "taking a partial derivative" as an action of a (linear) operator, i.e. a linear mapping, say from $C^1(G)$ to $C(G)$, $G \subset \mathbb{R}^n$ open, thus for $1 \leq j \leq n$ we consider

$$\frac{\partial}{\partial x_j} : C^1(G) \to C(G), \ f \mapsto \frac{\partial f}{\partial x_j}. \tag{5.16}$$

If $G = \mathbb{R}^n$ we can also consider the translation τ_a, $a \in \mathbb{R}^n$, defined by $\tau_a : \mathbb{R}^n \to \mathbb{R}^n$, $\tau_a(x) = x + a$. For $f \in C^1(\mathbb{R}^n)$ we find

$$\frac{\partial}{\partial x_j}(f \circ \tau_a)(x) = \frac{\partial}{\partial x_j}f(x+a) = \left(\frac{\partial f}{\partial x_j}\right)(x+a) = \left(\frac{\partial f}{\partial x_j}\circ \tau_a\right)(x),$$

or with the translation operator $\tau_{a,op} : C(\mathbb{R}^n) \to C(\mathbb{R}^n)$, $\tau_{a,op}(f) = f \circ \tau_a$, note that $\tau_{a,op}$ maps $C^1(\mathbb{R}^n)$ into $C^1(\mathbb{R}^n)$, we have

$$\left(\frac{\partial}{\partial x_j}\circ \tau_{a,op}\right)f = \left(\tau_{a,op}\circ\frac{\partial}{\partial x_j}\right)f, \tag{5.17}$$

i.e. partial derivatives are invariant under translations.

Introducing for two linear operators A, B the **commutator** $[A, B] = AB - BA$ (whenever it is defined), we can rewrite (5.17) as $\left[\frac{\partial}{\partial x_j}, \tau_{a,op}\right] = 0$ which holds on $C^1(\mathbb{R}^n)$.

Now if $G \subset \mathbb{R}^n$ is open then $\tau_a(G) = a + G$ is open too and $f \in C^1(G)$ induces a function $\tau_{a,op}f \in C^1(\tau_a(G))$. In particular if $x_0 \in G$ then $0 \in \tau_{-x_0}(G)$ and instead of studying $f \in C^1(G)$ at $x_0 \in G$ we may study $\tau_{-x_0,op}f \in C^1(\tau_{-x_0}(G))$ at 0. Clearly we can extend these considerations to higher order partial derivatives.

The next result gives a sufficient condition allowing us to interchange the order of higher order partial derivatives.

Theorem 5.18. *For $f \in C^2(G)$, $G \subset \mathbb{R}^n$ open, the following holds for all $j, l = 1, \ldots, n$ and $x \in G$*

$$\frac{\partial^2 f(x)}{\partial x_l \partial x_j} = \frac{\partial^2 f(x)}{\partial x_j \partial x_l}. \tag{5.18}$$

Proof. It is sufficient to prove (5.18) for a function of two variables and, having Example 5.17 in mind, we may assume that $x = 0 \in G \subset \mathbb{R}^2$. Thus we need to prove for an open set $G \subset \mathbb{R}^2$ with $0 \in G$ that $f \in C^2(G)$ implies

$$\frac{\partial^2 f}{\partial y \partial x}(0,0) = \frac{\partial^2 f}{\partial x \partial y}(0,0), \tag{5.19}$$

or by introducing a more convenient notation

$$D_2 D_1 f(0,0) = D_1 D_2 f(0,0). \tag{5.20}$$

Since G is open there exists $\delta > 0$ such that

$$\{(x,y) \in \mathbb{R}^2 \,|\, |x| < \delta \text{ and } |y| < \delta\} \subset G$$

holds. For $|y| < \delta$ let $F_y : (-\delta, \delta) \to \mathbb{R}$ be the function

$$F_y(x) = f(x,y) - f(x,0).$$

By the mean value theorem, in the formulation of Corollary I.22.6, there is ξ, $|\xi| \le |x|$, such that

$$F_y(x) - F_y(0) = F_y'(\xi)x.$$

Further we have

$$F_y'(\xi) = D_1 f(\xi, y) - D_1 f(\xi, 0).$$

The mean value theorem applied to $y \mapsto D_1 f(\xi, y)$ yields the existence of η, $|\eta| \leq |y|$, such that

$$D_1 f(\xi, y) - D_1 f(\xi, 0) = D_2 D_1 f(\xi, \eta) y.$$

Thus we have

$$f(x, y) - f(x, 0) - f(0, y) + f(0, 0) = D_2 D_1 f(\xi, \eta) xy. \tag{5.21}$$

Now consider $G_x : (-\delta, \delta) \to \mathbb{R}$ with

$$G_x(y) = f(x, y) - f(0, y), |x| < \delta.$$

As before we find $\tilde{\xi}$ and $\tilde{\eta}$ such that $|\tilde{\xi}| \leq |x|$, $|\eta| \leq |y|$ and

$$G_x'(y) - G_x(0) = G_x'(\tilde{\eta}) y$$

as well as

$$G_x'(\tilde{\eta}) = D_2 f(x, \tilde{\eta}) - D_2 f(0, \tilde{\eta}) = D_1 D_2 f(\tilde{\xi}, \tilde{\eta}) xy,$$

which gives

$$f(x, y) - f(0, y) - f(x, 0) + f(0, 0) = D_1 D_2 f(\tilde{\xi}, \tilde{\eta}) xy. \tag{5.22}$$

From (5.21) and (5.22) we derive for $x \neq 0$ and $y \neq 0$ that

$$D_2 D_1 f(\xi, \eta) = D_1 D_2 f(\tilde{\xi}, \tilde{\eta})$$

with (ξ, η) and $(\tilde{\xi}, \tilde{\eta})$ depending on (x, y). As $(x, y) \to (0, 0)$ it follows however that $(\xi, \eta) \to (0, 0)$ and $(\tilde{\xi}, \tilde{\eta}) \to (0, 0)$, thus the continuity of $D_2 D_1 f$ and $D_1 D_2 f$, respectively, yields

$$D_2 D_1 f(0, 0) = D_1 D_2 f(0, 0),$$

and the theorem is proved. $\qquad\qquad\qquad\qquad\qquad\qquad\qquad\qquad \square$

An immediate consequence of Theorem 5.18 is

Corollary 5.19. *Let $G \subset \mathbb{R}^n$ be open and $f : G \to \mathbb{R}$ be k-times continuously partial differentiable, i.e. $f \in C^k(G)$. For any permutation π of the numbers $1, \ldots, k$ we have*

$$D_{i_k} \cdots D_{i_2} D_{i_1} f = D_{i_{\pi(k)}} \cdots D_{i_{\pi(2)}} D_{i_{\pi(1)}} f \tag{5.23}$$

with $1 \le i_k \le n$.

Some of our current notations are too complicated to handle more difficult situations, for example finding Leibniz's rule for higher order partial derivatives. For this reason we now introduce the **multi-index notation** which will turn out to be very helpful since it often makes formulae in higher dimensional calculus look completely similar to their counterparts in one-dimensional calculus.

A **multi-index** α is an element in $\mathbb{N}_0^n$, i.e. $\alpha = (\alpha_1, \ldots, \alpha_n)$ and $\alpha_j \in \mathbb{N}_0$. For $\alpha, \beta \in \mathbb{N}_0^n$ we define

$$\alpha + \beta := (\alpha_1 + \beta_1, \ldots, \alpha_n + \beta_n), \tag{5.24}$$

$$|\alpha| := \alpha_1 + \ldots + \alpha_n, \tag{5.25}$$

$$\alpha! = \alpha_1! \cdot \ldots \cdot \alpha_n!, \tag{5.26}$$

$$\alpha \le \beta \text{ if } \alpha_j \le \beta_j \text{ for all } 1 \le j \le n, \tag{5.27}$$

$$\alpha - \beta = (\alpha_1 - \beta_1, \ldots, \alpha_n - \beta_n) \text{ if } \beta \le \alpha, \tag{5.28}$$

$$\binom{\alpha}{\beta} := \binom{\alpha_1}{\beta_1} \cdot \ldots \cdot \binom{\alpha_n}{\beta_n}. \tag{5.29}$$

Further, if $x \in \mathbb{R}^n$ and $\alpha \in \mathbb{N}_0^n$ we set

$$x^\alpha = x_1^{\alpha_1} \cdot \ldots \cdot x_n^{\alpha_n}. \tag{5.30}$$

Now we can write a polynomial in n-variables $x_1, \ldots, x_n \in \mathbb{R}$ in a rather simple way:

$$p(x) = \sum_{|\alpha| \le m} a_\alpha x^\alpha$$

with $a_\alpha \in \mathbb{R}$ and $x = (x_1, \ldots, x_n)$.

Lemma 5.20 (Binomial theorem for vectors). *For $x, y \in \mathbb{R}^n$ and $\alpha \in \mathbb{N}_0^n$ we have*

$$(x + y)^\alpha = \sum_{\beta \le \alpha} \binom{\alpha}{\beta} x^\beta y^{\alpha - \beta}. \tag{5.31}$$

Proof. By definition we have

$$(x+y)^\alpha = (x_1 + y_1)^{\alpha_1} \cdot \ldots \cdot (x_n + y_n)^{\alpha_n},$$

and the binomial yields

$$(x_1 + y_1)^{\alpha_1} \cdot \ldots \cdot (x_n + y_n)^{\alpha_n} = \sum_{\beta_1 \leq \alpha_1} \binom{\alpha_1}{\beta_1} x_1^{\beta_1} y_1^{\alpha_1 - \beta_1} \cdot \ldots \cdot \sum_{\beta_n \leq \alpha_n} \binom{\alpha_n}{\beta_n} x_n^{\beta_n} y_n^{\alpha_n - \beta_n}$$

$$= \sum_{\beta \leq \alpha} \binom{\alpha}{\beta} x^\beta y^{\alpha - \beta}.$$

$\square$

For $\alpha \in \mathbb{N}_0^n$, $|\alpha| \leq k$, and $f \in C^k(G)$, $G \subset \mathbb{R}^n$ open, we write

$$D^\alpha f := \partial^\alpha f := D_1^{\alpha_1} \cdot \ldots \cdot D_n^{\alpha_n} f := \frac{\partial^{|\alpha|} f}{\partial x_1^{\alpha_1} \cdots \partial x_n^{\alpha_n}} \tag{5.32}$$

where $D_j^l = \left(\frac{\partial}{\partial x_j} \right)^l = \frac{\partial^l}{\partial x_j^l}$. With this notation Leibniz's rule for higher order derivatives becomes "simple":

Lemma 5.21. *For $\alpha \in \mathbb{N}_0^n$, $|\alpha| \leq k$, and $f, g \in C^k(G)$ we have*

$$\partial^\alpha (f \cdot g) = \sum_{\beta \leq \alpha} \binom{\alpha}{\beta} \partial^\beta f \partial^{\alpha - \beta} g. \tag{5.33}$$

Proof. Since $|\alpha| \leq k$ and $f, g \in C^k(G)$ we can interchange the order of all partial derivatives that appear in formula (5.33). Using Leibniz's rule for functions of one variable, see Remark 5.6 and Corollary I.21.12, we find

$$D^\alpha(f \cdot g) = D_{x_n}^\alpha \cdots D_{x_1}^{\alpha_1}(f \cdot g) = D_{x_n}^{\alpha_n} \cdots D_{x_2}^{\alpha_2} \left(\sum_{\beta_1 \leq \alpha_1} \binom{\alpha_1}{\beta_1} D_{x_1}^{\beta_1} f D^{\alpha_1 - \beta_1} g \right)$$

$$= D_{x_n}^{\alpha_n} \cdots D_{x_3}^{\alpha_3} \left(\sum_{\beta_2 \leq \alpha_2} \sum_{\beta_1 \leq \alpha_1} \binom{\alpha_2}{\beta_2} \binom{\alpha_1}{\beta_1} D^{\beta_2} D^{\beta_1} f D^{\alpha_2 - \beta_2} D^{\alpha_1 - \beta_1} g \right)$$

$$= \cdots = \sum_{\beta_n \leq \alpha_n} \cdots \sum_{\beta_1 \leq \alpha_1} \binom{\alpha_n}{\beta_n} \cdots \binom{\alpha_1}{\beta_1} D^{\beta_n} \ldots D^{\beta_1} f D^{\alpha_n - \beta_n} \ldots D^{\alpha_1 - \beta_1} g$$

$$= \sum_{\beta \leq \alpha} \binom{\alpha}{\beta} D^\beta f D^{\alpha - \beta} g.$$

$\square$

In Chapter I.21, we gave without proof the **Faà di Bruno formula**, i.e. formula (I.21.13), to calculate higher order derivatives of the composition of two real-valued functions of one real variable. A useful extension to the case $f \circ g$ where $f : \mathbb{R} \to \mathbb{R}$ and $g : \mathbb{R}^n \to \mathbb{R}$ is the formula

$$\partial^\alpha (f \circ g) = \sum_{j=1}^{|\alpha|} f^{(j)}(g(\cdot)) \sum \frac{\alpha!}{\delta_\beta! \delta_\gamma! \ldots \delta_\omega!} \cdot \left(\frac{\partial^\beta g}{\beta!} \right)^{\delta_\beta} \cdot \ldots \cdot \left(\frac{\partial^\omega g}{\omega!} \right)^{\delta_\omega} \quad (5.34)$$

with the second sum running over all pairwise different multi-indices $0 \neq \beta, \gamma, \ldots, \omega \in \mathbb{N}_0^n$ and all $\delta_\beta, \delta_\gamma, \ldots, \delta_\omega \in \mathbb{N}$ such that $\delta_\beta \beta + \delta_\gamma \gamma + \ldots + \delta_\omega \omega = \alpha$ and $\delta_\beta + \delta_\gamma + \ldots + \delta_\omega = j$. We refer to L. E. Fraenkel [21] for more details. In particular we have for C^∞-functions $u, v, w : \mathbb{R}^n \to \mathbb{R}$, $u > 0$, $v \neq 0$, and for $\gamma \in \mathbb{N}_0^n$, $|\gamma| = l$,

$$\partial^\gamma \ln u = \sum_{\gamma^1 + \cdots + \gamma^l = \gamma} c'_{\{\gamma^j\}} \prod_{j=1}^{l} \frac{\partial^{\gamma^j} u}{u}, \quad \gamma \neq 0, \quad (5.35)$$

$$\partial^\gamma \left(\frac{1}{v} \right) = \frac{1}{v} \sum_{\gamma^1 + \cdots + \gamma^l = \gamma} c''_{\{\gamma^j\}} \prod_{j=1}^{l} \frac{\partial^{\gamma^j} v}{v}, \quad (5.36)$$

and

$$\partial^\gamma (e^\omega) = e^\omega \sum_{\substack{\gamma^1 + \cdots + \gamma^{l'} = \gamma \\ l' = 1, \ldots, |\gamma|}} c'''_{\{\gamma^j\}} \prod_{j=1}^{l'} \partial^{\gamma^j} \omega, \quad (5.37)$$

where the summations are taken over all choices of $\gamma^1, \ldots, \gamma^l \in \mathbb{N}_0^n$ and $\gamma^1, \ldots, \gamma^{l'} \in \mathbb{N}_0^n$, respectively, that add up to γ. The constants $c'_{\{\gamma^j\}}, c''_{\{\gamma^j\}}$ and $c'''_{\{\gamma^j\}}$ depend on these multi-indices.

Problems

Problems 1-5 are meant to be for practice in calculating partial derivatives. Instead of creating elaborate examples we prefer to work with solutions to certain partial differential equations.

1. Prove that $h(t, x) := (4\pi t)^{-\frac{n}{2}} e^{-\frac{||x||^2}{4t}}$ satisfies in $(0, \infty) \times \mathbb{R}^n$ the **heat equation** $\frac{\partial u(t,x)}{\partial t} - \Delta u(t, x) = 0$ where $\Delta_n = \frac{\partial^2}{\partial x_1^2} + \cdots + \frac{\partial^2}{\partial x_n^2}$ is the **Laplace operator** in $\mathbb{R}^n$.

2. For $r > 0$ and $\varphi \in (0, 2\pi)$ prove that

$$\left(\frac{1}{r} \frac{\partial}{\partial r} \left(r \frac{\partial}{\partial r} \right) + \frac{1}{r^2} \frac{\partial^2}{\partial \varphi^2} \right) \left(\sum_{k=0}^{N} c_k r^k \cos k\varphi \right) = 0$$

for $c_k \in \mathbb{R}$ and $N \in \mathbb{N}$.

3. Define the differential operator

$$L = \left(\frac{1}{\sin \vartheta} \frac{\partial}{\partial \vartheta} \left(\sin \vartheta \frac{\partial}{\partial \vartheta} \right) + \frac{1}{\sin^2 \vartheta} \frac{\partial^2}{\partial \varphi^2} \right)$$

acting on functions defined on $(0, \pi) \times (0, 2\pi)$. Prove that $Y_{1,1}(\vartheta, \varphi) :=$ $\sqrt{\frac{3}{8\pi}} \sin \vartheta \cos \varphi$ is an eigenfunction with eigenvalue -2 of L, i.e. $LY_{1,1} = -2Y_{1,1}$.

4. Prove that Δ_n is rotationally invariant, i.e. for all $T \in O(n)$ and $u \in C^2(\mathbb{R}^n)$ we have

$$(\Delta_n(u \circ T))(x) = (\Delta_n u)(Tx).$$

5. Consider the function $f : \mathbb{R}^2 \to \mathbb{R}$ defined by

$$f(x, y) = \begin{cases} \frac{x^2(y-2)^2}{x^6+(y-2)^6}, & (x, y) \neq (0, 2), \\ 0, & (x, y) = (0, 2). \end{cases}$$

Prove that

$$f_x(x, y) = \begin{cases} 2x(y - 2)^2 \frac{(y-2)^6 - 2x^6}{(x^6+(y-2)^6)^2}, & (x, y) \neq (0, 2) \\ 0, & (x, y) = (0, 2) \end{cases}$$

and

$$f_y(x, y) = \begin{cases} 2x^2(y - 2) \frac{x^6 - 2(y-2)^6}{(x^6+(y-2)^6)^2}, & (x, y) \neq (0, 2) \\ 0, & (x, y) = (0, 2). \end{cases}$$

Is f continuous at $(0, 2)$?

6. The function $g : B_1(0) \to \mathbb{R}$ is defined by

$$g(x, y) = \begin{cases} xy \ln \left(\ln \frac{1}{\sqrt{x^2+y^2}} \right), & (x, y) \neq (0, 0) \\ 0, & (x, y) = (0, 0). \end{cases}$$

Find f_x, f_y, f_{xx} and f_{yy}, and prove that at $(0,0)$ f_{xy} and f_{yx} do not exist.

7. For $1 \leq k \leq N$ let $v_k : G \to \mathbb{R}$, $G \subset \mathbb{R}^n$ open, have the partial derivative $\frac{\partial v_k}{\partial x_j}$ for $1 \leq j \leq n$. Find $\frac{\partial}{\partial x_j} \prod_{k=1}^{N} v_k(x)$.

8. Let $H : (0, \infty) \times (0, 2\pi) \to \mathbb{R}$ be defined with the help of $f : \mathbb{R}^2 \to \mathbb{R}$ by
$$H(\rho, \varphi) := f(x, y) = f(\rho \cos \varphi, \rho \sin \varphi).$$
Prove that
$$\left(\frac{\partial f}{\partial x} \right) + \left(\frac{\partial f}{\partial y} \right)^2 = \left(\frac{\partial H}{\partial \rho} \right)^2 + \left(\frac{1}{\rho} \frac{\partial H}{\partial \varphi} \right)^2.$$

9. Let $f : \mathbb{R}^2 \to \mathbb{R}$, $f(x, y) = \cos^2 x + y^4 - 1$, and let $g : (0, \pi) \to \mathbb{R}$, $g(x) = \sqrt{\sin x}$. Verify that $f(x, g(x)) = 0$ for all $x \in (0, \pi)$ and verify that
$$g'(x) = -\frac{\frac{\partial f}{\partial x}(x, g(x))}{\frac{\partial f}{\partial y}(x, g(x))}.$$

10. For $x \in \mathbb{R}^n, 1 \leq j \leq n, m \in \mathbb{N}_0$, prove the existence of a constant $c = c_{k,m,j}$ such that
$$\left| \frac{\partial^m}{\partial x_j^m} \frac{1}{(1 + ||x||^2)^{\frac{k}{2}}} \right| \leq c_{k,m,j} \frac{1}{(1 + ||x||^2)^{\frac{k+m}{2}}}.$$

11. The following exercises will help to develop some familiarity in using the multi-index notation:

a) For $x \in \mathbb{R}^3$ and $\alpha = (2, 1, 2)$ find $\partial^\alpha e^{-||x||^2}$;

b) For $x \in \mathbb{R}^n, \alpha \in \mathbb{N}_0^n, 1 \leq k, l, \leq n, l \neq k$, find
$$\frac{\partial^2}{\partial x_k \partial x_l} e^{x^\alpha};$$

c) For the polynomial $p(x) = \sum_{|\alpha| \leq m} c_\alpha x^\alpha, c_\alpha \in \mathbb{R}, x \in \mathbb{R}^n$, prove that $\partial^\beta p(x) = 0$ for $|\beta| > m$;

d) Prove for $a \in \mathbb{R}^n, x \in \mathbb{R}^n$ and $\gamma \in \mathbb{N}_0^n$ the estimate
$$|\partial^\gamma \cos \langle a, x \rangle| \leq |a^\gamma|.$$

6   The Differential of a Mapping

In one dimension the classical motivation for introducing derivatives is the tangent problem, but it turns out that for a general function $f : (a, b) \to \mathbb{R}$ we can define the tangent at a point $(x_0, f(x_0))$ of its graph only **after** we have defined the derivative of f at x_0. However, we can transfer the geometric idea of finding a tangent to an algebraic-analytic idea, namely to approximate $f(x) - f(x_0)$ in a neighbourhood of x_0 by a linear mapping. Once this is achieved we can return to the geometric problem. The idea of linear approximation is also possible in higher dimensions.

Definition 6.1. A. *Let $G \subset \mathbb{R}^n$ be an open set and $f : G \to \mathbb{R}^m$ a mapping. We call f **differentiable at** $x_0 \in G$ if a linear mapping $A : \mathbb{R}^n \to \mathbb{R}^m$ (depending in general on x_0) exists such that in a neighbourhood $U(x_0) \subset G$ of x_0 the following holds with $x_0 + y \in U(x_0)$*

$$f(x_0 + y) - f(x_0) = Ay + \varphi_{x_0}(y) \tag{6.1}$$

where with $U(0) := -x_0 + U(x_0)$ the mapping $\varphi_{x_0} : U(0) \to \mathbb{R}^m$ satisfies

$$\lim_{y \to 0} \frac{\varphi_{x_0}(y)}{\|y\|} = 0. \tag{6.2}$$

*We call A the **differential of** f at x_0.*
B. *If $f : G \to \mathbb{R}^m$ is differentiable at all points $x \in G$ we call f **differentiable in** **G**.*

If $f : G \to \mathbb{R}^m$ is differentiable at $x_0 \in G$ we often write for its differential at x_0

$$d_{x_0} f = A \quad \text{with } A \text{ from (6.1)}. \tag{6.3}$$

Since $d_{x_0} f : \mathbb{R}^n \to \mathbb{R}^m$ is a linear mapping we can apply it to $y \in \mathbb{R}^n$ for which we write $(d_{x_0} f)y$. When f is differentiable in G we obtain a new mapping

$$
\begin{aligned}
df &: G \to L(\mathbb{R}^n, \mathbb{R}^m) \\
x &\mapsto d_x f : \mathbb{R}^n \to \mathbb{R}^m
\end{aligned} \tag{6.4}
$$

where $L(\mathbb{R}^n, \mathbb{R}^m)$ is the vector space of all linear mappings from $\mathbb{R}^n$ to $\mathbb{R}^m$. Later on, so that no confusion arises, we will also use df instead of $d_x f$. Fixing in $\mathbb{R}^n$ and $\mathbb{R}^m$ as usual the canonical basis we can identify $L(\mathbb{R}^n, \mathbb{R}^m)$ with the $m \times n$-dimensional vector space of all $m \times n$ matrices (m rows and

n columns) which we denote by $M(m, n; \mathbb{R})$. Therefore we can identify $d_x f$ with the matrix function

$$x \mapsto d_x f = \begin{pmatrix} a_{11}(x) & \cdots & a_{1n}(x) \\ \vdots & & \vdots \\ a_{m1}(x) & \cdots & a_{mn}(x) \end{pmatrix} \tag{6.5}$$

and the task is to determine the coefficients of this matrix function in terms of f.

For $n = m = 1$ Definition 6.1 coincides of course with Definition I.21.2 when taking Theorem I.21.3 into account. In particular for $f : G \to \mathbb{R}$, $G \subset \mathbb{R}$ open, we find

$$d_x f = f'(x), \tag{6.6}$$

and the real number $f'(x)$, $x \in G$ fixed, represents the 1×1-matix in (6.5). To get some ideas for the general case we recall that $f : G \to \mathbb{R}^m$, $G \subset \mathbb{R}^n$, has components $f_l : G \to \mathbb{R}$, $l = 1, \ldots, m$, so $f = (f_1, \ldots, f_m)$, and φ_{x_0} has components $\varphi_{x_0,l} : G \to \mathbb{R}$, $l = 1, \ldots, m$, i.e. $\varphi_{x_0} = (\varphi_{x_0,1}, \ldots, \varphi_{x_0,n})$. Therefore (6.1) stands for m equations

$$f_l(x_0 + y) - f_l(x_0) = \sum_{j=1}^{n} a_{lj}(x_0) y_j + \varphi_{x_0,l}(y) \tag{6.7}$$

with $A = (a_{lj}(x_0))_{\substack{l=1,\ldots,m \\ j=1,\ldots,n}}$ and $y = (y_1, \ldots, y_n)$. Note that $\frac{\varphi_{x_0}(y)}{\|y\|}$ tends to $0 \in \mathbb{R}^m$ if and only if for all $1 \le l \le m$ the terms $\frac{\varphi_{x_0,l}(y)}{\|y\|}$ tend to $0 \in \mathbb{R}$. Moreover, if f is differentiable in G we find

$$f_l(x + y) - f_l(x) = \sum_{j=1}^{n} a_{lj}(x) y_j + \varphi_{x,l}(y)$$

in a neighbourhood $U(x)$ of x and $\varphi_{x,l} : U_x(0) \to \mathbb{R}$, $U_x(0) = -x + U(x)$. Since $d_{x_0} f$ is locally an approximation of f by a linear mapping, the expectation is that if $f : \mathbb{R}^n \to \mathbb{R}^m$ is itself a linear mapping then it must itself be its best linear approximation. Thus for a linear mapping we shall expect $d_x f = f$ for all $x \in \mathbb{R}^n$, or when working with the canonical basis in $\mathbb{R}^n$ and $\mathbb{R}^m$ and when representing f as corresponding matrix $A : \mathbb{R}^n \to \mathbb{R}^m$ we expect

$$dA = A.$$

Thus we get

Example 6.2. A. For $A \in M(m, n; \mathbb{R})$ interpreted as a linear mapping $A : \mathbb{R}^n \to \mathbb{R}^m$ we have $dA = A$. Indeed, we just have to choose for all $x \in \mathbb{R}^n$ the function $\varphi_x : \mathbb{R}^n \to \mathbb{R}^m$ as $\varphi_x(y) = 0$ for all $y \in \mathbb{R}^n$ and we find

$$A(x + y) - Ax = Ay$$

by the linearity of A.

B. Let $C = (c_{jl})_{j,l=1,\ldots,n}$ be a symmetric $n \times n$ matrix with real coefficients, i.e. $c_{jl} = c_{lj} \in \mathbb{R}$. Consider the function $f : \mathbb{R}^n \to \mathbb{R}$, $f(x) = \sum_{j,l=1}^n c_{jl} x_j x_l$.

With $x = \begin{pmatrix} x_1 \\ \vdots \\ x_n \end{pmatrix}$ and $x^t = (x_1, \ldots, x_n)$ (recall our convention when using matrix notation) we may write

$$f(x) = x^t C x = \sum_{j,l=1}^n c_{jl} x_j x_l.$$

For $x, y \in \mathbb{R}^n$ we now find

$$
\begin{aligned}
f(x + y) &= (x + y)^t C (x + y) \\
&= x^t C x + y^t C x + x^t C y + y^t C y \\
&= x^t C x + 2(Cx)^t y + y^t C y,
\end{aligned}
$$

and $d_x f : \mathbb{R}^n \to \mathbb{R}$, $(d_x f) y := 2(Cx)^t y$ is a linear mapping (depending on x). Furthermore, for $\varphi : \mathbb{R}^n \to \mathbb{R}$ defined as $\varphi(y) := y^t C y$ we have the estimate

$$|\varphi(y)| = \left| \sum_{j,l=1}^n c_{jl} y_j y_l \right| \leq \gamma \|y\|^2$$

implying that $\lim_{y \to 0} \frac{\varphi(y)}{\|y\|} = 0$. Thus we find

$$f(x + y) - f(x) = 2(Cx)^t y + \varphi(y)$$

with the linear mapping $y \mapsto 2(Cx)^t y$ and φ satisfying $\lim_{y \to 0} \frac{\varphi(y)}{\|y\|} = 0$ implying that f is differentiable with differential $d_x f = 2(Cx)^t$. Note that

$$\frac{\partial}{\partial x_k} f(x) = \frac{\partial}{\partial x_k} \sum_{j,l=1}^n c_{jl} x_j x_l = 2(Cx)_k^t$$

and we find that for f the differential at x is given by

$$\mathrm{d}_x f = \left(\frac{\partial f}{\partial x_1}(x), \cdots, \frac{\partial f}{\partial x_n}(x) \right) : \mathbb{R}^n \to \mathbb{R}$$

and

$$(\mathrm{d}_x f)y = \sum_{j=1}^{n} \frac{\partial f}{\partial x_j}(x)y_j. \tag{6.8}$$

With Example 6.2.B in mind the following result is not too surprising:

Theorem 6.3. *Let $G \subset \mathbb{R}^n$ be an open set and $f : G \to \mathbb{R}^m$ a mapping differentiable at $x \in G$, i.e.*

$$f(x+y) - f(x) = (\mathrm{d}_x f)y + \varphi_x(y)$$

with $\lim_{y \to 0} \frac{\varphi_x(y)}{\|y\|} = 0$. Then f is continuous at x, all components $f_j : G \to \mathbb{R}$, $1 \le j \le m$, of f have all first order partial derivatives at x, and moreover the differential of $\mathrm{d}_x f$ is given by the matrix $\left(\frac{\partial f_j}{\partial x_l}(x) \right)_{\substack{j=1,\dots,m \\ l=1,\dots n}}$, i.e.

$$(\mathrm{d}_x f)y = \begin{pmatrix} \frac{\partial f_1}{\partial x_1}(x) & \cdots & \frac{\partial f_1}{\partial x_n}(x) \\ \vdots & & \vdots \\ \frac{\partial f_m}{\partial x_1}(x) & \cdots & \frac{\partial f_m}{\partial x_n}(x) \end{pmatrix} \begin{pmatrix} y_1 \\ \vdots \\ y_n \end{pmatrix}. \tag{6.9}$$

Proof. Since f is differentiable at x there exists a linear mapping in $L(\mathbb{R}^n, \mathbb{R}^m)$ represented by a matrix $A \in M(m,n)$ and a function φ_x such that

$$f(x+y) - f(x) = Ay + \varphi_x(y)$$

and $\lim_{y \to 0} \frac{\varphi_x(y)}{\|y\|} = 0$. From Theorem 2.30 we deduce that $\|Ay\| \le \|A\| \, \|y\|$ and $\lim_{y \to 0} \frac{\varphi_x(y)}{\|y\|} = 0$ also implies $\lim_{y \to 0} \varphi_x(y) = 0$ which together yields

$$\lim_{y \to 0} f(x+y) = f(x),$$

i.e. f is continuous at x. With $A = \mathrm{d}_x f$ (more precisely $A(x) \in M(m,n;\mathbb{R})$, $A(x) = \mathrm{d}_x f$) we find

$$f_j(x+y) - f_j(x) = \sum_{l=1}^{n} a_{jl}(x)y_l + \varphi_{x,j}(y), \quad j = 1, \dots, m,$$

96

and in particular with $y = he_l$, $h \in \mathbb{R}$ and $|h|$ small such that $x + he_l \in G$, we have

$$f_j(x + he_l) - f_j(x) = ha_{jl}(x) + \varphi_{x,j}(he_l)$$

implying, recall $\|he_l\| = |h|$, that

$$\frac{\partial f_j}{\partial x_l}(x) = \lim_{h \to 0} \frac{f_j(x + he_l) - f_j(x)}{h} = a_{jl}(x) + \lim_{h \to 0} \frac{\varphi(he_j)}{h} = a_{jl}(x),$$

i.e. we find for $A = \mathrm{d}_x f$

$$\mathrm{d}_x f = \begin{pmatrix} \frac{\partial f_1}{\partial x_1}(x) & \cdots & \frac{\partial f_1}{\partial x_n}(x) \\ \vdots & & \vdots \\ \frac{\partial f_m}{\partial x_1}(x) & \cdots & \frac{\partial f_m}{\partial x_n}(x) \end{pmatrix}$$

and the theorem is proven.   □

Definition 6.4. *Let $f : G \to \mathbb{R}^m$, $G \subset \mathbb{R}^n$ open, be a mapping with components $f = (f_1, \ldots, f_m)$. If for $x \in G$ all partial derivatives $\frac{\partial f_j}{\partial x_l}(x)$, $1 \leq j \leq m$, $1 \leq l \leq n$, exist we call the matrix*

$$J_f(x) = \left(\frac{\partial f_j}{\partial x_l}(x) \right)_{\substack{j=1,\ldots,m \\ l=1,\ldots,n}} \tag{6.10}$$

*the **Jacobi matrix** of f at x. If $n = m$ we call $\det J_f(x)$ the **Jacobi determinant** of f at x.*

By Theorem 6.3, if $f : G \to \mathbb{R}^m$, $G \subset \mathbb{R}^n$ open, is differentiable at $x \in G$ then its differential at x is given by the Jacobi matrix of f at x. We are interested in a converse statement, i.e. that the existence of all partial derivatives $\frac{\partial f_j}{\partial x_l}(x)$ implies the differentiability of f at x. However we know that some caution is needed: the existence of all first order partial derivatives does not imply the continuity of a function, but by Theorem 6.3 a differentiable function is continuous.

Theorem 6.5. *Let $G \subset \mathbb{R}^n$ be an open set and $f : G \to \mathbb{R}^m$ a mapping such that for all its components $f_j : G \to \mathbb{R}$, $1 \leq j \leq m$, all partial derivatives $\frac{\partial f_j}{\partial x_l}(x)$, $j = 1, \ldots, m$, $l = 1, \ldots, n$, exist and are continuous at $x \in G$. Then f is differentiable at x.*

Proof. Since G is open there exists $\delta > 0$ such that $B_\delta(x) \subset G$. Take $\xi = (\xi_1, \ldots, \xi_n)$, $|\xi| < \delta$, and for $0 \le l \le n$ consider

$$z^{(l)} = x + \sum_{v=1}^{l} \xi_v e_v,$$

i.e. $z^{(0)} = x$ and $z^{(n)} = x + \xi$. By the mean value theorem we find

$$f_j(z^{(l)}) - f_j(z^{(l-1)}) = D_l f_j(y^{(l)})\xi_l, \quad l = 1, \ldots, n,$$

where $y^{(l)} = z^{(l-1)} + \theta_l \xi_l e_l$, $0 \le \theta_l \le 1$. Therefore we have

$$f_j(x + \xi) - f_j(x) = \sum_{l=1}^{n} D_l f_j(y^{(l)})\xi_l$$

and finally

$$f(x + \xi) - f(x) = \begin{pmatrix} \sum_{l=1}^{n} D_l f_1(y^{(l)})\xi_l \\ \vdots \\ \sum_{l=1}^{n} D_l f_m(y^{(l)})\xi_l \end{pmatrix}.$$

We now set $a_{lj} := \frac{\partial f_j(x)}{\partial x_l} = D_l f_j(x)$ and

$$\varphi_j(\xi) = \sum_{l=1}^{n} (D_l f_j(y^{(l)}) - a_{lj})\xi_l, \quad 1 \le j \le m.$$

Since $x \mapsto \frac{\partial f_j(x)}{\partial x_l}$ is continuous at x it follows that

$$\lim_{\xi \to 0} ((D_l f_j)(y^{(l)}) - a_{lj}) = 0, \quad 1 \le j \le m,$$

implying for $1 \le j \le m$ that $\lim_{\xi \to 0} \frac{\varphi_j(l\xi)}{\|\xi\|} = 0$, hence $\lim_{\xi \to 0} \frac{\varphi(\xi)}{\|\xi\|} = 0$ and the theorem is proven. $\qquad\square$

Example 6.6. Consider the mapping $S : (0, \infty) \times (0, \pi) \times (0, 2\pi) \to \mathbb{R}^3$ defined by $S(r, \vartheta, \varphi) = (r \sin \vartheta \cos \varphi, r \sin \vartheta \sin \varphi, r \cos \vartheta)$. With $S = (S_1, S_2, S_3)$ we find

$$J(r, \vartheta, \varphi) = d_{(r,\vartheta,\varphi)} S = \begin{pmatrix} \frac{\partial S_1}{\partial r}(r, \vartheta, \varphi) & \frac{\partial S_1}{\partial \vartheta}(r, \vartheta, \varphi) & \frac{\partial S_1}{\partial \varphi}(r, \vartheta, \varphi) \\ \frac{\partial S_2}{\partial r}(r, \vartheta, \varphi) & \frac{\partial S_2}{\partial \vartheta}(r, \vartheta, \varphi) & \frac{\partial S_2}{\partial \varphi}(r, \vartheta, \varphi) \\ \frac{\partial S_3}{\partial r}(r, \vartheta, \varphi) & \frac{\partial S_3}{\partial \vartheta}(r, \vartheta, \varphi) & \frac{\partial S_3}{\partial \varphi}(r, \vartheta, \varphi) \end{pmatrix}$$

$$= \begin{pmatrix} \sin \vartheta \cos \varphi & r \cos \vartheta \sin \varphi & -r \sin \vartheta \sin \varphi \\ \sin \vartheta \sin \varphi & r \cos \vartheta \sin \varphi & r \sin \vartheta \cos \varphi \\ \cos \vartheta & -r \sin \vartheta & 0 \end{pmatrix}.$$

Moreover the Jacobi determinant of S is given by

$$\det J_S(r, \vartheta, \varphi) = \det \begin{pmatrix} \sin\vartheta\cos\varphi & r\cos\vartheta\sin\varphi & -r\sin\vartheta\sin\varphi \\ \sin\vartheta\sin\varphi & r\cos\vartheta\sin\varphi & r\sin\vartheta\cos\varphi \\ \cos\vartheta & -r\sin\vartheta & 0 \end{pmatrix}.$$

$$= r^2 \sin\vartheta.$$

In Chapter 12 we will return to S when studying orthogonal curvilinear coordinates, especially spherical coordinates.

Corollary 6.7. **A.** *If for $f : G \to \mathbb{R}^m$, $G \subset \mathbb{R}^n$ open, all partial derivatives $\frac{\partial f_j}{\partial x_l}$, $1 \leq j \leq m$, $1 \leq l \leq n$, are continuous then f is continuous in G.*
B. *We have the implications*

i) $f \in C^1(G)$ implies f is differentiable in G;

ii) f is differentiable in G implies all partial derivatives $\frac{\partial f_j}{\partial x_l}$ exist.

In general neither the converse of i) nor of ii) holds. Indeed, Example I.21.15 gives a function, namely

$$f(x) = \begin{cases} x^2 \sin\frac{1}{x}, & x \neq 0 \\ 0, & x = 0 \end{cases}$$

which is differentiable on $\mathbb{R}$, but $f'(0)$ is not continuous at $x = 0$, which implies that the converse of i) cannot hold.
Furthermore, we know that

$$f(x, y) := \begin{cases} \frac{xy}{x^2+y^2}, & (x, y) \neq (0, 0) \\ 0, & (x, y) = (0, 0) \end{cases}$$

has for all $(x, y) \in \mathbb{R}^2$ the partial derivative $f_x(x, y)$ and $f_y(x, y)$, and in particular $f_x(0, 0) = f_y(0, 0) = 0$, see Example 5.8, but we know by Example 4.1 that f is not continuous at $(0, 0)$, hence by Theorem 6.3 the function f cannot be differentiable at $(0, 0)$.

Corollary 6.8. *Differentiability is a linear operation, i.e. if $f, g : G \to \mathbb{R}^m$, $G \subset \mathbb{R}^n$ open, are differentiable at $x = 0$ (or in G) then for all $\lambda, \mu \in \mathbb{R}$ the mapping $\lambda f + \mu g : G \to \mathbb{R}^m$ is differentiable at x (in G) and we have*

$$d_x(\lambda f + \mu g) = \lambda d_x f + \mu d_x g. \tag{6.11}$$

Proof. Since $\frac{\partial}{\partial x_l}(\lambda f_j + \mu g_j) = \lambda\frac{\partial}{\partial x_l}f_j + \mu\frac{\partial}{\partial x_l}g_j$ the result follows from Theorem 6.3. $\qquad\square$

Next we want to prove the **chain rule** for the differential of the composition of two differentiable mappings.

Theorem 6.9. *Let $U \subset \mathbb{R}^n$ and $V \subset \mathbb{R}^m$ be open sets and $g : U \to \mathbb{R}^m$, $f : V \to \mathbb{R}^k$ mappings such that $g(U) \subset V$. Suppose that g is differentiable at $x \in U$ and f is differentiable at $y := g(x)$. Then $f \circ g : U \to \mathbb{R}^k$ is differentiable in x and we have*

$$d(f \circ g)(x) = (df)(g(x)) \circ dg(x). \tag{6.12}$$

(Note: $dg(x) : \mathbb{R}^n \to \mathbb{R}^m$, $(df)(g(x)) : \mathbb{R}^m \to \mathbb{R}^k$, hence

$$df(g(x)) \circ dg(x) : \mathbb{R}^n \to \mathbb{R}^k \tag{6.13}$$

is the composition of two linear mappings.)

Proof. Let $A := dg(x)$ and $B := df(y)$. We have to show that $d(f \circ g)(x) = BA$. By assumption the following hold

$$g(x + \xi) = g(x) + A\xi + \varphi(\xi)$$

$$f(y + \eta) = f(y) + B\eta + \psi(\eta)$$

where

$$\lim_{\xi \to 0} \frac{\varphi(\xi)}{\|\xi\|} = 0, \quad \lim_{\eta \to 0} \frac{\psi(\eta)}{\|\eta\|} = 0.$$

For $\eta := g(x + \xi) - g(x) = A\xi + \varphi(\xi)$ we now find

$$\begin{aligned}
(f \circ g)(x + \xi) &= f(g(x) + \eta) \\
&= f(g(x)) + BA\xi + B\varphi(\xi) + \psi(A\xi + \varphi(\xi)) \\
&= (f \circ g)(x) + BA\xi + \chi(\xi),
\end{aligned}$$

where

$$\chi(\xi) := B\varphi(\xi) + \psi(A\xi + \varphi(\xi)).$$

We have to prove that

$$\lim_{\xi \to 0} \frac{\chi(\xi)}{\|\xi\|} = 0.$$

We know that
$$\lim_{\xi \to 0} \frac{\varphi(\xi)}{\|\xi\|} = 0, \text{ and } \lim_{\xi \to 0} \frac{B\varphi(\xi)}{\|\xi\|} = 0.$$

Further there is a constant $K > 0$ such that $\|\varphi(\xi)\| \le K \|\xi\|$ for all ξ, $\|\xi\|$ sufficiently small. Since $\lim_{\eta \to 0} \frac{\psi(\eta)}{\|\varphi\|} = 0$ it follows that $\psi(\eta) = \|\eta\| \psi_1(\eta)$ where $\lim_{\eta \to 0} \psi_1(\eta) = 0$. Thus for $\|\xi\|$ sufficiently small we get

$$\|\psi(A\xi + \varphi(\xi))\| \le (C_{A,2,2} + K) \|\xi\| \|\psi_1(A\xi + \varphi(\xi))\|,$$

or
$$\lim_{\xi \to 0} \frac{\psi(A\xi + \varphi(\xi))}{\|\xi\|} = 0,$$

since $A\xi + \varphi(\xi) \to 0$ as $\xi \to 0$, i.e.

$$\lim_{\xi \to 0} \frac{\chi(\xi)}{\|\xi\|} = 0$$

which proves the theorem.   □

Corollary 6.10. *Let $U \subset \mathbb{R}^n$ and $V \subset \mathbb{R}^m$ be open and*

$$f : V \to \mathbb{R}$$
$$y \mapsto f(y)$$

and

$$g : U \to \mathbb{R}^m$$

$$x \mapsto g(x) = \begin{pmatrix} g_1(x) \\ \vdots \\ g_m(x) \end{pmatrix}$$

be two differentiable mappings such that $g(U) \subset V$. Then $h := f \circ g : U \to \mathbb{R}$ has first order partial derivatives and for $i = 1, \ldots, n$ the following holds

$$\frac{\partial h}{\partial x_i} = \sum_{j=1}^{n} \frac{\partial f}{\partial y_j}(g_1(x), \ldots, g_m(x)) \frac{\partial g_j}{\partial x_i}(x_1, \ldots, x_n). \tag{6.14}$$

Proof. The Jacobi matrices of h, g and f are given by

$$dh(x) = \left(\frac{\partial h}{\partial x_1}(x), \cdots, \frac{\partial h}{\partial x_n}(x) \right),$$

$$(df)(g(x)) = \left(\frac{\partial f}{\partial y_1}(g(x)), \cdots, \frac{\partial f}{\partial y_m}(g(x)) \right)$$

and

$$dg(x) = \begin{pmatrix} \frac{\partial g_1}{\partial x_1}(x) & \cdots & \frac{\partial g_1}{\partial x_n}(x) \\ \vdots & & \vdots \\ \frac{\partial g_m}{\partial x_1}(x) & \cdots & \frac{\partial g_m}{\partial x_n}(x) \end{pmatrix}.$$

Now the corollary follows from

$$dh(x) = (df)(g(x)) \circ dg(x).$$

$\square$

Definition 6.11. *A function* $F : \mathbb{R}^n \setminus \{0\} \to \mathbb{R}$ *satisfying with some* $m \in \mathbb{R}$

$$F(\lambda x_1, \dots, \lambda x_n) = \lambda^m F(x_1, \dots, x_n) \tag{6.15}$$

for all $\lambda > 0$ *and* $x = (x_1, \dots, x_n) \in \mathbb{R}^n \setminus \{0\}$ *is called a **homogeneous function of degree** m.*

Example 6.12. A. Let $m \in \mathbb{N}$ then

$$p(x) = \sum_{|\alpha|=m} a_\alpha x^\alpha$$

with $\alpha \in \mathbb{N}_0^n$, $a_\alpha \in \mathbb{R}$ and $x \in \mathbb{R}^n$ is a homogeneous polynomial of degree m.
B. Let $h : \mathbb{R} \to \mathbb{R}$ be a function and define $f : \mathbb{R} \setminus \{0\} \times \mathbb{R} \to \mathbb{R}$ by $f(x,y) = h\left(\frac{y}{x}\right)$. Then h is homogeneous of degree 0: for $\lambda \neq 0$ we have

$$f(\lambda x, \lambda y) = h\left(\frac{\lambda y}{\lambda x}\right) = h\left(\frac{y}{x}\right) = f(x,y) = \lambda^0 f(x,y).$$

(Note that in this example we must exclude $\lambda = 0$ or we need to amend our definition of a homogeneous function.)
C. If $f_1 : \mathbb{R}^n \setminus \{0\} \to \mathbb{R}$ is homogeneous of degree m_1 and if $f_2 : \mathbb{R}^n \setminus \{0\} \to \mathbb{R}$ is homogeneous of degree m_2 then $f_1 \cdot f_2 : \mathbb{R}^n \setminus \{0\} \to \mathbb{R}$ is homogeneous of degree $m_1 + m_2$:

$$\begin{aligned} (f_1 \cdot f_2)(\lambda x_1, \dots, \lambda x_n) &= f_1(\lambda x_1, \dots, \lambda x_n) f_2(\lambda x_1, \dots, \lambda x_n) \\ &= \lambda^{m_1} f_1(x_1, \dots, x_n) \lambda^{m_2} f_2(x_1, \dots, x_n) \\ &= \lambda^{m_1+m_2} (f_1 \cdot f_2)(x_1, \dots, x_n). \end{aligned}$$

Proposition 6.13 (Euler's theorem on homogeneous functions). *For a differentiable function $F : \mathbb{R}^n \setminus \{0\} \to \mathbb{R}$ homogeneous of degree m the relation*

$$\sum_{j=1}^{n} x_j \frac{\partial F}{\partial x_j}(x) = mF(x) \qquad (6.16)$$

holds.

Proof. We want to apply Corollary 6.10. For this we define the mapping $g : \mathbb{R} \times \mathbb{R}^n \to \mathbb{R}^n$, $(\lambda, x) \mapsto y = g(\lambda, x) := \lambda x$ and now we consider $h : \mathbb{R} \times \mathbb{R}^n \to \mathbb{R}$, $h := F \circ g$. From (6.14) we derive

$$\frac{\partial h}{\partial \lambda}(\lambda, x) = \sum_{j=1}^{n} \frac{\partial g_j}{\partial \lambda}(\lambda, x) \frac{\partial F}{\partial y_j}(y)$$

$$= \sum_{j=1}^{n} x_j \frac{\partial F}{\partial y_j}(y)$$

Since $h(\lambda, x) = F(\lambda x) = \lambda^m F(x)$ it follows further that

$$\frac{\partial h}{\partial \lambda}(\lambda, x) = m\lambda^{m-1} F(x)$$

and for $\lambda = 1$ we arrive at

$$\sum_{j=1}^{n} x_j \frac{\partial F}{\partial y_j}(y) = \sum_{j=1}^{n} x_j \frac{\partial F}{\partial x_j}(x) = mF(x).$$

$\square$

A partial derivative of a given function can be interpreted as a one-dimensional derivative in the direction of a coordinate axis. We want to study derivatives in an arbitrary direction.

Definition 6.14. *Let $G \subset \mathbb{R}^n$ be an open set and $f : G \to \mathbb{R}$ be a function. Further let $x \in G$ and $v \in \mathbb{R}^n$, $\|v\| = 1$, a vector. The **directional derivative** of f at x in the direction of v is defined as*

$$D_v f(x) := \frac{d}{dt} f(x + tv)\Big|_{t=0} := \lim_{t \to 0} \frac{f(x + tv) - f(x)}{t}. \qquad (6.17)$$

Clearly, for $v = e_j$ we find

$$D_v = D_{e_j} = \frac{\partial}{\partial x_j},$$

i.e. we recover the partial derivative with respect to x_j.

In order to relate directional derivatives to partial derivatives (as well as for other reasons which we will encounter soon) it is helpful to introduce the gradient of a scalar valued function.

Definition 6.15. *For a differentiable function $\varphi : G \to \mathbb{R}$, $G \subset \mathbb{R}^n$ open, the **gradient** is defined as the mapping*

$$\operatorname{grad} \varphi : G \to \mathbb{R}^n, \ (\operatorname{grad} \varphi)(x) = \left(\frac{\partial \varphi}{\partial x_1}(x), \cdots, \frac{\partial \varphi}{\partial x_n}(x) \right). \tag{6.18}$$

Remark 6.16. For defining $\operatorname{grad} \varphi$ it is sufficient to assume the existence of all first order partial derivatives. Further we will write often $\operatorname{grad} \varphi(x)$ for $(\operatorname{grad} \varphi)(x)$, however we try to avoid the notation $\nabla \varphi(x)$ for $(\operatorname{grad} \varphi)(x)$, where ∇ denotes the **nabla-operator**.

Theorem 6.17. *Let $g \subset \mathbb{R}^n$ be an open set and $f : G \to \mathbb{R}$ a continuously differentiable function, i.e. f is differentiable and $x \mapsto \mathrm{d}f(x)$ is continuous. Then we have for every $x \in G$ and $v \in \mathbb{R}^n$, $v = (v_1, \ldots, v_n)$, with $\|v\| = 1$ the formula*

$$D_v f(x) = \sum_{j=1}^{n} v_j \frac{\partial f(x)}{\partial x_j} = \langle v, \operatorname{grad} f(x) \rangle. \tag{6.19}$$

Proof. Let $g : \mathbb{R} \to \mathbb{R}^n$ be defined by

$$g(t) := x + tv = (x_1 + tv_1, \ldots, x_n + tv_n).$$

For $\epsilon > 0$ sufficiently small we have $g((-\epsilon, \epsilon)) \subset G$ and it follows for $h := f \circ g : (-\epsilon, \epsilon) \to \mathbb{R}$ that

$$D_v f(x) = \frac{d}{dt} f(x + tv) \Big|_{t=0} = \frac{dh}{dt}(0).$$

The chain rule implies however

$$\frac{dh(t)}{dt} = \sum_{j=1}^{n} \frac{\partial f}{\partial x_j}(g(t)) \frac{dg_j(t)}{dt},$$

and since $\frac{dg_j(t)}{dt} = \frac{d}{dt}(x_j + tv_j) = v_j$, we finally arrive at

$$\frac{dh}{dt}(0) = \sum_{j=1}^{n} \frac{\partial f}{\partial x_j}(x)v_j.$$

$\square$

Example 6.18. A. From Example 5.7.C we deduce for $h : \mathbb{R}^3 \setminus \{0\} \to \mathbb{R}$, $h(x) = \frac{1}{\|x\|}$, that $\operatorname{grad} h(x) = \frac{-x}{\|x\|^3}$.

B. Let $h : \mathbb{R} \to \mathbb{R}$ be a continuously differentiable function and define $f : \mathbb{R}^2 \to \mathbb{R}$ by $f(x,t) := h(x - ct)$, $c > 0$. The vector $v = \left(\frac{c}{(1+c^2)^{\frac{1}{2}}}, \frac{1}{(1+c^2)^{\frac{1}{2}}} \right)$ has length 1 and the directional derivative of f in the direction of v is given by

$$D_v f(x,t) = \frac{c}{(1+c^2)^{\frac{1}{2}}} \frac{\partial f}{\partial x}(x,t) + \frac{1}{(1+c^2)^{\frac{1}{2}}} \frac{\partial f}{\partial t}(x,t)$$

$$= \frac{c}{(1+c^2)^{\frac{1}{2}}} h'(x - ct) - \frac{c}{(1+c^2)^{\frac{1}{2}}} h'(x - ct) = 0,$$

which means that $f(x,t) = h(x - ct)$ is constant in the direction of v. If we write $v = \frac{c}{(1+c^2)^{\frac{1}{2}}} \left(1, \frac{1}{c}\right)$ the result becomes a bit more clear: by $x - ct = \kappa =$ constant a straight line is given in the plane $\mathbb{R}^2$ determined by the equation $\left\langle \begin{pmatrix} x \\ t \end{pmatrix}, \begin{pmatrix} 1 \\ -c \end{pmatrix} \right\rangle - \kappa = 0$ and the vector $\begin{pmatrix} 1 \\ -c \end{pmatrix}$ is orthogonal to $\begin{pmatrix} 1 \\ \frac{1}{c} \end{pmatrix}$ since $\left\langle \begin{pmatrix} 1 \\ -c \end{pmatrix}, \begin{pmatrix} 1 \\ \frac{1}{c} \end{pmatrix} \right\rangle = 1 - 1 = 0.$

C. For the directional derivative $D_v f$ the estimate

$$|D_v f(x)| \le \|\operatorname{grad} f(x)\| \tag{6.20}$$

holds where as usual $\|\cdot\|$ denotes the Euclidean norm in $\mathbb{R}^n$. We derive (6.20) from (6.19) by applying the Cauchy-Schwarz inequality

$$|D_v f(x)| = |\langle v, \operatorname{grad} f(x)\rangle| \le \|v\| \|\operatorname{grad} f(x)\| = \|\operatorname{grad} f(x)\|$$

since $\|v\| = 1$.

We can now prove a variant of the **mean value theorem**.

Theorem 6.19. *Let $G \subset \mathbb{R}^n$ be an open set and for $x, y \in G$ we assume that the line segment connecting x and y is entirely contained in G. If $f \in C^1(G)$ then there exists $\vartheta = \vartheta_{x,y} \in (0, 1)$ such that*

$$f(y) - f(x) = \langle \operatorname{grad} f(x + \vartheta(y - x)), y - x \rangle. \tag{6.21}$$

Proof. We define the function $g : [0, 1] \to \mathbb{R}$, $g(t) = f(x + t(y - x))$. This function is continuously differentiable on $[0, 1]$ and by the one-dimensional mean value theorem, Corollary I.22.6, we know the existence of ϑ such that

$$g(1) - g(0) = g'(\vartheta).$$

Now, $g(1) = f(y)$ and $g(0) = f(x)$. Furthermore, the chain rule in the form of Corollary 6.10 yields

$$g'(t) = \sum_{j=1}^{n} \frac{\partial f}{\partial x_j}(x + t(y - x))(y - x) = \langle \operatorname{grad} f(x + \vartheta(y - x)), y - x \rangle$$

implying the theorem. $\qquad\square$

In view of Theorem 6.19 we give

Definition 6.20. *A set $K \subset \mathbb{R}^n$ is called **convex** if for every $x, y \in K$ it follows that the line segment connecting x and y belongs entirely to K, i.e. $x, y \in K$ implies $\{z = \lambda x + (1 - \lambda)y | \lambda \in [0, 1]\} \subset K$.*

We will see more applications of the gradient when turning our attention to geometry as well as dealing with vector calculus. We want to close the chapter by looking at "higher order" differentials.

Let $f : G \to \mathbb{R}^m$, once more $G \subset \mathbb{R}^n$ open, be a differentiable mapping. Thus for every $x \in G$ there exists the differential $d_x f$ and we can consider the mapping $x \mapsto d_x f$. This is a mapping from G to $L(\mathbb{R}^n, \mathbb{R}^m)$ or when introducing the canonical basis in $\mathbb{R}^n$ and $\mathbb{R}^m$, we can consider this as a mapping from G to $M(m, n; \mathbb{R})$ which we can identify with $\mathbb{R}^{mn}$. Thus the differential $d_x(df)$ must be a linear mapping from $\mathbb{R}^n$ to $L(\mathbb{R}^n, \mathbb{R}^m)$ (or from $\mathbb{R}^n$ to $M(m, n; \mathbb{R})$), i.e. $d_x f(df) \in L(\mathbb{R}^n, L(\mathbb{R}^n, \mathbb{R}^m))$.

Problems

Problems 1-5 are designed for practice in calculating Jacobi matrices and Jacobi determinants. Instead of creating artificial examples we prefer to consider coordinate transformations which we will take up in Chapter 12. Some problems consist of the calculus part of a geometric problem which we will return to later.

1. Let $F : [0, \infty) \times [0, 2\pi) \times \mathbb{R} \to \mathbb{R}^3$ be given by
   $F(\rho, \varphi, z) := (\rho \cos \varphi, \rho \sin \varphi, z)$. Find $J_F(\rho, \varphi, z)$ and $\det J_F(\rho, \varphi, z)$.

2. Consider the mapping $G : \mathbb{R} \times [0, \infty) \times \mathbb{R} \to \mathbb{R}^3$ defined by $G(u, v, z) := (\frac{1}{2}(u^2 - v^2), uv, z)$ and calculate $\det J_G(u, v, z)$.

3. The mapping $H : [0, \infty) \times [0, \infty) \times [0, 2\pi) \to \mathbb{R}^n$ is given by $H(u, v, \varphi) := (uv \cos \varphi, uv \sin \varphi, \frac{1}{2}(u^2 - v^2))$. Find $J_H(u, v, \varphi)$ and its determinant.

4. Find the Jacobi determinant of $K : [0, \infty) \times [0, \pi] \times [0, 2\pi) \to \mathbb{R}^3$ with $K(\xi, \eta, \varphi) := (\sinh \xi \sin \eta \cos \varphi, \sinh \xi \sin \eta \sin \varphi, \cosh \xi \cos \eta)$.

5. Spherical coordinates on $\mathbb{R}^3$ are given by the mapping $S : [0, \infty) \times [0, \pi] \times [0, 2\pi) \to \mathbb{R}^3$ where $S(r, \vartheta, \varphi) := (r \sin \vartheta \cos \varphi, r \sin \vartheta \sin \varphi, r \cos \vartheta)$. For $r = 2$ and $\varphi = \frac{\pi}{4}$ find a linear approximation to $S(2, \vartheta, \frac{\pi}{4})$ for $\left| \vartheta - \frac{\pi}{2} \right| < \epsilon$.

6. The vector space $M(m, n; \mathbb{R})$ of all $m \times n$ matrices can be identified by $\mathbb{R}^{nm}$, hence we can handle matrix functions $L : G \to M(m, n; \mathbb{R}), G \subset \mathbb{R}^k$,

$$L(x) = \begin{pmatrix} a_{11}(x) \ldots a_{1n}(x) \\ \cdot \\ \cdot \\ \cdot \\ a_{m1}(x) \ldots a_{mn}(x) \end{pmatrix}.$$

   For $f : G \to \mathbb{R}^m$, $G \subset \mathbb{R}^n$, with Jacobi matrix at $x \in G$ given by

$$A(x) := J_f(x) = \left(\frac{\partial f_j}{\partial x_k}(x) \right)_{\substack{j=1,\ldots,m \\ k=1,\ldots,n}}$$

   we can consider the mapping $A : J_f : G \to M(m, n; \mathbb{R})$. Suppose that $f_j \in C^2(G)$, $1 \leq j \leq m$. Find the Jacobi matrix $J_A(x) \in M(mn, n; \mathbb{R})$. **Hint**: first identify $M(m, n; \mathbb{R})$ with $\mathbb{R}^{mn}$.

7. For $(r, \varphi) \in (0, \infty) \times [0, 2\pi)$ define $U(r, \varphi) = \begin{pmatrix} r \cos \varphi & -r \sin \varphi \\ r \sin \varphi & r \cos \varphi \end{pmatrix}$ and consider the mapping

$$U : (0, \infty) \times [0, 2\pi) \to M(2, \mathbb{R}) \cong \mathbb{R}^4$$
$$(r, \varphi) \mapsto U(r, \varphi).$$

Find the differential of U at (r_0, φ_0).

8. Consider $f : \mathbb{R}^3 \to \mathbb{R}^2, f(x, y, z) = \begin{pmatrix} (x^2 + y^2 + z^2 + 1)^{\frac{1}{2}} \\ x + y + z \end{pmatrix}$ and $g :$ $\mathbb{R}^2 \to \mathbb{R}^3, g(p, q) = \begin{pmatrix} p + q \\ p \cdot q + 2p \\ 0 \end{pmatrix}$. Find the Jacobi matrix of $f \circ g :$ $\mathbb{R}^2 \to \mathbb{R}^2$ and $g \circ f : \mathbb{R}^3 \to \mathbb{R}^3$.

9. Let $f : G \to \mathbb{R}^k, G \subset \mathbb{R}^n$, be a differentiable function and let $\mathrm{pr}_j :$ $\mathbb{R}^k \to \mathbb{R}, 1 \leq j \leq k$, be the jth coordinate projection. Use the chain rule to find $d(\mathrm{pr}_j \circ f)(x), x \in G$.

10. Let $g : G \to \mathbb{R}, G \subset \mathbb{R}^n$ open, be a C^1-function and $g(x) \neq 0$ for all $x \in G$. Verify that $\mathrm{grad} \left(\frac{1}{g} \right)(x) = -\frac{1}{g^2(x)} \mathrm{grad} g(x)$.

11. We call $f : \mathbb{R}^n \to \mathbb{R}$ rotationally invariant if $f(x_1, \ldots, x_n) = g(r), r = (x_1 + \cdots + x_n^2)^{\frac{1}{2}}$ holds for some $g : [0, \infty) \to \mathbb{R}$, see Example 5.15. Suppose that g is a C^2-function homogeneous of degree $\alpha \in \mathbb{R}$. Is $\frac{\partial^2 f}{\partial x_j \partial x_k}$ a homogeneous function?
Hint: first prove that if $g : [0, \infty) \to \mathbb{R}$ is homogeneous of degree $\alpha > 0$ then g' is homogeneous of degree $\alpha - 1$.

12. Let $f : \mathbb{R}^2 \to \mathbb{R}, f(x, y) = h(x^2 + y^2), h : [0, \infty) \to \mathbb{R}$ is a C^2-function. Let $(x_0, y_0) \in \partial B_r(0), r > 0$, and consider the tangential direction $\vec{t} = \begin{pmatrix} -\frac{y_0}{r^2} \\ \frac{x_0}{r^2} \end{pmatrix}$ to $\partial B_r(0)$ at $(x_0, y_0) \in \partial B_r(0)$. Prove that the directional derivative of f in the direction $\vec{t}$ at (x_0, y_0) is zero, i.e. $\frac{df}{dt}(x_0, y_0) = 0$.

13. Let $f : \mathbb{R}^k \to \mathbb{R}$ be a C^1-function and let $v \in \mathbb{R}^k, ||v|| = 1$. Suppose that $\mathrm{grad} f(x_0) \neq 0$ and prove that

$$\sup_{||v||=1} \frac{\partial f(x_0)}{\partial v} = \frac{\partial f(x_0)}{\partial w} = ||\mathrm{grad} f(x_0)||,$$

where $w = \frac{\operatorname{grad} f(x_0)}{||\operatorname{grad} f(x_0)||}$.

14.   a) Use the mean value theorem to prove that if $G \subset \mathbb{R}^n$ is an open convex set and $g : G \to \mathbb{R}$ is a C^1-function with $\operatorname{grad} g(x) = 0$ for all $x \in G$, then f is constant.

        b) Let $G \subset \mathbb{R}^n$ be an open convex set and $f : G \to \mathbb{R}$ a C^1-function such that $\sup_{x \in G} ||\operatorname{grad} f(x)|| \leq M < \infty$. Prove that f is Lipschitz continuous, i.e. for some $L > 0$ the following holds for $x, y \in G$,

$$|f(x) - f(y)| \leq L||x - y||.$$

7 Curves in $\mathbb{R}^n$

So far we have extended ideas such as continuity and differentiability from real-valued functions of one real variable to mappings $f : G \to \mathbb{R}^m$, $G \subset \mathbb{R}^n$ an open set. In this and the next chapter we want to apply these new concepts to geometry. We would like to mention that the book [11] by M. DoCarmo is a nice introduction to these geometric topics, and we have used some examples treated in [33]. We start this chapter with curves, more precisely with curves in $\mathbb{R}^n$, but particular emphasis will be given to curves in the plane ($n = 2$) and curves in $\mathbb{R}^3$.

In order to get a better understanding of connectivity we have introduced in Chapter 4, see Definition 4.11, the concept of a path as a continuous mapping $\gamma : [a, b] \to G$, $G \subset \mathbb{R}^n$. The idea of pathwise connected sets extends to general topological spaces and it is worth keeping the name "path" in this context. A curve as defined below is a path, but often we will add regularity assumptions; moreover in the geometric context the name curve is more common.

Definition 7.1. *A **parametric curve** in $\mathbb{R}^n$ is a continuous mapping $\gamma : I \to \mathbb{R}$ where $I \subset \mathbb{R}$ is an interval. By definition the **trace** of a curve γ is its range, i.e.* $\mathrm{tr}(\gamma) := \gamma(I) \subset \mathbb{R}^n$.

Example 7.2. Every continuous function $f : I \to \mathbb{R}$ gives rise to a curve in $\mathbb{R}^2$ by $\gamma : I \to \mathbb{R}^2$, $\gamma(t) = (t, f(t))$. The trace of γ is just the graph of f, $\mathrm{tr}(\gamma) = \Gamma(f)$.

In order to study curves we need to better understand the analysis of functions $f : I \to \mathbb{R}^n$, and it even makes sense to look at curves in $M(m, n; \mathbb{R}) \cong \mathbb{R}^{mn}$.

We remind the reader that for as long as we do not make explicit use of matrix operations we do not distinguish for a vector $v \in \mathbb{R}^n$ between $v = (v_1, \dots, v_n)$ and $v = \begin{pmatrix} v_1 \\ \vdots \\ v_n \end{pmatrix}$. Let $I \subset \mathbb{R}$ be an open interval and $f : I \to \mathbb{R}^n$, $f(t) = (f_1(t), \dots, f_n(t))$, as well as $A : I \to M(m, n; \mathbb{R})$, $A(t) = (a_{jl}(t))_{\substack{j=1,\dots,m \\ l=1,\dots,n}}$, be mappings with components $f_j : I \to \mathbb{R}$ and $a_{jl} : I \to \mathbb{R}$, respectively. The mappings f and A are continuous if and only if all their components are continuous functions. Moreover, there is only one natural way to define the derivative of f and A, respectively, namely by

$$f'(t) := \frac{df}{dt}(t) := \begin{pmatrix} f_1(t) \\ \vdots \\ f_n(t) \end{pmatrix} \tag{7.1}$$

and

$$A'(t) := \frac{dA}{dt}(t) := \begin{pmatrix} a'_{11}(t) & \cdots & a'_{1n}(t) \\ \vdots & & \vdots \\ a'_{m1}(t) & \cdots & a'_{mn}(t) \end{pmatrix}, \tag{7.2}$$

and this coincides of course with our general Definition 6.1.

How to define higher order derivatives of f (or A) is now obvious. Furthermore we can integrate f and A over any compact interval $[a, b] \subset I$, provided all components are continuous, and we get

$$\int_a^b f(t)\, dt = \begin{pmatrix} \int_a^b f_1(t)\, dt \\ \vdots \\ \int_a^b f_n(t)\, dt \end{pmatrix} \in \mathbb{R}^n \tag{7.3}$$

as well as

$$\int_a^b A(t)\, dt \begin{pmatrix} \int_a^b a_{11}(t)\, dt & \cdots & \int_a^b a_{1n}(t)\, dt \\ \vdots & & \vdots \\ \int_a^b a_{m1}(t)\, dt & \cdots & \int_a^b a_{mn}(t)\, dt \end{pmatrix} \in M(m, n; \mathbb{R}). \tag{7.4}$$

Since all operations are reduced to operations in the components we find immediately under appropriate assumptions on the components that the fundamental theorem holds:

$$f(b) - f(a) = \int_a^b f'(t)\, dt \tag{7.5}$$

or

$$\begin{pmatrix} f_1(b) - f_1(a) \\ \vdots \\ f_n(b) - f_n(a) \end{pmatrix} = \begin{pmatrix} \int_a^b f'_1(t)\, dt \\ \vdots \\ \int_a^b f'_n(t)\, dt \end{pmatrix}. \tag{7.6}$$

112

A further consequence is for $r_1, r_2 \in I$

$$f(r_1 + r_2) - f(r_1) = \int_{r_1}^{r_1+r_2} f'(s) \, \mathrm{d}s = \left(\int_0^1 f'(r_1 + tr_2) \, \mathrm{d}t \right) r_2. \qquad (7.7)$$

In addition we have

Lemma 7.3. *For a continuous mapping $f : [a, b] \to \mathbb{R}^n$ the following holds*

$$\left\| \int_a^b f(t) \, \mathrm{d}t \right\| \leq \int_a^b \| f(t) \| \, \mathrm{d}t. \qquad (7.8)$$

Proof. For $u := \int_a^b f(t) \, \mathrm{d}t \in \mathbb{R}^n$ we put $K := \|u\|$. Using the Cauchy-Schwarz inequality it follows that

$$K^2 = \sum_{j=1}^n u_j^2 = \sum_{j=1}^n \int_a^b f_j(t) \, \mathrm{d}t \, u_j$$

$$= \int_a^b \left(\sum_{j=1}^n f_j(t) u_j \right) \mathrm{d}t$$

$$\leq \int_a^b \| f(t) \| \, \| u \| \, \mathrm{d}t = K \int_a^b \| f(t) \| \, \mathrm{d}t,$$

and when dividing by K we arrive at

$$K = \left\| \int_a^b f(t) \, \mathrm{d}t \right\| \leq \int_a^b \| f(t) \| \, \mathrm{d}t.$$

$\square$

With these preparations we can now start to investigate parametric curves and their traces as geometric objects. In this chapter we are only interested in **local properties** of a curve, i.e. properties of its trace in a "neighbourhood" of a point $\gamma(t_0)$, and we want to use calculus methods for our investigations. For this $\gamma : I \to \mathbb{R}^n$ needs to have a certain number of derivatives. It is custom to denote in geometry (and in mechanics) the derivative with respect to the curve parameter with a dot and we will adopt this custom, thus we will write

$$\dot{\gamma}(t) := \gamma'(t), \quad \ddot{\gamma} := \gamma''(t), \text{ etc.} \qquad (7.9)$$

However we also use $\gamma^{(l)}(t)$ for $\frac{d^l \gamma(t)}{dt^l}$.

We will need some simple operations for vectors and vector-valued functions in $\mathbb{R}^3$, for more details we refer to Appendix I. For $x, y \in \mathbb{R}^3$ the **scalar product** in $\mathbb{R}^3$ is given as usual by

$$\langle x, y \rangle = x_1 y_1 + x_2 y_2 + x_3 y_3 \tag{7.10}$$

and the **cross product** by

$$x \times y = (x_2 y_3 - x_3 y_2, x_3 y_1 - x_1 y_3, x_1 y_2 - x_2 y_1). \tag{7.11}$$

If $\langle x, y \rangle = 0$ for non-zero vectors $x, y \in \mathbb{R}^3$ we call x and y **orthogonal** and for this we write $x \perp y$. It is easy to see that $(x \times y) \perp x$ and $(x \times y) \perp y$. Thus, if x and y are independent then the vectors x, y and $x \times y$ form a basis of $\mathbb{R}^3$. More properties of the scalar product and the cross product are discussed in Appendix I. For C^1-functions $x, y : I \to \mathbb{R}^3$ we find

$$\frac{d}{dt}\langle x(t), y(t) \rangle = \langle \dot{x}(t), y(t) \rangle + \langle x(t), \dot{y}(t) \rangle, \tag{7.12}$$

in particular

$$\frac{d}{dt}\langle x(t), x(t) \rangle = \frac{d}{dt}\|\dot{x}(t)\|^2 = 2\langle \dot{x}(t), x(t) \rangle. \tag{7.13}$$

Furthermore

$$\frac{d}{dt}(x(t) \times y(t)) = \dot{x}(t) \times y(t) + x(t) \times \dot{y}(t). \tag{7.14}$$

We also refer to Problem 3.

Definition 7.4. *A parametric curve $\gamma : I \to \mathbb{R}^n$ is called a C^k-curve or of class C^k, $0 \le k \le \infty$, if $\gamma|_{\mathring{I}}$ is k-times continuously differentiable and the one-sided derivatives $\gamma^{(l)}(a)$ and $\gamma^{(l)}(b)$, $0 \le l \le k$, exist in the sense that $\lim_{\substack{t>a \\ t \to a}} \gamma^{(l)}(t) = \gamma^{(l)}(a)$ and $\lim_{\substack{t<b \\ t \to b}} \gamma^{(l)}(t) = \gamma^{(l)}(b)$, if I has the end points $a < b$.*

Remark 7.5. If $I' \subset \mathbb{R}$ is open and $\gamma : I' \to \mathbb{R}^n$ is k-times continuously differentiable, then for every closed interval $I \subset I'$ the parametric curve $\gamma|_I$ belongs to the class C'^k.

Example 7.6. A. If $f : I \to \mathbb{R}$ is k-times continuously differentiable then $\gamma : I \to \mathbb{R}^2$, $\gamma(t) = (t, f(t))$, is a C^k-curve.

B. Let $a, b \in \mathbb{R}^n$ and define $\gamma : [0, 1] \to \mathbb{R}^n$ as $\gamma(t) := at + b$. The trace of γ is the line segment connecting $\gamma(0) = a$ with $\gamma(1) = a + b$. In general, if $x, y \in \mathbb{R}^n$, $x \neq y$, are two points we can consider the line segment connecting x with y as the trace of the parametric curve $\gamma : [0, 1] \to \mathbb{R}^n$, $\gamma(t) = (y-x)t+x$, see Figure 7.1.

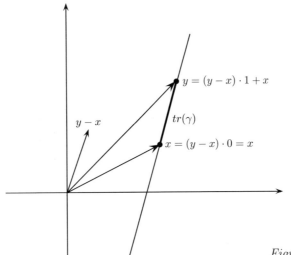

$$y = (y - x) \cdot 1 + x$$

$$y - x$$

$$\text{tr}(\gamma)$$

$$x = (y - x) \cdot 0 = x$$

Figure 7.1

C. Consider $\gamma : [0, \pi] \to \mathbb{R}^2$, $\gamma(t) = (\alpha \cos t, \beta \sin t)$ with $\alpha > 0$ and $\beta > 0$. Clearly γ is a C^∞-curve and its trace is the **ellipse**

$$\mathcal{E} = \left\{ (x, y) \in \mathbb{R}^2 \,\middle|\, \frac{x^2}{\alpha^2} + \frac{y^2}{\beta^2} = 1 \right\}.$$

Indeed, for $(x, y) = (\alpha \cos t, \beta \sin t)$ we find

$$\frac{x^2}{\alpha^2} + \frac{y^2}{\beta^2} = \frac{\alpha^2 \cos^2 t}{\alpha^2} + \frac{\beta^2 \sin^2 t}{\beta^2} = \cos^2 t + \sin^2 t = 1,$$

so $\text{tr}(\gamma) \subset \mathcal{E}$. We leave it as an exercise to prove the converse, i.e. $\mathcal{E} \subset \text{tr}(\gamma)$, see Problem 4. The trace of γ is indicated in Figure 7.2.

115

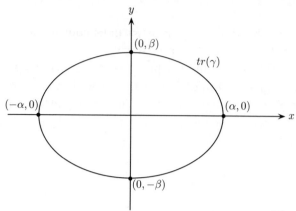

Figure 7.2

Note that $(\alpha, 0) = (\alpha \cos 0, \beta \sin 0) = (\alpha \cos 2\pi, \beta \sin 2\pi)$.

Example 7.7. The **circular helix** is given as the trace of $\gamma : \mathbb{R} \to \mathbb{R}^3$, $\gamma(t) = (r \cos t, r \sin t, ht)$, $r > 0$ and $h > 0$, see Figure 7.3. It is a C^∞-curve and its trace is a helix on the circular cylinder $Z = \{(x, y, z) \in \mathbb{R}^3 \mid x^2 + y^2 = r^2\}$ with height $2\pi h$.

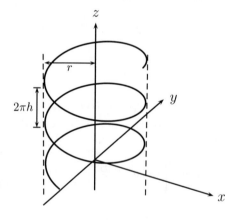

Figure 7.3

Example 7.8. The trace of the C^∞-curve $\gamma : \mathbb{R} \to \mathbb{R}^2$, $\gamma(t) = (t^3, t^2)$ is given in Figure 7.4, and at $\gamma(0) = (0, 0)$ the trace looks "singular". This curve is the (upper branch) of **Neil's parabola**.

116

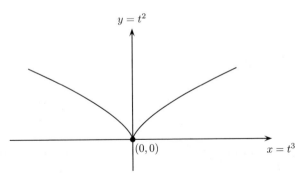

$y = t^2$

$(0,0)$

$x = t^3$

Figure 7.4

Definition 7.9. *Let $\gamma : I \to \mathbb{R}^n$ be a C^k-curve, $k \geq 1$. We call $\dot{\gamma}(t) = \frac{d\gamma}{dt}(t)$, $t \in I$, the **velocity vector** of γ at t.*

Remark 7.10. A. In the case where the end points of I belongs to I we will take the one-sided derivative to define $\dot{\gamma}(a)$ or $\dot{\gamma}(b)$ if $a < b$ are the end points.
B. Note that $\dot{\gamma}(t)$ is related to t and not to $\gamma(t) \in \mathrm{tr}(\gamma)$. In particular γ can be very smooth (class C^∞) while $\mathrm{tr}(\gamma)$ may not look like a smooth geometric object, see Example 7.7. Moreover, if for $t_1 \neq t_2$, $t_1, t_2 \in I$, we have $\gamma(t_1) = \gamma(t_2)$ we may have two different velocity vectors attached to $p \in \mathrm{tr}(\gamma)$, $p = \gamma(t_1) = \gamma(t_2)$, see Example 7.11.

Example 7.11. The trace of the C^∞-curve $\gamma; \mathbb{R} \to \mathbb{R}^2$, $\gamma(t) = (t^3 - 4t, t^2 - 4)$, is given in Figure 7.5. At $p = (0,0) = \gamma(2) = \gamma(-2)$ we have the two velocity vectors $\dot{\gamma}(2) = (8,4)$ and $\dot{\gamma}(-2) = (8,-4)$.

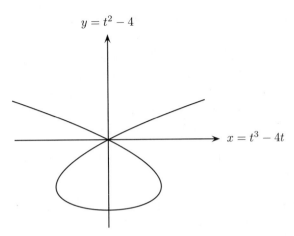

$y = t^2 - 4$

$x = t^3 - 4t$

Figure 7.5

117

These observations lead to

Definition 7.12. *Let $\gamma : I \to \mathbb{R}^n$ be a parametric curve belonging to the class C^k, $k \geq 1$. **A.** We call γ **regular** if $\dot{\gamma}(t) \neq 0 \in \mathbb{R}^n$ for all $t \in I$. A point with $\dot{\gamma}(t) \neq 0$ is called a **regular point** and if $\dot{\gamma}(t) = 0$ we call $\gamma(t)$ a **singular point**.*
***B.** Let $p \in \gamma(I) = \mathrm{tr}(\gamma)$. We call p an **m-fold point** of γ, $m = 2, \ldots, \infty$, if for $t_1 < t_2 < \cdots < t_m$, $t_j \in I$, we have $\gamma(t_j) = p$. A two-fold point is called a **double point**.*
***C.** A parametric curve $\gamma : [a, b] \to \mathbb{R}^n$ is called **closed** if with $\omega := b - a > 0$ there exists $\gamma_\omega : \mathbb{R} \to \mathbb{R}^n$ such that $\gamma_\omega|_{[a,b]} = \gamma$ and if $\gamma_\omega(t + \omega) = \gamma_\omega(t)$ for all $t \in \mathbb{R}$.*
***D.** A closed parametric curve $\gamma : [a, b] \to \mathbb{R}^n$ is called **simple closed** if $\gamma|_{[a,b)}$ is injective.*

Example 7.13. A. Since for $f \in C^1(I)$ we find for $\gamma(t) = (t, f(t))$ that $\dot{\gamma}(t) = \frac{d}{dt}(t, f(t)) = (1, \dot{f}(t)) \neq (0, 0)$ for all $t \in I$, the graph of a C^1-function always gives rise to a regular, parametric curve of class C^1.
B. It is easy to check that an ellipse, and hence a circle, as well as line segments are traces of regular, parametric curves as is the circular helix in Example 7.7.
C. The curve in Example 7.8 has a singular point for $t = 0$ since $\dot{\gamma}(t) = (6t^2, 2t)$ which is equal to $(0, 0)$ for $t = 0$.
D. The curve in Example 7.11 is regular. Indeed, $\dot{\gamma}(t) = (3t^2 - 4, 2t - 4)$ and in order to be equal to 0 for the same t_0 we must have $3t_0^2 = 4$ and $t_0 = 2$ which is impossible. However the point $p = \gamma(2) = \gamma(-2)$ is a double point of $\gamma(I)$.
E. The ellipse $t \overset{\gamma}{\mapsto} (\alpha \cos t, \beta \sin t)$ is interesting. If we take the domain of γ to be the interval $I = [0, 2\pi]$, then γ is simply closed with $\gamma(0) = \gamma(2\pi) = (\alpha, 0)$. But if we take the domain $I = [0, 2\pi k]$, $k \in \mathbb{N}$ and $k \geq 2$, the curve γ is still closed but all points not equal to $(\alpha, 0)$ are k-fold points while $(\alpha, 0)$ is a $(k + 1)$-fold point. This is due to the periodicity of cos and sin.

Now we relate the velocity vector to geometry by introducing the tangent line of γ at $\gamma(t)$.

Definition 7.14. *Let $\gamma : I \to \mathbb{R}^n$ be a regular parametric C^k-curve, $k \geq 1$. For $t_0 \in I$ fixed the mapping $g_{t_0} : \mathbb{R} \to \mathbb{R}^n$, $g_{t_0}(s) = \dot{\gamma}(t_0)s + \gamma(t_0)$ is called the **tangent line** to γ at t_0.*

Remark 7.15. Note that g_{t_0} itself is a parametric C^∞-curve the trace of which is a straight line which is called the **tangent** or again the **tangent line** to $\mathrm{tr}(\gamma)$ at $\gamma(t_0)$. The velocity vector of $g(t_0)$ is for all $s \in \mathbb{R}$ given by $\dot{g}_{t_0}(s) = \dot{\gamma}(t_0)$. Hence the tangent line is a parametric curve with the property that for $s = 0$ it passes through $\gamma(t_0)$ and has the same velocity vector as γ at t_0.

Example 7.16. Consider the graph of a function $f : I \to \mathbb{R}$. The corresponding curve is given by $\gamma(t) = (t, f(t))$ and hence its tangent line at t_0 is

$$g_{t_0}(s) = (1, (1, f'(t_0)))s + (t_0, f(t_0)),$$

see Figure 7.6

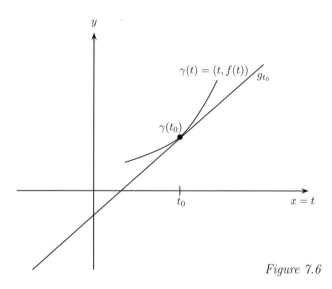

Figure 7.6

Example 7.17. Let $\gamma : \mathbb{R} \to \mathbb{R}^2$, $\gamma(t) = (t^3 - 4t, t^2 - 4)$ be as in Example 7.11. With $\dot{\gamma}(t) = (3t^2 - 4, 2t)$ the tangent line to γ at $t = -2$ is given by $g_{-2}(s) = \begin{pmatrix} 4 \\ -4 \end{pmatrix} s$ and the tangent line to γ at $t = 2$ is given by $g_2(s) = \begin{pmatrix} 4 \\ 4 \end{pmatrix} s$.

Both lines pass through the point $\gamma(-2) = \gamma(2) = (0, 0)$, see Figure 7.7. This example shows that we can speak about a unique tangent line to γ at $t = -2$ as well as $t = 2$, however we cannot speak about a unique tangent line to γ at the point $(0, 0) = \gamma(2) = \gamma(-2)$.

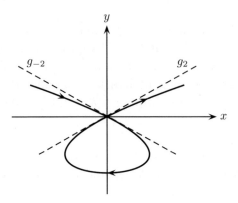

Figure 7.7

Example 7.18. Consider the circle S^1 as the trace of the parametric curve $\gamma : [0, 2\pi] \to \mathbb{R}^2$, $t \mapsto (\cos t, \sin t)$. This is a simply closed curve with $(1, 0) = \gamma(0) = \gamma(2\pi)$, we find that the tangent line to γ at $t = 0$ and the tangent line to γ at $t = 2\pi$ are identical. Hence there are cases where the tangent lines at m-fold points coincide, namely whenever $\dot{\gamma}(t_1) = \cdots = \dot{\gamma}(t_m)$ for the m-fold point $p = \gamma(t_1) = \cdots \gamma(t_m)$, and in such a case we can define the tangent line to the trace of γ at $p \in \text{tr}(\gamma)$.

Let $\gamma : I \to \mathbb{R}^2$ be a C^1-curve and $t_0 \in \overset{\circ}{I}$. The parametric curve $n_{t_0} : \mathbb{R} \to \mathbb{R}^2$ defined by $n_{t_0}(s) = \begin{pmatrix} -\dot{\gamma}_2(t_0) \\ \dot{\gamma}_1(t_0) \end{pmatrix} s + \gamma(t_0)$ is a straight line passing through $\gamma(t_0)$ and the following holds

$$\left\langle \begin{pmatrix} \dot{\gamma}_1(t_0) \\ \dot{\gamma}_2(t_0) \end{pmatrix}, \begin{pmatrix} -\dot{\gamma}_2(t_0) \\ \dot{\gamma}_1(t_0) \end{pmatrix} \right\rangle = -\dot{\gamma}_1(t_0)\dot{\gamma}_2(t_0) + \dot{\gamma}_2(t_0)\dot{\gamma}_1(t_0) = 0$$

i.e. the trace of n_{t_0} intersects the trace of g_{t_0} at $\gamma(t_0)$ and these lines are orthogonal. Moreover we have

$$\det \begin{pmatrix} \dot{\gamma}_1(t_0) & -\dot{\gamma}_2(t_0) \\ \dot{\gamma}_2(t_0) & \dot{\gamma}_1(t_0) \end{pmatrix} = \dot{\gamma}_1^2(t_0) + \dot{\gamma}_2^2(t_0) > 0,$$

i.e. the basis $\left\{ \begin{pmatrix} \dot{\gamma}_1(t_0) \\ \dot{\gamma}_2(t_0) \end{pmatrix}, \begin{pmatrix} -\dot{\gamma}_2(t_0) \\ \dot{\gamma}_1(t_0) \end{pmatrix} \right\}$ is positively orientated.

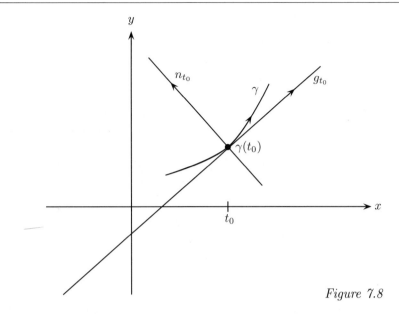

Figure 7.8

As Figure 7.8 suggests, for studying γ in a neighbourhood of $\gamma(t_0)$ we may introduce a new coordinate system with origin at $\gamma(t_0)$, abscissa being the tangent line to γ at t_0 and ordinate being the normal line to γ at t_0. We will take up this approach in our chapters on differential geometry.

Definition 7.19. *Let $\gamma : I \to \mathbb{R}^2$ be a regular parametric C^1-curve and $t_0 \in \mathring{I}$. We call*

$$n_{t_0} : \mathbb{R} \to \mathbb{R}^2, \quad n_{t_0}(s) = \begin{pmatrix} -\dot{\gamma}_2(t_0) \\ \dot{\gamma}_1(t_0) \end{pmatrix} s + \gamma(t_0) \tag{7.15}$$

*the **normal line** to γ at t_0.*

Definition 7.20. *Let $\gamma : I \to \mathbb{R}^2$ be a regular parametric C^1-curve. For $s_0 \in \mathring{I}$ we call $\vec{t}(s_0) := \frac{\dot{\gamma}(s_0)}{\|\dot{\gamma}(s_0)\|}$ the **tangent vector** to γ at s_0. The **normal vector** $\vec{n}(s_0)$ to γ at s_0 is the uniquely determined vector satisfying*

$$\|\vec{n}(s_0)\| = 1, \quad \langle \vec{t}(s_0), \vec{n}(s_0) \rangle = 0, \quad \det(\vec{t}(s_0), \vec{n}(s_0)) > 0, \tag{7.16}$$

in fact we find

$$\vec{n}(s_0) = \frac{1}{\|\dot{\gamma}(s_0)\|} \begin{pmatrix} -\dot{\gamma}_2(s_0) \\ \dot{\gamma}_1(s_0) \end{pmatrix}.$$

121

Remark 7.21. When starting with our chapters on differential geometry we will call (for good reasons) $\vec{t}(s_0)$ the **pre-tangent vector** and $\vec{n}(s_0)$ the **pre-normal vector**.

Example 7.22. Let $f : I \to \mathbb{R}$ be a C^1-function and $\gamma : I \to \mathbb{R}^2$, $\gamma(t) = (t, f(t))$ be the associated parametric curve. The tangent vector to γ at $s_0 \in \mathring{I}$ is

$$\vec{t}(s_0) = \frac{1}{\sqrt{1 + f'(s_0)}} \begin{pmatrix} 1 \\ f'(s_0) \end{pmatrix} \qquad (7.17)$$

and the normal vector to γ at $s_0 \in \mathring{I}$ is

$$\vec{n}(s_0) = \frac{1}{\sqrt{1 + f'(s_0)}} \begin{pmatrix} -f'(s_0) \\ 1 \end{pmatrix}.$$

Using the tangent vector and the normal vector we can represent the tangent line and the normal line as

$$\tilde{g}_{t_0}(\tilde{s}) = \vec{t}(t_0)\tilde{s} + \gamma(t_0) \ \text{ and } \ \tilde{n}_{t_0}(\tilde{s}) = \vec{n}(t_0)\tilde{s} + \gamma(t_0). \qquad (7.18)$$

We arrive at (7.18) from g_{t_0} and n_{t_0}, respectively, by formally introducing the new curve parameter $\tilde{s} = \|\dot{\gamma}(t_0)\| s$. Before continuing our geometric considerations it is helpful to first study general parameter transformations or change of parameter.

Definition 7.23. *Let $\gamma_j : I_j \to \mathbb{R}^n$, $j = 1, 2$, be two parametric curves belonging to the class C^k, $k \geq 1$, and $\varphi : I_2 \to I_1$ a bijective C^k-function with differentiable inverse. If $\gamma_2 = \gamma_1 \circ \varphi$ we say that γ_2 is obtained from γ_1 by a* **change of parameter** *or the* **parameter transformation** φ. *If $\dot{\varphi} > 0$ we call φ* **orientation preserving**.

Remark 7.24. A. Obviously, if γ_2 is obtained from γ_1 by a change of parameter then $\mathrm{tr}(\gamma_2) = \mathrm{tr}(\gamma_1)$.
B. In order to understand the meaning of orientation suppose that $I_1 = [a, b]$ and $I_2 = [\alpha, \beta]$. If $\dot{\varphi} > 0$ then $\varphi(\alpha) = a$ and $\varphi(\beta) = b$ and therefore the initial points and the terminal points of γ_2 and γ_1 are the same, however if $\dot{\gamma} < 0$ then $\varphi(\alpha) = b$ and $\varphi(\beta) = a$ we "run" through the trace in the opposite direction.

An interesting question is whether for a regular curve $\gamma : I \to \mathbb{R}^n$ we can always find a parametrization such that $\|\dot{\gamma}(s)\| = 1$ for all $s \in I$, i.e. in this case (for $n = 2$) the velocity vector is always the tangent vector. Consider the regular parametric curve $\gamma : [a, b] \to \mathbb{R}^n$ belonging to the class C^1. With

$$l_\gamma := \int_a^b \|\dot{\gamma}(r)\| \, \mathrm{d}r \tag{7.19}$$

we define the mapping

$$s : [a, b] \to [0, l_\gamma]$$
$$t \mapsto s(t) := \int_a^t \|\dot{\gamma}(r)\| \, \mathrm{d}r. \tag{7.20}$$

Since $\dot{s}(t) = \|\dot{\gamma}(t)\| > 0$, recall $\dot{\gamma}(r) \neq 0$ since γ is regular, the function s is strictly increasing and further with $\varphi : [0, l_\gamma] \to [a, b]$, $\varphi := s^{-1}$, we find

$$\|(\gamma \circ \varphi)^\cdot(s)\| = \|\dot{\gamma}(\varphi(s))\| \, |\dot{\varphi}(s)| = \|\dot{\gamma}(t)\| \frac{1}{\|\dot{\gamma}(t)\|} = 1.$$

Thus we have proved

Theorem 7.25. *Let $\gamma : [a, b] \to \mathbb{R}^n$ be a regular parametric curve in the class C^1 and define $\varphi : [0, l_\gamma] \to [a, b]$ as the inverse to the function $s : [a, b] \to [0, l_\gamma]$, $s(t) := \int_a^t \|\dot{\gamma}(r)\| \, \mathrm{d}r$. Then the inverse function $\varphi := s^{-1} : [0, l_\gamma] \to [a, b]$ is an orientation preserving parameter transformation and $\gamma \circ \varphi : [0, l_\gamma] \to \mathbb{R}^n$ is a parametric curve the velocity vector of which has always length 1, i.e. $\|(\gamma \circ \varphi)^\cdot(t)\| = 1$ for all $\varphi \in [0, l_\gamma]$.*

Remark 7.26. Theorem 7.25 can be extended to the case where $[a, b]$ is replaced by any interval $I \subset \mathbb{R}$.

For reasons we will understand later, see Chapter 15, it makes sense to give

Definition 7.27. *Let $\gamma : I \to \mathbb{R}^n$ be a regular parametric curve. We call γ **parametrized with respect to the arc length** if $\|\dot{\gamma}(t)\| = 1$ for all $t \in I$. Moreover with $I = [a, b]$ we call $l_\gamma := \int_a^b \|\dot{\gamma}(t)\| \mathrm{d}t$ the **length** of γ.*

The statement of Theorem 7.25 (or that of Remark 7.26) is that every regular parametric curve admits a parametrization with respect to the arc length and that the definition of the length is independent of the parametrization.

Example 7.28. The circle $\gamma : [0, 2\pi] \to \mathbb{R}^2$, $\gamma(t) = R(\cos t, \sin t)$, with centre $(0,0)$ and radius $R > 0$ is in general not parametrized with respect to the arc length: $\dot{\gamma}(t) = R(-\sin t, \cos t)$, hence $\|\dot{\gamma}(t)\| = R$. However by $\eta : [0, 2\pi R] \to \mathbb{R}^2$, $\eta(t) = R\left(\cos\frac{t}{R}, \sin\frac{t}{R}\right)$, a parametrization with respect to the arc length is given:

$$\dot{\eta}(t) = R\left(-\frac{1}{R}\sin\frac{t}{R}, \frac{1}{R}\cos\frac{t}{R}\right) \text{ and } \|\dot{\eta}(t)\| = \left\|\left(-\sin\frac{t}{R}, \cos\frac{t}{R}\right)\right\| = 1.$$

If $\gamma : I \to \mathbb{R}^n$ is a regular curve parametrized with respect to the arc length we write $\vec{t}(s) := \dot{\gamma}(s)$, i.e. $\|\vec{t}(s)\| = 1$ for all $s \in I$, and hence

$$1 = \langle \dot{\gamma}(s), \dot{\gamma}(s) \rangle = \langle \vec{t}(s), \vec{t}(s) \rangle$$

for all s, which implies for a C^2-curve

$$0 = \frac{d}{ds}1 = \frac{d}{ds}\langle \dot{\gamma}(s), \dot{\gamma}(s) \rangle = \frac{d}{ds}\sum_{j=1}^{n} \dot{\gamma}_j^2(s)$$

$$= 2\sum_{j=1}^{n} \dot{\gamma}_j(s)\ddot{\gamma}(s) = 2\langle \dot{\gamma}(s), \ddot{\gamma}(s) \rangle,$$

i.e. the vector $\ddot{\gamma}(s)$ is orthogonal to $\dot{\gamma}(s) = \vec{t}(s)$.

Definition 7.29. *Let $\gamma : I \to \mathbb{R}^n$ be a regular C^2-curve parametrized with respect to the arc length. We call $\vec{t}(s) := \dot{\gamma}(s)$ the **tangent vector** to γ at s and if $\ddot{\gamma}(s) \neq 0$*

$$\vec{n}(s) := \frac{\ddot{\gamma}(s)}{\|\ddot{\gamma}(s)\|} \tag{7.21}$$

*the **normal vector** to γ at s.*

(Again, as mentioned in Remark 7.21 we will later on understand why pre-tangent vector and pre-normal vector for $\vec{t}(s)$ and $\vec{n}(s)$, respectively, are better names).

In Problem 11 we will see that for $n = 2$ the definitions of $\vec{n}(s)$ given in Definition 7.20 and in Definition 7.29 coincide.

Here we encounter, maybe for the first time, the fact that while analysis allows us to handle all dimensions in the same way, geometry sometimes makes a distinction between dimensions. In the situation of Definition 7.29

the normal vector is not defined if $\ddot{\gamma}(s) = 0$. For $n = 2$ however, there are only two independent directions, thus even if $\ddot{\gamma}(s) = 0$ there is a unique straight line passing through $\gamma(s)$ which is orthogonal to the tangent line to γ at s. Thus when fixing the orientation we can for $n = 2$ define a unique normal vector to γ at s even in the case $\ddot{\gamma}(s) = 0$. For $n = 3$ this causes difficulties. (See also the following example part B.)

Example 7.30. A. Consider the circle $S^1 = \{(x, y) \in \mathbb{R}^2 \mid x^2 + y^2 = 1\}$ as the trace of the parametric curve $\gamma : [0, 2\pi] \to \mathbb{R}^2$, $\gamma(t) = (\cos t, \sin t)$. Since $\dot{\gamma}(t) = (-\sin t, \cos t)$ we find $\|\dot{\gamma}(t)\| = (\sin^2 t + \cos^2 t)^{\frac{1}{2}} = 1$, i.e. γ is parametrized with respect to the arc length and for all $t \in [0, 2\pi]$ we have $\vec{t}(t) = (-\sin t, \cos t)$. Moreover, $\ddot{\gamma}(t) = (-\cos t, -\sin t)$, hence $\|\ddot{\gamma}(t)\| = 1$, we find the normal vector $\vec{n}(t) = (-\cos t, -\sin t)$. A simple calculation gives that we have indeed

$$\langle \vec{t}(t), \vec{n}(t) \rangle = \left\langle \begin{pmatrix} -\sin t \\ \cos t \end{pmatrix}, \begin{pmatrix} -\cos t \\ -\sin t \end{pmatrix} \right\rangle = 0$$

and in addition

$$\det \begin{pmatrix} -\sin t & -\cos t \\ \cos t & -\sin t \end{pmatrix} = \sin^2 t + \cos^2 t = 1 > 0.$$

B. For the straight line $h : \mathbb{R} \to \mathbb{R}^n$, $h(s) = \vec{a}s + \vec{b}$ with $\|a\| = 1$ we find $\dot{h}(s) = \vec{a}$ and therefore $\left\| \dot{h}(s) \right\| = 1$, i.e. $\vec{t}(s) = \vec{a}$ for all $s \in \mathbb{R}$, and we find that h is parametrized with respect to the arc length. Further we find for all $s \in \mathbb{R}$ that $\ddot{h}(s) = 0$ and therefore the normal vector in the sense of Definition 7.29 is never defined and indeed it is difficult to justify any preference for calling a specific straight line passing through $\gamma(s)$ and belonging to the plane orthogonal to $\dot{\gamma}(s)$ the normal line to γ at s. However in the case $n = 2$ we still may introduce a well defined normal line and hence a normal vector. Note that the condition $\ddot{h}(s) = 0$ implies $\ddot{h}_j(s) = 0$ for all $1 \leq j \leq n$ and therefore $h_j(s) = \alpha_j s + \beta_j$ for some $\alpha_j, \beta_j \in \mathbb{R}$. Consequently $h(s) = \vec{a}s + \vec{b}$ with $\vec{a} = (\alpha_1, \ldots, \alpha_n)$ and $\vec{b} = (\beta_1, \ldots, \beta_n)$.

We now want to study further geometric properties of (families of) parametric curves in $\mathbb{R}^2$ and $\mathbb{R}^3$.

Let $\gamma_j : I_j \to \mathbb{R}^2$ be two regular parametric C^2-curves each having at each point a well defined unique tangent line. We suppose that for some $t_0 \in \mathring{I}_1$

and $s_0 \in \mathring{I}_2$ we have $\gamma_1(t_0) = \gamma_2(s_0)$, i.e. γ_1 and γ_2 intersect at the point $p := \gamma_1(t_0) = \gamma_2(s_0) \in \mathbb{R}^2$. By our assumption the curves have a tangent vector at t_0 and s_0, respectively, which are given by $\frac{\dot{\gamma}_1(t_0)}{\|\dot{\gamma}_1(t_0)\|}$ and $\frac{\dot{\gamma}_2(s_0)}{\|\dot{\gamma}_2(2_0)\|}$. By

$$\cos \alpha = \frac{|\langle \dot{\gamma}_1(t_0), \dot{\gamma}_2(s_0) \rangle|}{\|\dot{\gamma}_1(t_0)\| \, \|\dot{\gamma}_2(s_0)\|}, \quad 0 \leq \alpha \leq \frac{\pi}{2} \tag{7.22}$$

an angle α is well defined.

Definition 7.31. *Suppose that $\gamma_j : I_j \to \mathbb{R}^2$ intersect at $p = \gamma_1(t_0) = \gamma_2(s_0)$, $t_0 \in \mathring{I}_1$ and $s_0 \in \mathring{I}_2$. Suppose further that both curves have at p a uniquely defined tangent vector. Then we say that γ_1 and γ_2 intersect in p under the angle $\alpha \in \left[0, \frac{\pi}{2}\right]$, α given by (7.22).*

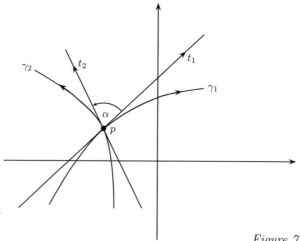

Figure 7.9

Remark 7.32. Our definition restricts α to the range $0 \leq \alpha \leq \frac{\pi}{2}$ and does not take into account orientation, i.e. we do not distinguish between $\sphericalangle(\gamma_1(t_0), \gamma_2(s_0))$ and $\sphericalangle(\gamma_2(s_0), \gamma_1(t_0))$. This simplification is sufficient for our current purpose.

Definition 7.33. *Two regular parametric C^1-curves each admitting a unique tangent line at $p = \gamma_1(t_0) = \gamma_2(s_0)$ are said to be **orthogonal** at p if $\cos \alpha = 0$, i.e. $\alpha = \frac{\pi}{2}$. If however $\cos \alpha = 1$, i.e. $\alpha = 0$, the curves are said to be the **tangent** at p.*

Example 7.34. A. Consider two C^1-functions $f, h : I \to \mathbb{R}$ and for $t_0 \in \overset{\circ}{I}$ suppose that $f(t_0) = h(t_0)$. Thus their graphs and hence the corresponding curves $\gamma_f(t) = \begin{pmatrix} t \\ f(t) \end{pmatrix}$ and $\gamma_h(t) = \begin{pmatrix} t \\ h(t) \end{pmatrix}$, $t \in I$, intersect at $p = \begin{pmatrix} t_0 \\ f(t_0) \end{pmatrix} = \begin{pmatrix} t_0 \\ h(t_0) \end{pmatrix}$. Their tangent vectors at p are given by

$$\frac{\dot{\gamma}_f(t_0)}{\|\dot{\gamma}_f(t_0)\|} = \frac{1}{\sqrt{1 + f'(t_0)^2}} \begin{pmatrix} 1 \\ f'(t_0) \end{pmatrix}$$

and

$$\frac{\dot{\gamma}_h(t_0)}{\|\dot{\gamma}_h(t_0)\|} = \frac{1}{\sqrt{1 + h'(t_0)^2}} \begin{pmatrix} 1 \\ h'(t_0) \end{pmatrix}$$

implying that

$$\cos \alpha = \frac{1}{\sqrt{1 + f'(t_0)^2}\sqrt{1 + h'(t_0)^2}} |1 + f'(t_0)h'(t_0)|.$$

If we take for example $I = (0, 2)$ and $f(t) = t^2$, $h(t) = \frac{1}{t}$ we find $f'(1) = 2$ and $h'(1) = -1$ and consequently

$$\cos \alpha = \frac{1}{2}|1 - 2| = \frac{1}{\sqrt{10}},$$

i.e. the parabola $t \mapsto t^2$ and the hyperbola $t \mapsto \frac{1}{t}$ intersect at the point $(1, 1)$ under the angle α, $\cos \alpha = \frac{1}{\sqrt{10}}$. Note: the diagram below is not to scale.

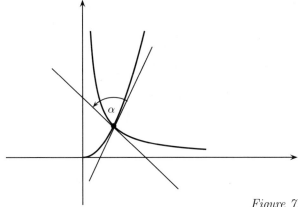

Figure 7.10

B. Suppose that $\gamma_1(t_0) = \gamma_2(s_0)$ as well as $\dot{\gamma}_1(t_0) = \dot{\gamma}_2(s_0)$. In this case we have

$$\cos\alpha = \frac{|\langle \dot{\gamma}_1(t_0), \dot{\gamma}_2(s_0)\rangle|}{\|\dot{\gamma}_1(t_0)\| \, \|\dot{\gamma}_2(s_0)\|} = \frac{|\langle \dot{\gamma}_1(t_0), \dot{\gamma}_1(t_0)\rangle|}{\|\dot{\gamma}_1(t_0)\|^2} = 1$$

and it follows that $\alpha = 0$. For example the two circles $\gamma_1 : [0, 2\pi] \to \mathbb{R}^2$, $\gamma_1(t) = (\cos t, \sin t)$ and $\gamma_2 : [0, 2\pi] \to \mathbb{R}^2$, $\gamma_2(s) = (\sqrt{2} + \cos s, \sqrt{2} + \sin s)$ intersect at $p = \left(\frac{1}{2}\sqrt{2}, \frac{1}{2}\sqrt{2}\right) = \gamma_1\left(\frac{\pi}{4}\right) = \gamma_2\left(\frac{5\pi}{4}\right)$ and we find

$$\dot{\gamma}_1\left(\frac{\pi}{4}\right) = \left(-\sin\frac{\pi}{4}, \cos\frac{\pi}{4}\right) = \left(\frac{1}{2}\sqrt{2}, \frac{1}{2}\sqrt{2}\right)$$

as well as

$$\dot{\gamma}_2\left(\frac{5\pi}{4}\right) = \left(-\sin\frac{5\pi}{4}, \cos\frac{5\pi}{4}\right) = \left(\frac{1}{2}\sqrt{2}, -\frac{1}{2}\sqrt{2}\right)$$

implying

$$\left|\left\langle \dot{\gamma}_1\left(\frac{\pi}{4}\right), \dot{\gamma}_2\left(\frac{5\pi}{4}\right)\right\rangle\right| = \left|-\frac{1}{2} - \frac{1}{2}\right| = 1,$$

and therefore we have $\cos\alpha = 1$ or $\alpha = 0$.

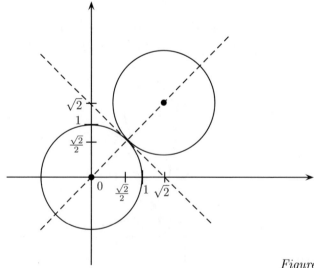

Figure 7.11

The next example prepares us for the introduction of new coordinates, namely polar coordinates in the plane, or more generally the introduction of curvilinear coordinates in $\mathbb{R}^n$.

In the plane $\mathbb{R}^2$ we consider the circles $C_\rho \subset \mathbb{R}^2$ with centre $(0,0) \in \mathbb{R}^2$ and radius $\rho > 0$, and in addition we consider the rays R_φ, $\varphi \in [0, 2\pi)$, where R_φ is the trace of $r_\varphi : (0, \infty) \to \mathbb{R}^2$, $r_\varphi(t) = \begin{pmatrix} \cos \varphi \\ \sin \varphi \end{pmatrix} t$, see Figure 7.12.

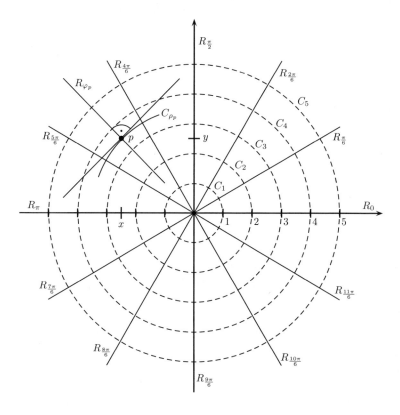

Figure 7.12

First we note that for every $\rho > 0$ and $\varphi \in [0, 2\pi)$ the intersection $C_\rho \cap R_\varphi$ consists of one and only one point, namely $p = \rho(\cos \varphi, \sin \varphi)$. Moreover, at p the circle C_ρ and the ray R_φ are orthogonal as we can either calculate (see Problem 12) or we know this fact from elementary geometry. Thus every point $p \in \mathbb{R}^2 \setminus \{0\}$ has a unique representation $p = (\rho \cos \varphi, \rho \sin \varphi)$ $(= (x, y))$ and we may define a mapping $\Phi : \mathbb{R}_+ \setminus \{0\} \times [0, 2\pi) \to \mathbb{R}^2 \setminus \{0\}$ which is bijective and differentiable. The relations between (ρ, φ) and (x, y) are

$$x = \rho \cos \varphi, \quad y = \rho \sin \varphi \tag{7.23}$$

$$\rho = (x^2 + y^2)^{\frac{1}{2}}, \quad \tan \varphi = \frac{y}{x}, \quad \cos \varphi \neq 0, \quad \cot \varphi = \frac{x}{y}, \quad \sin \varphi \neq 0. \tag{7.24}$$

While we have a bijective mapping Φ, for the inverse we cannot find a single expression. Further, for $(x, y) = (0, 0)$ we cannot find a unique (ρ, φ) such that (7.23) and (7.24) hold. Since

$$\frac{\partial x}{\partial \rho} = \cos \varphi, \quad \frac{\partial x}{\partial \varphi} = -\rho \sin \varphi, \quad \frac{\partial y}{\partial \rho} = \sin \varphi, \quad \frac{\partial y}{\partial \varphi} = \rho \cos \varphi$$

we find for the Jacobi matrix of Φ

$$J_\Phi(\rho, \varphi) = \begin{pmatrix} \cos \varphi & -\rho \sin \varphi \\ \sin \varphi & \rho \cos \varphi \end{pmatrix} \tag{7.25}$$

which gives the Jacobi determinant

$$\det J_\Phi(\rho, \varphi) = \rho. \tag{7.26}$$

What we have discovered has however far reaching consequences. We can cover the plane (without the origin) by two families F_1 and F_2 of regular curves having the following properties:

1. For $p \in \mathbb{R}^2 \setminus \{0\}$ there exists exactly one $\xi_p \in F_1$ and exactly one $\eta_p \in F_2$ with $p \in \operatorname{tr}(\xi_p) \cap \operatorname{tr}(\eta_p)$, i.e. p is in the intersection of exactly one pair of curves $(\xi_p, \eta_p) \in F_1 \times F_2$.

2. Whenever $\xi \in F_1$ and $\eta \in F_2$ intersect, the intersection contains only one point and at this point ξ and η are orthogonal.

This is the basic idea behind **orthogonal curvilinear coordinates** and this will be picked up in Chapter 12. The coordinates (ρ, φ) are called (plane) **polar coordinates** and we will also use them when treating complex-valued functions of a complex variable.

Polar coordinates allow us to discuss some interesting curves in the plane in a more geometric manner.

Example 7.35. A. The curve $\gamma : [0, 2\pi] \to \mathbb{R}^2$, $\gamma(t) = (re^{\eta t} \cos t, re^{\eta t} \sin t)$, $\eta < 0 < r$ has as its trace a segment of the **logorithmic spiral**, see Figure 7.13. With polar coordinates ρ and φ we find

$$\gamma(\varphi) = \rho(\varphi) \begin{pmatrix} \cos \varphi \\ \sin \varphi \end{pmatrix}, \quad \rho(\varphi) = re^{\eta \varphi}, \quad \varphi \in [0, 2\pi].$$

B. A segment of the **Archimedian spiral** can be described by

$$\gamma(\varphi) = \rho(\varphi) \begin{pmatrix} \cos \varphi \\ \sin \varphi \end{pmatrix} = \alpha \varphi \begin{pmatrix} \cos \varphi \\ \sin \varphi \end{pmatrix}, \quad \varphi \in [0, 2\pi], \quad \alpha > 0,$$

see Figure 7.14.

Note that in both cases we have a curve given in polar coordinates

$$\gamma(\varphi) = \rho(\varphi) \begin{pmatrix} \cos \varphi \\ \sin \varphi \end{pmatrix}, \quad \rho : [0, 2\pi] \to \mathbb{R}_+ \setminus \{0\}, \quad \varphi \in [0, 2\pi].$$

But often we can extend ρ to a larger domain, so we may consider curves $\gamma(\varphi) = \rho(\varphi) \begin{pmatrix} \cos \varphi \\ \sin \varphi \end{pmatrix}$, $\rho : I \to \mathbb{R}_+ \setminus \{0\}$, $I \subset \mathbb{R}$. In some cases even $I = \mathbb{R}$ is possible.

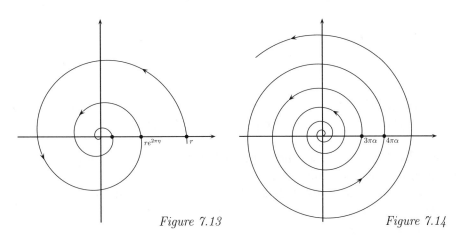

Figure 7.13   Figure 7.14

Next we want to discuss curvature for plane curves and we start with

Definition 7.36. *Let* $\gamma : I \to \mathbb{R}^2$ *be a regular parametric* C^2*-curve parametrized with respect to the arc length, i.e. with tangent vector* $\vec{t}(t) =$

131

$\dot{\gamma}(t)$, $\|\dot{\gamma}(t)\| = 1$ for all t. We call the non-negative number

$$\kappa(t) := \left\|\dot{\vec{t}}(t)\right\| = \|\ddot{\gamma}(t)\| \tag{7.27}$$

the **curvature** of γ at t and

$$\rho(t) := \frac{1}{\kappa(t)} \tag{7.28}$$

is called the **radius of curvature** (of γ at t).

Remark 7.37. A. If $\kappa(t_0) = 0$ we have $\rho(t_0) = \infty$.
B. If γ is not parametrized with respect to the arc length one can prove (see Proposition 7.42 and Remark 7.43) that the curvature is given by

$$\kappa(t) = \frac{|\dot{\gamma}_1(t)\ddot{\gamma}_2(t) - \dot{\gamma}_2(t)\ddot{\gamma}_1(t)|}{\|\dot{\gamma}(t)\|^3}. \tag{7.29}$$

With Remark 7.37.A we can deduce

Example 7.38. If $f : (a, b) \to \mathbb{R}$ is a C^2-function then the curvature of $\gamma : (a, b) \to \mathbb{R}^2$, $t \mapsto (t, f(t))$, is given by

$$\kappa(t) = \frac{|f''(t)|}{(1 + f'(t)^2)^{\frac{3}{2}}} \tag{7.30}$$

which follows from (7.29) and the fact that $\dot{\gamma}_1(t) = 1$, $\ddot{\gamma}_1(t) = 0$, $\dot{\gamma}_2(t) = f'(t)$ and $\ddot{\gamma}_2(t) = f''(t)$. Also recall our considerations in Chapter I.22.

Example 7.39. For a straight line the curvature is of course 0 and for a circle with centre $a = (a_1, a_2)$ and radius $r > 0$ we find with $\gamma : [0, 2\pi r] \to \mathbb{R}^2$, $\gamma(t) = r\left(a_1 + \cos\frac{t}{r}, a_2 + \sin\frac{t}{r}\right)$ that $\dot{\gamma}(t) = \left(-\sin\frac{t}{r}, \cos\frac{t}{r}\right)$ implying $\|\dot{\gamma}(t)\| = 1$ for all $t \in [0, 2\pi r]$, i.e. γ is parametrized with respect to the arc length, and it follows for the curvature

$$\kappa(t) = \|\ddot{\gamma}(t)\| = \left\|\left(-\frac{1}{r}\cos\frac{t}{r}, -\frac{1}{r}\sin\frac{t}{r}\right)\right\| = \frac{1}{r} \tag{7.31}$$

which gives the radius of curvature $\rho(t) = r$, also compare with Chapter I.22.

132

Example 7.40. In the case where a plane curve is given with respect to polar coordinates, i.e. $\gamma(\varphi) = \rho(\varphi) \begin{pmatrix} \cos \varphi \\ \sin \varphi \end{pmatrix}$ we find

$$|\dot{\gamma}(\varphi)| = (\dot{\rho}^2(\varphi) + \rho^2(\varphi))^{\frac{1}{2}},$$
$$\dot{\gamma}_1(\varphi) = \dot{\rho}(\varphi) \cos \varphi - \rho(\varphi) \sin \varphi,$$
$$\ddot{\gamma}_1(\varphi) = \ddot{\rho}(\varphi) \cos \varphi - 2\dot{\rho}(\varphi) - \rho(\varphi) \cos \varphi,$$
$$\dot{\gamma}_2(\varphi) = \dot{\rho}(\varphi) \sin \varphi + \rho(\varphi) \cos \varphi,$$
$$\ddot{\gamma}_2(\varphi) = \ddot{\rho}(\varphi) \sin \varphi + 2\dot{\rho}(\varphi) \cos \varphi - \rho(\varphi) \sin \varphi,$$

implying

$$\kappa(\varphi) = \frac{|2\dot{\rho}(\varphi) - \rho(\varphi)\ddot{\rho}(\varphi) + \rho^2(\varphi)|}{(\dot{\rho}^2(\varphi) + \rho^2(\varphi))^{\frac{3}{2}}}. \tag{7.32}$$

For the Archimedian spiral, i.e. $\rho(\varphi) = \alpha\varphi$, we find

$$\kappa(\varphi) = \frac{2 + \varphi^2}{\alpha(1 + \varphi^2)^{\frac{3}{2}}}$$

and for the logarithmic spiral, i.e. $\rho(\varphi) = e^{\eta\varphi}$ we get

$$\kappa(\varphi) = \frac{e^{-\eta\varphi}}{\sqrt{1 + \eta^2}}.$$

Now we turn our attention to space curves, i.e. parametric curves $\gamma : I \to \mathbb{R}^3$. In the following, if not stated otherwise, $\gamma : I \to \mathbb{R}^3$ is a regular C^3-curve which is parametrized with respect to the arc length. Thus the tangent vector of γ at t_0 is given by $\vec{t}(t_0) = \dot{\gamma}(t_0)$, $\|\dot{\gamma}(t_0)\| = 1$, the **curvature** of γ at t_0 is $\kappa(t_0) = \left\|\dot{\vec{t}}(t_0)\right\| = \|\ddot{\gamma}(t_0)\|$, and if $\ddot{\gamma}(t_0) \neq 0$ the normal vector to γ at t_0 is given by $\vec{n}(t_0) = \frac{\ddot{\gamma}(t_0)}{\|\ddot{\gamma}(t_0)\|}$ or $\ddot{\gamma}(t_0) = \kappa(t_0)\vec{n}(t_0)$, provided $\ddot{\gamma}(t_0) \neq 0$. We already know that if $\kappa(t) = 0$ for all $t \in I$ then γ is a line segment and the converse is trivial. Moreover, if $\|\ddot{\gamma}(t_0)\| \neq 0$ we know that $\vec{t}(t_0)$ and $\vec{n}(t_0)$ are orthogonal, hence they span a plane in $\mathbb{R}^3$. The plane

$$\mathcal{E}_{\gamma(t_0)} := \left\{ x \in \mathbb{R}^3 \,\middle|\, x = \gamma(t_0) + \lambda\vec{t}(t_0) + \mu\vec{n}(t_0), \, \lambda, \mu \in \mathbb{R} \right\} \tag{7.33}$$

is called the **osculating plane** to γ at t_0. Let us assume that $\ddot{\gamma}(t) \neq 0$ for all $t \in I$, hence the osculating plane is always defined. Since we are working in $\mathbb{R}^3$ we need in addition to $\vec{t}(t_0)$ and $\vec{n}(t_0)$ a further, independent direction in order to span $\mathbb{R}^3$. For this we introduce the **bi-normal vector** $\vec{b}(t_0)$ to γ at t_0 as the uniquely determined unit vector in $\mathbb{R}^3$ such that $\vec{t}(t_0) \perp \vec{b}(t_0)$, $\vec{n}(t_0) \perp \vec{b}(t_0)$, hence $\vec{b}(t_0) \perp \mathcal{E}_{\gamma(t_0)} - \gamma(t_0)$, and $\det(\vec{t}(t_0), \vec{n}(t_0), \vec{b}(t_0)) = 1$, i.e.

$$\vec{b}(t_0) = \vec{t}(t_0) \times \vec{n}(t_0). \tag{7.34}$$

For the derivative of $\vec{b}$ we find

$$\begin{aligned}
\dot{\vec{b}}(t) = (\vec{t}(t) \times \vec{n}(t))\dot{} &= \dot{\vec{t}}(t) \times \vec{n}(t) + \vec{t}(t) \times \dot{\vec{n}}(t) \\
&= \kappa(t)\vec{n}(t) \times \vec{n}(t) + \vec{t}(t) \times \dot{\vec{n}}(t) \\
&= \vec{t}(t) \times \dot{\vec{n}}(t). \tag{7.35}
\end{aligned}$$

Moreover, since $\langle \vec{n}(t), \vec{n}(t) \rangle = 1$ it follows that

$$0 = \frac{d}{dt}\langle \vec{n}(t), \vec{n}(t) \rangle = 2\langle \vec{n}(t), \dot{\vec{n}}(t) \rangle, \tag{7.36}$$

i.e. $\vec{n}(t) \perp \dot{\vec{n}}(t)$ and (7.35) combined with (7.36) yields the existence of $\tau(t)$ such that

$$\dot{\vec{b}}(t) = \tau(t)\vec{n}(t). \tag{7.37}$$

Definition 7.41. *The function $\tau : I \to \mathbb{R}$ defined by (7.37) is called the* **torsion** *of γ.*

We already know that if $\kappa(t) = 0$ for all $t \in I$ then γ is a line segment. Further if $\tau(t) = 0$ for all $t \in I$ then $\vec{b}(t) = \vec{b}_0$ is a constant vector by (7.37) which implies

$$\frac{d}{dt}\langle \gamma(t), \vec{b}_0 \rangle = \langle \vec{t}(t), \vec{b}_0 \rangle = 0$$

and therefore γ remains in the osculation plane $\mathcal{E}_{\gamma(t_0)} = \mathcal{E}_{\gamma(t)}$ for all $t \in I$. Later on in our treatise we will prove that γ is essentially determined by its curvature κ and torsion τ. The proof will be dependent on a discussion of the **Frenet-Serret formulae**

$$\begin{aligned}
\dot{\vec{t}}(s) &= \kappa(s)\vec{n}(s) \\
\dot{\vec{n}}(s) &= -\kappa(s)\vec{t}(s) - \tau(s)\vec{b}(s) \\
\dot{\vec{b}}(s) &= \tau(s)\vec{n}(s),
\end{aligned} \tag{7.38}$$

where the first and third equations are already known to us and the second follows from

$$\frac{d}{ds}\vec{n}(s) = \frac{d}{ds}(\vec{b}(s) \times \vec{t}(s)) = \dot{\vec{b}}(s) \times \vec{t}(s) + \vec{b}(s) \times \dot{\vec{t}}(s)$$
$$= \tau(s)\vec{n}(s) \times \vec{t}(s) + \kappa\vec{b}(s) \times \vec{n}(s)$$
$$= -\tau(s)\vec{b}(s) - \kappa(s)\vec{t}(s),$$

where we used that $\{\vec{t}(s), \vec{n}(s), \vec{b}(s)\}$ is a orthogonal basis such that $\det(\vec{t}(s), \vec{n}(s), \vec{b}(s)) = 1$.

Finally we state a formula for the curvature of γ if γ is not parametrized with respect to the arc length, a proof is provided in Problem 9a).

Proposition 7.42. *The curvature of a regular parametric C^3-curve $\gamma : I \to \mathbb{R}^3$ is given by*

$$\kappa(t) = \frac{|\dot{\gamma}(t) \times \ddot{\gamma}(t)|}{|\dot{\gamma}(t)|^3}. \tag{7.39}$$

Remark 7.43. If we consider a plane curve as a curve in $\mathbb{R}^3$ with $\gamma_3(t) = 0$ for all $t \in I$, then (7.39) reduces to (7.29).

Problems

1. Let $f : (a, b) \to \mathbb{R}^k$ be a function and $t_0 \in (a, b)$. Prove that f is differentiable at t_0 if and only if

$$\lim_{h \to 0} \frac{f(t_0 + h) - f(t_0)}{h}$$

exists.

2. Let $A : (0, 1) \to M(m, n; \mathbb{R})$ and $u : (0, 1) \to \mathbb{R}^n$ be differentiable.

   a) Find

   $$\frac{d}{dt}(A(t)u(t)).$$

b) For $0 < a < b < 1$ prove

$$\left\|\int_a^b A(t)u(t)dt\right\| \le \left(\int_a^b \sum_{k=1}^m \sum_{l=1}^n |a_{kl}(t)|^2 dt\right)^{\frac{1}{2}} \left(\int_a^b \|u(t)\|^2 dt\right)^{\frac{1}{2}}$$

$$\le (mn(b-a))^{\frac{1}{2}} \left(\max_{\substack{1\le k\le m\\1\le l\le n}} \|a_{kl}\|_{\infty,[a,b]}\right) \left(\int_a^b \|u(t)\|^2 dt\right)^{\frac{1}{2}},$$

where $\|\cdot\|$ denotes the Euclidean norm (on the left hand side in $\mathbb{R}^m$ and on the right hand side in $\mathbb{R}^n$) and $\|a_{kl}\|_{\infty,[a,b]} = \sup_{t\in[a,b]} |a_{kl}(t)|$.

Hint: compare with the proof of Theorem 2.30

3.   a) For C^1-functions $u, v : I \to \mathbb{R}^n$ prove

$$\frac{d}{dt}\langle u(t), v(t)\rangle = \langle\frac{du(t)}{dt}, v(t)\rangle + \langle u(t), \frac{dv(t)}{dt}\rangle.$$

b) Prove Leibniz's rule for the cross product, i.e.

$$\frac{d}{dt}(\alpha(t) \times \beta(t)) = \frac{d\alpha(t)}{dt} \times \beta(t) + \alpha(t) \times \frac{d\beta(t)}{dt}$$

for C^1-function $\alpha, \beta : I \to \mathbb{R}^3$.

4.   a) Let $\gamma : [0, 2\pi] \to \mathbb{R}^2$, $f(t) = (\alpha \cos t, \beta \sin t), \alpha > 0, \beta > 0$, be a parametrization of the ellipse. Prove that

$$\operatorname{tr}(\gamma) = \mathcal{E} = \left\{(x, y) \in \mathbb{R}^2 \,\middle|\, \frac{x^2}{\alpha^2} + \frac{y^2}{\beta^2} = 1\right\}.$$

b) Prove that $\operatorname{tr}(\gamma) \subset S^1 = \{(x, y) \in \mathbb{R}^2 | x^2 + y^2 = 1\}$ where $\gamma : \mathbb{R} \to \mathbb{R}^2, \gamma(t) = \left(\frac{1-t^2}{1+t^2}, \frac{2t}{1+t^2}\right)$.

c) Let $\psi : \mathbb{R} \to \mathbb{R}^3$ be given by $\psi(t) = (3t \cos t, 3t \sin t, 5t)$ and show that $\operatorname{tr}(\psi) \subset \left\{(x, y, z) \in \mathbb{R}^3 \,\middle|\, \frac{x^2}{9} + \frac{y^2}{9} - \frac{z^2}{25} = 0\right\}$.

5.   a) Consider $\varphi : [0, \infty) \to \mathbb{R}^2, \varphi(t) = (at - b \sin t, a - b \cos t)$ where $0 < b \le a$. Find the velocity vector $\dot\varphi(t), t \in [0, \infty)$. If $b < a$, does φ have singular points? Prove that for $b = a$ the curve φ has singular points. Sketch the curve φ for $t \in [0, 6\pi]$ and $a = b = 2$.

b) For $\gamma_j : \mathbb{R} \to \mathbb{R}^2$, $j = 1, 2$, find all regular, singular and multiple points, here $\gamma_1(t) = (t^5, t^4)$ and $\gamma_2(t) = (t^2 + t - 2, t^3 + t^2 - 2t)$.

6.   a) Find the tangent line to the logarithmic spiral $\gamma : \mathbb{R} \to \mathbb{R}^2$, $\gamma(t) = (e^{at} \cos t, e^{at} \sin t), a > 0$, at the point $t_1 = \frac{\pi}{2}$.

b) Consider the circular helix $\gamma : \mathbb{R} \to \mathbb{R}^3, \gamma(t) = (r \cos t, r \sin t, bt)$, $r > 0$ and $b > 0$. Find the tangent line to γ at $t_0 \in \mathbb{R}$ and prove that the tangent line at t_0 and at $t_0 + 2\pi k, k \in \mathbb{Z}$ are parallel.

7.   a) For the circle $x^2 + y^2 = r^2$ find a representation $\gamma : [0, 2\pi] \to \mathbb{R}^2$ such that γ is simply closed and prove that all normal lines to the circle intersect at one point.

b) Let $\gamma : [0, 2\pi] \to \mathbb{R}^2$, $\gamma(\varphi) = \rho(\varphi) \begin{pmatrix} \cos \varphi \\ \sin \varphi \end{pmatrix}$ be a plane curve in polar coordinates with a C^1-function ρ. Find the normal line to γ at $\varphi_0 \in (0, 2\pi)$.

8.   a) Prove that the curve $\gamma : \mathbb{R} \to \mathbb{R}^3$ given by

$$\gamma(s) = \left(\frac{1}{2}(s + \sqrt{s^2 + 1}), \frac{1}{2}(s + \sqrt{s^2 + 1})^{-1}, \frac{1}{2}\sqrt{2}(\ln(s + \sqrt{s^2 + 1})) \right)$$

is parametrized with respect to the arc length.
Hint: it is easier to write $\gamma(s) = \gamma(h(s)) = (\frac{1}{2}h(s), \frac{1}{2h(s)}, \frac{\sqrt{2}}{2} \ln h(s))$.

b) For the straight line segment $\gamma : [-1, 1] \to \mathbb{R}^2$ given by $\gamma(t) = \begin{pmatrix} 2 \\ 4 \end{pmatrix}t + \begin{pmatrix} 2 \\ 0 \end{pmatrix}$ find a parametrization with respect to the arc length.

9.   a) Prove that the curvature of a regular parametric C^3-curve is given by

$$\kappa(t) = \frac{|\dot{\gamma}(t) \times \ddot{\gamma}(t)|}{|\dot{\gamma}(t)|^3}.$$

This is the content of Proposition 7.42 and the main point is that γ is not necessarily parametrized with respect to the arc length.

b) Let $\gamma : I \to \mathbb{R}^3$ be a regular C^3-curve parametrized with respect to the arc length and with non-vanishing curvature. Prove that its torsion is given by

$$\tau(s) = \frac{-\langle \dot{\gamma}(s) \times \ddot{\gamma}(s), \dddot{\gamma}(s) \rangle}{\kappa^2(s)}.$$

Hint: express $\dot{\gamma}(s), \ddot{\gamma}(s)$ and $\dddot{\gamma}(s)$ with the help of $\vec{t}(s), \vec{n}(s)$ and $\vec{b}(s)$ and make use of orthogonality relations.

137

10.   a) Given the circular helix $\gamma : \mathbb{R} \to \mathbb{R}^3, \gamma(t) = (r \cos t, r \sin t, ht)$, $r > 0$ and $h > 0$. Parametrize γ with respect to the arc length.

b) Given the following part of a circular helix:

$$\gamma : [0, 4\pi] \to \mathbb{R}^3, \gamma(t) = (3 \cos t, 3 \sin t, 4t).$$

Let $\varphi : \tilde{I} \to [0, 4\pi]$ such that $\tilde{\gamma} := \gamma \circ \varphi : \tilde{I} \to \mathbb{R}^3$ is parametrized with respect to the arc length. Find φ (hence $\tilde{I}$) and $\tilde{\gamma} : \tilde{I} \to \mathbb{R}^3$. For $\tilde{\gamma}$ find the curvature and torsion (use Problem 9), as well as the osculating plane at $\tilde{\gamma}(s_0)$, $s_0 \in I$.

11. In two dimensions we apparently have two definitions of the normal vector to a regular C^2-curve, Definition 7.20 and Definition 7.29 for $n = 2$. Prove that both definitions coincide.

12. Consider on $\mathbb{R}^2$ the ray $R_{\rho_0} : [0, \infty) \to \mathbb{R}^2, R_{\rho_0}(\rho) = \rho \binom{\cos \varphi_0}{\sin \varphi_0}$ for some fixed $\varphi_0 \in [0, 2\pi)$, and the circle $C_{\rho_0} : [0, 2\pi] \to \mathbb{R}^2, C_{\rho_0}(\varphi) = \rho_0 \binom{\cos \varphi}{\sin \varphi}$ for some fixed $\rho_0 > 0$. Prove that $R_{\varphi_0} \cap C_{\rho_0}$ consists of one point and both curves are orthogonal at the intersection point.

13. Since $\cos \frac{\pi}{4} = \sin \frac{\pi}{4} = \frac{1}{2}\sqrt{2}$ the functions $g : \mathbb{R} \to \mathbb{R}, g(x) = \cos x$, and $h : \mathbb{R} \to \mathbb{R}, h(x) = \sin x$, intersect at the point $(\frac{\pi}{4}, \frac{1}{2}\sqrt{2})$. Find the angle under which these two curves intersect.

8 Surfaces in $\mathbb{R}^3$. A First Encounter

We have seen that every (differentiable) function $f : I \to \mathbb{R}$, $I \subset \mathbb{R}$ being an interval, gives rise to a parametric curve $\gamma : I \to \mathbb{R}^2$, $\gamma(t) = (t, f(t))$, the trace of which is just the graph of f. A different point of view is that we "deform" the line segment $I \times \{0\} \subset \mathbb{R}^2$ and obtain a curve, and in fact this is an interpretation we can apply to every parametric curve in $\mathbb{R}^n$.

We may look at a surface in $\mathbb{R}^3$ first as a graph of a function $f : G \to \mathbb{R}$, $G \subset \mathbb{R}^2$, and then more generally as a "deformation" of $G \times \{0\} \subset \mathbb{R}^3$. Instead of I being an interval we now require $G \subset \mathbb{R}^2$ to be a region, i.e. a non-empty open and connected subset of $\mathbb{R}^2$, see Definition 3.30.D. Studying graphs of functions as surfaces allows two complementary approaches: we may use geometry to study the function, but we can also employ properties of the functions to study geometry. We start with a lengthy example: the open upper half sphere S_+^2.

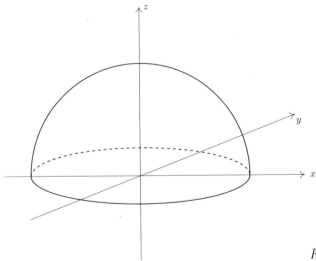

Figure 8.1

On the open unit ball $B_1(0) \subset \mathbb{R}^2$ we define the function $f : B_1(0) \to \mathbb{R}$, $f(x, y) = \sqrt{1 - x^2 - y^2}$. The graph of f is by definition $S_+^2 := \Gamma(f) = \left\{ (x, y, z) \in \mathbb{R}^3 \,\middle|\, (x, y) \in B_1(0) \text{ and } z = f(x, y) = \sqrt{1 - x^2 - y^2} \right\}$, see Figure 8.1, recall $S^2 = \left\{ (x, y, z) \in \mathbb{R}^3 \,\middle|\, x^2 + y^2 + z^2 = 1 \right\}$. (Note that $S^2 = \partial B_1^{(3)}(0)$

where $B_1^{(3)}(0)$ is the unit ball in $\mathbb{R}^3$.) In $B_1(0) \subset \mathbb{R}^2$ we can consider certain curves and lift these curves onto $\Gamma(f)$. For example we may look at circles $\omega_\rho : [0, 2\pi] \to B_1(0)$, $\omega_\rho(\varphi) = \begin{pmatrix} \rho\cos\varphi \\ \rho\sin\varphi \end{pmatrix}$, $0 < \rho < 1$ fixed. The image of ω_ρ under f is a subset of $\Gamma(f) \subset \mathbb{R}^3$ namely the trace of the curve
$$\gamma_\rho(\varphi) = \begin{pmatrix} \rho\cos\varphi \\ \rho\sin\varphi \\ \sqrt{1-\rho^2} \end{pmatrix}, \quad \varphi \in [0, 2\pi].$$ Thus $(x, y, z) \in \operatorname{tr}(\gamma_\rho)$ if and only if
$x = \rho\cos\varphi$, $y = \rho\sin\varphi$, $z = \sqrt{1 - \rho^2\cos^2\varphi - \rho^2\sin^2\varphi} = \sqrt{1-\rho^2}$. In other words $\operatorname{tr}(\gamma_\rho)$ is a circle with centre $(0, 0, \sqrt{1-\rho^2})$ and radius ρ, see Figure 8.2.

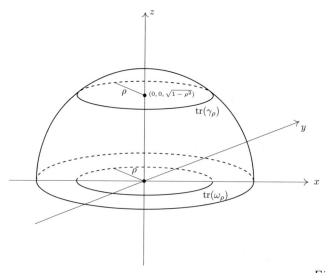

Figure 8.2

The ray segment $R_\gamma : (0, 1) \to B_1(0) \subset \mathbb{R}^2$, $\varphi \in [0, 2\pi]$ fixed, $R_\gamma(\rho) = \begin{pmatrix} \rho\cos\varphi \\ \rho\sin\varphi \end{pmatrix}$ induces the curve $\eta_\varphi(\rho) = \begin{pmatrix} \rho\cos\varphi \\ \rho\sin\varphi \\ \sqrt{1-\rho^2} \end{pmatrix}$ whose trace is a quarter circle starting at $(0, 0, 1)$ and ending at $(\cos\varphi, \sin\varphi, 0)$ with the two "endpoints" excluded, see Figure 8.3.

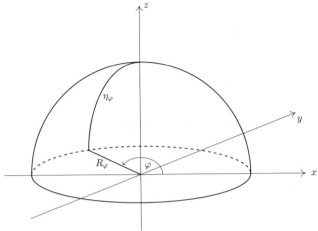

Figure 8.3

When we consider the diameter of $B_1(0)$ as a line segment, i.e. $D_\varphi : (-1, 1) \to B_1(0)$, $D_\varphi(r) = \begin{pmatrix} r \cos\varphi \\ r \sin\varphi \end{pmatrix}$, $\varphi \in [0, 2\pi]$ fixed, the corresponding curve on $\Gamma(f)$ is of course the (open) semi-circle $\zeta_\varphi(r) = \begin{pmatrix} r \cos\varphi \\ r \sin\varphi \\ \sqrt{1 - \rho^2} \end{pmatrix}$ connecting $(-\cos\varphi, -\sin\varphi, 0)$ with $(\cos\varphi, \sin\varphi, 0)$, see Figure 8.4.

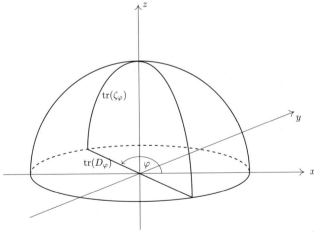

Figure 8.4

For $(x_0, y_0) = (\rho_0 \cos \varphi_0, \rho_0 \sin \varphi_0) \in B_1(0)$ the circle w_{ρ_0} and the ray segment R_{φ_0} (more precisely their traces) intersect at the point (x_0, y_0) and consequently the traces of γ_{ρ_0} and η_{φ_0} intersect at $(x_0, y_0, f(x_0, y_0)) = \left(\rho_0 \cos \varphi_0, \rho_0 \sin \varphi_0, \sqrt{1 - \rho_0^2} \right)$. Clearly w_{ρ_0} and R_{φ_0} are orthogonal at their intersection point: classical geometry teaches that the "radius" and the "circle line" are orthogonal curves. The tangent line to γ_{ρ_0} at $\left(\rho_0 \cos \varphi_0, \rho_0 \sin \varphi_0, \sqrt{1 - \rho_0^2} \right)$ is given by

$$g_{\gamma_{\rho_0}(\varphi_0)}(s) = \begin{pmatrix} -\rho_0 \sin \varphi_0 \\ \rho_0 \cos \varphi_0 \\ 0 \end{pmatrix} s + \begin{pmatrix} \rho_0 \cos \varphi_0 \\ \rho_0 \sin \varphi_0 \\ \sqrt{1 - \rho_0^2} \end{pmatrix}$$

and the tangent line to η_{φ_0} at $\left(\rho_0 \cos \varphi_0, \rho_0 \sin \varphi_0, \sqrt{1 - \rho_0^2} \right)$ is given by

$$h_{\eta_{\varphi_0}(\rho_0)}(t) = \begin{pmatrix} \cos \varphi_0 \\ \sin \varphi_0 \\ \frac{-\rho_0}{\sqrt{1 - \rho_0^2}} \end{pmatrix} t + \begin{pmatrix} \rho_0 \cos \varphi_0 \\ \rho_0 \sin \varphi_0 \\ \sqrt{1 - \rho_0^2} \end{pmatrix}$$

and at their intersection point $\left(\rho_0 \cos \varphi_0, \rho_0 \sin \varphi_0, \sqrt{1 - \rho_0^2} \right)$ they are orthogonal too:

$$\left\langle \begin{pmatrix} -\rho_0 \sin \varphi_0 \\ \rho_0 \cos \varphi_0 \\ 0 \end{pmatrix}, \begin{pmatrix} \cos \varphi_0 \\ \sin \varphi_0 \\ \frac{-\rho_0}{\sqrt{1 - \rho_0^2}} \end{pmatrix} \right\rangle = 0.$$

Moreover, if we take any point $(x_0, y_0, z_0) \in \Gamma(f)$ it is of the form $\left(\rho_0 \cos \varphi_0, \rho_0 \sin \varphi_0, \sqrt{1 - \rho_0^2} \right)$ for some ρ_0 and φ_0, and projecting all points with $\rho_0 = $ constant onto the plane $z = 0$ we will obtain the circle with radius ρ_0 and centre $(0, 0, 0)$, while projecting all points with $\varphi_0 = $ constant we obtain the ray segment connecting the origin $(0, 0, 0)$ with the point $(\cos \varphi_0, \sin \varphi_0, 0)$ (both "end points" being excluded), see Figures 8.5 and 8.6.

142

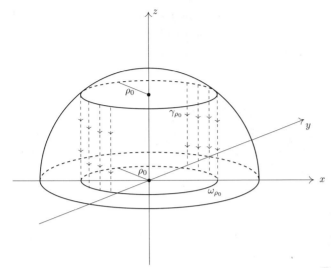

Figure 8.5

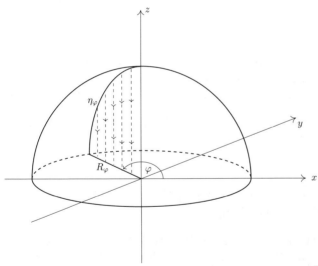

Figure 8.6

In this example there is a one-to-one correspondence between the points on S_+^2 and their projections onto $B_1(0) \times \{0\} \subset \mathbb{R}^3$. We can use the families of curves on S_+^2, $\rho = $ constant, $\rho \in (0,1)$, and $\varphi = $ constant, $\varphi \in [0, 2\pi]$, i.e. γ_ρ and η_φ, as coordinate lines on $S_+^2 \setminus \{(0,0,1)\}$: every point on $S_+^2 \setminus \{(0,0,1)\}$ is uniquely determined as the intersection point of exactly one curve of the family γ_ρ, $\rho \in (0,1)$, and exactly one curve of the family η_φ, $\varphi \in [0, 2\pi]$. Note that when allowing $\rho = 0$ we will have covered all points on S_+^2, but we have to give up the bijectivity of the mapping $(x, y) \mapsto (\rho \cos \varphi, \rho \sin \varphi)$.

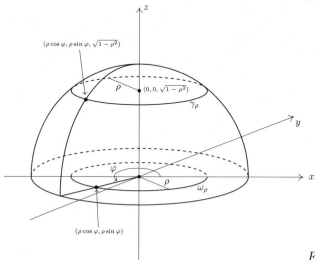

Figure 8.7

With this example in mind, the main question is: what is typical for graphs of functions and what is special in this example? We start by introducing level lines, even better level sets of a function.

Definition 8.1. *Let $G \subset \mathbb{R}^2$ be a region and $f : G \to \mathbb{R}$ be a continuous function. For $c \in \mathbb{R}$ we call*

$$f^{-1}(c) := \{(x, y) \in G \mid f(x, y) = c\} \tag{8.1}$$

*the **level set** of f to the level c. A subset of $f^{-1}(c)$ which is the trace of a (parametric) curve is called a **level line** of f.*

Remark 8.2. A. If $c \notin f(G)$ then $f^{-1}(c) = \emptyset$. Otherwise, since f is continuous and $\{c\}$ is closed the set $f^{-1}(c)$ is a closed subset in $\mathbb{R}^2$.
B. For $c_1 \neq c_2$ it always follows that $f^{-1}(c_1) \cap f^{-1}(c_2) = \emptyset$, i.e. level sets for two different "levels" will never intersect.
C. Since G is by assumption a region, in particular G is connected, $f(G)$ is an interval and every point $(x, y) \in G$ belongs to exactly one level set $f^{-1}(c)$, $c \in I = f(G)$.

Example 8.3. A. For $f : B_1(0) \to \mathbb{R}$, $f(x, y) = \sqrt{1 - x^2 - y^2}$, the level sets are the following:

(i) $f^{-1}(c) = \emptyset$ if $c \in (-\infty, 0] \cup (1, \infty]$;

(ii) $f^{-1}(c) = \{(x, y) \in \mathbb{R}^2 \,|\, x^2 + y^2 = 1 - c^2\}$ if $c \in (0, 1)$;

(iii) $f^{-1}(1) = \{(0, 0)\}$.

B. Consider the function $g : \mathbb{R}^2 \to \mathbb{R}$, $g(x, y) = (x^2 + y^2)^2 - (x^2 + y^2) + 1$. The level sets of this function can be obtained in the following way: for all $(x, y) \in \mathbb{R}^2$ such that $x^2 + y^2 = r^2$ the function has the same value namely $r^4 - r^2 + 1$. Moreover, when r varies in $[0, \infty)$ the set $\{(x, y) \in \mathbb{R}^2 \,|\, x^2 + y^2 = r^2,\ r \in [0, \infty)\}$ is equal to $\mathbb{R}^2$. Thus we can reduce our discussion to an investigation of $h : [0, \infty) \to \mathbb{R}$, $h(r) = r^4 - r^2 + 1$. If $h(r) = c$ then the level set of $f^{-1}(c)$ consists of all $(x, y) \in \mathbb{R}^2$ such that $x^2 + y^2 = r^2$. The function h has a local maximum at $r_0 = 0$ with $h(r_0) = 1$, it has a local minimum at $r_1 = \frac{1}{2}\sqrt{2}$ with $h(r_1) = \frac{3}{4}$, and as $r \to \infty$ it follows that $h(r) \to \infty$.

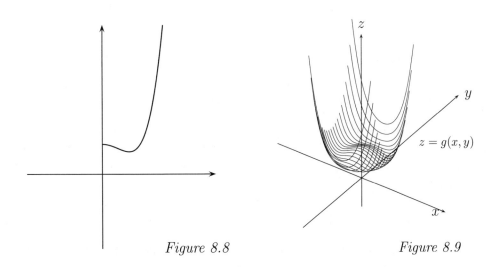

Figure 8.8   Figure 8.9

If $c < \frac{3}{4}$ we find $f^{-1}(c) = \emptyset$ and

$$f^{-1}\left(\frac{3}{4}\right) = \left\{(x, y) \in \mathbb{R}^2 \,\Big|\, h\left((x^2 + y^2)^{\frac{1}{2}}\right) = \frac{3}{4}\right\} = \left\{(x, y) \in \mathbb{R}^2 \,\Big|\, x^2 + y^2 = \frac{1}{2}\right\},$$

which is a circle with radius $\frac{1}{2}\sqrt{2}$ and centre $(0, 0)$. For $\frac{3}{4} < c < 1$ the

145

equation $h(r) = c$ has two solutions,

$$r_s = \sqrt{\frac{1}{2} - \sqrt{c - \frac{3}{4}}} \quad \text{and} \quad r_l = \sqrt{\frac{1}{2} + \sqrt{c - \frac{3}{4}}}$$

and therefore we find

$$f^{-1}(c) = \left\{ (x,y) \in \mathbb{R}^2 \,\Big|\, x^2 + y^2 = \frac{1}{2} - \sqrt{c - \frac{3}{4}} \right\} \cup \left\{ (x,y) \in \mathbb{R}^2 \,\Big|\, x^2 + y^2 = \frac{1}{2} + \sqrt{c - \frac{3}{4}} \right\},$$

i.e. $f^{-1}(c)$ for this range of c consists of two circles with centre $(0,0)$ and with radius r_s and r_l, respectively. If $c = 1$ then

$$f^{-1}(1) = \{(0,0)\} \cup \left\{ (x,y) \in \mathbb{R}^2 \,\big|\, x^2 + y^2 = 1 \right\},$$

and for $c > 1$ we have the level sets

$$f^{-1}(c) = \left\{ (x,y) \in \mathbb{R}^2 \,\Big|\, x^2 + y^2 = \frac{1}{2} + \sqrt{c - \frac{3}{4}} \right\}$$

which are circles with centre $(0,0)$ and radius $\sqrt{\frac{1}{2} + \sqrt{c - \frac{3}{4}}}$. Thus, level sets may be nice curves, may be disconnected sets, each component consisting of a nice curve, may be the union of a nice curve with a point etc.

Given the level sets of a function $f : G \to \mathbb{R}$, $G \subset \mathbb{R}^2$, and the value on each level set we can of course recover the function and its graph: if c is in the range of f, i.e. $c \in \mathrm{ran}(f)$, and if $N(c) = f^{-1}(c) \subset G$ is the corresponding level set then

$$\Gamma(f) = \left\{ (z,c) \,\big|\, z \in N(c) \text{ and } c \in \mathrm{ran}(f) \right\}.$$

This observation is used to produce geographical maps. For a small patch of land we can assume that the surface of the Earth is well approximated by a plane. The landscape is then uniquely determined when knowing the altitude for every point of this patch (say with sea level being normalised to 0).

We will return to level sets in Chapter 10 when dealing with implicitly defined functions and in Chapter 12 when handling curvilinear coordinates. The idea to approximate locally the surface of the Earth by a plane points of course to the question of the existence of a tangent plane to graphs of functions $f : G \to \mathbb{R}$. Indirectly we encountered this problem earlier, namely when looking at parametric curves $\gamma : [a, b] \to \mathbb{R}^3$ whose trace $\mathrm{tr}(\gamma)$ are contained in the graph of a given function.

Let $f : G \to \mathbb{R}$, $G \subset \mathbb{R}^2$ a region, be a C^1-function and consider the graph $\Gamma(f) = \{(x, y, f(x, y)) \mid (x, y) \in G\} \subset \mathbb{R}^3$. For $(x_0, y_0) \in G$ we can find $a > 0$ and $b > 0$ such that $(x_0 - a, x_0 + a) \times (y_0 - b, y_0 + b) \subset G$. Now consider the two families of curves $\xi_{y_0} : (x_0 - a, x_0 + a) \to \Gamma(f) \subset \mathbb{R}^3$, $\xi_{x_0}(x) = (x, y_0, f(x, y_0))$ and $\eta_{x_0} : (y_0 - b, y_0 + b) \to \Gamma(f) \subset \mathbb{R}^3$, $\eta_{x_0}(y) = (x_0, y, f(x_0, y))$. The velocity vector of ξ_{y_0} at x_0 is given by $\begin{pmatrix} 1 \\ 0 \\ \frac{\partial f}{\partial x}(x_0, y_0) \end{pmatrix}$ and the velocity vector of η_{x_0} at y_0 is given by $\begin{pmatrix} 0 \\ 1 \\ \frac{\partial f}{\partial y}(x_0, y_0) \end{pmatrix}$. Let us change our point of view. We can consider the mapping $F : G \to \mathbb{R}^3$, $(x, y) \mapsto (F_1(x, y), F_2(x, y), F_3(x, y)) = (x, y, f(x, y))$. Clearly $F(G) = \Gamma(f)$. The differential of F at the point $(x_0, y_0) \in G$ is given by the Jacobi matrix

$$(\mathrm{d}F)_{(x_0, y_0)} = J_F(x_0, y_0) = \begin{pmatrix} \frac{\partial F_1}{\partial x}(x_0, y_0) & \frac{\partial F_1}{\partial y}(x_0, y_0) \\ \frac{\partial F_2}{\partial x}(x_0, y_0) & \frac{\partial F_2}{\partial y}(x_0, y_0) \\ \frac{\partial F_3}{\partial x}(x_0, y_0) & \frac{\partial F_3}{\partial y}(x_0, y_0) \end{pmatrix}$$

$$= \begin{pmatrix} 1 & 0 \\ 0 & 1 \\ \frac{\partial f}{\partial x}(x_0, y_0) & \frac{\partial f}{\partial y}(x_0, y_0) \end{pmatrix}. \qquad (8.2)$$

Thus the Jacobi matrix of F at (x_0, y_0) is the matrix built up by the two vectors $\begin{pmatrix} 1 \\ 0 \\ \frac{\partial f}{\partial x}(x_0, y_0) \end{pmatrix}$ and $\begin{pmatrix} 0 \\ 1 \\ \frac{\partial f}{\partial y}(x_0, y_0) \end{pmatrix}$. Since the Jacobi matrix gives the linear approximation of F we shall consider the plane

$$T_{(x_0, y_0, f(x_0, y_0))}(\Gamma(f)) := \left\{ \begin{pmatrix} x_0 \\ y_0 \\ f(x_0, y_0) \end{pmatrix} + \lambda \begin{pmatrix} 1 \\ 0 \\ \frac{\partial f}{\partial x}(x_0 y_0) \end{pmatrix} + \mu \begin{pmatrix} 0 \\ 1 \\ \frac{\partial f}{\partial y}(x_0, y_0) \end{pmatrix} \middle| \lambda, \mu \in \mathbb{R} \right\} \qquad (8.3)$$

as the **tangent plane** to $\Gamma(f)$ at $\begin{pmatrix} x_0 \\ y_0 \\ f(x_0, y_0) \end{pmatrix}$.

Note that $\begin{pmatrix} 1 \\ 0 \\ \frac{\partial f}{\partial x}(x_0, y_0) \end{pmatrix}$ and $\begin{pmatrix} 0 \\ 1 \\ \frac{\partial f}{\partial y}(x_0, y_0) \end{pmatrix}$ are independent whatever f is

and hence they always span a plane. Thus every graph $\Gamma(f)$ of a C^1-function $f : G \to \mathbb{R}$ has at every point $(x_0, y_0, f(x_0, y_0))$ a tangent plane. Of course we can now normalise these two vectors to

$$\vec{t}_1(x_0, y_0) = \frac{1}{\left(1 + \left(\frac{\partial f}{\partial x}(x_0, y_0)\right)^2\right)^{\frac{1}{2}}} \begin{pmatrix} 1 \\ 0 \\ \frac{\partial f}{\partial x}(x_0, y_0) \end{pmatrix} \tag{8.4}$$

and

$$\vec{t}_2(x_0, y_0) = \frac{1}{\left(1 + \left(\frac{\partial f}{\partial y}(x_0, y_0)\right)^2\right)^{\frac{1}{2}}} \begin{pmatrix} 0 \\ 1 \\ \frac{\partial f}{\partial y}(x_0, y_0) \end{pmatrix}. \tag{8.5}$$

Moreover the vector

$$\vec{n}(x_0, y_0) := \frac{\begin{pmatrix} 1 \\ 0 \\ f_x(x_0, y_0) \end{pmatrix} \times \begin{pmatrix} 0 \\ 1 \\ f_y(x_0, y_0) \end{pmatrix}}{\left\| \begin{pmatrix} 1 \\ 0 \\ f_x(x_0, y_0) \end{pmatrix} \times \begin{pmatrix} 0 \\ 1 \\ f_y(x_0, y_0) \end{pmatrix} \right\|} \tag{8.6}$$

is a unit vector orthogonal to $\vec{t}_1(x_0, y_0)$ and $\vec{t}_2(x_0, y_0)$, i.e. it is orthogonal to the plane $T_{(x_0, y_0, f(x_0, y_0))}(\Gamma(f)) - \begin{pmatrix} x_0 \\ y_0 \\ f(x_0, y_0) \end{pmatrix} = \mathrm{span}(\vec{t}_1(x_0, y_0), \vec{t}_2(x_0, y_0))$.

We define the **normal line** to $\Gamma(f)$ at $(x_0, y_0, f(x_0, y_0))$ as the line

$$\tilde{N}_{(x_0, y_0, f(x_0, y_0))}(\Gamma(f)) = \left\{ \begin{pmatrix} x_0 \\ y_0 \\ f(x_0, y_0) \end{pmatrix} + s\vec{n}(x_0, y_0) \,\middle|\, s \in \mathbb{R} \right\}. \tag{8.7}$$

Hence at every point $(x_0, y_0, f(x_0, y_0))$ we may introduce a new coordinate system with origin $(x_0, y_0, f(x_0, y_0))$ and basis $\vec{t}_1(x_0, y_0)$, $\vec{t}_2(x_0, y_0)$ and $\vec{n}(x_0, y_0)$.

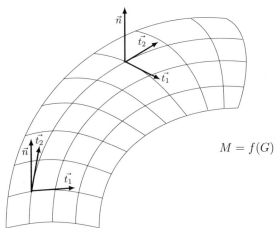

Figure 8.10

The considerations made above lead to the idea to introduce and study more general surfaces than graphs of functions. In the following $G \subset \mathbb{R}^2$ is a region and typical points in G are denoted by (u, v).

Definition 8.4. A. *A **parametric surface** in $\mathbb{R}^3$ is a C^1-mapping $f : G \to \mathbb{R}^3$ for which the differential (Jacobi matrix) is at all points $(u, v) \in G$ injective, i.e. $(\mathrm{d}f)_{(u,v)} = J_f(u, v) : \mathbb{R}^2 \to \mathbb{R}^3$ is injective. We call $M := f(G) \subset \mathbb{R}^3$ the **trace** of f.*
B. *Assume in addition that f is injective. The vectors $f_u(u, v)$ and $f_v(u, v)$ are called **tangent vectors** to $M = f(G)$ at $p = f(u, v)$ and $N(u, v) := f_u(u, v) \times f_v(u, v)$ is the **normal vector** corresponding to M at p. The plane $f(u, v) + \mathrm{span}\{f_u(u, v), f_v(u, v)\} \subset \mathbb{R}^3$ is called the **tangent plane** to M at p and $f(u, v) + \{\lambda N(u, v) | \lambda \in \mathbb{R}\} \subset \mathbb{R}^3$ is called the **normal line** to M at p.*

Clearly the normal line to M at p is orthogonal to the tangent plane to M at p since $f_u \times f_v$ is orthogonal to f_u and f_v. The normalised vectors

$$\vec{t}_1(u, v) := \frac{f_u(u, v)}{||f_u(u, v)||} \quad \text{and} \quad \vec{t}_2(u, v) = \frac{f_v(u, v)}{||f_v(u, v)||} \tag{8.8}$$

are called **unit tangent vectors** to M at $p = f(u, v)$. Note that the injectivity of $df_{(u,v)}$ implies that $||f_u|| \neq 0$ and $||f_v|| \neq 0$. The normalised vector

$$\vec{n}(u, v) := \frac{f_u(u, v) \times f_v(u, v)}{||f_u(u, v) \times f_v(u, v)||} \tag{8.9}$$

is often called the **unit normal vector** to M at p. We have

$$\det(\vec{t}_1, \vec{t}_2, \vec{n}) > 0 \qquad (8.10)$$

for all $(u, v) \in G$ where df is injective. The triple $\{\vec{t}_1(u, v), \vec{t}_2(u, v), \vec{n}(u, v)\}$ is called the **Gaussian frame** to M at $p = f(u, v)$.

Remark 8.5. Although it is not always necessary, we want to agree that f itself is injective, and by this we avoid double points or self-intersections of M.

Example 8.6. A. For an open set $G \subset \mathbb{R}^2$ and $h : G \to \mathbb{R}$ a C^1-function we can consider the parametric surface $f : G \to \mathbb{R}^3$, $f(u, v) = \begin{pmatrix} u \\ v \\ h(u, v) \end{pmatrix}$, which is regular. Further we have

$$\vec{t}_1 = \frac{1}{\sqrt{1 + h_u^2}} \begin{pmatrix} 1 \\ 0 \\ h_u \end{pmatrix}, \vec{t}_2 = \frac{1}{\sqrt{1 + h_v^2}} \begin{pmatrix} 0 \\ 1 \\ h_v \end{pmatrix}, \vec{n} = \frac{1}{\sqrt{1 + h_u^2 + h_v^2}} \begin{pmatrix} -h_u \\ -h_v \\ 1 \end{pmatrix}.$$
$$(8.11)$$

B. Let $g = B_1(0) = \{(u, v) \in \mathbb{R}^2 \mid u^2 + v^2 < 1\}$ and $f : G \to \mathbb{R}^3$ be the mapping $f(u, v) = (u, v, \sqrt{1 - u^2 - v^2})$. Since

$$(df)_{(u,v)} = \begin{pmatrix} 1 & 0 \\ 0 & 1 \\ -\frac{u}{\sqrt{1-u^2-v^2}} & -\frac{v}{\sqrt{1-u^2-v^2}} \end{pmatrix},$$

$(d f)_{(u,v)}$ is injective and the trace M is the (open) upper half-sphere.

C. Let $G = (0, \pi) \times (0, 2\pi)$ and $g : G \to \mathbb{R}^3$ be given by

$$g(u, v) = (\sin u \cos v, \sin u \sin v, \cos u),$$

which implies

$$(dg)_{(u,v)} = \begin{pmatrix} \cos u \cos v & -\sin u \sin v \\ \cos u \sin v & \sin u \cos v \\ -\sin u & 0 \end{pmatrix},$$

and since

$$\left\langle \begin{pmatrix} \cos u \cos v \\ \cos u \sin v \\ -\sin u \end{pmatrix}, \begin{pmatrix} -\sin u \sin v \\ \sin u \cos v \\ 0 \end{pmatrix} \right\rangle = 0$$

these vectors are orthogonal, hence independent and $(dg)_{(u,v)}$ is for all $(u,v) \in G$ injective. Furthermore with $g(u,v) = (x,y,z)$ we have

$$x^2 + y^2 + z^2 = \sin^2 u \cos^2 v + \sin^2 u \sin^2 v + \cos^2 u = 1$$

and we find immediately that $f(G) \subset S^2$, i.e. the trace M of f is a subset of the unit sphere in $\mathbb{R}^3$. When studying spherical polar coordinates we will look at M in more detail.

D. This example serves to understand the two dimensional torus as a parametric surface in $\mathbb{R}^n$. The torus and the topological sum of tori are fundamental to understand concepts such as the genus of a topological space or the meaning of a topological group. Let $a > b > 0$ and $g : [0, 2\pi] \times [0, 2\pi] \to \mathbb{R}^3$ be given by

$$g(u,v) = ((a + b \cos u) \cos v, (a + b \cos u) \sin v, b \sin u).$$

We claim that $g|_{(0,2\pi) \times (0,2\pi)}$ is a parametric surface. First we find

$$(dg)_{(u,v)} = \begin{pmatrix} b \sin u \cos v & -(a + b \cos u) \sin v \\ -b \sin u \sin v & (a + b \cos u) \cos v \\ b \cos u & 0 \end{pmatrix}.$$

and if $\cos u \neq 0$ then the vectors g_u and g_v are independent. But for $\cos u = 0$ it follows that $\sin u \in \{-1, 1\}$ and we have to check the independence of $\begin{pmatrix} \pm b \cos v \\ \pm b \sin v \\ 0 \end{pmatrix}$ and $\begin{pmatrix} -a \sin v \\ a \cos v \\ 0 \end{pmatrix}$ and since the scalar product of these two vectors is 0 it follows that $(dg)_{(u,v)}$ is always injective, i.e. g is a parametric surface. In order to get some idea about the trace of g we start with $u = 0$ and $u = \pi$. For $u = 0$ we find

$$g(0, v) = ((a + b) \cos v, (a + b) \sin v, 0)$$

which is a circle in the $x - y$-plane with centre $(0, 0, 0)$ and radius $a + b$. For $u = \pi$ we find

$$g(\pi, v) = ((a - b) \cos v, (a - b) \sin v, 0)$$

which gives the circle in the $x - y$-plane with centre $(0, 0, 0)$ and radius $a - b$.

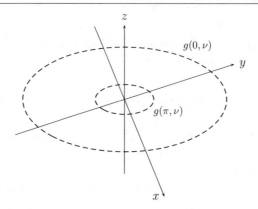

Figure 8.11

In general, for $u_0 \in [0, 2\pi]$ we find that

$$g(u_0, v) = ((a + b\cos u_0)\cos v, (a + b\cos u_0)\sin v, b\sin u_0)$$

and with $x = (a + b\cos u_0)\cos v$, $y = (a + b\cos u_0)\sin v$, $z = b\sin u_0$ it follows that

$$
\begin{aligned}
x^2 + y^2 + (z - b\sin u_0)^2 &= ((a + b\cos u_0)\cos v)^2 + ((a + b\cos u_0)\sin v)^2 \\
&\quad + (b\sin u_0 - b\sin u_0)^2 \\
&= (a + b\cos u_0)^2\cos^2 v + (a + b\cos u_0)^2\sin^2 v \\
&= (a + b\cos u_0)^2.
\end{aligned}
$$

Thus $g(u_0, v)$ gives a circle with centre $(0, 0, b\sin u_0)$ and radius $a + b\cos u_0$.

The union of all these circles gives $M = \operatorname{tr} g$ and this is the **torus**:

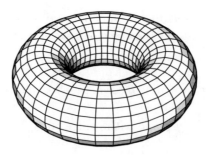

Figure 8.12

152

Example 8.7. First have a look at Figure 8.13.

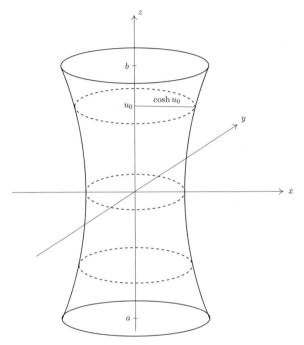

Figure 8.13

This surface has rotational symmetry and is in fact obtained by rotating a certain curve around the z-axis. This observation leads to a new class of parametric surfaces. A **surface of revolution** is given by

$$f : (a, b) \times (0, 2\pi) \to \mathbb{R}^3$$
$$f(u, v) = (h(u) \cos v, h(u) \sin v, k(u)) \tag{8.12}$$

with C^2-functions $h, k : (a, b) \to \mathbb{R}$, $h(u) > 0$. For $\mathrm{d}f$ we find

$$(\mathrm{d}f)_{(u,v)} = \begin{pmatrix} h'(u) \cos v & -h(u) \sin v \\ h'(u) \sin v & h(u) \cos v \\ k'(u) & 0 \end{pmatrix}$$

and since

$$\left\langle \begin{pmatrix} h'(u) \cos v \\ h'(u) \sin v \\ k'(u) \end{pmatrix}, \begin{pmatrix} -h(u) \sin v \\ h(u) \cos v \\ 0 \end{pmatrix} \right\rangle = 0,$$

these vectors are independent as long as not both $h'(u) = 0$ and $k'(u) = 0$, i.e. $(df)_{(u,v)}$ is injective. If we set

$$x = h(u_0) \cos v, \quad y = h(u_0) \sin v, \quad z = k(u_0)$$

we find

$$x^2 + y^2 + (z - k(u_0))^2 = h^2(u_0) \cos^2 v + h^2(u_0) \sin^2 v = h^2(u_0),$$

i.e. we obtain a circle with centre $(0, 0, k(u_0))$ and radius $h(u_0)$. Thus the trace of f is obtained by rotating the graph of h around the z-axis. In particular if $k(u) = u$ we rotate around the z-axis without any distortion on the axis. For example $h(u) = \cosh u$ and $k(u) = u$ gives the surface in Figure 8.13.

If we choose $h(u) = u$ and $k(u) = u$ as well as $(a, b) = \mathbb{R}_+ \setminus \{0\}$ we get a surface on the upper circular cone

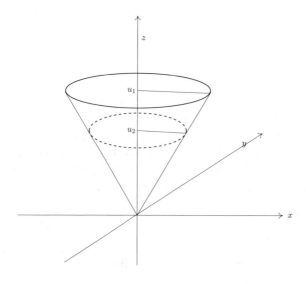

Figure 8.14

Remark 8.8. When discussing the parametrization of S^2, the torus, a surface of revolution we need (in order to keep within our definition) to take an open set G as our parameter domain and hence we do not obtain the full sphere, torus etc., as the trace. The periodicity of sin and cos however allows us to extend the domain of f in such a way that we obtain for the trace the "closed" surface, i.e. we may consider parametrization defined on $[a, b] \times [0, 2\pi]$ etc.

Let $f : G \to \mathbb{R}^3$ be a parametric surface. If $\omega : [a, b] \to G$ is a regular parametric curve we can lift ω to $M = f(G)$ by defining $\gamma : [a, b] \to \mathbb{R}^3$, $\gamma = f \circ \omega$. For the velocity vector of γ we find

$$\dot{\gamma}(t) = \frac{d}{dt}\gamma(t) = \frac{d}{dt}f(\omega(t)) = \frac{\partial f}{\partial x}(\omega(t))\dot{\omega}_1(t) + \frac{\partial f}{\partial y}(\omega(t))\dot{\omega}_2(t)$$

and it follows that $\dot{\gamma}(t) \in \text{span}\left(\frac{\partial f}{\partial x}(\omega(t)), \frac{\partial f}{\partial y}(\omega(t))\right)$, i.e. the tangent line to γ at t_0 lies in the tangent plane to f at $\omega(t_0)$.

Some curves of special interest are coordinate lines. For simplicity assume $G = (a, b) \times (c, d)$. The curves $\xi_{y_0}; (a, b) \to G$, $\xi_{y_0}(x) = (x, y_0)$, $y_0 \in (c, d)$ and $\eta_{x_0} : (c, d) \to G$, $\eta_{x_0}(y) = (x_0, y)$ are line segments parallel to the coordinate axis and of course regular curves with $\dot{\xi}_{y_0}(x) = (1, 0)$ and $\dot{\eta}_{x_0}(x_0, y) = (0, 1)$.

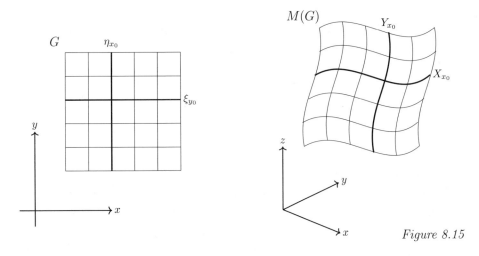

Figure 8.15

The lifted curves $X_{y_0}(x) := f(\xi_{y_0}(x))$ and $Y_{x_0}(y) := f(\eta_{x_0}(y))$ are the **coordinates lines**. Thus we may think to use (locally) coordinate lines as "coordinates" or a "frame" on M. We will return to this idea several times later on.

A final remark: we can often characterise "surfaces" by equations, say the sphere S^2 is given by the set of all solutions of the equation $x^2 + y^2 + z^2 = 1$, $(x, y, z) \in \mathbb{R}^3$. More generally, given a function $F : \Omega \to \mathbb{R}$, $\Omega \subset \mathbb{R}^3$, and we may consider the set M of all $(x, y, z) \in \Omega$ such that $F(x, y, z) = 0$. A natural question is whether $M \subset \mathbb{R}^3$ is a graph of a function $f : I \to \mathbb{R}$, or a parametric surface $h : G \to \mathbb{R}^3$, or at least locally of this type.

155

Problems

1. Let $h : (0,1) \to \mathbb{R}$ be a C^1-function. Consider the graph $\Gamma(h)$ as a subset of the $x - z$-plane in $\mathbb{R}^3$. Find a parametrization of the surface S in $\mathbb{R}^3$ obtained by translating $\Gamma(h)$ along the y-axis.

2. Find the level sets of the graph of the function $f : \mathbb{R}^2 \to \mathbb{R}$, $f(x,y) = \sqrt{r^2 - \frac{x^2}{a^2} - \frac{y^2}{b^2}}$.

3.   a) Find the tangent plane and the normal line at $(2,2,9) \in \mathbb{R}^3$ to the surface with the trace $P = \{(x,y,z) \in \mathbb{R}^3 | z = x^2 + y^2 + 1\}$.

   b) Find the tangent plane and the normal line to the surface with trace $H = \{(x,y,z) \in \mathbb{R}^3 | z = x^2 - y^2\}$ for the points with $x_0 = y_0$ and with $x_0 = -y_0$.

4.   a) Let $f : (a,b) \times (0, 2\pi) \to \mathbb{R}^3$, $f(u,v) = (h(u)\cos v, h(u)\sin v, k(u))$ be a surface of revolution where $h, k : (a,b) \to \mathbb{R}$, $h(u) > 0$, are two C^2-functions. For $(u_0, v_0) \in (a,b) \times (0, 2\pi)$ find the tangent plane and the normal line to $M = f((a,b) \times (0, 2\pi))$ at $p = f(u_0, v_0)$.

   b) Use the result of part a) to find the tangent plane and normal line for the cone as drafted in Figure 8.14, i.e. to the surface of revolution with $(a,b) = (0, \infty)$ and $h(u) = k(u) = u$.

5. The **catenoid** is the surface of revolution generated by $h : \mathbb{R} \to \mathbb{R}$, $h(u) = \cosh u$, and $k : \mathbb{R} \to \mathbb{R}$, $k(u) = u$. Sketch the trace of f. Consider the coordinate lines $u \mapsto f(u, v_0)$ and $v \mapsto f(u_0, v)$ and find the angle of their intersection at $f(u_0, v_0)$.

6. Consider the mapping $S : \mathbb{R}^2 \to \mathbb{R}^3$ given by

$$S(u,v) = \left(\frac{4u}{u^2 + v^2 + 4}, \frac{4v}{u^2 + v^2 + 4}, \frac{2(u^2 + v^2)}{u^2 + v^2 + 4} \right).$$

For which values of $(u,v) \in \mathbb{R}^2$ is this a parametric surface? Prove that $S(\mathbb{R}^2) \subset \partial B_1((0,0,1))$, i.e. the image of $\mathbb{R}^2$ under S is contained in the sphere with radius 1 and centre $(0,0,1) \in \mathbb{R}^3$.

9 Taylor Formula and Local Extreme Values

Let $f : G \to \mathbb{R}$, $G \subset \mathbb{R}^2$ open, be a C^k-function, $k \geq 1$. Its range is a subset of $\mathbb{R}$, hence we may ask for points $x, y \in G$ such that

$$f(x) = \sup_{z \in G} f(z), \quad f(y) = \inf_{z \in G} f(z). \qquad (9.1)$$

From our investigations in one dimension we know that it is reasonable to first study the existence of **local extreme values** of f, i.e. try to find points $x, y \in G$ such that

$$f(x) = \sup_{z \in U(x)} f(z), \quad f(y) = \inf_{z \in U(y)} f(z) \qquad (9.2)$$

for (small) neighbourhoods $U(x)$ and $U(y)$ of x and y, respectively. We expect at such points $\Gamma(f)$ to have a horizontal tangent plane, i.e. a tangent plane being parallel to the plane spanned by e_1 and e_2 in $\mathbb{R}^3$.

A complementary approach is to approximate f locally, i.e. in a neighbourhood of a point x using the differential of f at x. If $f \in C^1(G)$ and $x \in G$ we have for $z \in B_\rho(x)$, $B_\rho(x) \subset G$, by Theorem 6.3

$$f(x + z) = f(x) + (d_x f)z + \varphi_x(z) \qquad (9.3)$$

with $\lim_{z \to 0} \frac{\varphi_x(z)}{\|z\|} = 0$. For $\|z\|$ small we may "neglect" $\varphi_x(z)$, more precisely we may achieve $\|\varphi_x(z)\| \leq \epsilon \|z\|$ for $\|z\| \leq \eta$, and therefore if $x + z$ is close to x we have

$$f(x + z) \approx f(x) + (d_x f)(z).$$

Since $d_x(f)$ is a linear mapping for $\|z\| \leq \eta$ the term $(d_x f)(z)$ must change its sign except $d_x f = 0$. Thus in order for f to have an extreme value at x we would expect $d_x f = 0$. For $f \in C^1(G)$ we know that $d_x f = \begin{pmatrix} \frac{\partial f}{\partial x_1}(x) \\ \frac{\partial f}{\partial x_2}(x) \end{pmatrix}$ and the tangent plane to $\Gamma(f)$ at $x \in G$ is given by

$$\left\{ \begin{pmatrix} x_1 \\ x_2 \\ f(x_1, x_2) \end{pmatrix} + \lambda \begin{pmatrix} 1 \\ 0 \\ \frac{\partial f}{\partial x_1}(x_1, x_2) \end{pmatrix} + \mu \begin{pmatrix} 0 \\ 1 \\ \frac{\partial f}{\partial x_2}(x_1, x_2) \end{pmatrix} \,\middle|\, \lambda, \mu \in \mathbb{R} \right\}$$

which is indeed parallel to the plane spanned by e_1 and e_2 in $\mathbb{R}^3$ if and only if $\frac{\partial f}{\partial x_1}(x_1, x_2) = \frac{\partial f}{\partial x_2}(x_1, x_2) = 0$. In order to study the behaviour of f

157

in a neighbourhood of a point $x \in G$, for example with the aim to draw conclusions about local extreme values, we need more terms in (9.3), i.e. a better approximation (as we already know from the one-dimensional case). This leads to the question whether we can attain an analogue to Taylor's formula for functions $f : G \to \mathbb{R}$, $G \subset \mathbb{R}^n$. We start with some auxiliary results. In $\mathbb{R}^n$ we will always choose the basis $\{e_1, \ldots, e_n\}$ and we will always use the Euclidean norm $\|x\| = \left(\sum_{j=1}^n x_j^2 \right)^{\frac{1}{2}}$. Moreover, we now have to use matrix operations and therefore we need to be more strict in making a distinction between row and column vectors. Vectors (points) in $G \subset \mathbb{R}^n$ will now be column vectors $\begin{pmatrix} x_1 \\ \vdots \\ x_n \end{pmatrix}$ and consequently row vectors can be viewed as the transpose of column vectors $x^t = \begin{pmatrix} x_1 \\ \vdots \\ x_n \end{pmatrix}^t = (x_1, \ldots, x_n)$. Let $f : G \to \mathbb{R}$ be a differentiable function. Its differential $d_x f$ at $x \in G$ is a linear mapping from $\mathbb{R}^n$ to $\mathbb{R}$ represented by its gradient

$$d_x f = (\mathrm{grad} f)(x) = \left(\frac{\partial f}{\partial x_1}(x), \cdots, \frac{\partial f}{\partial x_n} \right) \tag{9.4}$$

and

$$(d_x f)(z) = ((\mathrm{grad} f)(x))(z) \tag{9.5}$$

$$= \left(\frac{\partial f}{\partial x_1}(x), \cdots, \frac{\partial f}{\partial x_n}(x) \right) \begin{pmatrix} z_1 \\ \vdots \\ z_n \end{pmatrix}$$

$$= \sum_{j=1}^n \frac{\partial f}{\partial x_j}(x) z_j$$

$$= \langle (\mathrm{grad} f(x))^t, z \rangle.$$

Thus we may rewrite (9.3) as

$$f(x + z) = f(x) + ((\mathrm{grad} f)(x)) z + \varphi_x(z)$$
$$= f(x) + \langle (\mathrm{grad} f(x))^t, z \rangle + \varphi_x(z).$$

Our aim is to replace $\varphi_x(z)$ by a quadratic form in z and a term $\psi_x(z)$ such that $\lim_{\|z\| \to 0} \frac{\psi_x(z)}{\|z\|^2} = 0$. For this we need some better understanding of how to estimate linear mappings and Jacobi matrices.

Let A be a $m \times n$ matrix, $A = (a_{kl})_{\substack{k=1,\ldots,m \\ l=1,\ldots,n}}$. For $x \in \mathbb{R}^n$ we find using the Cauchy-Schwarz inequality (and with $\|y\|_{(k)}$ denoting for the moment the Euclidean norm of $y \in \mathbb{R}^k$)

$$\|Ax\|_{(m)} = \left\| \begin{pmatrix} (Ax)_1 \\ \vdots \\ (Ax)_m \end{pmatrix} \right\|_{(m)}$$

$$= \left(\sum_{k=1}^{m} \left(\sum_{l=1}^{n} a_{kl} x_l \right)^2 \right)^{\frac{1}{2}}$$

$$\leq \left(\sum_{k=1}^{m} \left(\sum_{l=1}^{n} a_{kl}^2 \right) \left(\sum_{l=1}^{n} x_l^2 \right) \right)^{\frac{1}{2}}$$

$$= \left(\sum_{k=1}^{m} \sum_{l=1}^{n} a_{kl}^2 \right)^{\frac{1}{2}} \|x\|_{(n)},$$

and with

$$\|A\| := \left(\sum_{k=1}^{m} \sum_{l=1}^{n} a_{kl}^2 \right)^{\frac{1}{2}} \tag{9.6}$$

we have

$$\|Ax\|_{(m)} \leq \|A\| \, \|x\|_{(n)}. \tag{9.7}$$

Define

$$|||A||| := \inf \left\{ c > 0 \mid \|Ax\|_{(m)} \leq c \, \|x\|_{(n)} \right\}. \tag{9.8}$$

Since $\frac{\|Ax\|_{(m)}}{\|x\|_{(n)}} \leq |||A|||$ we find that

$$\tau := \sup \left\{ \frac{\|Ax\|_{(m)}}{\|x\|_{(n)}} \,\middle|\, x \in \mathbb{R}^n \setminus \{0\} \right\} \leq |||A|||.$$

However, for all $x \in \mathbb{R}^n$

$$\|Ax\|_{(m)} \leq \tau \, \|x\|_{(n)},$$

and therefore

$$|||A||| = \sup\left\{ \frac{\|Ax\|_{(m)}}{\|x\|_{(m)}} \,\middle|\, x \in \mathbb{R}^n \setminus \{0\} \right\}. \qquad (9.9)$$

From (9.7) and (9.8) we find immediately that

$$|||A||| \le \|A\|,$$

while (9.7) and (9.9) yield

$$\|A\| \le |||A|||,$$

i.e. we have proved

$$\|A\| = |||A||| = \inf\left\{ c > 0 \,\middle|\, \|Ax\|_{(m)} \le c\,\|x\|_{(n)} \right\}. \qquad (9.10)$$

(From now on we will again write $\|x\|$ for $\|x\|_{(k)}$, $x \in \mathbb{R}^k$.)
Preparing the Taylor formula in $\mathbb{R}^n$ we prove

Theorem 9.1. *Let $G \subset \mathbb{R}^n$ be open and $f : G \to \mathbb{R}^m$ a continuously differentiable mapping. Let $x \in G$ and $\xi \in \mathbb{R}^n$ a vector such that the straight line segment $x + t\xi$, $0 \le t \le 1$, belongs completely to G. It follows that*

$$f(x + \xi) - f(x) = \left(\int_0^1 (\mathrm{d}f)(x + t\xi)\,\mathrm{d}t \right)\xi. \qquad (9.11)$$

Proof. Consider the components $f_i : G \to \mathbb{R}$, $1 \le i \le m$, of f and the functions $g_i(t) := f_i(x + t\xi)$. It follows that

$$f_i(x + \xi) - f_i(x) = g_i(1) - g_i(0) = \int_0^1 g_i'(t)\,\mathrm{d}t$$

$$= \int_0^1 \left(\sum_{j=1}^n \frac{\partial f_i}{\partial x_j}(x + t\xi)\xi_i \right)\mathrm{d}t$$

$$= \sum_{j=1}^n \left(\int_0^1 \frac{\partial f_i}{\partial x_j}(x + t\xi)\,\mathrm{d}t \right)\xi_j.$$

But the components of $\mathrm{d}f$ are the functions $\frac{\partial f_i}{\partial x_j}$ and the theorem is proved. $\qquad \square$

160

Corollary 9.2. *Under the assumptions of Theorem 9.1 suppose that*

$$M := \sup_{0 \le t \le 1} \|(df)(x + t\xi)\| < \infty.$$

Then we find that

$$\|f(x + \xi) - f(x)\| \le M \|\xi\|. \tag{9.12}$$

Proof. Combining Theorem 9.1 and Lemma 7.3 we arrive at

$$
\begin{aligned}
\|f(x + \xi) - f(x)\| &= \left\| \int_0^1 (df)(x + t\xi)\xi \, dt \right\| \\
&\le \int_0^1 \|df(x + t\xi)\| \, \|\xi\| \, dt \\
&\le M \|\xi\|.
\end{aligned}
$$

$\square$

The one-dimensional Taylor formula, Theorem I.29.11, gives a representation of the values $g(x)$ of a C^{k+1}-function $g : (-\eta + c, c + \eta) \to \mathbb{R}$ at $x \in (-\eta + c, c + \eta)$ by using the values $g^{(l)}(c)$, $0 \le l \le k$, and the values of $g^{(k+1)}$ on the line segment connecting x and c. Let $f \in C^{k+1}(G)$, $G \subset \mathbb{R}^n$ open, and suppose that $x, z \in G$ as well as the line segment connecting x and z belongs to G. We can consider the function of one real variable defined by

$$
\begin{aligned}
g &: [0, 1] \to \mathbb{R} \\
&t \mapsto g(t) := f(x + ty), \ y = z - x,
\end{aligned} \tag{9.13}
$$

and we may apply to g one-dimensional results in order to find a Taylor formula for f. In the following we will use multi-index notation as introduced in Chapter 5.

Proposition 9.3. *Let $f \in C^k(G)$, $G \subset \mathbb{R}^n$ open, and assume that with $x, z \in G$ the line segment connecting x and z belongs to G too. With $y := z - x$ it follows that g as defined in (9.13) is an element in $C^k([0, 1])$ and we have*

$$\frac{d^k g(t)}{dt^k} = \sum_{|\alpha| = k} \frac{k!}{\alpha!}(D^\alpha f)(x + ty)y^\alpha. \tag{9.14}$$

Proof. For $k = 1$ the chain rule yields

$$\frac{dg(t)}{dt} = \frac{d}{dt} f(x_1 + ty_1, \ldots, x_n + ty_n) = \sum_{j=1}^{n} (D_j f)(x + ty) y_j$$

with $D_j = \frac{\partial}{\partial x_j}$. Now we want to proceed by induction. We assume first that for $l \leq k - 1$ the following holds

$$\frac{d^l}{dt^l} g(t) = \sum_{i_1, \ldots, i_l = 1}^{n} (D_{i_l} \cdot \ldots \cdot D_{i_1} f)(x + ty) y_{i_1} \cdot \ldots \cdot y_{i_l}$$

and we find

$$\frac{d^{l+1}}{dt^{l+1}} g(t) = \frac{d}{dt} \left(\sum_{i_1, \ldots, i_n = 1}^{n} (D_{i_l} \cdot \ldots \cdot D_{i_1} f)(x + ty) y_{i_1} \cdot \ldots \cdot y_{i_l} \right)$$

$$= \sum_{j=1}^{n} D_j \left(\sum_{i_1, \ldots, i_l = 1}^{n} (D_{i_l} \ldots D_{i_1} f)(x + ty) y_{i_1} \cdot \ldots \cdot y_{i_l} \right)$$

$$= \sum_{i_1, \ldots, i_{l+1} = 1}^{n} (D_{i_{l+1}} \cdot \ldots \cdot D_{i_1} f)(x + ty) y_{i_1} \cdot \ldots \cdot y_{i_{l+1}}.$$

Thus we have proved

$$\frac{d^k g(t)}{dt^k} = \sum_{i_1, \ldots, i_k = 1}^{n} (D_{i_k} \cdot \ldots D_{i_1} f)(x + ty) y_{i_1} \cdot \ldots \cdot y_{i_k}. \tag{9.15}$$

Denote by α_j the number of occurrences of the index j, $1 \leq j \leq n$, among the indices $i_1, \ldots, i_k$ in one of the terms of (9.15). It follows that

$$(D_{i_k} \cdot \ldots \cdot D_{i_1} f)(x + ty) y_{i_1} \cdot \ldots \cdot y_{i_k} = (D_1^{\alpha_1} \cdot \ldots \cdot D_n^{\alpha_n} f)(x + ty) y_1^{\alpha_1} \cdot \ldots \cdot y_n^{\alpha_n}.$$

The number of k-tuples $(i_1, \ldots, i_k)$ of numbers $1 \leq i_l \leq n$, where j, $1 \leq j \leq n$, occurs exactly α_j-times, $\alpha_1 + \cdots + \alpha_n = k$, is given by

$$\frac{k!}{\alpha_1! \cdot \ldots \cdot \alpha_n!} = \frac{k!}{\alpha!}.$$

Therefore we arrive at

$$\frac{d^k g(t)}{dt^k} = \sum_{i_1,\ldots,i_k=1}^{k} (D_{i_k} \cdot \ldots \cdot D_{i_1} f)(x + ty) y_{i_1} \cdot \ldots \cdot y_{i_k}$$

$$= \sum_{|\alpha|=k} \frac{k!}{\alpha!} (D^\alpha f)(x + ty) y^\alpha.$$

□

Now we can prove a first version of **Taylor's formula** in higher dimensions.

Theorem 9.4. *Let $G \subset \mathbb{R}^n$ be an open set, $x, z \in G$ and suppose that the line segment connecting x and z belongs to G. For $f \in C^{k+1}(G)$ the following holds*

$$f(z) = \sum_{|\alpha|\le k} \frac{D^\alpha f(x)}{\alpha!}(z-x)^\alpha + \sum_{|\alpha|=k+1} \frac{(D^\alpha f)(x + \vartheta(z-x))}{\alpha!}(z-x)^\alpha \quad (9.16)$$

for some $\vartheta \in [0, 1]$.

Proof. Consider the function $g : [0,1] \to \mathbb{R}$, $g(t) := f(x + t(z - x))$. The function g is $(k+1)$-times continuously differentiable and the one-dimensional Taylor formula with Lagrange form of the remainder term, see Theorem I.29.14, yields the existence of $\vartheta \in [0, 1]$ such that

$$f(z) = g(1) = \sum_{m=0}^{k} \frac{g^{(m)}(0)}{m!} + \frac{g^{(k+1)}(\vartheta)}{(k+1)!}.$$

By Proposition 9.3 it follows for $m = 0, \ldots, k$ that

$$\frac{g^{(m)}(0)}{m!} = \sum_{|\alpha|=m} \frac{D^\alpha f(x)}{\alpha!}(z - x)^\alpha$$

and

$$\frac{g^{(k+1)}(\vartheta)}{(k+1)!} = \sum_{|\alpha|=k+1} \frac{(D^\alpha f)(x + \vartheta(z - x))}{\alpha!}(z - x)^\alpha$$

implying (9.16).

□

As in the one-dimensional case we call $\sum_{|\alpha|\leq k}\frac{D^\alpha f(x)}{\alpha!}(z-x)^\alpha$ the **Taylor polynomial** of f of order k about x.

Corollary 9.5. *Let $G \subset \mathbb{R}^n$ be an open set and $B_\delta(x) \subset G$. Further let $f \in C^k(G)$. For all $y \in B_\delta(0)$ we have*

$$f(x+y) = \sum_{|\alpha|\leq k} \frac{D^\alpha f(x)}{\alpha!} y^\alpha + \varphi_x(y), \tag{9.17}$$

where $\varphi_x : B_\delta(0) \to \mathbb{R}$ is a function such that $\varphi_x(0) = 0$ and

$$\lim_{y\to 0} \frac{\varphi_x(y)}{\|y\|^k} = 0. \tag{9.18}$$

Proof. By Theorem 9.4 there exists $\vartheta \in [0,1]$ such that

$$f(x+y) = \sum_{|\alpha|\leq k-1} \frac{D^\alpha f(x)}{\alpha!} y^\alpha + \sum_{|\alpha|=k} \frac{(D^\alpha f)(x+\vartheta y)}{\alpha!} y^\alpha$$

$$= \sum_{|\alpha|\leq k} \frac{D^\alpha f(x)}{\alpha!} y^\alpha + \sum_{|\alpha|=k} r_\alpha(y) y^\alpha,$$

with

$$r_\alpha(y) := \frac{D^\alpha f(x+\vartheta y) - D^\alpha f(x)}{\alpha!}, \qquad |\alpha| = k.$$

The continuity of $D^\alpha f$, $|\alpha| = k$, implies that

$$\lim_{y\to 0} r_\alpha(y) = 0.$$

Now we define

$$\varphi_x(y) := \sum_{|\alpha|=k} r_\alpha(y) y^\alpha \tag{9.19}$$

and observe

$$\frac{|\varphi_x(y)|}{\|y\|^k} \leq \sum_{|\alpha|=k} |r_\alpha(y)| \frac{|y^\alpha|}{\|y\|^k}$$

$$= \sum_{|\alpha|=k} |r_\alpha(y)| \frac{|y_1^{\alpha_1}|}{\|y\|^{\alpha_1}} \cdots \frac{|y_n^{\alpha_n}|}{\|y\|^{\alpha_n}}$$

$$\leq \sum_{|\alpha|=k} |r_\alpha(y)|,$$

164

implying that

$$\lim_{y \to 0} \frac{\varphi_x(y)}{\|y\|^k} = 0.$$

$\square$

We might ask to extend Taylor's formula to Taylor's series for an arbitrarily often differentiable function $f : G \to \mathbb{R}, G \subset \mathbb{R}$. However, this is much more sophisticated than it seems at a first glance - it is the domain of convergence which for $G \subset \mathbb{R}^n$, $n \geq 2$, causes some surprises and problems. We will return to this later.

We now return to the problem of finding local extreme values of $f : G \to \mathbb{R}$ and for this we want to use Corollary 9.5 since it allows us to compare locally the values of f. We pay particular attention to the terms up to order 2 and it is helpful to introduce a special notation for the matrix with entries being the second order partial derivatives of f.

Definition 9.6. *For a function $f \in C^2(G)$, $G \subset \mathbb{R}^n$ open, the matrix*

$$(\mathrm{Hess} f)(x) := \left(\frac{\partial^2 f}{\partial x_k \partial x_l}(x) \right)_{k,l=1,\dots,n} \tag{9.20}$$

*is called the **Hesse matrix** of f at x.*

Since the second partial derivatives of f are by assumption continuous we have

$$\frac{\partial^2 f}{\partial x_k \partial x_l}(x) = \frac{\partial^2 f}{\partial x_l \partial x_k}(x),$$

i.e. $(\mathrm{Hess} f)(x)$ is a symmetric matrix. Combining this new definition with Corollary 9.5 we obtain

Corollary 9.7. *If $f \in C^2(G)$, $G \subset \mathbb{R}^n$ open, and if $B_\delta(x) \subset G$ then we have for all $y \in B_\delta(0)$*

$$f(x+y) = f(x) + (\mathrm{grad} f(x))y + \frac{1}{2}y^t(\mathrm{Hess} f(x))y + \varphi_x(y) \tag{9.21}$$

with

$$\lim_{y \to 0} \frac{\varphi_x(y)}{\|y\|^2} = 0. \tag{9.22}$$

165

Remark 9.8. Instead of working with $y \in B_\delta(0)$ we may restrict ourselves to the case where the line segment connecting x and $x + y$ belongs to G.

Definition 9.9. *Let $f : G \to \mathbb{R}$, $G \subset \mathbb{R}^n$, be a function. We say that f has a **local maximum (minimum)** at $x \in \overset{\circ}{G}$ if there is a neighbourhood $U(x) \subset G$ of x and*

$$f(x) \geq f(y) \quad (f(x) \leq f(y)) \quad \text{for all } y \in U(x). \tag{9.23}$$

*If in (9.23) equality holds only for $x = y$ we say that f has an **isolated local maximum (minimum)** at $x \in \overset{\circ}{G}$. By definition a **local extreme value** is either a local maximum or a local minimum.*

The following necessary condition for f having a local extreme value at x is completely analogous to the one-dimensional case and should not come as a surprise to us.

Theorem 9.10. *If a function $f : G \to \mathbb{R}$, $G \subset \mathbb{R}^n$ open, has all first order partial derivatives in G and a local extreme value at $x \in G$, then we must have that*

$$(\operatorname{grad} f)(x) = 0, \tag{9.24}$$

i.e. $\frac{\partial f}{\partial x_j}(x) = 0$ for $1 \leq j \leq n$.

Proof. For $j = 1, \ldots, n$ we consider the functions

$$s \mapsto g_j(s) := f(x + se_j)$$

defined on some interval $(-\epsilon, \epsilon)$ such that $x + se_j \in G$ for $|s| < \epsilon$, and of course these differentiable functions have a local extreme value for $s = 0$ because $g_j(0) = f(x)$. Therefore it follows that

$$g_j'(0) = (D_j f)(x) = 0 \quad \text{for } j = 1, \ldots, n,$$

i.e. $\operatorname{grad} f(x) = 0$. $\qquad \square$

We call a point $x_0 \in G$ a **critical point** or a **stationary point** of $f : G \to \mathbb{R}$, $G \subset \mathbb{R}^n$, if $(\operatorname{grad}) f(x_0) = 0$. This notion will be generalised later on to mappings $f : G \to \mathbb{R}^m$.

Before turning to sufficient criteria for local maxima or minima, respectively, we need a better understanding of symmetric $n \times n$ matrices. For more details we refer to Appendix I.

Let $A = (a_{kl})_{k,l=1,\ldots,n}$ be a real symmetric $n \times n$ matrix, i.e. $a_{kl} = a_{lk}$ for all $k, l = 1, \ldots, n$, or equivalently $A^t = A$ where A^t denotes the transpose of the matrix A. The eigenvalues of A are real and A is equivalent to a diagonal matrix. If we denote by $\lambda_1, \ldots, \lambda_n$ the eigenvalues of A each counted according to its multiplicity, there exists an <u>orthogonal</u> matrix U such that

$$\begin{pmatrix} \lambda_1 & & 0 \\ & \ddots & \\ 0 & & \lambda_n \end{pmatrix} = U A U^t, \tag{9.25}$$

recall that $U^{-1} = U^t$ for an orthogonal matrix U.

Moreover the column vectors of U form an orthogonal basis of $\mathbb{R}^n$, i.e. if

$$U = (u_1, \ldots, u_n) = \begin{pmatrix} u_{11} & \cdots & u_{1n} \\ \vdots & & \vdots \\ u_{n1} & \cdots & u_{nn} \end{pmatrix}$$

then the following holds

$$\langle u_j, u_l \rangle = \delta_{jl},$$

and the vectors u_j, $j = 1, \ldots, n$ are eigenvectors of A. (Note that some care is needed in the notation due to the multiplicity of eigenvalues, see Appendix I.) Thus on $\mathbb{R}^n$ we can introduce instead of the canonical basis $\{e_1, \ldots, e_n\}$ the basis $\{u_1, \ldots, u_n\}$ which we obtain from $u_j = U e_j$. Given a symmetric $n \times n$ matrix A we can associate to it a **bilinear form**

$$\beta_A(x, y) = \langle Ax, y \rangle = \sum_{k,l=1}^{n} a_{kl} x_k y_l. \tag{9.26}$$

The symmetry of A implies

$$\langle Ax, y \rangle = \langle x, A^t y \rangle = \langle x, Ay \rangle$$

or

$$\beta_A(x, y) = \beta_A(y, x). \tag{9.27}$$

In addition the linearity of A (interpret as mapping $A : \mathbb{R}^n \to \mathbb{R}^n$) implies the bilinearity of β_A:

$$\beta_A(\lambda x + \mu y, z) = \lambda \beta_A(x, z) + \mu \beta_A(y, z)$$

as well as

$$\beta_A(x, \lambda y + \mu z) = \lambda \beta_A(x, y) + \mu \beta_A(x, z)$$

for $x, y, z \in \mathbb{R}^n$ and $\lambda, \mu \in \mathbb{R}$. With respect to the basis $\{u_1, \dots, u_n\}$ we find with $\xi = Ux$, $\eta = Uy$

$$\begin{aligned}
\beta_A(x, y) &= \langle Ax, y \rangle = \langle AU^{-1}\xi, U^{-1}\eta \rangle \\
&= \langle UAU^{-1}\xi, \eta \rangle = \langle UAU^t\xi, \eta \rangle \\
&= \left\langle \begin{pmatrix} \lambda_1 & & 0 \\ & \ddots & \\ 0 & & \lambda_n \end{pmatrix} \xi, \eta \right\rangle \\
&= \sum_{j=1}^{n} \lambda_j \xi_j \eta_j =: \tilde{\beta}_A(\xi, \eta).
\end{aligned}$$

In particular we obtain for the associated **quadratic form**

$$\beta_A := \beta_A(x, x) \tag{9.28}$$

that

$$\tilde{\beta}_A(\xi) = \sum_{j=1}^{n} \lambda_j \xi_j^2. \tag{9.29}$$

Definition 9.11. *Let A be a symmetric matrix with associated bilinear form β_A. **A.** We call β_A and A **positive definite** if*

$$\beta_A(x) = \sum_{k,l=1}^{n} a_{kl} x_k x_l > 0 \tag{9.30}$$

for all $x \in \mathbb{R}^n$.
*B. If $\beta_A(x) \geq 0$ for all $x \in \mathbb{R}^n$ we call β_A and A **positive semi-definite**.*
*C. If $\beta_A < 0$ ($\beta_A \leq 0$) for all $x \in \mathbb{R}^n$ we call β_A and A **negative definite** (**negative semi-definite**).*
*D. If $\beta_A > 0$ for some $x \in \mathbb{R}^n$ and $\beta_A(y) < 0$ for some $y \in \mathbb{R}^n$, then we call β_A and A **indefinite**.*

The following result summarises criteria for A (or β_A) being positive definite. Recall that if $A = (a_{kl})_{k,l=1,\dots,n}$ is an $n \times n$ matrix the **principal minors** of A are the determinants of the matrices $(a_{kl})_{k,l=1,\dots,m}$, $1 \leq m \leq n$.

Theorem 9.12. *The following are equivalent*

   i) A is positive definite;

   ii) all eigenvalues of A are strictly positive;

   iii) all principal minors of A are strictly positive;

   iv) there exists $B \in G((n, \mathbb{R}))$ such that $A = B^t B$;

   v) A is positive semi-definite and $\det A \neq 0$;

   vi) A is invertible and A^{-1} is positive definite.

For a discussion of this theorem and related topics we refer to Appendix I. A word of caution: if A is positive definite then $-A$ is negative definite and vice versa. Now consider $A = \begin{pmatrix} -1 & 0 \\ 0 & -1 \end{pmatrix}$ which is of course negative definite, but $\det A = 1 > 0$.

Lemma 9.13. *Let A be a positive definite matrix. For the associated bilinear form the following holds for all $x \in \mathbb{R}^n$*

$$\beta_A(x) \geq \alpha \, \|x\|^2 \tag{9.31}$$

where

$$\alpha := \inf \left\{ \beta_A(x) \,\middle|\, x \in S^{n-1} \right\} > 0. \tag{9.32}$$

(Recall: $S^{n-1} = \{x \in \mathbb{R}^n \mid \|x\| = 1\}$)

Proof. First we note that S^{n-1} is compact and that $x \mapsto \beta_A(x)$ is continuous. Hence β_A attains its infimum α on S^{n-1} and since $\beta_A(x) > 0$ for all $x \in \mathbb{R}^n$ it follows that $\alpha > 0$. Now let $x \in \mathbb{R}^n \setminus \{0\}$ and $\lambda := \frac{1}{\|x\|}$. Since $y := \lambda x \in S^{n-1}$ we find $\beta_A(y) \geq \alpha$ implying

$$\alpha \leq \beta_A(y) = \beta_A(\lambda x) = \lambda^2 \beta_A(x) = \frac{1}{\|x\|^2 \beta_A(x)},$$

or

$$\beta_A(x) \geq \alpha \, \|x\|^2 \quad \text{for } x \in \mathbb{R}^n \setminus \{0\}.$$

But for $x = 0$ the inequality is trivial. $\qquad\square$

Now we can prove a sufficient criterion for the existence of local extreme values.

Theorem 9.14. *Let $f \in C^2(G)$, $G \subset \mathbb{R}^n$ open, and suppose that $\operatorname{grad} f(x_0) = 0$ for some $x_0 \in G$.*
A. *If $(\operatorname{Hess} f)(x_0)$ is positive definite then f has an isolated local minimum at x_0.*
B. *If $(\operatorname{Hess} f)(x_0)$ is negative definite then f has an isolated local maximum at x_0.*
C. *If $(\operatorname{Hess} f)(x_0)$ is indefinite then f has no local extreme values at x_0.*

Proof. **A.** We apply Corollary 9.7 and find in a neighbourhood of x_0, $V(x_0) \subset G$, and $y \in -x_0 + V(x_0)$

$$f(x_0 + y) = f(x) = \frac{1}{2} y^t (\operatorname{Hess} f(x_0)) y + \varphi_{x_0}(y) \tag{9.33}$$

where we used that $\operatorname{grad} f(x_0) = 0$. Since $\lim_{y \to 0} \frac{\varphi_{x_0}(y)}{\|y\|^2} = 0$ for $\epsilon > 0$ there exists $\delta > 0$ such that $\|y\| < \delta$ implies $|\varphi_{x_0}(y)| \le \epsilon \|y\|^2$. With $\alpha := \inf \left\{ \langle \operatorname{Hess} f(x_0) x, x \rangle \,\middle|\, x \in S^{n-1} \right\}$ we find by Lemma 9.13 that

$$\langle \operatorname{Hess} f(x_0) y, y \rangle \ge \alpha \|y\|^2 . \tag{9.34}$$

If we now choose δ such that $|\varphi_{x_0}(y)| \le \frac{\alpha}{4} \|y\|^2$ and combine (9.33) and (9.34) we arrive at

$$f(x_0 + y) \ge f(x_0) + \frac{1}{2} \alpha \|y\|^2 - \frac{\alpha}{4} \|y\|^2 = f(x_0) + \frac{\alpha}{4} \|y\|^2 \tag{9.35}$$

for $\|y\| < \delta$ and $x_0 + y \in V(x_0)$, i.e. $B_\delta(x_0) \subset V$ and $y \in B_\delta(0)$. However (9.35) implies $f(x_0 + y) > f(x_0)$ for all $y \in B_\delta(0) \setminus \{0\}$ or $f(x) > f(x_0)$ for all $x \in B_\delta(x_0) \setminus \{x_0\}$ proving part A.
B. This follows from part A by considering $-f$ instead of f.
C. Suppose that $\operatorname{Hess} f(x_0)$ is indefinite. We have to show that in any neighbourhood $V(x_0) \subset G$ of x_0 there exists points $y_1, y_2 \in V(x_0)$ such that

$$f(y_2) < f(x_0) < f(y_1).$$

Since $\operatorname{Hess} f(x_0)$ is indefinite there exists $x \in \mathbb{R}^n \setminus \{0\}$ such that $\gamma := \langle \operatorname{Hess} f(x_0) y, y \rangle > 0$. For $|t|$ sufficiently small it follows from (9.33) that

$$f(x_0 + tx) = f(x_0) + \frac{1}{2} \langle \operatorname{Hess} f(x_0)(tx), tx \rangle + \varphi_{x_0}(tx)$$

$$= f(x_0) + \frac{\gamma}{2} t^2 + \varphi_{x_0}(tx),$$

170

and choosing $|t|$ small enough, say $|t| < \delta_1$, we have $|\varphi_{x_0}(tx)| \le \frac{\alpha}{4}t^2$, which implies

$$f(x_0 + tx) > f(x_0) \qquad \text{for } 0 < |t| < \delta_1.$$

In addition we have the existence of $z \in \mathbb{R}^n \backslash \{0\}$ such that $\eta := \langle \operatorname{Hess} f(x_0)z, z\rangle < 0$ and by the same type of reasoning as before we find for $|s|$ sufficiently small, say $|s| < \delta_2$, that

$$f(x_0 + sz) < f(x_0) \qquad \text{for } 0 < |s| < \delta_2,$$

implying the assertion. $\qquad\qquad\square$

Theorem 9.15. *For $f \in C^2(G)$, $G \subset \mathbb{R}^n$ open, to have a local minimum (maximum) at $x_0 \in G$ it is necessary that $(\operatorname{Hess} f)(x_0)$ is positive semi-definite (negative semi-definite), i.e.*

$$\beta_{(\operatorname{Hess} f)(x_0)}(x) \ge 0 \qquad (\beta_{(\operatorname{Hess} f)(x_0)}(x) \le 0) \qquad \text{for all } x \in \mathbb{R}^n.$$

Proof. We discuss the case of a local minimum, the second case goes analogously. We know that $\operatorname{grad} f(x_0) = 0$. Further let $B_r(x_0) \subset G$ such that $f(x) \ge f(x_0)$ for all $x \in B_r(x_0)$. Using the Taylor formula for $x = x_0 + h$, $|h| < r$, we find

$$0 \le f(x_0 + h) - f(x_0) = \frac{1}{2}\langle (\operatorname{Hess} f)(x_0 + \vartheta h), h\rangle$$

for some $\vartheta \in (0,1)$. With $h = sx$ we find with s small

$$0 \le \frac{s^2}{2}\langle (\operatorname{Hess} f)(x_0 + \vartheta sx)x, x\rangle,$$

or

$$0 \le \langle (\operatorname{Hess} f)(x_0 + \vartheta sx)x, x\rangle.$$

Letting s tend to zero gives

$$0 \le \langle (\operatorname{Hess} f)(x_0)x, x\rangle.$$

$\qquad\qquad\square$

Example 9.16. Let A be a positive definite matrix and β_A the corresponding quadratic form, i.e. $\beta_A : \mathbb{R}^n \to \mathbb{R}$ is given by $\beta_A(x) = \langle Ax, x\rangle$. From Lemma 9.13 we know that $\beta_A(x) \ge \alpha \|x\|^2$, $\alpha = \inf \{\beta_A(x) \,|\, x \in S^{n-1}\} > 0$, implying that β_A has a local isolated, in fact global minimum for $x_0 = 0$. We now derive the same result using our sufficient criterion, i.e. Theorem 9.14. First we note that with the Kronecker symbol δ_{ij}

$$\frac{\partial}{\partial x_j}\beta_A(x) = \frac{\partial}{\partial x_j}\sum_{k,l=1}^{n} a_{kl}x_k x_l = \sum_{k,l=1}^{n} a_{kl}\frac{\partial}{\partial x_j}(x_k x_l)$$

$$= 2\sum_{k,l=1}^{n} a_{kl}x_k\frac{\partial x_l}{\partial x_j} = 2\sum_{k,l=1}^{n} a_{kl}x_k\delta_{lj}$$

$$= 2\sum_{k=1}^{n} a_{kj}x_k = 2\sum_{k=1}^{n} a_{jk}x_k,$$

i.e.

$$\operatorname{grad}\beta_A(x) = 2Ax.$$

The condition $\operatorname{grad}\beta_A(x_0) = 0$ is equivalent to $2Ax_0 = 0$, but as a positive definite matrix A is invertible only if $x_0 = 0$ satisfies the necessary condition $\operatorname{grad}\beta_A(x_0) = 0$. Furthermore we find

$$\frac{\partial^2}{\partial x_i \partial x_j}\beta_A(x) = 2\sum_{k=1}^{n} a_{kj}\frac{\partial x_k}{\partial x_i} = 2\sum_{k=1}^{n} a_{kj}\delta_{ki} = 2a_{ij},$$

which yields for all $x \in \mathbb{R}^n$

$$\operatorname{Hess}\beta_A(x) = 2A.$$

Since A is by assumption positive definite, hence $2A$ is positive definite, it follows indeed that β_A has a local minimum at $x_0 = 0$.

Example 9.17. Consider $g : \mathbb{R}^2 \to \mathbb{R}$, $g(x,y) = c + x^2 - y^2$. Since $\operatorname{grad} g(x,y) = (2x, -2y)$ it follows that only for $(x,y) = (0,0)$ the gradient of g vanishes. The Hesse matrix of g is given by

$$(\operatorname{Hess} g)(x,y) = \begin{pmatrix} 2 & 0 \\ 0 & -2 \end{pmatrix} \quad \text{for all } (x,y) \in \mathbb{R}^2$$

and this matrix is indefinite. Thus g has no local isolated extreme values at $(0,0)$.

Example 9.18. For the function $h : \mathbb{R}^2 \to \mathbb{R}$, $h(x, y) = x^3 + y^3 - 4xy$ we find $\frac{\partial h}{\partial x}(x, y) = 3x^2 - 4y$ and $\frac{\partial h}{\partial y}(x, y) = 3y^2 - 4x$. The system $\frac{\partial h}{\partial x}(x, y) = 3x^2 - 4y = 0$ and $\frac{\partial h}{\partial y}(x, y) = 3y^2 - 4x = 0$ has the solutions $(x_1, y_1) = (0, 0)$ and $(x_2, y_2) = \left(\frac{4}{3}, \frac{4}{3}\right)$. Furthermore we find

$$\frac{\partial^2 h}{\partial x^2}(x, y) = 6x, \qquad \frac{\partial^2 h}{\partial y^2}(x, y) = 6y, \qquad \frac{\partial^2 h}{\partial x \partial y}(x, y) = -4,$$

giving as Hesse matrix

$$(\text{Hess } h)(x, y) = \begin{pmatrix} 6x & -4 \\ -4 & 6y \end{pmatrix}.$$

For $(x_2, y_2) = \left(\frac{4}{3}, \frac{4}{3}\right)$ we obtain

$$(\text{Hess } h)\left(\frac{4}{3}, \frac{4}{3}\right) = \begin{pmatrix} 8 & -4 \\ -4 & 8 \end{pmatrix}.$$

The leading principal minors are 8 and $\det \begin{pmatrix} 8 & -4 \\ -4 & 8 \end{pmatrix} = 48$, and hence $(\text{Hess } h)\left(\frac{4}{3}, \frac{4}{3}\right)$ is positive definite implying that h has a local minimum at $\left(\frac{4}{3}, \frac{4}{3}\right)$. For the second critial value $(x_1, y_1) = (0, 0)$ we obtain

$$(\text{Hess } h)(0, 0) = \begin{pmatrix} 0 & -4 \\ -4 & 0 \end{pmatrix}.$$

and since the leading minors are 0 and -16, we know that $(\text{Hess } h)(0, 0)$ is not positive definite, nor is it negative definite since $\begin{pmatrix} 0 & 4 \\ 4 & 0 \end{pmatrix}$ is not positive definite by the same type of argument. We claim that $(\text{Hess } h)(0, 0)$ is indefinite. The corresponding quadratic form is given by

$$\left\langle \begin{pmatrix} 0 & -4 \\ -4 & 0 \end{pmatrix} \begin{pmatrix} \xi \\ \eta \end{pmatrix}, \begin{pmatrix} \xi \\ \eta \end{pmatrix} \right\rangle = \left\langle \begin{pmatrix} -4\eta \\ -4\xi \end{pmatrix}, \begin{pmatrix} \xi \\ \eta \end{pmatrix} \right\rangle = -8\xi\eta$$

and for $\xi = -\eta$ this quadrartic form is strictly positive if $\eta \neq 0$, but for $\xi = \eta$ this quadratic form is strictly negative if $\eta \neq 0$, hence it is indefinite and h has no local extreme value at $(0, 0)$.

Example 9.19. Consider $f : \mathbb{R}^2 \to \mathbb{R}$, $f(x,y) = 4x^2 + y^2 - 4xy + 2$. The gradient of f is given by

$$(\operatorname{grad} f)(x, y) = (8x - 4y, 2y - 4x)$$

and the critical points (x_0, y_0) satisfying $(\operatorname{grad} f)(x_0, y_0) = (0, 0)$ are all points on the straight line $8x = 4y$ or $y = 2x$. For the Hesse matrix we find

$$(\operatorname{Hess} f)(x, y) = \begin{pmatrix} 8 & -4 \\ -4 & 2 \end{pmatrix}$$

with the leading principal minors 8 and $\det \begin{pmatrix} 8 & -4 \\ -4 & 2 \end{pmatrix} = 0$, thus

$(\operatorname{Hess} f)(x, y)$ is for no point in $\mathbb{R}^2$ positive definite. However we easily find that $f(x, y) = (2x - y)^2 + 2 \geq 2$ and for all points on the line $y = 2x$ the function f attains its minimum. But we cannot find this result using Theorem 9.14.

Let $A = (a_{kl})_{k,l=1,\dots,n}$ be an $n \times n$ matrix. Its **trace** is by definition the sum of its diagonal elements

$$\operatorname{tr}(A) = \sum_{k=1}^{n} a_{kk}. \tag{9.36}$$

It can be shown that $\operatorname{tr}(A)$ is independent of the chosen basis to represent A, i.e. $\operatorname{tr}(A)$ is a number determined by the linear mapping induced by A, not by the representation of this linear mapping. In particular for a symmetric matrix A the trace is equal to the sum of all its eigenvalues, also see Appendix I. This implies in particular that the trace of a positive definite (semi-definite) matrix is positive (non-negative) and the trace of a negative definite (semi-definite) matrix is negative (non-positive).

Now let $u \in C^2(G)$, $G \subset \mathbb{R}^n$ open and bounded, and assume that u has a continuous extension to $\bar{G}$. The Hesse matrix of u is the matrix

$$\begin{pmatrix} \frac{\partial^2 u}{\partial x_1^2} & \frac{\partial^2 u}{\partial x_1 x_2} & \cdots & \frac{\partial^2 u}{\partial x_1 x_n} \\ \frac{\partial^2 u}{\partial x_2 x_1} & \frac{\partial^2 u}{\partial x_2^2} & \cdots & \frac{\partial^2 u}{\partial x_2 x_n} \\ \vdots & & \ddots & \vdots \\ \frac{\partial^2 u}{\partial x_n x_n} & \cdots & \cdots & \frac{\partial^2 u}{\partial x_n^2} \end{pmatrix}$$

and for its trace we find

$$\operatorname{tr}(\operatorname{Hess} u)(x) = \sum_{j=1}^{n} \frac{\partial^2 u(x)}{\partial x_j^2} =: \Delta_n u(x),$$

where Δ_n denotes the n-dimensional **Laplace operator**. A function $u \in C^2(G)$ is called **harmonic** if $\Delta_n u = 0$ in G. Let $u \in C^2(G) \cap C(\bar{G})$ be harmonic in G, $\bar{G} \subset \mathbb{R}^n$ compact. As a continuous function on the compact set $\bar{G}$ the function u attains its maximum in $\bar{G}$. We claim a first **maximum principle for harmonic functions**:

Proposition 9.20. *If $u \in C^2(G) \cap C(\bar{G})$ is harmonic in the bounded open set G then u attains its maximum on ∂G, i.e.*

$$\max_{x \in \bar{G}} u(x) = \max_{u \in \partial G} u(x). \tag{9.37}$$

Proof. Since $\Delta_n v(x) = 2n > 0$ for $v(x) = |x|^2$ we deduce that $w = u + \epsilon v$, $\epsilon > 0$, satisfies $\Delta_n w(x) > 0$ for all $x \in G$. Suppose now that for $x_0 \in G$ the function w attains a maximum. By Theorem 9.15 it follows that $(\text{Hess } w)(x_0)$ must be negative semi-definite, hence its trace must be non-positive, or

$$\Delta_n w(x_0) = \text{tr}(\text{Hess } w)(x_0) \leq 0,$$

which contradicts $\Delta_n w(x) > 0$ for $x \in G$. Consequently the continuous functions $w \in C(\bar{G})$ must attain its maximum on ∂G which implies for $G \subset B_R(0)$ and all $\epsilon > 0$

$$w(x) < \max_{y \in \partial G} w(y) \leq \max_{y \in \partial G} u(x) + \epsilon R^2.$$

For $\epsilon \to 0$ we obtain now

$$u(x) \leq \max_{y \in \partial G} u(y)$$

implying

$$\max_{x \in \bar{G}} u(x) = \max_{y \in \partial G} u(y).$$

$\square$

(The reader should note that some, but not all theoretical considerations in this chapter were influenced by the presentation in O. Forster [19].)

Problems

1. Let A be a symmetric $n \times n$ matrix and suppose that for the corresponding bilinear form β_A we have that $\beta_A(x, x) \geq 0$ for all $x \in B_\epsilon(0) \setminus \{0\}$ with some $\epsilon > 0$. Prove that A is positive semi-definite and that when $\beta_A(x, x) > 0$ for all $x \in B_\epsilon(0) \setminus \{0\}$ the matrix A is positive definite.

2. Let $A \in M(n; \mathbb{R})$ be a symmetric matrix with eigenvalues $\lambda_1 \leq \lambda_2 \leq \cdots \leq \lambda_n$ counted according to their multiplicity. Prove that A is positive definite if $\lambda_1 > 0$ and positive semi-definite if $\lambda_1 \geq 0$. Formulate similar criteria for negative definite and negative semi-definite matrices. Further prove that A is indefinite if and only if $\lambda_1 < 0$ and $\lambda_n > 0$.

3. Let $|| \cdot ||$ be a norm on $\mathbb{R}^n$ (not necessarily the Euclidean norm). On $M(n; \mathbb{R})$ define

$$|||A||| := \inf\{c > 0 | \, ||Ax|| \leq c||x||\}.$$

a) Prove that $||| \cdot |||$ is a norm on $M(n; \mathbb{R})$ which also satisfies $|||AB||| \leq |||A||| \, |||B|||$. Such a norm on $M(n; \mathbb{R})$ is called the **matrix norm** on $M(n; \mathbb{R})$ with respect to $|| \cdot ||$. Show that

$$|||A||| = \sup_{x \neq 0} \frac{||Ax||}{||x||}.$$

b) Find the matrix norm with resect to the maximum norm $||x||_\infty = \max_{1 \leq j \leq n} |x_j|$ on $\mathbb{R}^n$.

4. Let $G \subset \mathbb{R}^m$ be an open set and $A : G \to M(n; \mathbb{R})$ a continuous function. Prove that if for some $x_0 \in G$ the matrix $A(x_0)$ is positive definite, then there exists $r > 0$ and $\delta > 0$ such that on $B_r(x_0) \subset G$ we have $\beta_{A(x)}(\xi, \xi) \geq \delta ||\xi||^2$ for all $\xi \in \mathbb{R}^n$.

5. Use the mean value theorem to prove that the function $g : (\mathbb{R}_+ \setminus \{0\}) \times (\mathbb{R}_+ \setminus \{0\}) \to \mathbb{R}, g(x, y) = \arctan \frac{y}{x} + \arctan \frac{x}{y}$, is a constant.

6. Let $G \subset \mathbb{R}^2$ be a convex set and $f : G \to \mathbb{R}$ a differentiable function with the property that $\frac{\partial f}{\partial x}(x, y) = 0$ for all $(x, y) \in G$. Prove that f is independent of x, i.e. $f(x, y) = h(y)$ with some suitable function h.

7. a) Find the Taylor polynomial of order 2 about $(0,0)$ for the function $f : \mathbb{R}^2 \to \mathbb{R}$, $f(x,y) = e^{xy}\cos(x+y)$.

b) Find the Taylor polynomial of order 3 about $(1,1)$ of the function $g : (\mathbb{R}_+ \setminus \{0\}) \times (\mathbb{R}_+ \setminus \{0\}) \to \mathbb{R}, g(x,y) = \ln xy$.

8. Show that for $x \in B_\epsilon(0) \subset \mathbb{R}^2$ we have

$$\left| \cos x_1 \cos x_2 - 1 + \frac{1}{2}(x_1^2 + x_2^2) \right| \leq \frac{4}{3}\epsilon^3.$$

9. Use the Taylor formula to show for all $k \in \mathbb{N}$ and $x = (x_1 + \cdots + x_n)^k = \sum_{|\alpha|=k} \frac{k!}{\alpha!} x^\alpha$.

10. Let $A \in M(n; \mathbb{R})$ be a symmetric matrix and $0 \neq x_0 \in \mathbb{R}^n$ be a critical point of $g(x) := \frac{\beta_A(x,x)}{||x||^2}$. Prove that x_0 is an eigenvector of A to the eigenvalue $\lambda := \frac{\beta_A(x_0,x_0)}{||x_0||^2}$.

11. For the following functions find the Hesse matrix

a) $f(x_1, x_2, x_3) = (x_3 + \sqrt{1 + x_1^2}) - x_2 x_3$,

b) $g(x,y) = \det \begin{pmatrix} g_{11}(x) & g_{12}(y) \\ g_{21}(y) & g_{22}(x) \end{pmatrix}$, $\quad g_{12}(y) = g_{21}(y)$, where g_{11}, g_{22}, $g_{12} : \mathbb{R} \to \mathbb{R}$ are C^2-functions.

12. For the following functions find their isolated extreme values

a) $g(x,y) = xye^{-(x+y)}$, $x, y \in \mathbb{R}$;

b) $h(u,v) = u^2 - v^2 - (u^2 + v^2)^2$, $u, v \in \mathbb{R}$.

13. Let $\xi^1, \ldots, \xi^k$ be k fixed points and consider on $\mathbb{R}^n$ the function $f(x) = \sum_{l=1}^k ||x - \xi^l||^2$. Prove that f attains its minimum at the midpoint $\xi := \frac{1}{k}\sum_{l=1}^k \xi^l$.

14. Let $f : \mathbb{R}^n \to \mathbb{R}$ be a rotational invariant function, i.e. $f(x_1, \ldots, x_n) = g(||x||)$ with a suitable function $g : \mathbb{R} \to \mathbb{R}$. Suppose that f and g are C^2-functions. Prove that for $x \neq 0$ the Hesse matrix of f is given by

$$(\text{Hess} f)(x) = \left(g''(r)\frac{x_k x_l}{r^2} + g'(r)\left(\frac{r^2\delta_{kl} - x_k x_l}{r^3}\right)\right)_{k,l=1,\ldots,n},$$

where $r = (x_1^2 + \cdots + x_n^2)^{\frac{1}{2}} = ||x||$. Deduce that f is harmonic in $\mathbb{R}^n \setminus \{0\}$ if

$$g''(r) + \frac{n-1}{r}g'(r) = 0.$$

15. Let $f : \mathbb{R}^n \to \mathbb{R}$ be a C^2-function and assume that $(\mathrm{Hess}\, f)(x)$ is for all $x \in \mathbb{R}^n$ positive definite. Show that f can have at most one critical point.

10 Implicit Functions and the Inverse Mapping Theorem

Let us start with the following problem: can we describe the circle in the plane as the graph of one (several) function(s). We denote by C_r the circle with centre $0 = (0,0) \in \mathbb{R}^2$ and radius $r > 0$, see Figure 10.1.

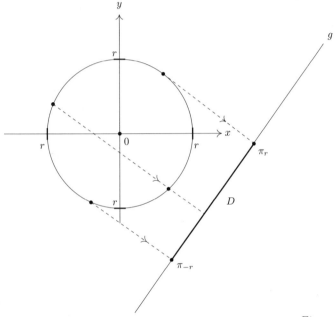

Figure 10.1

We have different possibilities to describe the circle as a set, for example

$$C_r := \left\{ (x, y) \in \mathbb{R}^2 \,\big|\, x^2 + y^2 = r^2 \right\} \tag{10.1}$$

or

$$C_r := \operatorname{tr}(\gamma), \ \gamma : [0, 2\pi] \to \mathbb{R}^2, \ \gamma(\varphi) := (r \cos \varphi, r \sin \varphi). \tag{10.2}$$

However C_r is not a graph $\Gamma(f)$ of a function $f : D \to \mathbb{R}$, $D \subset \mathbb{R}$. Indeed, suppose D is a subset of a straight line $g \subset \mathbb{R}^2$, see Figure 10.1. In order to obtain C_r as the graph of a function defined on D, D must be the orthogonal projection of C_r onto g, and then we take g as the new abscissa and after fixing a point $0 \in g$ we take the line orthogonal to g passing through 0 as

179

the new ordinate. But it is clear that with the exception of the two boundary points π_{-r} and π_r of D, every point on D corresponds to two points on C_r.

However, a closer look at the situation shows how we can locally describe C_r as the graph of a function, locally means in a (small) neighbourhood of a point $(x_0, y_0) \in C_r$, see Figure 10.2.

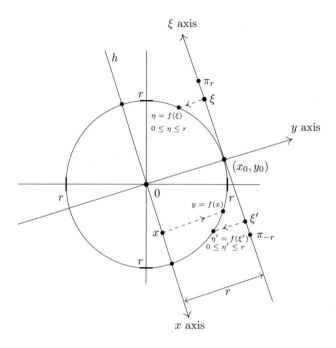

Figure 10.2

We take the tangent g to C_r passing through (x_0, y_0) as the new abscissa, (x_0, y_0) as the new origin and the ordinate n is the line orthogonal to g passing through (x_0, y_0). The new coordinates we denote by (ξ, η), hence g becomes the ξ-axis, n becomes the η-axis, and in the (ξ, η)-system the origin (x_0, y_0) has coordinates $(0, 0)$. The orthogonal projection $(\xi, \eta) \mapsto \eta$ is denoted by pr_2. It follows that $D := \mathrm{pr}_2(C_r)$ is the interval $[\pi_{-r}, \pi_r]$ on the ξ-axis and it has length $2r$. We now define $f : [\pi_{-r}, \pi_r] \to \mathbb{R}$ by

$$f(\xi) = \eta, \, \eta \le r \quad \text{and} \quad \eta \in \mathrm{pr}_2^{-1}(\{\xi\}),$$

where $\mathrm{pr}_2^{-1}(A)$ is the pre-image of A under pr_2. For $\xi \in (\pi_{-r}, \pi_r)$ we know that $\mathrm{pr}_2^{-1}(\{\xi\})$ consists always of two points (ξ, η) and $(\xi, \tilde{\eta})$ with $\eta < r$ but

180

$\tilde{\eta} > r$. Thus in a neighbourhood of (x_0, y_0), after introducing an appropriate new coordinate system, we can represent C_r as the graph of a function. For the circle this construction was rather complicated. It is much simpler to replace g by h, the straight line parallel to g and passing through the centre of C_r, and then taking the centre as the new origin. Denoting the new coordinates by (X, Y) we find that $Y = F(X) = \sqrt{r^2 - X^2}$ locally as a describing function in these coordinates . In particular if we choose $(x_0, y_0) = (0, r)$ we can use the original coordinate system to find as a locally describing function $y = k(x) = \sqrt{r^2 - x^2}$, and for $(x_0, y_0) = (r, 0)$ we find $x = l(y) = \sqrt{r^2 - y^2}$, see Figure 10.3 and Figure 10.4.

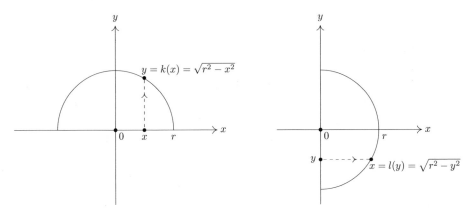

Figure 10.3   Figure 10.4

Note that as long as we can define a tangent to a curve in $\mathbb{R}^2$, the first construction can always be applied, while the second approach, or a similar one only works easily for a few special curves.

Instead of a curve in the plane we may consider a surface in $\mathbb{R}^3$, for example the sphere S^2, and may ask whether locally, i.e. in a neighbourhood of a point $(x, y, z) \in S^2$, the sphere has a representation as the graph of a function. A way to formulate the problem in $\mathbb{R}^n$ goes as follows:
Let $f : G \to \mathbb{R}$, $G \subset \mathbb{R}^{n+1}$, be a function and consider the set

$$M_0(f) := \left\{ (x_1, \ldots, x_{n+1}) \in G \,\middle|\, f(x_1, \ldots, x_{n+1}) = 0 \right\}. \tag{10.3}$$

Now we ask whether it is possible to find for $x_0 = (x_1^0, \ldots, x_{n+1}^0) \in M_0(f)$ a neighbourhood $U(x_0) \subset G \subset \mathbb{R}^{n+1}$ and a function $h : \tilde{G} \to \mathbb{R}$, $\tilde{G} \subset \mathbb{R}^n$, such that

$$M_0(f) \cap U(x_0) = \Gamma(h) = \left\{ (y, h(y)) \,\middle|\, y \in \tilde{G} \right\}. \tag{10.4}$$

We may turn this problem into a different direction. Suppose that we pay particular attention to $x_k =: z$ in the defining equation of $M_0(f)$, and we may ask whether we can express z as a function of $(x_1, \ldots, x_{k-1}, x_{k+1}, \ldots, x_{n+1})$. Thus we want to solve

$$f(x_1, \ldots, x_{k-1}, z, x_{k+1}, \ldots, x_{n+1}) = 0 \tag{10.5}$$

with the help of a function g, i.e.

$$z = g(x_1, \ldots, x_{k-1}, x_{k+1}, \ldots, x_{n+1}).$$

Example 10.1. We want to solve $x^2 + y^2 = r^2$ for $y \in \mathbb{R}$. Clearly, for $|x| > r$ there is no solution, for $x = r$ the solution is $y = 0$ as it is for $x = -r$. However, for $-r < x < r$ we have two solutions $y = \sqrt{r^2 - x^2}$ and $y = -\sqrt{r^2 - x^2}$. But locally around (x, y_0) there is only one solution, either $y_0(x) = \sqrt{r^2 - x^2}$ or $y_0(x) = -\sqrt{r^2 - x^2}$. Moreover, these two expressions can be interpreted as functions defined on $[-r, r]$, each given locally, i.e. close to (x, y_0), $y_0 > 0$ $(y_0 < 0)$, the solution.

We can extend the problem further: On $G \subset \mathbb{R}^n$ let $f_j : G \to \mathbb{R}$, $j = 1, \ldots, m$, be given functions. We want to find all solutions to the system of equations

$$\begin{aligned} f_1(x) &= f_1(x_1, \ldots, x_n) &= 0 \\ &\vdots &\vdots &\qquad \vdots \\ f_m(x) &= f_m(x_1, \ldots, x_n) &= 0. \end{aligned} \tag{10.6}$$

This is rather ambitious, so we may ask to find local solutions but what does this mean now?

The following problem is well known (systems of linear equations):

$$\begin{aligned} a_{11}x_1 + \ldots + a_{1n}x_n &= 0 \\ &\vdots \\ a_{m1}x_1 + \ldots + a_{mn}x_n &= 0, \end{aligned} \tag{10.7}$$

where $a_{jl} \in \mathbb{R}$ and we take now $G = \mathbb{R}^n$ and

$$f_j(x_1, \ldots, x_n) = a_{j1}x_1 + \ldots + a_{jn}x_n.$$

Each equation $f_j(x_1, \ldots, x_n) = 0$ describes a plane in $\mathbb{R}^n$ and the solution to (10.7) is the intersection of these planes. Other problems we may encounter are (also) of geometric nature, for example finding the intersection of a sphere with a plane, which transforms in $\mathbb{R}^3$ with

$$S = \left\{ (x, y, z) \in \mathbb{R}^3 \mid (x - x_0)^2 + (y - y_0)^2 + (z - z_0)^2 = r^2 \right\}$$

and

$$E = \left\{ (x, y, z) \in \mathbb{R}^3 \mid ax + by + cz = d \right\}$$

to the system of equations

$$(x - x_0)^2 + (y - y_0)^2 + (z - z_0)^2 = r^2$$
$$ax + by + cz = d. \tag{10.8}$$

Of course we would like to describe the solution geometrically, say locally as the graph of a function.

As an attempt to understand the problem better, we ask the following: suppose that $f : G \to \mathbb{R}$, $G \subset \mathbb{R}^2$, is a function, and in addition let $g : I \to \mathbb{R}$, $I \subset \mathbb{R}$ being an interval, be a function such that $\Gamma(g) \subset G$ and $f(x, g(x)) = 0$ for all $x \in I$. Assume further that f and g are differentiable. Can we find the derivative of g using the partial derivatives of f? Thus, if we can solve locally $f(x, y) = 0$ with a function $y = g(x)$, we try to find the derivative of g using the partial derivatives of f.
The solution goes as follows: since $0 = f(x, g(x))$ for all $x \in I$ the chain rule yields

$$0 = \frac{d}{dx} f(x, g(x)) = \left(\frac{\partial f}{\partial x} \right) (x, g(x)) + \left(\frac{\partial f}{\partial y} \right) (x, g(x)) g'(x) \tag{10.9}$$

and if

$$\left(\frac{\partial f}{\partial y} \right) (x, g(x)) \neq 0 \tag{10.10}$$

we find

$$g'(x) = -\frac{\left(\frac{\partial f}{\partial x} \right) (x, g(x))}{\left(\frac{\partial f}{\partial y} \right) (x, g(x))}. \tag{10.11}$$

Thus, once we know that g exists, we can find its derivative.

Example 10.2. Consider on $\mathbb{R}^2$ the function $f(x,y) = \frac{x^2}{a^2} + \frac{y^2}{b^2} - 1$. The set $\{f(x,y) = 0\}$ is an ellipse, of course we assume $a > 0$, $b > 0$. The partial derivatives of f are given by

$$\left(\frac{\partial f}{\partial x}\right)(x,y) = \frac{2x}{a^2} \quad , \quad \left(\frac{\partial f}{\partial y}\right)(x,y) = \frac{2y}{b^2}. \tag{10.12}$$

For $y \neq 0$ we have $\frac{\partial f}{\partial y}(x,y) \neq 0$. Now let $g : \left(-\frac{a}{2}, \frac{a}{2}\right) \to \mathbb{R}$, $g(x) = \frac{b}{a}\sqrt{a^2 - x^2}$. It follows that

$$f(x, g(x)) = \frac{x^2}{a^2} + \frac{b^2(a^2 - x^2)}{a^2 b^2} - 1 = \frac{x^2}{a^2} - \frac{x^2}{a^2} + 1 - 1 = 0$$

and therefore

$$g'(x) = -\frac{\frac{2x}{a^2}}{\frac{2g(x)}{b^2}} = -\frac{b^2 x}{a^2 g(x)} = -\frac{bx}{a\sqrt{a^2 - x^2}},$$

which of course coincides with $\frac{d}{dx}\left(\frac{b}{a}\sqrt{a^2 - x^2}\right)$.

Returning to our general considerations, we may ask whether (10.10) already implies the existence of g. This is in principle the case, but we first want to extend (10.11) to higher dimensions. We follow the considerations of O. Forster [19].

Consider the open balls $B_{r_1}(x_0) = \{x \in \mathbb{R}^n \mid \|x - x_0\| < r_1\}$ and $B_{r_2}(y_0) = \{y \in \mathbb{R}^m \mid \|y - y_0\| < r_2\}$. Note that we use $\|.\|$ in $\mathbb{R}^n$ and $\mathbb{R}^m$ as the symbol for the Euclidean norm. For $(x,y) = (x_1, \ldots, x_n, y_1, \ldots, y_m) \in \mathbb{R}^{n+m}$ we write $\|(x,y)\|$ for $\|(x_1, \ldots, x_n, y_1, \ldots, y_n)\|$, i.e. the Euclidean norm in $\mathbb{R}^{n+m}$. Consider $f : B_{r_1}(x_0) \times B_{r_2}(y_0) \to \mathbb{R}^m$ with components $f_j : B_{r_1}(x_0) \times B_{r_2}(y_0) \to \mathbb{R}$, $j = 1, \ldots, m$. Assume that $d_{(x_0, y_0)} f$ exists, i.e. f is differentiable at (x_0, y_0). Then all partial derivatives $\frac{\partial f_j}{\partial x_k}(x_0, y_0)$ and $\frac{\partial f_j}{\partial y_l}(x_0, y_0)$ exist for $1 \leq j \leq m$, $1 \leq k \leq n$ and $1 \leq l \leq m$, and therefore we can form the matrices

$$\frac{\partial f}{\partial x}(x_0, y_0) := \begin{pmatrix} \frac{\partial f_1}{\partial x_1}(x_0, y_0) & \cdots & \frac{\partial f_1}{\partial x_n}(x_0, y_0) \\ \vdots & & \vdots \\ \frac{\partial f_m}{\partial x_1}(x_0, y_0) & \cdots & \frac{\partial f_m}{\partial x_n}(x_0, y_0) \end{pmatrix} \in M(m, n; \mathbb{R}) \tag{10.13}$$

and

$$\left(\begin{array}{ccc} \frac{\partial f_1}{\partial y_1}(x_0, y_0) & \cdots & \frac{\partial f_1}{\partial y_m}(x_0, y_0) \\ \vdots & & \vdots \\ \frac{\partial f_m}{\partial y_1}(x_0, y_0) & \cdots & \frac{\partial f_m}{\partial y_m}(x_0, y_0) \end{array} \right) \in M(m; \mathbb{R}). \tag{10.14}$$

Our goal is to find $\mathrm{d}_{x_0} g$ if g is a mapping such that $f(x, g(x)) = 0$ holds for all $x \in B_{r_1}(x_0)$ and $y_0 = g(x_0)$. Such a mapping g must map $B_{r_1}(x_0)$ into $B_{r_2}(y_0)$ i.e. $g : B_{r_1}(x_0) \to B_{r_2}(y_0)$, it will have components $g_j : B_{r_1}(x_0) \to \mathbb{R}$, and if $\mathrm{d}_{x_0} g$ exists we will have

$$\mathrm{d}_{x_0} g = \left(\begin{array}{ccc} \frac{\partial g_1}{\partial x_1}(x_0, y_0) & \cdots & \frac{\partial g_1}{\partial x_n}(x_0, y_0) \\ \vdots & & \vdots \\ \frac{\partial g_m}{\partial x_1}(x_0, y_0) & \cdots & \frac{\partial g_m}{\partial x_n}(x_0, y_0) \end{array} \right) \in M(m, n; \mathbb{R}). \tag{10.15}$$

Note that for $n = m = 1$ under appropriate conditions we recover, as expected, the situation leading to (10.11).

Theorem 10.3. *Let $f : B_{r_1}(x_0) \times B_{r_2}(y_0) \to \mathbb{R}^m$ be differentiable at $(x_0, y_0) \in B_{r_1}(x_0) \times B_{r_2}(y_0) \subset \mathbb{R}^n \times \mathbb{R}^m$ and assume that $f(x_0, y_0) = 0$. Suppose that $\frac{\partial f}{\partial y}(x_0, y_0)$ given by (10.14) is invertible and that $g : B_{r_1}(x_0) \to \mathbb{R}^m$ is a continuous mapping such that $g(B_{r_1}(x_0)) \subset B_{r_2}(y_0)$ and $g(x_0) = y_0$, as well as $f(x, g(x)) = 0$ for all $x \in B_{r_1}(x_0)$. Then g is differentiable at x_0 and $\mathrm{d}_{x_0} g$ is given by*

$$\mathrm{d}_{x_0} g = - \left(\frac{\partial f}{\partial y}(x_0, y_0) \right)^{-1} \frac{\partial f}{\partial x}(x_0, y_0). \tag{10.16}$$

Proof. First we note that $\left(\frac{\partial f}{\partial y}(x_0, y_0) \right)^{-1} \in M(m; \mathbb{R})$ and $\frac{\partial f}{\partial x}(x_0, y_0) \in M(m, n; \mathbb{R})$, hence $\left(\frac{\partial f}{\partial y} \right)^{-1} \frac{\partial f}{\partial x}(x_0, y_0)$ is a well defined element in $M(m, n; \mathbb{R})$. Next we observe that the differentiation is translation invariant and therefore we may assume $x_0 = 0 \in \mathbb{R}^n$ and $y_0 = 0 \in \mathbb{R}^m$, i.e. $(x_0, y_0) = (0, 0) \in \mathbb{R}^{n+m}$. We now set

$$A := \frac{\partial f}{\partial x}(0, 0) \in M(m, n; \mathbb{R})$$

and

$$B := \frac{\partial f}{\partial y}(0, 0) \in M(m; \mathbb{R}).$$

The differential of f at $(0,0) \in \mathbb{R}^{n+m}$ is now the $m \times (m+n)$ matrix

$$
\mathrm{d}_{(0,0)} f = \begin{pmatrix} \frac{\partial f_1}{\partial x_1}(0,0) & \cdots & \frac{\partial f_1}{\partial x_n}(0,0) & \frac{\partial f_1}{\partial y_1}(0,0) & \cdots & \frac{\partial f_1}{\partial y_m}(0,0) \\ \vdots & & & & & \vdots \\ \frac{\partial f_m}{\partial x_1}(0,0) & \cdots & \frac{\partial f_m}{\partial x_n}(0,0) & \frac{\partial f_m}{\partial y_1}(0,0) & \cdots & \frac{\partial f_m}{\partial y_m}(0,0) \end{pmatrix}
$$

or written as a block matrix

$$
\mathrm{d}_{(0,0)} f = (A, B), \tag{10.17}
$$

which yields with $f(0,0) = 0$ that

$$
f(x,y) = (\mathrm{d}_{(0,0)}f) \begin{pmatrix} x \\ y \end{pmatrix} + \varphi(x,y) = Ax + By + \varphi(x,y) \tag{10.18}
$$

where $\varphi : B_{r_1}(0) \times B_{r_2}(0) \to \mathbb{R}^m$ satisfies

$$
\lim_{\|(x,y)\| \to 0} \frac{\varphi(x,y)}{\|(x,y)\|} = 0. \tag{10.19}
$$

By assumption we have $f(x, g(x)) = 0$ for all $x \in B_{r_1}(0)$ implying by (10.18) that

$$
0 = Ax + Bg(x) + \varphi(x, g(x)),
$$

or

$$
g(x) = -B^{-1}Ax - B^{-1}\varphi(x, g(x)). \tag{10.20}
$$

The result follows once we can prove that for some δ_1, $0 < \delta_1 < r_1$, and some $c_0 > 0$ the following holds

$$
\|g(x)\| \leq c_0 \|x\| \qquad \text{for all} \qquad x \in \overline{B_{\delta_1}(0)}. \tag{10.21}
$$

Indeed, assume (10.21) holds and define

$$
\psi(x) := -B^{-1}\varphi(x, g(x)). \tag{10.22}
$$

Since $-B^{-1}$ is continuous as a linear mapping from $\mathbb{R}^m$ to $\mathbb{R}^m$, we deduce from (10.19) first that for $\eta > 0$ there exists $\delta_2 > 0$ such that $\|(x,y)\| < \delta_2$ implies

$$
\|\varphi(x,y)\| \leq \eta \|(x,y)\| = \eta \left(\|x\|^2 + \|y\|^2\right)^{\frac{1}{2}}.
$$

186

Further, (10.21) implies now

$$\|\varphi(x, g(x))\| \leq \eta \left(\|x\|^2 + \|y\|^2 \right)^{\frac{1}{2}}$$
$$\leq \eta \left(\|x\|^2 + c_0^2 \|x\|^2 \right)^{\frac{1}{2}}$$
$$= \eta(1 + c_0) \|x\|$$

provided $\|x\| < \delta_1$. Thus, given $\varepsilon > 0$ we can find some $0 < \delta < r_1$ such that $x \in B_\delta(x)$ implies

$$\|\varphi(x, g(x))\| \leq \varepsilon \|x\|,$$

or

$$\lim_{\|x\| \to 0} \frac{\varphi(x, g(x))}{\|x\|} = 0, \tag{10.23}$$

and therefore

$$\lim_{\|x\| \to 0} \frac{\psi(x)}{\|x\|} = \lim_{\|x\| \to \infty} \frac{-B^{-1}\varphi(x, g(x))}{\|x\|} = 0.$$

Hence g is differentiable at $x_0 = 0$ and has differential

$$d_0 g = -B^{-1} A = - \left(\frac{\partial f}{\partial y} \right)^{-1} (0, 0) \frac{\partial f}{\partial x}(0, 0).$$

It remains to prove (10.21). In the following we denote by $\|C\|$ the norm of a matrix C, i.e. $\|C\| = \left(\sum c_{kl}^2 \right)^{\frac{1}{2}}$. We set

$$\gamma_1 := \|B^{-1} A\| \qquad \text{and} \qquad \gamma_2 := \|B^{-1}\|.$$

Condition (10.19) implies the existence of $\rho_1 \in (0, r_1)$ and $\rho_2 \in (0, r_2)$ such that $\|x\| < \rho_1$ and $\|y\| < \rho_2$ will imply

$$\|\varphi(x, y)\| \leq \frac{1}{2\gamma_2} \|(x, y)\| \leq \frac{1}{2\gamma_2}(\|x\| + \|y\|).$$

The continuity of g yields the existence of $0 < \delta_1 \leq \rho_1$ such that $x \in B_{\delta_1}(0)$ implies $\|g(x)\| \leq \rho_2$, note that $g(x_0) = y_0 = 0$, and therefore it follows for $x \in B_{\delta_1}(0)$

$$\|\varphi(x, g(x))\| \leq \frac{1}{2\gamma_2}(\|x\| + \|g(x)\|).$$

187

Now we use (10.20) to find for $x \in B_{\delta_1}(0)$

$$\|g(x)\| \le \gamma_1 \|x\| + \gamma_2 \|\varphi(x, g(x))\|$$
$$\le \left(\gamma_1 + \frac{1}{2}\right) \|x\| + \frac{1}{2} \|g(x)\|$$

or

$$\|g(x)\| \le (2\gamma_1 + 1) \|x\|$$

implying (10.21) with $c_0 = 2\gamma_1 + 1$ and the theorem is proven. $\qquad\square$

Remark 10.4. For $n = m = 1$ we recover (10.11).

Let us return to our original question and stay for a moment in the one-dimensional case. Thus we ask whether we can solve the equation

$$f(x, z) = 0 \tag{10.24}$$

by a function $z = g(x)$, i.e. solutions shall be given as a function of x. Suppose we know such a function g exists, is differentiable and $g(0) = 0$ (for simplicity). From (10.11) we know that

$$g'(x) = -\left(\frac{\partial f}{\partial y}\right)^{-1} (x, g(x)) \left(\frac{\partial f}{\partial x}\right) (x, g(x)), \tag{10.25}$$

provided $\left(\frac{\partial f}{\partial y}\right)^{-1} (x, g(x))$ exists. Locally, i.e. in a neighbourhood of $(0,0)$ we may work with $\left(\frac{\partial f}{\partial y}\right)^{-1} (0,0)$ as an approximation of $\left(\frac{\partial f}{\partial y}\right)^{-1} (x, g(x))$ and consider for a function h such that $f(x, h(x)) = 0$ and $h(0) = 0$ the equation

$$h'(x) = -\left(\frac{\partial f}{\partial y}\right)^{-1} (0,0) \left(\frac{\partial f}{\partial x}\right) (x, h(x)). \tag{10.26}$$

Integration approximately yields

$$h(x) = -\left(\frac{\partial f}{\partial y}\right)^{-1} (0,0) f(x, h(x)) \tag{10.27}$$

and with

$$G(x, h(x)) := h(x) - \left(\frac{\partial f}{\partial y}\right)^{-1} (0,0) f(x, h(x)) \tag{10.28}$$

we get

$$G(x, h(x)) = h(x) \qquad \text{if and only if} \qquad f(x, h(x)) = 0. \qquad (10.29)$$

Let us try to find an iteration to obtain a function h satisfying $G(x, h(x)) = h(x)$. For some h_0 ($h_0(x) = 0$ for all x is allowed) set $h_1 := G(x, h_0(x))$ and now we continue with

$$h_{\nu+1}(x) := G(x, h_\nu(x)). \qquad (10.30)$$

If we have for such a sequence of functions $(h_\nu)_{\nu \in \mathbb{N}}$ that $h_\nu \to h$ and $G(x, h_\nu(x)) \to G(x, h(x))$ in some sense, then (10.30) implies

$$h(x) = G(x, h(x)), \qquad (10.31)$$

and if in addition there is only one h satisfying (10.31) then we can indeed solve $f(x, y) = 0$ by the function h, i.e. $f(x, y) = 0$ if and only if $y = h(x)$. Of course, all this needs a justification, but before doing this we note that although the "integration" leading to (10.27) is only a one-dimensional argument, we can start in the higher dimensional case (using the notation of Theorem 10.3 and its proof) with

$$G(x, h(x)) := h(x) - \left(\frac{\partial f}{\partial y}\right)^{-1} (x_0, y_0) f(x, h(x)), \qquad (10.32)$$

and the (formal) iteration $h_{\nu+1}(x) := G(x, h_\nu(x))$ will lead to the same idea how to obtain possibly a solution to $f(x, y) = 0$ in the form of a mapping $y = g(x)$.

We want to prepare a part of the proof of our main result, the implicit function theorem, in a separate heuristic consideration. Let $G : \mathbb{R}^n \times \mathbb{R}^m \to \mathbb{R}^n$ be a mapping and suppose that $h : V_n \to V_m$, $V_n \subset \mathbb{R}^n$, $V_m \subset \mathbb{R}^m$, is a function such that $G(x, h(x)) \in \mathbb{R}^m$ is well defined. Consider the norm

$$\|h\|_\infty := \sup_{x \in V_n} \|h(x)\|$$

and assume that

$$\|G(x, y) - G(x, z)\| \le \kappa \|y - z\|$$

for all $x \in V_n$ and all $y, z \in \mathbb{R}^m$, or $y, z \in V_m$. For the iteration

$$h_{\nu+1}(x) = G(x, h_\nu(x)) \qquad (10.33)$$

189

this implies that

$$h_{\nu+1}(x) - h_\nu(x) = G(x, h_\nu(x)) - G(x, h_{\nu-1}(x)),$$

or

$$\|h_{\nu+1} - h_\nu\|_\infty \leq \kappa \|h_\nu - h_{\nu-1}\|_\infty.$$

With

$$h_N = \sum_{\nu=0}^{N-1}(h_{\nu+1} - h_\nu), \qquad h_0 = 0,$$

we now find

$$\|h_N\|_\infty \leq \sum_{\nu=0}^{N-1} \|h_{\nu+1} - h_\nu\|_\infty$$

$$\leq \sum_{\nu=0}^{N-1} \kappa^\nu \|h_1 - h_0\|_\infty = \left(\sum_{\nu=0}^{N-1} \kappa^\nu\right) \|h_1\|_\infty,$$

and if $\kappa < 1$ we can apply a variant of the Weierstrass test, compare Theorem I.29.1, to $\left(\sum_{\nu=0}^{N-1}(h_{\nu+1} - h_\nu)\right)_{N\in\mathbb{N}}$, i.e. to the sequence $(h_N)_{N\in\mathbb{N}}$. Thus $(h_N)_{N\in\mathbb{N}}$ converges uniformly to some function h. Passing now in (10.33) to the limit gives

$$h(x) = G(x, h(x)),$$

i.e. $f(x, h(x)) = 0$.

With these preparations in mind we eventually prove the **implicit function theorem** where we follow the main ideas in O. Forster [19] again.

Theorem 10.5. *Let $U_1 \subset \mathbb{R}^n$ and $U_2 \subset \mathbb{R}^m$ be two open sets and $f : U_1 \times U_2 \to \mathbb{R}^m$ be a continuously differentiable mapping. For $(x_0, y_0) \in U_1 \times U_2$ suppose that $f(x_0, y_0) = 0$ and that $\left(\frac{\partial f}{\partial y}\right)(x_0, y_0) \in M(m; \mathbb{R})$ has the inverse $\left(\frac{\partial f}{\partial y}\right)^{-1}(x_0, y_0)$. Then there exist open neighbourhoods $V_1 \subset U_1$ of x_0 and $V_2 \subset U_2$ of y_0 and a continuous mapping $g : V_1 \to V_2$ such that*

$$f(x, g(x)) = 0 \qquad\qquad (10.34)$$

holds for all $x \in V_1$. Moreover, if (10.34) holds for some $(x, y) \in V_1 \times V_2$ then $y = g(x)$.

Remark 10.6. A. The function g is said to be implicitly defined by the equation $f(x, y) = 0$. **B.** As our examples in the beginning of this Chapter show, in general we cannot expect g to be defined on all of U_1, i.e. the result is a local one in the sense that for every point the assertion holds (only) in a (small) neighbourhood. **C.** For the point (x_0, y_0) we can immediately apply Theorem 10.3 to calculate $\mathrm{d}_{x_0} g$. Now, if $\left(\frac{\partial f}{\partial y}\right)(x_0, y_0)$ is invertible, it is invertible in a neighbourhood $W_1 \times W_2$ of (x_0, y_0). This follows from the fact that if $\det\left(\frac{\partial f}{\partial y}\right)(x_0, y_0) \neq 0$, then $\det \frac{\partial f}{\partial y}$ must be not equal to zero in a neighbourhood of (x_0, y_0). Hence formula (10.16) holds for $\mathrm{d}g$ in a neighbourhood $W_1 \times W_2 \subset V_1 \times V_2$ of (x_0, y_0).

Proof of Theorem 10.5. First we note as in the proof of Theorem 10.3 that we may assume $(x_0, y_0) = (0, 0) \in \mathbb{R}^n \times \mathbb{R}^m$, and then we introduce once more the notation

$$A := \left(\frac{\partial f}{\partial x}\right)(0, 0) \in M(m, n; \mathbb{R}) \qquad \text{and} \qquad B := \left(\frac{\partial f}{\partial y}\right) \in M(m; \mathbb{R}).$$

Now we define as in our preparation the mapping

$$G : U_1 \times U_2 \to \mathbb{R}^m \tag{10.35}$$
$$G(x, y) := y - B^{-1} f(x, y).$$

We note that

$$\left(\frac{\partial G}{\partial y}\right)(x, y) = \mathrm{id}_{\mathbb{R}^m} - B^{-1}\left(\frac{\partial f}{\partial y}\right)(x, y) \tag{10.36}$$

and therefore

$$\left(\frac{\partial G}{\partial y}\right)(0, 0) = \mathrm{id}_{\mathbb{R}^m} - B^{-1}\left(\frac{\partial f}{\partial y}\right)(0, 0) = \mathrm{id}_{\mathbb{R}^m} - B^{-1}B = 0_{\mathbb{R}^m} \tag{10.37}$$

where $0_{\mathbb{R}^m}$ is the zero element in $M(m; \mathbb{R})$. By assumption $\frac{\partial G}{\partial y}$ is continuous implying the existence of neighbourhoods $W_1 \subset U_1$ of $0 \in \mathbb{R}^n$ and $W_2 \subset U_2$ of $0 \in \mathbb{R}^m$ such that

$$\left\|\left(\frac{\partial G}{\partial y}\right)(x, y)\right\| \leq \frac{1}{2} \qquad \text{for all} \qquad (x, y) \in W_1 \times W_2. \tag{10.38}$$

We pick $r > 0$ such that

$$V_2 := B_r(0) \subset W_2 \,(\subset \mathbb{R}^m).$$

Since $G(0,0) = 0$, recall $f(0,0) = 0$, the continuity of G implies the existence of a neighbourhood $V_1 \subset W_1$ such that

$$\rho := \sup_{x \in V_1} \|G(x,0)\| < \frac{r}{2}. \qquad (10.39)$$

(Problem 7 asks to verify this argument.)

As we know that $f(x,y) = 0$ is equivalent to $G(x,y) = y$ we are now investigating the latter equation. First we prove that for $x \in V_1$ there exists at most one $y \in V_2$ such that $G(x,y) = y$. Indeed if $y_1 = G(x,y_1)$ and $y_2 = G(x,y_2)$ then

$$y_1 - y_2 = G(x,y_1) - G(x,y_2)$$

and an application of the mean value theorem in the form of Corollary 9.2 implies by (10.38)

$$\|y_1 - y_2\| = \|G(x,y_1) - G(x,y_2)\| \leq \frac{1}{2}\|y_1 - y_2\| \qquad (10.40)$$

which yields $y_1 = y_2$. Eventually we prove the existence of $g : V_1 \to V_2$ such that $f(x, g(x)) = 0$, $x \in V_1$. We set $g_0 : V_1 \to V_2$, $g_0(x) = 0$ for all $x \in V_1$ and now we define the iteration

$$g_{\nu+1}(x) := G(x, g_\nu(x)). \qquad (10.41)$$

Note that by (10.39) we find

$$\|g_1\|_{\infty, V_1} := \sup_{x \in V_1} \|g_1(x)\| = \rho. \qquad (10.42)$$

Suppose that we can use (10.40) to show

$$\|g_{\nu+1} - g_\nu\|_{\infty, V_1} \leq 2^{-\nu}\rho. \qquad (10.43)$$

In this case the series $\sum_{\nu=0}^{\infty}(g_{\nu+1} - g_\nu)$ would have the majorant $\sum_{\nu=0}^{\infty} 2^{-\nu}\rho = 2\rho$ and the M-test, Theorem I.29.1, implied the uniform convergence to a limit function

$$g := \lim_{N \to \infty} g_N = \lim_{N \to \infty} \sum_{\nu=0}^{N-1}(g_{\nu+1} - g_\nu) = \sum_{\nu=0}^{\infty}(g_{\nu+1} - g_\nu),$$

and $g : V_1 \to \mathbb{R}^m$ is continuous and satisfies $\|g\|_{\infty, V_1} \leq 2\rho < r$, i.e. $g(V_1) \subset V_2$. Now we may pass in (10.41) to the limit to obtain for all $x \in V_1$

$$g(x) = \lim_{N \to \infty} g_{N+1}(x) = \lim_{N \to \infty} G(x, g_N(x)) = G(x, g(x)),$$

i.e. $f(x, g(x)) = 0$.

It remains to prove (10.43) which we will do by induction. For $\nu = 0$ this is just (10.42). With

$$g_N := \sum_{\nu=0}^{N-1} (g_{\nu+1} - g_\nu)$$

we have under the assumption of (10.43) that

$$\|g_N\|_{\infty, V_1} \leq \sum_{\nu=0}^{N-1} 2^{-\nu} \rho = 2\rho < r,$$

i.e. $g_N(V_1) \subset V_2$ and $g_{\nu+1} := G(x, g_\nu)$ is well defined. Furthermore, since

$$g_{\nu+1}(x) - g_\nu(x) = G(x, g_\nu(x)) - G(x, g(x)_{\nu-1})$$

we find again by the mean value theorem in the form of Corollary 9.2 that

$$\|g_{\nu+1}(x) - g_\nu(x)\| \leq \frac{1}{2} \|g_\nu(x) - g_{\nu-1}(x)\|$$

or

$$\|g_{\nu+1} - g_\nu\|_{\infty, V_1} \leq \frac{1}{2} \|g_\nu - g_{\nu-1}\|_{\infty, V_1}$$

and therefore (10.43), since $\|g_1 - g_0\|_{\infty, V_1} = \|g_1\|_{\infty, V_1} = \rho$. Finally we observe that the uniqueness result also implies that if $f(x, y) = 0$ and $(x, y) \in V_1 \times V_2$ then $y = g(x)$. $\qquad\square$

Although we started our considerations with the problem to solve (systems of) equation(s), the implicit function theorem, Theorem 10.5, is hiding this aspect, while the chosen formulation will become very useful within more geometric investigations. For this reason we want to give a more "equation solving" formulation of Theorem 10.5.

Let $f_j \in C^1(U_1 \times U_2)$, $f_j : U_1 \times U_2 \to \mathbb{R}$, $1 \leq j \leq m$, and $U_1 \times U_2 \subset \mathbb{R}^n \times \mathbb{R}^m$, and consider the system of (non-) linear equations

$$f_1(x_1, \ldots, x_n, x_{n+1}, \ldots, x_{n+m}) = 0$$
$$\vdots \tag{10.44}$$
$$f_m(x_1, \ldots, x_n, x_{n+1}, \ldots, x_{n+m}) = 0.$$

Suppose that for some point $x^0 := (x_1^0, \ldots, x_n^0, x_{n+1}^0, \ldots, x_{n+m}^0) \in U_1 \times U_2$ the following holds

$$\det\left(\left(\frac{\partial f_j}{\partial x_l}\right)(x^0)\right)_{\substack{j=1,\ldots,m \\ l=n+1,\ldots,n+m}} \neq 0.$$

Then there exists $\rho > 0$ such that with $x^{01} = (x_1^{01}, \ldots, x_n^{01})$

$$B_\rho(x^{01}) := \left\{ \tilde{x} \in \mathbb{R}^n \mid \sum_{j=1}^n |\tilde{x}_j - x_j^{01}|^2 < \rho^2 \right\} \subset U_1,$$

and there exists continuous functions $g_j : B_\rho(x^{01}) \to \mathbb{R}$, $j = 1, \ldots, m$, such that for all $\tilde{x} \in B_\rho(x^{01})$ we have

$$f_1(\tilde{x}_1, \ldots, \tilde{x}_n, g_1(\tilde{x}_1, \ldots, \tilde{x}_n), \ldots, g_m(\tilde{x}_1, \ldots, \tilde{x}_n)) = 0$$
$$\vdots \tag{10.45}$$
$$f_m(\tilde{x}_1, \ldots, \tilde{x}_n, g_1(\tilde{x}_1, \ldots, \tilde{x}_n), \ldots, g_m(\tilde{x}_1, \ldots, \tilde{x}_n)) = 0.$$

Now we want to discuss several applications of the implicit function theorem. First however we want to discuss in more detail graphs of functions as a set of zeroes of equations. We first handle the case $n = 2$. Let $U \subset \mathbb{R}^2$ be open and $f \in C^1(U)$. Suppose that $f(x_0, y_0) = 0$ and $\left(\frac{\partial f}{\partial y}\right)(x_0, y_0) \neq 0$ hold for $(x_0, y_0) \in U$. Then we can find $\rho > 0$ such that $(x_0 - \rho, x_0 + \rho) \times \{y\} \subset U$ and a function $g : (x_0 - \rho, x_0 + \rho) \to \mathbb{R}$ such that $f(x, g(x)) = 0$ for all $x \in (x_0 - \rho, x_0 + \rho)$. With $\Gamma(g) = \{(x, g(x)) \mid x \in (x_0 - \rho, x_0 + \rho)\}$ we have $(x, y) \in \Gamma(g)$, $x \in (x_0 - \rho, x_0 + \rho)$ if and only if

$$(x, y) \in Z_f := \left\{ (z_1, z_2) \in U \mid f(z_1, z_2) = 0 \text{ and } z_1 \in (x_0 - \rho, x_0 + \rho) \right\}.$$

Example 10.7. We return to our introductory example with $f(x, y) = x^2 + y^2 - r^2 = 0$ where $f : \mathbb{R}^2 \to \mathbb{R}$ is obviously a C^1-function. There are natural bounds for x and y, namely $-r \leq x \leq r$ and $-r \leq y \leq r$. Moreover we find that $\frac{\partial f}{\partial y}(x, y) = 2y \neq 0$ for all $y \neq 0$. Thus for $(x_0, y_0) \in [-r, r] \times [-r, r]$ with $y_0 \neq 0$ we can find a neighbourhood U_1 of x_0, say $(x_0 - \rho, x_0 + \rho)$, $0 < \rho < r$, and a function $g : U_1 \to \mathbb{R}$ such that for $x \in U_1$ the following holds

$$x^2 + g^2(x) - r^2 = 0 \qquad \text{or} \qquad g^2(x) = r^2 - x^2.$$

Hence, if $y_0 > 0$ then $y = g_+(x) = \sqrt{r^2 - x^2}$ and if $y_0 < 0$ then $y = g_-(x) = -\sqrt{r^2 - x^2}$. However if $y_0 = 0$ and $x_0^2 + y_0^2 - r^2 = 0$, then we observe that $\left(\frac{\partial f}{\partial x}\right)(x_0, 0) = 2x_0 \neq 0$. Changing the roles of x and y we can now solve

$$f(x, y) = 0, \quad y_0 = 0,$$

with the help of a function $h_\pm : U_2 \to \mathbb{R}$, $y \mapsto h_\pm(y)$ where $h_\pm(y) = \pm\sqrt{r^2 - y^2}$ with the sign depending on whether $x > 0$ or $x < 0$. Eventually we find that we can extend $g_\pm$ and $h_\pm$ to $(-r, r)$ and then we arrive at

$$\partial B_r(0) = \left\{ (x, y) \in \mathbb{R}^2 \mid x^2 + y^2 - r^2 = 0 \right\} = \Gamma(g_+) \cup \Gamma(g_-) \cup \Gamma(h_+) \cup \Gamma(h_-).$$
$$(10.46)$$

Thus while $\partial B_r(0)$ is not the graph of one function, it is the union of finitely many graphs. (The very careful reader will notice a certain notational problem in (10.46) which is however easy to resolve along the lines of the following considerations.)

The following considerations will link graphs of functions to the set of zeroes of functions. It is a bit more technical, but we include it here since this is its natural place and it will be very helpful when dealing with differentiable manifolds defined by equations later (in Volume VI, Part 16). The student may skip this part during a first reading.

Let $X \subset \mathbb{R}^{n+1}$ be an open set and $f : X \to \mathbb{R}$ a C^1-function. Let $x^0 = (x_1^0, \ldots, x_{n+1}^0) \in X$ be a zero of f, i.e. $f(x_1^0, \ldots, x_{n+1}^0) = 0$ and assume for some j, $1 \leq j \leq n + 1$, that $\frac{\partial f}{\partial x_j}(x_1^0, \ldots, x_{n+1}^0) \neq 0$. Let $r > 0$ such that $\prod_{l=1}^{n+1}(x_l^0 - r, x_l^0 + r) \subset X$ and define

$$Y_{j,1} := \begin{cases} \emptyset & j = 1 \\ \prod_{l=1}^{j-1}(x_l^0 - r, x_l^0 + r), \end{cases} \qquad Y_{j,2} := \begin{cases} \prod_{l=1}^{j-1}(x_l^0 - r, x_l^0 + r) \\ \emptyset \quad \text{if } j = 1 \end{cases}$$

as well as

$$Y_j := \begin{cases} Y_{n+1,1}, & j = n+1 \\ Y_{j,1} \times Y_{j,2}, & 2 \le j \le n \\ Y_{1,2}, & j = 1 \end{cases}.$$

The implicit function theorem yields for r small enough the existence of g_j : $Y_j \to \mathbb{R}$, $Y_{j,1} \times g_j(Y_j) \times Y_{j,2} \subset X \left(g_1(Y_1) \times Y_{n+1,2} \text{ for } j = 1, Y_{n+1,1} \times g_{n+1}(Y_{n+1}) \right.$ for $j = n+1 \left. \right)$ such that

$$f(x_1, \ldots, x_{j-1}, g_j(x_1, \ldots, x_{j-1}, x_{j+1}, \ldots, x_{n+1}), x_{j+1}, \ldots, x_{n+1}) = 0$$

for all $(x_1, \ldots, x_{j-1}, x_{j+1}, \ldots x_{n+1}) \in Y_j$. Next we define

$$\tilde{\Gamma}(g_j) := \{(x_1, \ldots, x_{n+1}) \in Y_{j,1} \times \mathbb{R} \times Y_{j,2} \,|\, x_j = g(x_1, \ldots, x_{j-1}, x_{j+1}, \ldots, x_{n+1})\}$$

and introduce $\pi_j : \mathbb{R}^{n+1} \to \mathbb{R}^{n+1}$, $(x_1, \ldots, x_{n+1}) \mapsto (x_1, \ldots, x_{j-1}, x_{j+1}, \ldots, x_{n+1}, x_j)$ which induces a mapping $P_j : \tilde{\Gamma}(g_j) \to \Gamma(g_j)$ by $(x_1, \ldots, x_{j-1}, y_j, x_{j+1}, \ldots, x_{n+1}) \mapsto (x_1, \ldots, x_{j-1}, x_{j+1}, \ldots, x_{n+1}, y_j)$. Clearly P_j is bijective. In a neighbourhood of $x^0 = (x_1, \ldots, x^0_{n+1})$ the set of zeroes of f coincides with $\tilde{\Gamma}(g_j)$ and hence can be mapped bijectively onto the graph of a function.

Example 10.8. We consider the **hyperboloid**

$$H_1 = \left\{ (x, y, z) \in \mathbb{R}^3 \,\middle|\, x^2 + y^2 - z^2 = 1 \right\},$$

see Figure 10.5. We can introduce the function $f : \mathbb{R}^3 \to \mathbb{R}$, $f(x, y, z) = x^2 + y^2 - z^2 - 1$, and find that H_1 is the set of zeroes of f. Further we have

$$\frac{\partial f}{\partial z} = -2z \ne 0 \quad \text{for} \quad z \ne 0$$

and therefore we find for $(x_0, y_0, z_0) \in H_1$, $z_0 \ne 0$, the existence of a function $g : U_1 \to \mathbb{R}$, U_1 being a neighbourhood of (x_0, y_0), such that $f(x, y, g(x, y)) = 0$ for all $x, y \in U_1$. Of course, depending on whether $z_0 > 0$ or $z_0 < 0$ we have $g(x, y) = \sqrt{x^2 + y^2 - 1}$ or $g(x, y) = -\sqrt{x^2 + y^2 - 1}$. From our previous consideration we now see that locally around a point $(x_0, y_0, z_0) \in H_1$, $z_0 \ne 0$, we can represent H_1 as a graph of a function. Since $\frac{\partial f}{\partial x} = 2x$ and $\frac{\partial f}{\partial y} = 2y$, the condition $x^2 + y^2 - z^2 = 1$ implies that for every $(x, y, z) \in H_1$ at least one of the partial derivatives $\frac{\partial f}{\partial x}, \frac{\partial f}{\partial y}, \frac{\partial f}{\partial z}$ at this point (hence at any point) is not zero and consequently we can think of H_1 as the union of overlapping graphs of functions $g_j : U_j \to \mathbb{R}$, $U_j \subset \mathbb{R}^2$.

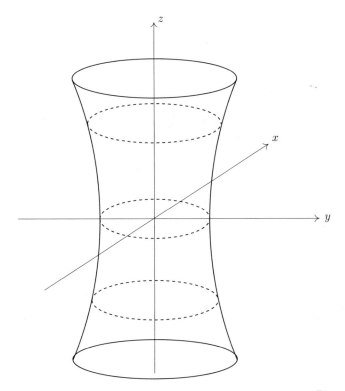

Figure 10.5

Example 10.9. Let $G \subset \mathbb{R}^n$ be open and $f : G \to \mathbb{R}$ be a C^1-function such that $\operatorname{grad} f(x) \neq 0$ for all $x \in G$. The level sets $N_c(f)$ of f are given by the equation

$$f(x) = f(x_1, \ldots, x_n) = c, \quad x \in G, \quad c \in \mathbb{R},$$

i.e. a level set is a subset of G consisting of all points where f attains the value c, i.e.

$$N_c(f) = \left\{ x \in G \,\middle|\, f(x_1, \ldots, x_n) = c \right\}.$$

If $N_c(f) \neq \emptyset$ and $x^0 \in N_c(f)$, then the gradient condition implies that for at least one j, $1 \leq j \leq n$, we have $\frac{\partial f}{\partial x_j}(x^0) \neq 0$. Consequently, by the implicit function theorem there exists an open set $U \subset \mathbb{R}^{n-1}$ and a function $g_j : U \to \mathbb{R}$ such that in a neighbourhood V of $x^0 \in G$ we can represent $N_c(f)$ as graph of the function g_j. Assuming $j = n$, we thus can write in a

197

neighbourhood V of x^0

$$N_c(f) = \left\{ (x_1, \ldots, x_{n-1}, g_n(x_1, \ldots, x_{n-1})) \,\middle|\, g_n : B_r(x_1^0, \ldots, x_{n-1}^0) \to \mathbb{R} \right\}$$

with $g(x_1^0, \ldots, x_{n-1}^0) = x_n^0$.

We want to consider systems of equations and see how we may apply the implicit function theorem. Our first example handles linear systems of equations and recovers results from linear algebra.

Example 10.10. Let $a_{kl} \in \mathbb{R}$, $k, l \in 1, \ldots, n$, and consider the linear system

$$x_1 = a_{11}y_1 + \cdots + a_{1n}y_n$$

$$\vdots$$

$$x_n = a_{n1}y_1 + \cdots + a_{nn}y_n,$$

or with $f_j : \mathbb{R}^n \times \mathbb{R}^n \to \mathbb{R}$,

$$f_j(y_1, \ldots, y_n, x_1, \ldots, x_n) = a_{j1}y_1 + \cdots + a_{jn}y_n - x_j,$$

we have

$$f_j(y_1, \ldots, y_n, x_1, \ldots, x_n) = 0 \qquad \text{for} \qquad j = 1, \ldots, n,$$

which we can even rewrite as

$$f(y_1, \ldots, y_n, x_1, \ldots, x_n) = 0, \qquad f = (f_1, \ldots, f_n).$$

We are seeking solutions $y_j = y_j(x_1, \ldots, x_n)$ and therefore we look at the Jacobi matrix $\frac{\partial f}{\partial y}$ or

$$\frac{\partial f_j}{\partial y_l} = a_{jl}$$

which is independent of $(y_1, \ldots, y_n, x_1, \ldots, x_n)$ and invertible if and only if $\det(a_{jl}) \neq 0$. In this case we find

$$y_j = g_j(x_1, \ldots, x_n)$$

for example in explicit form using Cramer's rule, see Appendix I.

Of course this result is not surprising, but maybe gives some illustration.

198

Example 10.11. We want to solve the system

$$x^2 + u^3 v^3 + y^2 = \frac{1}{2} \tag{10.47}$$
$$-x^2 u^2 + y^2 v^2 = 9$$

for $x > 1$ and $y > 1$. In other words we try to find $x = h_1(u, v)$ and $y = h_2(u, v)$ such that

$$h_1^2(u, v) + u^3 v^3 + h_2^2(u, v) = \frac{1}{2}$$
$$-h_1^2(u, v) u^2 + h_2^2(u, v) v^2 = 9$$

holds for functions $h_1 > 1$ and $h_2 > 1$. We rewrite (10.47) as

$$\left. \begin{array}{c} f_1(x, y, u, v) = 0 \\ f_2(x, y, u, v) = 0 \end{array} \right\} \tag{10.48}$$

and find

$$\begin{pmatrix} \frac{\partial f_1}{\partial u} & \frac{\partial f_2}{\partial u} \\ \frac{\partial f_1}{\partial v} & \frac{\partial f_2}{\partial v} \end{pmatrix} = \begin{pmatrix} 3u^2 v^3 & -2ux^2 \\ 3u^3 v^2 & 2vy^2 \end{pmatrix}$$

and further

$$\det \begin{pmatrix} 3u^2 v^3 & -2ux^2 \\ 3u^3 v^2 & 2vy^2 \end{pmatrix} = 6u^2 v^2 (v^2 y^2 + u^2 x^2)$$

since by assumption $x > 1$ and $y > 1$, this determinant vanishes for either $u = 0$ or $v = 0$ and the implicit function theorem tells us that locally for points (x, y, u, v), $x > 1$, $y > 1$, $u \neq 0$ and $v \neq 0$, we can solve (10.46) with the help of functions $x = h_1(u, v)$ and $y = h_2(u, v)$.

In one dimension we know how to find the derivative of the inverse of a given differentiable function f provided the derivative of f is not zero, see Theorem I.21.11. Since bijective real-valued functions of a real variable are necessarily strictly monotone, the existence of an inverse function seems to be independent of the way we calculate its derivative. However strictly monotone functions which are continuously differentiable have with the exception of isolated points non-zero derivatives. Indeed, if $f'(x_0) \neq 0$ then $f'(x) \neq 0$ in a neighbourhood of x_0 and the formula (compare (I.21.12))

$$(f^{-1})'(y_0) = \frac{1}{f'(f^{-1}(y_0))}, \quad f(x_0) = y_0, \tag{10.49}$$

does not only allow us to calculate $(f^{-1})'$, but shows the significance of the condition $f'(x_0) \neq 0$. In higher dimensions we face two problems. First of all, if $g : G \to \mathbb{R}^m$, $G \subset \mathbb{R}^n$, is a mapping, we need a condition for g being invertible, and then we want of course to calculate the derivative of the inverse. We are restricting ourselves to continuously differentiable mappings. Since every linear mapping $A : \mathbb{R}^n \to \mathbb{R}^m$ is continuously differentiable with differential (at x) given by $\mathrm{d}_x A = A$, from linear algebra we deduce that A can have only an inverse if $n = m$. Now, let $g : G \to \mathbb{R}^n$, $G \subset \mathbb{R}^n$, be continuously differentiable and $h : H \to \mathbb{R}^n$, $H \subset \mathbb{R}^n$, a further continuously differentiable mapping such that

$$h \circ g = \mathrm{id}_n \quad \text{or} \quad h(g(x)) = x \quad \text{for} \quad x \in G,$$

here id_n denote the identity mapping on $\mathbb{R}^n$. The chain rule yields

$$\mathrm{d}_{g(x)} h \circ \mathrm{d}_x g = \mathrm{id}_n$$

and if $\mathrm{d}_x g$ is invertible, then

$$\mathrm{d}_{g(x)} h = (\mathrm{d}_x g)^{-1}. \tag{10.50}$$

But in this case we clearly have $h = g^{-1}$ (at least close to $x \in G$ where $\mathrm{d}_x g$ is invertible). Hence (10.50) reads as

$$\mathrm{d}_y g^{-1} = (\mathrm{d}_x g)^{-1}, \quad y = g(x), \tag{10.51}$$

or

$$\left(\mathrm{d}\, g^{-1} \right)(y) = \left(\mathrm{d}\, g \right)^{-1} \left(g^{-1}(y) \right), \tag{10.52}$$

which reduces for $n = 1$ to (10.49). This calculation suggests that the condition $\mathrm{d}_x g$ being invertible implies the (local) existence of g^{-1} and its differentiability with (10.52) to hold.

Theorem 10.12 (Inverse mapping theorem). *Let $G \subset \mathbb{R}^n$ be an open set and $g : G \to \mathbb{R}^n$ be a continuously differentiable mapping and assume that for $x_0 \in G$ the differential $\mathrm{d}_{x_0} g$ is invertible, i.e. the Jacobi matrix $J_g(x_0)$ of g at x_0 has an inverse or equivalently $\det J_g(x_0) \neq 0$. Then there exists an open neighbourhood $U_1 \subset G$ of x_0 and an open neighbourhood $U_2 \subset \mathbb{R}^n$ of $y_0 := g(x_0)$ such that $g|_{U_1}$ is invertible with continuously differentiable inverse $g^{-1} : U_2 \to U_1$ and*

$$\left(\mathrm{d}\, g^{-1} \right)(y_0) = \left(\mathrm{d}\, g \right)^{-1} \left(g(y_0) \right) \tag{10.53}$$

holds.

Proof. The basic idea is to use the implicit function theorem. For this define $f : \mathbb{R}^n \times G \to \mathbb{R}^n$, $f(y, z) = y - g(z)$. For $z = x_0$ and $y_0 = g(x_0)$ we find $f(y_0, x_0) = 0$. Moreover we have $\frac{\partial f}{\partial z}(y_0, x_0) = d_{x_0} g (= d\,g(x_0))$ where we used the notation of the proof of Theorem 10.5. Since by assumption $d_{x_0} g$ is invertible the implicit function theorem yields the existence of an open neighbourhood U_2 of y_0, an open neighbourhood $\tilde{U}_1$ of x_0, $\tilde{U}_1 \subset G$, and a mapping $h : U_2 \to \tilde{U}_1$ such that

$$f(y, h(y)) = y - g(h(y)) \qquad \text{for all} \qquad y \in U_2.$$

Without loss of generality we may assume that h is continuously differentiable (otherwise we need to shrink U_2). The set

$$U_1 := \tilde{U}_1 \cap f^{-1}(U_2) = \tilde{U}_1 \cap \left\{ z \in \tilde{U}_1 \,\big|\, f(z) \in U_2 \right\}$$

is open since $\tilde{U}_1$ is open and $f^{-1}(U_2)$ is open as a pre-image of an open set, further $x_0 \in U_1$. By construction $g : U_1 \to U_2$ is bijective with inverse h. Now (10.53) follows by the chain rule as in our preceding considerations. $\square$

Example 10.13. Consider on $\mathbb{R}^2$ the mapping $f : \mathbb{R}^2 \to \mathbb{R}^2$ defined by

$$f(x_1, x_2) = \begin{pmatrix} f_1(x_1, x_2) \\ f_2(x_1, x_2) \end{pmatrix} = \begin{pmatrix} x_1^4 \\ x_1 + x_2 \end{pmatrix}.$$

The Jacobi matrix is given by

$$\begin{pmatrix} \frac{\partial f_1}{\partial x_1}(x_1, x_2) & \frac{\partial f_1}{\partial x_2}(x_1, x_2) \\ \frac{\partial f_2}{\partial x_1}(x_1, x_2) & \frac{\partial f_2}{\partial x_2}(x_1, x_2) \end{pmatrix} = \begin{pmatrix} 4x_1^3 & 0 \\ 1 & 1 \end{pmatrix}$$

and $\det \begin{pmatrix} 4x_1^3 & 0 \\ 1 & 1 \end{pmatrix} = 4x_1^3 \neq 0$ for all $(x_1, x_2) \in \mathbb{R}^2$ such that $x_1 \neq 0$. If $x_1 > 0$ then we can solve

$$y_1 = x_1^4$$
$$y_2 = x_1 + x_2$$

by $x_1 = \sqrt[4]{y_1}$ and $x_2 = y_2 - \sqrt[4]{y_1}$, and if $x_2 < 0$ the solution is $x_1 = -\sqrt[4]{y_1}$, $x_2 = y_2 + \sqrt[4]{y_1}$. Thus we find for $h(y_1, y_2) = \begin{pmatrix} \sqrt[4]{y_1} \\ y_2 - \sqrt[4]{y_1} \end{pmatrix}$ that

$$(h \circ f)(x_1, x_2) = \begin{pmatrix} \sqrt[4]{f_1(x_1, x_2)} \\ f_2(x_1, x_2) - \sqrt[4]{f_1(x_1, x_2)} \end{pmatrix} = \begin{pmatrix} \sqrt[4]{x_1^4} \\ x_1 + x_2 - \sqrt[4]{x_1^4} \end{pmatrix} = \begin{pmatrix} x_1 \\ x_2 \end{pmatrix},$$

i.e. $h = f^{-1}$ for $x_1 > 0$, and we leave it as an exercise to verify the case $x_1 < 0$.

In Chapter 12, when dealing with curvilinear coordinates we will encounter many further applications of the implicit and the inverse function theorem.

Problems

1. Given in $\mathbb{R}^2$ the function $f(x,y) = \frac{x^4}{16} + \frac{y^2}{9} - 25$ find a function $g : D(g) \to \mathbb{R}_+$ with the largest possible domain $D(g) \subset \mathbb{R}$ such that $f(x, g(x)) = 0$ holds for all $x \in D(g)$.

2. For $t \in \mathbb{R}$ consider the function $z(t) = x^2(t)y^3(t)$ where x and y are implicitly given by $x^4 + y = t$ and $x^2 + y^2 = t^2$. Prove that for $x(t) \neq 0$ and $4x^2(t)y(t) \neq 1$ we have

$$z'(t) = \frac{y(t)(2y(t) - 2t - 3x^2(t)y(t) + 12x^4(t)y(t)t)}{4x^2(t)y(t) - 1}.$$

3. Show that for some $\epsilon > 0$ there exist functions $f, g : (-\epsilon, \epsilon) \to \mathbb{R}_+ \setminus \{0\}$ such that solutions to

$$x^2 + y^2 - 2z^2 = 0 \quad \text{and} \quad 2x^2 + y^2 + z^2 = 4$$

are given in a neighbourhood of $y = 0$ by $x = f(y)$ and $z = g(y)$.

4. Let $f : \mathbb{R}^4 \to \mathbb{R}^2$ have the two components

$$f_1(x, u) = f_1(x_1, x_2, u_1, u_2) = u_1^2 - u_2 - 3x_1 - x_2$$

and

$$f_2(x, u) = f_2(x_1, x_2, u_1, u_2) = u_1 - 2u_2^2 - x_1 + 2x_2,$$

where we use the notation $(x, u) = (x_1, x_2, u_1, u_2) \in \mathbb{R}^4$, i.e. $x = (x_1, x_2) \in \mathbb{R}^2$ and $u = (u_1, u_2) \in \mathbb{R}^2$. Now consider the equation $f(x, u) = 0$. Prove that for $(x, u) \in \mathbb{R}^4$ such that $u_1 \cdot u_2 \neq \frac{1}{8}$ the 2×2 matrix $\frac{\partial f}{\partial u}(x, u)$ is invertible. Now use the implicit function theorem to establish the existence of open balls $B_{\rho_1}(x_0) \subset \mathbb{R}^2$ and $B_{\rho_2}(u_0)$, $u_{01} \cdot u_{02} \neq \frac{1}{8}$, and the existence of a differentiable mapping $g : B_{\rho_1}(x_0) \to B_{\rho_2}(u_0)$ such that for $x \in B_{\rho_1}(x_0)$ we have $f(x, g(x)) = 0$. Find $(dg)(x)$ for $x \in B_{\rho_1}(x_0)$.

5. Let $G := \{x \in \mathbb{R}^3 | x_1 + x_2 + x_3 \neq -1\}$ and consider $f : G \to \mathbb{R}^3$ defined by

$$f(x) = \left(\frac{x_1}{1 + x_1 + x_2 + x_3}, \frac{x_2}{1 + x_1 + x_2 + x_3}, \frac{x_3}{1 + x_1 + x_2 + x_3} \right).$$

Prove that f is injective, determine $f(G)$ and $f^{-1} : f(G) \to G$. Further find $d(f^{-1})$.

6. Let $f : G \to \mathbb{R}$, $G \subset \mathbb{R}^2$ open, be a C^1-function and $\gamma : I \to G$ a C^1-curve. We call γ or its trace a **gradient line** of f or $\Gamma(f)$ (or the trace of f) if for a strictly positive function $\kappa : I \to \mathbb{R}$ the following holds

$$\dot{\gamma}(t) = \kappa(t)(\mathrm{grad} f)(\gamma(t)), \ t \in I.$$

Prove that if a gradient line of f and a level line, i.e. a curve defined by $f(x, y) = c$, intersect, then they are orthogonal at the intersection point.

7. Justify statement (10.39).

11   Further Applications of the Derivatives

In this chapter we want to discuss further applications of (partial) derivatives. These depend on both the extra "variability" and the more challenging geometry that we have in higher dimensions. Indeed, the critical reader may have spotted that we have not (yet) transferred all one-dimensional results to higher dimensions, for example the Taylor series has not been discussed. A function of several variables can be interpreted as a representation of a quantity depending on certain parameters, say the temperature of a gas depending on volume and pressure, and our interest is to study the change of such a quantity under the change of the parameters. For example we might be interested in extreme values or a "tendency" to approach an equilibrium. Sometimes we may ask for extreme values under certain constraints, for example we might be interested in the body with largest volume under the assumption that its surface has a fixed area. We may know that the quantity is subjected to invariances, i.e. shows some symmetries and thus we may ask whether these symmetries can be used for a more simple description. Think of a quantity depending on three spatial coordinates (the parameters) but we know that this quantity is invariant under rotation. Instead of a function of three variables a function of one variable should be sufficient for our mathematical description. We are used to geometric interpretations, functions have graphs, they define level sets in their domains, etc. We may think at a more geometric and intuitive representation of a function. In these and many more examples partial derivatives or the differential turn out to be helpful tools and in this chapter we want to discuss some of these applications of derivatives with more to follow in Chapters 12 and 13. The two topics we want to discuss in this chapter are extreme values under constraints, in particular under constraints which are not "solvable", and envelopes of families of curves. We start our investigations by looking at extreme value problems under constraints. Here is an example.

Example 11.1. Find the dimensions of a rectangular box open at the top such that the volume of the box is $500m^3$ and the surface area becomes minimal.

We put one vertex of the box into the origin of our coordinate system and denote the other end point of the edges starting at the origin by x, y and z and we assume them to lie on the non-negative axes, see Figure 11.1.

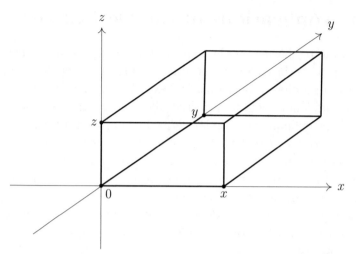

Figure 11.1

The volume V of the box is $V = xyz$ and by assumption we have $xyz = 500$. The surface area of the box (open at the top) is given by

$$\tilde{S}(x, y, z) = xy + 2yz + 2xz.$$

Using $z = \frac{500}{xy}$ we find

$$\tilde{S}(x, y, z) = S(x, y) = xy + \frac{1000}{x} + \frac{1000}{y},$$

and hence we can eliminate the constraint which leads to the eliminating of one variable. Now we can apply the methods developed in Chapter 9 to find a minimum of S:

$$\frac{\partial S}{\partial x}(x, y) = y - \frac{1000}{x^2} = 0$$

$$\frac{\partial S}{\partial y}(x, y) = x - \frac{1000}{y^2} = 0$$

are the necessary conditions, and they lead to $x^2 y = 1000$ and $xy^2 = 1000$, implying $\frac{x}{y} = 1$ or $x = y$, which yields $x^3 = y^3 = 1000$ or $x = y = \sqrt[3]{1000} = 10$, and consequently $z = \frac{500}{10 \cdot 10} = 5$. Thus we expect a local extreme value of S at $(10, 10)$. The Hessian of S is given by

$$\begin{pmatrix} \frac{\partial^2 S}{\partial x^2}(x, y) & \frac{\partial^2 S}{\partial x \partial y}(x, y) \\ \frac{\partial^2 S}{\partial y \partial x}(x, y) & \frac{\partial^2 S}{\partial y^2}(x, y) \end{pmatrix} = \begin{pmatrix} \frac{2000}{x^3} & 1 \\ 1 & \frac{2000}{y^3} \end{pmatrix},$$

and for $x = y = 10$ it follows that

$$\text{Hess}(S)(10, 10) = \begin{pmatrix} 2 & 1 \\ 1 & 2 \end{pmatrix}$$

with the eigenvalues $\lambda_1 = 3$ and $\lambda_2 = 1$. Hence $\text{Hess}(S)(10, 10)$ is positive definite and S has a minimum at $(10, 10)$, consequently $\tilde{S}$ has under the constraint $xyz = 500$ a minimum at $(10, 10, 5)$ and the surface has the area of 300m².

In Example 11.1 we could solve the equation given the constraint $(xyz = 500)$ to eliminate one variable (we have chosen to eliminate z) and then we would have obtained an extreme value problem without a constraint which we then can handle with the theory developed in Chapter 9. In general however we cannot expect to solve a constrained extreme value problem along that line.

We may even pose the following more general problem:

Problem 11.2. Find (isolated) extreme values for $f : G \to \mathbb{R}$, $G \subset \mathbb{R}^n$, under the constraints $g_l(x) = 0$, $x \in G$, $l = 1, \ldots, m$, where $g_l : G \to \mathbb{R}$ are suitable functions.

Suppose that in Problem 11.2 we have $m = 1$ but we cannot eliminate one variable by solving $g_1(x_1, \ldots, x_n) = 0$ globally, say

$$x_j = h(x_1, \ldots, x_{j-1}, x_{j+1}, \ldots, x_n),$$

and then try to find unrestricted extreme values for

$$\tilde{f}(x_1, \ldots, x_{j-1}, x_{j+1}, \ldots, x_n)$$
$$= f(x_1, \ldots, x_{j-1}, h(x_1, \ldots, x_{j-1}, x_{j+1}, \ldots, x_n), , x_{j+1}, \ldots, x_n).$$

We then face a new challenge since we cannot apply our known results from Chapter 9.

If we have no constraints and if f has an isolated extreme value at $x^0 = (x_1^0, \ldots, x_n^0)$ under the assumption that f is differentiable we may find x^0 by solving the system

$$\frac{\partial f}{\partial x_j}(x^0) = 0, \qquad j = 1, \ldots, n.$$

When we have the constraints $g_l(x) = 0$, $l = 1, \ldots, k$, we have k more equations to satisfy and following J. L. Lagrange we introduce k additional variables $\lambda = (\lambda_1, \ldots, \lambda_k)$ called **Lagrange multipliers**, and then we can consider the new problem:

Find the extreme values for

$$F(x, \lambda) = f(x) + \sum_{l=1}^{k} \lambda_l g_l(x)$$

under the constraints

$$g_l(x) = 0, \qquad l = 1, \ldots, k.$$

Now if F has an extreme value x^0 under these constraints then f also has an extreme value at x^0. We now find the following necessary conditions

$$\begin{cases} \frac{\partial F}{\partial x_j}(x^0, \lambda^0) = \frac{\partial f}{\partial x_j}(x_0) + \sum_{l=1}^{k} \lambda_l \frac{\partial g_l}{\partial x_l}(x^0) = 0, & j = 1, \ldots, n \\ g_m(x^0) = 0, & m = 1, \ldots, k, \end{cases} \tag{11.1}$$

so we have $n + k$ equations for $n + k$ unknowns. The following result is known as **Lagrange's multiplier rule**. For its proof we follow H. Heuser [25], note that this is a proof where we need to distinguish more carefully between column and row vectors.

Theorem 11.3. *Let $G \subset \mathbb{R}^n$ be an open set and $f : G \to \mathbb{R}$ as well as $g = \begin{pmatrix} g_1 \\ \vdots \\ g_k \end{pmatrix} : G \to \mathbb{R}^k$, $k < n$, continuously differentiable mappings. Suppose that f has a local extreme value at $x^0 = (x_1^0, \ldots, x_n^0) \in G$ under the constraints $g_l(x) = 0$ for $l = 1, \ldots, k$ and $x \in G$. Suppose that the $k \times n$ matrix*

$$(d_x g)(x^0) = J_g(x^0) = \begin{pmatrix} \frac{\partial g_1}{\partial x_1}(x^0) & \cdots & \frac{\partial g_1}{\partial x_n}(x^0) \\ \vdots & & \vdots \\ \frac{\partial g_k}{\partial x_1}(x^0) & \cdots & \frac{\partial g_k}{\partial x_n}(x^0) \end{pmatrix} \tag{11.2}$$

has a $k \times k$ minor with non-vanishing determinant. Then there exist $\lambda_1, \ldots, \lambda_k \in \mathbb{R}$ such that

$$\mathrm{grad}\left(f + \sum_{l=1}^{k} \lambda_l g_l \right)(x^0) = 0 \tag{11.3}$$

holds, i.e.

$$
\begin{cases}
\frac{\partial f}{\partial x_1}(x^0) + \lambda_1 \frac{\partial g_1}{\partial x_1}(x^0) + \cdots + \lambda_k \frac{\partial g_k}{\partial x_1}(x^0) = 0 \\
\vdots \\
\frac{\partial f}{\partial x_n}(x^0) + \lambda_1 \frac{\partial g_1}{\partial x_n}(x^0) + \cdots + \lambda_k \frac{\partial g_k}{\partial x_n}(x^0) = 0.
\end{cases}
\tag{11.4}
$$

Proof. If necessary after a change of the enumeration of the variables we may assume

$$
\det
\begin{pmatrix}
\frac{\partial g_1}{\partial x_1}(x^0) & \cdots & \frac{\partial g_1}{\partial x_k}(x^0) \\
\vdots & & \vdots \\
\frac{\partial g_k}{\partial x_1}(x^0) & \cdots & \frac{\partial g_k}{\partial x_k}(x^0)
\end{pmatrix}
\neq 0.
\tag{11.5}
$$

Furthermore we introduce the following notation

$$
y = \begin{pmatrix} x_1 \\ \vdots \\ x_k \end{pmatrix}, \quad
z = \begin{pmatrix} x_{k+1} \\ \vdots \\ x_n \end{pmatrix}, \quad
\xi = \begin{pmatrix} x_1^0 \\ \vdots \\ x_n^0 \end{pmatrix}, \quad
\eta = \begin{pmatrix} x_1^0 \\ \vdots \\ x_k^0 \end{pmatrix}, \quad
\zeta = \begin{pmatrix} x_{k+1}^0 \\ \vdots \\ x_n^0 \end{pmatrix}
$$

and we now write $f(y,z)$ for $f(x)$, as well as $g_l(y,z)$ for $g_l(x)$ with the obvious meaning that $f(\xi) = f(\eta, \zeta)$ and $g_l(\xi) = g_l(\eta, \zeta)$. By definition

$$
\xi \in M := \{ x \in G \mid g(x) = 0 \}
$$

and in particular $g(\xi) = \begin{pmatrix} g_1(\xi) \\ \vdots \\ g_k(\xi) \end{pmatrix} = \begin{pmatrix} g_1(\eta, \zeta) \\ \vdots \\ g_k(\eta, \zeta) \end{pmatrix} = \begin{pmatrix} 0 \\ \vdots \\ 0 \end{pmatrix}$, and by (11.4) the

matrix

$$
\frac{\partial g}{\partial y}(\eta, \zeta) =
\begin{pmatrix}
\frac{\partial g_1}{\partial x_1}(\eta, \zeta) & \cdots & \frac{\partial g_1}{\partial x_k}(\eta, \zeta) \\
\vdots & & \vdots \\
\frac{\partial g_k}{\partial x_1}(\eta, \zeta) & \cdots & \frac{\partial g_k}{\partial x_k}(\eta, \zeta)
\end{pmatrix}
\tag{11.6}
$$

is invertible. Thus we are in a situation where we can apply the implicit function theorem, Theorem 10.5. Hence there exists an open neighbourhood $\tilde{U} \subset \mathbb{R}^{n-k}$ of ζ and a continuously differentiable function $h : \tilde{U} \to \mathbb{R}^k$,

$h = \begin{pmatrix} h_1 \\ \vdots \\ h_k \end{pmatrix}$, such that $h(\zeta) = \eta$ and $g(h(z), z) = 0$ for all $z \in \tilde{U}$. Furthermore,

if necessary on a smaller neighbourhood $U \subset \tilde{U}$ of ζ the function

$$
\varphi(z) := f(h(z), z)
\tag{11.7}
$$

is differentiable and for the partial derivatives at ζ we find for $k+1 \le m < n$

$$\frac{\partial \varphi}{\partial x_m}(\zeta) = \frac{\partial f}{\partial x_1}(\xi)\frac{\partial h_1}{\partial x_m}(\zeta) + \cdots + \frac{\partial f}{\partial x_k}(\xi)\frac{\partial h_k}{\partial x_m}(\zeta) + \frac{\partial f}{\partial x_m}(\xi). \qquad (11.8)$$

With $\frac{\partial f}{\partial y}(\xi) = \left(\frac{\partial f}{\partial x_1}(\xi), \cdots, \frac{\partial f}{\partial x_k}(\xi)\right)$ and $\frac{\partial f}{\partial z}(\xi) = \left(\frac{\partial f}{\partial x_{k+1}}(\xi), \cdots, \frac{\partial f}{\partial x_n}(\xi)\right)$ we can write (11.8) as

$$(\mathrm{d}\varphi)(\zeta) = \frac{\partial f}{\partial y}(\xi)(\mathrm{d}\,h)(\zeta) + \frac{\partial f}{\partial z}(\xi) \qquad (11.9)$$

or more explicitly (in order to check the structure)

$$\left(\frac{\partial \varphi}{\partial x_{k+1}}(\zeta), \cdots, \frac{\partial \varphi}{\partial x_n}(\zeta)\right) = \left(\frac{\partial f}{\partial x_1}(\xi), \cdots, \frac{\partial f}{\partial x_k}(\xi)\right) \begin{pmatrix} \frac{\partial h_1}{\partial x_{k+1}}(\zeta) & \cdots & \frac{\partial h_1}{\partial x_n}(\zeta) \\ \vdots & & \vdots \\ \frac{\partial h_k}{\partial x_{k+1}}(\zeta) & \cdots & \frac{\partial h_k}{\partial x_n}(\zeta) \end{pmatrix}$$

$$+ \left(\frac{\partial f}{\partial x_{k+1}}(\xi), \cdots, \frac{\partial f}{\partial x_n}(\xi)\right). \qquad (11.10)$$

Since f has a local extreme value at ξ under the constraint $g(x) = 0$, it follows that φ has a local extreme value at ζ (without any constraints), hence $(\mathrm{d}\varphi)(\zeta) = 0$, or with (11.9) or (11.10)

$$\frac{\partial f}{\partial y}(\xi)(\mathrm{d}\,h)(\zeta) + \frac{\partial f}{\partial z}(\xi) = 0. \qquad (11.11)$$

We now want to exploit the constraints $g_l(x) = 0$ which we can rewrite as

$$\psi_l(x_{k+1}, \ldots, x_n) \qquad (11.12)$$
$$:= g_l(h_1(x_{k+1}, \ldots, x_n), \ldots, h_k(x_{k+1}, \ldots, x_n), x_{k+1}, \ldots, x_n) = 0$$

for all $(x_{k+1}, \ldots, x_n) \in U$ and $1 \le l \le k$. As we have (11.9) we find now for $l + 1, \ldots, k$

$$(\mathrm{d}\psi_l)(\zeta) = \frac{\partial g_l}{\partial y}(\xi)(\mathrm{d}\,h)(\zeta) + \frac{\partial g_l}{\partial z}(\xi) \qquad (11.13)$$

with $\frac{\partial g_l}{\partial y} = \left(\frac{\partial g_l}{\partial x_1}, \cdots, \frac{\partial g_l}{\partial x_k}\right)$ and $\frac{\partial g_l}{\partial z} = \left(\frac{\partial g_l}{\partial x_{k+1}}, \cdots, \frac{\partial g_l}{\partial x_n}\right)$. Since $(\mathrm{d}\psi_l)(\zeta) = 0$ we arrive at

$$\frac{\partial g_l}{\partial y}(\xi)(\mathrm{d}\,h)(\zeta) + \frac{\partial g_l}{\partial z}(\xi) = 0 \qquad \text{for} \qquad l = 1, \ldots, k,$$

or using matrix notation

$$\frac{\partial g}{\partial y}(\xi)(\mathrm{d}h)(\zeta) + \frac{\partial g}{\partial z}(\xi) = 0. \tag{11.14}$$

The invertibility of $\frac{\partial g}{\partial y}(\xi)$ yields

$$(\mathrm{d}h)(\zeta) = -\left(\frac{\partial g}{\partial y}(\xi)\right)^{-1} \frac{\partial g}{\partial z}(\xi) \tag{11.15}$$

and with (11.11) we find

$$-\frac{\partial f}{\partial y}(\xi)\left(\frac{\partial g}{\partial y}(\xi)\right)^{-1}\frac{\partial g}{\partial z}(\xi) + \left(\frac{\partial f}{\partial z}\right)(\xi) = 0. \tag{11.16}$$

We now set

$$L := (\lambda_1, \ldots, \lambda_k) := -\frac{\partial f}{\partial y}(\xi)\left(\frac{\partial g}{\partial y}(\xi)\right)^{-1}$$

and it follows

$$L\frac{\partial g}{\partial y}(\xi) = -\frac{\partial f}{\partial y}(\xi)$$

or

$$\frac{\partial f}{\partial y}(\xi) + L\frac{\partial g}{\partial y}(\xi) = 0, \tag{11.17}$$

but on the other hand with (11.15) we find

$$\frac{\partial f}{\partial z}(\xi) + L\frac{\partial g}{\partial z}(\xi) = 0. \tag{11.18}$$

However, (11.17) and (11.18) are just (11.3). $\qquad\square$

Remark 11.4. This is a very challenging proof and therefore it may take a number of attempts to fully understand it. In particular it may be better to return to the proof once the reader is more familiar with higher dimensional analysis.

Now we want to study how to use Theorem 10.3 to find (local) extreme values under constraints. It is important to be aware that the system (11.4) (or (11.3)) together with the constraints $g_l(x^0) = 0$, $l = 1, \ldots, k$, gives necessary conditions. Once we have $(x_1^0, \ldots, x_n^0) \in G$ and $(\lambda_1, \ldots, \lambda_k) \in \mathbb{R}^k$ satisfying these equations, we still need to check whether we are indeed dealing with a local extreme value satisfying the constraints, and when we do we need to find out whether we have a maximum or a minimum. In some cases we can use our criterion involving the Hessian, but in general sufficient criteria are more involved and not always "practical useful". We refer to M. Moskowitz and F. Paliogiannis [38], Theorem 3.9.9 for one of such result.

The following remark might be further helpful to understand the idea behind Lagrange's approach. Introducing the function

$$G(x, \mu) = f(x) + \mu_1 g_1(x) + \cdots + \mu_k g_k(x) \tag{11.19}$$

we find that (11.4) together with the constraints $g_l(x^0) = 0$ are obtained by requiring at an extreme value for G

$$\frac{\partial G}{\partial x_j}(x^0, \lambda) = 0, \quad 1 \le j \le k, \quad \frac{\partial G}{\partial \mu_l}(x^0, \lambda) = g_l(x^0) = 0, \quad l = 1, \ldots, k,$$

but these are the necessary conditions for G having a local extreme value at $(x^0, \lambda) \in G \times \mathbb{R}^k$.

Instead of developing a more sophisticated theory, we prefer to discuss some examples.

Example 11.5. We want to find local extreme values for $f : \mathbb{R}^2 \to \mathbb{R}$, $f(x, y) = xy$, under the constraint $g(x, y) = x^2 + y^2 - 1 = 0$. Note that the condition $g(x, y) = 0$ means that (x, y) lies on the circle with centre $(0, 0)$ and radius 1. We define

$$G(\lambda, x, y) = xy + \lambda(x^2 + y^2 - 1)$$

and find the three equations

$$\frac{\partial G}{\partial x}(\lambda, x, y) = y + 2\lambda x = 0 \tag{11.20}$$

$$\frac{\partial G}{\partial y}(\lambda, x, y) = x + 2\lambda y = 0 \tag{11.21}$$

$$g(x, y) = x^2 + y^2 - 1 = 0 \tag{11.22}$$

as necessary conditions. From (11.20) and (11.21) we deduce

$$\lambda = -\frac{y}{2x} \quad \text{and} \quad \lambda = -\frac{x}{2y}$$

implying $y^2 = x^2$ and since by (11.22) we must have that $x^2 + y^2 = 1$ it follows that $2x^2 = 1$ or $x^2 = \frac{1}{2}$. This yields four combinations for critical values $\left(\frac{1}{2}\sqrt{2}, \frac{1}{2}\sqrt{2}\right)$, $\left(-\frac{1}{2}\sqrt{2}, -\frac{1}{2}\sqrt{2}\right)$, $\left(\frac{1}{2}\sqrt{2}, -\frac{1}{2}\sqrt{2}\right)$ and $\left(-\frac{1}{2}\sqrt{2}, \frac{1}{2}\sqrt{2}\right)$. Note that f has a certain symmetry: $f(x, y) = f(-x, -y)$ and $f(x, -y) = f(-x, y)$. Since $f\left(\frac{1}{2}\sqrt{2}, \frac{1}{2}\sqrt{2}\right) = \frac{1}{2} > 0$ and $f\left(\frac{1}{2}\sqrt{2}, -\frac{1}{2}\sqrt{2}\right) = -\frac{1}{2} < 0$ we expect a maximum at $\left(\frac{1}{2}\sqrt{2}, \frac{1}{2}\sqrt{2}\right)$ and $\left(-\frac{1}{2}\sqrt{2}, -\frac{1}{2}\sqrt{2}\right)$ and a minimum at $\left(\frac{1}{2}\sqrt{2}, -\frac{1}{2}\sqrt{2}\right)$ and $\left(-\frac{1}{2}\sqrt{2}, \frac{1}{2}\sqrt{2}\right)$. For $\left(\frac{1}{2}\sqrt{2}, \frac{1}{2}\sqrt{2}\right)$ we can locally solve the equation $x^2 + y^2 = 1$ to get $y = \sqrt{1 - x^2}$ and consequently

$$f(x, y(x)) = h(x) = x\sqrt{1 - x^2}.$$

Since

$$h'(x) = \frac{1 - 2x^2}{\sqrt{1 - x^2}}$$

and

$$h''(x) = \frac{x(-3 + 2x^2)}{(1 - x^2)^{\frac{3}{2}}}$$

we find that $h'\left(\frac{1}{2}\sqrt{2}\right) = 0$ and $h''\left(\frac{1}{2}\sqrt{2}\right) = -4$, implying that at $\left(\frac{1}{2}\sqrt{2}, \frac{1}{2}\sqrt{2}\right)$ the function $f|_{\{(x,y)\,|\,x^2+y^2=1\}}$ has a local maximum. Analogously we can prove that $f|_{\{(x,y)\,|\,x^2+y^2=1\}}$ has at $\left(-\frac{1}{2}\sqrt{2}, \frac{1}{2}\sqrt{2}\right)$ a local minimum.

Example 11.6. We want to determine the local extreme values of $f : \mathbb{R}^3 \to \mathbb{R}$, $f(x, y, z) = 5x + y - 3z$ under the constraints $g_1(x, y, z) = x + y + z = 0$ and $g_2(x, y, z) = x^2 + y^2 + z^2 - 8 = 0$, i.e. we search for local extreme values of f on the intersection of the plane $x + y + z = 0$ and the circle $x^2 + y^2 + z^2 = 8$. With $F(\lambda_1, \lambda_2, x, y, z) = 5x + y - 3z + \lambda_1(x + y + z) + \lambda_2(x^2 + y^2 + z^2 - 8)$ we find the following five equations forming necessary conditions

$$\frac{\partial F}{\partial x}(\lambda_1, \lambda_2, x, y, z) = 5 + \lambda_1 + 2\lambda_2 x = 0, \tag{11.23}$$

$$\frac{\partial F}{\partial y}(\lambda_1, \lambda_2, x, y, z) = 1 + \lambda_1 + 2\lambda_2 y = 0, \tag{11.24}$$

$$\frac{\partial F}{\partial z}(\lambda_1, \lambda_2, x, y, z) = -3 + \lambda_1 + 2\lambda_2 z = 0 \tag{11.25}$$

$$x + y + z = 0, \tag{11.26}$$

$$x^2 + y^2 + z^2 - 8 = 0. \tag{11.27}$$

Adding (11.23) - (11.25) we arrive at

$$3 + 3\lambda_1 + 2\lambda_2(x + y + z) = 0$$

which together with (11.26) yields $3 + 3\lambda_1 = 0$ or $\lambda_1 = -1$. With this λ_1 in (11.23) and (11.24) we find

$$4 + 2\lambda_2 x = 0 \qquad \text{and} \qquad 2\lambda_2 y = 0.$$

The first of these equations implies $\lambda_2 \neq 0$ and $x \neq 0$ which now implies $y = 0$. Thus (11.26) and (11.27) read as

$$x + z = 0 \qquad \text{and} \qquad x^2 + z^2 = 8,$$

i.e. $z = -x$ and $2x^2 = 8$ or $x^2 = 4$, hence $x_{1,2} = \pm 2$. Thus local extreme values might be at $(2, 0, 2)$, $(2, 0, -2)$, $(-2, 0, 2)$ and $(-2, 0, -2)$. However $(2, 0, 2)$ and $(-2, 0, -2)$ do not satisfy (11.26). Thus only $(2, 0, -2)$ and $(-2, 0, 2)$ with corresponding values $f(2, 0, -2) = 16$ and $f(-2, 0, 2) = -16$ are candidates with a possible maximum at $(2, 0, -2)$ and a possible minimum at $(-2, 0, 2)$. But we still need to decide whether these are indeed local extreme values. For this we need further tools which we will first develop before returning to this example.

There are situations where we have additional a priori information which allows us to decide whether or not a function $f : G \to \mathbb{R}$, $G \subset \mathbb{R}^n$, has a local extreme value under constraints $g_1(x) = 0, \ldots, g_m(x) = 0$. A typical situation is given by

Lemma 11.7. Let $f : G \to \mathbb{R}$ and $g_l : G \to \mathbb{R}$, $1 \leq l \leq m$, be continuously differentiable functions and suppose that $\xi_1, \ldots, \xi_k \in G$ are determined as solutions to (11.4) under the constraints $g_l(x) = 0$, $1 \leq l \leq m$, with corresponding Lagrange multipliers. Let $f(\xi) = \max_{1 \leq j \leq k} f(\xi_j)$ and $f(\eta) := \min_{1 \leq j \leq k} f(\xi_j)$. If $M := \{x \in G \mid g_l(x) = 0, l = 1, \ldots, m\} = \bigcap_{l=1}^{m} \{x \in G \mid g_l(x) = 0\}$ is compact then f has a global maximum at ξ and a global minimum at η.

Proof. Since M is by assumption compact and f is continuous, the function $f|_M$ is a continuous function on a compact set and according to Theorem 3.27 it attains a global maximum and a global minimum which must be also a local maximum and local minimum, respectively. Hence the global maximum of $f|_M$ must be attained at ξ and the global minimum must be attained at η. $\square$

Now we can complete Example 11.6.

Example 11.8. (Example 11.6 continued) The constraints are given by the sets $\{(x,y,z) \in \mathbb{R}^3 \mid x^2 + y^2 + z^2 = 8\} \cap \{(x,y,z) \in \mathbb{R}^3 \mid x+y+z = 0\}$. The set $\{(x,y,z \in \mathbb{R}^3 \mid x^2 + y^2 + z^2 = 8\}$ is the circle with centre $(0,0,0)$ and radius $\sqrt{8}$, and hence it is compact. The second set is closed as it is the pre-image of the closed set $0 \in \mathbb{R}$ under the continuous mapping $g_1 : \mathbb{R}^3 \to \mathbb{R}$, $g_1(x,y,z) = x+y+z$. Since the intersection of a compact set with a closed set is compact, we are in the situation of Lemma 11.7, and hence f has a maximum at $(2,0,-2)$ and a minimum at $(-2,0,2)$.

A further typical problem is to determine the distance between two sets, for example to find a point on the graph of a function which has the smallest distance to a given set. In Example 3.28 we have already discussed the distance between two sets A, $K \subset X$ of a metric space (X,d) which is defined by

$$\mathrm{dist}(K,A) := \inf\{d(x,y) \mid x \in K, y \in A\}. \tag{11.28}$$

In particular we know that if A is closed, K compact and $A \cap K \neq \emptyset$ then $\mathrm{dist}(K,A) > 0$. We now claim

Lemma 11.9. Let $\emptyset \neq A \subset \mathbb{R}^n$ be closed and $\emptyset \neq K \subset \mathbb{R}^n$ be compact. Then there exist $x_0 \in K$ and $y_0 \in A$ such that

$$\|x_0 - y_0\| \leq \|x - y\| \qquad \text{for all} \quad x \in K \quad \text{and} \quad y \in A. \tag{11.29}$$

Proof. For $\delta := \mathrm{dist}(K,A)$ there exist sequences $(x_n)_{n \in \mathbb{N}}$, $x_n \in K$, and $(y_n)_{n \in \mathbb{N}}$, $y_n \in A$, such that $\lim_{n \to \infty} \|x_n - y_n\| = \delta$. Since K is compact $(x_n)_{n \in \mathbb{N}}$ must have a converging subsequence $(x_{n_l})_{l \in \mathbb{N}}$ with some limit x_0. The corresponding sequence $(y_n)_{n \in \mathbb{N}}$ is denoted by $(y_{n_l})_{l \in \mathbb{N}}$. We observe that

$$\|y_{n_l}\| \leq \|y_{n_l} - x_{n_l}\| + \|x_{n_l}\| \leq M$$

since $(y_{n_l} - x_{n_l})_{l \in \mathbb{N}}$ and $(x_{n_l})_{l \in \mathbb{N}}$ are convergent, hence bounded. By the Bolzano-Weierstrass theorem the sequence $(y_{n_l})_{l \in \mathbb{N}}$ has a convergent subsequence denoted by $(y'_k)_{k \in \mathbb{N}}$ with limit denoted by y_0 and we write $(x'_k)_{k \in \mathbb{N}}$ for the corresponding subsequence of $(x_{n_l})_{l \in \mathbb{N}}$ and this subsequence converges of course again to x_0. Thus we find $\lim_{k \to \infty} \|x'_k - y'_k\| = \delta$ as well as $\lim_{k \to \infty} \|x'_k - y'_k\| = \|x_0 - y_0\|$. The definition of δ now implies

$$\|x_0 - y_0\| \leq \|x - y\| \qquad \text{for all} \quad x \in K \quad \text{and} \quad y \in A.$$

$\square$

Example 11.10. Find the point in the plane $z = x + y$ which has smallest distance to the point $(\alpha, \beta, \gamma) \in \mathbb{R}^3$. The distance from (α, β, γ) to a point (x, y, z) is of course given by

$$\left((x - \alpha)^2 + (y - \beta)^2 + (z - \gamma)^2\right)^{\frac{1}{2}}$$

which attains its minimum (under constraints) as this term is bounded from below by 0. Moreover the minimum of $((x - \alpha)^2 + (y - \beta)^2 + (z - \gamma)^2)^{\frac{1}{2}}$ is the same as the minimum of $(x - \alpha)^2 + (y - \beta)^2 + (z - \gamma)^2$. Thus we have to find the minimum of

$$f(x, y, z) = (x - \alpha)^2 + (y - \beta)^2 + (z - \gamma)^2 \tag{11.30}$$

under the constraint

$$g(x, y, z) = x + y - z = 0. \tag{11.31}$$

Clearly, if $\alpha + \beta - \gamma = 0$ then the solution is (α, β, γ) itself. In general we find using Theorem 11.3 the necessary conditions

$$2(x - \alpha) + \lambda = 0, \tag{11.32}$$
$$2(y - \beta) + \lambda = 0, \tag{11.33}$$
$$2(z - \gamma) + \lambda = 0, \tag{11.34}$$
$$x + y - z = 0. \tag{11.35}$$

Subtracting (11.33) from (11.32) we find

$$x = y + \alpha - \beta,$$

adding (11.34) to (11.33) we get

$$y = -z + \beta + \gamma$$

which yields in (11.35)

$$z = \frac{2\gamma + \alpha + \beta}{3}$$

implying

$$y = \frac{2\beta - \alpha + \gamma}{3}, \qquad x = \frac{2\alpha - \beta + \gamma}{3}$$

and $\lambda = 2\left(\frac{\alpha + \beta - \gamma}{3}\right)$, but λ is not needed anymore.

Thus $\left(\frac{2\alpha-\beta+\gamma}{3}, \frac{2\beta-\alpha+\gamma}{3}, \frac{2\gamma+\alpha+\beta}{3}\right)$ is the only candidate for an extreme value. Further by Lemma 11.9 we know that there must be an extreme value, in fact a minimum, and this is obtained at that point. For the distance we find $\frac{|\gamma-\alpha-\beta|}{\sqrt{3}}$, in particular for $\gamma = \alpha + \beta$ the distance is as expected 0.

We want to discuss two further, maybe more theoretical applications showing how we can use the Lagrange multiplier method to prove theorems.

Theorem 11.11 (Hadamard's determinant estimate). *For n vectors*

$$x_1 = \begin{pmatrix} x_{11} \\ \vdots \\ x_{n1} \end{pmatrix}, \cdots, x_n = \begin{pmatrix} x_{1n} \\ \vdots \\ x_{nn} \end{pmatrix}$$

in $\mathbb{R}^n$ and the corresponding matrix

$$X := (x_1, \ldots, x_n) = \begin{pmatrix} x_{11} & \cdots & x_{1n} \\ \vdots & & \vdots \\ x_{n1} & \cdots & x_{nn} \end{pmatrix}$$

we have

$$|\det(X)| \leq \prod_{j=1}^{n} \|x_j\|. \tag{11.36}$$

In addition, if $x_j \neq 0$ for $1 \leq j \leq n$, then equality holds in (11.36) if and only if the vectors $x_1, \ldots, x_n$ are mutually orthogonal, i.e. $\langle x_j, x_k \rangle = 0$ for $j \neq k$.

Proof. We follow the sketch in S. Hildebrandt [26] and prove the result in several steps. We include the result here not only because it is interesting, but also to get the reader used to the fact that proofs can be very long and involved.

1. Define

$$M := \left\{ X \in GL(n; \mathbb{R}) \mid \|x_j\|^2 = 1, j = 1, \ldots, n \right\}.$$

Suppose that $|\det X| \leq 1$ for all $X \in M$ and that equality holds if and only if $X \in O(n)$. We then claim that (11.36) follows. Indeed, take any

$X \in M(n; \mathbb{R})$. If $\|x_j\| = 0$ for some j, $1 \le j \le n$, then $\det X = 0$ and the result is trivial. So we assume $\|x_j\| \ne 0$ for all $j = 1, \ldots, n$ and consider

$$\tilde{X} := \left(\frac{x_1}{\|x_1\|}, \ldots, \frac{x_n}{\|x_n\|} \right).$$

Again, if $\tilde{X} \notin GL(n; \mathbb{R})$ then $\det \tilde{X} = 0$ implying that

$$0 = \frac{1}{\prod_{j=1}^{n} \|x_j\|} \det(x_1, \ldots, x_n),$$

i.e. $\det X = 0$ and again (11.36) is trivial. If however $\tilde{X} \in GL(n; \mathbb{R})$ then $\tilde{X} \in M$ and by our assumption $|\det \tilde{X}| \le 1$ holds, but

$$\det \tilde{X} = \det \left(\frac{x_1}{\|x_1\|}, \cdots, \frac{x_n}{\|x_n\|} \right) = \frac{1}{\prod_{j=1}^{n} \|x_j\|} \det(x_1, \ldots, x_n)$$

$$= \frac{1}{\prod_{j=1}^{n} \|x_j\|} \det X$$

implying (11.36). Suppose that in (11.36) holds equality, then we must have $|\det \tilde{X}| = \left| \det \left(\frac{x_1}{\|x_1\|}, \ldots, \frac{x_n}{\|x_n\|} \right) \right| = 1$, hence $\tilde{X} \in O(n)$ by our assumptions. This implies $\left\langle \frac{x_j}{\|x_j\|}, \frac{x_k}{\|x_k\|} \right\rangle = \delta_{jk}$ and therefore $\langle x_j, x_k \rangle = 0$ if $j \ne k$.

2. We want to study properties of M, note first

$$M \subset M(n; \mathbb{R}) \cong \mathbb{R}^{n^2}.$$

Further, M is a set bounded in $\mathbb{R}^{n^2}$ since for $X \in M$

$$\|X\| = \sum_{k,l=1}^{n} x_{kl}^2 \le n.$$

However M is not closed, hence not compact in $\mathbb{R}^{n^2}$. To see this take a sequence $(a_k)_{k \in \mathbb{N}}$, $0 < a_k < 1$ such that $\lim_{k \to \infty} a_k = 1$ and consider

$$A_k = \begin{pmatrix} (1-a_k)^{\frac{1}{2}} & 0 & 0 & \cdots & 0 \\ a_k^{\frac{1}{2}} & 1 & 0 & \cdots & 0 \\ 0 & 0 & 1 & \cdots & 0 \\ \vdots & \vdots & \vdots & \ddots & \vdots \\ 0 & 0 & 0 & \cdots & 1 \end{pmatrix} = (x, e_2, e_3, \ldots, e_n).$$

218

Clearly $\|e_j\|^2 = 1$ and further $\|x\|^2 = 1 - a_k + a_k = 1$. Moreover we find $\det A_k = (1 - a_k)^{\frac{1}{2}} \neq 0$, so $A_k \in GL(n; \mathbb{R})$. But for $k \to \infty$ we find

$$\lim_{k \to \infty} A_k = \begin{pmatrix} 0 & 0 & 0 & \cdots & 0 \\ 1 & 1 & 0 & \cdots & 0 \\ 0 & 0 & 1 & \cdots & 0 \\ \vdots & \vdots & \vdots & \ddots & \vdots \\ 0 & 0 & 0 & \cdots & 1 \end{pmatrix} \notin GL(n; \mathbb{R}).$$

However $\overline{M}$ is compact and $O(n) \subset M$.

3. This part of the proof may help in getting used to the fact that "points" in an underlying domain of a function might be rather complex objects, for example matrices. Consider $\det : M \to \mathbb{R}$. Since $\overline{M}$ is compact and $\det : M \to \mathbb{R}$ is continuous there exists $X_0 = (x_1^0, \ldots, x_n^0) \in \overline{M}$ such that

$$m := \sup \left\{ \det X \mid X \in \overline{M} \right\} = \det X_0.$$

Note that for $X \in SO(n)$ we have $\det X = 1$, so $m \geq 1$ and therefore $\det X_0 > 0$ implying $X_0 \in GL(n; \mathbb{R})$, thus $X_0 \in M$. By Theorem 11.3 there exists n Lagrange multipliers $\lambda_1, \ldots, \lambda_n \in \mathbb{R}$ such that X_0 is a critical point for

$$F(X) := \det X + \lambda_1 \|x_1\|^2 + \cdots + \lambda_n \|x_n\|^2.$$

This F and the multipliers $\lambda_1, \ldots, \lambda_n$ refer to the problem: find the supremum of $\det X$, $X \in GL(n; \mathbb{R})$, subjected to the constraints $g_l(X) = g_l(x_1, \ldots, x_n) = \|x_l\|^2 = 1$, $l = 1, \ldots, n$, recall $x_j = \begin{pmatrix} x_{1j} \\ \vdots \\ x_{nj} \end{pmatrix} \in \mathbb{R}^n$. Now

we have for $F(X_0)$ the n^2 equations (11.4). To get an insight into this we note that

$$\det X = \det(x_1, \ldots, x_n)$$
$$= \sum_{l=1}^{n} (-1)^{l+j} x_{lj} \det X_{(lj)}$$

with

219

$$M_{lj} = \begin{pmatrix} x_{11} & \cdots & x_{1j} & \cdots & x_{1n} \\ & \vdots & & & \vdots \\ \overline{x_{l1}} & \cdots & \overline{x_{lj}} & \cdots & \overline{x_{ln}} \\ & \vdots & & & \vdots \\ x_{n1} & \cdots & x_{nj} & \cdots & x_{nn} \end{pmatrix}$$

being the minor corresponding to position lj. Thus we find

$$\frac{\partial}{\partial x_{kj}} \det X = \frac{\partial}{\partial x_{kj}} \sum_{l=1}^{n} (-1)^{l+j} x_{lj} \det X_{(lj)}$$

$$= (-1)^{k+j} 1 \det X_{(kj)}$$

$$= \sum_{l=1}^{n} (-1)^{l+j} (e_k)_l \det X_{lj}$$

$$= \det (x_1, \ldots, x_{j-1}, e_k, x_{j+1}, \ldots, x_n),$$

where e_k is as usual the kth unit vector in $\mathbb{R}^n$. On the other hand we have

$$\frac{\partial}{\partial x_{kj}} \left(\lambda_1 \|x_1\|^2 + \cdots + \lambda_n \|x_n\|^2 \right) = 2\lambda_j x_{kj} = 2\lambda_j \langle e_k, x_j \rangle.$$

If we write for $1 \leq j \leq n$

$$\nabla_{X_j} F(X) = \left(\frac{\partial}{\partial x_{1j}} F(X), \ldots, \frac{\partial}{\partial x_{nj}} F(X) \right)$$

we obtain (11.3) in the form

$$\nabla_{X_j} F(X_0) = 0 \qquad \text{for} \quad j = 1, \ldots, n$$

which stands, j being fixed, for the n equations

$$\det(x_1^0, \ldots, x_{j-1}^0, e_k, x_{j+1}^0, \ldots, x_n^0) + 2\lambda_j \langle e_k, x_j^0 \rangle = 0 \qquad (11.37)$$

and $k = 1, \ldots, n$. With $y := \sum_{l=1}^{n} y_l e_l$ we get by linearity

$$\det(x_1^0, \ldots, x_{j-1}^0, y, x_{j+1}^0, \ldots, x_n^0) + 2\lambda_j \langle y, x_j^0 \rangle = 0 \quad j = 1, \ldots, n. \quad (11.38)$$

Taking $y = x_j^0$ in the jth equation of (11.38) we find using $\|x_j^0\| = 1$ that

$$\det X_0 + 2\lambda_j \langle x_j^0, x_j^0 \rangle = \det X_0 + 2\lambda_j,$$

or

$$\lambda_1 = \lambda_2 = \cdots = \lambda_n = -\frac{1}{2} \det X_0 = -\frac{1}{2}m \neq 0.$$

This now leads to

$$\langle y, x_j^0 \rangle = \frac{1}{m} \det(x_1^0, \ldots, x_{j-1}^0, y, x_{j+1}^0, \ldots, x_n^0)$$

for $1 \leq j \leq n$, and for $y = x_k^0$, $k \neq j$, it follows that

$$\det(x_1^0, \ldots, x_{j-1}^0, x_k^0, x_{j+1}^0, \ldots, x_n^0) = 0$$

and therefore

$$\langle x_k^0, x_j^0 \rangle = 0 \qquad \text{for} \quad k \neq j,$$

and since $\langle x_j^0, x_j^0 \rangle = 1$ we eventually find that

$$\langle x_k^0, x_j^0 \rangle = \delta_{kj}$$

i.e. $X_0 \in O(n)$, but $\det X_0 = m \geq 1$, so $X_0 \in SO(n)$. Thus the maximiser of $\det X$ on M is in $SO(n)$. Analogously, for the minimiser X^1 we find $X^1 \in O(n) \backslash SO(n)$, and therefore

$$|\det X| \leq 1 \qquad \text{for} \quad X \in M,$$

and equality can occur only for $X \in O(n)$. $\qquad\qquad\qquad\qquad\qquad$ $\square$

Remark 11.12. Let $x_1, \ldots, x_n \in \mathbb{R}^n$ be independent and let $W(x_1, \ldots, x_n)$ be the parallelotope spanned by $\{x_1, \ldots, x_n\}$. The volume of $W(x_1, \ldots, x_n)$ is given by $\lambda^{(n)}(W(x_1, \ldots, x_n)) = (\det(x_1, \ldots, x_n))$ and therefore Hadamard's determinant estimates state that among all parallelotopes with corresponding sides of the same length the hyper-rectangle has the largest volume. For $n = 2$ this is the statement that among all parallelograms with corresponding sides of the same length the rectangle has the largest area.

In the problems at the end of the chapter we will discuss some further applications of the Lagrange multiplier theorem.

We now want to turn to some further geometric applications of partial derivatives. These are of interest on their own, but most of all they also have many implications in the theory of partial differential equations, and thus to mathematical physics, for example as Huygens' principle in the theory of wave propagation. We restrict ourselves here to the case of plane curves and the case of surfaces in $\mathbb{R}^3$. Our starting point are two examples.

Example 11.13. Consider the family of circles

$$C_r := \left\{ (x, y) \in \mathbb{R}^2 \, \middle| \, (x - 2r)^2 + y^2 = r^2 \right\}, \qquad r \in \mathbb{R}.$$

For $r \neq 0$ the circle C_r has centre $(2r, 0)$ and radius r, for $r = 0$ we deal with the degenerate case of one point, namely the origin, see Figure 11.2

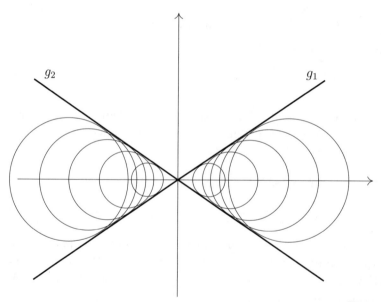

Figure 11.2

We are interested in the lines g_1 and g_2 which are in the some sense generated by the family C_r. A closer investigation reveals that each point (x_0, y_0) on g_1 or g_2 has the following two properties: each belongs to some circle C_{r_0}; and is tangent to this circle at (x_0, y_0).

Example 11.14. In the $x - y$ plane we consider all line segments having one end point in the set $[0, 1] \times \{0\}$ and the other in the set $\{0\} \times [0, 1]$, i.e. the end points belong to the unit interval on the x- axis and the y-axis, respectively, and each line segment has length 1, see Figure 11.3.

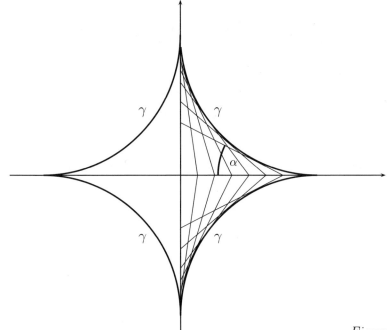

Figure 11.3

Again we conjecture that by this construction a curve γ is defined which seems to have four singular points at $(1, 0)$, $(0, 1)$, $(-1, 0)$ and $(0, -1)$. Every point on this curve except the four singular points belongs to one line segment (interpreted as a parametric curve) and at the common point (x_0, y_0) the line segment belongs to the tangent line of this new curve at (x_0, y_0). We may ask whether we can find the "equation" of this curve called the **astroid curve**. To answer this question we need some preparation.

We want to discuss the following: given a plane C^1-curve $\gamma : I \to \mathbb{R}^2$, $t \mapsto \gamma(t) = (\gamma_1(t), \gamma_2(t)) = (x(t), y(t))$. Suppose that its trace $\mathrm{tr}(\gamma)$ coincides with the set $\{(x, y) \in \mathbb{R}^2, \,|\, f(x, y) = 0\}$ for some suitable function $f : G \to \mathbb{R}^2$. Thus $f(\gamma_1(t), \gamma_2(t)) = 0$ for all $t \in I$. We ask whether we can find the tangent line and the normal line to γ at t_0 as the graph of straight lines given as functions $y = y(x)$ (or $x = x(y)$). Recall (compare Definition 7.14 and Definition 7.19) that the tangent line and normal line to γ at t_0 are given by

$$g_{t_0}(s) = \vec{t}(t_0)s + \gamma(t_0) = \begin{pmatrix} \dot{\gamma}_1(t_0) \\ \dot{\gamma}_2(t_0) \end{pmatrix} s + \begin{pmatrix} \gamma_1(t_0) \\ \gamma_2(t_0) \end{pmatrix} \tag{11.39}$$

and

$$n_{t_0}(s) = \vec{n}(t_0)s + \gamma(t_0) = \begin{pmatrix} -\dot{\gamma}_2(t_0) \\ \dot{\gamma}_1(t_0) \end{pmatrix} s + \begin{pmatrix} \gamma_1(t_0) \\ \gamma_2(t_0) \end{pmatrix}. \tag{11.40}$$

The crucial relation is

$$\langle \vec{t}(t_0), \vec{n}(t_0) \rangle = 0,$$

i.e. the orthogonality of the tangent and the normal vectors. In our situation we have the information that $f(\gamma_1(t), \gamma_2(t)) = 0$ for all $t \in I$ implying

$$0 = \frac{d}{dt} f(\gamma_1(t), \gamma_2(t)) = \dot{\gamma}_1(t) f_x(\gamma_1(t), \gamma_2(t)) + \dot{\gamma}_2(t) f_y(\gamma_1(t), \gamma_2(t))$$

with f_x and f_y denoting $\frac{\partial f}{\partial x}$ and $\frac{\partial f}{\partial y}$, respectively. The normal vector at t_0 is given by $\begin{pmatrix} f_x(\gamma_1(t_0), \gamma_2(t_0)) \\ f_y(\gamma_1(t_0), \gamma_2(t_0)) \end{pmatrix}$ and consequently the tangent vector is given by $\begin{pmatrix} -f_y(\gamma_1(t_0), \gamma_2(t_0)) \\ f_x(\gamma_1(t_0), \gamma_2(t_0)) \end{pmatrix}$ or any non-zero scalar multiplier thereof. Thus an equation for the tangent line and normal line, respectively, to $\text{tr}(\gamma)$ at t_0, $\gamma(t_0) = (x_0, y_0)$, are

$$f_x(x_0, y_0)(x - x_0) + f_y(x_0, y_0)(y - y_0) = 0 \tag{11.41}$$

and

$$f_x(x_0, y_0)(y - y_0) - f_y(x_0, y_0)(x - x_0) = 0. \tag{11.42}$$

We now consider a family M_φ of curves $\gamma : I \to \mathbb{R}^2$ in the (x, y)-plane each given by the equation

$$\varphi(x, y, c) = 0 \tag{11.43}$$

with a C^1 function $\varphi : G \times I \to \mathbb{R}, G \subset \mathbb{R}^2$ open. The traces of these curves form a set $M \subset \mathbb{R}^2$. We seek a (parametric) curve $\eta : I \to \mathbb{R}^2$ such that for every $c \in I$, $\eta(c) \in M_\varphi$, i.e. $\eta(c) = \gamma_c(t_c)$ for some $\gamma_c \in M_\varphi, t_c \in I$, and $\dot{\eta}(c) = \dot{\gamma}_c(t_c)$. Note that for $c_1 \neq c_2$ in general γ_{c_1} and γ_{c_2} will be different curves. If η exists then for all $c \in I$ we have

$$\varphi(\eta_1(c), \eta_2(c), c) = 0, \ c \in I \tag{11.44}$$

224

and

$$\frac{\partial \varphi}{\partial x}(\eta_1(c), \eta_2(c), c)\dot{\eta}_1(c) + \frac{\partial \varphi}{\partial y}(\eta_1(c), \eta_2(c), c)\dot{\eta}_2(c) = 0 \qquad (11.45)$$

where the first equation expresses the fact that $\eta(c) = \gamma_c(t_c)$, while the second equation states that the tangent vector $\binom{\dot{\eta}_1(c)}{\dot{\eta}_2(c)}$ to η at c is orthogonal to the normal vector to γ_c at t_c, i.e. $\binom{\dot{\eta}_1(c)}{\dot{\eta}_2(c)}$ is parallel to $\binom{-\varphi_y(\eta_1(c), \eta_2(c))}{\varphi_x(\eta_1(c), \eta_2(c))} = \binom{-\varphi_y(\gamma_{c,1}(t_c), \gamma_{c,2}(t_c))}{\varphi_x(\gamma_{c,1}(t_c), \gamma_{c,2}(t_c))}$, the tangent vector of γ_c at t_c. Differentiating (11.44) we obtain

$$0 = \frac{d}{dc}\varphi(\eta_1(c), \eta_2(c), c) = \frac{\partial \varphi}{\partial x}(\eta_1(c), \eta_2(c), c)\dot{\eta}_1(c) + \frac{\partial \varphi}{\partial y}(\eta_1(c), \eta_2(c), c)\dot{\gamma}_2(c)$$

$$+ \frac{\partial \varphi}{\partial c}(\eta_1(c), \eta_2(c), c), \qquad (11.46)$$

which implies by (11.45) that

$$\frac{\partial \varphi}{\partial c}(\eta_1(c), \eta_2(c), c) = 0, c \in I. \qquad (11.47)$$

Conversely, (11.45) and (11.47) give (11.46). But (11.46) yields

$$\varphi(\eta_1(c), \eta_2(c), c) = \text{constant},$$

and hence if (11.44) holds for one $c_0 \in I$ then it holds for all $c \in I$.

Definition 11.15. *Let M_φ be a family of plane curves defined by the equation $\varphi(x, y, c) = 0$ where $\varphi : G \times I \to \mathbb{R}, G \subset \mathbb{R}^2$ open and $I \subset \mathbb{R}$ an interval, is a C^1-mapping. The **envelope** $\mathcal{E}$ of the family M_φ is the set*

$$\mathcal{E} := \left\{ (x, y) \in G \mid \varphi(x, y, c) = 0 \text{ and } \frac{\partial \varphi}{\partial c}(x, y, c) = 0 \text{ for all } c \in I \right\}. \qquad (11.48)$$

The envelope of M_φ is by definition a point set in $G \subset \mathbb{R}^2$. We may ask whether $\mathcal{E}$ is at least locally the trace of a parametric curve.

A short inspection shows that we are exactly in the situation of the implicit function theorem, Theorem 10.5. We must choose $U_2 = G$, $U_1 = I$ (changing the order $I \times G$ to $G \times I$ does of course not matter) and $f = \binom{\varphi}{\varphi_c} : G \times I \to \mathbb{R}^2$. Thus in order to apply Theorem 10.5 we need to add the condition

$$\det \begin{pmatrix} \frac{\partial \varphi}{\partial x} & \frac{\partial \varphi}{\partial y} \\ \frac{\partial \varphi_c}{\partial x} & \frac{\partial \varphi_c}{\partial y} \end{pmatrix} (x_0, y_0, c_0) \neq 0. \qquad (11.49)$$

Theorem 11.16. *Let $\varphi : G \times I \to \mathbb{R}$ be a C^1-function and suppose that $\varphi_c :$ $G \times I \to \mathbb{R}$ is a C^1-function too, where $G \subset \mathbb{R}^2$ is open and $I \subset \mathbb{R}$ is an open interval. Further we assume that (11.49) holds for some $(x_0, y_0, c_0) \in G \times I$. Then there exists a mapping $\gamma : I_1 \to G$, $c_0 \in I_1 \subset I$ an open interval, such that for $c \in I_1$*

$$\varphi(\gamma_1(c), \gamma_2(c), c) = 0 \tag{11.50}$$

holds. If in addition

$$\varphi_{cc}(x_0, y_0, c_0) \neq 0 \tag{11.51}$$

then there exists an open interval $I_2 \subset I_1$, $c_0 \in I_2$, such that $\dot{\gamma}(c) \neq 0$ for $c \in I_2$, i.e. $\gamma : I_2 \to G$ is a regular parametric curve representing in a neighbourhood of c_0 the envelope of the family of curves given by the equation $\varphi(x, y, c) = 0$, $c \in I$.

Proof. The main statement follows as already explained from the implicit function theorem, Theorem 10.5. Hence we know that in some open interval $I_1 \subset I$, $c_0 \in I_1$,

$$\varphi(\gamma_1(c), \gamma_2(c), c) = 0 \tag{11.52}$$

and

$$\varphi_c(\gamma_1(c), \gamma_2(c), c) = 0 \tag{11.53}$$

holds. Differentiating both equations with respect to c yields

$$\varphi_x(\gamma_1(c), \gamma_2(c), c)\dot{\gamma}_1(c) + \varphi_y(\gamma_1(c), \gamma_2(c), c)\dot{\gamma}_2(c) + \varphi_c(\gamma_1(c), \gamma_2(c), c) = 0$$

and with (11.53)

$$\varphi_x(\gamma_1(c), \gamma_2(c), c)\dot{\gamma}_1(c) + \varphi_y(\gamma_1(c), \gamma_2(c), c)\dot{\gamma}_2(c) = 0 \tag{11.54}$$

and

$$\varphi_{cx}(\gamma_1(c), \gamma_2(c), c)\dot{\gamma}_1(c) + \varphi_{cy}(\gamma_1(c), \gamma_2(c), c)\dot{\gamma}_2(c) = -\varphi_{cc}(\gamma_1(c), \gamma_2(c), c). \tag{11.55}$$

We read (11.54) and (11.55) as a linear system for the two unknowns $\dot{\gamma}_1(c)$ and $\dot{\gamma}_2(c)$. By (11.49) and (11.51) we know that for c_0 this system has a (unique) non-trivial solution $\dot{\gamma}_1(c_0)$, $\dot{\gamma}_2(c_0)$, $\dot{\gamma}_1^2(c_0) + \dot{\gamma}_2^2(c_0) \neq 0$. By continuity we conclude the existence of an open interval $I_2 \subset I_1$, $c_0 \in I_2$ such that $\gamma : I_2 \to G$ is a regular curve. $\qquad\square$

While this result gives a satisfactory solution to the problem of the local representation of envelopes as parametric curves, it does not help much to calculate the envelope or to find a (globally) representing curve. However in some cases we may find the envelope explicitly by direct investigations.

Example 11.17. (Example 11.14 continued) With the notation in Figure 11.3 we find that a line segment in the family of curves generating the astroid curve belongs to a straight line

$$\frac{x}{\cos\alpha} + \frac{y}{\sin\alpha} - 1 = 0, \ \alpha \in [0, 2\pi],$$

which follows from the geometric definition of cos and sin and the fact that the corresponding triangle with right angle at the origin has hypotenuse of length 1. We take the angle α as a parameter, hence

$$\varphi(x, y, \alpha) = \frac{x}{\cos\alpha} + \frac{y}{\sin\alpha} - 1 = 0$$

and

$$\varphi_\alpha(x, y, \alpha) = \frac{x\sin\alpha}{\cos^2\alpha} - \frac{y\cos\alpha}{\sin^2\alpha} = 0$$

which leads to $x = \cos^3\alpha$ and $y = \sin^3\alpha$, and since $\cos^2\alpha + \sin^2\alpha = 1$ we find further

$$x^{\frac{2}{3}} + y^{\frac{2}{3}} = 1$$

for $0 \le x, y \le 1$ and by symmetry we find the other branches. As we are dealing with a parametric curve, to recover the full astroid curve we may take $\gamma : [0, 2\pi] \to \mathbb{R}^2$, $\gamma(\alpha) = \begin{pmatrix} \cos^3\alpha \\ \sin^3\alpha \end{pmatrix}$.

A short inspection of our treatment of envelopes of curves reveals that we can extend the considerations to surfaces.

So far we have encountered parametric surfaces, see Definition 8.4. In some cases we can represent a parametric surface $f : G_0 \to \mathbb{R}^3$, $G_0 \subset \mathbb{R}^2$, locally as the solution set of an equation

$$g(x, y, z) = 0 \tag{11.56}$$

with a suitable mapping $g : G \to \mathbb{R}$. For example for the sphere with centre $0 \in \mathbb{R}^3$ and radius $r > 0$ we have $g(x, y, z) = x^2 + y^2 + z^2 - r^2$, further if

$h : H \to \mathbb{R}$, $H \subset \mathbb{R}^2$, is a function then

$$\Gamma(h) := \{(x, y, z) \in H \times \mathbb{R} \mid z = h(x, y)\}$$
$$= \{(x, y, z) \in H \times \mathbb{R} \mid g(x, y, z) = h(x, y) - z = 0\}.$$

Thus we may look more generally at equations such as (11.56) and ask when its solution set is locally a (parametric) surface. If for example we find that for all $(x_0, y_0, z_0) \in \{(x, y, z) \in G \mid g(x, y, z) = 0\}$ we have $(\operatorname{grad} g)(x_0, y_0, z_0) \neq 0$, then the implicit function theorem allows us to locally represent $\{(x, y, z) \in G \mid g(x, y, z) = 0\}$ as the graph of functions of two variables, see Chapter 10. Thus, as a next step we may consider families of surfaces given by equations and indexed by a parameter.

Let $I \subset \mathbb{R}$ be an interval and for $c \in I$ let a surface S_c in $\mathbb{R}^3$ be given by $S_c = \{(x, y, z) \in \mathbb{R}^3 \mid f(x, y, z, c) = 0\}$ where $f : \mathbb{R}^3 \times I \to \mathbb{R}$ is a C^2-function. We define the **envelope** $\mathcal{E}$ of the family $(S_c)_{c \in I}$ as

$$\mathcal{E} := \{(x, y, z) \in \mathbb{R}^3 \mid f(x, y, z, c) = 0 \text{ and } f_c(x, y, z, c) = 0 \text{ for some } c \in I\}.$$
$$(11.57)$$

Once more the implicit function theorem shows that if

$$f_{cc}(x_0, y_0, z_0, c_0) \neq 0$$

for some $(x_0, y_0, z_0, c_0) \in \mathbb{R}^3 \times \overset{\circ}{I}$, then in a neighbourhood of (x_0, y_0, z_0) we find $c = \sigma(x, y, z)$ and the envelope is locally given as a solution set of

$$f(x, y, z, \sigma(x, y, z)) = 0.$$

A more detailed discussion of envelopes is given in S. Hildebrandt [26]. We also refer to the classical course of R. Courant [8].

Example 11.18. For $c \in \mathbb{R}$ consider the family of spheres with centres $(c, 0, 0)$ and radius 1,

$$S_c = \{(x, y, z) \in \mathbb{R}^3 \mid (x - c)^2 + y^2 + z^2 = 1\}.$$

With $f : \mathbb{R}^3 \times \mathbb{R} \to \mathbb{R}$, $f(x, y, z, c) = (x - c)^2 + y^2 + z^2 - 1$, and $f_c(x, y, z, c) = -2(x - c)$ we find for the envelope of $(S_c)_{c \in \mathbb{R}}$

$$\mathcal{E} = \{(x, y, z) \in \mathbb{R}^3 \mid (x - c)^2 + y^2 + z^2 - 1 = 0 \quad \text{and} \quad x - c = 0\}$$
$$= \{(x, y, z) \in \mathbb{R}^3 \mid y^2 + z^2 = 1\},$$

which is a circular cylinder with axis being the x-axis and with radius 1, see Figure 11.4

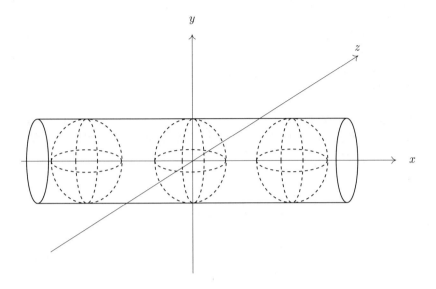

Figure 11.4

Problems

1. Prove that under the condition $xy = 1$ the function $f : (0, \infty) \times (0, \infty) \to \mathbb{R}, f(x, y) = \frac{x^p}{p} + \frac{y^q}{q}, p, q > 1$ and $\frac{1}{p} + \frac{1}{q} = 1$, attains its minimum at $(x_0, y_0) = (1, 1)$. Derive **Young's inequality** from this result, i.e. $|uv| \le \frac{|u|^p}{p} + \frac{|v|^q}{q}$ for all $u, v \in \mathbb{R}, \frac{1}{p} + \frac{1}{q} = 1, 1 < p, q$.

2.   a) Give an example of a non-convex but compact set $K \subset \mathbb{R}^2$ and a point $(\xi, \eta) \in K^{\complement}$ such that $\mathrm{dist}(K, (\xi, \eta))$ is attained at exactly two points $(x_1, y_1), (x_2, y_2) \in K$.

    b) Give an example of a non-convex, non-compact set $G \subset \mathbb{R}^2$ and a point $(\xi, \eta) \in G^{\complement}$ such that $(x, y) \in \partial G$ implies $||(x, y) - (\xi, \eta)|| = \mathrm{dist}(G, (\xi, \eta))$.

(In both cases an argument relying on elementary geometry is sufficient.)

3. Consider the cube $Q := [-1, 1]^3 \subset \mathbb{R}^3$ and the straight line $G := \{g(x) = (x, 0, 2)| x \in \mathbb{R}\} \subset \mathbb{R}^3$. Determine the trace of the curve

$\gamma : \mathbb{R} \to Q$ defined by $\gamma(x) \in Q$ and $||\gamma(x) - g(x)|| = \text{dist}(Q, g(x))$.

4. Prove that out of all triangles with given circumference length the equilateral triangle has the largest area.
   Hint: if the sides of the triangle ABC have length a, b, c and if $l = \frac{a+b+c}{2}$ then the area of ABC is given by $\sqrt{l(l-a)(l-b)(l-c)}$.

5. Find the minimum of $f(x) = x_1^k + \cdots + x_n^k$, $x_j > 0$, $n \in \mathbb{N}$, $k > 1$, under the constraint $g(x) = x_1 + \cdots + x_n - 1 = 0$. Use this result to derive for all $x_j > 0, j = 1, \ldots, n$, the estimate

$$\left(\frac{1}{n} \sum_{j=1}^{n} x_j \right)^k \leq \frac{1}{n} \sum_{j=1}^{n} x_j^k.$$

6. Find the local extreme values of $f(x, y, z) = x + 4y - 2z$ under the constraints $x + y - z = 0$ and $x^2 + y^2 + z^2 = 8$.

7. Let $x_1, \ldots, x_n, y_1, \ldots, y_n > 0$ such that $\sum_{j=1}^{n} x_j^2 = \sum_{j=1}^{n} y_j^2 = 1$. Find the maximum of $\sum_{j=1}^{n} x_j y_j$ under these constraints and use this result to give a further proof of the Cauchy-Schwarz inequality.

8. Let $A = (a_{kl})_{k,l=1,\ldots,n}$, $a_{kl} = a_{lk} \in \mathbb{R}$, be a symmetric matrix and $f(x) = \langle Ax, x \rangle = \sum_{k,l=1}^{n} a_{kl} x_k x_l$ the corresponding quadratic form. Prove that $\lambda_1 := \min_{x \in S^{n-1}} f(x)$ is an eigenvalue of A. Denote the corresponding eigenvector by ξ^1. Let $H(\xi^1) := \{x \in \mathbb{R}^n | \langle x, \xi^1 \rangle = 0\}$ be the orthonormal complement of the subspace spanned by ξ^1. Prove that $\lambda_2 := \min\{f(x) | x \in S^{n-1} \text{ and } x \in H(\xi^1)\}$ is a further eigenvalue of A.

9. a) Find the envelope of the family of parabolas given by $\psi(x, y; c) = y - (x - c)^5 = 0$.

   b) Let $\varphi : \mathbb{R}^2 \times [0, 2\pi] \to \mathbb{R}$ be the function $\varphi(x, y; \alpha) = x \sin \alpha + y \cos \alpha - 1$. The equation $\varphi(x, y, \alpha) = 0$ defines for a fixed α a straight line $g_\alpha \subset \mathbb{R}^2$. Find the envelope of the family $g_\alpha, \alpha \in [0, 2\pi]$, and sketch the situation.

10. Given the planes $E_c \subset \mathbb{R}^3$ defined for $c \in \mathbb{R}$ by the equation $(2+c)x + (3+c)y + (4+c)z - c^2 = 0$. Find the envelope of this family of planes.

12   Curvilinear Coordinates

Often curvilinear coordinates are treated on a very basic calculus level, in vector calculus or when transforming volume integrals. This makes it later on more difficult to see connections to local coordinates needed in the theory of differentiable manifolds when introducing charts or an atlas. Clearly these two considerations are closely related and for this reason our approach to curvilinear coordinates is a bit different. We first discuss (in the case of the plane) how to introduce coordinates starting with Cartesian coordinates. We emphasise that we need to take decisions: we have to choose an origin, an abscissa then we can construct the ordinate, and eventually we have to choose units of length on both axes. We will see that neither orthogonality nor dealing with straight lines is really necessary for introducing coordinates. Thus we arrive at the insight that in one and the same set, say a plane, we can have quite different types of coordinates, essentially generated by families of "coordinate lines". This leads to a new problem, namely how do different coordinate systems transform into each other. Closely related is the following question: can we map E, the plane, onto a subset of $\mathbb{R}^2$, say a rectangle $I_1 \times I_2$, in such a way that we can use the pairs $(x_1, x_2) \in I_1 \times I_2$ as coordinates on E? We are led to certain bijective mappings $h : I_1 \times I_2 \to \mathbb{R}^2$ or more generally $h : I_1 \times \cdots \times I_n \to \mathbb{R}^n$ which we want to consider as "coordinate mappings", or more precisely as "coordinatisation mappings". Since $I_1 \times \cdots \times I_n$ is a subset of $\mathbb{R}^n$ we may ask how do partial derivatives transform under such a change in coordinates? Finally we have a closer look at the (differential) geometric background: orthogonality and symmetry. In addition, many examples are discussed.

Consider the plane E and a point $p \in E$. How can we "locate" or "find" p in E? We may fix a point $\mathcal{O} \in E$ and call $\mathcal{O}$ the **origin**. Then we may choose a line X passing through $\mathcal{O}$, calling this line the **abscissa**, and eventually we construct the line Y orthogonal to X passing through $\mathcal{O}$, this line is called the **ordinate**. On each of these lines we fix a unit of length which need not be the same for X and Y. Now we can characterise p by its **coordinates** with respect to X and Y, see Figure 12.1.

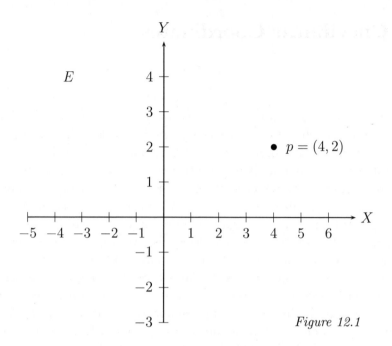

Figure 12.1

We may look at this situation differently. We first consider the family $\mathcal{F}_X$ of all lines parallel to a given line X. This family $\mathcal{F}_X$ covers the whole plane E. Then p belongs to exactly one of these lines, say $X_p \in \mathcal{F}_X$:

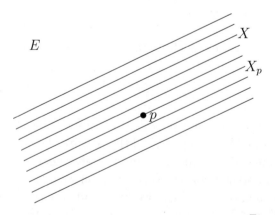

Figure 12.2

Next we want to locate p on X_p. For this we choose a second line not parallel to X and cover E with the family $\mathcal{F}_Y$ of all lines parallel to Y. Then p belongs exactly to one of these lines, say Y_p, and we can find p as an intersection of X_p and Y_p. Since X and Y are not parallel, p is uniquely determined:

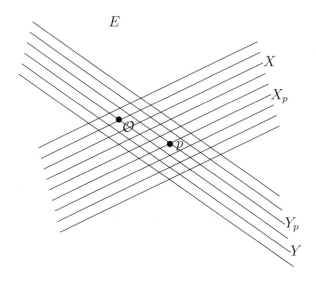

Figure 12.3

Note the fact that a point is uniquely determined as the intersection of two lines each coming from a certain family. This does not make use of "numbers", i.e. coordinates. It just uses classical Euclidean geometry. While in analysis we prefer to work with analytic or coordinate geometry, we should not forget about synthetic geometry or projective geometry.

At the moment we use some results from coordinate free geometry to introduce coordinates. Indeed, the procedure used to localise P in the consideration described in Figure 12.3 is a way to introduce coordinates: if we choose the intersection of X with Y as the origin $\mathcal{O}$, see Figure 12.3, and units of length on X and Y, respectively, we can label X_p as $(0, p_Y)$ and Y_p as $(p_X, 0)$ and find p as (p_X, p_Y). For X and Y being orthogonal we obtain the well known diagram:

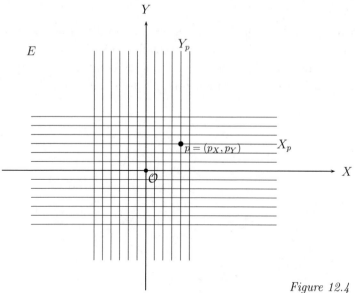

Figure 12.4

It turns out that we do not have to restrict ourselves to pairs of lines. For example we may consider the family of all circles $\mathcal{C}_{\mathcal{O}}$ with a common centre (= origin) $\mathcal{O}$ and the family of all rays $\mathcal{R}_{\mathcal{O}}$ starting at $\mathcal{O}$:

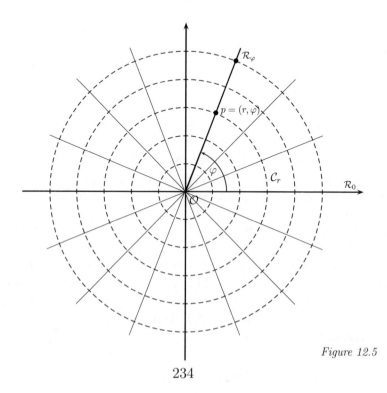

Figure 12.5

234

Next we fix one ray which we denote by $\mathcal{R}_0$. We can now label each ray $\mathcal{R}_\varphi$ by the angle φ, $0 \leq \varphi < 2\pi$, that it forms with the ray $\mathcal{R}_0$ (with $\mathcal{O}$ being the vertex). Moreover, if we choose on $\mathcal{R}_0$ a unit of length we can label each circle C_r of the family using this unit of length with r being the radius of C_r with respect to this unit. Every point p in the plane E with the exception of $\mathcal{O}$ is now uniquely characterised as the intersection of one ray $\mathcal{R}_\varphi$ and one circle C_r, i.e. by the pair $(r, \varphi) \in (0, \infty) \times [0, 2\pi)$. For the origin $\mathcal{O}$ we choose $r = 0$, but φ is not defined. We call (r, φ) the **polar coordinates** of p. If we look more closely at these examples we see that in each case we cover the plane E (maybe with the exception of one point) by two families $\mathcal{F}_1$ and $\mathcal{F}_2$ of curves having the following properties: if γ_1, $\gamma_2 \in \mathcal{F}_k$, $k = 1, 2$ fixed, then $\operatorname{tr}(\gamma_1) \cap \operatorname{tr}(\gamma_2) = \emptyset$, i.e. the traces of two curves belonging to the same family will never intersect, and moreover the union of all traces of curves belonging to such a family is the plane (or the plane with one point removed). In addition, for $\gamma_1 \in \mathcal{F}_1$ and $\gamma_2 \in \mathcal{F}_2$ the traces intersect in exactly one point. Once we can label the curves in $\mathcal{F}_k$ in a unique way we can characterise each point in E (or in E with one point removed) with the help of the labels of these curves. Note that in order to obtain such a characterisation we had to make several choices: to fix two non-parallel lines, or to fix an origin and a ray, or to fix two other families of curves with certain properties and in addition we had to fix units of length.

When replacing the plane E by an n-dimensional vector space V we may try to use n families $\mathcal{F}_k$, $1 \leq k \leq n$, of curves each having the property that any two distinct curves in a fixed family $\mathcal{F}_k$ have non-intersecting traces and that the union of all traces of curves belonging to a family $\mathcal{F}_k$ is V (or V one point (the same for all k) removed), i.e. V is covered by the traces of curves from $\mathcal{F}_k$. In addition we must require that $\bigcap_{k=1}^n \operatorname{tr}(\gamma_k)$, $\gamma_k \in \mathcal{F}_k$, $k = 1, \ldots, n$, consists of exactly one point. This is the basic idea behind curvilinear coordinates, although our rough sketch leaves several problems open, easily to be resolved when working in $\mathbb{R}^n$ which we will do in the following.

Let us stay for a while in a plane E which we identify now with $\mathbb{R}^2$ considered as a two-dimensional Euclidean space with standard orthogonal basis $\{e_1, e_2\}$. We may ask how to get a family of curves $\mathcal{F}$ with the properties required above. In principle we know an answer from Chapter 8: level lines. Let $u : \mathbb{R}^2 \to \mathbb{R}$ be a function with range $[0, \infty)$ and $f(0, 0) = 0$. Since u is defined on $\mathbb{R}^2$ every point $(x, y) \in \mathbb{R}^2$ belongs at least to one level line and obviously two level lines cannot intersect. Assuming that we can represent level lines as, say C^1-curves, we have constructed such a family $\mathcal{F}$. The condition that we can represent level lines as nice curves is necessary, for example we know that level sets need not be connected.

Example 12.1. Let $f : \mathbb{R}^2 \to \mathbb{R}$, $f(x,y) = \frac{x^2}{a^2} + \frac{y^2}{b^2}$, $a > b > 0$, and consider the level lines $f(x,y) = r^2$, note that since $f \geq 0$ we can represent every value in the range of f in a unique way by the square of a non-negative number $r \geq 0$. For $r = 0$ we have $x = y = 0$, i.e. this level line is a point, not a smooth curve. For $r > 0$ the level sets $N(r^2) = \left\{ (x,y) \in \mathbb{R}^2 \,|\, f(x,y) = \frac{x^2}{a^2} = \frac{y^2}{b^2} = r^2 \right\}$ are ellipses $\mathcal{E}_r$ with foci $(-r\sqrt{a^2 - b^2}, 0)$ and $(r\sqrt{a^2 - b^2}, 0)$, and eccentricity $\eta = \sqrt{1 - \frac{b^2}{a^2}}$. Of course for $a = b$ we obtain a circle with centre $(0,0)$ and radius r. Again we may consider as a second family of curves the rays $\mathcal{R}_\varphi$ starting at $(0,0)$ and labelled by the angle φ made with the positive x-axis, i.e. $\mathcal{R}_0$: in Figure 12.6 we have chosen $a = 5$ and $b = 4$ implying $\sqrt{a^2 - b^2} = 3$ and we consider the ellipse $\mathcal{E}_r$ the values $r_1 = 1$, $r_2 = \frac{1}{2}$ and $r_3 = \frac{1}{4}$, whereas φ is not specified.

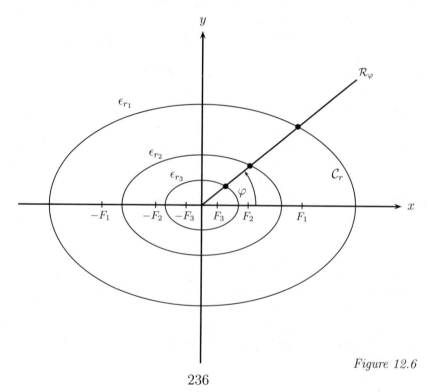

Figure 12.6

236

We want to determine the angle between the tangent to $\mathcal{E}_r$ and the ray $\mathcal{R}_{\varphi_0}$ at their intersection point. Given the parametrization $\gamma_r(\varphi) = \begin{pmatrix} ra\cos\varphi \\ rb\sin\varphi \end{pmatrix}$ we find that for r fixed $\mathrm{tr}(\gamma_r) = \mathcal{E}_r$ and the direction of the tangent line at $\gamma_r(\varphi_0)$ is given by $\begin{pmatrix} -ra\sin\varphi_0 \\ rb\cos\varphi_0 \end{pmatrix}$. On the other hand $\mathcal{R}_{\varphi_0}$ is given by the parametrizations $\rho(\tau) = \begin{pmatrix} \tau\cos\varphi_0 \\ \tau\sin\varphi_0 \end{pmatrix}$, $\tau > 0$, with tangent vectors $\begin{pmatrix} \cos\varphi_0 \\ \sin\varphi_0 \end{pmatrix}$. This yields for the cosine of the angle between these two tangent lines

$$\cos\alpha = \frac{-ra\sin\varphi_0\cos\varphi_0 + rb\cos\varphi_0\sin\varphi_0}{\sqrt{r^2a^2\sin^2\varphi_0 + r^2b^2\cos^2\varphi_0}}$$

and $\cos\alpha = 0$ if and only if

$$a\sin\varphi_0\cos\varphi_0 = b\cos\varphi_0\sin\varphi_0.$$

Clearly, for $\varphi_0 = 0, \frac{\pi}{2}, \pi$ and $\frac{3\pi}{2}$ this holds for all values of a and b, as it does hold for all values of φ_0 if $a = b$. If $\frac{b}{a} < 1$ we have no further solution, i.e. the tangent lines are not orthogonal. Still we can characterise every point $\{(x,y) \in \mathbb{R}^2 \mid (0,0)\}$ in a unique way using these ellipses and rays.

In $\mathbb{R}^2$ equipped with the canonical basis we have of course Cartesian coordinates x_1 and x_2 for each point. Thus the following question arises: why introduce new ways to characterise or locate points? In addition we need to understand the relation between two such characterisations. The first question can be answered by hinting to possible simplifications for example caused by symmetry.
Consider on $\mathbb{R}^2$ a function $f : \mathbb{R}^2 \to \mathbb{R}$ which is rotational invariant, i.e. $f(x) = f(U(x))$ for all $U \in SO(2)$ and all $x \in \mathbb{R}^2$. For such a function the value $f(x)$ depends only on the distance r of x to the origin $\mathcal{O} = (0,0)$, i.e. for some function $g : [0,\infty) \to \mathbb{R}$ we have that $f(x) = g(\|x\|)$, or using polar coordinates (r,φ) we find $f(x) = g(r)$, $r = \|x\|$. Clearly, a function of one variable is much easier to handle than a function of two variables. Note that we can extend these considerations when $\mathbb{R}^2$ is replaced by $\mathbb{R}^2\backslash\{0\}$ or an annulus $A_{R_1,R_2} = \{x \in \mathbb{R}^2 \mid R_1 < \|x\| < R_2\}$ or an even more general set as we will see below.

In order to understand the relation between two "characterisations" of points in a plane we change our point of view by switching from one plane E (or

$\mathbb{R}^2$) to two copies of the plane E and we assume that in E we can describe every point p by a pair of numbers $(a_1, a_2) \in I_1^{(1)} \times I_2^{(1)}$ and in addition by a pair of numbers $(\alpha_1, \alpha_2) \in I_1^{(2)} \times I_2^{(2)}$. Here we assume $I_j^{(k)}$ to be an interval, but we do not make any restriction on the interval to be open, closed or half-open or bounded. Hence we assume to have bijective mappings $h_1 : E \to I_1^{(1)} \times I_2^{(1)}$, $p \mapsto h_1(p) = (a_1(p), a_2(p))$, and $h_2 : E \to I_1^{(2)} \times I_2^{(2)}$, $p \mapsto h_2(p) = (\alpha_1(p), \alpha_2(p))$. We call such a mapping a **coordinate mapping** or a **global chart** and (a_1, a_2) (or (α_1, α_2)) the **coordinates** of p with respect to h_1 (or h_2). The two mappings

$$h_2 \circ h_1^{-1} : I_1^{(1)} \times I_2^{(1)} \to I_1^{(2)} \times I_2^{(2)}$$

and

$$h_1 \circ h_2^{-2} : I_1^{(2)} \times I_2^{(2)} \to I_1^{(1)} \times I_2^{(1)}$$

are called a **change of coordinates**. In particular taking $E = \mathbb{R}^2$ and in $\mathbb{R}^2$ the Cartesian coordinates (i.e. the coordinates with respect to the canonical basis) we may choose as h_1 the identity id_2 on $\mathbb{R}^2$. Clearly, nothing changes if we replace E by $E \backslash \{\mathcal{O}\}$ where $\mathcal{O} \in E$ is a fixed point and the domain and the range of h_j is changed accordingly.

Example 12.2. Consider the mapping $h : \mathbb{R}^2 \backslash \{0\} \to (0, \infty) \times [0, 2\pi)$ where $h(x, y) = (r, \varphi)$ with $r = (x^2 + y^2)^{\frac{1}{2}}$ and φ is the angle that the ray starting at $(0, 0)$ and passing through (x, y) makes with the positive x-axis. Note that φ is uniquely determined and h is indeed a bijective mapping with inverse given by $h^{-1} : (0, \infty) \times [0, 2\pi) \to \mathbb{R}^2 \backslash \{0\}$, $h^{-1}(r, \varphi) = (r \cos \varphi, r \sin \varphi)$. However we must be a bit careful when calculating φ from $x = r \cos \varphi$ and $y = r \sin \varphi$ as $\varphi = \arctan \frac{y}{x}$ due to the zeroes of $\cos \varphi$ and the periodicity of $\cos$ and $\sin$.

With Example 12.2 in mind we now can better understand the role of the families of curves $\mathcal{F}_j$, $j = 1, 2$, discussed above. If we fix in Example 12.2 $\varphi = \varphi_0$, then $r \mapsto (r \cos \varphi_0, r \sin \varphi_0)$ gives a ray in the direction determined by φ_0, whereas for $r = r_0$ fixed the trace of $\varphi \mapsto (r_0 \cos \varphi, r_0 \sin \varphi)$ is of course a circle with centre $(0, 0)$ and radius r_0. In general we obtain the families $\mathcal{F}_k$, $k = 1, 2$, as **coordinate lines** in the following way: If $h : E \to I_1 \times I_2$ (or $h : E \backslash \{0\} \to I_1 \times I_2$) then $\mathcal{F}_1$ is the family $\gamma_{t_2} := h^{-1}|_{I_1 \times \{t_2\}} \to E$, $t_2 \in I_2$, and $\mathcal{F}_2$ is the family of curves $\eta_{t_1} := h^{-1}|_{\{t_1\} \times I_2} \to E$, $t_1 \in I_1$. The traces of these curves intersect at the point with "coordinates" (t_1, t_2), i.e. $\gamma_{t_2}(t_1) = \eta_{t_1}(t_2) = h^{-1}(t_1, t_2)$. In the case $E = \mathbb{R}^2$ and if $\mathcal{F}_1, \mathcal{F}_2$ consist of differentiable curves we may ask whether the traces of η_{t_1} and γ_{t_2} are orthogonal at $p = h^{-1}(t_1, t_2)$ in the sense that these tangent lines at p are orthogonal.

It is often more convenient to start with h^{-1} than with h.

The above considerations lead to different, but related extensions. One extension to which the rest of this chapter is devoted is curvilinear coordinates induced essentially by symmetries. Here we are dealing with new systems of coordinates in $\mathbb{R}^n$ or $\mathbb{R}^n \backslash K$ where K is a certain (small) subset. These coordinates are globally, i.e. on $\mathbb{R}^n$ or $\mathbb{R}^n \backslash K$, defined as we have seen for example in the case of polar coordinates. Our interest is to study given functions on $\mathbb{R}^n$ in these coordinates with the aim to simplify their investigation by taking symmetries into account.

A second, far reaching extension will lead to the notion of a differentiable manifold and we will deal with this extension in Volume VI. The basic idea is to start with a Hausdorff space and a family of charts $(U_j, \varphi_j)_{j \in I}$ where $U_j \subset M$ is an open set, $\varphi_j : U_j \to \mathbb{R}^n$ is an injective mapping such that $\varphi_j : U_j \to \varphi(U_j)$ is a homeomorphism. In addition we assume that $M = \bigcup_{j \in I} U_j$. Charts induce on M, local coordinates, namely on U_j, by $\varphi_j^{-1} : \varphi_j(U_j) \to U_j$, recall $\varphi_j(U_j) \subset \mathbb{R}^n$. The mappings $\varphi_j \circ \varphi_l^{-1} : \varphi_l(U_j \cap U_l) \to \varphi_j(U_j \cap U_l)$ are assumed to be smooth and they determine locally a change of coordinates. In this sense we may look at curvilinear coordinates as being determined by one chart on $\mathbb{R}^n$ (or $\mathbb{R}^n \backslash K$).

Let us return once more to polar coordinates in the plane $\mathbb{R}^2$ defined by $x = r \cos \varphi$, $y = r \sin \varphi$, $r \geq 0$, $0 \leq \varphi < 2\pi$, with inverse written as $r = r(x, y)$ and $\varphi = \varphi(x, y)$, where $r = (x^2 + y^2)^{\frac{1}{2}}$ but some caution is needed with an expression for φ, $\varphi = \arctan \frac{y}{x}$ holds only in a subset. Suppose that $f : \mathbb{R}^2 \to \mathbb{R}$ is a smooth, say C^k-function with respect to Cartesian coordinates, i.e. x and y. We may consider the new function $g : [0, \infty) \times [0, 2\pi) \to \mathbb{R}$ defined by

$$g(r, \varphi) = f(x(r, \varphi), y(r, \varphi)) = f(r \cos \varphi, r \sin \varphi).$$

Now we want to find partial derivatives of g with respect to r or φ in terms of partial derivatives of f with respect to x and y. The following hold

$$\frac{\partial g}{\partial r}(r, \varphi) = \frac{\partial x}{\partial r}\left(\frac{\partial f}{\partial x}\right)(x(r, \varphi), y(r, \varphi)) + \frac{\partial y}{\partial r}\left(\frac{\partial f}{\partial y}\right)(x(r, \varphi), y(r, \varphi))$$

$$= \cos\varphi\left(\frac{\partial f}{\partial x}\right)(x(r, \varphi), y(r, \varphi)) + \sin\varphi\left(\frac{\partial f}{\partial y}\right)(x(r, \varphi), y(r, \varphi))$$

$$= \cos\varphi\left(\frac{\partial f}{\partial x}\right)(r\cos\varphi, r\sin\varphi) + \sin\varphi\left(\frac{\partial f}{\partial y}\right)(r\cos\varphi, r\sin\varphi),$$

and

$$\frac{\partial g}{\partial \varphi}(r, \varphi) = \frac{\partial x}{\partial \varphi}\left(\frac{\partial f}{\partial x}\right)(x(r, \varphi), y(r, \varphi)) + \frac{\partial y}{\partial \varphi}\left(\frac{\partial f}{\partial y}\right)(x(r, \varphi), y(r, \varphi))$$

$$= -r\sin\varphi\left(\frac{\partial f}{\partial x}\right)(r\cos\varphi, r\sin\varphi) + r\cos\varphi\left(\frac{\partial f}{\partial y}\right)(r\cos\varphi, r\sin\varphi).$$

Moreover, for the second partial derivatives we now find

$$\frac{\partial^2 g}{\partial r^2}(r, \varphi) = \frac{\partial}{\partial r}\left(\cos\varphi\left(\frac{\partial f}{\partial x}\right)(x(r, \varphi), y(r, \varphi)) + \sin\varphi\left(\frac{\partial f}{\partial y}\right)(x(r, \varphi), y(r, \varphi))\right)$$

$$= \cos\varphi\left(\frac{\partial x}{\partial r}\left(\frac{\partial^2 f}{\partial x^2}\right)(x(r, \varphi), y(r, \varphi)) + \frac{\partial y}{\partial r}\left(\frac{\partial^2 f}{\partial y^2}\right)(x(r, \varphi), y(r, \varphi))\right)$$

$$+ \sin\varphi\left(\frac{\partial x}{\partial r}\left(\frac{\partial^2 f}{\partial x \partial y}\right)(x(r, \varphi), y(r, \varphi)) + \frac{\partial y}{\partial r}\left(\frac{\partial^2 f}{\partial y^2}\right)(x(r, \varphi), y(r, \varphi))\right)$$

$$= \cos^2\varphi\left(\frac{\partial^2 f}{\partial x^2}\right)(r\cos\varphi, r\sin\varphi) + 2\cos\varphi\sin\varphi\left(\frac{\partial^2 f}{\partial x \partial y}\right)(r\cos\varphi, r\sin\varphi)$$

$$+ \sin^2\varphi\left(\frac{\partial^2 f}{\partial y^2}\right)(r\cos\varphi, r\sin\varphi),$$

$$\frac{\partial^2 g}{\partial \varphi^2}(r, \varphi) = \frac{\partial}{\partial \varphi}\left(-r\sin\varphi\left(\frac{\partial f}{\partial x}\right)(x(r, \varphi), y(r, \varphi)) + r\cos\varphi\left(\frac{\partial f}{\partial y}\right)(x(r, \varphi), y(r, \varphi))\right)$$

$$= -r\cos\varphi\left(\frac{\partial f}{\partial x}\right)(x(r, \varphi), y(r, \varphi)) - r\sin\varphi\frac{\partial}{\partial \varphi}\left(\frac{\partial f}{\partial x}\right)(x(r, \varphi), y(r, \varphi))$$

$$- r\sin\varphi\left(\frac{\partial f}{\partial y}\right)(x(r, \varphi), y(r, \varphi)) + r\cos\varphi\frac{\partial}{\partial \varphi}\left(\frac{\partial f}{\partial y}(x(r, \varphi), y(r, \varphi))\right)$$

$$= -r\cos\varphi\left(\frac{\partial f}{\partial x}\right)(x(r,\varphi),y(r,\varphi)) - r\sin\varphi\frac{\partial x}{\partial\varphi}\left(\frac{\partial^2 f}{\partial x^2}\right)(x(r,\varphi),y(r,\varphi))$$

$$-r\sin\varphi\frac{\partial y}{\partial\varphi}\left(\frac{\partial^2 f}{\partial x\partial y}\right)(x(r,\varphi),y(r,\varphi)) - r\sin\varphi\left(\frac{\partial f}{\partial y}\right)(x(r,\varphi),y(r,\varphi))$$

$$+r\cos\varphi\frac{\partial x}{\partial\varphi}\left(\frac{\partial^2 f}{\partial x\partial y}\right)(x(r,\varphi),y(r,\varphi)) - r\cos\varphi\frac{\partial y}{\partial\varphi}\left(\frac{\partial^2 f}{\partial y^2}\right)(x(r,\varphi),y(r,\varphi))$$

$$= -r\cos\varphi\left(\frac{\partial f}{\partial x}\right)(x(r,\varphi),y(r,\varphi)) + r^2\sin^2\varphi\left(\frac{\partial^2 f}{\partial x^2}\right)(x(r,\varphi),y(r,\varphi))$$

$$-r^2\sin\varphi\cos\varphi\left(\frac{\partial^2 f}{\partial x\partial y}\right)(x(r,\varphi),y(r,\varphi)) - r\sin\varphi\left(\frac{\partial f}{\partial y}\right)(x(r,\varphi),y(r,\varphi))$$

$$-r^2\sin\varphi\cos\varphi\left(\frac{\partial^2 f}{\partial x\partial y}\right)(x(r,\varphi),y(r,\varphi)) + r^2\cos^2\varphi\left(\frac{\partial^2 f}{\partial y^2}\right)(x(r,\varphi),y(r,\varphi))$$

$$= -r\cos\varphi\left(\frac{\partial f}{\partial x}\right)(r\cos\varphi,r\sin\varphi) - r\sin\varphi\left(\frac{\partial f}{\partial y}\right)(r\cos\varphi,r\sin\varphi)$$

$$+r^2\sin^2\varphi\left(\frac{\partial f}{\partial x^2}\right)(r\cos\varphi,r\sin\varphi) + r^2\cos^2\varphi\left(\frac{\partial^2 f}{\partial y^2}\right)(r\cos\varphi,r\sin\varphi)$$

$$-2r^2\cos\varphi\sin\varphi\left(\frac{\partial^2 f}{\partial x\partial y}\right)(r\cos\varphi,r\sin\varphi),$$

and we leave it as an exercise, see Problem 4, to find $\frac{\partial^2 g}{\partial r\partial\varphi}$. Next we form

$$r^2\frac{\partial^2 g}{\partial r^2}(r,\varphi) + r\frac{\partial g}{\partial r}(r,\varphi) + \frac{\partial^2 g}{\partial\varphi^2}(r,\varphi)$$

to find (while suppressing the arguments)

$$r^2\frac{\partial^2 g}{\partial r^2} + r\frac{\partial g}{\partial r} + \frac{\partial^2 g}{\partial\varphi^2} = r^2\cos^2\varphi\frac{\partial^2 f}{\partial x^2} + 2r^2\cos\varphi\sin\varphi\frac{\partial^2 f}{\partial x\partial y} + r^2\sin^2\varphi\frac{\partial^2 f}{\partial y^2}$$

$$+\cos\varphi\frac{\partial f}{\partial x} + r\sin\varphi\frac{\partial f}{\partial y} - r\cos\varphi\frac{\partial f}{\partial x} - r\sin\varphi\frac{\partial f}{\partial y}$$

$$+r^2\sin^2\varphi\frac{\partial^2 f}{\partial x^2} - 2r^2\cos\varphi\sin\varphi\frac{\partial^2 f}{\partial x\partial y} + r^2\cos^2\varphi\frac{\partial^2 f}{\partial y^2}$$

$$= r^2(\cos^2\varphi + \sin^2\varphi)\frac{\partial^2 f}{\partial x^2} + r^2(\sin^2\varphi + \cos^2\varphi)\frac{\partial^2 f}{\partial y^2}$$

$$= r^2\frac{\partial^2 f}{\partial x^2} + r^2\frac{\partial f}{\partial y^2},$$

241

which yields for $r > 0$

$$\Delta_2 f = \frac{\partial^2 f}{\partial x^2} + \frac{\partial^2 f}{\partial y^2} = \frac{\partial^2 g}{\partial r^2} + \frac{1}{r}\frac{\partial g}{\partial r} + \frac{1}{r^2}\frac{\partial^2 g}{\partial \varphi^2}. \tag{12.1}$$

In particular, if f is harmonic, i.e. $\Delta_2 f = 0$ (in $\mathbb{R}^2\backslash\{0\}$) and f is rotational invariant, i.e. $f(x, y) = g(r, \varphi) = h(r)$, then h must satisfy on $(0, \infty)$ the ordinary differential equation

$$\frac{d^2 h}{dr^2} + \frac{1}{r}\frac{dh}{dr} = 0, \tag{12.2}$$

which has two "independent" solutions, namely $h_1(r) = 1$ for all r, which is trivial, and $h_2(r) = \ln r$ which follows from $\frac{d}{dr}\ln r = \frac{1}{r}$, $\frac{d}{dr}\frac{1}{r} = -\frac{1}{r^2}$. Thus, introducing polar coordinates allows us to find (all) rotational symmetric harmonic functions in $\mathbb{R}^2\backslash\{0\}$; they are given by $r \mapsto \lambda_1 + \lambda_2 \ln r$. (A full justification requires a bit more knowledge about the equation (12.2) which we will provide in Volume IV.)

This example lets us imagine how useful coordinates which take symmetry into account are, but it also teaches us that we need a more systematic approach for our calculations and that certain (small) subsets might occur as singular points - in the above example the origin.

We now work in $\mathbb{R}^n$ with Cartesian coordinates, i.e. $x \in \mathbb{R}^n$ has the representation $x = (x_1, \ldots, x_n)$ where the x_j's are the coordinates with respect to the canonical basis $\{e_1, \ldots, e_n\}$. Further we use in $\mathbb{R}^n$ the Euclidean topology which gives us the standard scalar product and hence a notion of orthogonality. New global coordinates are introduced by the equation

$$x = x(\xi) = x(\xi_1, \ldots, \xi_n) = (x_1(\xi_1, \ldots, \xi_n), \ldots, x_n(\xi_1, \ldots, \xi_n)), \tag{12.3}$$

with $\xi = (\xi_1, \ldots, \xi_n) \in U \subset \mathbb{R}^n$, $U = I_1 \times \cdots \times I_n$ where I_j is an interval (open, closed, half-open, bounded, unbounded - all are allowed and possible). We assume that we can solve (12.3) for ξ, i.e.

$$\xi = \xi(x) = \xi(x_1, \ldots, x_n) = (\xi_1(x_1, \ldots, x_n), \ldots, \xi_n(x_1, \ldots, x_n)), \tag{12.4}$$

and that the mappings $\xi \mapsto x(\xi)$ as well as $x \mapsto \xi(x)$ are smooth, say of class C^k, $k \geq 1$. In other words we introduce a mapping $\Phi : U \to \mathbb{R}^n$ which we assume to be bijective with inverse $\Phi^{-1} : \mathbb{R}^n \to U$. We may later on have to reduce this assumption in order to allow coordinates being only uniquely defined for $\mathbb{R}^n \backslash K$, where K is a certain "small" subset of $\mathbb{R}^n$. The Jacobi matrix, i.e. the differential, of Φ at ξ is now

$$J_\Phi(\xi) = \left(\frac{\partial x_j}{\partial \xi_l}(\xi) \right)_{j,l=1,\dots,n} \tag{12.5}$$

and we assume further that for $\xi \in U$ we have

$$\det J_\Phi(\xi) \neq 0, \tag{12.6}$$

implying that the matrix $J_\Phi(\xi)$ has an inverse $J_\Phi^{-1}(x)$ and

$$J_{\Phi^{-1}}(x) = \left(\frac{\partial \xi_l}{\partial x_j}(x) \right)_{j,l=1,\dots,n} \tag{12.7}$$

must hold.

By $\xi_j \mapsto \gamma_j(\xi_j) := x(\xi_1, \dots, \xi_{j-1}, \xi_j, \xi_{j+1}, \dots, \xi_n)$, $\xi_1, \dots, \xi_{j-1}, \xi_{j+1}, \dots, \xi_n$ fixed, a curve $\gamma_j : I_j \to \mathbb{R}^n$ is given and we call this curve the jth coordinate line with respect to the coordinates ξ. For every $\xi = (\xi_1, \dots, \xi_n)$ we now obtain for $x = x(\xi_1, \dots, \xi_n)$ n tangent vectors $\dot\gamma_j(\xi_j) = \frac{\partial x}{\partial \xi_j}(\xi_1, \dots, \xi_n)$. We call the new coordinates **orthogonal curvilinear coordinates** if for every ξ, i.e. every x, the tangent vectors $\dot\gamma_j(\xi_j)$ are non-zero and mutually orthogonal, i.e. $\langle \dot\gamma_j(\xi_j), \dot\gamma_l(\xi_l) \rangle = \left\langle \frac{\partial x}{\partial \xi_j}(\xi), \frac{\partial x}{\partial \xi_l}(\xi) \right\rangle = 0$ for $j \neq l$. Of course we can now introduce normalised tangent vectors $\vec{t}_j(\xi) = \frac{\dot\gamma_j(\xi_j)}{\|\dot\gamma_j(\xi_j)\|} = \frac{1}{\left\| \frac{\partial x}{\partial \xi_j}(\xi) \right\|} \frac{\partial x}{\partial \xi_j}(\xi)$.

If $f : \mathbb{R}^n \to \mathbb{R}$ is a function, $x \mapsto f(x) = f(x_1, \dots, x_n)$, we can consider this function with respect to the coordinates ξ. We look at $g : U \to \mathbb{R}$, $\xi \mapsto g(\xi)$, where

$$g(\xi) = f(x(\xi)) = f(x_1(\xi_1, \dots, \xi_n), \dots, x_n(\xi_1, \dots, \xi_n)). \tag{12.8}$$

The chain rule yields

$$\frac{\partial g}{\partial \xi_j}(\xi) = \sum_{l=1}^n \frac{\partial x_l(\xi)}{\partial \xi_j} \left(\frac{\partial f}{\partial x_l} \right)(x(\xi)), \tag{12.9}$$

and consequently we find

$$\left(\frac{\partial f}{\partial x_l}\right)(x) = \sum_{j=1}^{n} \frac{\partial \xi_j}{\partial x_l} \left(\frac{\partial g}{\partial \xi_j}\right)(\xi(x)). \tag{12.10}$$

For second order partial derivatives we have

$$\frac{\partial^2 g}{\partial \xi_k \partial \xi_j}(\xi) = \frac{\partial}{\partial \xi_k} \sum_{l=1}^{n} \frac{\partial x_l(\xi)}{\partial \xi_j} \left(\frac{\partial f}{\partial x_l}\right)(x(\xi)) \tag{12.11}$$

$$= \sum_{l=1}^{n} \frac{\partial^2 x_l(\xi)}{\partial \xi_k \partial \xi_l} \left(\frac{\partial f}{\partial \xi_l}\right)(x(\xi)) + \sum_{m,l=1}^{n} \frac{\partial x_l(\xi)}{\partial \xi_j} \frac{\partial x_m(\xi)}{\partial \xi_k} \left(\frac{\partial^2 f}{\partial x_k \partial \xi_j}\right)(x(\xi)),$$

and

$$\left(\frac{\partial^2 f}{\partial x_k \partial x_l}\right)(x) = \sum_{j=1}^{n} \frac{\partial^2 \xi_j(x)}{\partial x_k \partial x_l} \left(\frac{\partial g}{\partial \xi_j}\right)(\xi(x)) \tag{12.12}$$

$$+ \sum_{m,j=1}^{n} \frac{\partial \xi_j(x)}{\partial x_l} \frac{\partial \xi_m(x)}{\partial x_k} \left(\frac{\partial^2 g}{\partial \xi_j \partial \xi_m}\right)(\xi(x)).$$

These expressions allow us to find for example the gradient or the Laplace operator of f in the new coordinates ξ.

When now discussing concrete examples we will have to pay attention to the following questions:

- Is Φ globally invertible or do we have to restrict Φ to a subset $\tilde{U}$ in order to get a globally invertible mapping $\Phi_{\tilde{U}} : \tilde{U} \to \mathbb{R}^n \backslash K$?

- What are the coordinate lines? Are the new coordinates orthogonal?

- Find the expressions for partial derivatives.

Example 12.3 (Polar coordinates). We have already discussed polar coordinates in $\mathbb{R}^2$ given by $x = (x_1, x_2) = (r \cos \varphi, r \sin \varphi)$, so $\Phi : [0, \infty) \times [0, 2\pi) \to \mathbb{R}^2$, $\Phi(r, \varphi) = (r \cos \varphi, r \sin \varphi)$. The coordinate lines are the rays starting at $x = 0$, $R_{\varphi_0}(r) = (r \cos \varphi_0, r \sin \varphi_0)$, and the circles $C_{r_0}(\varphi) = (r_0 \cos \varphi, r_0 \sin \varphi)$. We have $R_{\varphi_0} \perp C_{r_0}$ at the intersection point (r_0, φ_0), see Figure 12.5. The Jacobi matrix is given by $J_\Phi(r, \varphi) = \begin{pmatrix} \cos \varphi & -r \sin \varphi \\ \sin \varphi & r \cos \varphi \end{pmatrix}$

giving the Jacobi determinant $\det J_\Phi(r, \varphi) = r$ which is only zero for $r = 0$ and we have seen that for $r > 0$ and $\varphi \in [0, 2\pi)$ we can invert $x_1 = r \cos \varphi$, $x_2 = r \sin \varphi$ to get $r = (x_1^2 + x_2^2)^{\frac{1}{2}}$ and depending on φ we have $\frac{\sin \varphi}{\cos \varphi} = \frac{x_2}{x_1}$ or $\frac{\cos \varphi}{\sin \varphi} = \frac{x_1}{x_2}$, which we can write as $\sin \varphi = \frac{x_2}{r}$ or $\cos \varphi = \frac{x_1}{r}$. In (12.1) we have already given an expression for the Laplacian in polar coordinates, see (12.1).

Example 12.4 (Cylindrical coordinates). These are coordinates in $\mathbb{R}^3$ given by

$$x = (x_1, x_2, x_3) = (r \cos \varphi, r \sin \varphi, x_3) \tag{12.13}$$

for $r \geq 0$, $0 \leq \varphi < 2\pi$ and $x_3 \in \mathbb{R}$. The coordinate lines are $r \mapsto (r \cos \varphi_0, r \sin \varphi_0, x_3^0)$, i.e. rays in the plane $x_3 = x_3^0$ with starting point $(0, 0, x_3^0)$ lying on the x_3-axis (at x_3^0), circles $\varphi \mapsto (r_0 \cos \varphi, r_0 \sin \varphi, x_3^0)$ in the plane $x_3 = x_3^0$ with centre $(0, 0, x_3^0)$ and radius r_0, and lines $x_3 \mapsto (r_0 \cos \varphi_0, r_0 \sin \varphi_0, x_3)$ parallel to the x_3-axis passing through $(r_0 \cos \varphi_0, r_0 \sin \varphi_0, 0)$. Clearly a ray or a circle in the plane $x_3 = x_3^0$ is orthogonal to a line intersecting the ray or the circle and being parallel to the x_3-axis, and since rays and circles are orthogonal, the three coordinate lines are orthogonal at any point of intersection.

Since we are in $\mathbb{R}^3$ we may also consider **coordinate surfaces**

$$(r, \varphi) \mapsto (r \cos \varphi, r \sin \varphi, x_3^0)$$
$$(\varphi, x_3) \mapsto (r_0 \cos \varphi, r_0 \sin \varphi, x_3)$$
$$(r, x_3) \mapsto (r \cos \varphi_0, r \sin \varphi_0, x_3).$$

The first surface is a plane parallel to the x_1-x_2-plane passing through $(0, 0, x_3^0)$, the second surface is the boundary of the cylinder with central axis being the x_3-axis and radius r_0, the third surface is the half plane the boundary of which is the x_3-axis and whose projections onto the x_1-x_2-plane has angle φ_0 with the positive x_1-axis, see Figure 12.7. In Problem 1 of Chapter 6 we calculated the Jacobi matrix and the Jacobi determinant of $\Phi : [0, \infty) \times [0, 2\pi) \times \mathbb{R} \to \mathbb{R}$, $(r, \varphi, x_3) \mapsto (r \cos \varphi, r \sin \varphi, x_3)$ as

$$J_\Phi(r, \varphi, x_3) = \begin{pmatrix} \cos \varphi & -r \sin \varphi & 0 \\ \sin \varphi & r \sin \varphi & 0 \\ 0 & 0 & 1 \end{pmatrix} \tag{12.14}$$

and

$$\det J_\Phi(r, \varphi, x_3) = r \tag{12.15}$$

which is zero for $r = 0$. However, now we find

$$r \equiv (x_1^2 + x_2^2)^{\frac{1}{2}}, \quad \frac{\sin \varphi}{\cos \varphi} = \frac{x_2}{x_1} \quad \text{or} \quad \frac{\cos \varphi}{\sin \varphi} = \frac{x_1}{x_2}, \quad x_3 = x_3,$$

implying that the whole x_3-axis is singular for this coordinate transformation. The Laplacian in cylindrical coordinates reads for $f(x_1, x_2, x_3) = g(r, \varphi, x_3)$ as

$$\Delta_3 f(x_1, x_2, x_3) = \frac{\partial^2 g(r,\varphi,x_3)}{\partial r^2} + \frac{1}{r}\frac{\partial g(r,\varphi,x_3)}{\partial r} + \frac{1}{r^2}\frac{\partial^2 g(r,\varphi,x_3)}{\partial \varphi^2} + \frac{\partial^2 g(r,\varphi,x_3)}{\partial x_3^2}. \quad (12.16)$$

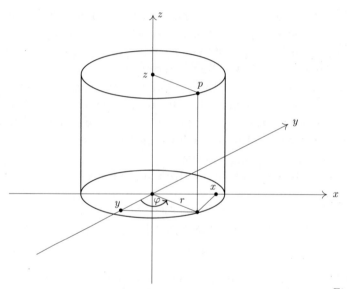

Figure 12.7

Example 12.5 (Spherical coordinates in $\mathbb{R}^3$). For $r \geq 0$, $0 \leq \varphi < 2\pi$ and $0 \leq \vartheta \leq \pi$ we define

$$x_1 = r \sin \vartheta \cos \varphi, \quad x_2 = r \sin \vartheta \sin \varphi, \quad x_3 = r \cos \vartheta.$$

The coordinate lines are

- $r \mapsto (r \sin \vartheta_0 \cos \vartheta_0, r \sin \vartheta_0 \sin \varphi_0, r \cos \vartheta_0)$,

- $\vartheta \mapsto (r_0 \sin \vartheta \cos \varphi_0, r_0 \sin \vartheta \sin \varphi_0, r_0 \cos \vartheta)$,

- $\varphi \mapsto (r_0 \sin \vartheta_0 \cos \varphi, r_0 \sin \vartheta_0 \sin \varphi, r_0 \cos \vartheta_0)$.

246

These are the following curves:

- rays starting at $0 \in \mathbb{R}^3$ and passing through $(\sin \vartheta_0 \cos \varphi_0, \sin \vartheta_0 \sin \varphi_0, \cos \vartheta_0)$ $\in S^2$;

- half circles with centre $0 \in \mathbb{R}^3$ and radius r_0 belonging to the plane spanned by $(0, 0, 1)$ and $(\cos \varphi_0, \sin \varphi_0, 0)$;

- circles with centre $(0, 0, r_0 \cos \vartheta_0)$ and radius $r_0 \cos \vartheta_0$.

Since

$$\frac{d}{dr} x(r, \vartheta_0, \varphi_0) = (\sin \vartheta_0 \cos \varphi_0, \sin \vartheta_0 \sin \varphi_0, \cos \vartheta_0)$$

$$\frac{d}{d\vartheta} x(r_0, \vartheta, \varphi_0) = (r_0 \cos \vartheta \cos \varphi_0, r_0 \cos \vartheta \sin \varphi_0, -r_0 \sin \vartheta)$$

$$\frac{d}{d\varphi} x(r_0, \vartheta_0, \varphi) = (-r_0 \sin \vartheta_0 \sin \varphi, r_0 \sin \vartheta_0 \cos \varphi, 0)$$

we find at the point $(r_0, \vartheta_0, \varphi_0)$

$$\left\langle \frac{d}{dr} x(r_0, \vartheta_0, \varphi_0), \frac{d}{d\vartheta} x(r_0, \vartheta_0, \varphi_0) \right\rangle = 0$$

$$\left\langle \frac{d}{d\vartheta} x(r_0, \vartheta_0, \varphi_0), \frac{d}{d\varphi} x(r_0, \vartheta_0, \varphi_0) \right\rangle = 0$$

$$\left\langle \frac{d}{d\varphi} x(r_0, \vartheta_0, \varphi_0), \frac{d}{dr} x(r_0, \vartheta_0, \varphi_0) \right\rangle = 0, \qquad (12.17)$$

which we leave as an exercise, see Problem 5. Hence spherical coordinates are orthogonal curvilinear coordinates.
The corresponding coordinate surfaces are

- $(r, \vartheta) \mapsto (r \sin \vartheta \cos \varphi_0, r \sin \vartheta \sin \varphi_0, r \cos \vartheta)$
  a half plane with boundary being the x_3-axis and with projection onto the x_1-x_2-plane being the ray $(r \cos \varphi_0, r \sin \varphi_0)$.

- $(\vartheta, \varphi) \mapsto (r_0 \sin \vartheta \cos \varphi, r_0 \sin \vartheta \sin \varphi, r_0 \cos \vartheta)$
  which is the sphere $\partial B_{r_0}(0)$;

- $(r, \varphi) \mapsto (r \sin \vartheta_0 \cos \varphi, r \sin \vartheta_0 \sin \varphi, r \cos \vartheta_0)$
  for $\vartheta = \frac{\pi}{2}$ this is the x_1-x_2-plane, for $0 \leq \vartheta < \frac{\pi}{2}$ this is a cone with

vertex $0 \in \mathbb{R}^3$ the positive x_3-axis as cone axis and the angle between the positive x_1-axis and the cone is $\frac{\pi}{2} - \vartheta$, for $\frac{\pi}{2} < \vartheta < \pi$ this is a cone with vertex $0 \in \mathbb{R}^3$ the negative x_3-axis as cone axis and the angle between the positive x_1-axis and the cone is $\vartheta - \frac{\pi}{2}$.

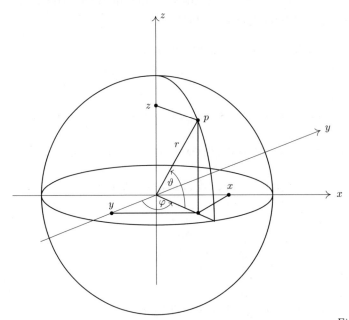

Figure 12.8

The Jaocbi matrix of $\Phi(r, \vartheta, \varphi) = (r \sin \vartheta \cos \varphi, r \sin \vartheta \sin \varphi, r \cos \vartheta)$ is

$$J_\Phi(r, \vartheta, \varphi) = \begin{pmatrix} \sin \vartheta \cos \varphi & r \cos \vartheta \cos \varphi & -r \sin \vartheta \sin \varphi \\ \sin \vartheta \sin \varphi & r \cos \vartheta \sin \varphi & r \sin \vartheta \cos \varphi \\ \cos \varphi & -r \sin \varphi & 0 \end{pmatrix} \tag{12.18}$$

which gives the Jacobi determinant

$$\det J_\Phi(r, \vartheta, \varphi) = r^2 \sin \vartheta \tag{12.19}$$

which is zero for $r = 0$ as well as for $\vartheta = 0$ or $\vartheta = \pi$, thus Φ has only a differentiable inverse on $(0, \infty) \times (0, \pi), [0, 2\pi)$, compare with Example 6.6.

248

The Laplace operator for $f(x_1, x_2, x_3) = g(r, \vartheta, \varphi)$ is given by

$$(\Delta_3 f)(x_1, x_2, x_3) = \frac{1}{r^2} \frac{\partial}{\partial r} \left(r^2 \frac{\partial g}{\partial r}(r, \vartheta, \varphi) \right) + \frac{1}{r^2 \sin \vartheta} \frac{\partial}{\partial \vartheta} \left(\sin \vartheta \frac{\partial g(r, \vartheta, \varphi)}{\partial \vartheta} \right)$$

$$+ \frac{1}{r^2 \sin \vartheta} \frac{\partial^2 g(r, \vartheta, \varphi)}{\partial \varphi^2} \qquad (12.20)$$

$$= \frac{\partial^2 g}{\partial r^2}(r, \vartheta, \varphi) + \frac{2}{r} \frac{\partial g}{\partial r}(r, \vartheta, \varphi) + \frac{1}{r^2} \frac{\partial^2 g(r, \vartheta, \varphi)}{\partial \vartheta^2} + \frac{\cos \vartheta}{r^2 \sin \vartheta} \frac{\partial g(r, \vartheta, \varphi)}{\partial \vartheta}$$

$$+ \frac{1}{r^2 \sin \vartheta} \frac{\partial^2 g(r, \vartheta, \varphi)}{\partial \varphi^2}.$$

We leave it as a further exercise, Problem 6, to find r, ϑ and φ as functions of x_1, x_2, x_3, however we refer also to Example 6.6.

Spherical coordinates can also be introduced in $\mathbb{R}^n$ which we want to discuss briefly.

Example 12.6 (Spherical coordinates in $\mathbb{R}^n$). We now want to introduce spherical coordinates in $\mathbb{R}^n$ and clearly we will recover the cases $n = 2$ and $n = 3$ as polar coordinates in $\mathbb{R}^2$ and spherical coordinates in $\mathbb{R}^3$. In our presentation we follow closely H. Triebel [46]. The idea is to start with one family of coordinate hypersurfaces, namely the spheres $\partial B_r(0) = \{rz \in \mathbb{R}^n \mid z \in S^{n-1}, r \geq 0\}$, i.e. our first new coordinate variable is the distance from $x = (x_1, \ldots, x_n) \in \mathbb{R}^n$ to the origin $0 \in \mathbb{R}^n$, i.e.

$$r = \|x\| = (x_1^2 + \cdots + x_n^2)^{\frac{1}{2}}.$$

The other $n - 1$ coordinates $\vartheta_1, \ldots, \vartheta_{n-1}$ we need only define for $\|x\| = 1$, i.e. for points on the unit sphere S^{n-1}. Let $x \in S^{n-1}$, i.e.

$$x_1^2 + \cdots + x_n^2 = 1. \qquad (12.21)$$

Then there exists a uniquely determined ϑ_1, $0 \leq \vartheta_1 \leq \pi$, such that $x_1 = \cos \vartheta_1$. Suppose that $|x_1| \neq 1$, otherwise we have $x_2 = \cdots = x_n = 0$. It follows from (12.21) that

$$x_2^2 + \cdots + x_n^2 = 1 - \cos^2 \vartheta_1 = \sin^2 \vartheta_1$$

or

$$\left(\frac{x_2}{\sin \vartheta_1} \right)^2 + \cdots + \left(\frac{x_n}{\sin \vartheta_1} \right)^2 = 1,$$

which implies the existence of ϑ_2, $0 \leq \vartheta_2 \leq \pi$, such that

$$\frac{x_2}{\sin \vartheta_1} = \cos \vartheta_2 \qquad \text{or} \qquad x_2 = \sin \vartheta_1 \cos \vartheta_2,$$

and the latter equality extends to $\vartheta_1 = 0$ or $\vartheta_1 = \pi$. Suppose now $0 < \vartheta_1, \vartheta_2 < \pi$. If follows that

$$\left(\frac{x_3}{\sin \vartheta_1}\right)^2 + \cdots + \left(\frac{x_n}{\sin \vartheta_1}\right)^2 = 1 - \cos^2 \vartheta_2 = \sin^2 \vartheta_2$$

or

$$\left(\frac{x_3}{\sin \vartheta_1 \sin \vartheta_2}\right)^2 + \cdots + \left(\frac{x_n}{\sin \vartheta_1 \sin \vartheta_2}\right)^2 = 1.$$

Thus, by an obvious (finite) induction we arrive at

$$x_1 = \cos \vartheta_1$$
$$x_2 = \sin \vartheta_1 \cos \vartheta_2$$
$$x_3 = \sin \vartheta_1 \sin \vartheta_2 \cos \vartheta_3$$
$$\vdots$$
$$x_{n-2} = \sin \vartheta_1 \sin \vartheta_2 \cdots \sin \vartheta_{n-3} \cos \vartheta_{n-2}$$

and

$$\left(\frac{x_{n-1}}{\sin \vartheta_1 \cdots \sin \vartheta_{n-2}}\right)^2 + \left(\frac{x_n}{\sin \vartheta_1 \cdots \sin \vartheta_{n-2}}\right)^2 = 1, \tag{12.22}$$

where we first assume the $0 < \vartheta_j < \pi$, $j = 1, \ldots, n - 2$. From (12.22) we deduce the existence of a unique ϑ_{n-1}, $0 \leq \vartheta_{n-1} < 2\pi$, such that

$$\frac{x_{n-1}}{\sin \vartheta_1 \cdots \sin \vartheta_{n-2}} = \cos \vartheta_{n-1}, \quad \frac{x_n}{\sin \vartheta_1 \cdots \sin \vartheta_{n-2}} = \sin \vartheta_{n-1}, \tag{12.23}$$

which yields finally for $x \in S^{n-1}$

$$\begin{cases} x_1 & = \cos \vartheta_1 \\ x_2 & = \sin \vartheta_1 \cos \vartheta_2 \\ \vdots \\ x_{n-1} & = \sin \vartheta_1 \cdots \sin \vartheta_{n-2} \cos \vartheta_{n-1} \\ x_n & = \sin \vartheta_1 \cdots \sin \vartheta_{n-2} \sin \vartheta_{n-1}, \end{cases} \tag{12.24}$$

which extends to $0 \leq \vartheta_j \leq \pi$, $j = 1, \ldots, n - 2$, and $0 \leq \vartheta_{n-1} < 2\pi$, and eventually, for $x \in \mathbb{R}^n$ with $\|x\| = r$ we arrive at

$$
\begin{cases}
x_1 & = r \cos \vartheta_1 \\
x_2 & = r \sin \vartheta_1 \cos \vartheta_2 \\
\vdots \\
x_{n-1} & = r \sin \vartheta_1 \cdots \sin \vartheta_{n-2} \cos \vartheta_{n-1} \\
x_n & = r \sin \vartheta_1 \cdots \sin \vartheta_{n-2} \sin \vartheta_{n-1},
\end{cases}
\tag{12.25}
$$

for $0 \le r < \infty$, $0 \le \vartheta_j \le \pi$ and $j = 1, \dots, n-2$, as well as $0 \le \vartheta_{n-1} < 2\pi$. In Problem 7 we will see that we recover for $n = 2$ and $n = 3$ the known formulae for polar coordinates and spherical coordinates in $\mathbb{R}^3$, respectively. Without proof we state that by (12.25) orthogonal coordinates are given, a proof can be found in [46]. The Jacobi determinant of the mapping $x = \Phi(r, \vartheta_1, \dots, \vartheta_{n-1})$ is

$$
\det \Phi(r, \vartheta_1, \dots, \vartheta_{n-1}) = r^{n-1} \sin^{n-2} \vartheta_1 \sin^{n-3} \vartheta_2 \cdots \sin \vartheta_{n-2}, \tag{12.26}
$$

which allows us to determine the set where Φ has no (differentiable) inverse. For the non-singular points the inversion formula is

$$
\begin{cases}
r = (x_1^2 + \cdots + x_n^2)^{\frac{1}{2}} \\
\vartheta_j = \frac{x_j}{r} = \frac{x_j}{(x_1^2 + \cdots + x_n^2)^{\frac{1}{2}}}, \quad j = 1, \dots, n-1.
\end{cases}
\tag{12.27}
$$

Finally we note for the Laplacian for $f(x_1, \dots, x_n) = g(r, \vartheta_1, \dots, \vartheta_{n-1})$

$$
\Delta_n f = \frac{1}{r^{n-1}} \frac{\partial}{\partial r} \left(r^{n-1} \frac{\partial g}{\partial r} \right) + \frac{1}{r^2 \sin^{n-2} \vartheta_1} \frac{\partial}{\partial \vartheta_1} \left(\sin^{n-2} \vartheta_1 \frac{\partial g}{\partial \vartheta_1} \right) \tag{12.28}
$$

$$
+ \frac{1}{r^2 \sin^2 \vartheta_1 \sin^{n-3} \vartheta_2} \frac{\partial}{\partial \vartheta_2} \left(\sin^{n-3} \vartheta_2 \frac{\partial g}{\partial \vartheta_2} \right) + \cdots +
$$

$$
+ \frac{1}{r^2 \sin^2 \vartheta_1 \cdots \sin^2 \vartheta_{n-3} \sin \vartheta_{n-2}} \frac{\partial}{\partial \vartheta_{n-2}} \left(\sin \vartheta_{n-2} \frac{\partial g}{\partial \vartheta_{n-2}} \right)
$$

$$
+ \frac{1}{r^2 \sin^2 \vartheta_1 \cdots \sin^2 \vartheta_{n-2}} \frac{\partial^2 g}{\partial \vartheta_{n-2}}.
$$

When f is rotational symmetric, i.e. $f(x) = g(r)$, we find

$$
\Delta_{n,\,\text{radial}}\, g = \frac{\partial^2 g}{\partial r^2} + \frac{n-1}{r} \frac{\partial g}{\partial r}. \tag{12.29}
$$

Further examples of orthogonal curvilinear coordinates can be found in Ph. M. Morse and H. Feshbach [37], in particular Chapter 5.1, and in M. R. Spiegel [44].

Let us return to symmetries and coordinate lines or coordinate hypersurfaces. We have already discussed the case of certain functions, say $f : \mathbb{R}^3 \to \mathbb{R}$, depending only on $r = (x_1^2 + x_2^2 + x_3^2)^{\frac{1}{2}}$, i.e. the distance to the origin. If $T \in SO(3)$ is a proper rotation then we find for all $x \in \mathbb{R}^3$ that $f(x) = f(Tx)$. Thus symmetry is reflected by invariance under a group of transformations (or under a group action) $T : \mathbb{R}^n \to \mathbb{R}^n$. Transformations leaving coordinate lines or surfaces invariant will leave the function invariant not depending on the complementary coordinates: in order for $(x_1, x_2, x_3) \mapsto f(x_1, x_2, x_3)$ to be invariant under all proper rotations f can only depend on r, not on ϑ or φ. To put it the other way round: if we are dealing with a problem (a function) which has a certain symmetry leaving certain families of lines or surfaces invariant, we are tempted to use these families as coordinate lines or surfaces, and if necessary, to complement these families to get new (curvilinear, orthogonal) coordinates. If these curves are given in $\mathbb{R}^n$ with respect to Cartesian coordinates by n equations

$$\xi_1(x_1, \ldots, x_n) = 0, \ldots, \xi_n(x_1, \ldots, x_n) = 0, \tag{12.30}$$

the implicit function theorem, Theorem 10.5, gives us a criteria to get at least locally new coordinates

$$x_1 = x_1(\xi_1, \ldots, \xi_n), \ldots, x_n = x_n(\xi_1, \ldots, \xi_n), \tag{12.31}$$

and we may try to find the largest domain for (12.31) to hold. Such a set of equations is for example (12.27). However, we already know that this is not always possible and we will see later in Volume VI that part of the problem is that no smooth mapping with smooth inverse will map compact sets such as circles onto non-compact sets such as straight lines, also see Theorem 3.26. In Chapter 21 we will return to orthogonal curvilinear coordinates when discussing the transformation theorem for volume integrals.

Problems

1.   a) Give a description of the closed set $A \subset \mathbb{R}^2$ with boundary ∂A as given below in terms of polar coordinates.

$$\partial A := \{(x, y) \in \mathbb{R}^2 \mid y = 0, 1 \leq x \leq 2\} \cup \{(x, y) \in \mathbb{R}^2 \mid x = 0, 1 \leq y \leq 2\}$$
$$\cup \{(x, y) \in \mathbb{R}^2 \mid x^2 + y^2 = 1, x \geq 0, y \geq 0\}$$
$$\cup \{(x, y) \in \mathbb{R}^2 \mid x^2 + y^2 = 4, x \geq 0, y \geq 0\}.$$

b) Consider the closed set $A \subset \mathbb{R}^3$ in the following figure and characterise this set with the help of cylindrical coordinates. (The grey shaded area is part of the boundary.)

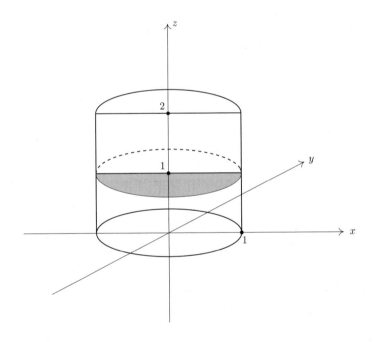

2. In the plane $\mathbb{R}^2$ consider the coordinates $x = \cosh u \cos \varphi$ and $y = \sinh u \sin \varphi$, $u \geq 0, 0 \leq \varphi < 2\pi$. For $u = 1, \frac{3}{2}, 2$ and $\varphi = 0, \frac{\pi}{6}, \frac{\pi}{4}, \frac{\pi}{3}, \frac{\pi}{3}, \frac{\pi}{2}, \frac{2\pi}{3}, \frac{3\pi}{4}, \frac{5\pi}{6}, \pi, \frac{7\pi}{6}, \frac{5\pi}{4}, \frac{4\pi}{3}, \frac{3\pi}{2}, \frac{5\pi}{3}, \frac{7\pi}{4}, \frac{11\pi}{6}$ draft the coordinate lines. Are these orthogonal coordinates?

(This problem follows closely M. R. Spiegel [44], p. 139.)

3. Consider in the plane $\mathbb{R}^2$ the partial differential equation $u_x + u_y = 0$. Prove that in the new coordinates $x = \xi + \eta$ and $y = \xi - \eta$ this equation takes the form $v_\xi = 0$. Now solve the initial value problem $u_x + u_y = 0$ in $\mathbb{R}^2$ and $u(x, 0) = f(x), f \in C^1(\mathbb{R})$, by first transforming it into a problem in the new coordinates.

   Hint: note that $v_\xi = 0$ means that v is independent of ξ, i.e. $v(\xi, \eta) = h(\eta)$ for some function $h : \mathbb{R} \to \mathbb{R}$.

4. Let $g(r, \varphi) = f(x(r, \varphi), y(r, \varphi)) = f(r \cos \varphi, r \sin \varphi)$ be given in polar coordinates. Find $\frac{\partial^2 g}{\partial r \partial \varphi}$ in terms of partial derivatives of $f : \mathbb{R}^2 \to \mathbb{R}$ assuming that f is a C^2 function.

5. In the case of spherical coordinates in $\mathbb{R}^3$ prove the orthogonality of the coordinate lines, i.e. verify (12.17).

6. Given spherical coordinates in $\mathbb{R}^3$, i.e. for $r \geq 0, 0 \leq \varphi < 2\pi$ and $0 \leq \vartheta \leq \pi$ we have $x_1 = r \sin \vartheta \cos \varphi, x_2 = r \sin \vartheta \sin \varphi, x_3 = r \cos \vartheta$. Find r, ϑ and φ in terms of x_1, x_2 and x_3.

7. Given spherical coordinates in $\mathbb{R}^n$, i.e. (12.25). Recover for $n = 2$ polar coordinates and for $n = 3$ the formula for spherical coordinates in $\mathbb{R}^3$ as introduced in Example 12.5.

8. Let $\gamma : [0, 1] \to \mathbb{R}^3$ be a regular C^1-curve and assume that $\text{tr}(\gamma) \subset S^2$. How can we use spherical coordinates in $\mathbb{R}^3$ to describe $\text{tr}(\gamma) \subset S^2$?

9. a) Let $u : \mathbb{R}^3 \to \mathbb{R}$ be a harmonic function, i.e. $\Delta_3 u = 0$. Suppose that with respect to spherical coordinates we have $u(x_1, x_2, x_3) = h(\vartheta, \varphi)$. Find a second order partial differential equation which h must satisfy.

   b) The three dimensional wave equation is given by $\left(\frac{\partial^2}{\partial t^2} - \Delta_3 \right) u(x_1, x_2, x_3, t) = 0$. A spherical wave is a solution to the wave equation depending only on t and $r = (x_1^2 + x_2^2 + x_3^2)^{\frac{1}{2}}$. Find a partial differential equation of second order for a spherical wave $v = v(r, t)$.

13 Convex Sets and Convex Functions in $\mathbb{R}^n$

In Chapter 23 of Volume I we discussed convex functions of one real variable, i.e. real-valued functions f defined on an interval $I \subset \mathbb{R}$ satisfying for all $x_1, x_2 \in I$ and all $\lambda \in [0,1]$

$$f(\lambda x_1 + (1-\lambda)x_2) \le \lambda f(x_1) + (1-\lambda)f(x_2). \tag{13.1}$$

Using the notion of convex sets as introduced in Chapter 6 we now want to investigate convex functions defined on a convex set in $\mathbb{R}^n$. We start with

Definition 13.1. A. *A set $K \subset \mathbb{R}^n$ is **convex** if for every two points $x_1, x_2 \in K$ the line segment joining x_1 and x_2 belongs to K, i.e. $x_1, x_2 \in K$ implies $\lambda x_1 + (1-\lambda)x_2 \in K$ for all $\lambda \in [0,1]$.*
***B.** A function $f : K \to \mathbb{R}$ defined on a convex set $K \subset \mathbb{R}^n$ is called a **convex function** if*

$$f(\lambda x_1 + (1-\lambda)x_2) \le \lambda f(x_1) + (1-\lambda)f(x_2) \tag{13.2}$$

*holds for all $x_1, x_2 \in K$ and $\lambda \in [0,1]$. If for $x_1 \ne x_2$ in (13.2) the strict inequality holds for all $\lambda \in (0,1)$, then f is called a **strictly convex function** on K.*
***C.** We call f (**strictly**) **concave** if $-f$ is (strictly) convex, i.e. if*

$$f(\lambda x_1 + (1-\lambda)x_2) \ge \lambda f(x_1) + (1-\lambda)f(x_2) \tag{13.3}$$

for all $x_1, x_2 \in K$ and $\lambda \in [0,1]$ with the strict inequality to hold for $x_1 \ne x_2$ and $\lambda \in (0,1)$ in the case of a strictly concave function.

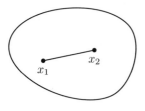

Convex Set

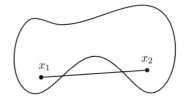

Non-convex Set

Figure 13.1

255

As in the case of convex functions of one variable we can prove (compare with Proposition I.23.8, Proposition I.23.10 and Problem 1 in Chapter I.23).

Proposition 13.2. A. *Let $K \in \mathbb{R}^n$ be a convex set and $f, g, f_n : K \to \mathbb{R}$, $n \in \mathbb{N}$, be convex functions. Then $f + g$ and αf, $\alpha \geq 0$, are convex functions too and if $F(x) := \lim_{n \to \infty} f_n(x)$ exists and is finite for every $x \in K$, then $F : K \to \mathbb{R}$ is convex.*
B. *Let $K \subset \mathbb{R}^n$ be a convex set and $J \neq \emptyset$ be an index set such that for each $j \in J$ a convex function $f_j : K \to \mathbb{R}$ is given. If*

$$g(x) := \sup \{ f_j(x) \mid j \in J \} < \infty \tag{13.4}$$

for each $x \in K$, then $g : K \to \mathbb{R}$ is a convex function.
C. Jensen's inequality *holds: let $f : K \to \mathbb{R}$ be a convex function. Then for every $m \in \mathbb{N}$, $m \geq 2$, and any choice of points $x_1, \ldots, x_m \in K$ and all $0 \leq \lambda_j \leq 1$, $j = 1, \ldots, m$, such that $\lambda_1 + \cdots + \lambda_m = 1$ it follows that*

$$f(\lambda_1 x_1 + \cdots + \lambda_m x_m) \leq \lambda_1 f(x_1) + \cdots + \lambda_m f(x_m). \tag{13.5}$$

Example 13.3. A. Let $K \subset \mathbb{R}^n$, $x_0 \in \mathbb{R}^n$ and $r > 0$. If K is convex then the sets $x_0 + K := \{ y \in \mathbb{R}^n \mid y = x_0 + z, \, z \in K \}$ and $rK := \{ y \in \mathbb{R}^n \mid y = rz, \, z \in K \}$ are convex too, hence $x_0 + rK$ is convex. Indeed, we find for $0 \leq \lambda \leq 1$ and $y_1 = x_0 + z_1$, $y_2 = x_0 + z_2$ with $z_1, z_2 \in K$ that

$$\lambda y_1 + (1 - \lambda) y_2 = \lambda x_0 + \lambda z_1 + (1 - \lambda) x_0 + (1 - \lambda) z_2$$
$$= x_0 + \lambda z_1 + (1 - \lambda) z_2 \in x_0 + K.$$

Moreover for $0 \leq \lambda \leq 1$ and $v_1 = r w_1$, $v_2 = r w_2$, $w_1, w_2 \in K$, we find

$$\lambda v_1 + (1 - \lambda) v_2 = \lambda r w_1 + (1 - \lambda) r w_2 = r(\lambda w_1 + (1 - \lambda) w_2) \in rK.$$

B. A closed half space $H_{a,c} = \{ x \in \mathbb{R}^n \mid \langle a, x \rangle \geq c \}$, $c \in \mathbb{R}$ and $a \in \mathbb{R}^n \backslash \{0\}$, is a convex set as is its interior $\overset{\circ}{H}_{a,c}$ and the boundary $\partial H_{a,c} = \{ x \in \mathbb{R}^n \mid \langle a, x \rangle = c \}$ which is a hyperplane in $\mathbb{R}^n$.

Definition 13.4. *For vectors $x_1, \ldots, x_k \in \mathbb{R}^n$, $k \in \mathbb{N}$ and $\lambda_j \geq 0$, $j = 1, \ldots, k$, such that $\lambda_1 + \cdots + \lambda_k = 1$ we call*

$$x := \lambda_1 x_1 + \cdots + \lambda_k x_k \tag{13.6}$$

*a **convex combination** of $x_1, \ldots, x_k$.*

Lemma 13.5. *A set $K \subset \mathbb{R}^n$ is convex if and only if every convex combination of its elements belongs to K.*

Proof. If every convex combination of elements of K belongs to K, with $k = 2$ we obtain the convexity of K. The converse goes by induction with an argument close to that of proving Jensen's inequality. If K is convex then for $k = 1$ and $k = 2$ the convex combinations of elements of K belong to K. Now assume that we know that all convex combinations of $k - 1$ elements of K belong to K and let $x := \lambda_1 x_1 + \cdots + \lambda_k x_k$, $x_j \in K$, be a convex combination. If $\lambda_k = 0$ by our hypothesis we know that $x \in K$ and if $\lambda_k = 1$ the result is trivial. For $0 < \lambda_k < 1$ we find $0 < \lambda_1 + \cdots + \lambda_{k-1} = 1 - \lambda_k < 1$ implying

$$\lambda_1 x_1 + \cdots + \lambda_k x_k = (1 - \lambda_k) \left(\frac{\lambda_1}{1 - \lambda_k} x_1 + \cdots + \frac{\lambda_{k-1}}{1 - \lambda_k} x_{k-1} \right) + \lambda_k x_k \in K$$

since $\frac{\lambda_1}{1-\lambda_k} + \cdots + \frac{\lambda_{k-1}}{1-\lambda_k} = 1$ and therefore $\frac{\lambda_1}{1-\lambda_k} x_1 + \cdots + \frac{\lambda_{k-1}}{1-\lambda_k} x_{k-1} \in K$. $\square$

Let $k \in \mathbb{N}$, $k \le n$ and $x_0, \ldots, x_n \in \mathbb{R}^n$ such that $x_1 - x_0, \ldots, x_k - x_0$ are linearly independent. The set of all convex combinations of $\{x_0, \ldots, x_k\}$ is called a **k-simplex** in $\mathbb{R}^n$ with vertices $x_0, \ldots, x_n$.

In order to get a better understanding of convex sets and convex functions we now show that these are quite a closely related notion. We need

Definition 13.6. *Let $f : G \to \mathbb{R}$, $G \subset \mathbb{R}^n$, be a function. The **epi-graph** epi(f) of f is by definition the set*

$$\text{epi}(f) := \{(x, y) \in G \times \mathbb{R} \,|\, y \ge f(x)\}. \tag{13.7}$$

If we consider epi(f) *as a subset of $\mathbb{R}^{n+1} = \mathbb{R}^n \times \mathbb{R}$, then G is the projection of* epi(f) *onto the first n variables, i.e. onto $\mathbb{R}^n$.*

Example 13.7. A. The epi-graph of $f : [-1, 1] \to \mathbb{R}$, $f(x) = x^2$, is the set $\{(x, y) \in [-1, 1] \times \mathbb{R} \,|\, y > x^2\}$, see Figure 13.2.
B. The epi-graph of $g : [-1, 1] \to \mathbb{R}$, $f(x) = -x^2$ is shown in Figure 13.3, the epi-graph of $\sin : \mathbb{R} \to \mathbb{R}$ is given in Figure 13.4.

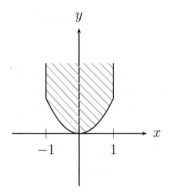

Figure 13.2

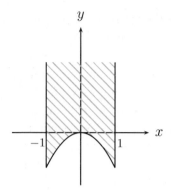

Figure 13.3

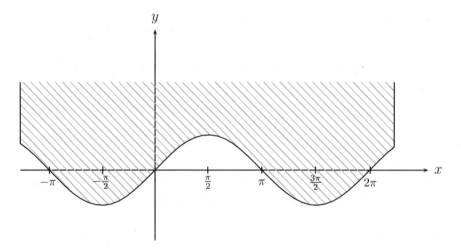

Figure 13.4

Theorem 13.8. *Let $K \subset \mathbb{R}^n$ and $f : K \to \mathbb{R}$ be a function. The function f is convex if and only if $\mathrm{epi}(f) \subset \mathbb{R}^n \times \mathbb{R}$ is a convex set.*

Proof. Suppose that f is convex which already implies by the definition of a convex function that $K \subset \mathbb{R}^n$ is convex. It follows further for $x_1, x_2 \in K$, $\lambda \in [0, 1]$, and $(x_1, z_1), (x_2, z_2) \in \mathrm{epi}(f)$ that

$$f(\lambda_1 x_1 + (1 - \lambda)x_2) \le \lambda f(x_1) + (1 - \lambda)f(x_2) \le \lambda z_1 + (1 - \lambda)z_2,$$

i.e. $((\lambda x_1 + (1 - \lambda)x_2), \lambda z_1 + (1 - \lambda)z_2) \in \mathrm{epi}(f)$, hence $\mathrm{epi}(f)$ is convex. Now suppose that $\mathrm{epi}(f)$ is convex. For $x_1, x_2 \in K$ the points $(x_1, f(x_1))$

258

and $(x_2, f(x_2))$ belong to epi(f) and therefore it follows from the convexity of epi(f) that for $\lambda \in [0, 1]$ also $(\lambda x_1 + (1-\lambda)x_2, \lambda f(x_1) + (1-\lambda)f(x_2)) \in$ epi(f) implying that K is convex and $f(\lambda x_1 + (1-\lambda)x_2) \leq \lambda f(x_1) + (1-\lambda)f(x_2)$, i.e. the convexity of f. $\qquad\square$

Let $f : \mathbb{R}^n \to [0, \infty)$ be a function. We may ask whether the set $\{x \in \mathbb{R}^n \,|\, f(x) \leq 1\}$ is convex. Replacing f by $\frac{1}{r}f$ we will obtain a criterion for the convexity of $\{x \in \mathbb{R}^n \,|\, f(x) \leq r\}$. The proof of the following result is taken from B. Simon [42].

Theorem 13.9. *If $f : \mathbb{R}^n \to [0, \infty)$ is a homogeneous function of degree 1, i.e. $f(\lambda x) = \lambda f(x)$ for all $x \in \mathbb{R}^n$ and $\lambda \geq 0$, then f is convex if and only if $\{x \in \mathbb{R}^n \,|\, f(x) \leq 1\}$ is convex and for the convexity of either f or $\{x \in \mathbb{R}^n \,|\, f(x) \leq 1\}$ it is necessary and sufficient that*

$$f(x_1 + x_2) \leq f(x_1) + f(x_2) \tag{13.8}$$

holds for all $x_1, x_2 \in \mathbb{R}^n$.

Proof. Suppose that f is convex. For the set $K := \{x \in \mathbb{R}^n \,|\, f(x) \leq 1\}$ we deduce with $0 \leq \lambda \leq 1$ and $x_1, x_2 \in K$ that

$$f(\lambda x_1 + (1 - \lambda)x_2) \leq \lambda f(x_1) + (1 - \lambda)f(x_2) \leq \lambda + (1 - \lambda) = 1,$$

i.e. $\lambda x_1 + (1 - \lambda)x_2 \in K$, hence K is convex.
Next we show that the convexity of K implies (13.8). First we assume that $f(x_1) \neq 0$ and $f(x_2) \neq 0$, $x_1, x_2 \in K$. We set $\lambda := \frac{f(x_1)}{f(x_1)+f(x_2)}$ which yields $1 - \lambda = \frac{f(x_2)}{f(x_1)+f(x_2)}$. We note that $\frac{x_j}{f(x_j)} \in K$ since by homogeneity $f\left(\frac{x_j}{f(x_j)}\right) = \frac{1}{f(x_j)}f(x_j) = 1$, $j = 1, 2$, and therefore the convexity of K implies

$$\lambda \frac{x_1}{f(x_1)} + (1 - \lambda)\frac{x_2}{f(x_2)} = \frac{f(x_1)}{f(x_1) + f(x_2)}\frac{x_1}{f(x_1)} + \frac{f(x_2)}{f(x_1) + f(x_2)}\frac{x_2}{f(x_2)}$$
$$= \frac{x_1 + x_2}{f(x_1) + f(x_2)}$$

which implies $\frac{x_1+x_2}{f(x_1)+f(x_2)} \in K$ or $\frac{f(x_1+x_2)}{f(x_1)+f(x_2)} \leq 1$ which is however (13.8). If $f(x_1) = 0$ and $f(x_2) = 1$ then for any $\vartheta > 0$ it follows that $\vartheta x_1, x_2 \in K$ implying with $\alpha_\vartheta := \frac{\vartheta}{1+\vartheta}$ and $\beta_\vartheta = \frac{1}{1+\vartheta}$ that $\alpha_\vartheta(x_1 + x_2) = \beta_\vartheta(\vartheta x_1) + (1 - \beta_\vartheta)x_2 \in K$, or $\frac{\vartheta}{1+\vartheta}f(x_1 + x_2) \leq 1$. The limit $\vartheta \to \infty$ gives $f(x_1 + x_2) \leq$

$1 = f(x_1) + f(x_2)$. If $f(x_1) = 0$ and $f(x_2) \neq 0$ we consider in the argument made above $\tilde{x}_1 = \frac{x_1}{f(x_2)}$ and $\tilde{x}_2 = \frac{x_2}{f(x_2)}$. Finally, if $f(x_1) = f(x_2) = 0$ then for any $\vartheta > 0$, $\vartheta x_1, \vartheta x_2 \in K$, so $\frac{1}{2}\vartheta(x_1 + x_2) \in K$ implying that $f(x_1 + x_2) \leq \frac{2}{\vartheta}$ and for $\vartheta \to \infty$ we arrive at $f(x_1 + x_2) = 0 = f(x_1) + f(x_2)$. Eventually we show that (13.8) implies convexity for a function homogeneous of degree 1: for $0 \leq \lambda \leq 1$ we find that

$$f(\lambda x_1 + (1-\lambda)x_2) \leq f(\lambda x_1) + f((1-\lambda)x_2) \leq \lambda f(x_1) + (1-\lambda)f(x_2).$$

$\square$

Example 13.10. A. For a norm $\|.\| : \mathbb{R}^n \to \mathbb{R}$ on $\mathbb{R}$ the balls $\overline{B_r(x_0)} \subset \mathbb{R}^n$ are convex. Since $\overline{B_r(x_0)} = \{x \in \mathbb{R}^n \,|\, \|x - x_0\| \leq r\} = x_0 + r\overline{B_1(0)}$, by Example 13.3.A we only need to prove the convexity of $\overline{B_1(0)}$. However a norm is homogeneous of degree 1 since $\|\lambda x\| = |\lambda| \, \|x\| = \lambda \, \|x\|$ for $\lambda > 0$, and a norm satisfies the triangle inequality $\|x + y\| \leq \|x\| + \|y\|$ which is (13.8) for $f(\cdot) = \|.\|$. Now Theorem 13.8 gives the result which, as the convexity of the open ball $B_r(x_0)$ can be easily proved directly, is not surprising.

B. For $p > 0$ the function $\rho_p : \mathbb{R}^n \to \mathbb{R}$, $\rho_p(x) = (|x_1|^p + \cdots + |x_n|^p)^{\frac{1}{p}}$ is homogeneous of degree 1. For $p \geq 1$ we know from Minkowski's inequality that $\rho_p(x+y) \leq \rho_p(x) + \rho_p(y)$ and indeed ρ_p is a norm. For $0 < p < 1$ we find for the unit vectors e_1, e_2 that $e_1, e_2 \in \{x \in \mathbb{R}^n \,|\, \rho_p(x) \leq 1\}$ since $\rho_p(e_j) = 1$. However for the convex combination $\frac{1}{2}e_1 + \frac{1}{2}e_2$ we have

$$\rho_p\left(\frac{1}{2}e_1 + \frac{1}{2}e_2\right) = \rho_p\left(\left(\frac{1}{2}, \frac{1}{2}, 0, \ldots, 0\right)\right) = \left(\left(\frac{1}{2}\right)^p + \left(\frac{1}{2}\right)^p\right)^{\frac{1}{p}}$$

$$= \left(2\left(\frac{1}{2}\right)^p\right)^{\frac{1}{p}} = \frac{1}{2}2^{\frac{1}{p}} = 2^{\frac{1}{p}-1}.$$

The condition $2^{\frac{1}{p}-1} \leq 1$ implies $1 - \frac{1}{p} \geq 0$ or $p \geq 1$. Thus for $0 < \rho < 1$ the set $\{x \in \mathbb{R}^n \,|\, \rho_p(x) \leq 1\}$ is not convex, hence (13.8) does not hold and therefore $\rho_p(x) = \left(\sum_{j=1}^n |x_j|^p\right)^{\frac{1}{p}}$ is not a norm on $\mathbb{R}^n$ for $0 < \rho < 1$.

C. We call $\rho : \mathbb{R}^n \to \mathbb{R}$ a **semi-norm** if $\rho(x) \geq 0$ for all $x \in \mathbb{R}^n$, $\rho(\lambda x) = |\lambda|\rho(x)$ for all $\lambda \in \mathbb{R}$, $x \in \mathbb{R}^n$, and $\rho(x + y) \leq \rho(x) + \rho(y)$. For every semi-norm the set $B_1^\rho(0) := \{x \in \mathbb{R}^n \,|\, \rho(x) \leq 1\}$ is convex and hence all sets $B_r^\rho(x_0) := x_0 + B_1^\rho(0)$ are convex.

Lemma 13.11. *For a family* $K_j \subset \mathbb{R}^n$, $j \in J$, *of convex sets the intersection* $\bigcap_{j \in J} K_j$ *is convex too.*

Proof. For $x_1, x_2 \in \bigcap_{j \in J} K_j$ it follows that $x_1, x_2 \in K_j$ for all $j \in J$. Hence for $0 \leq \lambda \leq 1$ we have $\lambda x_1 + (1 - \lambda)x_2 \in K_j$ for all $j \in J$ implying that $\lambda x_1 + (1 - \lambda)x_2 \in \bigcap_{j \in J} K_j$ is convex. $\square$

Since closed half spaces are convex we deduce that the intersection of finitely many closed half spaces is a closed convex set, called a **convex polyeder** or **polytop**.

Definition 13.12. A. *The* ***convex hull*** $\operatorname{conv}(A)$ *of* $A \subset \mathbb{R}^n$ *is defined by*

$$\operatorname{conv}(A) := \bigcap \{ K \subset \mathbb{R}^n \,|\, A \subset K \text{ and } K \text{ is convex} \}, \tag{13.9}$$

and hence $\operatorname{conv}(A)$ *is the smallest convex set containing* A.
B. *A convex set* $K \subset \mathbb{R}^n$ *is called a* ***closed convex set*** *if it is both closed and convex.*
C. *The* ***closed convex hull*** $\overline{\operatorname{conv}}(A)$ *of* $A \subset \mathbb{R}^n$ *is the closure of* $\operatorname{conv}(A)$.
D. *A closed half space* $H_{a,c}$ *is called a* ***supporting half space*** *of the non-empty, closed convex set* $K \neq \mathbb{R}^n$ *if* $K \subset H_{a,c}$ *and the hyperplane* $\partial H_{a,c}$ *contains at least one point of* K. *In this case* $\partial H_{a,c}$ *is called* ***supporting hyperplane***.

Remark 13.13. A. From Lemma 13.5 it follows that $\operatorname{conv}(A)$ is the set of all convex combinations of elements belonging to A.
B. From Lemma 13.11 we deduce that for a closed convex set $K \subset \mathbb{R}^n$ the following holds

$$K \subset \bigcap \{ H_{a,c} \,|\, H_{a,c} \text{ is a supporting half space of } K \}. \tag{13.10}$$

C. More properties of convex hulls are discussed in problems 6-8. In particular we will see that $\overline{\operatorname{conv}}(A)$ is the intersection of all closed convex sets containing A.

The following result is interesting in its own right, in particular since it does not use the finite dimensionality of $\mathbb{R}^n$ but only the fact that $\mathbb{R}^n$ is a complete scalar product space, i.e. a Hilbert space. However we prove it here in order to derive a better geometric understanding of closed convex sets using supporting half spaces, Theorem 13.16, and supporting hyperplanes, Theorem 13.19.

Theorem 13.14. *Let $K \subset \mathbb{R}^n$ be a closed convex set and $\xi \in \mathbb{R}^n$. Then there exists a unique element $x_0 = p(\xi, K)$, called the **metric projection** of ξ onto K, such that*

$$||\xi - x_0|| = \inf\{||\xi - \eta|| \,|\, \eta \in K\}. \tag{13.11}$$

The metric projection is characterised by

$$\langle x_0, \eta - x_0 \rangle \geq \langle \xi, \eta - x_0 \rangle \quad \text{for all} \ \ \eta \in K. \tag{13.12}$$

Moreover the mapping $\xi \mapsto p(\xi, K)$ from $\mathbb{R}^n$ to K is Lipschitz continuous with constant 1, i.e.

$$||p(\xi_1, K) - p(\xi_2, K)|| \leq ||\xi_1 - \xi_2|| \tag{13.13}$$

for all $\xi_1, \xi_2 \in \mathbb{R}^n$.

Proof. (Following D. Kinderlehrer and G. Stampacchia [29])
Let $(\eta_k)_{k \in \mathbb{N}}$, $\eta_k \in K$, be a minimising sequence, i.e.

$$\delta := \inf\{||\xi - \eta|| \,|\, \eta \in K\} = \lim_{k \to \infty} ||\xi - \eta_k||.$$

A direct calculation shows for $k, l \in \mathbb{N}$

$$||\eta_k - \eta_l||^2 = 2||\xi - \eta_k||^2 + 2||\xi - \eta_l||^2 - 4||\xi - \frac{1}{2}(\eta_k + \eta_l)||^2. \tag{13.14}$$

Moreover by convexity $\frac{1}{2}(\eta_k + \eta_l) \in K$ and therefore $\delta^2 \leq ||x - \frac{1}{2}(\eta_k - \eta_l)||^2$. It follows that

$$||\eta_k - \eta_l||^2 \leq 2||\xi - \eta_k||^2 + 2||\xi - \eta_l||^2 - 4\delta,$$

and since $(\eta_k)_{k \in \mathbb{N}}$ is a minimising sequence we deduce further that $\lim_{k,l \to \infty} ||\eta_k - \eta_l||^2 = 0$, i.e. $(\eta_k)_{k \in \mathbb{N}}$ is a Cauchy sequence, hence it has a limit $x_0 = \lim_{k \to \infty} \eta_k \in \mathbb{R}^n$ and since K is closed and $\eta_k \in K$ we find $x_0 \in K$ and

$$||\xi - x_0|| = \lim_{k \to \infty} ||\xi - \eta_k|| = \delta,$$

which proves the existence of $x_0 = p(\xi, K)$. For two minimisers $x_1, x_2 \in K$ we have using (13.14) (with η_k and η_l replaced by x_1 and x_2)

$$||x_1 - x_2||^2 = 2||\xi - x_1||^2 + 2||\xi - x_2||^2 - 4||\xi - \frac{1}{2}(x_1 + x_2)||^2$$
$$\leq 2\delta^2 + 2\delta^2 - 4\delta^2 = 0,$$

i.e. $x_1 = x_2$.

In order to see (13.12), let $\xi \in \mathbb{R}^n$ and $x_0 = p(\xi, K)$ the metric projection of ξ onto K. Using the convexity of K we find for $\eta \in K$ and $0 \leq \lambda \leq 1$ that $\lambda\eta + (1-\lambda)x_0 = x_0 + \lambda(\eta - x_0) \in K$, and by (13.11) the function $g : [0,1] \to \mathbb{R}$ defined by

$$g(\lambda) := ||\xi - x_0 - \lambda(\eta - x_0)||^2$$

has a minimum for $\lambda = 0$, so $g'(0) \geq 0$. However

$$g(\lambda) = ||\xi - x_0 - \lambda(\eta - x_0)||^2 = ||\xi - x_0||^2 - 2\lambda\langle\xi - x_0, \eta - x_0\rangle + \lambda^2||\eta - x_0||^2,$$

which yields $0 \leq g'(0) = - <\xi - x_0, \eta - x_0>$, or

$$< x_0, \eta - x_0 > = < \xi, \eta - x_0 >,$$

i.e. (13.12) holds. Now assume that (13.12) holds. Then it follows that

$$0 \leq \langle x_0 - \xi, (\eta - \xi) + (\xi - x_0)\rangle \leq -||\xi - x_0||^2 + \langle x_0 - \xi, \eta - \xi\rangle,$$

or by the Cauchy-Schwarz inequality

$$||x_0 - \xi||^2 \leq \langle x_0 - \xi, \eta - \xi\rangle \leq ||x_0 - \xi|| \, ||\eta - \xi||,$$

i.e.

$$||x_0 - \xi|| \leq ||\eta - \xi||$$

for all $\eta \in K$. Finally we show (13.13). Let $\xi_1, \xi_2 \in \mathbb{R}^n$ and $x_1 = p(\xi_1, K)$ and $x_2 = p(\xi_2, K)$. Since for all $\eta \in K$ we have

$$\langle x_1, \eta - x_1\rangle \geq \langle \xi_1, \eta - x_1\rangle \text{ and } \langle x_2, \eta - x_2\rangle \geq \langle \xi_2, \eta - x_2\rangle.$$

We may choose $\eta = x_2$ in the first inequality and $\eta = x_1$ in the second one to obtain by adding these inequalities

$$||x_1 - x_2||^2 = \langle x_1 - x_2, x_1 - x_2\rangle \leq \langle \xi_1 - \xi_2, x_1 - x_2\rangle \leq ||\xi_1 - \xi_2|| \, ||x_1 - x_2||,$$

or

$$||x_1 - x_2|| = ||p(\xi_1, K) - p(\xi_2, K)|| \leq ||\xi_1 - \xi_2||.$$

$\square$

Remark 13.15. Note that for $\xi \in \mathbb{R}^n \setminus K$ we have for a closed convex set $K \subset \mathbb{R}^n$ that $||\xi - p(\xi, K)|| = \text{dist}(\xi, K)$, and for $\xi \in K$ we have $p(\xi, K) = \xi$.

Theorem 13.16. *If $K \neq \emptyset$ is a closed convex set $K \subset \mathbb{R}^n$, $K \neq \mathbb{R}^n$, then equality holds in (13.10), i.e. K is the intersection of all its supporting half spaces.*

Proof. (Following St. Hildebrandt [26]). Denote the right hand side of (13.10) by $\tilde{K}$ which is a closed convex set and suppose that $\xi \in \tilde{K} \backslash K$. Since K is closed it follows by Theorem 13.14 the existence of $x_0 \in K$ such that

$$\|\xi - x_0\| \leq \|\xi - x\| \quad \text{for all } x \in K, \tag{13.15}$$

and therefore $\|\xi - x_0\| = \text{dist}(\xi, K)$. Let

$$H_0 := \left\{ x \in \mathbb{R}^n \,\middle|\, \langle x_0 - \xi, x - x_0 \rangle \geq 0 \right\} \quad \left(= H_{x_0 - \xi, \langle x_0, x_0 - \xi \rangle} \right).$$

Since $\langle x_0 - \xi, \xi - x_0 \rangle = -\|\xi - x_0\|^2 < 0$ it follows that $\xi \notin H_0$, however $x_0 \in H_0$. We will show that $K \subset H_0$ which will imply that H_0 is a supporting half space of K, hence $\tilde{K} \subset H_0$ and therefore $\xi \in H_0$ which is a contradiction. Now we prove that $K \subset H_0$. For $x \in K$ and $\lambda \in [0, 1]$ it follows by convexity that $\lambda x + (1 - \lambda)x_0 \in K$ and (13.15) yields for all $\lambda \in [0, 1]$ that

$$\|\xi - x_0\|^2 \leq \|\xi - (\lambda x + (1 - \lambda)x_0)\|^2 = \|\xi - x_0 - \lambda(x - x_0)\|^2$$

or

$$\|\xi - x_0\|^2 \leq \|\xi - x_0\|^2 - 2\lambda\langle \xi - x_0, x - x_0 \rangle + \lambda^2 \|x - x_0\|^2$$

implying that for all $\lambda \in (0, 1]$ that

$$\langle x_0 - \xi, x - x_0 \rangle + \frac{\lambda}{2} \|x - x_0\|^2 \geq 0$$

and for $\lambda \to 0$ we arrive at $\langle x_0 - \xi, x - x_0 \rangle \geq 0$ for all $x \in K$, i.e. $K \subset H_0$. $\square$

For $K \subset \mathbb{R}^n$ a closed convex set and $\xi \in \mathbb{R}^n \setminus K$ we define

$$u(\xi, K) := \frac{\xi - p(\xi, K)}{\text{dist}(\xi, K)} = \frac{\xi - p(\xi, K)}{\|\xi - p(\xi, K)\|}.$$

Lemma 13.17. *Let $\emptyset \neq K \subset \mathbb{R}^n, K \neq \mathbb{R}^n$, be a closed convex set and $\xi \in \mathbb{R}^n \setminus K$. The hyperplane H through $p(\xi, K)$ and orthogonal to $u(\xi, K)$ is a supporting hyperplane of K.*

Proof. Since $p(\xi, K) \in H$ and $p(\xi, K) \in K$ it follows that $H \cap K \neq \emptyset$. Denote by H^- the closed half space with boundary H not containing ξ, and assume for some $y \in K$ that $y \notin H^-$. Let z be on the line segment joining $p(\xi, K)$ with y and which is nearest to ξ, $||z - \xi|| = \inf\{||w - \xi|| \, | \, w = \lambda y + (1 - \lambda)p(\xi, K), \lambda \in [0,1]\}$. It follows that $||\xi - z|| < ||\xi - p(\xi, K)||$ contradicting the definition of $p(\xi, K)$ since $z \in K$. Hence $K \subset H^-$. $\qquad\square$

Lemma 13.18. *Let $K \subset \mathbb{R}^n$ be a compact convex set such that $K \subset B_r(w)$ and denote $\partial B_r(w)$ by S. It follows that $p(S, K) = \partial K$, i.e. the image of the sphere S under the metric projection onto K is the boundary of K.*

Proof. Clearly the projection of S onto K must be a subset of ∂K, i.e. $p(S, K) \subset \partial K$. Now we want to prove the converse inclusion. In the following we will denote by $R(\xi, K)$ the ray through $p(\xi, K)$ in the direction of $u(\xi, K)$, i.e.

$$R(\xi, K) := \{z = p(\xi, K) + \mu(\xi, K) \, | \, \mu \geq 0\}. \tag{13.16}$$

Let $x \in \partial K$ and for $k \in \mathbb{N}$ choose $x_k \in B_\rho(w) \setminus K$ such that $||x_k - x|| < \frac{1}{k}$. Since $p(x, K) = x$ we find by (13.13) that

$$||x - p(x_k, K)|| = ||p(x, K) - p(x_k, K)|| \leq ||x - x_k|| < \frac{1}{k}.$$

The ray $R(x_k, K)$ intersects S at some point y_k and $p(y_k, K) = p(x_k, K)$, as we will see in Problem 11. Thus $||x - p(y_k, K)|| < \frac{1}{k}$. Since $(y_k)_{k \in \mathbb{N}}$ is a sequence in S and S is compact in $\mathbb{R}^n$, a subsequence $(y_{kl})_{l \in \mathbb{N}}$ of $(y_k)_{k \in \mathbb{N}}$ converges to some limit $y \in S$. Since $\xi \mapsto p(\xi, K)$ is continuous and $\lim_{k \to \infty} p(y_k, K) = x$, we deduce that $x = p(y, K)$, i.e. $\partial K \subset p(S, K)$. $\qquad\square$

Now we can prove

Theorem 13.19. *Through every boundary point of a compact convex set $K \subset \mathbb{R}^n$ passes a supporting hyperplane, i.e. for all $x \in \partial K$ there exists a supporting hyperplane $\partial H_{a,c}$ such that $x \in \partial H_{a,c}$.*

Proof. Let $x \in \partial K$. By Lemma 13.18 there exists $\xi \in \mathbb{R}^n \setminus K$ such that $x = p(\xi, K)$. By Lemma 13.17 the hyperplane through $x = p(\xi, K)$ and orthogonal to $\xi - x$ supports K.

$\qquad\square$

For deriving Theorem 13.19 we made use of R. Schneider [41].

Definition 13.20. *Let $K \subset \mathbb{R}^n$ be a convex set. We call $z \in K$ an **extreme point** if $z = \lambda x + (1 - \lambda)y$ for some $\lambda \in [0, 1]$ it follows that $\lambda = 0$ or $\lambda = 1$. The set of all extreme points of K is denoted by $\mathrm{ext}(K)$.*

Remark 13.21. A point $z \in K$ is an extreme point of $K \in \mathbb{R}^n$ if it is not an interior point (with respect to the relative topology) of a line segment belonging to K.

Example 13.22. A. The extreme points of a triangle are of course the vertices as they are in the case of a rectangle. More generally, the extreme points of a closed hypercube in $\mathbb{R}^n$ are the 2^n vertices.
B. The extreme points of a closed ball $\overline{B_r(x_0)}$ are the boundary points, i.e. $\mathrm{ext}\left(\overline{B_r(x_0)}\right) = \partial B_r(x_0)$. Indeed, if $y_0 \notin \partial B_r(x_0)$ but $y_0 \in \overline{B_r(x_0)}$, it is an interior point of the line segment connecting x_0 with the boundary passing through y_0, so $\mathrm{ext}\left(\overline{B_r(x_0)}\right) \subset \partial B_r(x_0)$. On the other hand if $y_0 \in \partial B_r(x_0)$ is an interior point of a line segment which does not belong to the tangent plane to $\partial B_r(x_0)$ at y_0, then this line segment is a subset of a line which intersects the tangent plane to $\partial B_r(x_0)$ at y_0 in the point $y_0 \in \partial B_r(x_0)$. Hence this line segment cannot belong to $\overline{B_r(x_0)}$ as line segments belonging to a tangent plane to $\overline{B_r(x_0)}$ at y_0 do not belong to $\overline{B_r(x_0)}$.

Lemma 13.23. *If $x \in \mathrm{ext}(K)$ then $K \backslash \{x\}$ is convex.*

Proof. Suppose that $x \in \mathrm{ext}(K)$ and $K \backslash \{x\}$ is not convex. Then there exists $y, z \in K \backslash \{x\}$ such that the line segment joining y and z is not contained in $K \backslash \{x\}$, but by convexity this line segment belongs to K, thus x must belong to this line segment. Since $x \neq y$ and $x \neq z$ this point x is an interior point of the line segment connecting y and z which is a contradiction to x being an extreme point. $\qquad \square$

The following result is due to H. Minkowski and extends to infinite dimensional topological vector spaces where it is known as Krein-Milman theorem (and we will discuss this theorem in Volume V).

Theorem 13.24. *For a compact convex set $K \subset \mathbb{R}^n$ the set of extreme points $\mathrm{ext}(K)$ is the smallest subset of K satisfying*

$$\mathrm{conv}(\mathrm{ext}(K)) = K. \tag{13.17}$$

Proof. Lemma 13.23 implies the minimality of $\mathrm{ext}(K)$. It remains to prove (13.17) which we do by induction with respect to the dimension n, $K \subset \mathbb{R}^n$. The cases $n = 0$ and $n = 1$ are trivial; they refer to a point and an interval, respectively. Let $n > 1$, $K \subset \mathbb{R}^n$ and assume $\overset{\circ}{K} \neq \emptyset$ (otherwise consider $K \subset \mathbb{R}^k$ with k such that $\overset{\circ}{K} \neq \emptyset$ in $\mathbb{R}^k$). We assume that for $0, 1, \ldots, n-1$ equality (13.17) holds. Let $x \in K$. If $x \in \partial K$ then by Theorem 13.19 there exists a supporting hyperplane H passing through x and $K \cap H$ is a compact convex set which we can consider as a compact set in $\mathbb{R}^{n-1}$, hence $x \in \mathrm{conv}\,(\mathrm{ext}(K \cap H))$. Since $\mathrm{ext}(K \cap H) \subset \mathrm{ext}(K)$ we find $x \in \mathrm{conv}\,(\mathrm{ext}(K))$. If $x \notin \partial K$, i.e. $x \in \overset{\circ}{K}$, we can find $y, z \in \partial K$ such that $x \in \{\lambda y + (1-\lambda)z \mid 0 < \lambda < 1\}$, but $y, z \in \mathrm{conv}\,(\mathrm{ext}(K))$, so $x \in \mathrm{ext}\,(\mathrm{conv}(K))$. $\qquad\square$

As $\mathbb{R}^n$ or a halfspace shows, for non-compact sets Theorem 13.24 does not hold.

A first application of Minkowski's theorem allows us to locate the maximum of a continuous convex function defined on a compact convex set.

Theorem 13.25. *A continuous convex function $f : K \to \mathbb{R}$ defined on a compact convex set $K \subset \mathbb{R}^n$ attains its maximum at an extreme point of K.*

Proof. Let $x_0 \in K$ be such that $f(x_0) = \max_{x \in K} f(x)$. Note that K is compact and f is continuous, hence such $x_0 \in K$ exists. By Theorem 13.24 it follows that x_0 must be a convex combination of extreme points of K, i.e.

$$x_0 = \lambda_1 y_1 + \cdots + \lambda_l y_l, \ \ 0 < \lambda_j \leq 1, \ y_j \in \mathrm{ext}(K).$$

Now we apply Jensen's inequality, Proposition 13.2.C, to find

$$f(x_0) \leq \lambda_1 f(y_1) + \cdots + \lambda_l f(y_l) \leq (\lambda_1 + \cdots + \lambda_l) f(x_0) = f(x_0),$$

i.e. $f(x_0) = \lambda_1 f(y_1) + \cdots + \lambda_l f(y_l)$ which can only hold if $f(y_1) = \cdots = f(y_l) = f(x_0)$. $\qquad\square$

This result, which does not require any differentiability assumption has many applications in convex optimisation and hence in mathematical economics. This applies even more to the next one relating local minima of a convex function to its (global) minimum, for the one-dimensional case, compare with Problem 3 in Chapter I.23. For more details related to these questions we refer to P. M. Gruber [23].

Lemma 13.26. *Let $f : \mathbb{R}^n \to \mathbb{R}$ be a convex function and suppose that f has at $x_0 \in \mathbb{R}^n$ a local minimum, then $f(x_0) \leq f(x)$ for all $x \in \mathbb{R}^n$, i.e. f has a (global) minimum at x_0.*

Proof. Since f has a local minimum at x_0, there exists $\epsilon > 0$ such that $f(x_0) \leq f(x)$ for all $x \in \overline{B_\epsilon(x_0)}$. Take $x \in \overline{B_\epsilon(x_0)}^{\mathsf{c}}$, i.e. $\|x - x_0\| > \epsilon$ and consider

$$y := \frac{\epsilon}{\|x - x_0\|} x + \left(1 - \frac{\epsilon}{\|x - x_0\|} \right) x_0.$$

First note that

$$\|y - x_0\| = \left\| \frac{\epsilon x - \epsilon x_0}{\|x - x_0\|} \right\| = \epsilon$$

and therefore by Jensen's inequality

$$f(x_0) \leq f(y) = f\left(\frac{\epsilon}{\|x - x_0\|} x + \left(1 - \frac{\epsilon}{\|x - x_0\|} \right) x_0 \right)$$

$$\leq \frac{\epsilon}{\|x - x_0\|} f(x) + \left(1 - \frac{\epsilon}{\|x - x_0\|} \right) f(x_0)$$

which implies $f(x_0) \leq f(x)$ for all $x \in \mathbb{R}^n$. $\qquad \square$

We now turn to the question: how "regular" are convex functions? In the case $n = 1$ we know that on (a, b) a convex function $f : [a, b] \to \mathbb{R}$ is differentiable from the left and from the right (see Theorem I.23.5) and hence by Corollary I.23.6 it is on (a, b) continuous. First we prove the continuity of convex functions:

Theorem 13.27. *Let $K \subset \mathbb{R}^n$ be an open convex set and $f : K \to \mathbb{R}$ a convex function. Then f is continuous and for every compact subset $K' \subset K$ the function $f|_{K'}$ is Lipschitz continuous.*

Proof. Let $x_0 \in K$. Since K is open we can find a closed cube $\overline{Q} = I_1 \times \cdots \times I_n$, $I_j \subset \mathbb{R}$ intervals, such that for some $\rho > 0$

$$x_0 \in B_\rho(x_0) \subset \mathring{Q} \subset \overline{Q} \subset K.$$

The closed cube $\overline{Q}$ is a closed convex set with the set $\text{ext}(\overline{Q})$ consisting of its 2^n vertices $y_1, \ldots, y_{2^n}$. Hence by Theorem 13.24 every $x \in \overline{Q}$ admits a representation

$$x = \sum_{j=1}^{2^n} \lambda_j y_j \text{ with } \lambda_j \geq 0 \text{ and } \sum_{j=1}^{2^n} \lambda_j = 1. \tag{13.18}$$

Jensen's inequality now implies

$$f(x) \leq \sum_{j=1}^{2^n} \lambda_j f(y_j) \leq M := \max \left\{ f(y_j) \,|\, j = 1, \ldots, 2^n \right\}. \tag{13.19}$$

Let $z \in B_\rho(x_0)$, i.e. $z = x_0 + \eta e$ where $\eta \in [0, 1]$ and $\|e\| = \rho$. We may write $z = (1 - \eta)x_0 + \eta(x_0 + e)$ and the convexity of f yields

$$f(z) \leq (1 - \eta)f(x_0) + \eta f(x_0 + e), \tag{13.20}$$

or

$$f(z) - f(x_0) \leq \eta(M - f(x_0)). \tag{13.21}$$

Moreover we observe that

$$x_0 = \frac{1}{1 + \eta} z + \frac{\eta}{1 + \eta}(x_0 - e) \tag{13.22}$$

and again by convexity we find

$$f(x_0) \leq \frac{1}{1 + \eta} f(z) + \frac{\eta}{1 + \eta} f(x_0 - e) \tag{13.23}$$

which gives

$$f(x_0) \leq \frac{f(z) + \eta M}{1 + \eta} \tag{13.24}$$

or

$$f(x_0) - f(z) \leq \eta(M - f(x_0)). \tag{13.25}$$

From (13.21) and (13.25) we deduce

$$|f(z) - f(x_0)| \leq \eta(M - f(x_0)) \leq \frac{1}{\rho}(M - f(x_0)) \|z - x_0\| \tag{13.26}$$

for all $z \in B_\rho(x_0)$ and this implies the continuity of f at x_0. Now let $K' \subset K$ be a compact subset of K. Since $\operatorname{dist}(K', K^\complement) > 0$ there exists $\tau > 0$ such that the compact set $K'_\tau := \left\{ w = u + v \,|\, u \in K', v \in \overline{B_\tau(0)} \right\}$ is a subset of K, i.e. $K' \subset K'_\tau \subset K$. Since f is continuous $|f|$ attains on K'_τ a maximum M at some point belonging to K'_τ. For $x, y \in K'$, $x \neq y$, it follows that

$$z := y + \frac{\tau}{\|y - x\|}(y - x) \in K'_\tau \tag{13.27}$$

and with $\lambda := \frac{\|y-x\|}{\tau + \|y-x\|}$ we obtain $0 \leq \lambda \leq 1$ and

$$y = (1 - \lambda)x + \lambda z. \tag{13.28}$$

By convexity we find

$$f(y) \leq (1 - \lambda)f(x) + \lambda f(z)$$

or

$$f(y) - f(x) \leq \lambda(f(z) - f(x)) \leq 2M\frac{\|y - x\|}{\tau + \|y - x\|} \leq \frac{2M}{\tau}\|y - x\|. \tag{13.29}$$

Interchanging the roles of x and y we get

$$|f(y) - f(x)| \leq \frac{2M}{\tau}\|x - y\| = \kappa\|x - y\| \tag{13.30}$$

for all $x, y \in K'$, i.e. for every compact subset K' of K the function $f|_{K'}$ is Lipschitz constant κ depending only on f and K'. $\qquad\square$

Differentiability results for $f : K \to \mathbb{R}$, $K \subset \mathbb{R}^n$ convex and f convex, are much more involved and we refer to the monograph of R. Schneider [41] for such results. However, for differentiable convex functions we have the following characterisation.

Theorem 13.28. *Let $K \subset \mathbb{R}^n$ be an open convex set and $f : K \to \mathbb{R}$ a continuously differentiable function. The convexity of f is equivalent to either of the following conditions:*

$$f(y) - f(x) \geq \langle \operatorname{grad} f(x), y - x \rangle \text{ for all } x, y \in K, \tag{13.31}$$

and

$$\langle \operatorname{grad} f(y) - \operatorname{grad} f(x), y - x \rangle \geq 0 \text{ for all } x, y \in K. \tag{13.32}$$

In order to prove this theorem we need a result for convex functions defined on an interval.

Lemma 13.29. *Let $I \subset \mathbb{R}$ be a closed interval and $f : I \to \mathbb{R}$ a continuous function differentiable in $\mathring{I}$. If f' is increasing then f is convex.*

Proof. For $s, t \in \overset{\circ}{I}$, $s < t$, and $0 < \lambda < 1$ we deduce from the mean value theorem the existence of $\vartheta_1 \in [s, (1 - \lambda)s + \lambda t]$ and $\vartheta_2 \in [(1 - \lambda)s + \lambda t, t]$ such that

$$f'(\vartheta_1) = \frac{f((1 - \lambda)s + \lambda t) - f(s)}{\lambda(t - s)}$$

and

$$f'(\vartheta_2) = \frac{f(t) - f((1 - \lambda)s + \lambda t)}{(1 - \lambda)(t - s)}.$$

Since $f'(\vartheta_1) \le f'(\vartheta_2)$ we find

$$f((1 - \lambda)s + \lambda t) \le (1 - \lambda)f(s) + \lambda f(t),$$

i.e. the convexity of f. $\qquad\qquad\qquad\qquad\qquad\qquad\qquad\qquad\quad\square$

Proof of Theorem 13.24. Suppose that f is convex and let $x, y \in K$, $0 < \lambda < 1$. Since f is convex we have

$$f(x + \lambda(y - x)) = f((1 - \lambda)x + \lambda y) \le (1 - \lambda)f(x) + \lambda f(y)$$

and find

$$\frac{f(x + \lambda(y - x)) - f(x)}{\lambda} \le f(y) - f(x)$$

or

$$\frac{f(x+\lambda(y-x))-f(x)}{\lambda} - \langle \operatorname{grad} f(x), y - x \rangle \le f(y) - f(x) - \langle \operatorname{grad} f(x), y - x \rangle.$$
$$(13.33)$$

As λ tends to 0 the term $\frac{f(x+\lambda(y-x))-f(x)}{\lambda}$ converges to the directional derivative of f in the $y - x$ direction at the point x, which by Theorem 6.17 is given by $\langle \operatorname{grad} f(x), y - x \rangle$ implying (13.31). We may interchange in (13.31) the role of x and y to find

$$f(y) - f(x) \ge \langle \operatorname{grad} f(x), y - x \rangle$$

and

$$f(x) - f(y) \ge \langle \operatorname{grad} f(y), x - y \rangle = \langle - \operatorname{grad} f(y), y - x \rangle.$$

Adding these two inequalities we obtain

$$0 \ge \langle \operatorname{grad} f(x) - \operatorname{grad} f(y), x - y \rangle$$

or (13.32).
Now we prove that (13.32) implies the convexity of f. For this let $x, y \in K$ and $0 \leq \lambda \leq 1$. We consider the function

$$h(\lambda) := f(x + \lambda(y - x))$$

to find for $0 \leq \lambda_1 < \lambda_2 \leq 1$ by the chain rule

$$
\begin{aligned}
h'(\lambda_2) - h'(\lambda_1) &= \langle \operatorname{grad} f(x + \lambda_2(y - x)), y - x \rangle - \langle \operatorname{grad} f(x + \lambda_1(y - x)), y - x \rangle \\
&= \langle \operatorname{grad} f(x + \lambda_2(y - x)) - \operatorname{grad} f(x + \lambda_1(y - x)), y - x \rangle \\
&= \frac{1}{\lambda_2 - \lambda_1} \langle \operatorname{grad} f(x + \lambda_2(y - x)) - \operatorname{grad} f(x + \lambda_1(y - x)), x + \lambda_2(y - x) \\
&\quad - x - \lambda_1(y - x) \rangle \geq 0,
\end{aligned}
$$

i.e. the monotonicity of h' and by Lemma 13.29 the convexity of h. Since x and y are arbitrary by Lemma 13.29 we conclude that f is convex:

$$
\begin{aligned}
f(\lambda y - (1 - \lambda)x) = h(\lambda) = h(\lambda \cdot 1 + (1 - \lambda) \cdot 0) \\
\leq \lambda h(1) + (1 - \lambda)h(0) \\
= \lambda f(y) + (1 - \lambda)f(x).
\end{aligned}
$$

$\square$

Remark 13.30. Note that (13.32) is a type of monotonicity condition for $\operatorname{grad} f$. In particular if $f : I \to \mathbb{R}$, $I \subset \mathbb{R}$ an interval, then (13.32) becomes

$$(f'(y) - f'(x))(y - x) \geq 0$$

which implies for $y > x$ that $f'(y) \geq f'(x)$, i.e. f' is increasing. Combined with Lemma 13.29 this yields that a continuously differentiable function $f : I \to \mathbb{R}$ is convex if and only if f' is monotone increasing. In Problem 13 we will also give a geometric interpretation to 13.26.

A necessary and sufficient condition for the convexity of a C^2-function on an interval is, see Theorem I.23.2, that $f'' \geq 0$. An analogous result holds in $\mathbb{R}^n$.

Theorem 13.31. Let $K \subset \mathbb{R}^n$ be an open convex set and $f \in C^2(K)$. The function f is convex if and only if its Hesse matrix $\operatorname{Hess} f$ is positive semi-definite, i.e. for all $x \in K$ and all $\xi \in \mathbb{R}^n$

$$\langle \operatorname{Hess} f(x)\xi, \xi \rangle = \sum_{k,l=1}^{n} \frac{\partial^2 f}{\partial x_k \partial x_l}(x)\xi_k \xi_l \geq 0.$$

Proof. The result as well as its proof is closely related to the sufficient condition (and its proof) for a C^2-function to have a local minimum at a point x_0, see Theorem 9.14.

For $x \in K$ and $h \in \mathbb{R}^n$ such that $x + h \in K$ we find by Taylor's formula, Theorem 9.4, that

$$f(x + h) = f(x) + \langle \operatorname{grad} f(x), h \rangle + \frac{1}{2} \langle \operatorname{Hess} f(x + \vartheta h)h, h \rangle \qquad (13.34)$$

with some suitable $\vartheta \in (0, 1)$. If Hess f is positive semi-definite in K we find

$$f(x + h) - f(x) \geq \langle \operatorname{grad} f(x), h \rangle$$

or with $h = y - x$, i.e. $x + h = y \in K$, we obtain (13.31) implying the convexity of f. Now we assume that f is convex. Combining (13.34) with (13.31) we find that

$$\langle \operatorname{Hess} f(x + \vartheta h)h, h \rangle \geq 0 \qquad (13.35)$$

for all h such that $\|h\|$ is sufficiently small and $\vartheta \in (0, 1)$. Since for $h = \rho w$, $w \in S^{n-1}$ and $\rho > 0$, we find further

$$\rho^2 \langle \operatorname{Hess} f(x + \rho \vartheta w)w, w \rangle \geq 0,$$

or $\langle \operatorname{Hess} f(x + \rho \vartheta w)w, w \rangle \geq 0$ and for $\rho \to 0$ we obtain $\langle \operatorname{Hess} f(x)w, w \rangle \geq 0$ for all $w \in S^{n-1}$ and all $x \in K$ which implies the positive definiteness of Hess f on K. $\qquad \square$

We want to introduce a very useful tool for handling problems related to convexity, namely the **Legendre transform** or the **conjugate function** of a convex function. Here we restrict ourselves to finite-valued convex functions and soon even to smooth, i.e. C^1- or even C^2-functions. When discussing convexity in topological vector spaces (in Volume V) we will allow more general functions, while in Volume IV when dealing with partial differential equations of first order or Hamiltonian systems, we will always work with C^2-functions.

Definition 13.32. *A function $f : \mathbb{R}^n \to \mathbb{R}$ is called **coercive** if*

$$\lim_{|x| \to \infty} \frac{f(x)}{\|x\|} = \infty. \qquad (13.36)$$

Let f be a continuous coercive function. From (13.36) it follows for $\kappa > 0$ that the existence of $R \geq 0$ such that $\|x\| \geq R$ implies $f(x) \geq \kappa \|x\|$ and the continuity of f yields that $f(x) \geq -\delta$, $\delta \geq 0$, on $B_R(0)$. Hence we find for a continuous coercive function, by Theorem 13.27, in particular for a convex coercive function, that for $\kappa > 0$ there exists $\delta \geq 0$ such that

$$f(x) \geq \kappa \|x\| - \delta \text{ for all } x \in \mathbb{R}^n. \tag{13.37}$$

Definition 13.33. *Let* $f : \mathbb{R}^n \to \mathbb{R}$ *be a coercive convex function. Its* **Legendre transform** $f^* : \mathbb{R}^n \to \mathbb{R}$ *is defined by*

$$f^*(y) := \sup_{x \in \mathbb{R}^n} (\langle x, y \rangle - f(x)). \tag{13.38}$$

In the situation of Definition 13.33 we find with $\kappa := 1 + \|y\|$ in (13.37) that for all $x \in \mathbb{R}^n$

$$\langle x, y \rangle - f(x) \leq \|x\| \, \|y\| - f(x)$$
$$\leq \|x\| \, \|y\| - \|x\| \, (\|y\| + 1) + \delta = - \|x\| + \delta.$$

For $\|x\| > \delta + f(0)$ this yields

$$\langle x, y \rangle - f(x) < -f(0)$$

implying the existence of some $x_0 = x_0(y) \in B_{\delta + f(0)}(0)$ such that

$$\langle x, y \rangle - f(x) \leq \sup_{x \in \mathbb{R}^n} (\langle x, y \rangle - f(x)) = \langle x_0, y_0 \rangle - f(x_0).$$

Theorem 13.34. *Let* $f : \mathbb{R}^n \to \mathbb{R}$ *be a coercive convex function. For* $y \in \mathbb{R}^n$ *there exists* $x_0(y) \in \mathbb{R}^n$ *such that*

$$f(x) - f(x_0(y)) \geq \langle y, x - x_0 \rangle \text{ for all } x \in \mathbb{R}^n, \tag{13.39}$$

and

$$f^*(y) = \langle x_0(y), y \rangle - f(x_0(y)). \tag{13.40}$$

In particular $f^*(y) < \infty$, f^* *is coercive and convex. Moreover we have* $(f^*)^* = f$.

Proof. (Following B. Simon [42]) From the preceding considerations we already know that (13.39) holds. In particular $f^*(y)$ is finite and since f^* is the supremum of convex functions, namely the functions $y \mapsto \langle x, y \rangle - f(x)$, f^* is convex which implies by Theorem 13.27 the continuity of f^*. From (13.38) we conclude further

$$f^*(y) + f(x) \geq \langle x, y \rangle \tag{13.41}$$

and therefore with $x = \frac{\rho y}{\|y\|}$, $\rho > 0$,

$$f^*(y) \geq \rho \|y\| - f\left(\frac{\rho y}{\|y\|}\right). \tag{13.42}$$

Since f is bounded on compact sets we have

$$\left| f\left(\frac{\rho y}{\|y\|}\right) \right| \leq \max_{z \in B_\rho(0)} |f(z)|$$

implying by (13.42) that

$$\liminf_{\|y\| \to \infty} \frac{f^*(y)}{\|y\|} \geq \rho,$$

and since $\rho > 0$ was arbitrary we deduce that $\lim_{\|y\| \to \infty} \frac{f^*(y)}{\|y\|} = \infty$, i.e. f^* is coercive. Thus f^* is convex and coercive in $\mathbb{R}^n$, hence $(f^*)^*$ exists and is a convex, coercive function on $\mathbb{R}^n$. Using (13.41) we find

$$f(x) \geq \langle x, y \rangle - f^*(y)$$

which yields

$$f(x) \geq \sup_{x \in \mathbb{R}^n} (\langle x, y \rangle - f^*(y)) = (f^*)^*(x). \tag{13.43}$$

On the other hand epi(f) is convex and the intersection of epi(f) with $\overline{B_1((x_0, f(x_0)))}$ is a compact convex set with $(x_0, f(x_0))$ being a boundary point. By Theorem 13.19 there exists a supporting hyperplane through $(x_0, f(x_0))$, i.e. there exists y_0 such that

$$f(x) - f(x_0) \geq \langle y_0, x - x_0 \rangle, \tag{13.44}$$

which implies

$$\sup_{x \in \mathbb{R}^n} (\langle x, y_0 \rangle - f(x)) = \langle x_0, y_0 \rangle - f(x_0) = f^*(y_0)$$

and therefore

$$f(x_0) = \langle x_0, y_0 \rangle - f^*(y_0) \leq \sup_{y_0 \in \mathbb{R}^n} (\langle x_0, y_0 \rangle - f^*(y_0)) = (f^*)^*(x_0)$$

and with (13.43) the result follows. $\qquad\square$

Suppose that $f : \mathbb{R}^n \to \mathbb{R}$ is a coercive and convex C^1-function. In this case

$$f^*(y) = \sup_{x \in \mathbb{R}^n} (\langle x, y \rangle - f(x)) = \max_{x \in \mathbb{R}^n} (\langle x, y \rangle - f(x)) = \langle x_0, y \rangle - f(x_0)$$

and hence for $y \in \mathbb{R}^n$ fixed $x \mapsto \langle x, y \rangle - f(x)$ has also a local extreme value at x_0 implying that the gradient of $x \mapsto \langle x, y \rangle - f(x)$ must vanish at x_0 or

$$y = (\operatorname{grad} f)(x_0). \tag{13.45}$$

We can read (13.45) as a mapping $x_0 \mapsto y(x_0)$ and try to find the inverse mapping, say $x_0 = z(y)$. If this is always possible we find

$$f^*(y) = \langle z(y), y \rangle - f(z(y)). \tag{13.46}$$

Here are some examples

Example 13.35. A. Consider $f : \mathbb{R} \to \mathbb{R}$, $f(x) = \frac{1}{2}x^2$. It follows that $y = f'(x) = x$ and hence $y = z(y) = x$ implying

$$f^*(y) = y^2 - \frac{1}{2}y^2 = \frac{1}{2}y^2 = f(y).$$

B. Let $A \in M(n; \mathbb{R})$ be a positive definite, symmetric matrix and define $f(x) := \frac{1}{2}\langle Ax, x \rangle$. It follows that

$$y = (\operatorname{grad} f)(x) = Ax \text{ or } x = A^{-1}y.$$

Therefore we find

$$f^*(y) = \langle y, A^{-1}y \rangle - \frac{1}{2}\langle A^{-1}y, AA^{-1}y \rangle = \frac{1}{2}\langle y, A^{-1}y \rangle.$$

C. For $k \in \mathbb{N}$ consider $f(x) = \frac{1}{2k}x^{2k}$. We find that $y = f'(x) = x^{2k-1}$ which yields $x = y^{\frac{1}{2k-1}} = |y|^{\frac{1}{2k-1}}$ for $x \geq 0$ and $x = -|y|^{\frac{1}{2k-1}}$ for $x < 0$. This however implies

$$f^*(y) = |y|^{\frac{1}{2k-1}}|y| - \frac{1}{2k}|y|^{\frac{2k}{2k-1}} = \frac{2k-1}{2k}|y|^{\frac{2k}{2k-1}}.$$

D. Let $\kappa > 0$ and consider $f(x) = |x|^{1+\kappa}$. This function is convex and coercive as long as $\kappa > 0$ and we find

$$y = (1+\kappa)|x|^\kappa \text{ for } x \geq 0 \text{ and } y = -(1+\kappa)|x|^\kappa \text{ for } x < 0.$$

For the Legendre transform of f we now deduce

$$f^*(y) + \frac{|y|^{\frac{1}{\kappa}+1}}{(1+\kappa)^{\frac{1}{\kappa}}} - \left|\frac{|y|^{\frac{1}{\kappa}}}{(1+\kappa)^{\frac{1}{\kappa}}}\right|^{1+\kappa} = c(\kappa)|y|^{\frac{1+k}{\kappa}}$$

where $c(\kappa) = \frac{1}{(1+\kappa)^{\frac{1}{\kappa}}}\left(1 - \frac{1}{1+\kappa}\right)$. Note that in this example as well as in Examples A and C the exponent p of f and q of f^* are conjugate, i.e. $\frac{1}{p} + \frac{1}{q} = 1$.

E. The function $f : \mathbb{R} \to \mathbb{R}$, $x \mapsto |x|$, is convex but not coercive. For f^* we find

$$f^*(y) = \sup_{x \in \mathbb{R}} (\langle x, y\rangle - f(x)) = \sup_{x \in \mathbb{R}} (xy - |x|).$$

For $|y| > 1$ it is obvious that $f^*(y) = +\infty$, while for $|y| \leq 1$ it follows that $f^*(y) = 0$. Thus f^* is no longer real-valued but $f^* : \mathbb{R} \to \mathbb{R} \cup \{\infty\}$, however epi($f^*$) is convex in $\mathbb{R} \times (\mathbb{R} \cup \{\infty\})$ for a convex set $K \subset \mathbb{R}^n$. This will be done in Volume V.

A further example, especially in higher dimensions is given in Problem 16.

We now want to provide for $n = 1$ a geometric interpretation of the Legendre transform.

Let $f : \mathbb{R} \to \mathbb{R}$ be a coercive convex C^2-function such that $f''(x) > 0$ for all $x \in \mathbb{R}$. In this case f' is strictly increasing and hence $(f')^{-1}$ exists. Since with f the function $x \mapsto f(x + x_0) - f(x_0)$ has the same properties we may assume for simplicity that f has its unique minimum at $x_0 = 0$. For $p \in \mathbb{R}$ we consider the straight line $y = g_p(x) = px$, see Figure 13.5 below. We may ask for the largest "distance" in the y-direction between f and g, more precisely for the maximum of $g_p(x) - f(x)$ (note that we do not take the absolute value of $g_p(x) - f(x)$, so if $z_p > 0$ is the second point where g_p intersects f, then for $x < 0$ or $x > z_p$ this "distance" becomes negative). So we are looking for a maximum of $g_p - f$ in $(0, z_p)$, hence we have to look at $g_p'(x) - f'(x) = 0$ or $p = f'(x)$, i.e. $x = (f')^{-1}(p)$ is a critical value and since $f'' > 0$, we have indeed an isolated global maximum at $x = (f')^{-1}(p)$.

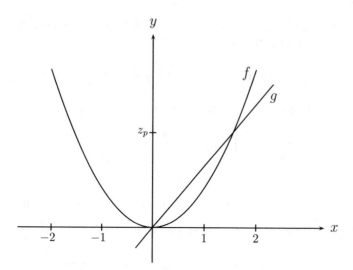

Figure 13.5

The value at this maximum is given by

$$\sup_{x \in \mathbb{R}} (xp - f(x)) = p(f')^{-1}(p) - f\left((f')^{-1}(p)\right), \qquad (13.47)$$

but $\sup_{x \in \mathbb{R}}(xp - f(x))$ is nothing but $f^*(p)$.
Next we consider the family of lines given by

$$\varphi(x, y, p) = y - px + f(p)$$

which we view as a one-parameter family, p being the parameter, in the x-y plane. This family determines an envelope

$$\mathcal{E} = \left\{ (x, y) \in \mathbb{R}^2 \,\middle|\, \varphi(x, y, p) = 0, \ \frac{\partial \varphi}{\partial p}(x, y, p) = 0 \right\}$$

since $\frac{\partial^2 \varphi}{\partial p^2}(x, y, p) = f''(p) \neq 0$ for all p, see Theorem 11.16. The two equations we have are

$$y = px - f(p) = 0 \text{ and } x = f'(p)$$

and since f' is invertible we find $p = (f')^{-1}(x)$ and hence

$$y = (f')^{-1}(x)x - f\left((f')^{-1}(x)\right)$$

and comparing with (13.47) we see that the Legendre transform of f admits an interpretation as an envelope. For more details we refer to V. I. Arnold [5].

278

Problems

1. Let $f : [0, \infty) \to \mathbb{R}$ be a convex increasing function and $|| \cdot ||$ a norm on $\mathbb{R}^n$. Prove that $g(x) := f(||x||)$ is a convex function on $\mathbb{R}^n$.

2.   a) Prove the following variant of Jensen's inequality: let $K \subset \mathbb{R}^n$ be a convex set and $f : K \to \mathbb{R}$ a convex function. Then for $\kappa_j \geq 0$ and $x_j \in K$, $j = 1, \ldots, m$, we have

$$f \left(\frac{1}{\kappa} \sum_{j=1}^{m} \kappa_j x_j \right) \leq \frac{1}{\kappa} \sum_{j=1}^{m} \kappa_j f(x_j)$$

where $\kappa = \sum_{j=1}^{m} \kappa_j$.

   b) Use Jensen's inequality to prove for $x_j \geq 0$ and $\lambda_j \geq 0$, $\sum_{j=1}^{n} \lambda_j = 1$, that

$$x_1^{\lambda_1} \cdot \ldots \cdot x_n^{\lambda_n} \leq \lambda_1 x_1 + \cdots + \lambda_n x_n$$

and deduce once more the geometric-arithmetic mean inequality.

3. For $K \subset \mathbb{R}^n$ convex and $f : K \to \mathbb{R}$ convex prove that if $K_c := \{x \in K \mid f(x) < c\} \neq \emptyset$ then K_c is convex.

4. Let $\psi : \mathbb{R}^{n_1} \times \mathbb{R}^{n_2} \to \mathbb{R}$, $(\xi, \eta) \mapsto \psi(\xi, \eta) = ||\xi||^\alpha + ||\beta||^\beta$ where $|| \cdot ||$ denotes the Euclidean norm in $\mathbb{R}^{n_j}$. For $0 < \alpha < 1$ and $0 < \beta < 1$ prove that

$$\operatorname{conv} \left\{ (\xi, \eta) \in \mathbb{R}^{n_1} \times \mathbb{R}^{n_2} \, \left| \, (||\xi||^\alpha + ||\eta||^\beta)^{\frac{1}{2}} < \rho \right. \right\}$$

$$\subset \left\{ (\xi, \eta) \in \mathbb{R}^{n_1} \times \mathbb{R}^{n_2} \, \left| \, (||\xi||^\alpha + ||\eta||^\beta)^{\frac{1}{2}} < 2^{\frac{1-(\alpha \wedge \beta)}{2}} \rho \right. \right\}$$

where $\alpha \wedge \beta = \min(\alpha, \beta)$.

5. Let $|| \cdot ||$ be a norm on $\mathbb{R}^n$ and $K \subset \mathbb{R}^n$ a non-empty convex set. Show that the function $g : \mathbb{R}^n \to \mathbb{R}$, $g(x) := \inf \{||x - y|| \mid y \in K\}$ is convex.

6.   a) Let $A, B \subset \mathbb{R}^n$ be convex sets. Is $A \cup B$ convex?

   b) For two convex sets $A, B \subset \mathbb{R}^n$ prove that

$$\operatorname{conv}(A \cup B) = \bigcup_{\lambda \in [0,1]} (\lambda A + (1 - \lambda)B).$$

7. a) Prove that the closure of a convex set is convex.

b) For $A \subset \mathbb{R}^n$ show that

$$\overline{\mathrm{conv}}(A) = \bigcap \{K \mid A \subset K, \, K \subset \mathbb{R}^n \text{ is a closed convex set}\}.$$

8. Let $\emptyset \neq K \subset \mathbb{R}^n$ be convex and for $r > 0$ define $K_r := \bigcup_{z \in K} B_r(z)$. Show that $K_r \subset \mathbb{R}^n$ is an open convex set.

9. a) Draft a $0-, 1-, 2-$ and $3-$simplex.

b) Given the circle with centre $0 \in \mathbb{R}^2$ and radius r. Consider the regular polygon with vertices $p_j = (r \cos \varphi_j, r \sin \varphi_j)$ where $j = 0, \ldots, n-1$ and $\varphi_j = \frac{2\pi j}{n}$. Prove by using "obvious geometric reasoning" that this regular polygon is the boundary of a compact convex set P and find the extreme points of P.

10. Given a system of linear inequalities in $\mathbb{R}^n$:

$$a_{j1}x_1 + \cdots + a_{jn}x_n \leq b_j, \quad j \in J, a_{jk} \in \mathbb{R}.$$

Prove that the solution set of this system is a convex set.

11. Let $K \subset \mathbb{R}^n$ be a compact convex set and $\xi \in \mathbb{R}^n \setminus K$. Denote by $R(\xi, K)$ the ray (13.16). Prove that for the metric projection $p(\xi, K) = p(y, K)$ for every $y \in R(\xi, K)$.

12. Let $K \subset \mathbb{R}^n$ be a convex set and $u : \mathbb{R}^n \to \mathbb{R}$ linear. For $x_0 \in K$ define $\xi := u(p) := \min\{u(x) \mid x \in K\}$ and $K' = K \cap \{x \in \mathbb{R}^n \mid u(x) = \xi\}$. Now show that $p \in \mathrm{ext}(K')$ implies $p \in \mathrm{ext}(K)$.

13.* Use Problem 11 to prove: given the system

$$a_{j1}x_1 + \cdots + a_{jn}x_n + b_j \geq 0, \quad j = 1, \ldots, m,$$

and assume that the set K of all solutions is compact in $\mathbb{R}^n$. Then K has at least one extreme point.

14. Give a geometric interpretation for (13.31), at least for $n = 1$ and $n = 2$.

15. Prove that if $f : K \to \mathbb{R}, K \subset \mathbb{R}^n$ convex, is a strictly convex function then in (13.31) and (13.32) the strict inequality holds.

16.   Use Problem 14 to show that if $f : \mathbb{R}^n \to \mathbb{R}$ is a strictly convex coercive C^1-function then f attains a unique minimum.

17.   Prove that the function $g : \mathbb{R}^2 \to \mathbb{R}$, $x \mapsto g(x) = \frac{2}{3}(1 + ||x||^2)^{\frac{3}{4}} - \frac{2}{3}$ is coercive and convex, and find the Legendre transform of g.

14 Spaces of Continuous Functions as Banach Spaces

We have seen that on certain infinite dimensional vector spaces we can introduce a norm, hence a metric, and hence we can study convergence and continuity on these spaces. For example we have considered the space $C(I) := \{f : I \to \mathbb{R} \mid f \text{ is continuous}\}$ where $I \in \mathbb{R}$ is an interval. If I is compact we may choose on $C(I)$ the sup-norm, i.e. $\|f\|_\infty := \sup_{x \in I} |f(x)|$. Now we may ask whether $C(I)$ is complete, whether we can find dense (with respect to $\|.\|_\infty$) subsets of $C(I)$, or whether we can characterise compact subsets of $C(I)$, etc.

We start with a rather general result for bounded functions on a set $X \neq \emptyset$. By $\mathcal{M}_b(X; \mathbb{R})$ we denote the vector space of all bounded functions $u : X \to \mathbb{R}$, i.e.

$$\mathcal{M}_b(X; \mathbb{R}) := \{u : X \to \mathbb{R} \mid \|u\|_\infty < \infty\} \tag{14.1}$$

where as usual we write

$$\|u\|_\infty := \sup_{x \in X} |u(x)|. \tag{14.2}$$

By Example 1.9 we know that $\mathcal{M}_b(X; \mathbb{R})$ is a normed space. We claim

Theorem 14.1. *The normed space $(\mathcal{M}_b(X; \mathbb{R}), \|.\|_\infty)$ is complete, i.e. a Banach space.*

Proof. We have to prove that if $(u_\nu)_{\nu \in \mathbb{N}}$, $u_\nu \in \mathcal{M}_b(X; \mathbb{R})$, is a Cauchy sequence then it has a limit $u \in \mathcal{M}_b(X; \mathbb{R})$, i.e. there exists $u \in \mathcal{M}_b(X; \mathbb{R})$ such that $\lim_{\nu \to \infty} \|u_\nu - u\|_\infty = 0$. Since $(u_\nu)_{\nu \in \mathbb{N}}$ is a Cauchy sequence, for $\epsilon > 0$ there exists $N_0 = N_0(\epsilon) \in \mathbb{N}$ such that $n, m \geq N_0(\epsilon)$ implies

$$\|u_n - u_m\|_\infty = \sup_{x \in X} |u_n(x) - u_m(x)| < \epsilon.$$

In particular we have for all $x \in X$ and $n, m \geq N_0(\epsilon)$ that

$$|u_n(x) - u_m(x)| < \epsilon. \tag{14.3}$$

This however means that for every $x \in X$ the sequence $(u_\nu(x))_{\nu \in \mathbb{N}}$ is a Cauchy sequence in $\mathbb{R}$ and therefore it has a limit

$$u(x) := \lim_{\nu \to \infty} u_\nu(x). \tag{14.4}$$

Thus by (14.4) we have defined a function $u : X \to \mathbb{R}$. Moreover, for $n \geq N_0(\epsilon)$ we may pass in (14.3) to the limit as m tends to infinity to arrive at

$$|u_n(x) - u(x)| \leq \epsilon \tag{14.5}$$

for all $x \in X$ and $n \geq N_0(\epsilon)$. This in turn yields that

$$\|u_n - u\|_\infty = \sup_{x \in X} |u_n(x) - u(x)| \leq \epsilon$$

for all $n \geq N_0(\epsilon)$ telling us that $(u_\nu)_{\nu \in \mathbb{N}}$ converges to u in the norm $\|.\|_\infty$. In addition we have

$$|u(x)| \leq |u_{N_0}(x) - u(x)| + |u_{N_0}(x)| \leq \epsilon + \|u_{N_0}\|_\infty$$

implying that $\|u\|_\infty = \sup_{x \in X} |u(x)|$ is bounded and hence $u \in \mathcal{M}_b(X; \mathbb{R})$. $\square$

If X is a metric space (in fact we may assume X to be a topological space) then we can consider the bounded and continuous functions $f : X \to \mathbb{R}$. For this vector space we will write $C_b(X; \mathbb{R})$ or just $C_b(X)$. We know by Theorem 2.32 that the uniform limit, i.e. the limit with respect to the sup-norm $\|.\|_\infty$, of continuous functions is continuous. Combined with Theorem 14.1 this gives

Theorem 14.2. *For a metric space (X, d) the space $C_b(X)$ is a Banach space.*

We now want to find "nice" dense subsets of $(C_b(K; \mathbb{R}), \|.\|_\infty)$ where K is a compact metric space. In particular the case $K \subset \mathbb{R}^n$, K compact, is covered. Recall that a subset Y in a metric space Z is dense if $\overline{Y} = Z$. Since $C_b(K)$ is a normed space, a subset $Y \subset C_b(K)$ will be dense if we can approximate every continuous function $u : K \to \mathbb{R}$ uniformly, i.e. with respect to $\|.\|_\infty$, by a sequence $(u_\nu)_{\nu \in \mathbb{N}}$, $u_\nu \in Y$. Approximating continuous functions is of central importance in numerical analysis, but it has also a lot of theoretical applications. For this reason we have spent time with approximation problems, in a first reading a student may skip this chapter. We start with the classical **Weierstrass approximation theorem**.

Theorem 14.3. *The set of all restrictions of polynomials to $[0, 1]$ is dense in $(C([0, 1]; \mathbb{R}), \|.\|_\infty)$.*

Remark 14.4. A. The above theorem claims that for every continuous function $u : [0, 1] \to \mathbb{R}$ there exists a sequence of polynomials $p_n : \mathbb{R} \to \mathbb{R}$ such that $\lim_{n \to \infty} \|p_n|_{[0,1]} - u\|_\infty = 0$, i.e. the functions $p_n|_{[0,1]}$ converge uniformly to u. Equivalently, for every $u \in C([0, 1])$ and $\epsilon > 0$ there exists a polynomial $p_\epsilon : \mathbb{R} \to \mathbb{R}$ such that $\|p_\epsilon|_{[0,1]} - u\|_\infty < \epsilon$.

B. Let $[a, b] \subset \mathbb{R}$ be any compact interval. The mapping $\varphi : [0, 1] \to [a, b]$, $\varphi(t) = (b - a)t + a$, is continuous and has a continuous inverse. For $u \in C([a, b]; \mathbb{R})$ it follows that $v := u \circ \varphi \in C([0, 1]; \mathbb{R})$ and

$$\sup_{s \in [a,b]} |u(s)| = \sup_{t \in [0,1]} |v(t)| = \sup_{t \in [0,1]} |(u \circ \varphi)(t)|.$$

This implies, see Problem 2, that Theorem 14.3 extends to $(C_b([a, b]; \mathbb{R}), \|.\|_\infty)$.

We will deduce the Weierstrass theorem from **Korovkin's theorem**:

Theorem 14.5. *Let $L_n : C([0, 1]) \to C([0, 1])$, $n \in \mathbb{N}$, be a sequence of bounded linear operators which are positivity preserving, i.e. $\|L_n f\|_\infty \le c_n \|f\|_\infty$ and $f \ge 0$ implies $L_n f \ge 0$. For $j = 0, 1, 2$ we denote by $f_j \in C([0, 1])$ the functions defined by $f_0(x) = 1$, $f_1(x) = x$ and $f_2(x) = x^2$. If $\lim_{n \to \infty} \|L_n f_j - f_j\|_\infty = 0$ for $j = 0, 1, 2$, then $\lim_{n \to \infty} \|L_n f - f\|_\infty = 0$ for all $f \in C([0, 1])$.*

Proof. The beautiful fact is that we need to check the convergence only for 3 special and rather simple functions and we will get convergence for all functions.

We know that $f \in C([0, 1])$ is bounded and uniformly continuous, i.e. $|f(x)| \le M$ for all $x \in [0, 1]$ and for $\epsilon > 0$ there exists $\delta > 0$ such that for all $x, y \in [0, 1]$

$$|x - y| < \sqrt{\delta} \text{ implies } |f(x) - f(y)| < \epsilon,$$

or equivalently

$$(x - y)^2 < \delta \text{ implies } |f(x) - f(y)| < \epsilon.$$

For $(x - y)^2 \ge \delta$ we note that $1 \le \frac{(x-y)^2}{\delta}$ and hence

$$|f(x) - f(y)| \le |f(x)| + |f(y)| \le \frac{2M}{\delta}(x - y)^2$$

for these x and y, implying for all $x, y \in [0, 1]$ that

$$|f(x) - f(y)| \le \epsilon + \frac{2M}{\delta}(x - y)^2. \tag{14.6}$$

Now we start to re-interpret (14.6). For all $y \in [0, 1]$ we have

$$|f(\cdot) - f(y)| \leq \epsilon f_0(\cdot) + \frac{2M}{\delta}(f_1(\cdot) - y)^2. \tag{14.7}$$

Next we have a closer look at the the operators L_n. They are linear and positivity preserving, hence they are monotone in the sense that $f \geq g$, i.e. $f(x) \geq g(x)$ for all $x \in [0, 1]$, implies $L_n f \geq L_n g$. Indeed, $f \geq g$ implies $f - g \geq 0$, i.e. $L_n f - L_n g = L_n(f - g) \geq 0$ or $L_n f \geq L_n g$. Since $-|f| \leq f \leq |f|$ we deduce that $-L_n|f| \leq L_n f \leq L_n|f|$, or $|L_n f| \leq L_n|f|$, and for the constant function $x \mapsto f(y)$ it follows that $L_n(f(y)) = L_n(f(y)f_0(\cdot)) = f(y)L_n f_0$. With these considerations (14.7) yields

$$|(L_n f)(x) - f(y)(L_n f_0)(x)|$$
$$\leq \epsilon(L_n f_0)(x) + \frac{2M}{\delta}\left((L_n f_2)(x) - 2y(L_n f_1)(x) + y^2(L_n f_0)(x)\right),$$

and for $y = x$ we find for all $x \in [0, 1]$

$$|(L_n f)(x) - f(x)L_n(f_0)(x)| \tag{14.8}$$
$$\leq \epsilon(L_n f_0)(x) + \frac{2M}{\delta}\left((L_n f_2)(x) - 2x(L_n f_1)(x) + x^2(L_n f_0)(x)\right).$$

Again, we re-interpret the result by reading (14.8) now as an estimate for functions, recall $x = f_1(x)$ and $x^2 = f_2(x)$, so we arrive at

$$|L_n f - f L_n f_0| \leq \epsilon L_n f_0 + \frac{2M}{\delta}(L_n f_2 - 2f_1 L_n f_1 + f_2 L_n f_0), \tag{14.9}$$

which gives further

$$|L_n f - f| \leq |L_n f - f L_n f_0| + |f L_n f_0 - f|$$
$$\leq \epsilon L_n f_0 + \frac{2M}{\delta}(L_n f_2 - 2f_1 L_n f_1 + f_2 L_n f_0) + |f L_n f_0 - f|.$$

The uniform convergence of $L_n f_j$ to f_j, $j = 0, 1, 2$ implies now

$$\|L_n f - f\|_\infty \leq \epsilon f_0 + \frac{2M}{\delta}(f_2 - 2f_1 f_1 + f_2 f_0) + |f f_0 - f|,$$

but $f_1 \cdot f_1 = f_2$, $f_2 \cdot f_0 = f_2$ and $f \cdot f_0 = f$ implying with $f_0 = 1$ that $\|L_n f - f\|_\infty \leq \epsilon$ proving the theorem. $\qquad \square$

Proof of Theorem 14.3. We will apply Korovkin's result and the operators L_n we define with the help of the **Bernstein polynomials** p_n

$$p_n(f)(x) := \sum_{k=0}^{n} \binom{n}{k} f\left(\frac{k}{n}\right) x^k (1-x)^{n-k}, \tag{14.10}$$

as $L_n f := p_n(f)|_{[0,1]}$. Clearly L_n is linear and positivity preserving and since

$$|L_n f(x)| = \left| p_n|_{[0,1]}(f)(x) \right| \leq \sum_{k=0}^{n} \binom{n}{k} \left| f\left(\frac{k}{n}\right) \right| x^k (1-x)^{n-k}$$

$$\leq \|f\|_\infty \sum_{k=0}^{n} \binom{n}{k} x^k (1-x)^{n-k} = \|f\|_\infty$$

it follows that $L_n : C([0,1]) \to C([0,1])$ is bounded, note that in the last steps we used that $\sum_{k=0}^{n} \binom{n}{k} x^k (1-x^k)^{n-k} = (x + (1-x))^n = 1$. Further, for $j = 0, 1, 2$ we find

$$p_n(f_0)(x) = 1 = f_0(x),$$
$$p_n(f_1)(x) = f_1(x),$$
$$p_n(f_2)(x) = \frac{1}{n} f_0(x) + \frac{n-1}{n} f_2(x),$$

implying on $[0,1]$ the uniform convergence of $L_n f_j$ to f_j, $j = 0, 1, 2$, and the Korovkin theorem implies now the uniform convergence of the polynomials $p_n(f)|_{[0,1]}$ to f and the Weierstrass theorem is proved. □

Remark 14.6. Our proof is taken from H. Bauer [7].

The Weierstrass theorem was analysed more carefully by M. Stone and finally extended to the Stone-Weierstrass theorem below. In our presentation we follow H. Bauer [6] who however was much influenced by J. Dieudonné, compare [10]. We need

Definition 14.7. *Let X be a set and $\mathcal{F}$ be a family of real-valued functions $f : X \to \mathbb{R}$. **A.** We say that $\mathcal{F}$ **separates points** in X if for every $x, y \in X$, $x \neq y$, there exists $f \in \mathcal{F}$ such that $f(x) \neq f(y)$. **B.** If for every $x, y \in X$, $x \neq y$, there exist $f, g \in \mathcal{F}$ such that $f(x)g(y) \neq f(y)g(x)$ we call $\mathcal{F}$ **cross separating** for points in X.*

Remark 14.8. If $\mathcal{F}$ is cross separating for points in X then F is separating points. Indeed, let $x, y \in X$, $x \neq y$. Then there exist $f, g \in \mathcal{F}$ such that $f(x)g(y) \neq f(y)g(x)$. If $f(x) = f(y)$ then $g(x) \neq g(y)$ and if $g(x) = g(y)$, then $f(x) \neq f(y)$. Moreover, if X has more than one point, then for every $x \in X$ there exists $f \in \mathcal{F}$ such that $f(x) \neq 0$. In addition, a certain converse statement holds: if $\mathcal{F}$ is separating points and if the constant function $x \mapsto 1$ belongs to $\mathcal{F}$, then $\mathcal{F}$ is also cross separating for points in X, see Problem 1.

Now we can prove **Stone's approximation theorem**:

Theorem 14.9. *Let K be a compact metric space and $H \subset C(K)$ a linear subspace. Suppose that H is cross separating for points in K and that $f \in H$ implies $|f| \in H$. Then H is dense in $C(K)$, i.e. for every $u \in C(K)$ and $\epsilon > 0$ there exists $h_\epsilon \in H$ such that $\|u - h_\epsilon\|_\infty < \epsilon$, or, equivalently, we can approximate $u \in C(K)$ uniformly by a sequence $(h_\nu)_{\nu \in \mathbb{N}}$, $h_\nu \in H$.*

Proof. First we claim that for $x, y \in K$, $x \neq y$, and $\alpha, \beta \in \mathbb{R}$ there exists $h \in H$ such that $h(x) = \alpha$ and $h(y) = \beta$. We can find $u, v \in H$ such that $u(x)v(y) \neq u(y)v(x)$ and therefore h defined by

$$h := \frac{\alpha v(y) - \beta v(x)}{u(x)v(y) - u(y)v(x)} u - \frac{\alpha u(y) - \beta u(x)}{u(x)v(y) - u(y)v(x)} v$$

belongs to H and $h(x) = \alpha$ as well as $h(y) = \beta$.
Now let $f \in C(K)$ and $\epsilon > 0$ be given. For $x, y \in K$, $x \neq y$, choose $h_{x,y} \in H$ such that

$$h_{x,y}(x) = f(x) \text{ and } h_{x,y}(y) = f(y),$$

note that by our preceding consideration such a function $h_{x,y}$ exists. The set

$$U_{x,y} := \{z \in K \,|\, h_{x,y}(z) < f(z) + \epsilon\}$$

is an open neighbourhood of x and for $y \in K$ fixed we have

$$K = \bigcup_{x \in K} U_{x,y},$$

i.e. $(U_{x,y})_{x \in K}$ is an open covering of K. Since K is compact there exists finitely many $x_1, \ldots, x_N \in K$ such that

$$K = \bigcup_{l=1}^{N} U_{x_l,y}.$$

We define
$$h_y := \inf \{ h_{x_1,y}, \ldots, h_{x_N,y} \},$$
i.e. $h_y : K \to \mathbb{R}$, $h_y(z) := \inf \{ h_{x_1,y}(z), \ldots, h_{x_N,y}(z) \}$. For $z \in K$ we find j, $1 \leq j \leq N$, such that $z \in U_{x_j,y}$, implying that
$$h_y(z) = h_{x_j,y}(z) \leq f(z) + \epsilon,$$
i.e. the inequality
$$h_y(z) \leq f(z) + \epsilon \tag{14.11}$$
holds for all $z \in K$. Since $h_y(z) = f(y)$ we deduce that
$$V_y := \{ z \in K \,|\, h_y(z) > f(z) - \epsilon \}$$
is an open neighbourhood of y. Again we find that
$$K = \bigcup_{y \in K} V_y,$$
i.e. $(V_y)_{y \in K}$ is an open covering of K implying the existence of $y_1, \ldots, y_M$ such that
$$K = \bigcup_{j=1}^{M} V_{y_j}.$$
We define
$$h := \sup \{ h_{y_1}, \ldots, h_{y_M} \}$$
and we find
$$f(z) - \epsilon < h(z) < f(z) + \epsilon$$
implying that
$$\| f - h \|_\infty < \epsilon.$$
Since $g \in H$ implies $|g| \in H$ it follows for $g_1, g_2 \in H$ that $\sup\{g_1, g_2\} = \frac{1}{2}(g_1 + g_2) + \frac{1}{2}|g_1 - g_2|$ and $\inf\{g_1, g_2\} = \frac{1}{2}(g_1 + g_2) - \frac{1}{2}|g_1 - g_2|$ belong to H, and hence h_y and h belong to H and the theorem is proven. $\qquad\square$

We want to replace the conditions H being a subspace and $f \in H$ implies $|f| \in H$ by a seemingly more algebraic one, namely we want to consider instead of H a sub-algebra of $C(K)$, i.e. a subspace A with the property that for $h, g \in A$ the product $f \cdot g$ belongs to A too. First we note that the closure of a sub-algebra $A \subset C(K)$ is again a sub-algebra.

Proposition 14.10. *Let $A \subset C(K)$ be a sub-algebra and $\bar{A}$ its closure in $C(K)$, i.e. with respect to the norm $\|.\|_\infty$. Then $\bar{A}$ is a sub-algebra of $C(K)$ and $f \in \bar{A}$ implies $|f| \in \bar{A}$.*

Proof. We give the proof for the case where the constants belong to A, in Appendix II we will handle the general case. Let $g, h \in \bar{A}$ and $(g_\nu)_{\nu \in \mathbb{N}}$, $(h_\nu)_{\nu \in \mathbb{N}}$ be sequences in A such that $\lim_{\nu \to \infty} \|g_\nu - g\|_\infty = 0$ and $\lim_{\nu \to \infty} \|h_\nu - h\|_\infty = 0$. The properties of the limit immediately give

$$\lim_{\nu \to \infty} \|(\lambda g_\nu + \mu h_\nu) - (\lambda g + \mu h)\|_\infty = 0$$

as well as

$$\lim_{\nu \to \infty} \|g_\nu h_\nu - gh\|_\infty = 0,$$

so $\bar{A}$ is a sub-algebra of $C(K)$. Now we prove that $f \in \bar{A}$ implies $|f| \in \bar{A}$. Let $(f_\nu)_{\nu \in \mathbb{N}}$, $f_\nu \in A$, such that $\lim_{\nu \to \infty} \|f_\nu - f\|_\infty = 0$. The converse triangle inequality yields

$$\left\| |f| - |f_\nu| \right\|_\infty \leq \|f - f_\nu\|_\infty,$$

hence we need to prove that $f \in A$ implies $|f| \in \bar{A}$. If f is identically zero, nothing is to prove, otherwise $\|f\|_\infty \neq 0$ and we may consider instead of f the function $\frac{f}{\|f\|_\infty}$, so we may assume without loss of generality that $\|f\|_\infty \leq 1$. By the Weierstrass theorem (applied to the interval $[-1, 1]$, see Problem 2) we can find a sequence $(p_n)_{n \in \mathbb{N}}$ of polynomials such that this sequence approximates uniformly on $[-1, 1]$ the function $|.|$. This however implies the uniform convergence of $(p_n \circ f)_{n \in \mathbb{N}}$ to $|f|$, note that $\frac{|f|}{\|f\|_\infty} \leq 1$, i.e. $f(x) = y \in [-1, 1]$, and therefore

$$\left| p_n(f(x)) - |f(x)| \right| = \left| p_n(y) - |y| \right| \leq \sup_{z \in [-1,1]} \left| p_n(z) - |z| \right|.$$

Once we know that $p_n(f) \in A$ the result follows. Under the extra assumption that the constants belong to A this condition holds. For the general case we need the condition that $p_n(0) = 0$ and such an approximating sequence of polynomials converging uniformly on $[-1, 1]$ to $|.|$ will be constructed in Appendix II. $\square$

As a consequence of Stone's theorem and Proposition 14.10 we have the **Stone-Weierstrass theorem**

290

Theorem 14.11. *Let K be a compact metric space and $A \subset C(K)$ a sub-algebra which is cross separating for points in K. Then A is dense in $C(K)$ with respect to the norm $\|.\|_\infty$.*

Proof. Since $\bar{A}$ satisfies all conditions of Stone's theorem the result follows immediately. ☐

Of course, we can now recover the Weierstrass theorem, but note that our proof (without Appendix II) made use of the Weierstrass theorem. Thus only if we replace the final part of our argument of the proof of Proposition 14.10 by Appendix II we get an independent proof of the Weierstrass theorem.

Example 14.12. In the Weierstrass theorem we cannot replace the compact interval by a non-compact interval. In particular for $\mathbb{R}$ we find that $u :$
$$\mathbb{R} \to \mathbb{R}, \; u(x) = \begin{cases} 1 - |x| & , |x| \leq 1 \\ 0 & , |x| > 1 \end{cases} \text{ is bounded and continuous. If } p \text{ was}$$
a polynomial on $\mathbb{R}$ such that $|p(x) - u(x)| < \frac{1}{2}$ for all $x \in \mathbb{R}$, it followed for $|x| > 1$ that $|p(x)| \leq \frac{1}{2} + |u(x)| \leq \frac{1}{2}$. Hence p must be a constant, i.e. $p(x) = \alpha$ for all $x \in \mathbb{R}$ and $\alpha < \frac{1}{2}$. But $|u(0) - p(0)| = |1 - \alpha| > \frac{1}{2}$ which is a contradiction.

It is common in textbooks to add as a further application of the Stone-Weierstrass theorem the following result:

Example 14.13. Consider on $\mathbb{R}$ the space of all continuous 2π-periodic functions

$$C_{\text{per}} := C_{\text{per}}([0, 2\pi]; \mathbb{R}) := \left\{ u \in C(\mathbb{R}) \, \middle| \, u(x) = u(x + 2\pi) \right\}$$

on which we take the norm

$$\|u\|_\infty := \sup_{x \in [0, 2\pi]} |u(x)|.$$

The span of the functions $1, \cos k(\cdot)$ and $\sin k(\cdot)$, $k \in \mathbb{N}$, forms an algebra since

$$\cos kx \cos lx = \frac{1}{2} \left(\cos(k - l)x + \cos(k + l)x \right)$$

$$\cos kx \sin lx = \frac{1}{2} \left(\sin(l - k)x + \sin(k + l)x \right)$$

$$\sin kx \sin lx = \frac{1}{2} \left(\cos(k - l)x - \cos(k + l)x \right).$$

Denote this algebra by $\mathrm{Trig}[0, 2\pi]$. Now, $\mathrm{Trig}[0, 2\pi]$ can never be point separating on $[0, 2\pi]$ due to the periodicity, hence all $u \in \mathrm{Trig}[0, 2\pi]$, in fact all $u \in C_{\mathrm{per}}$ have at 0 and 2π the same value. However on $[0, 2\pi)$ the sub-algebra $\mathrm{Trig}[0, 2\pi] \subset C_{\mathrm{per}}$ contains the constant functions using the injectivity of $\cos$ on $[0, \pi]$ and $[\pi, 2\pi]$, as well as the fact that $\sin$ is positive in $(0, \pi)$ and negative on $(\pi, 2\pi)$, it is easy to see that $\mathrm{Trig}[0, 2\pi]$ is cross separating for points in $[0, 2\pi)$. But $[0, 2\pi)$ is **not** compact, so we cannot apply the Stone-Weierstrass directly. There are two (essentially identical) ways to overcome the problem and in Volume III in our chapter on Fourier analysis we will discuss them in detail.

The first approach is to consider first $C(S^1, \mathbb{C})$, $S^1 \subset \mathbb{R}^2 \cong \mathbb{C}$, and extend (in a rather straightforward way) the results to complex-valued functions. Instead of $\mathrm{Trig}[0, 2\pi]$ we then use the span of the functions $\varphi \mapsto e^{ik\varphi}$, $k \in \mathbb{Z}$. This is a point separating sub-algebra of $C(S^1, \mathbb{C})$ containing the constants, hence it is cross separating for points in S^1 and consequently dense in $C(S^1, \mathbb{C})$. Now switching to real and imaginary parts and using the periodicity of $\varphi \mapsto e^{ik\varphi}$ we can deduce that $\mathrm{Trig}[0, 2\pi]$ is dense in C_{per}. The other approach is to start with the fact that we cannot distinct 0 and 2π using 2π-periodic functions, thus we may identify them and consider the corresponding (topological) identification space, which turns out to be homeomorphic to S^1.

Let (X, d) be a metric space (or even consider a topological space $(X, \mathcal{O})$) and let $u : X \to \mathbb{R}$ be a function. We define the **support** of u, $\mathrm{supp}\, u$, as the closure of the set on which u does not vanish, i.e.

$$\mathrm{supp}\, u = \overline{\{x \in X \mid u(x) \neq 0\}}. \tag{14.12}$$

By definition $\mathrm{supp}\, u$ is closed and it is the complement of the largest open set in which u vanishes. If $\mathrm{supp}\, u$ is compact we say that u has **compact support**. The space of all continuous functions with compact support we denote by $C_0(X)$. If $X = K$ is compact, then we clearly have $C_0(K) = C_b(K)$, however for non-compact X we have $C_0(X) \subset C_b(X)$, but $C_0(X) \neq C_b(X)$. Let us restrict to the case $X = \mathbb{R}^n$ and on $\mathbb{R}^n$ we use as always the Euclidean norm. First we claim

Lemma 14.14. *For $x_0 \in \mathbb{R}^n$ and $r > 0$ the function $u : \mathbb{R}^n \to \mathbb{R}$ defined by*

$$u(x) := \begin{cases} \exp\left(\frac{-1}{r^2 - \|x - x_0\|^2}\right) & , \|x - x_0\| < r \\ 0 & , \|x - x_0\| \geq r \end{cases}$$

belongs to $C_0(\mathbb{R}^n) \cap C^\infty(\mathbb{R}^n)$ *and* $\operatorname{supp} u = \overline{B_r(x_0)}$.

Proof. That $u \in C^\infty(\mathbb{R}^n)$ follows essentially from Problem 5 in Chapter I.9 and the support property is trivial. $\qquad\square$

Further we note

Lemma 14.15. *If* $u : \mathbb{R}^n \to \mathbb{R}$ *has a partial derivative* $\frac{\partial u}{\partial x_j}$, *then* $\operatorname{supp} \frac{\partial u}{\partial x_j} \subset$ $\operatorname{supp} u$.

Proof. Let $y \in (\operatorname{supp} u)^\complement$. Then there exists an open ball $B_\rho(y) \subset (\operatorname{supp} u)^\complement$ and in $B_\rho(y)$ we have $\frac{\partial u}{\partial x_j} = 0$, thus $\frac{\partial u}{\partial x_j} = 0$ in $(\operatorname{supp} u)^\complement$ implying $\operatorname{supp} \frac{\partial u}{\partial x_j} \subset$ $\operatorname{supp} u$. $\qquad\square$

From Lemma 14.15 we deduce for every linear partial differential operator with continuous coefficients, i.e.

$$L(x, D) = \sum_{|\alpha| \leq m} a_\alpha(x) D^\alpha,$$

where $L(x, D)$ is defined on $C^m(\mathbb{R}^n)$ (or $C^m(G)$, $G \subset \mathbb{R}^n$ open) by

$$L(x, D)u(x) = \sum_{|\alpha| \leq m} a_\alpha(x) D^\alpha u(x),$$

that
$$\operatorname{supp} L(x, D)u \subset \operatorname{supp} u, \tag{14.13}$$

i.e. differential operators do not increase the support of a function. Operators with such a property are called local:

Definition 14.16. *Let* $A : D(A) \to C(\mathbb{R}^n)$ *be a linear mapping defined as the subspace* $D(A) \subset C(\mathbb{R}^n)$. *If for all* $u \in D(A)$ *the following holds*

$$\operatorname{supp} Au \subset \operatorname{supp} u \tag{14.14}$$

we call A *a **local operator**.*

Thus differential operators are local operators.

Next we want to determine the closure of $C_0(\mathbb{R}^n)$ in $C_b(\mathbb{R}^n)$.

Definition 14.17. *We say that* $u : \mathbb{R}^n \to \mathbb{R}$ *vanishes at infinity if for every* $\epsilon > 0$ *there exists* $R = R(\epsilon) \geq 0$ *such that* $\|x\| \geq R(\epsilon)$ *implies* $|u(x)| \leq \epsilon$.

Remark 14.18. Equivalent to this definition is that for every $\epsilon > 0$ there exists a compact set $K \subset \mathbb{R}^n$ such that $x \in K^{\complement}$ implies $|u(x)| < \epsilon$. This formulation allows us to extend the notion "vanishing at infinity" to locally compact spaces.

Lemma 14.19. *The uniform limit* u *of a sequence* $(u_\nu)_{\nu \in \mathbb{N}}$, $u_\nu \in C_0(\mathbb{R}^n)$, *vanishes at infinity.*

Proof. Let $\epsilon > 0$ and $N(\epsilon) \in \mathbb{N}$ such that $\nu \geq N(\epsilon)$ implies that $\|u - u_\nu\|_\infty < \epsilon$. In particular we find

$$|u(x)| \leq |u(x) - u_{N(\epsilon)}(x)| + |u_{N(\epsilon)}|$$

and for $x \in (\operatorname{supp} u_{N(\epsilon)})^{\complement}$ it follows that

$$|u(x)| \leq |u(x) - u_{N(\epsilon)}(x)| \leq \left\| u - u_{N(\epsilon)} \right\|_\infty < \epsilon.$$

$\square$

Theorem 14.20. *The closure of* $C_0(\mathbb{R}^n)$ *in* $C_b(\mathbb{R}^n)$ *is the space of all continuous functions* $u : \mathbb{R}^n \to \mathbb{R}$ *vanishing at infinity.*

Proof. We already know that every element in the closure of $C_0(\mathbb{R}^n)$ must vanish at infinity. It remains to show that if $u : \mathbb{R}^n \to \mathbb{R}$ is continuous and vanishes at infinity, then we can find a sequence $(u_\nu)_{\nu \in \mathbb{N}}$, $u_\nu \in C_0(\mathbb{R}^n)$, converging uniformly to u. For $\nu \in \mathbb{N}$ define $\varphi_\nu : \mathbb{R}^n \to \mathbb{R}$ by

$$\varphi_\nu(x) := \begin{cases} 1 & , \|x\| \leq \nu \\ -\frac{1}{\nu}\|x\| + 2 & , \nu \leq \|x\| < 2\nu \\ 0 & , \|x\| \geq 2\nu. \end{cases}$$

The functions $u_\nu := u\varphi_\nu$ are continuous and have compact support, $\operatorname{supp} u_\nu \subset \overline{B_{2\nu}(0)}$. Since u vanishes at infinity, for $\epsilon > 0$ we can find $R(\epsilon)$ such that $|u(x)| < \frac{\epsilon}{2}$ for $x \in B_{R(\epsilon)}^{\complement}(0)$. With $\nu \geq N = N(\epsilon) \geq R(\epsilon)$ it follows for $x \in \mathbb{R}^n$

$$|u(x) - u_\nu(x)| = |u(x)||1 - \varphi_\nu(x)| < \epsilon,$$

indeed, if $x \in B_{R(\epsilon)}^{\complement}(0)$ then $|u(x)||1 - \varphi_\nu(x)| \leq 2|u(x)|$, and if $x \in B_{R(\epsilon)}(0)$ then $1 - \varphi_\nu(x) = 0$ since $\varphi_\nu|_{B_\nu(0)} = 1$. Thus $\|u - u_\nu\|_\infty \leq \epsilon$ for $\nu \geq N(\epsilon)$ implying the result. $\square$

Definition 14.21. *We denote by $C_\infty(\mathbb{R}^n)$ or $C_\infty(\mathbb{R}^n; \mathbb{R})$ the space of all continuous functions $u : \mathbb{R}^n \to \mathbb{R}$ vanishing at infinity.*

From our previous considerations we conclude

Theorem 14.22. *The space $(C_\infty(\mathbb{R}^n), \|.\|_\infty)$ is a Banach space with $C_0(\mathbb{R}^n)$ being dense.*

Thus we have $C_0(\mathbb{R}^n) \subset C_\infty(\mathbb{R}^n) \subset C_b(\mathbb{R}^n)$ and these are proper inclusions. Note that Theorem 14.22 holds for more general spaces, and we will pick this up in later parts. For reference purpose we introduce now the notation

$$C_0^k(G) := C_0(G) \cap C^k(G) \tag{14.15}$$

for $k \in \mathbb{N} \cup \{\infty\}$ and $G \subset \mathbb{R}^n$ open.

Finally in this chapter we want to characterise compact sets of $C(K)$ where K is a compact metric space. We will follow closely F. Hirzebruch and W. Scharlau [27]. Let (X, d) be a metric space. We recall the notion of a relative compact subset of X as a set whose closure is compact and that of a totally bounded set $Y \subset X$ as a set with the property that for every $\epsilon > 0$ there exists a finite covering of Y with open sets of diameter less than ϵ, see Definition 3.15. By Theorem 3.18 we know that X is compact if it is totally bounded and complete. Moreover we have

Lemma 14.23. *Let (X, d) be a metric space. A subset $Y \subset X$ is totally bounded if and only if its closure $\overline{Y}$ is totally bounded. Moreover, a subset $Y \subset X$ is relative compact if and only if it is totally bounded and its closure is complete.*

Proof. If the set Y is totally bounded, given $\epsilon > 0$ there exist $x_1, \ldots, x_N \in Y$ such that $Y \subset \bigcup_{j=1}^N B_{\frac{\epsilon}{2}}(x_j)$ which implies that

$$\overline{Y} \subset \overline{\bigcup_{j=1}^N B_{\frac{\epsilon}{2}}(x_j)} \subset \bigcup_{j=1}^N B_\epsilon(x_j),$$

hence $\overline{Y}$ is totally bounded. If $\overline{Y}$ is totally bounded we find for some $\epsilon > 0$ points $y_1, \ldots, y_m \in \overline{Y}$ such that $\overline{Y} \subset \bigcup_{l=1}^M B_{\frac{\epsilon}{2}}(y_l)$. Let $x_l \in Y$, $1 \leq l \leq M$, such that $d(x_l, y_l) < \frac{\epsilon}{2}$. Then $B_{\frac{\epsilon}{2}}(y_l) \subset B_\epsilon(x_l)$ and $Y \subset \bigcup_{l=1}^M B_\epsilon(x_l)$. This proves the first part of the lemma. The second assertion follows now from the first part together with Theorem 3.18. $\qquad\square$

To proceed further we need

Definition 14.24. *We call a subset $H \subset C(K)$* ***equi-continuous*** *if for every $x \in K$ and $\epsilon > 0$ there exists a neighbourhood $U(x)$ of x such that $|f(y) - f(x)| < \epsilon$ for all $f \in H$ and all $y \in U(x)$.*

Now we can prove a first version of the **Theorem of Arzela and Ascoli**:

Theorem 14.25. *Let K be a compact metric space. A subset $H \subset C(K; \mathbb{R})$ is relative compact if and only if H is equi-continuous and for all $x \in K$ the set $H(x) := \{ f(x) \,|\, f \in H \} \subset \mathbb{R}$ is relative compact in $\mathbb{R}$. (Recall that on $C(K)$ we take the sup-norm $\|.\|_\infty$.)*

Proof. We first prove that if H is relative compact then it must be equi-continuous and $H(x)$ must be for every $x \in K$ relative compact in $\mathbb{R}$. We start with proving that H is equi-continuous. Since H is relative compact it is totally bounded, hence for $\epsilon > 0$ there exists $f_1, \ldots, f_M \in H$ such that for every $f \in H$ and some $1 \leq j \leq M$ it follows that

$$\|f - f_j\|_\infty < \frac{\epsilon}{3}.$$

For $x \in K$ fixed we define the neighbourhood $U(x)$ of x

$$U(x) := \left\{ y \in K \,\Big|\, |f_j(y) - f_j(x)| < \frac{\epsilon}{3} \text{ for all } j = 1, \ldots, M \right\}.$$

It follows for $y \in U(x)$ and $f \in H$ with $j = j(f)$ suitable

$$|f(y) - f(x)| \leq |f(y) - f_j(y)| + |f_j(y) - f_j(x)| + |f_j(x) - f(x)| < \epsilon$$

which proves that H is equi-continuous. Next we show that for every $x \in K$ the set $H(x)$ is relative compact in $\mathbb{R}$. Since H is totally bounded it follows that $H(x)$ must be totally bounded and hence bounded. Since a bounded subset of $\mathbb{R}$ is relative compact we have proved the first direction. Now we want to prove that if H is equi-continuous and for all $x \in K$ the set $H(x)$ is relative compact then $H \subset C(K)$ is relative compact. First we show that $\overline{H}$ is complete. For this let $(f_\nu)_{\nu \in \mathbb{N}}$, $f_\nu \in H$, be a Cauchy sequence. Since $H(x)$ is relative compact it follows that $(f_\nu(x))_{\nu \in \mathbb{N}}$ is a Cauchy sequence in $\mathbb{R}$, hence is convergent. Denote the limit of $(f_\nu(x))_{\nu \in \mathbb{N}}$ by $g(x)$. Thus by $g : K \to \mathbb{R}$, $g(x) := \lim_{\nu \to \infty} f_\nu(x)$, a function g is defined. We claim that g is continuous. Let $x \in K$ and $\epsilon > 0$. The equi-continuity of H implies the existence of an open neighbourhood $U(x)$ of x such that for all $y \in U(x)$ and all f_ν, $\nu \in \mathbb{N}$,

$$|f_\nu(x) - f_\nu(y)| < \frac{\epsilon}{3}$$

holds. This implies for all $y \in U(x)$ and ν_0 sufficiently large (ν_0 may depend on y) that

$$|g(y) - g(x)| \leq |g(y) - f_{\nu_0}(y)| + |f_{\nu_0}(y) - f_{\nu_0}(x)| + |f_{\nu_0}(x) - g(x)| < \epsilon$$

and hence the pointwise continuity of the function g. Thus the sequence $(f_\nu)_{\nu \in \mathbb{N}}$ converges pointwise to a continuous function g. We want to prove that this is indeed a uniform limit.

Now we will use that K is compact. Hence elements in $C(K)$, and therefore elements in H, are bounded and uniformly continuous. The equi-continuity of H implies now the existence of open balls $B_{r_1}(x_1), \ldots, B_{r_N}(x_N)$ such that

$$K = \bigcup_{j=1}^{N} B_{r_j}(x_j)$$

and for all $y \in B_{r_j}(x_j)$ and all $f \in H$

$$|f(y) - f(x_j)| < \frac{\epsilon}{3} \text{ as well as } |g(y) - g(x_j)| < \frac{\epsilon}{3}$$

holds. The estimates for g use the uniform continuity of g on K.

Further we know the existence of $\nu_0 \in \mathbb{N}$ such that $\nu \geq \nu_0$ implies for $j = 1, \ldots, N$

$$|f_\nu(x_j) - g(x_j)| < \frac{\epsilon}{3}.$$

Together we find for $y \in K$, say $y \in B_{r_{j_0}}(x_0)$, and $\nu > \nu_0$ that

$$|f_\nu(y) - g(y)| \leq |f_\nu(y) - f_\nu(x_{j_0})| + |f_\nu(x_{j_0}) - g(x_{j_0})| + |g(x_{j_0}) - g(y)| < \epsilon$$

which yields the uniform convergence of $(f_\nu)_{\nu \in \mathbb{N}}$ to g.

Finally we will prove that H is totally bounded which will imply that H is relative compact in $C(K)$. For this let $B_{r_j}(x_j)$, $j = 1, \ldots, N$, be as above. We know that $H(x_j)$ is relative compact and hence $Y := \bigcup_{j=1}^{N} H(x_j)$ is relative compact, hence totally bounded. Therefore we can cover Y with a finite number M of balls $B_{\frac{\epsilon}{6}}(y_l)$, i.e. $Y \subset \bigcup_{l=1}^{M} B_{\frac{\epsilon}{6}}(y_l)$. Denote by Φ the set of all mappings $\varphi : \{1, \ldots, N\} \to \{1, \ldots, M\}$. For $\varphi \in \Phi$ we define

$$G_\varphi := \left\{ f \in H \,\middle|\, |f(x_j) - y_{\varphi(j)}| < \frac{\epsilon}{6} \text{ for } j = 1, \ldots, N \right\}.$$

It follows that $H = \bigcup_{\varphi \in \Phi} G_\varphi$. Since Φ is a finite set we have shown that H is totally bounded if we can prove that for $f, g \in G_\varphi$ it follows that $\|f - g\|_\infty < \epsilon$. However for $y \in K$, say $y \in B_{r_{j_0}}(x_{j_0})$, and $f, g \in G_\varphi$ we find

$$|f(y) - g(y)| \le |f(y) - f(x_{j_0})| + |f(x_{j_0}) - y_{\varphi(j_0)}|$$
$$+ |y_{\varphi(j_0)} - g(x_{j_0})| + |g(x_{j_0}) - g(y)| < \frac{\epsilon}{3} + \frac{\epsilon}{6} + \frac{\epsilon}{6} + \frac{\epsilon}{3} = \epsilon,$$

and the result follows. $\qquad\qquad\square$

Remark 14.26. We have not used much the fact that we are dealing with real-valued mappings. In fact with a few small modifications, see [27], the proof holds when $C(K)$ is replaced by $C(E; X)$ where E is a topological space and (X, d) is a metric space. Here $C(E; X)$ stands for all continuous mappings $u : E \to X$, and is not necessarily a vector space.

Corollary 14.27. *A set $H \subset C(K)$ is relative compact if and only if H is equi-continuous and H is bounded in $C(K)$, i.e. $\sup_{f \in H} \|f\|_\infty \le M < \infty$.*

Proof. If H is bounded in $C(K)$ then all sets $H(x)$ are bounded in $\mathbb{R}$, hence relative compact. Conversely, if all the sets $H(x)$ are relative compact, hence bounded, the equi-continuity of H and the compactness of K yields the boundedness of H. $\qquad\qquad\square$

Example 14.28. On $C([0, 1])$ we consider the family $H = \{f_n \mid n \in \mathbb{N}\}$ where $f_n(x) = \frac{x^4}{x^4 + (1 - nx)^4}$. Since $|f_n(x)| \le 1$ we find $\sup_{f_n \in H} \|f_n\|_\infty = 1$. Moreover, for all $x \in [0, 1]$ we have $\lim_{n \to \infty} f_n(x) = 0$, but the convergence is not uniform since $f_n\left(\frac{1}{n}\right) = 1$. Thus H cannot be equi-continuous and uniform boundedness of H is not sufficient for relative compactness in $C([0, 1])$ (or $C(K)$, K compact).

Example 14.29. Let H be a family of continuous mappings $f : [a, b] \to \mathbb{R}$ such that $\sup_{f \in H} \|f\|_\infty \le M < \infty$ and all elements in H have on (a, b) a derivative which extends to a continuous function f' on $[a, b]$ and in addition

$\sup_{f \in H} \|f'\|_\infty \leq L < \infty$. We claim that H is equi-continuous. Indeed, the mean value theorem implies

$$|f(x) - f(y)| = |f'(\xi)||x - y| \leq L|x - y|$$

where L is independent of f. Hence for $x \in [a, b]$ fixed and $\epsilon > 0$ given we choose $\delta := \frac{\epsilon}{L}$ to find for $U(x) := B_\delta(x)$ that $y \in B_\delta(x)$ implies

$$|f(y) - f(x)| \leq L|x - y| < L \cdot \delta = \epsilon.$$

Hence H satisfies the conditions of the Arzela-Ascoli theorem.

We want to discuss some further applications of the notion of equi-continuity, for which we follow mainly H. Heuser [25] and [27].

Definition 14.30. *Let (X, d) be a metric space and $\mathcal{F}$ a family of functions $f : X \to \mathbb{R}$. We call $\mathcal{F}$ **uniformly equi-continuous** if for every $\epsilon > 0$ there exists $\delta > 0$ such that for all $f \in \mathcal{F}$ and all $x, y \in X$ the condition $d(x, y) < \delta$ implies $|f(x) - f(y)| < \epsilon$.*

Example 14.31. Let $[a, b] \subset \mathbb{R}$ be a compact interval and (X, d) a compact metric space. Let $f : [a, b] \times X \to \mathbb{R}$ be a continuous function. This function is on the compact space uniformly continuous and therefore, for every $\epsilon > 0$ there exists $\delta > 0$ such that $d(x, x_0) < \delta$ implies $|f(y, x) - f(y, x_0)| < \epsilon$. This means however that the family

$$\left\{ g_y : X \to \mathbb{R} \,\middle|\, y \in [a, b] \text{ and } g_y(x) = f(x, y) \right\}$$

is a uniformly equi-continuous family (and a subset of $C([a, b])$).

Theorem 14.32 (Continuity of parameter dependent integrals). *Let $[a, b] \subset \mathbb{R}$ be compact and (X, d) be a compact metric space. Further let $f : [a, b] \times X \to \mathbb{R}$ be a continuous function. Then for every $x \in X$ the integral*

$$F(x) := \int_a^b f(y, x) \, dy \tag{14.16}$$

exists and $F : X \to \mathbb{R}$ is a continuous function.

Proof. Since $f(\cdot, x) : [a, b] \to \mathbb{R}$ is continuous the integral is well defined for every $x \in X$. Now, by Example 14.31 we know that the family

$\{f(y, \cdot) : X \to \mathbb{R} \mid y \in [a, b]\}$ is a uniformly equi-continuous family of functions. It follows that for every $\epsilon > 0$ there exists $\delta > 0$ such that for all $y \in [a, b]$ and all $x, x_0 \in X$ the condition $d(x, x_0) < \delta$ implies $|f(y, x) - f(y, x_0)| < \frac{\epsilon}{b-a}$. This yields however

$$|F(x) - F(x_0)| = \left| \int_a^b (f(y, x) - f(y, x_0)) \, dy \right|$$

$$\leq \int_a^b |f(y, x) - f(y, x_0)| \, dy < \frac{\epsilon}{b - a} \cdot (b - a) = \epsilon,$$

proving the continuity of F. $\qquad\square$

Along these lines it is also possible to derive a differentiability result for F. We prefer a slightly different approach following O. Forster [19]. We start with

Lemma 14.33. Let $I, J \subset \mathbb{R}$ be two compact intervals and

$$f : I \times J \to \mathbb{R}$$
$$(x, y) \mapsto f(x, y)$$

be a continuous function which has a continuous partial derivative with respect to the second variable. Further let $c \in J$ and $(y_k)_{k \in \mathbb{N}}$, $y_k \in J$, be a sequence such that $y_k \neq c$ for all $k \in \mathbb{N}$ and $\lim_{k \to \infty} = c$. Define

$$\tilde{F}_k(x) := \frac{f(x, y_k) - f(x, c)}{y_k - c}, \quad k \in \mathbb{N},$$

and

$$\tilde{F}(x) = \frac{\partial f}{\partial y}(x, c).$$

It follows that $(\tilde{F}_k)_{k \in \mathbb{N}}$ converges uniformly on I to $\tilde{F}$.

Proof. Let $\epsilon > 0$. The continuous function $D_2 f : I \times J \to \mathbb{R}$ is on the compact set $I \times J$ uniformly continuous, hence there is $\delta > 0$ such that for $y, y' \in J$, $|y - y'| < \delta$ implies

$$|D_2 f(x, y) - D_2(x, y')| < \epsilon.$$

Now the mean value theorem yields that for each $k \in \mathbb{N}$ there is η_k lying between c and y_k such that

$$\tilde{F}_k(x) = D_2 f(x, \eta_k).$$

Take $N \in \mathbb{N}$ sufficiently large such that $|c - y_k| < \delta$ for all $k \geq N$. Then it follows that $|c - \eta_k| < \delta$ for all $k \in \mathbb{N}$ and further

$$|\tilde{F}(x) - \tilde{F}_k(x)| = |D_2 f(x, c) - D_2 f(x, \eta_k)| < \epsilon$$

for all $x \in I$, $k \geq N$, and the lemma is proved. $\square$

Theorem 14.34 (Differentiation of parameter dependent integrals).
Let $I, J \subset \mathbb{R}$ be compact intervals and $f : I \times J \to \mathbb{R}$ a continuous function which has a continuous partial derivative with respect to the second variable. For $y \in J$ set

$$\varphi(y) := \int_I f(x, y) \, dx. \tag{14.17}$$

Then the function $\varphi : J \to \mathbb{R}$ is continuously differentiable and

$$\frac{d\varphi}{dy}(y) = \int_I \frac{\partial f(x, y)}{\partial y} \, dx \tag{14.18}$$

holds.

Proof. Let $c \in J$ and $(y_k)_{k \in \mathbb{N}}$, $y_k \in J$, a sequence satisfying $\lim_{k \to \infty} y_k = c$ and $y_k \neq c$ for all k. Further let $\tilde{F}_k$ and $\tilde{F}$ be as in Lemma 14.33. Since $\tilde{F}_k$ converges uniformly to $\tilde{F}$ we find

$$\lim_{k \to \infty} \frac{\varphi(y_k) - \varphi(c)}{y_k - c} = \lim_{k \to \infty} \int_I \frac{f(x, y_k) - f(x, c)}{y_k - c} \, dx$$
$$= \int_I \frac{\partial f}{\partial y}(x, c) \, dx.$$

Hence $\varphi'(c)$ exists and for every $c \in J$ we have

$$\varphi'(c) = \int_I \frac{\partial f}{\partial y}(x, c) \, dx.$$

The continuity of $\frac{\partial f}{\partial y}$ on the compact set $I \times J$ implies also the continuity of φ'. $\square$

Of course we can iterate the result of Theorem 14.34 if f has higher order partial derivatives with respect to y:

Corollary 14.35. *If in the situation of Theorem 14.34 the function f has with respect to the second variable all continuous partial derivatives $\frac{\partial^l}{\partial y^l} f(x, y)$, $l = 1, \ldots, k$, then φ is a C^k-function and for $1 \leq l \leq k$ the following holds*

$$\frac{d^l \varphi}{dy^l}(y) = \int_I \frac{\partial^l f(x, y)}{\partial y^l} dx. \tag{14.19}$$

We want to use Corollary 14.35 to derive an integral representation for Bessel functions. In Problem 7 in Chapter I.29 we introduced for $l \in \mathbb{N}_0$ the lth Bessel function J_l as the solution of the ordinary differential equation

$$x^2 J_l''(x) + x J_l'(x) + (x^2 - l^2) J_l(x) = 0 \tag{14.20}$$

by giving a power series representation which converges on all $\mathbb{R}$:

$$J_l(x) = \sum_{n=0}^{\infty} \frac{(-1)^n x^{l+2n}}{2^{l+2n} n! (n+l)!}. \tag{14.21}$$

Note that the power series defines a unique function, but we do not yet have uniqueness conditions for solutions to (14.20). We claim that J_l admits the representation

$$J_l(x) = \frac{1}{\pi} v_l(x) := \frac{1}{\pi} \int_0^{\pi} \cos(l\vartheta - x \sin \vartheta) d\vartheta, \tag{14.22}$$

see M. Abramowitz and J. A. Stegun [2], 9.1.21, where further integral representations are given. Here we prove only that v_l solves the Bessel differential equation. For this we apply Corollary 14.35 noting that the function under the integral, i.e.
$(x, \vartheta) \mapsto \cos(l\vartheta - x \sin \vartheta)$, is on every compact set $[-R, R] \times [0, 2\pi]$ a C^∞-function. Therefore we find

$$\frac{d}{dx} v_l(x) = \frac{d}{dx} \int_0^{\pi} \cos(l\vartheta - x \sin \vartheta) d\vartheta$$

$$= \int_0^{\pi} \sin(l\vartheta - x \sin \vartheta) \sin \vartheta d\vartheta$$

and

$$\frac{d^2}{dx^2}v_l(x) = \frac{d}{dx}\int_0^\pi \sin(l\vartheta - x\sin\vartheta)\sin\vartheta d\vartheta$$

$$= \int_0^\pi \frac{d}{dx}(\sin(l\vartheta - x\sin\vartheta))\sin\vartheta d\vartheta$$

$$= -\int_0^\pi \cos(l\vartheta - x\sin\vartheta)\sin^2\vartheta d\vartheta,$$

which yields

$$x^2\frac{d^2 v_l(x)}{dx^2} + x\frac{dv_l(x)}{dx} + (x^2 - l^2)v(x)$$

$$= -\int_0^\pi ((x^2\sin^2\vartheta + l^2 - x^2)\cos(l\vartheta - x\sin\vartheta) - x\sin\vartheta\sin(l\vartheta - x\sin\vartheta))d\vartheta.$$

Since

$$\frac{d}{dx}((l + x\cos\vartheta)\sin(l\vartheta - x\sin\vartheta))$$

$$= (-x\sin\vartheta)\sin(l\vartheta - x\sin\vartheta) + (l + x\cos\vartheta)(l - x\cos\vartheta)\cos(l\vartheta - x\sin\vartheta)$$

$$= (-x\sin\vartheta)\sin(l\vartheta - x\sin\vartheta) + (l^2 - x^2 - x^2\sin^2\vartheta)\cos(l\vartheta - x\sin\vartheta)$$

we find that

$$-\int_0^\pi ((x^2\sin^2\vartheta + l^2 - x^2)\cos(l\vartheta - x\sin\vartheta) - x\sin\vartheta\sin(l\vartheta - x\sin\vartheta))d\vartheta$$

$$= -(l + \cos\vartheta)\sin(l\vartheta - x\sin\vartheta)|_0^\pi = 0.$$

Thus, once we can prove, see Volume IV, that J_l and $\frac{1}{\pi}v_l$ satisfy in addition the uniqueness criterion for (14.20), we have a complete justification of (14.22). With similar arguments one can deduce, compare with Abramowitz and Stegun [2], 9.1.20

$$J_l(x) = \frac{x^l}{2^l \pi^{\frac{1}{2}}\Gamma(l + \frac{1}{2})}\int_0^\pi \cos(x\cos\vartheta)\sin^{2l}\vartheta d\vartheta. \tag{14.23}$$

We refer to Problems 10 and 11 for further applications of Theorem 14.34.

Problems

1. Let $\mathcal{F}$ be a vector space of real-valued functions $g : X \to \mathbb{R}$ where (X, d) is a compact space. Suppose that the function $e : X \to \mathbb{R}$, $e(x) = 1$ for all $x \in X$, belongs to $\mathcal{F}$ and that $\mathcal{F}$ separates points. Show that F is cross separating.

2. Prove that the set of all restrictions of polynomials to $[a, b]$ is dense in $(C([a, b]), \|\cdot\|_\infty)$.

3. Let $f : [-1, 1] \to \mathbb{R}$ be a continuous function vanishing at 0, i.e. $f(0) = 0$. Prove that there exists a sequence of polynomials $(p_k)_{k\in\mathbb{N}}$ such that $\left(p_k|_{[-1,1]}\right)_{k\in\mathbb{N}}$ converges on $[-1, 1]$ uniformly to f and $p_k(0) = 0$ for all $k \in \mathbb{N}$.

4. Let $I \subset \mathbb{R}$ be a compact interval and consider the Banach space $(C(I), \|\cdot\|_\infty)$. Prove that there exists a countable subset $M \subset C(I)$ such that for every $f \in C(I)$ the following holds: for every $\epsilon > 0$ there exists $g_\epsilon \in M$ such that $\|f - g_\epsilon\|_\infty < \epsilon$, i.e. $C(I)$ contains a countable subset, hence it is separable.
   Hint: consider polynomials with rational coefficients.

5. Let $f : [0, 1] \to \mathbb{R}$ be a continuous function such that $\int_0^1 f(x)x^k dx = 0$ for all $k \in \mathbb{N}_0$. Prove that f is identically zero. Now deduce that if for two continuous functions $f, g : [0, 1] \to \mathbb{R}$ we have $\int_0^1 f(x)x^k dx = \int_0^1 g(x)x^k dx$ for all $k \in \mathbb{N}_0$ then $f = g$.

6.   a) Let X be a compact metric space and $A, B \subset X$ be two closed disjoint sets, i.e. $A \cap B = \emptyset$. Prove the existence of $h : X \to [0, 1]$ such that $h|_A = 0$ and $h|_B = 1$.
   Hint: consider $x \mapsto \frac{\text{dist}(x,A)}{\text{dist}(x,A)+\text{dist}(x,B)}$.

   b) Let (X, d_X) and (Y, d_Y) be two compact metric spaces. We define the **algebraic tensor product** $C(X) \otimes_a C(Y)$ of $C(X)$ and $C(Y)$ as the space of all functions $h : X \times Y \to \mathbb{R}$ with the representation $h(x, y) = \sum_{j=1}^m f_j(x)g_j(y), m \in \mathbb{N}$, where $f_j \in C(X)$ and $g_j \in C(Y)$. Prove that $C(X) \otimes_a C(Y)$ is dense in $C(X \times Y)$.

7.   a) Let $G \subset \mathbb{R}^n$ be an open set. Find $\text{supp}\,\chi_G$ where χ_G denotes as usual the characteristic function of G.

b) Give an example of a C^∞-function g defined on $\mathbb{R}$ with supp $g = \bigcup_{j=1}^{\infty} \overline{B_{\frac{1}{4}}(j)}$. Does g have a compact support?

8. Consider the set $K := \{u \in C^1([a,b]) \mid ||u||_\infty + ||u'||_\infty \leq 1\}$ as a subset of $(C([a,b]), ||\cdot||_\infty)$ and prove that K is relative compact in $C([a,b])$.

9. Let $(f_n)_{n\in\mathbb{N}}$, $f_n \in C([a,b])$, be a uniformly bounded sequence, i.e. $\sup_{n\in\mathbb{N}} ||f_n||_\infty \leq M < \infty$. Prove that the sequence $(F_n)_{n\in\mathbb{N}}$, $F_n(x) := \int_a^x f_n(t)dt$, $x \in [a,b]$, is relative compact in $C([a,b])$.

10. For two continuous functions $u, v : \mathbb{R} \to \mathbb{R}$ with compact supports, supp $u = K_u$, supp $v = K_v$, define the **convolution** $u*v$ as the function $u * v : \mathbb{R} \to \mathbb{R}$ defined by $(u * v)(x) := \int_\mathbb{R} u(x-y)v(y)dy$. Prove that $u * v \in C(\mathbb{R})$ and find supp$(u * v)$.

11. Let $u : \mathbb{R} \to \mathbb{R}$ be a C^1-function such that both $|u|$ and $|u'|$ have finite (improper) integrals, i.e. $\int_\mathbb{R} |u(x)|dx < \infty$ and $\int_\mathbb{R} |u'(x)|dx < \infty$. Prove that $\hat{u} : \mathbb{R} \to \mathbb{R}$ defined by

$$\hat{u}(\xi) = \int_\mathbb{R} \cos(x\xi)u(x)dx$$

is a continuous function vanishing at infinity.
Hint: prove that $(1 + |\xi|)|\hat{u}(\xi)| \leq M_u < \infty$ for some $M_u \in \mathbb{R}$ independent of $\xi \in \mathbb{R}$. For this you need to prove that if $|u|$ and $|u'|$ are integrable over $\mathbb{R}$ then $\lim_{R\to\infty} u(R) = \lim_{R\to\infty} u(-R) = 0$.

15 Line Integrals

In Chapter 7 we defined the length of a regular parametric C^1-curve $\gamma : [\alpha, \beta] \to \mathbb{R}^n$ by

$$l_\gamma := \int_\alpha^\beta \|\dot\gamma(t)\| \, dt \tag{15.1}$$

with $\dot\gamma(t) = \frac{d\gamma}{dt}(t)$, compare with (7.19). In some sense this is an ad hoc definition and needs some justification. Moreover, given $\mathrm{tr}(\gamma) \subset \mathbb{R}^n$ we may consider a vector field $f : \mathrm{tr}(\gamma) \to \mathbb{R}^n$ and ask whether we can define its integral on $\mathrm{tr}(\gamma)$. Both problems we will address in this chapter. In order to simplify notation we will in the following often write just γ for $\mathrm{tr}(\gamma)$.

Given a continuous curve $\gamma : [\alpha, \beta] \to \mathbb{R}^n$. We may ask how to define its length. If $\sigma : [\alpha, \beta] \to \mathbb{R}^n$, $\sigma(t) = at + b$ is a line segment then the length of $\mathrm{tr}(\gamma)$ (or just γ) is the distance from $\sigma(\beta)$ to $\sigma(\alpha)$, i.e.

$$l_\sigma = \|\sigma(\beta) - \sigma(\alpha)\| = \|a\beta + b - a\alpha - b\| = (\beta - \alpha) \|a\|. \tag{15.2}$$

Since $\dot\sigma(t) = a$ we find that

$$\int_\alpha^\beta \|\dot\sigma(t)\| \, dt = \int_\alpha^\beta \|a\| \, dt = \|a\| (\beta - \alpha)$$

confirming (15.1). Having in mind how we have introduced the Riemann integral when trying to calculate the area under the graph of a function, we may try to approximate (the trace of) a curve by a polygon and consider the length of the polygon as an approximation of the length of the curve, see Figure 15.1.

Thus we choose a partition $\alpha = t_0 < t_1 < \ldots < t_m = \beta$ of $[\alpha, \beta]$ and consider the polygon p with vertices $\gamma(t_0), \gamma(t_1), \ldots, \gamma(t_m)$ which has length

$$l_p = \sum_{k=0}^{m-1} l_{\gamma(t_k)\gamma(t_{k+1})}$$

where $l_{\gamma(t_k)\gamma(t_{k+1})}$ denotes the length of the line segment connecting $\gamma(t_k)$ with $\gamma(t_{k+1})$.

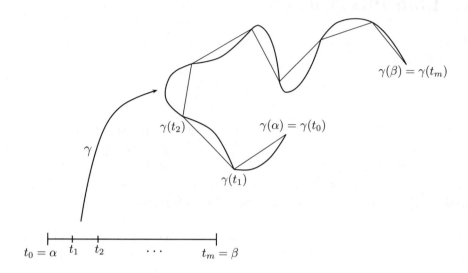

Figure 15.1

Of course, this length is given by

$$l_{\gamma(t_k)\gamma(t_{k+1})} = \|\gamma(t_{k+1}) - \gamma(t_k)\|$$

and therefore

$$l_p = \sum_{k=0}^{m-1} \|\gamma(t_{k+1}) - \gamma(t_k)\| . \tag{15.3}$$

Let us recall some notation from Chapter I.25. If $[\alpha, \beta] \subset \mathbb{R}$, $\alpha, \beta \in \mathbb{R}$, is an interval we write for a partition $\alpha = t_0 < t_1 < \ldots < t_m = \beta$ of $[\alpha, \beta]$ also $Z = Z(t_0, \ldots, t_m)$ with the understanding $t_0 = \alpha$ and $t_m = \beta$. The **joint partition** of $Z_1(t_0, \ldots, t_m)$ and $Z_2(s_0, \ldots, s_k)$ of $[\alpha, \beta]$ is the partition

$$Z(r_0, \ldots, r_l) := Z_1(t_0, \ldots, t_m) \cup Z_2(s_0, \ldots, s_k)$$

where $r_0 = t_0 = s_0 = \alpha$ and $r_l = t_m = s_k = \beta$.

Definition 15.1. *Let* $\gamma : [\alpha, \beta] \to \mathbb{R}^n$, $\alpha, \beta \in \mathbb{R}$, *be a continuous parametric curve and* $Z = Z(t_0, \ldots, t_m)$ *a partition of* $[\alpha, \beta]$.
A. *We call*

$$V_Z(\gamma) := \sum_{k=0}^{m-1} \|\gamma(t_{k+1}) - \gamma(t_k)\| \tag{15.4}$$

the **Z-variation** of γ.

B. *The* **total variation** *(or* **1-variation***) of γ is defined by*

$$V(\gamma) := \sup_{Z} V_Z(\gamma) \tag{15.5}$$

where the supremum is taken over all partitions of $[\alpha, \beta]$.

C. *We call γ* **rectifiable** *if $V(\gamma) < \infty$ and in this case we define the length l_γ of γ to be $V(\gamma)$, i.e.*

$$l_\gamma := V(\gamma). \tag{15.6}$$

Proposition 15.2. *A continuous parametric curve $\gamma : [\alpha, \beta] \to \mathbb{R}^n$, $\gamma = (\gamma_1, \ldots, \gamma_n)$, is rectifiable if and only if for all $1 \leq k \leq n$ the function $\gamma_k : [\alpha, \beta] \to \mathbb{R}^n$ is of bounded variation.*

Proof. For $a = (a_1, \ldots, a_n) \in \mathbb{R}^n$ the estimates

$$|a_k| \leq \|a\| = \left(\sum_{l=1}^{n} |a_l|^2\right)^{\frac{1}{2}} \leq \sum_{l=1}^{n} |a_l|$$

holds for all $k = 1, \ldots, n$. For a partition $Z(t_0, \ldots, t_m)$ of $[\alpha, \beta]$ we find now

$$|\gamma_k(t_{j+1}) - \gamma_k(t_j)| \leq \left(\sum_{l=1}^{n} |\gamma_l(t_{j+1}) - \gamma_l(t_j)|^2\right)^{\frac{1}{2}} \leq \sum_{l=1}^{n} |\gamma_l(t_{j+1}) - \gamma_l(t_j)|$$

implying

$$\sum_{j=0}^{m-1} |\gamma_k(t_{j+1}) - \gamma_k(t_j)| \leq \sum_{j=0}^{m-1} \left(\sum_{l=1}^{n} |\gamma_l(t_{j+1}) - \gamma_l(t_j)|^2\right)^{\frac{1}{2}} \leq \sum_{j=0}^{m-1} \sum_{l=1}^{n} |\gamma_l(t_{j+1}) - \gamma_l(t_j)|$$

or

$$V_Z(\gamma_k) \leq V_Z(\gamma) \leq \sum_{l=1}^{n} V_Z(\gamma_l).$$

Thus, if $V(\gamma)$ is finite then $V(\gamma_k)$, $k = 1, \ldots, n$, is finite and if all $V(\gamma_k)$ are finite then $V(\gamma)$ is finite and the proposition follows. $\square$

Note that for $n = 1$ these definitions coincide with those given in I.32.8, and recall Example I.32.13 where we have given an example of a continuous function not being of finite total variation, i.e. a continuous parametric curve need not be rectifiable.

309

Lemma 15.3. *Let $Z(t_0, \ldots, t_m)$ be a partition of $[\alpha, \beta]$ and for some j_0 let $t_{j_0} < t < t_{j_0+1}$. Denote by $Z_t = Z_t(t_0, \ldots, t_{j_0}, t, t_{j_0+1}, \ldots, t_m)$ the partition $\alpha = t_0 < t_1 < \ldots < t_{j_0} < t < t_{j_0+1} < \ldots < t_m = \beta$ of $[\alpha, \beta]$ and let $\gamma : [\alpha, \beta] \to \mathbb{R}^n$ be a continuous parametric curve. Then the following holds*

$$V_Z(\gamma) \leq V_{Z_t}(\gamma). \tag{15.7}$$

Proof. By the triangle inequality we have

$$V_Z(\gamma) = \sum_{l=0}^{m-1} \|\gamma(t_{l+1}) - \gamma(t_l)\|$$

$$= \sum_{l=0}^{j_0-1} \|\gamma(t_{l+1}) - \gamma(t_l)\| + \|\gamma(t_{j_0+1}) - \gamma(t_{j_0})\| + \sum_{l=j_0+1}^{m-1} \|\gamma(t_{l+1}) - \gamma(t_l)\|$$

$$\leq \sum_{l=0}^{j_0-1} \|\gamma(t_{l+1}) - \gamma(t_l)\| + \|\gamma(t_{j_0+1}) - \gamma(t)\| + \|\gamma(t) - \gamma(t_{j_0})\|$$

$$+ \sum_{l=j_0+1}^{m-1} \|\gamma(t_{l+1}) - \gamma(t_l)\|$$

$$= V_{Z_t}(\gamma).$$

$\square$

Lemma 15.4. *Let $\gamma : [\alpha, \beta] \to \mathbb{R}^n$ be a rectifiable curve and $\alpha < \eta < \beta$. Then the curves $\gamma|_{[\alpha,\eta]}$ and $\gamma|_{[\eta,\beta]}$ are also rectifiable and the following holds*

$$V\left(\gamma|_{[\alpha,\eta]}\right) + V\left(\gamma|_{[\eta,\beta]}\right) = V(\gamma). \tag{15.8}$$

Proof. Let $Z_1(t_0, \ldots, t_k)$ and $Z_2(t_k, \ldots, t_m)$ be partitions of $[\alpha, \eta]$ and $[\eta, \beta]$, respectively, hence $t_0 = \alpha$, $t_k = \eta$, $t_m = \beta$. Their union $Z_1(t_0, \ldots, t_k) \cup Z_2(t_k, \ldots, t_m)$ is a partition of $[\alpha, \beta]$ and

$$V_{Z_1}\left(\gamma|_{[\alpha,\eta]}\right) + V_{Z_2}\left(\gamma|_{[\eta,\beta]}\right) = V_{Z_1 \cup Z_2}(\gamma) \leq V(\gamma)$$

which implies that $V\left(\gamma|_{[\alpha,\eta]}\right)$ and $V\left(\gamma|_{[\eta,\beta]}\right)$ are finite, i.e. $\gamma|_{[\alpha,\eta]}$ and $\gamma|_{[\eta,\beta]}$ are rectifiable and

$$V\left(\gamma|_{[\alpha,\eta]}\right) + V\left(\gamma|_{[\eta,\beta]}\right) \leq V(\gamma)$$

holds. On the other hand, if $Z(s_0, \ldots, s_m)$ is a partition of $[\alpha, \beta]$ then $Z_1'(s_0, \ldots, \eta)$ and $Z_2'(\eta, \ldots, s_m)$ are partitions of $[\alpha, \eta]$ and $[\eta, \beta]$, respectively and $Z_1' \cup Z_2' = Z_\eta$, see (15.7). Thus we find

$$V_Z(\gamma) \le V_{Z_\eta}(\gamma) = V_{Z_1'}\left(\gamma|_{[\alpha,\eta]}\right) + V_{Z_2'}\left(\gamma|_{[\eta,\beta]}\right) \le V\left(\gamma|_{[\alpha,\eta]}\right) + V\left(\gamma|_{[\eta,\beta]}\right)$$

which yields

$$V(\gamma) \le V\left(\gamma|_{[\alpha,\eta]}\right) + V\left(\gamma|_{[\eta,\beta]}\right)$$

and the result follows. $\qquad\square$

Let $\gamma_j : [\alpha_j, \beta_j] \to \mathbb{R}^n$, $j = 1, 2$, $\alpha_2 = \beta_1$, be two continuous parametric curves such that $\gamma_1(\beta_1) = \gamma_2(\alpha_2)$. We define the sum $\gamma_1 \oplus \gamma_2 : [\alpha, \beta] \to \mathbb{R}^n$, $\alpha = \alpha_1$ and $\beta = \beta_2$ by

$$(\gamma_1 \oplus \gamma_2)(t) := \begin{cases} \gamma_1(t) & , t \in [\alpha_1, \beta_1] \\ \gamma_2(t) & , t \in [\alpha_2, \beta_2]. \end{cases} \tag{15.9}$$

Clearly $\gamma_1 \oplus \gamma_2$ is a continuous parametric curve. We claim

Lemma 15.5. *Let γ_1 and γ_2 be as above and suppose that they are rectifiable. Then $\gamma_1 \oplus \gamma_2$ is rectifiable and we have*

$$V(\gamma_1 \oplus \gamma_2) = V(\gamma_1) + V(\gamma_2). \tag{15.10}$$

Proof. Let $Z(t_0, \ldots, t_m)$ be a partition of $[\alpha, \beta] = [\alpha_1, \beta_2]$ and consider $Z_{\beta_1} = Z(t_0, \ldots, t_m) \cup Z(t_0, \beta_1, t_m)$, recall $t_0 = \alpha$ and $t_m = \beta$. It follows that

$$V_Z(\gamma_1 \oplus \gamma_2) \le V_{Z_{\beta_1}}(\gamma_1 \oplus \gamma_2)$$
$$= V_{Z(t_0,\ldots,\beta_1)}(\gamma_1) + V_{Z(\beta_1,\ldots,t_m)}(\gamma_2)$$
$$\le V(\gamma_1) + V(\gamma_2)$$

implying that $\gamma_1 \oplus \gamma_2$ is rectifiable and Lemma 15.4 gives now (15.10). $\quad\square$

Clearly, Lemma 15.4 and Lemma 15.5 extend to finite decompositions $\gamma = \gamma_1 \oplus \cdots \oplus \gamma_N$.

As in the one-dimensional case we can prove that Lipschitz continuity implies rectifiability, i.e. finite variation.

Theorem 15.6. *If $\gamma : [\alpha, \beta] \to \mathbb{R}^n$, $\alpha, \beta \in \mathbb{R}$, is a Lipschitz continuous curve, i.e. for some $\kappa \geq 0$ we have for all $s, t \in [\alpha, \beta]$ the estimate $\|\gamma(s) - \gamma(t)\| \leq \kappa |s - t|$, then γ is rectifiable and*

$$V(\gamma) \leq \kappa(\beta - \alpha). \tag{15.11}$$

Proof. Let $Z(t_0, \ldots, t_m)$ be a partition of $[\alpha, \beta]$. It follows that

$$V_Z(\gamma) = \sum_{k=0}^{m-1} \|\gamma(t_{k+1}) - \gamma(t_k)\|$$

$$\leq \sum_{k=0}^{m-1} \kappa(t_{k+1} - t_k) = \kappa(\beta - \alpha)$$

implying

$$l_\gamma = V(\gamma) = \sup_Z V_Z(\gamma) \leq \kappa(\beta - \alpha) < \infty.$$

$\square$

Before we establish (15.1), i.e. proving for C^1-curves that they are rectifiable and (15.1) holds, we want to extend the class of C^1-curves to piecewise C^1-curves.

Definition 15.7. *A continuous parametric curve $\gamma : [\alpha, \beta] \to \mathbb{R}^n$, $\alpha, \beta \in \mathbb{R}$, is said to be piecewise of class C^1 or a **piecewise continuously differentiable** curve if there exists a partition $Z(t_0, \ldots, t_m)$ of $[\alpha, \beta]$ such that with $I_j := [t_j, t_{j+1}]$, $j = 0, \ldots, m - 1$, the curves $\gamma|_{I_j}$ belong to the class C^1.*

Recall that $\gamma : [\alpha, \beta] \to \mathbb{R}^n$ is differentiable if $\gamma|_{(\alpha, \beta)}$ is differentiable and the one-sided derivatives $\dot{\gamma}_+(\alpha) := \dot{\gamma}(\alpha + 0) := \lim_{\substack{t \to \alpha \\ t > \alpha}} \frac{\gamma(t) - \gamma(\alpha)}{t - \alpha}$ and $\dot{\gamma}_-(\beta) := \dot{\gamma}(\beta - 0) := \lim_{\substack{t \to \beta \\ t < \beta}} \frac{\gamma(t) - \gamma(\beta)}{t - \beta}$ exist.

Note that if γ is piecewise continuously differentiable and t_j is a point of the partition used in Definition 15.7 then $\dot{\gamma}(t_j)$ need not exist. However every C^1-curve is also a piecewise differentiable curve.

Example 15.8. A. Every polygon is a piecewise continuously differentiable curve. **B.** Let $\gamma_j : [\alpha_j, \beta_j] \to \mathbb{R}^n$, $j = 1, 2$, $\beta_1 = \alpha_2$, be two piecewise continuously differentiable curves such that $\gamma_1(\beta_1) = \gamma_2(\alpha_2)$. Then $\gamma_1 \oplus \gamma_2$

is piecewise continuously differentiable. Clearly, $\gamma_1 \oplus \gamma_2$ is continuous and if $Z_1(t_0, \ldots, t_m)$ is a partition of $[\alpha_1, \beta_1]$ as well $Z_2(t_m, \ldots, t_k)$ is a partition of $[\alpha_2, \beta_2]$, $\beta_1 = \alpha_2$, such that $\gamma_1|_{[t_j,t_{j+1}]}$, $0 \le j \le m - 1$, and $\gamma_2|_{[t_l,t_{l+1}]}$, $m \le l \le k - 1$, are C^1-curves, then $Z(t_0, \ldots, t_k) := Z_1(t_0, \ldots, t_m) \cup Z_2(t_m, \ldots, t_k)$ is a partition of $[\alpha_1, \beta_2]$ and with respect to this partition $\gamma_1 \oplus \gamma_2$ is a piecewise continuously differentiable C^1-curve, see also Figure 15.2

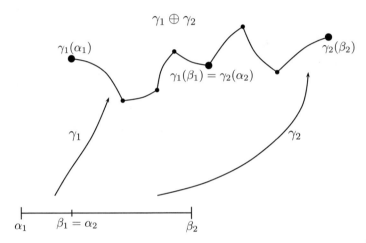

$$\gamma_1 \oplus \gamma_2$$
$$\gamma_1(\alpha_1)$$
$$\gamma_2(\beta_2)$$
$$\gamma_1(\beta_1) = \gamma_2(\alpha_2)$$
$$\gamma_1 \qquad \gamma_2$$
$$\alpha_1 \qquad \beta_1 = \alpha_2 \qquad \beta_2$$

Figure 15.2

Corollary 15.9. *A piecewise continuously differentiable curve $\gamma : I \to \mathbb{R}^n$ is rectifiable.*

Proof. First we note that by the mean value theorem every C^1-curve $\gamma : I \to \mathbb{R}^n$ is Lipschitz continuous, recall that I is compact. Thus by Lemma 15.4 a C^1-curve is rectifiable. Obviously the result is proved once we have proved it for a continuous curve $\gamma : [\alpha, \beta] \to \mathbb{R}^n$ such that with some $\eta_0 \in (\alpha, \beta)$ the curves $\gamma|_{[\alpha,\eta_0]}$ and $\gamma|_{[\eta_0,\beta]}$ are C^1-curves. Now let $Z(t_0, \ldots, t_m)$ be a partition of $[\alpha, \beta]$ and for some $0 \le j_0 \le m - 1$ assume $t_{j_0} < \eta_0 \le t_{j_0+1}$. It follows that

$$V_Z(\gamma) = V_{Z(t_0,\ldots,t_m)}(\gamma) \le V_{Z(t_0,\ldots,t_{j_0},\eta_0,t_{j_0+1},\ldots,t_m)}(\gamma)$$
$$= V_{Z(t_0,\ldots,\eta_0)}\left(\gamma|_{[t_0,\eta_0]}\right) + V_{Z(\eta_0,\ldots,t_m)}\left(\gamma|_{[\eta_0,t_m]}\right)$$
$$= V\left(\gamma|_{[\alpha,\eta_0]}\right) + V\left(\gamma|_{[\eta_0,t_m]}\right) < \infty$$

implying the result. $\qquad\square$

A central result for Riemann integrals is Theorem I.25.24 dealing with the convergence of Riemann sums to the Riemann integral of an integrable function. We need some preparation to prove a similar result for l_γ and we will mainly follow W. Rinow [39]. In the following $\gamma : [\alpha, \beta] \to \mathbb{R}^n$, $\alpha, \beta \in \mathbb{R}$, is a rectifiable curve, hence

$$V(\gamma) = \sup_Z V_Z(\gamma) < \infty$$

where the supremum is taken over all partitions of $[\alpha, \beta]$. Since γ is by assumption continuous, hence uniformly continuous on the compact interval $[\alpha, \beta]$ and since for every interval $[\eta, \xi] \subset [\alpha, \beta]$ the curve $\gamma|_{[\eta, \xi]}$ is rectifiable too, we conclude that for every $\epsilon > 0$ there exists $\delta > 0$ such that if $[\eta, \xi] \subset [\alpha, \beta]$ and $\xi - \eta < \delta$ then $\|\gamma(\xi) - \gamma(\eta)\| < \epsilon$.

Lemma 15.10. *Let* $Z_0(t_0, \ldots, t_{m_0})$ *and* $Z(s_0, \ldots, s_m)$ *be two partitions of* $[\alpha, \beta]$ *and*

$$\eta := \max_{0 \leq j \leq m-1} \mathrm{diam}\left(\gamma([s_j, s_{j+1}])\right).$$

Then we have

$$V_{Z_0}(\gamma) \leq V_Z(\gamma) + 2(m_0 - 1)\eta. \tag{15.12}$$

Proof. Let $t_l \in \{t_0, \ldots, t_{m_0}\}$. Suppose $t_l \in (s_j, s_{j+1})$ for some $0 \leq j \leq m-1$. For Z_{t_l}, see (15.17) for this notation, we find

$$V_{Z_{t_l}}(\gamma) - V_Z(\gamma) = \|\gamma(t) - \gamma(s_j)\| + \|\gamma(t) - \gamma(s_{j+1})\| - \|\gamma(s_{j+1}) - \gamma(s_j)\|$$
$$\leq \|\gamma(t) - \gamma(s_j)\| + \|\gamma(s_{j+1}) - \gamma(t)\| \leq 2\eta.$$

In the case where $t_l \in \{s_0, \ldots, s_m\}$ we find that $Z_{t_l} = Z$ and consequently $V_{Z_{t_l}}(\gamma) - V_Z(\gamma) \leq 2\eta$ too. By adding step by step the points of Z_0 to Z we obtain partitions $Z_{t_1}, Z_{t_1 t_2} := (Z_{t_1})_{t_2}, \ldots, Z_{t_1 \ldots t_{m_0 - 1}}$ such that

$$V_{Z_{t_1 \ldots t_{k+1}}}(\gamma) - V_{Z_{t_1 \ldots t_k}}(\gamma) \leq 2\eta.$$

Clearly $Z_0 \subset Z_{t_1 \ldots t_{m_0 - 1}}$ and therefore we obtain

$$V_{Z_0}(\gamma) \leq V_{Z_{t_1 \ldots t_{m_0 - 1}}}(\gamma) \leq V_Z(\gamma) + 2(m_0 - 1)\eta.$$

$\square$

Theorem 15.11. *Let* $\gamma : [\alpha, \beta] \to \mathbb{R}^n$ *be a rectifiable curve. Then for every* $\epsilon > 0$ *there exists* $\delta > 0$ *such that for every partition* Z *of* $[\alpha, \beta]$ *with mesh size less than* δ *it follows that* $V(\gamma) - V_Z(\gamma) < \epsilon$. *In particular if* $Z^{(\nu)}$, $\nu \in \mathbb{N}$, *is a sequence of partitions with mesh sizes* $\eta^{(\nu)}$ *and* $\lim_{\nu \to \infty} \eta^{(\nu)} = 0$, *then* $\lim_{\nu \to \infty} V_{Z^{(\nu)}}(\gamma) = V(\gamma) = l_\gamma$.

Proof. Let $\epsilon > 0$ be given and $Z_0(s_0, \ldots, s_m)$ be such that

$$V_{Z_0}(\gamma) > V(\gamma) - \frac{\epsilon}{2}. \tag{15.13}$$

The uniform continuity of γ implies the existence of $\delta > 0$ such that $r_1, r_2 \in [\alpha, \beta]$ and $|r_1 - r_2| < \delta$ implies that

$$\|\gamma(r_2) - \gamma(r_1)\| < \frac{\epsilon}{4(m_0 - 1)}.$$

Let $Z(t_0, \ldots, t_k)$ be a partition of $[\alpha, \beta]$ with mesh size less that δ, hence

$$\|\gamma(t_{l+1}) - \gamma(t_l)\| < \frac{\epsilon}{4(m_0 - 1)}$$

for all $0 \le l \le k - 1$, implying that

$$\eta := \max_{0 \le l \le m} \operatorname{diam} \gamma([t_l, t_{l+1}]) < \frac{\epsilon}{4(m_0 - 1)}.$$

Now Lemma 15.10 yields

$$V_{Z_0}(\gamma) \le V_Z(\gamma) + 2(m_0 - 1)\eta < V_Z(\gamma) + \frac{\epsilon}{2}$$

which together with (15.13) implies $V(\gamma) < V_Z(\gamma) + \epsilon$ and the theorem is proved. $\square$

Theorem 15.12. *For a* C^1-*curve* $\gamma : [\alpha, \beta] \to \mathbb{R}^n$ *the following holds*

$$l_\gamma = V(\gamma) = \int_\alpha^\beta \|\dot{\gamma}(t)\| \, dt. \tag{15.14}$$

Proof. Let $Z(t_0, \ldots, t_m)$ be a partition of $[\alpha, \beta]$. By the fundamental theorem of calculus in form of (7.5) we have

$$\gamma(t_j) - \gamma(t_{j-1}) = \int_{t_{j-1}}^{t_j} \dot{\gamma}(r) \, dr$$

implying by the triangle inequality, for vector-valued integrals, see Lemma 7.3,

$$V_Z(\gamma) = \sum_{j=0}^{m-1} \|\gamma(t_j) - \gamma(t_{j-1})\| \leq \sum_{j=0}^{m-1} \int_{t_{j-1}}^{t_j} \|\dot{\gamma}(r)\| \, dr = \int_{\alpha}^{\beta} \|\dot{\gamma}(r)\| \, dr,$$

hence

$$l_\gamma = V(\gamma) \leq \int_{\alpha}^{\beta} \|\dot{\gamma}(r)\| \, dr. \tag{15.15}$$

To prove the equality we note first that $\gamma|_{[\eta,\xi]}$ is again a C^1-curve for every interval $[\eta, \xi] \subset [\alpha, \beta]$ and its length is $l_{\gamma|_{[\eta,\xi]}} = V\left(\gamma|_{[\eta,\xi]}\right)$. We add the convention $l_{\gamma|_{[\alpha,\alpha]}} = 0$. It follows from Lemma 15.4 for $\alpha \leq t < t+h \leq \beta$ that

$$l_{\gamma|_{[t,t+h]}} = l_{\gamma|_{[\alpha,t+h]}} - l_{\gamma|_{[\alpha,t]}}$$

and (15.15) applied to $\gamma|_{[t,t+h]}$ yields

$$\frac{1}{h}\left|l_{\gamma|_{[\alpha,t+h]}} - l_{\gamma|_{[\alpha,t]}}\right| \leq \frac{1}{h}\int_t^{t+h} \|\dot{\gamma}(r)\| \, dr. \tag{15.16}$$

Since $\|\gamma(t+h) - \gamma(t)\| \leq l_{\gamma|_{[\alpha,t+h]}} - l_{\gamma|_{[\alpha,t]}}$ we find

$$\left\|\frac{1}{h}\int_t^{t+h} \dot{\gamma}(r) \, dr\right\| \leq \frac{1}{h}\left|l_{\gamma|_{[\alpha,t+h]}} - l_{\gamma|_{[\alpha,t]}}\right| \tag{15.17}$$

which implies with (15.16)

$$\left\|\frac{1}{h}\int_t^{t+h} \dot{\gamma}(r) \, dr\right\| \leq \frac{1}{h}\left|l_{\gamma|_{[\alpha,t+h]}} - l_{\gamma|_{[\alpha,t]}}\right| \leq \frac{1}{h}\int_t^{t+h} \|\dot{\gamma}(r)\| \, dr$$

for all $\alpha \leq t < t+h \leq \beta$. For $h \to 0$ we have

$$\lim_{h\to 0}\left\|\frac{1}{h}\int_t^{t+h} \dot{\gamma}(r) \, dr\right\| = \lim_{h\to 0}\int_t^{t+h} \|\dot{\gamma}(r)\| \, dr = 0$$

and therefore $t \mapsto l_{\gamma|_{[\alpha,t]}}$ is differentiable with $\left(l_{\gamma|_{[\alpha,t]}}\right)^{\cdot} = \|\dot{\gamma}(t)\|$. Hence for all $\alpha \leq t < \beta$ we have

$$l_{\gamma|_{[\alpha,t]}} = \int_{\alpha}^{t} \|\dot{\gamma}(r)\| \, dr$$

as well as (15.15)

$$0 \le l_\gamma - l_{\gamma|_{[\alpha,t]}} = l_{\gamma|_{[\alpha,\beta]}} - l_{\gamma|_{[\alpha,t]}} \le \int_t^b \|\dot{\gamma}(r)\| \, dr$$

implying for $t \to \beta$, $t < \beta$ that

$$l_\gamma = l_{\gamma|_{[\alpha,\beta]}} = \int_\alpha^\beta \|\dot{\gamma}(r)\| \, dr.$$

□

Corollary 15.13. Let $\gamma : [\alpha, \beta] \to \mathbb{R}^n$ be a piecewise continuously differentiable curve, i.e. with $\alpha = t_0 < t_1 < \ldots < t_m = \beta$ the curves $\gamma|_{[t_j, t_{j+1}]}$, $j = 0, \ldots, m-1$, are C^1-curves. Then γ is rectifiable and

$$l_\gamma = \sum_{j=0}^{m-1} \int_{t_j}^{t_{j+1}} \|\dot{\gamma}(r)\| \, dr. \tag{15.18}$$

With Theorem 15.12 we have now a more geometric justification of (15.1) as well as for Definition 7.27. Moreover it follows once more that l_γ is invariant under a change of parameter.

We have dealt here with parametric curves, i.e. certain mappings $\gamma : [\alpha, \beta] \to \mathbb{R}^n$. The geometric interesting objects are their traces. An important question is: When is a certain subset $\Gamma \subset \mathbb{R}^n$ trace of a parametric curve $\gamma : [\alpha, \beta] \to \mathbb{R}^n$, i.e. $\text{tr}(\gamma) = \Gamma$. Such a curve we would call a **parametrization** of Γ. Further, when Γ admits a parametrization we cannot expect it to be unique. Then we may ask whether Γ can have one parametrization which is rectifiable and another one which is not rectifiable - the answer is no. These questions refer to a topic called (geometric) measure theory which we will start to study in Volume III, namely in Part 6.

Let $\gamma : [\alpha, \beta] \to \mathbb{R}^n$ be a rectifiable curve, hence $\text{tr}(\gamma)$ is a subset of $\mathbb{R}^n$ on which we may define functions $f : \text{tr}(\gamma) \to \mathbb{R}$ and we may try to define an integral of f over $\text{tr}(\gamma)$. A similar problem can be posed for a function defined on a parametric surface or even more generally on a differentiable manifold. In Part 4 we will start to address this problem. For curves a different kind of integral turns out to be of utmost importance: line integrals of vector fields.

Definition 15.14. *Let $G \subset \mathbb{R}^n$ be a open set. A mapping $X : G \to \mathbb{R}^n$ is called a **vector field** on G. If X has components $X_j \in C^k(G)$ we call X a C^k-vector field.*

If X is a vector field on G and $p \in G$ we will later on often write X_p for $X(p)$.

Let $\gamma : [\alpha, \beta] \to \mathbb{R}^n$ be a parametric curve such that $\mathrm{tr}(\gamma) \subset G$ and let X be a vector field on G. We can consider now the vector field $X_{\gamma(t)}$, $t \in [\alpha, \beta]$, on γ (or better $\mathrm{tr}(\gamma)$) and we can form the expression $\langle X_{\gamma(t)}, \dot{\gamma}(t) \rangle$ which defines a function from $[\alpha, \beta]$ to $\mathbb{R}$.

Definition 15.15. *Let $\gamma : [\alpha, \beta] \to \mathbb{R}^n$ be a parametric curve, $\mathrm{tr}(\gamma) \subset G$, $G \subset \mathbb{R}^n$ open, and let $X : G \to \mathbb{R}^n$ be a vector field. Suppose further that $\dot{\gamma}(t)$ exists for all $t \in [\alpha, \beta]$. If the mapping $t \mapsto \langle X_{\gamma(t)}, \dot{\gamma}(t) \rangle$ is Riemann integrable we call*

$$\int_\gamma X_p \cdot \mathrm{d}p := \int_\gamma X(p)\, \mathrm{d}p := \int_\alpha^\beta \langle X_{\gamma(t)}, \dot{\gamma}(t) \rangle\, \mathrm{d}t = \int_\alpha^\beta \langle X(\gamma(t)), \dot{\gamma}(t) \rangle\, \mathrm{d}t \tag{15.19}$$

*the **line integral** of X along γ.*

First we note that in this definition we have not asked for any specific conditions on X and γ, but an integrability condition on a "combined" term, namely $\langle X(\gamma(t)), \dot{\gamma}(t) \rangle$. Clearly, if X is a continuous vector field and γ a C^1-curve then the line integral $\int_\gamma X_p\, \mathrm{d}p$ is well defined since in this case $t \mapsto \langle X_{\gamma(t)}, \dot{\gamma}(t) \rangle$ is a continuous function on the compact interval $[\alpha, \beta]$.

Proposition 15.16. *Let $X, Y : G \to \mathbb{R}^n$, $G \subset \mathbb{R}^n$ open, be vector fields and $\gamma : [\alpha, \beta] \to \mathbb{R}^n$ a parametric curve with $\mathrm{tr}(\gamma) \subset G$ and for which $\dot{\gamma}(t)$ exist for all $t \in [\alpha, \beta]$.*
A. *If X and Y are integrable along γ then $\lambda X + \mu Y$, $\lambda, \mu \in \mathbb{R}$, is also integrable along γ and*

$$\int_\gamma (\lambda X + \mu Y)(p)\, \mathrm{d}p = \lambda \int_\gamma X(p)\, \mathrm{d}p + \mu \int_\gamma Y(p)\, \mathrm{d}p \tag{15.20}$$

holds. **B.** *If X is integrable and bounded on γ in the sense that*

$$\|X\|_{\infty, \gamma} := \sup_{p \in \mathrm{tr}(\gamma)} \|X(p)\| < \infty,$$

then we have

$$\left| \int_\gamma X(p)\, \mathrm{d}p \right| \leq \|X\|_{\infty, \gamma}\, l_\gamma. \tag{15.21}$$

C. If $\gamma = \gamma_1 \oplus \gamma_2$ and X is integrable along γ then $X|_{\gamma_j}$ is integrable along γ_j, $j = 1, 2$, and

$$\int_{\gamma} X(p) \, \mathrm{d}p = \int_{\gamma_1} X(p) \, \mathrm{d}p + \int_{\gamma_2} X(p) \, \mathrm{d}p. \tag{15.22}$$

Proof. **A.** Since

$$\langle \lambda X(\gamma(t)) + \mu Y(\gamma(t)), \dot{\gamma}(t) \rangle = \lambda \langle X(\gamma(t)), \dot{\gamma}(t) \rangle + \mu \langle Y(\gamma(t)), \dot{\gamma}(t) \rangle$$

the linearity of the Riemann integral (on $\mathbb{R}^n$) implies (15.20).
B. By the Cauchy-Schwarz inequality we find

$$|\langle X(\gamma(t)), \dot{\gamma}(t) \rangle| \leq \|X\|_{\infty, \gamma} \|\dot{\gamma}(t)\|$$

which yields

$$\left| \int_{\gamma} X(p) \, \mathrm{d}p \right| = \left| \int_{\alpha}^{\beta} \langle X(\gamma(t)), \dot{\gamma}(t) \rangle \, \mathrm{d}t \right|$$

$$\leq \int_{\alpha}^{\beta} \|X(\gamma(t))\| \, \|\dot{\gamma}(t)\| \, \mathrm{d}t \leq \|X\|_{\infty, \gamma} \int_{\alpha}^{\beta} \|\dot{\gamma}(t)\| \, \mathrm{d}t$$

$$= \|X\|_{\infty, \gamma} \, l_{\gamma}.$$

C. Let $\eta \in (\alpha, \beta)$ and $\gamma_1 : [\alpha, \eta] \to G$, $\gamma_1 = \gamma|_{[\alpha, \eta]}$ as well as $\gamma_2 : [\eta, \beta] \to G$, $\gamma_2 = \gamma|_{[\eta, \beta]}$, thus $\gamma = \gamma_1 \oplus \gamma_2$. Then we find

$$\int_{\gamma} X(p) \, \mathrm{d}p = \int_{\alpha}^{\beta} \langle X(\gamma(t)), \dot{\gamma}(t) \rangle \, \mathrm{d}t$$

$$= \int_{\alpha}^{\eta} \langle X(\gamma(t)), \dot{\gamma}(t) \rangle \, \mathrm{d}t + \int_{\eta}^{\beta} \langle X(\gamma(t)), \dot{\gamma}(t) \rangle \, \mathrm{d}t$$

$$= \int_{\gamma_1} X|_{\gamma_1}(p) \, \mathrm{d}p + \int_{\gamma_2} X|_{\gamma_2}(p) \, \mathrm{d}p = \int_{\gamma_1} X(p) \, \mathrm{d}p + \int_{\gamma_2} X(p) \, \mathrm{d}p.$$

$\square$

Remark 15.17. A. We can sharpen part C to the following: let $\gamma_1 : [\alpha, \eta] \to G$ and $\gamma_2 : [\eta, \beta] \to G$ be two curves such that $\dot{\gamma}_j(t)$ exists for all t in the respective domains and assume $\gamma_1(\eta) = \gamma_2(\eta)$. Then we can define for $\gamma := \gamma_1 \oplus \gamma_2$ the two integrals

$$\int_{\alpha}^{\eta} \langle X(\gamma_1(t)), \dot{\gamma}_1(t) \rangle \, \mathrm{d}t \quad \text{and} \quad \int_{\eta}^{\beta} \langle X(\gamma_2(t)), \dot{\gamma}_2(t) \rangle \, \mathrm{d}t$$

for a vector field $X : G \to \mathbb{R}^n$ integrable along γ_1 and γ_2. With $\gamma := \gamma_1 \oplus \gamma_2$ we have on $[\alpha, \eta]$ that $\langle X(\gamma_1(t)), \dot{\gamma}_1(t) \rangle = \langle X(\gamma(t)), \dot{\gamma}(t) \rangle$ and on $(\eta, \beta]$ we find $\langle X(\gamma_2(t)), \dot{\gamma}_2(t) \rangle = \langle X(\gamma(t)), \dot{\gamma}(t) \rangle$ and under our assumptions these functions are integrable on $[\alpha, \eta]$ and $[\eta, \beta]$ respectively. Since the single point $t = \eta$ does not contribute to the Riemann integral, recall $X(\gamma(t))$ is bounded on $\mathrm{tr}(\gamma)$, as $\dot{\gamma}_j(\eta)$, $j = 1, 2$, is finite, it follows that X is integrable along γ and (15.22) holds.

B. If $X = (X_1, \ldots, X_n)$ is integrable along γ, then often one finds the notation

$$\int_\gamma X_1(p) \, \mathrm{d}p_1 + \cdots + X_n(p) \, \mathrm{d}p_n \tag{15.23}$$

for the integral $\int_\gamma X(p) \, \mathrm{d}p$.

We now want to assume that γ is at least a C^1-curve and that the vector fields are continuous in a neighbourhood of $\mathrm{tr}(\gamma)$. First we discuss how line integrals behave under a change of parameter, compare Definition 7.23.

Theorem 15.18. *Let $\gamma : [\alpha, \beta] \to \mathbb{R}^n$ be a C^1-curve and $\varphi : [\alpha', \beta'] \to [\alpha, \beta]$ a bijection which together with its inverse is C^1 and assume $\dot{\varphi} > 0$, i.e. φ is an orientation preserving change of parameter for γ. Denote by $\gamma' : [\alpha', \beta'] \to \mathbb{R}^n$ the curve $\gamma'(s) := (\gamma \circ \varphi)(s)$. For a continuous vector field $X : G \to \mathbb{R}^n$ where G is an open neighbourhood of $\mathrm{tr}(\gamma)$ we have*

$$\int_\gamma X(p) \, \mathrm{d}p = \int_{\gamma'} X(p') \, \mathrm{d}p'. \tag{15.24}$$

Proof. By the change of variable formula for Riemann integrals, see Theorem I.13.7, we have

$$\int_\gamma X(p) \, \mathrm{d}p = \int_\alpha^\beta \langle X(\gamma(t)), \dot{\gamma}(t) \rangle \, \mathrm{d}t = \int_{\alpha'}^{\beta'} \langle X(\varphi(s)), \dot{\gamma}(\varphi(s)) \rangle \, \dot{\varphi}(s) \, \mathrm{d}s$$

$$= \int_{\alpha'}^{\beta'} \langle X(\gamma'(s)), \dot{\gamma}'(s) \rangle \, \mathrm{d}s = \int_{\gamma'} X(p') \, \mathrm{d}p'.$$

$\square$

Remark 15.19. If in Theorem 15.18 we have $\dot{\varphi} < 0$, i.e. the change of parameter alternates the orientation of γ, then (15.23) becomes

$$\int_\gamma X(p) \, \mathrm{d}p = - \int_{\gamma'} X(p') \, \mathrm{d}p'. \tag{15.25}$$

In particular for $\varphi(s) = -s$ we find with $-\gamma$ being the curve $-\gamma : [-\beta, -\alpha] \to \mathbb{R}^n$, $(-\gamma)(t) = \gamma(-t)$, that

$$\int_\gamma X(p)\, \mathrm{d}p = -\int_{-\gamma} X(p)\, \mathrm{d}p. \tag{15.26}$$

Example 15.20. In this example we discuss some line integrals in the plane $\mathbb{R}^2$. Points will be $p \in \mathbb{R}^2$ with components (x, y) and vector fields $X : G \to \mathbb{R}^2$, $G \subset \mathbb{R}^2$, will be written as $X(x, y) = \begin{pmatrix} g_1(x, y) \\ g_2(x, y) \end{pmatrix}$.

A. We want to find the line integral of $X(x, y) = (y^2, -x^2)$ along the curve $\gamma(t) = (2\cos t, 2\sin t)$, $t \in [0, \pi]$, the trace of which is the half circle with centre 0 and radius 2:

$$\int_\gamma X(p)\, \mathrm{d}p = \int_0^\pi \langle X(\gamma(t)), \dot{\gamma}(t) \rangle\, \mathrm{d}t$$

$$= \int_0^\pi \left\langle \begin{pmatrix} g_1(2\cos t, 2\sin t) \\ g_2(2\cos t, 2\sin t) \end{pmatrix}, \begin{pmatrix} -2\sin t \\ 2\cos t \end{pmatrix} \right\rangle\, \mathrm{d}t$$

$$= 8 \int_0^\pi (-\cos^2 t \sin t + \sin^2 t \cos t)\, \mathrm{d}t$$

$$= 8 \left(\frac{1}{3}\cos^3 t + \frac{1}{3}\sin^3 t \right) \Big|_0^\pi = -\frac{16}{3}.$$

B. Along the arc $\gamma : \left[\frac{\pi}{4}, \frac{\pi}{3}\right] \to \mathbb{R}^2$ of the cycloid given by $\gamma(t) = \begin{pmatrix} t - \sin t \\ 1 - \cos t \end{pmatrix}$

we want to find the line integral of the vector field $X(x, y) = \begin{pmatrix} \frac{x}{y} \\ \frac{1}{y-1} \end{pmatrix}$. Since

$1 - \cos t \neq 0$ and $1 - \cos t - 1 = -\cos t \neq 0$ holds for $t \in \left[\frac{\pi}{4}, \frac{\pi}{3}\right]$, the vector field is continuous in an open neighbourhood of $\mathrm{tr}(\gamma)$. Now we get

$$\int_\gamma X(p)\,\mathrm{d}p = \int_{\frac{\pi}{4}}^{\frac{\pi}{3}} \langle X(\gamma(t)), \dot\gamma(t)\rangle\,\mathrm{d}t$$

$$= \int_{\frac{\pi}{4}}^{\frac{\pi}{3}} \left\langle \begin{pmatrix} \frac{t-\sin t}{1-\cos t} \\ \frac{1}{-\cos t} \end{pmatrix}, \begin{pmatrix} 1-\cos t \\ \sin t \end{pmatrix} \right\rangle\,\mathrm{d}t$$

$$= \int_{\frac{\pi}{4}}^{\frac{\pi}{3}} \left((t-\sin t) - \frac{\sin t}{\cos t} \right)\,\mathrm{d}t$$

$$= \left(\frac{1}{2}t^2 + \cos t + \ln\cos t \right)\Bigg|_{\frac{\pi}{4}}^{\frac{\pi}{3}}$$

$$= \frac{7\pi^2}{288} + \frac{1}{2}(1-\sqrt{2}) - \ln\sqrt{2}.$$

C. Let $X(x,y) = \begin{pmatrix} -y \\ x \end{pmatrix}$. We want to use this vector field to discover some interesting phenomena. First we consider the two curves $\gamma_1 : [0, 2\pi] \to \mathbb{R}^2$, $\gamma_1(t) = \begin{pmatrix} \cos t \\ \sin t \end{pmatrix}$ and $\gamma_2 : [0, 4\pi] \to \mathbb{R}^2$, $\gamma_2(t) = \begin{pmatrix} \cos t \\ \sin t \end{pmatrix}$. The trace of both curves is S^1, the difference is that γ_1 is simply closed while all points of γ_2 are double points, compare Definition 7.12. We find

$$\int_{\gamma_1} X(p)\,\mathrm{d}p = \int_0^{2\pi} \left\langle \begin{pmatrix} -\sin t \\ \cos t \end{pmatrix}, \begin{pmatrix} -\sin t \\ \cos t \end{pmatrix} \right\rangle\,\mathrm{d}t = \int_0^{2\pi} 1\,\mathrm{d}t = 2\pi$$

and

$$\int_{\gamma_2} X(p)\,\mathrm{d}p = \int_0^{4\pi} \left\langle \begin{pmatrix} -\sin t \\ \cos t \end{pmatrix}, \begin{pmatrix} -\sin t \\ \cos t \end{pmatrix} \right\rangle\,\mathrm{d}t = \int_0^{4\pi} 1\,\mathrm{d}t = 4\pi.$$

Thus even though γ_1 and γ_2 have the same trace we obtain different values for the line integrals.

Now we integrate X along the curve $\gamma_1|_{[0,\pi]}$ which is the half circle of radius 1 connecting $(1,0)$ with $(-1,0)$. Using our previous calculation we find

$$\int_{\gamma_1|_{[0,\pi]}} X(p)\,\mathrm{d}p = \int_0^\pi 1\,\mathrm{d}t = \pi.$$

Next we choose $\gamma_3 : [-1, 1] \to \mathbb{R}^2$, $\gamma(t) = (-t, 0)$ which is the line segment connecting $(1, 0)$ with $(-1, 0)$. It follows that

$$\int_{\gamma_3} X(p) \, \mathrm{d}p = \int_{-1}^{1} \left\langle \begin{pmatrix} 0 \\ -t \end{pmatrix}, \begin{pmatrix} -1 \\ 0 \end{pmatrix} \right\rangle \, \mathrm{d}t = 0.$$

So we find that although $\gamma_1|_{[0,\pi]}$ and γ_3 have the same initial and terminal points, namely $(1, 0)$ and $(-1, 0)$, the line integrals have different values, i.e. line integrals along curves which connect two fixed points do not only depend on these two points.

Having the considerations of Example 15.20.C in mind the following question arises: when does a line integral for a given vector field depend only on the end points but not on the particular curve connecting these points? To get a better understanding of this we start with

Definition 15.21. *A vector field* $X : G \to \mathbb{R}^n$, $G \subset \mathbb{R}^n$ *open, is called a* **gradient field** *if* $\operatorname{grad} \varphi = X$ *for some function* $\varphi : G \to \mathbb{R}$. *The function* φ *is called a* **potential** *or* **potential function** *of the vector field* X. *(Obviously* φ *needs to have all first order partial derivatives.)*

Theorem 15.22. *Let* $X : G \to \mathbb{R}^n$ *be a continuous gradient field and* $p_1, p_2 \in G$. *For every* C^1-*curve* γ *connecting* p_1 *and* p_2 *with* $\operatorname{tr}(\gamma) \subset G$ *the following holds*

$$\int_{\gamma} X(p) \, \mathrm{d}p = \varphi(p_2) - \varphi(p_1). \tag{15.27}$$

If $\gamma : [\alpha, \beta] \to \mathbb{R}^n$, $\gamma(\alpha) = p_1$ *and* $\gamma(\beta) = p_2$, *then* (15.27) *becomes*

$$\int_{\gamma} X(p) \, \mathrm{d}p = \varphi(\gamma(\beta)) - \varphi(\gamma(\alpha)). \tag{15.28}$$

Proof of Theorem 15.22. The chain rule yields

$$\int_{\gamma} X(p) \, \mathrm{d}p = \int_{\alpha}^{\beta} \langle X(\gamma(t)), \dot{\gamma}(t) \rangle \, \mathrm{d}t$$

$$= \int_{\alpha}^{\beta} \langle (\operatorname{grad} \varphi)(\gamma(t)), \dot{\gamma}(t) \rangle \, \mathrm{d}t$$

$$= \int_{\alpha}^{\beta} \frac{d}{dt} \left(\varphi(\gamma(t)) \right) \, \mathrm{d}t = \varphi(\gamma(\beta)) - \varphi(\gamma(t)).$$

$\square$

The next result gives a type of converse to Theorem 15.22.

Definition 15.23. *Let $X : G \to \mathbb{R}^n$ be a continuous vector field on the pathwise connected open set $G \subset \mathbb{R}^n$. We call the line integral $\int_\gamma X(p)\,\mathrm{d}p$* **path independent** *in G if for every C^1-curve $\gamma : [\alpha, \beta] \to G$, $\gamma(\alpha) = a$, $\gamma(\beta) = b$, the value of $\int_\gamma X(p)\,\mathrm{d}p$ depends only on a and b.*

If X is path independent in G and $a, b \in G$ we write $\int_a^b X(p)\,\mathrm{d}p$ for the line integral(s) $\int_\gamma X(p)\,\mathrm{d}p$ where γ is a C^1-curve with initial point a and terminal point b.

Theorem 15.24. *Let $X : G \to \mathbb{R}^n$ be a continuous vector field on the pathwise connected open set $G \subset \mathbb{R}^n$ such that its line integral is path independent in G. Then X is a gradient field. Moreover for every $a \in G$ fixed the function $\varphi : G \to \mathbb{R}$*

$$\varphi(x) := \int_a^x X(p)\,\mathrm{d}p \tag{15.29}$$

is a C^1-function and

$$\operatorname{grad} \varphi(p) = X(p) \ \text{ for all } p \in G. \tag{15.30}$$

Remark 15.25. A. We can interpret Theorem 15.24 as a version of the fundamental theorem of calculus to which it reduces for $n = 1$.
B. Clearly, Definition 15.23 and (the proof of) Theorem 15.24 need G to be pathwise connected.

Proof of Theorem 15.24. Let $X : G \to \mathbb{R}^n$ be a continuous vector field with path independent line integrals in G and $\xi \in G$. For $\epsilon > 0$ such that $B_\epsilon(\xi) \subset G$ and $h \in B_\epsilon(\xi)$ consider the line segment $\gamma : [0, 1] \to G$, $\gamma(t) = \xi + th$, connecting ξ with $\xi + h$. Since the line integral of X is path independent, by Proposition 15.16 we find

$$\int_a^{\xi+h} X(p)\,\mathrm{d}p - \int_a^\xi X(p)\,\mathrm{d}p = \int_\gamma X(p)\,\mathrm{d}p$$

which implies, note that $\dot{\gamma}(t) = h$,

$$\frac{1}{\|h\|}\left|\varphi(\xi + h) - \varphi(\xi) - \langle X(\xi), h\rangle\right| = \frac{1}{\|h\|}\left|\int_\gamma X(p)\,\mathrm{d}p - \int_\gamma X(\xi)\,\mathrm{d}p\right|$$

$$= \frac{1}{\|h\|}\left|\int_\gamma (X(p) - X(\xi))\,\mathrm{d}p\right|$$

$$\leq \frac{1}{\|h\|}\|X(\cdot) - X(\xi)\|_{\infty,\gamma}\,\|h\|$$

$$= \|X(\cdot) - X(\xi)\|_{\infty,\gamma}\,.$$

The continuity of X on G implies that $\|X(\cdot) - X(\xi)\|_{\infty,\gamma}$ tends to 0 as h tends to 0, hence

$$\varphi(\xi + h) - \varphi(\xi) = \langle X(\xi), h\rangle + \psi(\xi, h)$$

where

$$\lim_{\|h\|\to 0}\frac{\psi(\xi, h)}{\|h\|} = 0.$$

Thus φ is differentiable at ξ with $\varphi'(\xi) = X(\xi)$. □

Corollary 15.26. *Let $X : G \to \mathbb{R}^n$ be a continuous vector field admitting a potential function $\varphi : G \to \mathbb{R}$. If $\gamma : [\alpha, \beta] \to G$ is a closed C^1-curve then*

$$\int_\gamma X(p)\,\mathrm{d}p = 0.$$

Proof. We only have to note that

$$\int_\gamma X(p)\,\mathrm{d}p = \varphi(\gamma(\beta)) - \varphi(\gamma(\alpha))$$

and $\gamma(\alpha) = \gamma(\beta)$. □

A necessary condition for a C^1-vector field $X : G \to \mathbb{R}^n$ to be a gradient field is

Proposition 15.27. *Suppose that $X : G \to \mathbb{R}^n$, $X = (X_1, \ldots, X_n)$, is a C^1-vector field. Necessary for X to be a gradient field is the* **integrability condition**

$$\frac{\partial X_j}{\partial p_k} = \frac{\partial X_k}{\partial p_j} \quad \text{for } 1 \leq j, k \leq n. \tag{15.31}$$

Proof. If X is a C^1-vector field with potential function φ, $X(p) = \operatorname{grad} \varphi(p)$, then $\varphi : G \to \mathbb{R}$ is a C^2-function and therefore

$$\frac{\partial X_j}{\partial p_k} = \frac{\partial^2 \varphi}{\partial p_k \partial p_j} = \frac{\partial^2 \varphi}{\partial p_j \partial p_k} = \frac{\partial X_k}{\partial p_j}.$$

$\square$

It turns out that the integrability condition (15.31) is by no means a sufficient condition. In fact topological constraints on G are of great importance and we will return to the integrability condition in Chapter 26 and in later volumes.

We want to return to Definition 15.15 which looks as ad hoc as (15.1). Thus we may ask whether we can give a more "geometric" definition of a line integral of a vector field. A natural starting point would be to look for a continuous vector field $X : G \to \mathbb{R}^n$ and a continuous parametric curve $\gamma : [\alpha, \beta] \to G$ at a type of Riemann sum

$$V_Z(X) := \sum_{j=0}^{m-1} \langle X(\gamma(t_j)), \gamma(t_{j+1}) - \gamma(t_j) \rangle \tag{15.32}$$

for a partition $Z(t_0, \ldots, t_m)$ of $[\alpha, \beta]$. We observe that (15.32) means

$$V_Z(X) = \sum_{k=1}^{n} \sum_{j=0}^{m-1} X_k(\gamma(t_j))(\gamma_k(t_{j-1}) - \gamma_k(t_j)) = \sum_{k=1}^{n} V_Z(X_k), \tag{15.33}$$

where

$$V_Z(X_k) = \sum_{j=0}^{m-1} (X_k \circ \gamma)(t_j)(\gamma(t_{j+1}) - \gamma(t_j)). \tag{15.34}$$

However

$$V_Z(X_k) = \sum_{j=0}^{m-1} (X_k \circ \gamma)(t_j) \frac{\gamma_k(t_{j+1}) - \gamma(t_j)}{t_{j+1} - t_j} (t_{j+1} - t_j) \tag{15.35}$$

and we claim that for γ being continuously differentiable we can approximate $V_Z(X_k)$ by the Riemann sum

$$R_Z(X_k) = \sum_{j=0}^{m-1} (X_k \circ \gamma)(t_j) \dot{\gamma}_k(t_j)(t_{j+1} - t_j) \tag{15.36}$$

of the continuous function $t \mapsto (X_k \circ \gamma)(t)\dot{\gamma}_k(t)$. Indeed, by the mean value theorem there exists $\vartheta_{k,j} \in (0,1)$ such that

$$\gamma_k(t_{j+1}) - \gamma_k(t_j) = \dot{\gamma}_k(t_j + \vartheta_{k,j}t_{j+1})(t_{j+1} - t_j) \tag{15.37}$$

implying that

$$V_Z(X_k) = \sum_{j=0}^{m-1} (X_k \circ \gamma)(t_j)\dot{\gamma}_k(t_j + \vartheta_{k,j}t_{j+1})(t_{j+1} - t_j). \tag{15.38}$$

It follows that

$$|R_Z(X_k) - V_Z(X_k)| \leq \sum_{j=0}^{m-1} \|X_k\|_{\infty,\gamma} |\dot{\gamma}_k(t_j) - \dot{\gamma})_k(t_j + \vartheta_{k,j}t_{j+1})|(t_{j+1} - t_j). \tag{15.39}$$

The continuity of $\dot{\gamma}_k$ on the compact set $[\alpha, \beta]$ implies that for every $\eta > 0$ there exists $\delta > 0$ such that $|s - t| < \delta$ implies $|\dot{\gamma}_k(t) - \dot{\gamma}_k(s)| < \eta$. Therefore, given $\eta > 0$, there exists $\delta > 0$ such that for every partition $Z(t_0, \ldots, t_m)$ of $[\alpha, \beta]$ with mesh size less than δ it follows that

$$|R_Z(X_k) - V_Z(X_k)| \leq \|X_k\|_{\infty,\gamma} (\beta - \alpha)\eta. \tag{15.40}$$

If $(Z^{(\nu)})_{\nu \in \mathbb{N}}$ is a sequence of partitions of $[\alpha, \beta]$ with mesh sizes $(\delta^{(\nu)})_{\nu \in \mathbb{N}}$ converging to 0 we find that

$$\int_\alpha^\beta (X_k \circ \gamma)(t)\dot{\gamma}(t)\,dt = \lim_{\nu \to \infty} R_{Z^{(\nu)}}(X_k) = \lim_{\nu \to \infty} V_{Z^{(\nu)}}(X_k). \tag{15.41}$$

Hence we have proved

Theorem 15.28. *For a continuous vector field $X : G \to \mathbb{R}^n$, $G \subset \mathbb{R}^n$ open, and a C^1-curve $\gamma : [\alpha, \beta] \to G$ we have*

$$\int_\gamma X(p)\,dp = \lim_{\nu \to \infty} \sum_{k=1}^n V_{Z^{(\nu)}}(X_k) \tag{15.42}$$

where $(Z^{(\nu)})_{\nu \in \mathbb{N}}$ is any sequence of partitions of $[\alpha, \beta]$ with mesh sizes $(\delta^{(\nu)})_{\nu \in \mathbb{N}}$ converging to 0.

The question now emerges whether we can take (15.32) as a starting point to define a line integral for a continuous vector field and a rectifiable, not necessarily C^1 curve. We will return to this problem when discussing a more general theory of integration in Volume III Part 6.

Finally we want to discuss the question how to integrate a real-valued function defined on the trace of a rectifiable curve. Assume $\gamma : [\alpha, \beta] \to \mathbb{R}^n$ is a rectifiable curve and $f : \mathrm{tr}(\gamma) \to \mathbb{R}$ is a continuous function. If $s : [\alpha, \beta] \to \mathbb{R}$ is the length function of γ, i.e. $s(t) = l_\gamma|_{[0,t]}$, the natural approach is to consider variation sums of the type

$$V_\gamma(f, Z) = \sum_{j=1}^{M-1} f(\gamma(t_{j+1}))(s(t_{j+1}) - s(t_j))$$

where $Z(t_0, \ldots, t_M)$ is a partition of $[\alpha, \beta]$. Indeed, using the standard procedure to define a Riemann-type integral with the help of a variational sum, we may define

$$\int_\gamma f(x)\, \mathrm{d}s(x) := \sup_Z V_\gamma(f, Z)$$

where the supremum is again taken over all partitions of $[\alpha, \beta]$. When γ is a C^1-curve we may expect that we can replace $s(t_{j+1}) - s(t_j)$ by $\|\dot{\gamma}(t_{j+1})\|\,(t_{j+1} - t_j)$ as a "good" approximation and hence

$$V_\gamma(f, Z) \sim \sum_{j=1}^{M-1} f(\gamma(t_{j+1}))\, \|\dot{\gamma}(t_{j+1})\|\,(t_{j+1} - t_j)$$

which will give

$$\int_\gamma f(x)\, \mathrm{d}s(x) = \int_\alpha^\beta f(\gamma(t))\, \|\dot{\gamma}(t)\|\, \mathrm{d}t. \tag{15.43}$$

When discussing Stieltjes integrals in Volume III we will give a thorough justification of (15.43) but it is helpful to already have (15.43) at our disposal.

Problems

1. For the segment of a cylindrical helix given by $\gamma : [0, 4\pi] \to \mathbb{R}^3$, $\gamma(t) = (r\cos t, r\sin t, ht)$, $h > 0, r > 0$, find V_{Z_N} where for $N \in \mathbb{N}$ the points $t_j, j = 0, \ldots, N$ are $t_j = \frac{4\pi j}{N}$. Now pass to the limit as $N \to \infty$, i.e. calculate $\lim_{N \to \infty} V_{Z_N}(\gamma)$.

328

2. Consider the plane curve $\gamma : [\alpha, \beta] \to \mathbb{R}^2$, $\gamma(t) = \binom{t}{f(t)}$, where $f : [\alpha, \beta] \to \mathbb{R}$ is a continuous function. Prove that γ is rectifiable if and only if f is of bounded variation.

3. Let $\gamma : [\alpha, \beta] \to \mathbb{R}^n$ be a continuous curve and assume that for $j = 1, \ldots, n$ each of the components $\gamma_j : [\alpha, \beta] \to \mathbb{R}$ is a monotone function. Is γ rectifiable? Does the result hold even in the case when each γ_j is only piecewise monotone?

4. Consider the function $g : [0, 12] \to \mathbb{R}$ defined by

$$g(x) := \begin{cases} \frac{1}{8}x, & x \in [0, 1] \\ \frac{1}{8}x^2, & x \in [1, 4] \\ 2 + \sqrt{16 - (8 - x)^2}, & x \in [4, 12]. \end{cases}$$

Draft the graph of g and prove that the curve $\gamma : [0, 12] \to \mathbb{R}^2$, $\gamma(t) = \binom{t}{g(t)}$ is piecewise continuously differentiable, hence rectifiable. Now find its length l_γ.

5. Given the curve $\gamma_1 : [\frac{\pi}{4}, \frac{5\pi}{4}] \to \mathbb{R}^2$, $\gamma_1(t) = \binom{5\cos t}{5\sin t}$, find a curve γ_2 the trace of which is a line segment such that $\gamma_1 \oplus \gamma_2$ becomes a piecewise continuously differentiable, simply closed curve.

6. Find the line integral $\int_\gamma X_p dp$ for the following vector fields and curves:

a) $X_p = \binom{e^{p_1}}{e^{p_2}}$, $\gamma : [0, 1] \to \mathbb{R}^2$, $\gamma(t) = (t, \sqrt{t})$;

b) $X(z) = \binom{z_1 z_2}{z_2 e^{z_1}}$, γ is the simply closed polygon with vertices $(0, 0), (3, 0), (3, 2), (0, 2)$ and again $(0, 0)$.

7. Consider the vector field $V(x) = V(x_1, x_2) = \binom{x_2}{x_2 - x_1}$ on $\mathbb{R}^2$ and the two curves γ_1 and γ_2 defined as follows: γ_1 is the polygon with vertices $(0, 0), (0, \frac{1}{2}), (\frac{1}{2}, \frac{1}{2}), (\frac{1}{2}, 1)$ and $(1, 1)$, whereas $\gamma_2(t) = (t, t^2), t \in [0, 1]$. Note that both curves have the same initial point $(0, 0)$ and terminal point $(1, 1)$. Find the two line integrals $\int_{\gamma_1} V(x) dx$ and $\int_{\gamma_2} V(x) dx$.

8. Prove that the following vector fields are gradient fields:

a) $X : \mathbb{R}^2 \to \mathbb{R}^2$, $(x, y) \mapsto \binom{y}{x - y}$;

b) $Z : \mathbb{R}^n \setminus \{0\} \to \mathbb{R}^n$, $Z(x) = f(r)x, r = ||x||$.

329

9. Check whether the integrability conditions are satisfied for the following vector fields:

a) $U(x, y) = \left(\frac{y}{x^2+y^2}, \frac{-x}{x^2+y^2} \right), (x, y) \in \mathbb{R}^2 \setminus \{0\}$;

b) $W(x, y, z) = \left(xy, \frac{x^2+\ln z}{2}, \frac{y+z}{2z} \right), x, y \in \mathbb{R}, z > 0$;

c) $F(x) = -\frac{x}{||x||^\alpha}, x \in \mathbb{R}^k \setminus \{0\}, \alpha \in \mathbb{N}$.

10. a) Integrate the function $f : \mathbb{R}^3 \to \mathbb{R}, f(x, y, z) = xyz$ along the part of the cylindrical helix given by $\gamma : [0, 6\pi] \to (5\cos t, 5\sin t, 10t)$.

b) Let $\gamma : [0, 1] \to \mathbb{R}^2$ be the curve $\gamma(t) = \left(\begin{smallmatrix} \ln(1+t^2) \\ 2\arctan t - t + 3 \end{smallmatrix} \right)$ and $h : \mathbb{R}^2 \to \mathbb{R}$ the function $h(x, y) = ye^{-x}$. Find the integral of h along γ. (This problem is taken from G. M. Fichtenholz [14].)

Part 4: Integration of Functions of Several Variables

16 Towards Volume Integrals in the Sense of Riemann

This chapter will give an idea of how to define volume integrals for functions defined on a compact subset of $\mathbb{R}^n$. We want to point out problems and ways to resolve them. In some sense we could call this chapter "Motivations for the Connaisseur" and some students might prefer to start with Chapter 17 (which is possible). However we also invite these students to return to this chapter for a second reading to try to pick up some of the ideas why we sometimes need rather complicated definitions and most of all proofs.

We want to define an integral for a function $f : G \to \mathbb{R}$, $G \subset \mathbb{R}^n$. For $n = 1$ we expect that our definition will be consistent with our previous one, as we expect to maintain certain properties of the Riemann integral for functions defined on a bounded interval, e.g. linearity, preserving positivity, set additivity, etc. In particular, in the case where $G \subset \mathbb{R}^n$ is compact we would like to interpret the integral of a non-negative function $f : G \to \mathbb{R}$ as volume of the body $V = \{(x, z) \in \mathbb{R}^{n+1} \mid x \in G, 0 \le z \le f(x)\}$ which has the boundary $\partial V = G \times \{0\} \cup \{(x, z) \mid x \in \partial G, 0 \le z \le f(x)\} \cup \Gamma(f)$, where as usual $\Gamma(f) = \{(x, f(x)) \mid x \in G\}$ is the graph of f.

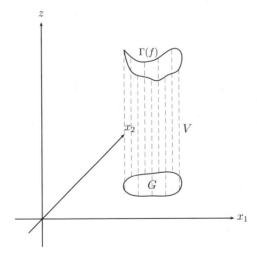

Figure 16.1

Recollecting the definition of the Riemann integral in one dimension we may have the following idea:

First we introduce a partition of G, say of finitely many sets $G_j \subset G$, $j = 1, \ldots, N$, then we define step functions f with respect to such a partition by requiring $f|_{G_j} = c_j \in \mathbb{R}$. Now we form the sum $\sum_{j=1}^{N} c_j \mathrm{vol}_n(G_j)$ and consider this as the integral of f. For a more general function $f : G \to \mathbb{R}$ we now investigate lower and upper sums as well as integrals and try to prove for a larger class of functions that the supremum of lower integrals coincides with the infimum of upper integrals.

The problems start early: how to find a good class of partitions and even more troublesome: what is $\mathrm{vol}_n(G_j)$? For hyper-rectangles $G = \bigtimes_{j=1}^{n}[a_j, b_j] \subset \mathbb{R}^n$, $a_j, b_j \in \mathbb{R}$, $a_j < b_j$, this is easily defined by

$$\mathrm{vol}_n(G) = \prod_{j=1}^{n}(b_j - a_j), \tag{16.1}$$

but with hyper-rectangles we cannot construct partitions of sets such as balls, intersections of ellipsoids with cylinders etc., or arbitrary compact sets in $\mathbb{R}^n$. The same applies when we replace hyper-rectangles by n-dimensional simplices. Thus, not only the definition of the integral itself needs an approximation process, but the definition of the volume of, say a compact set $G \subset \mathbb{R}^n$, may need some type of approximation and it is by no means clear that for every set we can find such an approximation. It is advantageous to use H. Lebesgue's approach to integration to first work out the problem of assigning volume to sets, i.e. to introduce measurable sets, and then to give an appropriate definition of an integral. We will discuss his theory in Volume III, Part 6. The surprise is that in Lebesgue's approach we cannot define for all sets the volume or the Lebesgue measure $\lambda^{(n)}(G)$ as long as we want $\lambda^{(n)}$ to have "good" properties, thus there are sets in $\mathbb{R}^n$ which do not have a well-defined volume, i.e. Lebesgue measure.

In this chapter we want to discuss certain more classical ideas on how to approach the problem of defining the volume of (compact) sets in $\mathbb{R}^n$ and on how this may lead to a consistent definition of an integral.

Based on earlier considerations of J. Kepler and G. Galilei, both being influenced by Archimedes, B. Cavalieri proposed the following principle to find the volume of a solid body by comparing sections through solid bodies: let

V_1 and V_2 be two solid bodies with ground surface on the plane E_1 and top surface on the plane E_2, E_1 and E_2 being parallel and having distance $h > 0$, see Figure 16.2.

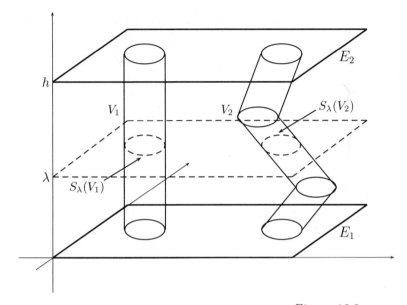

Figure 16.2

If all parallel sections $S_\lambda(V_1)$ and $S_\lambda(V_2)$ of V_1 and V_2 on the level λ, $0 \leq \lambda \leq h$, have the same surface area, then the two solid bodies V_1 and V_2 have the same volume. This is known as **Cavalieri's principle**.

We want to investigate ideas around Cavalieri's principle in order to get some hints into how we shall develop a theory of integration in higher dimensions.

Let us discuss this principle first for parallelograms. For two independent vectors a and b in the plane, $a, b \in \mathbb{R}^2$, the parallelogram spanned by a and b is the set

$$P(a,b) := \left\{ x \in \mathbb{R}^2 \,\middle|\, x = \lambda a + \mu b \,\middle|\, 0 \leq \lambda, \mu \leq 1 \right\}. \qquad (16.2)$$

335

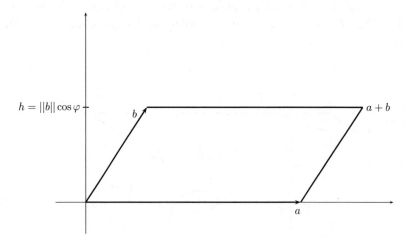

Figure 16.3

A parallelogram $P(a,c)$ with the same base side a, the same height h and all sections $S_\lambda(P(a,c))$, $0 \leq \lambda \leq h$, having the same "surface area", i.e. length as the sections $S_\lambda(P(a,b))$ are line segments, is given by $P(a,c) = P(a,b+\alpha a)$, $\alpha \in \mathbb{R}$, see Figure 16.4.

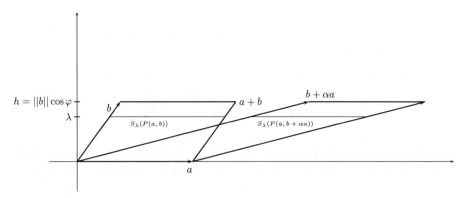

Figure 16.4

In $\mathbb{R}^n$ we define a **parallelotop** $P(a_1, \ldots, a_n) \subset \mathbb{R}^n$ spanned by n independent vectors $a_1, \ldots, a_n \in \mathbb{R}^n$ as

$$P(a_1, \ldots, a_n) := \left\{ x \in \mathbb{R}^n \,\middle|\, x = \sum_{j=1}^{n} \lambda_j a_j, \, \lambda_j \in [0,1] \right\} \qquad (16.3)$$

and the preceding considerations suggest that the volume $\mathrm{vol}_n(P(a_1, \ldots, a_n))$ of an n-dimensional parallelotop satisfies

$$\mathrm{vol}_n(P(a_1, \ldots, a_k + \alpha a_l, \ldots, a_n)) = \mathrm{vol}_n(P(a_1, \ldots, a_n)) \tag{16.4}$$

for $k \neq l$ and $\alpha \in \mathbb{R}$. In addition we expect that for $\alpha \in \mathbb{R}$

$$\mathrm{vol}_n(P(a_1, \ldots, a_{k-1}, \alpha a_k, a_{k+1}, \ldots, a_n)) = |\alpha| \mathrm{vol}_n(P(a_1, \ldots, a_n)) \tag{16.5}$$

which means that if we dilate one edge of the parallelotop the volume grows accordingly. Finally, the volume should be non-negative and translation invariant, i.e.

$$\mathrm{vol}_n(c + P(a_1, \ldots, a_n)) = \mathrm{vol}_n(P(a_1, \ldots, a_n)) \tag{16.6}$$

for all $c \in \mathbb{R}^n$, and we should have a normalisation, say

$$\mathrm{vol}_n(P(e_1, \ldots, e_n)) = 1. \tag{16.7}$$

It is easy to see, compare with Problem 3, that the unique volume form satisfying all these properties is given by

$$\mathrm{vol}_n(P(a_1, \ldots, a_n)) := |\det(a_1, \ldots, a_n)| \tag{16.8}$$

where $A = (a_1, \ldots, a_n)$ is the matrix with column vectors a_j, $1 \leq j \leq n$. Thus we can define the volume of parallelotops $P(a_1, \ldots, a_n)$ by (16.8). We now determine the change of the volume of a parallelotop under linear mappings.

Proposition 16.1. *Let $P(a_1, \ldots, a_n) \subset \mathbb{R}^n$ be a parallelotop and $T \in GL(n; \mathbb{R})$. Then $T(P(a_1, \ldots, a_n))$ is a parallelotop with volume*

$$\mathrm{vol}_n(T(P(a_1, \ldots, a_n))) = |\det T| \mathrm{vol}_n(P(a_1, \ldots, a_n)). \tag{16.9}$$

Proof. First we note that

$$T(P(a_1, \ldots, a_n)) = \left\{ Tx \,\Big|\, x = \sum_{j=1}^{n} \lambda_j a_j, 0 \leq \lambda_j \leq 1 \right\}$$

$$= \left\{ y \,\Big|\, y = \sum_{j=1}^{n} \lambda_j T a_j, 0 \leq \lambda_j \leq 1 \right\},$$

and since $T \in GL(n; \mathbb{R})$ the set $\{Ta_1, \ldots, Ta_n\}$ is linearly independent, i.e. $T(P(a_1, \ldots, a_n))$ is a parallelotop with volume

$$\text{vol}_n(T(P(a_1, \ldots, a_n))) = |\det(Ta_1, \ldots, Ta_n)|.$$

However the matrix $(Ta_1, \ldots, Ta_n)$ is the product of T with A, $A = (a_1, \ldots, a_n)$, and therefore we have

$$\begin{aligned}
\text{vol}_n(T(P(a_1, \ldots, a_n))) &= |\det(TA)| \\
&= |\det T||\det A| = |\det T|\, \text{vol}_n(P(a_1, \ldots, a_n)).
\end{aligned}$$

$\square$

Remark 16.2. Starting with the unit cube $P(e_1, \ldots, e_n)$ we find that every parallelotop $P(a_1, \ldots, a_n)$ is the image of $P(e_1, \ldots, e_n)$ under an element of $GL(n; \mathbb{R})$, namely the mapping defined by $Ae_j = a_j$.

We are interested in the change of the volume of "nice" sets under differentiable mappings. Since differentiable mappings can be locally approximated by their differential or Jacobi matrix, we should expect the Jacobi determinant to enter our considerations when dealing with such changes.

So far Cavalieri's principle leads to the insight that the determinant is a "volume form" and will have to play an important role when trying to define and to determine volumes of solid bodies in $\mathbb{R}^n$ and their behaviour under transformations. We can still get more from Cavalieri's principle, in fact it can guide us to some ideas about integrals over solid bodies. Note that in Cavalieri's time the notion of an integral did not exist.

We expect the volume of a body not only to be translation invariant but also invariant under rotations. Hence instead of comparing the volumes of two bodies between two parallel horizontal "planes" (lines, hyperspaces) we may switch to vertical parallel lines which will allow us to work in the more convenient setting of functions defined on a subset of the absissa rather than on the ordinate. We will restrict our consideration to "volumes in the plane", i.e. we apply Cavalieri's principle to sets in the plane which we still call bodies or volumes and the sections are lines having a length to which we still refer as area - just to have it easier when discussing the n-dimensional case. So, for $V \subset \mathbb{R}^2$ and a rectangle $R \subset \mathbb{R}^2$ Cavalieri's principle corresponds to Figure 16.5.

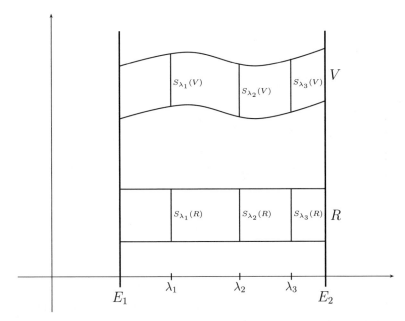

Figure 16.5

When for all $\lambda \in [a, b]$ the sections $S_\lambda(V)$ and $S_\lambda(R)$ have the same area (i.e. length) then V and R will have the same volume (i.e. area). We may introduce in Figure 16.5 Cartesian coordinates, see Figure 16.6, and consider V as a set being bounded from below and above by two functions $f_1, f_2 : [a, b] \to \mathbb{R}$.

This step can be viewed as switching from "classical" geometric methods to a method relying on calculus or analysis.

Suppose f_1 and f_2 are integrable. Then we find

$$\mathrm{vol}_2(V) = \int_a^b f_2(x)\,\mathrm{d}x - \int_a^b f_1(x)\,\mathrm{d}x = \int_a^b (f_2(x) - f_1(x))\,\mathrm{d}x$$
$$= \int_a^b (d - c)\,\mathrm{d}x = (b - a)(d - c)$$

as we expect.

339

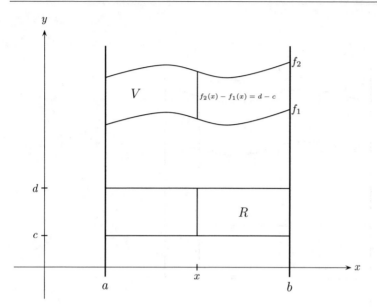

Figure 16.6

It is the appearance of the integral which gives us a hint to find maybe a partition of V into subsets which we can define a volume for and use this partition to introduce an integral for functions defined on V.
Consider (a new set) $V \subset \mathbb{R}^2$ in Figure 16.7.

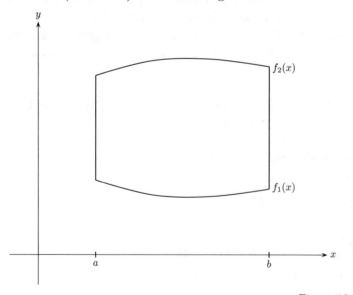

Figure 16.7

340

The set V is given by

$$V = \left\{ (x, y) \in \mathbb{R}^2 \mid x \in [a, b], f_1(x) \le y \le f_2(x) \right\}$$

for two functions $f_1, f_2 : [a, b] \to \mathbb{R}$. We do not assume anymore that $f_2(x) - f_1(x)$ is a constant independent of x, but we do assume that $f_1(x) < f_2(x)$ for all $x \in [a, b]$.

We now choose a partition $Z(x_0, \ldots, x_N)$ of $[a, b]$, recall $x_0 = a$, $x_N = b$, and in addition we define the functions $F_k : [a, b] \to \mathbb{R}$, $0 \le k \le M$, by $F_k = f_1(x) + \frac{k(f_2(x) - f_1(x))}{M}$. First we note that

$$F_{k+1}(x) - F_k(x) = \frac{f_2(x) - f_1(x)}{M} > 0.$$

Now with $y_{jk} := F_k(x_j)$, $0 \le j \le N$, $0 \le k \le M$, we get a partition of V into sets

$$V_{jk} = \left\{ (x, y) \in \mathbb{R}^2 \mid x_j \le x \le x_{j+1}, F_k(x) \le y < F_{k+1}(x) \right\}$$

for $0 \le j \le N$, $0 \le k \le M$, to which we must add the sets $\Gamma(f_2)$ as well as $\{b\} \times [f_1(b), f_2(b)]$. For notational purpose we set

$$V_{N+1\,M} = \{b\} \times [f_1(b), f_2(b))$$
$$V_{N\,M+1} = \Gamma(f_2) \backslash \{(b, f_2(b))\}$$
$$V_{N+1\,M+1} = \{(b, f_2(b))\},$$

so that $V_{jk} \cap V_{lm} = \emptyset$ for $(j, k) \ne (l, m)$ and $\bigcup_{j=0}^{N+1} \bigcup_{k=0}^{M+1} V_{jk} = V$.

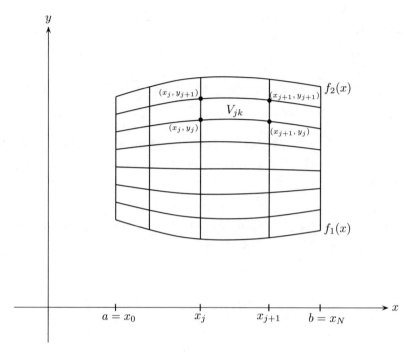

Figure 16.8

For the following we assume that $V_{N+1\,M}$, $V_{N\,M+1}$ and $V_{N+1\,M+1}$ to have volume 0 and we will not need to take care on these sets when discussing the volume of V or integrals for functions defined on V.

We can look at Figure 16.8 from two different points of view. It may serve us to illustrate once more Cavalieri's principle (for a set in the plane) but it may also lead to ideas how to construct partitions of sets in the plane. Both views are however related.

Introducing the sets

$$W(Z;j) := \bigcup_{k=0}^{M} V_{jk} \tag{16.10}$$

and

$$U(Z;k); = \bigcup_{j=0}^{N} V_{jk} \tag{16.11}$$

we expect

$$\text{vol}_2(V) = \sum_{j=0}^{N} \text{vol}_2(W(Z;j))$$

and

$$\text{vol}_2(V) = \sum_{k=0}^{M} \text{vol}_2(U(Z;k)).$$

Let $(Z_\nu)_{\nu \in \mathbb{N}}$ be a sequence of partitions of $[a, b]$ with $\text{mesh}(Z_\nu)$ tending to zero and assume that x_{j_0} is a point in any of these partitions. Intuitively we expect that for $\nu \to \infty$ the sets $W(Z_\nu; j_0)$ "converge" to the x_{j_0}-section of V, i.e. to the set $S_{x_{j_0}}(V) = \{(x_{j_0}, y) \in \mathbb{R}^2 \mid f_1(x_{j_0}) \le y \le f_2(x_{j_0})\}$, which suggests that

$$\text{vol}_2(V) = \int_a^b \text{area}(S_x(V)) \, dx$$

since

$$\sum_{j=0}^{N(\nu)} \text{vol}_2(W(Z_\nu, j)) \approx \sum_{j=0}^{N(\nu)} \text{area}(S_{x_j}(V))(x_{j+1} - x_j).$$

Now we turn to the problem of defining for $h : V \to \mathbb{R}$ an integral (in the Riemannian sense). We assume for the moment that V is compact and h is continuous, hence uniformly continuous on V. Having in mind that uniformly continuous functions are "very well behaved" we argue the following in a rather heuristic way hoping to provide rigorous proofs along the ideas developed later on.

We want to investigate Riemann type sums

$$\sum_{j=0}^{N} \sum_{k=0}^{M} h(\xi_{jk}, \eta_{jk}) \text{vol}_2(V_{jk}), \quad (\xi_{jk}, \eta_{jk}) \in V_{jk}, \tag{16.12}$$

where by our assumptions on f_1 and f_2 we know in principle how to define and calculate $\text{vol}_2(V_{jk})$. Note that we use the integral defined in dimension 1 to define areas (volumes) in dimension 2, and then we move on to define integrals on plane domains, i.e. it looks like an iteration process. For $\text{mesh}(Z)$ small and M large we may replace $\text{vol}_2(V_{jk})$ by $(x_{j+1} - x_j)(y_{jk+1} - y_{jk})$ which gives

$$\sum_{j=0}^{N} \sum_{k=0}^{M} h(\xi_{jk}, \eta_j) \text{vol}_2(V_{jk}) \approx \sum_{j=0}^{N} \sum_{k=0}^{M} h(\xi_{jk}, \eta_{jk})(x_{j+1} - x_j)(y_{jk+1} - y_{jk})$$

$$\tag{16.13}$$

343

and we write the right hand side in (16.13) as

$$\sum_{j=0}^{N}\left(\sum_{k=0}^{M} h(\xi_{jk},\eta_{jk})(y_{jk+1}-y_{jk})\right)(x_{j+1}-x_j). \qquad (16.14)$$

If now for $M \to \infty$ the points ξ_{jk} converge to some point ξ_j, for $M \to \infty$ the expression (16.14) should converge to

$$\sum_{j=0}^{N}\int_{f_1(x_j)}^{f_2(x_j)} h(\xi_j,y)\,\mathrm{d}y(x_{j+1}-x_j), \quad \xi_j \in (x_j,x_{j+1}), \qquad (16.15)$$

and for $N \to \infty$ such that $\mathrm{mesh}(Z) \to 0$ we should obtain for the expression in (16.15)

$$\int_a^b\left(\int_{f_1(x)}^{f_2(x)} h(x,y)\,\mathrm{d}y\right)\mathrm{d}x. \qquad (16.16)$$

Of course much is left open, on the other hand, for $h(x,y) = 1$ we already find what we expect

$$\mathrm{vol}_2(V) = \int_a^b\left(\int_{f_1(x)}^{f_2(x)} 1\,\mathrm{d}y\right)\mathrm{d}x = \int_a^b (f_2(x)-f_1(x))\,\mathrm{d}x,$$

and when V is a rectangle i.e. $f_1(x) = c$ and $f_2(x) = d$ are two constant functions, our arguments are well justified. The crucial step is to justify that by (16.13) a "good" approximation is given.

If we assume that for rectangles our argument holds, then we may try to avoid "curved" boundaries as much as possible but our construction of the subsets V_{jk} introduces a lot of new "curved" boundaries although it has the advantage to be in the spirit of Cavalieri's principle. So we may try to work with partitions of V in axes-parallel rectangles as much as possible and only to treat the boundary ∂V differently, see Figure 16.9.

344

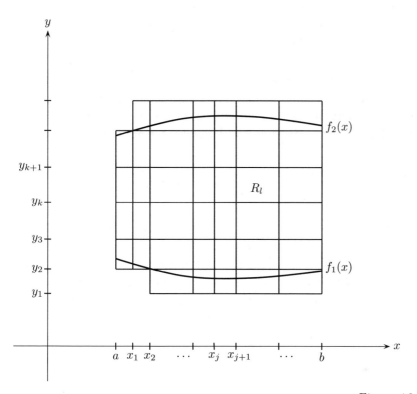

Figure 16.9

Let $Z(R_l,\ l = 1, \ldots, M)$ be a covering of V by closed axes-parallel rectangles such that $R_l \cap V \neq \emptyset$ for all $l = 1, \ldots, M$ and decompose $Z(R_l,\ l = 1, \ldots, M)$ into Z_1 and Z_2 where in Z_1 we collect those R_l which do not intersect $\Gamma(f_1)$ and $\Gamma(f_2)$. A corresponding Riemann sum for $h : V \to \mathbb{R}$ would have a decomposition

$$\sum_{l=1}^{M} h(\xi_l, \eta_l)\mathrm{vol}_2(R_l) = \sum_{Z_1} h(\xi_l, \eta_l)\mathrm{vol}_2(R_l) + \sum_{Z_2} h(\xi_k, \eta_k)\mathrm{vol}_2(R_k)$$

$$= S_1 + S_2,$$

and the second sum we split into two sums $S_{2\,\text{lower}}$ and $S_{2\,\text{upper}}$, where $S_{2\,\text{lower}}$ takes the rectangles into account covering $\Gamma(f_1)$ and $S_{2\,\text{upper}}$ takes the rectangles into account covering $\Gamma(f_2)$, note that for the mesh size of $Z(R_l,\ l = 1, \ldots, M)$ sufficiently small we can always achieve that $S_{2\,\text{lower}}$ and $S_{2\,\text{upper}}$

have no common term, recall $f_1(x) < f_2(x)$ for $x \in [a,b]$ and both functions are continuous. A convenient mesh size would be $\max_{1 \le l \le M} \operatorname{diam}(R_l)$. In principle we can handle the first sum since $\operatorname{vol}_2(R_l)$ is the volume of a rectangle, hence defined and known. The problem is S_2, we handle in more detail $S := S_{2\,\text{lower}}$. First we must note that for R_k contributing to S we have in general that $R_k \cap V^{\complement}$ has a non-empty interior, so a strictly positive volume (if defined) since it contains open rectangles or balls with positive area/volume. Thus R_k adds "volume" to V. A better approximation would be

$$\tilde{S} = \sum_{Z_{2\,\text{lower}}} h(\xi_k, \eta_k)\operatorname{vol}_2\left(R_k \cap \{(x,y) \mid x \in \operatorname{pr}_1(R_k),\, y \ge f(x)\}\right),$$

where pr_1 denotes the projection to the first coordinate. But the problem is to define $\operatorname{vol}_2\left(R_k \cap \{(x,y) \mid x \in \operatorname{pr}_1(R_k),\, y \ge f(x)\}\right)$, so we must stay with S. We want S to converge to 0 if h is continuous, hence bounded on V, recall V is compact. This means, if we long for a result for a larger class of functions h, that for a sequence of partitions $Z^{(\nu)}$ with $\operatorname{mesh}(Z^\nu) \to 0$, we obtain

$$\lim_{\nu \to \infty} \sum_{R_k \in Z_{2\,\text{lower}}} \operatorname{vol}_n(R_k) = 0,$$

see Figure 16.10.

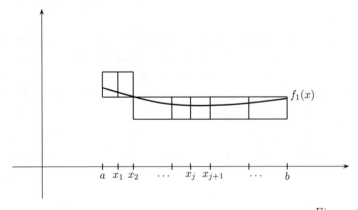

Figure 16.10

Suppose that we can approximate f_1 uniformly by a sequence of step functions, i.e. for $\epsilon > 0$ given we find a step function $T_N : [a,b] \to \mathbb{R}$ such that $\|T_N - f_1\|_\infty < \epsilon$, or $-\epsilon + T_N < f_1 < T_N + \epsilon$. Suppose that $T_N|_{I_\mu} = c_\mu$,

$\mu = 1, \ldots, K$, $I_\mu \subset [a, b]$ is an interval. Then we cover $\Gamma(f_1)$ with axes-parallel rectangles $H_\mu = \overline{I}_\mu \times [-\epsilon + c_\mu, c_\mu + \epsilon]$ and the total volume of $\bigcup_{\mu=1}^{K} H_\mu$ is $2\epsilon(b-a)$. Hence we should be able to control S, i.e. $S_{2\,\text{lower}}$ as well as $S_{2\,\text{upper}}$. The important insight is: it seems that it is the boundary of V which determines whether we can define a reasonable integral for $h : V \to \mathbb{R}$.

Here are two examples of "ugly" boundaries. The first is general: suppose a portion of ∂V is a non-rectifiable curve, then the above condition cannot hold. As a second concrete set take the set

$$A := \left\{ (x, y) \in \mathbb{R}^2 \,\middle|\, 0 \le x \le 1, \, 0 \le y \le 1 + D(x) \right\} \qquad (16.17)$$

where $D(x) = \begin{cases} 1 & , x \in \mathbb{Q} \cap [0, 1] \\ 0 & , x \in [0, 1] \text{ and } x \notin \mathbb{Q} \end{cases}$, i.e. D is the Dirichlet function of the interval $[0, 1]$. The boundary of ∂A of A is given, see Problem 4, by

$$\partial A = [0, 1] \times [1, 2] \cup \{0, 1\} \times [0, 1] \cup [0, 1] \times \{0\}. \qquad (16.18)$$

Thus $\text{vol}_2(\overline{A}) = 2$ and $\text{vol}_2(\partial A) \ge 1$. Of course A is not compact and the example does not fit exactly to our conditions, but it shows a very important fact: boundaries of sets can be very "strange".

In conclusion, we can draft a programme to develop an integration theory in $\mathbb{R}^n$ (we choose in the following $n = 2$):

- study iterated integrals $\int_a^b \left(\int_c^d f(x, y) \, dy \right) dx$;

- relate iterated integrals to "volume integrals" $\int_V f \, dz$, $V = [a, b] \times [c, d]$, $z = (x, y)$, and these integrals we want to define with the help of Riemann sums using rectangular partitions;

- classify compact sets for which we can "control" the boundary by certain coverings;

- construct an integral using Riemann sums for functions defined on compact sets the boundaries of which we can control by coverings;

- study this integral.

In light of this programme, it seems that the idea to use partitions which are more related to the geometry of V, see Figure 16.8, was an unnecessary detour. But it was not as the following heuristic discussion shows.

Consider the compact set $V_{R,r} := \left(\overline{B_R(0)} \backslash B_r(0)\right) \cap \{(x,y) \in \mathbb{R}^2 \mid x \geq 0, y \geq 0\}$, where $B_\rho(0) = \{(x,y) \in \mathbb{R}^2 \mid x^2 + y^2 < \rho^2\}$ and $0 < r < R$, see Figure 16.11.

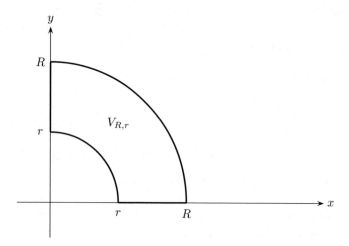

Figure 16.11

Another way to describe $V_{R,r}$ uses polar coordinates

$$V_{R,r} = \left\{ (\rho \cos \varphi, \rho \sin \varphi) \in \mathbb{R}^2 \mid r \leq \rho \leq R, 0 \leq \varphi \leq \frac{\pi}{2} \right\},$$

i.e.

$$V_{R,r} = \Phi\left([r,R] \times \left[0, \frac{\pi}{2}\right]\right)$$

where $\Phi(\rho, \varphi) = \begin{pmatrix} \rho \cos \varphi \\ \rho \sin \varphi \end{pmatrix}$. Thus $V_{R,r}$ is the image of a closed rectangle under Φ. The differential of Φ is given by $d\Phi_{(\rho,\varphi)} = \begin{pmatrix} \cos \varphi & -\rho \sin \varphi \\ \sin \varphi & \rho \cos \varphi \end{pmatrix}$ and $\det(d\Phi_{(\rho,\varphi)}) = \rho \neq 0$ for $(\rho, \varphi) \in [r, R] \times [0, \frac{\pi}{2}]$. Let us take a partition Z of $(\rho, \varphi) \in [r, R] \times [0, \frac{\pi}{2}]$ into rectangles $[\rho_j, \rho_{j+1}] \times [\varphi_k, \varphi_{k+1}]$ such that $\max(\rho_{j+1} - \rho_j)$ and $\max(\varphi_{k+1} - \varphi_k)$ are "small". This induces a partition of V into sets V_{jk}, see Figure 16.12.

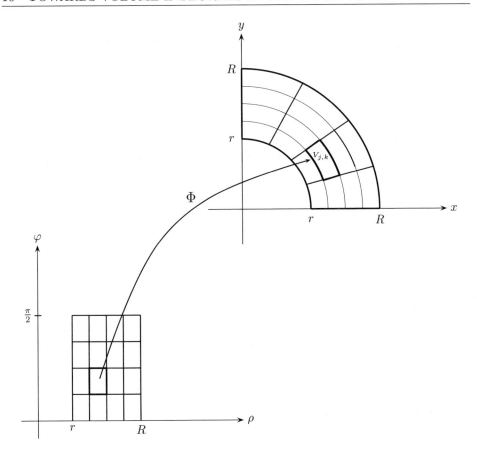

Figure 16.12

We may now replace $V_{jk} = \Phi\left([\rho_j, \rho_{j+1}] \times [\varphi_k, \varphi_{k+1}]\right)$ by $d\Phi([\rho_j, \rho_{j+1}] \times [\varphi_k, \varphi_{k+1}])$, i.e. we replace Φ by its linear approximation $d\Phi$ and this set has volume

$$|\det \Phi_{(\rho_j, \varphi_k)}| \text{vol}_2\left([\rho_j, \rho_{j+1}] \times [\varphi_k, \varphi_{k+1}]\right) = \rho_j(\rho_{j+1} - \rho_j)(\varphi_{k+1} - \varphi_k),$$

compare Proposition 16.1. If $h : V_{R,r} \to \mathbb{R}$ is continuous, then the Riemann sum

$$\sum h(x_j, y_j) \text{vol}_2(V_{jk}) \tag{16.19}$$

should be approximated by

$$\sum g(\rho_j, \varphi_j)\rho_j(\rho_{j+1} - \rho_j)(\varphi_{k+1} - \varphi_k) \tag{16.20}$$

where $g(\rho, \varphi) = h(\rho \cos \varphi, \rho \sin \varphi)$. However (16.20) is a Riemann sum for $(\rho, \varphi) \mapsto g(\rho, \varphi) \rho$ over a rectangle which we expect to converge to

$$\int_r^R \left(\int_0^{\frac{\pi}{2}} g(\rho, \varphi) \rho \, d\varphi \right) d\rho = \int_0^{\frac{\pi}{2}} \left(\int_r^R g(\rho, \varphi) \rho \, d\rho \right) d\varphi. \qquad (16.21)$$

Thus sometimes transformations may in fact lead to integrals defined on rectangles:

$$\int_V h(z) \, dz = \int_r^R \left(\int_0^{\frac{\pi}{2}} g(\rho, \varphi) \rho \, d\varphi \right) d\rho = \int_0^{\frac{\pi}{2}} \left(\int_r^R g(\rho, \varphi) \rho \, d\rho \right) d\varphi \qquad (16.22)$$

where the left hand side stands for the integral we expect to define with the help of (16.19) and in the integrals on the right hand side the Jacobi determinant of the transformation enters.

So transformations of sets fit to our programme which we now try to establish in $\mathbb{R}^n$, better for certain compact subsets of $\mathbb{R}^n$, and a suitable class of functions.

Problems

1.　a) Let $a, b, c \in \mathbb{R}^3$ be three independent vectors spanning a parallelotop $P(a, b, c)$ in $\mathbb{R}^3$. Prove that

$$\text{vol}_3(P(a, b, c)) = |\langle a \times b, c \rangle| = |\det(a, b, c)|.$$

　　b) Let $a, b \in \mathbb{R}^2$ be two independent vectors in the plane. Use the result of part a) to find a formula for $\text{vol}_2(P(a, b))$.

Hint: consider the vectors $\tilde{a} = (a_1, a_2, 0)$ and $\tilde{b} = (b_1, b_2, 0)$.

2. Find the image of the parallelotop $P(a, b, c) \subset \mathbb{R}^3$, $a = (1, 0, 2), b = (0, 3, 2), c = (3, 1, 0)$, under the linear mapping given by the matrix $\begin{pmatrix} 1 & 0 & 2 \\ 0 & 4 & 1 \\ -3 & 5 & 0 \end{pmatrix}$ and calculate $\text{vol}_3(T(P(a, b, c)))$.

3. Consider the mapping $|\det| : \mathbb{R}^n \times \cdots \times \mathbb{R}^n \to [0, \infty)$ defined by

$$|\det(a_1, \ldots, a_n)| = \left| \det \begin{pmatrix} a_{11} & \cdots & a_{1n} \\ \vdots & & \\ a_{n1} & \cdots & a_{nn} \end{pmatrix} \right|$$

and prove that this mapping defines a normalised volume form on $\mathbb{R}^n$, i.e. (16.4)-(16.8) hold for $|\det|$ replacing vol_n.
Hint: use Remark A.I.12.B as a characterisation of a determinant form.

4. Consider the set $A : \{(x, y) \in \mathbb{R}^2 | 0 \le x \le 1, 0 \le y \le 1 + D(x)\}$ where D is the Dirichlet function of $[0, 1]$, see (16.17). Find ∂A.

5. In the plane $\mathbb{R}^2$ consider the triangle ABC with vertices $(-1, 0), (1, 0)$, $(0, 1)$. Prove that for every $\epsilon > 0$ we can find $N = N(\epsilon)$ squares $Q_j(\epsilon)$, $1 \le j \le N(\epsilon)$, with sides parallel to the coordinate axes such that

$$\mathrm{vol}_2 \left(\bigcup_{j=1}^{N(\epsilon)} Q_j(\epsilon) \right) \le \epsilon.$$

17 Parameter Dependent and Iterated Integrals

Sometimes when dealing with functions of several variables we may interpret and handle some of the variables in a different way to the others. For example when discussing families of curves and their envelopes we considered functions $\varphi : G \times I \to \mathbb{R}$, $G \subset \mathbb{R}^2$ open, $I \subset \mathbb{R}$ an interval, and we assumed that for $\alpha \in I$ fixed the equation $\varphi(x, y, \alpha) = 0$ determines a curve in $\mathbb{R}^2$. Thus the variable α is a label for a curve while (x, y) is a point on a curve. Moreover, we can interpret a function $f : [a, b] \times [c, d] \to \mathbb{R}$ quite differently, namely that for every $y \in [c, d]$ a function $g_y = f(\cdot, y)$ defined on $[a, b]$ is given, i.e. y is used to label a specific function g_y within a given family of functions. Operations on these parameters (labels) such as α or y may lead to new objects such as envelopes, or as seen in Theorem 14.32, if $f : [a, b] \times [c, d] \to \mathbb{R}$ is continuous, integration with respect to the variable y (the label or the parameter) gives a new continuous function

$$F(x) := \int_c^d f(x, y) \, \mathrm{d}y. \tag{17.1}$$

Depending on properties of f this function may have further properties, for example if $\frac{\partial f}{\partial x}$ exists on $[a, b] \times [c, d]$ and is continuous then F is differentiable and (compare with Theorem 14.34)

$$\frac{dF}{dx}(x) = \int_c^d \frac{\partial f}{\partial x}(x, y) \, \mathrm{d}y. \tag{17.2}$$

We want to pick up these results and try to get a better understanding of functions defined as F in (17.1). A first question which arises after having already established continuity and differentiability is whether such a function is integrable. We first handle the case of two variables, the extension to n-variables is then straightforward.

Let $[a, b]$, $[c, d] \subset \mathbb{R}$ be two compact intervals and $f : [a, b] \times [c, d] \to \mathbb{R}$ be a continuous function. The function

$$x \mapsto F(x) := \int_c^d f(x, y) \, \mathrm{d}y \tag{17.3}$$

353

is by Theorem 14.32 well defined and continuous on $[a,b]$ hence integrable and we can form

$$\int_a^b F(x)\,dx = \int_a^b \left(\int_c^d f(x,y)\,dy \right) dx, \tag{17.4}$$

which we can call an **iterated integral** of f over $[a,b] \times [c,d]$. Of course, the situation is completely symmetric in the two variables x and y, so we may also consider the function

$$y \mapsto G(y) := \int_a^b f(x,y)\,dx \tag{17.5}$$

which is again continuous and we can form the other iterated integral of f, namely

$$\int_c^d G(y)\,dy = \int_c^d \left(\int_a^b f(x,y)\,dx \right) dy. \tag{17.6}$$

Clearly, F and G are different functions. However we may expect that the right hand sides of (17.4) and (17.6) coincide.

This is the content of the following result about interchanging the order of integration in iterated integrals.

Theorem 17.1. *Let $[a,b]$, $[c,d] \subset \mathbb{R}$ be compact intervals and $f : [a,b] \times [c,d] \to \mathbb{R}$ be a continuous function. Then we have*

$$\int_c^d \left(\int_a^b f(x,y)\,dx \right) dy = \int_a^b \left(\int_c^d f(x,y)\,dy \right) dx. \tag{17.7}$$

Proof. We define $g : [c,d] \to \mathbb{R}$ by

$$g(y) := \int_a^b \left(\int_c^y f(x,t)\,dt \right) dx.$$

It follows that $g(c) = 0$ and by Theorem 14.34 we find

$$g'(y) = \int_a^b \frac{\partial}{\partial y} \left(\int_c^y f(x,t)\,dt \right) dx = \int_a^b f(x,y)\,dy$$

which implies

$$\int_c^d \left(\int_a^b f(x,y)\,dx \right) dy = \int_c^d g'(y)\,dy = g(d) = \int_a^b \left(\int_c^d f(x,y)\,dy \right) dx.$$

$\square$

In Problem 1 and Problem 2 we extend Theorem 14.32 and Theorem 14.34 to the n-dimensional situation (which are straightforward results) and then we have all details proved to justify

Corollary 17.2. *For* $1 \leq j \leq n$ *let* $I_j \subset \mathbb{R}$ *be a compact interval and* $f : I_1 \times \cdots \times I_n \to \mathbb{R}$ *be a continuous function. Then for every permutation* $\sigma \in S(n)$ *the iterated integrals*

$$\int_{I_{\sigma(1)}} \left(\int_{I_{\sigma(2)}} \cdots \left(\int_{I_{\sigma(n)}} f(x_1, \ldots, x_n) \, dx_{\sigma(n)} \right) \cdots dx_{\sigma(2)} \right) dx_{\sigma(1)} \qquad (17.8)$$

are all defined and equal to each other.

Remark 17.3. Note that so far we only have the notion of the Riemann integral for functions of one real variable and neither Theorem 17.1 nor Corollary 17.2 uses any integration theory in higher dimensions. Often authors call Theorem 17.1 or Corollary 17.2 **Fubini's theorem**, but this is incorrect. We will discuss Fubini's result in Volume III Part 6 in more detail. It refers to the Lebesgue integration theory and has a more complex content than just the statement that certain iterated Riemann integrals are equal.

Let $f : [a, b] \times [c, d] \to \mathbb{R}$ be a continuous function and for $j = 1, 2$ let $\varphi_j : [a, b] \to [c, d]$, $\varphi_1(x) \leq \varphi_2(x)$, and $\psi_j : [c, d] \to [a, b]$, $\psi_1(y) \leq \psi_2(y)$, continuous functions too. We may consider the two functions

$$h_1(x) := \int_{\varphi_1(x)}^{\varphi_2(x)} f(x, y) \, dy \qquad (17.9)$$

and

$$h_2(y) = \int_{\psi_1(y)}^{\psi_2(y)} f(x, y) \, dx. \qquad (17.10)$$

Clearly, if f has with respect to x the same properties as with respect to y, and if the functions φ_j have similar properties as the functions ψ_j, h_1 and h_2 have the same properties and therefore we study only h_1.

Theorem 17.4. *For continuous functions* $\varphi_j : [a, b] \to [c, d]$, $j = 1, 2$, *the function* $h_1 : [a, b] \to \mathbb{R}$ *is continuous. Moreover, if* φ_j, $j = 1, 2$, *are differentiable and* f *has a continuous partial derivative* $\frac{\partial f}{\partial x}$ *on* $[a, b] \times [c, d]$ *then* h_1 *is differentiable on* $[a, b]$ *and*

$$\frac{d}{dx} h_1(x) = \int_{\varphi_1(x)}^{\varphi_2(x)} \frac{\partial f}{\partial x}(x, y) \, dy + \frac{d\varphi_2}{dx}(x) f(x, \varphi_2(x)) - \frac{d\varphi_1}{dx}(x) f(x, \varphi_1(x)) \, (17.11)$$

holds.

Proof. We want to first prove that for every $x_0 \in [a, b]$ the function h_1 is continuous. For this we write

$$h_1(x) = \int_{\varphi_1(x_0)}^{\varphi_2(x_0)} f(x, y)\, dy + \int_{\varphi_2(x_0)}^{\varphi_2(x)} f(x, y)\, dy - \int_{\varphi_1(x_0)}^{\varphi_1(x)} f(x, y)\, dy. \quad (17.12)$$

By Theorem 14.32 the first term is a continuous function, so given $\epsilon > 0$ there exists $\delta_1 > 0$ such that $|x - x_0| < \delta_1$, $x \in [a, b]$, implies that

$$\left| \int_{\varphi_1(x_0)}^{\varphi_2(x_0)} (f(x, y) - f(x_0, y))\, dy \right| < \frac{\epsilon}{3}.$$

Since f is bounded on the compact set $[a, b] \times [c, d]$ we have further

$$\left| \int_{\varphi_2(x_0)}^{\varphi_2(x)} f(x, y)\, dy \right| \leq \|f\|_\infty |\varphi_2(x) - \varphi_2(x_0)|$$

and

$$\left| \int_{\varphi_1(x_0)}^{\varphi_1(x)} f(x, y)\, dy \right| \leq \|f\|_\infty |\varphi_1(x) - \varphi_1(x_0)|.$$

From the continuity of φ_j we deduce the existence of $\delta_2 > 0$ such that $|x - x_0| < \delta_2$, $x \in [a, b]$, implies

$$\|f\|_\infty |\varphi_j(x) - \varphi_j(x_0)| < \frac{\epsilon}{3}.$$

Thus for $\delta = \min\{\delta_1, \delta_2\}$ we find with $|x - x_0| < \delta$, $x \in [a, b]$, that

$|h_1(x) - h_1(x_0)|$

$$= \left| \int_{\varphi_1(x_0)}^{\varphi_2(x_0)} (f(x, y) - f(x_0, y))\, dy + \int_{\varphi_2(x_0)}^{\varphi_2(x)} f(x, y)\, dy - \int_{\varphi_1(x_0)}^{\varphi_1(x)} f(x, y)\, dy \right|$$

$$\leq \left| \int_{\varphi_1(x_0)}^{\varphi_2(x_0)} (f(x, y) - f(x_0, y))\, dy \right| + \|f\|_\infty |\varphi_2(x) - \varphi_2(x_0)| + \|f\|_\infty |\varphi_1(x) - \varphi_1(x_0)|$$

$$< \frac{\epsilon}{3} + \frac{\epsilon}{3} + \frac{\epsilon}{3},$$

where we used that $\int_{\varphi_2(x_0)}^{\varphi_2(x_0)} f(x, y)\, dy = \int_{\varphi_1(x_0)}^{\varphi_1(x_0)} f(x, y)\, dy = 0$.
In order to prove that h_1 is differentiable if φ_j, $j = 1, 2$, are, we prove that

h_1 is differentiable for every $x_0 \in [a, b]$ and start with (17.12). By Theorem 14.34 we know that under the assumption that $\frac{\partial f}{\partial x}$ exists and is continuous on $[a, b] \times [c, d]$, the function $x \mapsto \int_{\varphi_1(x_0)}^{\varphi_2(x_0)} f(x, y)\, dy$ is differentiable with derivative $x \mapsto \int_{\varphi_1(x_0)}^{\varphi_2(x_0)} \frac{\partial f}{\partial x}(x, y)\, dy$. We apply the mean value theorem to the second integral in (17.12) to find

$$\frac{1}{x - x_0} \left(\int_{\varphi_2(x_0)}^{\varphi_2(x)} f(x, y)\, dy - \int_{\varphi_2(x_0)}^{\varphi_2(x_0)} f(x, y)\, dy \right) = \frac{1}{x - x_0} \int_{\varphi_2(x_0)}^{\varphi_2(x)} f(x, y)\, dy$$

$$= \frac{\varphi_2(x) - \varphi_2(x_0)}{x - x_0} f(x, \overline{\eta})$$

with $\overline{\eta}$ between $\varphi_2(x)$ and $\varphi_2(x_0)$. Letting x tend to x_0 we obtain

$$\lim_{x \to x_0} \frac{1}{x - x_0} \int_{\varphi_2(x_0)}^{\varphi_2(x)} f(x, y)\, dy = \frac{\partial \varphi_2}{\partial x}(x_0) f(x_0, \varphi_2(x_0))$$

where we used that f is continuous and φ_2 is differentiable. With the same type of argument we find that

$$\lim_{x \to x_0} \frac{1}{x - x_0} \int_{\varphi_1(x_0)}^{\varphi_1(x)} f(x, y)\, dy = \frac{\partial \varphi_1}{\partial x}(x_0) f(x_0, \varphi_1(x_0)),$$

implying the differentiability of h_1 and (17.11). $\qquad\square$

We gain more generality in Theorem 17.4 if we assume that f is defined on $[a, b] \times \mathbb{R}$ and is continuous on $[a, b] \times [\min \varphi_1, \max \varphi_2]$ or having in addition a continuous partial derivative $\frac{\partial f}{\partial x}$ on $[a, b] \times [\min \varphi_1, \max \varphi_2]$. The question is whether we want to start with a fixed domain $[a, b] \times [c, d]$ and restrict φ_j accordingly, or whether we want to start with a domain $\{(x, y)\,|\, x \in [a, b], \varphi_1(x) \le y \le \varphi_2(x)\}$ and now pose the appropriate conditions on f. For a moment let us take a second point of view and assume that $f : \mathbb{R}^2 \to \mathbb{R}$ is given as are $\varphi_j : [a, b] \to \mathbb{R}$. If $f|_{[a,b] \times [\min \varphi_1, \max \varphi_2]}$ is continuous, by Theorem 17.4 we know that

$$x \mapsto \int_{\varphi_1(x)}^{\varphi_2(x)} f(x, y)\, dy$$

is a continuous function, so the iterated integral

$$\int_a^b \left(\int_{\varphi_1(x)}^{\varphi_2(x)} f(x, y)\, dy \right) dx \tag{17.13}$$

is defined. For $f \equiv 1$ we find of course that

$$\int_a^b \left(\int_{\varphi_1(x)}^{\varphi_2(x)} 1 \, dy \right) dx = \int_a^b (\varphi_2(x) - \varphi_1(x)) \, dx$$

gives the area of the set V bounded by the lines $x = a$, $x = b$ and the graphs of the functions φ_1 and φ_2.

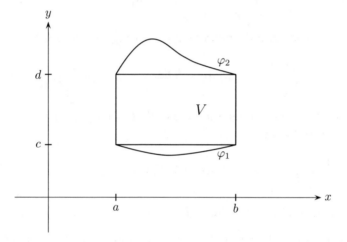

Figure 17.1

We may look at (17.13) as a first possibility to define an integral for a continuous function $f : V \to \mathbb{R}$ depending only on the one-dimensional Riemann integral. We can go even a step further, recall that f is at the moment defined on all of $\mathbb{R}^2$. Let $\psi_j : [c, d] \to \mathbb{R}$ be continuous functions such that $\psi_1(y) < \psi_2(y)$ and assume $\varphi_1(x) \leq c$, $d \leq \varphi_2(x)$, $\psi_1(y) \leq a$, $b \leq \psi_2(y)$ as well as for simplicity $\varphi_1(a) = \varphi_1(b) = c$, $\varphi_2(a) = \varphi_2(b) = d$, $\psi_1(c) = \psi_1(d) = a$, $\psi_2(c) = \psi_2(d) = b$. Now we define

$$W := \left\{ (x, y) \in \mathbb{R}^2 \, \middle| \, \varphi_1(x) \leq y \leq \varphi_2(x), \psi_1(y) \leq x \leq \psi_2(y) \right\},$$

see Figure 17.2.

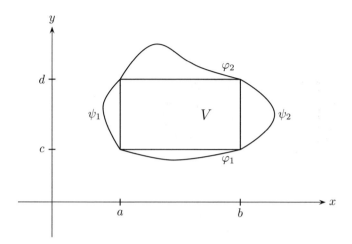

Figure 17.2

We can form two iterated integrals

$$\int_a^b \left(\int_{\varphi_1(x)}^{\varphi_2(x)} f(x,y)\,\mathrm{d}y \right) \mathrm{d}x \quad \text{and} \quad \int_c^d \left(\int_{\psi_1(y)}^{\psi_2(y)} f(x,y)\,\mathrm{d}x \right) \mathrm{d}y. \qquad (17.14)$$

For $f(x,y) = 1$, the first integral gives the area of the set

$$V := \left\{ (x,y) \in \mathbb{R}^2 \,\middle|\, x \in [a,b],\ \varphi_1(x) \le y \le \varphi_2(x) \right\}$$

see Figure 17.1 and the second integral gives the area of the set

$$U := \left\{ (x,y) \in \mathbb{R}^2 \,\middle|\, y \in [c,d],\ \psi_1(y) \le x \le \psi_2(y) \right\},$$

see Figure 17.3, and we conclude that the area of W should be

$$\int_a^b \left(\int_{\varphi_1(x)}^{\varphi_2(x)} 1\,\mathrm{d}y \right) \mathrm{d}x + \int_c^d \left(\int_{\varphi_1(y)}^{\varphi_2(y)} 1\,\mathrm{d}x \right) \mathrm{d}y - \int_a^b \left(\int_c^d 1\,\mathrm{d}y \right) \mathrm{d}x.$$

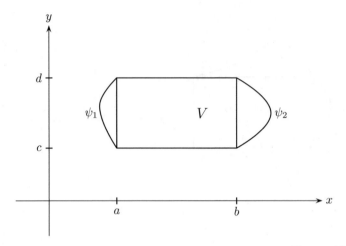

Figure 17.3

In fact we may even try to define

$$\int_W f := \int_a^b \left(\int_{\varphi_1(x)}^{\varphi_2(x)} f(x,y)\,dy \right) dx + \int_c^d \left(\int_{\varphi_1(y)}^{\varphi_2(y)} f(x,y)\,dx \right) dy - \int_a^b \left(\int_c^d f(x,y)\,dy \right) dx. \qquad (17.15)$$

At least for some functions f and domains W in $\mathbb{R}^2$ we could define integrals $\int_W f$ in this way, extensions to $\mathbb{R}^n$ seem to be possible. We can rewrite (17.15). Consider

$$\int_a^b \left(\int_{\varphi_1(x)}^{\varphi_2(x)} f(x,y)\,dy \right) dx - \int_a^b \left(\int_c^d f(x,y)\,dx \right) dy$$

$$= \int_a^b \left(\int_{\varphi_1(x)}^{\varphi_2(x)} f(x,y)\,dy - \int_c^d f(x,y)\,dy \right) dx$$

$$= \int_a^b \left(\int_{\min \varphi_1}^{\max \varphi_2} \chi_{Q(x)\setminus[c,d]}(y) f(x,y)\,dy \right) dx,$$

where

$$\chi_{Q(x)\setminus[c,d]}(y) = \begin{cases} 1, & \varphi_1(x) \le y \le \varphi_2(x),\ y \notin [c,d] \\ 0, & \text{otherwise.} \end{cases}$$

The advantage of this representation is that we integrate over a fixed rectangle $[a,b] \times [\min \varphi_1, \max \varphi_2]$, but the price to pay is that the function

$(x,y) \mapsto \chi_{Q(x)\setminus[c,d]}(y)f(x,y)$ is in general not continuous. But still we may follow up this idea and replace the right hand side of (17.15) by

$$\int_{\min\psi_1}^{\max\varphi_2} \left(\int_{\min\varphi_1}^{\max\varphi_2} \chi_W(x,y)f(x,y)\,\mathrm{d}y \right) \mathrm{d}x \qquad (17.16)$$

where

$$\chi_W(x,y) = \begin{cases} 1, & (x,y) \in W \\ 0, & \text{otherwise.} \end{cases}$$

This is an iterated integral over a rectangle for a function which is in general discontinuous and we do not even know whether this integral is equal to

$$\int_{\min\varphi_1}^{\max\varphi_2} \left(\int_{\min\psi_1}^{\max\psi_2} \chi_W(x,y)f(x,y)\,\mathrm{d}x \right) \mathrm{d}y.$$

However a strategy to define integrals $\int_W f$, $W \subset \mathbb{R}^n$, $f : W \to \mathbb{R}$, emerges: once we can define integrals over hyper-rectangles (rectangles in $\mathbb{R}^2$) and we can identify them with iterated integrals then we should be able to integrate certain classes of functions f defined on more general compact sets $G \subset \mathbb{R}^n$: find a hyper-rectangle K such that $G \subset K$ and extend f to be zero on $K\setminus G$, now integrate the extended function over G. Our next chapter will be devoted to volume integrals in the sense of Riemann for functions defined on hyper-rectangles, i.e. we will start with the outlined programme.

We want to close this chapter with some extensions and applications of Theorem 17.4. First we want to consider higher derivatives of h_1 defined by (17.9). Formula (17.11) suggests that for this we need of course higher order differentiability of φ_1 and φ_2, but we also need higher order partial derivatives of f with respect to x and we now need partial derivatives of f with respect to y.

Differentiating in (17.11) we find (provided all partial derivatives needed exist and $\frac{\partial^2 f}{\partial x^2}$ is continuous)

$$\frac{d^2 h_1}{dx^2}(x) = \frac{d}{dx}\left(\int_{\varphi_1(x)}^{\varphi_2(x)} \frac{\partial f}{\partial x}(x,y)\,dy + \frac{d\varphi_2}{dx}(x) f(x,\varphi_2(x)) - \frac{d\varphi_1}{dx}(x) f(x,\varphi_1(x)) \right)$$

$$= \int_{\varphi_1(x)}^{\varphi_2(x)} \frac{\partial^2 f}{\partial x^2}(x,y)\,dy + \frac{d\varphi_2}{dx}(x)\left(\frac{\partial f}{\partial x}\right)(x,\varphi_2(x)) - \frac{d\varphi_1}{dx}(x)\left(\frac{\partial f}{\partial x}\right)(x,\varphi_1(x))$$

$$+ \frac{d^2\varphi_2}{dx^2}(x) f(x,\varphi_2(x)) + \frac{d\varphi_2}{dx}(x)\left(\left(\frac{\partial f}{\partial x}\right)(x,\varphi_2(x)) + \frac{d\varphi_2}{dx}(x)\left(\frac{\partial f}{\partial y}\right)(x,\varphi_2(x)) \right)$$

$$- \frac{d^2\varphi_1}{dx^2}(x) f(x,\varphi_1(x)) - \frac{d\varphi_1}{dx}(x)\left(\left(\frac{\partial f}{\partial x}\right)(x,\varphi_1(x)) + \frac{d\varphi_1}{dx}(x)\left(\frac{\partial f}{\partial y}\right)(x,\varphi_1(x)) \right).$$

$$(17.17)$$

This suggests that for calculating the kth-derivative of h_1 and requiring $\frac{d^k h_1}{dx^k}$ to be continuous we should assume φ_1, φ_2 to be C^k-functions, f to be a C^{k-1}-function and $\frac{\partial^k f}{\partial x^k}$ to exist and to be continuous.

The next extension of (17.11) is with respect to functions of several variables: We are interested in

$$h(x_1,\ldots,x_k) = \int_{\varphi_1(x_1,\ldots,x_k)}^{\varphi_2(x_1,\ldots,x_k)} f(x_1,\ldots,x_k,y)\,dy \qquad (17.18)$$

and might even think to include several iterated integrals if f depends on $x_1,\ldots,x_k$ and $y_1,\ldots,y_l$. We look for partial derivatives of h in (17.18) and this leads to an iteration

$$\partial^\alpha h(x_1,\ldots,x_k) = \partial_k^{\alpha_k}\cdots\partial_1^{\alpha_1}\left(\int_{\varphi_1(x_1,\ldots,x_k)}^{\varphi_2(x_1,\ldots,x_k)} f(x_1,\ldots,x_k,y)\,dy \right)$$

with $\alpha \in \mathbb{N}_0^k$, φ_1, φ_2 being $C^{|\alpha|}$-functions on the domain G where $(x_1,\ldots,x_k)$ is coming from and f being a $C^{|\alpha|}$-function on $G \times [a,b]$, $y \in [a,b]$, and $a \le \varphi_1(x_1,\ldots,x_k) \le \varphi_2(x_1,\ldots,x_k) \le b$. The exact formula for $\partial^\alpha h$ needs some combinatorics to count all terms entering into the calculation and we leave this to the reader. We end the chapter with one example illustrating that parameter dependent integrals are often important tools for investigating partial differential equations.

Example 17.5. Let $f \in C([0,2\pi])$, $f(0) = f(2\pi)$, and for $0 < R_1 \le r < R_2 < 1$, $0 \le \varphi < 2\pi$, define the **Poisson integral** for the disc by

$$u(r,\varphi) := \frac{1}{2\pi} \int_0^{2\pi} \frac{1-r^2}{1 - 2r\cos(\varphi-\tau)+r^2} f(\tau)\,d\tau.$$

We claim that $w(x,y) = w(r\cos\varphi, r\sin\varphi) := u(r,\varphi)$ is in the annulus $B_{R_2}(0)\backslash\overline{B_{R_1}(0)}$ harmonic, i.e. satisfies $\Delta_2 w(x,y) = 0$. First we note that for $0 \le r \le R_2 < 1$ we have

$$0 < (1-r\cos(\varphi-\tau))^2 = 1 - 2r\cos(\varphi-\tau) + r^2\cos^2(\varphi-\tau) \le 1 - 2r\cos(\varphi-\tau) + r^2,$$

i.e. $1 - 2r\cos(\varphi - \tau) + r^2 \ne 0$. We use the Laplacian in polar coordinates, compare (12.1), and we aim to prove that

$$\left(\frac{\partial^2}{\partial r^2} + \frac{1}{r}\frac{\partial}{\partial r} + \frac{1}{r^2}\frac{\partial^2}{\partial\varphi^2}\right)\left(\frac{1}{2\pi}\int_0^{2\pi}\frac{1-r^2}{1-2r\cos(\varphi-\tau)+r^2}f(\tau)\,d\tau\right) = 0.$$

So we want to do the differentiations under the integral sign and aim to prove that

$$\left(\frac{\partial^2}{\partial r^2} + \frac{1}{r}\frac{\partial}{\partial r} + \frac{1}{r^2}\frac{\partial^2}{\partial\varphi^2}\right)\left(\frac{1-r^2}{1-2r\cos(\varphi-\tau)+r^2}\right) = 0.$$

Although it is possible to work out all partial derivatives, we want to use a little trick to do the calculations. This trick we will understand better when dealing with Fourier series in Volume III Part 8. Since for $0 < R_1 \le r \le R_2 < 1$ and all $\varphi, \tau \in [0, 2\pi]$ the series $\frac{1}{2} + \sum_{k=1}^\infty r^k\cos k(\varphi-\tau)$ converges uniformly as do all series obtained when taking partial derivatives with respect to r or to φ (or both), we find that

$$\frac{\partial^\alpha}{\partial r^{\alpha_1}\partial\varphi^{\alpha_2}}\left(\frac{1}{2} + \sum_{k=1}^\infty r^k\cos k(\varphi-\tau)\right)f(\tau)$$

is for all $\alpha \in \mathbb{N}_0^2$ a continuous function, in fact we have

$$\frac{\partial}{\partial r}\left(\frac{1}{2} + \sum_{k=1}^\infty r^k\cos k(\varphi-\tau)\right)f(\tau) = \sum_{k=1}^\infty kr^{k-1}\cos k(\varphi-\tau)f(\tau)$$

$$\frac{\partial^2}{\partial r^2}\left(\frac{1}{2} + \sum_{k=1}^\infty r^k\cos k(\varphi-\tau)\right)f(\tau) = \sum_{k=2}^\infty k(k-1)r^{k-2}\cos k(\varphi-\tau)f(\tau)$$

$$\frac{\partial}{\partial\varphi}\left(\frac{1}{2} + \sum_{k=1}^\infty \cos k(\varphi-\tau)\right)f(\tau) = -\sum_{k=1}^\infty kr^k\sin k(\varphi-\tau)f(\tau)$$

$$\frac{\partial^2}{\partial\varphi^2}\left(\frac{1}{2} + \sum_{k=1}^\infty r^k\cos k(\varphi-\tau)\right)f(\tau) = -\sum_{k=1}^\infty k^2r^k\cos k(\varphi-\tau)f(\tau)$$

and for $0 < R_1 \leq r \leq R_2 < 1$, $\varphi \in [0, 2\pi)$, we obtain

$$\left(\frac{\partial^2}{\partial r^2} + \frac{1}{r} \frac{\partial}{\partial r} + \frac{1}{r^2} \frac{\partial^2}{\partial \varphi^2} \right) \left(\left(\frac{1}{2} + \sum_{k=1}^{\infty} r^k \cos k(\varphi - \tau) \right) f(\tau) \right) = 0.$$

Now we prove

$$\frac{1}{2\pi} \int_0^{2\pi} \frac{1 - r^2}{1 - 2r \cos(\varphi - \tau) + r^2} f(\tau) \, d\tau = \frac{1}{\pi} \int_0^{2\pi} \left(\frac{1}{2} + \sum_{k=1}^{\infty} r^k \cos k(\varphi - \tau) \right) f(\tau) \, d\tau$$

which will imply the result since we have already justified

$$\left(\frac{\partial^2}{\partial r^2} + \frac{1}{r} \frac{\partial r}{\partial r} + \frac{1}{r^2} \frac{\partial^2}{\partial \varphi^2} \right) \frac{1}{\pi} \int_0^{2\pi} \left(\frac{1}{2} + \sum_{k=1}^{\infty} r^k \cos k(\varphi - \tau) \right) f(\tau) \, d\tau$$

$$= \frac{1}{\pi} \int_0^{2\pi} \left(\frac{\partial^2}{\partial r^2} + \frac{1}{r} \frac{\partial}{\partial r} + \frac{1}{r^2} \frac{\partial^2}{\partial \varphi^2} \right) \left(\frac{1}{2} + \sum_{k=1}^{\infty} r^k \cos k(\varphi - \tau) \right) f(\tau) \, d\tau = 0.$$

We observe that

$$\frac{1}{2} + \sum_{k=1}^{\infty} r^k \cos k(\varphi - \tau) = \frac{1}{2} \sum_{k=1}^{\infty} r^k \left(e^{ik(\varphi - \tau)} + e^{-ik(\varphi - \tau)} \right)$$

$$= \frac{1}{2} \left(1 + \sum_{k=1}^{\infty} \left((re^{i(\varphi - \tau)})^k + (re^{-i(\varphi - \tau)})^k \right) \right)$$

$$= \frac{1}{2} \left(1 + \frac{re^{i(\varphi - \tau)}}{1 - re^{i(\varphi - \tau)}} + \frac{re^{-i(\varphi - \tau)}}{1 - re^{-i(\varphi - \tau)}} \right)$$

$$= \frac{1}{2} \left(\frac{1 - r^2}{1 - 2r \cos(\varphi - \tau) + r^2} \right),$$

where we used that $\cos \vartheta = \frac{e^{i\vartheta} + e^{-i\vartheta}}{2}$, $\vartheta \in \mathbb{R}$, and for $a \in \mathbb{C}$, $|a| < 1$, we have $\sum_{k=0}^{\infty} a^k = \frac{1}{1-a}$, or $\sum_{k=1}^{\infty} a^k = \frac{a}{1-a}$, which we briefly will discuss in Problem 6.

Problems

1. Let $K \subset \mathbb{R}^n$ be a compact set and (X, d) be a compact metric space. Further let $f : K \times X \to \mathbb{R}$ be a continuous function. Denote by

$\text{pr}_j : \mathbb{R}^n \to \mathbb{R}$ the jth coordinate projection and by $\tilde{\text{pr}}_j : \mathbb{R}^n \to \mathbb{R}^{n-1}$ the projection defined for $z \in \mathbb{R}^n$ by $\tilde{\text{pr}}_j(z) = \tilde{\text{pr}}_j(z_1, \dots, z_n) = (z_1, \dots, z_{j-1}, z_{j+1}, \dots, z_n)$. Let $[a, b] \subset \text{pr}_j(K)$ and define

$$F(y_1, \dots, y_{j-1}, y_{j+1}, \dots, y_n, x) := \int_a^b f(y_1, \dots, y_n, x)\,dy_j.$$

Prove that $F : \tilde{\text{pr}}_j(K) \times X \to \mathbb{R}$ defines a continuous function.

2.   a) Let $I_1, \dots, I_n$ and $J_1, \dots, J_m$ be compact intervals and $f : I_1 \times \cdots \times I_n \times J_1 \times \cdots \times J_m \to \mathbb{R}$ be a continuous function which has the continuous partial derivatives $\frac{\partial f}{\partial y_j}$. Consider the function

$$\varphi(x_1, \dots, x_{k-1}, x_{k+1}, \dots, x_n, y_1, \dots, y_m) := \int_{I_k} f(x_1, \dots, x_n, y_1, \dots, y_n)\,dx_k.$$

Prove that $\varphi : I_1 \times \cdots \times I_{k-1} \times I_{k+1} \times \cdots \times I_n \times J_1 \times \cdots \times J_m \to \mathbb{R}$ has a continuous partial derivative with respect to y_j and the following holds

$$\frac{\partial \varphi}{\partial y_j}(x_1, \dots, x_{k-1}, x_{k+1}, \dots, x_n, y_1, \dots, y_m) = \int_{I_k} \frac{\partial f}{\partial y_j}(x_1, \dots, x_n, y_1, \dots, y_m)\,dx_k.$$

b) Indicate how you would extend Corollary 14.35 to higher order partial derivatives in light of part a).

3. Use Problems 1 and 2 to sketch the proof of Corollary 17.2.

4. For a continuously differentiable function $\varphi : [a, \infty) \to \mathbb{R}$ show that for $y > 0$

$$\int_0^{\varphi(y)} x^y \ln x\,dx = \frac{\varphi(y)^y}{(1+y)^2}\left((1+y)\varphi(y) \ln \varphi(y) - \varphi(y)\right).$$

Now derive the formula that for $\alpha > 0$

$$\int_0^1 x^\alpha \ln x\,dx = -\frac{1}{(1+\alpha)^2}.$$

Hint: consider the integral $\int_0^{\varphi(y)} x^y\,dx$ and its derivative.

5. Let $I_1, \ldots, I_k, J$ be compact intervals and consider functions $\varphi_\nu : I_1 \times \cdots \times I_k \to \mathbb{R}$, $\nu = 1, 2, \varphi_1 \leq \varphi_2$, and $h : I_1 \times \cdots \times I_k \times J \to \mathbb{R}$. Find conditions for $\varphi_\nu, \nu = 1, 2$, and h such that

$$\frac{\partial^2}{\partial x_j \partial x_l} \int_{\varphi_1(x_1,\ldots,x_k)}^{\varphi_2(x_1,\ldots,x_k)} h(x_1, \ldots, x_k, y) dy$$

exists, is a continuous function and can be calculated by a formula analogous to (17.17).

(There is no need to give an "$\epsilon - \delta$ proof" of the result, but a convincing argument relying on the results of this chapter.)

6. Use the power series expansion for exp, cos and sin which we assume to hold also for all $z \in \mathbb{C}$ to verify

$$e^{i\varphi} = \cos \varphi + i \sin \varphi, \phi \in \mathbb{R} \quad \text{with} \quad i^2 = -1.$$

18 Volume Integrals on Hyper-Rectangles

In this chapter we will define and investigate the Riemann integral for functions with domain being a hyper-rectangle - as analogously introduced for functions defined on a compact interval. A central tool for defining the Riemann integral of functions of one real variable was the algebra of all step functions $T[a, b]$, see Definition I.25.2. For a step function $\varphi : [a, b] \to \mathbb{R}$ there exists a partition $Z(x_0, \ldots, x_N)$ of $[a, b]$ such that $\varphi|_{(x_j, x_{j+1})} = c_j$, $0 \leq j \leq N - 1$, where $c_j \in \mathbb{R}$. No values at the points x_j of the partition are prescribed. In fact when constructing certain step functions we sometimes make use of the fact that their values at a finite number of points, the points of the underlying partition, do not matter. One may think of the definition of the integral (see below) or the approximation of continuous functions by step functions, Theorem I.25.6. This resonates well with another observation, namely with the fact that the definition of the length of a bounded interval $I \subset \mathbb{R}$ with end points $a < b$ is the same for $[a, b]$, $(a, b]$, $[a, b)$ and (a, b),

$$l(I) = b - a. \tag{18.1}$$

We can put this differently. The closed interval $[a, b]$ is, with respect to inclusion, the largest of all intervals with end points $a < b$ and has by definition length $b - a$. Removing one or two end points from $[a, b]$ does not change its length. Moreover, if $Z(x_0, \ldots, x_n)$ is a partition of $[a, b]$, the length of $\bigcup_{j=0}^{N-1} [x_j, x_{j+1}]$, $\bigcup_{j=0}^{N-1} (x_j, x_{j+1}]$, $\bigcup_{j=0}^{N-1} [x_j, x_{j+1})$ and $\bigcup_{j=1}^{N-1} (x_j, x_{j+1})$ is the same and equal to that of $[a, b]$, i.e. equal to $b - a$. Hence for calculating the length of an interval $[a, b]$ it does not matter whether a finite number of points is missing as in $\bigcup_{j=0}^{N-1} (x_j, x_{j+1})$ or a finite number of points is double counted as in $\bigcup_{j=0}^{N-1} [x_j, x_{j+1}]$.

The integral for a step function $\varphi \in T[a, b]$ is defined by

$$\int_a^b \varphi(x) \, \mathrm{d}x = \sum_{j=0}^{N-1} c_j (x_{j+1} - x_j), \tag{18.2}$$

if $Z(x_0, \ldots, x_N)$ is the underlying partition of $[a, b]$, see Definition I.25.7, and we see that φ need not be specified at the points x_j of the partition. In other words, the values of φ at a set of points which do not contribute when calculating the length of $[a, b]$ do not matter. We now try to extend these observations to hyper-rectangles in $\mathbb{R}^n$.

Definition 18.1. A. *A compact n-dimensional axes-parallel hyper-rectangle K is a set of the type*

$$K := \underset{j=1}{\overset{n}{\times}}[a_j, b_j], \quad a_j < b_j, \; j = 1, \ldots, n, \tag{18.3}$$

*and it is often called a **non-degenerate compact cell** in $\mathbb{R}^n$.*
B. *We call $\times_{j=1}^n [a_j, b_j]$ a **degenerate compact cell** of dimension $n - l$ in $\mathbb{R}^n$ if for a set $J \subset \{1, \ldots, n\}$ of $l \in \mathbb{N}$ elements we have $a_j = b_j$ for $j \in J$ and $a_j < b_j$ for $j \in \{1, \ldots, n\} \backslash J$.*

The set of all non-degenerate and degenerate cells in $\mathbb{R}^n$ is denoted by $\mathscr{R}_n$ and is called the set of all **compact cells** in $\mathbb{R}^n$.

Definition 18.2. *The n-dimensional **volume** of $K \in \mathscr{R}_n$, $K = \times_{j=1}^n [a_j, b_j]$, is defined as*

$$\mathrm{vol}_n(K) := \prod_{j=1}^{n}(b_j - a_j). \tag{18.4}$$

Note that for non-degenerate compact cells K we have $\mathrm{vol}_n(K) > 0$, where as $\mathrm{vol}_n(K) = 0$ if and only if K is a degenerate compact cell of $\mathbb{R}^n$.

We want to introduce partitions of compact cells into compact (sub-)cells. Since the aim is to define an integral for a suitable class of functions, partitions fit for purpose are sufficient, i.e. we do not long for the most general family of partitions. While Figure 18.1 shows a reasonable partition, it is not practical to work with. However the partition in Figure 18.2 is. This is a partition induced by partitions of the intervals forming the cell. Since we can pass from a general partition as in Figure 18.1 to a refined one as in Figure 18.2, see Figure 18.3, it turns out that partitions of cells induced by partitions of the defining intervals are sufficient for us to consider and we do not handle others.

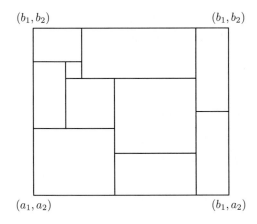

(b_1, b_2) (b_1, b_2)

(a_1, a_2) (b_1, a_2)

Figure 18.1

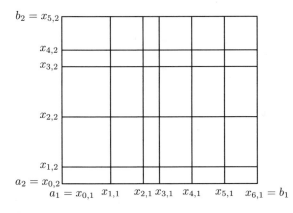

Figure 18.2

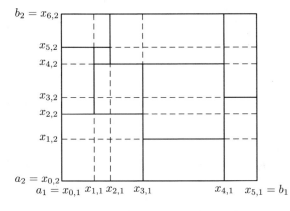

Figure 18.3

369

Let us introduce some notation to deal with such partitions. For $K = \times_{j=1}^{n}[a_j, b_j]$, $a_j < b_j$, we start with partitions $Z_j(x_{0,j}, \ldots, x_{N_j,j})$ of $[a_j, b_j]$, $j = 1, \ldots, n$, where as usual $a_j = x_{0,j}$ and $b_j = x_{N_j,j}$. The corresponding partition of K is $Z(Z_1, \ldots, Z_n)$. We set

$$A_j := \{0, 1, \ldots, N_j\} \times \{j\}, \tag{18.5}$$

$$A_Z := \overset{n}{\underset{j=1}{\times}} A_j \tag{18.6}$$

and $\alpha = (\alpha_1, \ldots, \alpha_n) \in A_Z$ has the typical structure

$$\alpha = (\alpha_1, \ldots, \alpha_n) = ((k_1, 1), \ldots, (k_n, n)), \quad \alpha_j = (k_j, j), k_j \in \{0, 1, \ldots, N_j\} \tag{18.7}$$

and with $x_{\alpha_j} = x_{k_j,j} \in Z_j(x_{0,j}, \ldots, x_{N_j,j}) \subset \mathbb{R}$ we define

$$x_\alpha := (x_{\alpha_1}, \ldots, x_{\alpha_n}) \in \mathbb{R}^n. \tag{18.8}$$

Now we find

$$Z(Z_1, \ldots, Z_n) = \{x_\alpha \in K \mid \alpha \in A_Z\} \tag{18.9}$$

and $y \in Z$ if $y = (x_{\alpha_1}, \ldots, x_{\alpha_n}) = (x_{k_1,1}, \ldots, x_{k_n,n})$, $\alpha_j = (k_j, j)$. For $\alpha_j \in A_j$, $\alpha_j = (k_j, j)$, we set for $k_j \geq 1$

$$I_{\alpha_j} := [x_{k_j-1,j}, x_{k_j,j}] \tag{18.10}$$

and

$$I_{\alpha_j} := \emptyset \quad \text{for} \quad k_j = 0.$$

Finally with

$$K_\alpha := \overset{n}{\underset{j=1}{\times}} I_{\alpha_j}, \quad \alpha = (\alpha_1, \ldots, \alpha_n) \tag{18.11}$$

we find

$$K = \bigcup_{\alpha \in A_Z} K_\alpha \tag{18.12}$$

and $\overset{\circ}{K}_\alpha \cap \overset{\circ}{K}_\beta = \emptyset$ for $\alpha, \beta \in A_Z$ where as usual $\overset{\circ}{K}_\alpha$ denotes the interior of K_α.

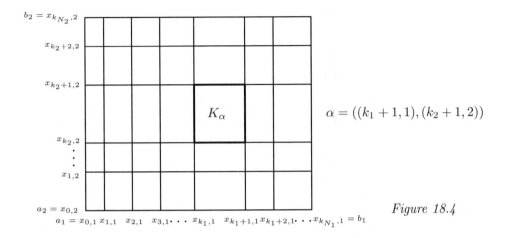

Figure 18.4

Definition 18.3. *Let $K = \bigtimes_{j=1}^{n}[a_j, b_j]$ be a non-degenerate compact cell and for $j = 1, \ldots, n$, partitions $Z_j(x_{0,j}, \ldots, x_{N_j,j})$ of $[a_j, b_j]$ given. We call $Z = Z(Z_1, \ldots, Z_n)$ the partition of K generated by the partitions Z_j, $1 \leq j \leq n$. The **mesh size** or width of Z is*

$$\mathrm{mesh}(Z) := \max_{\alpha \in A_Z}(\mathrm{diam}\, K_\alpha) \tag{18.13}$$

with K_α as in (18.11).

Since $\mathrm{diam}\left(\bigtimes_{j=1}^{n}[c_j, d_j]\right) = \left(\sum_{j=1}^{n}(d_j - c_j)^2\right)^{\frac{1}{2}}$ we have

$$\mathrm{mesh}(Z) = \max_{\alpha \in A_Z}\left(\sum_{j=0}^{n}(x_{k_j,j} - x_{k_j-1,j})^2\right)^{\frac{1}{2}} \tag{18.14}$$

where $\alpha = (\alpha_1, \ldots, \alpha_n)$ and $\alpha_j = (k_j, j)$.
Note that while $K = \bigcup_{\alpha \in A_Z} K_\alpha$ holds, in general we do not have $K_\alpha \cap K_\beta = \emptyset$ for $\alpha \neq \beta$. However for $\alpha \neq \beta$ this intersection is either empty or an $(n-m)$-dimensional degenerate hyper-rectangle in $\mathbb{R}^n$ with $m \geq 1$ and consequently we have

$$\mathrm{vol}_n(K_\alpha \cap K_\beta) = 0 \quad \text{for} \quad \alpha \neq \beta, \, \alpha, \beta \in A_Z.$$

Furthermore we have

$$\text{vol}_n(K) = \prod_{j=1}^{n}(b_j - a_j)$$

$$= \prod_{j=1}^{n}(x_{N_j,j} - x_{N_j-1,j} + \cdots + x_{1,j} - x_{0,j})$$

$$= \prod_{j=1}^{n}\sum_{k_j=1}^{N_j}(x_{k_j,j} - x_{k_j-1,j})$$

$$= \sum_{\alpha \in A_z} \text{vol}_n(K_\alpha).$$

Definition 18.4. *We call* $Z'_j(t_{0,j}, \ldots, t_{M_j,j})$ *a* **refinement** *of the partition* $Z_j(x_{0,j}, \ldots, x_{N_j,j})$ *of* $[a_j, b_j]$ *if every point of* Z'_j *belongs to* Z_j, *and we say that* $Z' = Z'(Z'_1, \ldots, Z'_n)$ *is a refinement of* $Z = Z(Z_1, \ldots, Z_n)$ *if every* Z'_j *is a refinement of* Z_j.

We are now prepared to extend the notion of step function to higher dimensions.

Definition 18.5. *Let* $K := \bigtimes_{j=1}^{n}[a_j, b_j]$ *be a non-degenerate compact cell in* $\mathbb{R}^n$ *and* $\varphi : K \to \mathbb{R}$ *a bounded function. We call* φ *a* **step function** *on* K *if there exists a partition* $Z = Z(Z_1, \ldots, Z_n)$ *of* K, $K = \bigcup_{\alpha \in A_Z} K_\alpha$, *such that* $\varphi|_{\mathring{K}_\alpha} = c_\alpha \in \mathbb{R}$, *i.e.* φ *is on* $\mathring{K}_\alpha$ *a constant. The set of all step functions on* K *is denoted by* $T(K)$.

Remark 18.6. A. No assumption on the values of φ on ∂K_α, $\alpha \in A_Z$, is made except that φ is bounded on K.
B. Note that $T(K)$ is defined with the help of partitions of K into cells. Of course, step functions can be defined with respect to more general partitions, for example the one indicated in Figure 18.1. However we do not need this generalisation for our theory.
C. As in the one-dimensional case it is possible to prove that $T(K)$ is an algebra, i.e. $\varphi, \psi \in T(K)$ and $\alpha, \beta \in \mathbb{R}$ imply $\alpha\varphi + \beta\psi \in T(K)$ and $\varphi \cdot \psi \in T(K)$, see Problem 2.

For a step function $\varphi \in T(K)$ we define now its integral as

$$\int_K \varphi(x)\,dx := \sum_{\alpha \in A_z} c_\alpha \text{vol}_n(K_\alpha). \tag{18.15}$$

In order to prove basic properties of this integral we have to introduce for two partitions $Z(Z_1, \ldots, Z_n) = \{x_\alpha \in K \mid \alpha \in A_Z\}$ and $Z'(Z_1', \ldots, Z_n') = \{y_\beta \in K \mid \beta \in A_{Z'}\}$ their **joint refinement** $Z^* := Z \cup Z' := \{x_\alpha \in K \mid \alpha \in A_Z\} \cup \{y_\beta \in K \mid \beta \in A_{Z'}\}$ for which we write also $\{x_\gamma \in K \mid \gamma \in A_{Z^*}\}$.

Proposition 18.7. A. *The definition of the integral* (18.15) *is independent of the partition used to represent* $\varphi \in T(K)$.

B. *On* $T(K)$ *the integral is linear, i.e. for* $\varphi, \psi \in T(K)$ *and* $\alpha, \beta \in \mathbb{R}$ *the following holds*

$$\int_K (\alpha\varphi + \beta\psi)(x)\, \mathrm{d}x = \alpha \int_K \varphi(x)\, \mathrm{d}x + \beta \int_K \psi(x)\, \mathrm{d}x. \tag{18.16}$$

C. *The integral is positivity preserving, i.e. for* $\varphi \in T(K)$ *and* $\varphi \geq 0$ *it follows that*

$$\int_K \varphi(x)\, \mathrm{d}x \geq 0, \tag{18.17}$$

which also yields the monotonicity of the integral, i.e. $\varphi \leq \psi$ *implies* $\int_K \varphi(x)\, \mathrm{d}x \leq \int_K \psi(x)\, \mathrm{d}x.$

Proof. The main observation is as in one dimension, namely that passing to a refinement of $Z(Z_1, \ldots, Z_n)$ does not change the value of the integral. So let $Z = Z(Z_1, \ldots, Z_n) = \{x_\alpha \in K \mid \alpha \in A_Z\}$ be a partition of K and $Z'(Z_1', \ldots, Z_n') = \{y_\beta \in K \mid \beta \in A_{Z'}\}$ a refinement of Z. We have to prove that

$$\sum_{\alpha \in A_Z} c_\alpha \mathrm{vol}_n(K_\alpha) = \sum_{\beta \in A_{Z'}} c_\beta' \mathrm{vol}_n(K_\beta').$$

Since Z' is a refinement of Z it follows that for every $\beta \in A_{Z'}$ there exists $\alpha \in A_Z$ such that $K_\beta' \subset K_\alpha$, and moreover for $\alpha \in A_Z$ there exists $\beta_1^{(\alpha)}, \ldots, \beta_{M_\alpha}^{(\alpha)} \in A_{Z'}$ such that $K_\alpha = \bigcup_{j=1}^{M_\alpha} K_{\beta_j}(\alpha)$, as well as $c_\alpha = c_{\beta_j}'(\alpha)$ for $1 \leq j \leq M_\alpha$. If we denote the integral (18.15) with respect to partition Z for a moment by $Z - \int_K \varphi$, it follows that

$$Z - \int_K \varphi(x)\, \mathrm{d}x = \sum_{\alpha \in A_Z} c_\alpha \mathrm{vol}_n(K_\alpha)$$

$$= \sum_{\alpha \in A_Z} \sum_{j=1}^{M_\alpha} c_{\beta_j(\alpha)}' \mathrm{vol}_n(K_{\beta_j^{(\alpha)}})$$

$$= \sum_{\beta \in A_{Z'}} c_\beta' \mathrm{vol}_n(K_\beta) = Z' - \int_K \varphi(x)\, \mathrm{d}x.$$

Now if φ has two representations with partitions Z and Z' we deduce that the integral of φ with respect to each of these partitions is equal to the integral of φ with respect to the joint partition Z^*, hence they are equal. From this part B follows easily. If φ and ψ are represented with respect to the partitions Z_φ and Z_ψ we represent both with respect to the joint partition $Z_\varphi \cup Z_\psi$ and then we use the linearity of finite sums. Part C is trivial since $\varphi \geq 0$ implies $\varphi \geq 0$ on $\mathring{K}_\alpha$, i.e. $c_\alpha \geq 0$, and therefore the sum in (18.15) consists of non-negative terms, whereas the monotonicity follows from the fact that $\varphi \leq \psi$ is equivalent to $0 \leq \psi - \varphi$ and the linearity of the integral. $\qquad \square$

Example 18.8. We have

$$\int_K 1 \, \mathrm{d}x = \mathrm{vol}_n(K)$$

and further for any partition of K

$$\int_K \varphi(x) \, \mathrm{d}x = \int_{\bigcup_{\alpha \in A_Z} K_\alpha} \varphi(x) \, \mathrm{d}x = \sum_{\alpha \in A_Z} \int_{K_\alpha} \varphi(x) \, \mathrm{d}x, \qquad (18.18)$$

which we interpret as the first step to prove set-additivity of the integral.

We are now starting to extend the integral to more general functions.

Definition 18.9. *Let* $K = \bigtimes_{j=1}^{n} [a_j, b_j] \subset \mathbb{R}^n$ *be a non-degenerate compact cell and* $f : K \to \mathbb{R}$ *a bounded function. For a partition* $Z = Z(Z_1, \ldots, Z_n)$ *of* K, $Z = \{x_\alpha \in K \mid \alpha \in A_Z\}$, *we choose points* $\xi_\alpha \in \mathring{K}_\alpha$ *and we set* $\xi := \{\xi_\alpha \mid \alpha \in A_Z\}$. *The term*

$$R(f, Z, \xi) := \sum_{\alpha \in A_Z} f(\xi_\alpha) \mathrm{vol}_n(K_\alpha) \qquad (18.19)$$

*is called the **Riemann sum** of* f *with respect to the partition* Z *and the points* ξ.

We can read (18.19) in the following way: given f, Z and ξ, we introduce a step function $\varphi_{f,Z,\xi}$ by $\varphi_{f,Z,\xi}|_{\mathring{K}_\alpha} = \xi_\alpha$ and on $K \backslash \bigcup_{\alpha \in A_Z} \mathring{K}_\alpha$ we allow $\varphi_{f,Z,\xi}$ to take any value however we require that $\varphi_{f,Z,\xi}$ is bounded on K. Then $R(f, Z, \xi)$ is the integral of the step function $\varphi_{f,Z,\xi}$. The natural question is: suppose we can approximate f by step functions, for example by functions of the type $\varphi_{f,Z,\xi}$, does this allow us to define in the limit an integral for f?

We follow the ideas from the theory in one dimension and introduce lower and upper integrals. Let $f : K \to \mathbb{R}$ be a bounded function. The **lower integral** of f is defined by

$$\int_* f(x)\,\mathrm{d}x = \int_{*K} f(x)\,\mathrm{d}x := \sup\left\{\int_K \varphi(x)\,\mathrm{d}x \,\middle|\, \varphi \in T(K) \text{ and } \varphi \le f\right\} \quad (18.20)$$

and the **upper integral** of f is defined by

$$\int^* f(x)\,\mathrm{d}x = \int_K^* f(x)\,\mathrm{d}x := \inf\left\{\int_K \psi(x)\,\mathrm{d}x \,\middle|\, \psi \in T(K) \text{ and } f \le \psi\right\}. \quad (18.21)$$

Proposition 18.10. *Let $f, g : K \to \mathbb{R}$ be bounded functions and $\lambda \ge 0$. For the lower and upper integral we have*

$$\int^* (f+g)(x)\,\mathrm{d}x \le \int^* f(x)\,\mathrm{d}x + \int^* g(x)\,\mathrm{d}x \quad (18.22)$$

and

$$\int^* (\lambda f)(x)\,\mathrm{d}x = \lambda \int^* f(x)\,\mathrm{d}x \quad (18.23)$$

as well as

$$\int_* (f+g)(x)\,\mathrm{d}x \ge \int_* f(x)\,\mathrm{d}x + \int_* g(x)\,\mathrm{d}x \quad (18.24)$$

and

$$\int_* (\lambda f)\,\mathrm{d}x = \lambda \int_* f(x)\,\mathrm{d}x. \quad (18.25)$$

In addition, for $\mu < 0$ we have

$$\int^* (\mu f)(x)\,\mathrm{d}x = \mu \int_* f(x)\,\mathrm{d}x \quad \text{and} \quad \int_* (\mu f)(x)\,\mathrm{d}x = \mu \int^* f(x)\,\mathrm{d}x. \quad (18.26)$$

$\square$

The proof of Proposition 18.10 is identical to those of Theorem I.25.13 and Corollary I.25.14 and we recommend working through those proofs again to adapt them to the new situation.

Of central importance now is

Definition 18.11. *Let* $K := \underset{j=1}{\overset{n}{\times}}[a_j, b_j]$ *be a non-degenerate compact cell in* $\mathbb{R}^n$ *and* $f : K \to \mathbb{R}$ *a bounded function. We call* f ***Riemann integrable*** *over* K *if*

$$\int_* f(x)\,\mathrm{d}x = \int^* f(x)\,\mathrm{d}x \tag{18.27}$$

and in this case we write

$$\int_K f(x)\,\mathrm{d}x = \int_* f(x)\,\mathrm{d}x = \int^* f(x)\,\mathrm{d}x. \tag{18.28}$$

We call $\int_K f(x)\,\mathrm{d}x$ *the* ***Riemann integral*** *of* f *over* K.
(Other notations are $\int_K f\,\mathrm{d}x$ *or* $\int f\,\mathrm{d}x$ *if from the context it is clear what the underlying set* K *is.)*

Obviously we have

Corollary 18.12. *Every step function* $\varphi \in T(K)$ *is Riemann integrable and the Riemann integral of* φ *is given by* (18.15).

Our basic integrability criterion in one dimension, Theorem I.25.15, extends to the new situation:

Theorem 18.13. *The bounded function* $f : K \to \mathbb{R}$ *defined on a non-degenerate compact cell in* $\mathbb{R}^n$ *is Riemann integrable if and only if for* $\epsilon > 0$ *there exist step functions* $\varphi, \psi \in T(K)$ *such that* $\varphi \leq f \leq \psi$ *and*

$$\int_K (\psi(x) - \varphi(x))\,\mathrm{d}x \leq \epsilon. \tag{18.29}$$

Proof. From our definitions we have $\int_* f(x)\,\mathrm{d}x \leq \int^* f(x)\,\mathrm{d}x$ and that for $\epsilon > 0$ there exist $\varphi, \psi \in T(K)$ such that $\varphi \leq f \leq \psi$ as well as

$$\int^* f(x)\,\mathrm{d}x \leq \int \psi(x)\,\mathrm{d}x + \frac{\epsilon}{2} \quad \text{and} \quad -\frac{\epsilon}{2} + \int \varphi(x)\,\mathrm{d}x \leq \int_* f(x)\,\mathrm{d}x.$$

Thus, if $\int^* f(x)\,\mathrm{d}x = \int_* f(x)\,\mathrm{d}x$ it follows that $\int(\psi(x) - \varphi(x))\,\mathrm{d}x < \epsilon$. On the other hand we have

$$\int^* f(x)\,\mathrm{d}x \leq \int \psi(x)\,\mathrm{d}x \quad \text{and} \quad \int_* f(x)\,\mathrm{d}x \geq \int \varphi(x)\,\mathrm{d}x$$

or

$$0 \leq \int^* f(x)\,\mathrm{d}x - \int_* f(x)\,\mathrm{d}x \leq \int (\psi(x) - \varphi(x))\,\mathrm{d}x,$$

implying that (18.29) yields the integrability of f. $\qquad\square$

The following result extends Theorem I.25.6 to K and will imply the Riemann integrability of continuous functions $f : K \to \mathbb{R}$.

Theorem 18.14. *Let* $K = \bigtimes_{j=1}^{n} [a_j, b_j]$ *be a non-degenerate compact cell of* $\mathbb{R}^n$ *and* $f : K \to \mathbb{R}$ *a continuous function. For* $\epsilon > 0$ *given there exist step functions* $\varphi, \psi \in T(K)$ *such that for all* $x \in K$

$$\varphi(x) \leq f(x) \leq \psi(x) \tag{18.30}$$

and

$$|\varphi(x) - \psi(x)| = \psi(x) - \varphi(x) \leq \epsilon. \tag{18.31}$$

Proof. We sketch the proof along the lines of the proof of Theorem I.25.6. Since K is compact f is uniformly continuous on K. Hence for $\epsilon > 0$ there exists $\delta > 0$ such that $|x_j - y_j| < \delta$, $j = 1, \ldots, n$, $x, y \in K$, $x = (x_1, \ldots, x_n)$, $y = (y_1, \ldots, y_n)$, implies $|f(x) - f(y)| < \frac{\epsilon}{2}$. We now divide each interval $[a_j, b_j]$ into N_j intervals of equal length less than δ, i.e. $\frac{b_j - a_j}{N_j} < \delta$. This gives for $j = 1, \ldots, n$ a partition $Z_j(x_{0,j}, \ldots, x_{N_j,j})$ of $[a_j, b_j]$ which induces a partition $Z = Z(Z_1, \ldots, Z_n) = \{x_\alpha \mid \alpha \in A_Z\}$ of K. We now define the two step functions

$$\varphi(x) := \begin{cases} c'_\alpha = f(x_\alpha) - \frac{\epsilon}{2}, & x \in \mathring{K}_\alpha \\ f(x), & x \in K \backslash \bigcup_{\alpha \in A_Z} \mathring{K}_\alpha \end{cases}$$

and

$$\psi(x) := \begin{cases} c_\alpha := f(x_\alpha) + \frac{\epsilon}{2}, & x \in \mathring{K}_\alpha \\ f(x), & x \in K \backslash \bigcup_{\alpha \in A_Z} \mathring{K}_\alpha. \end{cases}$$

From the definition follows

$$|\varphi(x) - \psi(x)| \leq \epsilon \quad \text{for all} \quad x \in K.$$

Further, for $x \in K \backslash \bigcup_{\alpha \in A_Z} \mathring{K}_\alpha$ we have $\varphi(x) = \psi(x) = f(x)$, and for $x \in \mathring{K}_\alpha$ it follows that $|x_j - (x_\alpha)_j| < \delta$ and therefore

$$-\frac{\epsilon}{2} < f(x) - f(x_\alpha) < \frac{\epsilon}{2}$$

or

$$\varphi(x) = c'_\alpha = f(x_\alpha) - \frac{\epsilon}{2} < f(x) < f(x_\alpha) + \frac{\epsilon}{2} = c_\alpha = \psi(x),$$

i.e. $\varphi(x) \leq f(x) \leq \psi(x)$ for all $x \in K$. $\square$

Corollary 18.15. *A continuous function $f : K \to \mathbb{R}$ defined on a non-degenerate compact cell in $\mathbb{R}^n$ is Riemann integrable.*

Proof. Given $\epsilon > 0$ by Theorem 18.14 we can find step functions $\varphi, \psi \in T(K)$ such that $\varphi \le f \le \psi$ and $\psi(x) - \varphi(x) < \frac{\epsilon}{\mathrm{vol}_n(K)}$ implying

$$\int_K (\psi(x) - \varphi(x)) \, \mathrm{d}x \le \int_K \frac{\epsilon}{\mathrm{vol}_n(K)} \, \mathrm{d}x \le \epsilon.$$

$\square$

Since we have prepared all auxiliary results we can prove, along the lines as we did in Chapter I.25, the following results. We omit the proofs urging the student to revisit the corresponding proofs in Volume I and make the obvious adaptations.

Theorem 18.16. *The set of all Riemann integrable functions $f : K \to \mathbb{R}$ defined on a non-degenerate compact cell of $\mathbb{R}^n$ forms a real vector space and in addition for Riemann integrable functions $f, g : K \to \mathbb{R}$ we have*

$$f_+, f_- \text{ are Riemann integrable;} \tag{18.32}$$

$$|f|^p \text{ is Riemann integrable for } p \ge 1; \tag{18.33}$$

$$f \cdot g \text{ is Riemann integrable;} \tag{18.34}$$

$$f \le g \text{ implies } \int_K f(x) \, \mathrm{d}x \le \int_K g(x) \, \mathrm{d}x, \tag{18.35}$$

in particular the Riemann integral is positivity preserving.

Let $f : K \to \mathbb{R}^n$ be a continuous function. Since K is compact, f attains its minimum $m := \min f(K)$ and maximum $M := \max f(K)$. Moreover, K is connected, so $f(K)$ is connected, hence $f(K) = [m, M]$. Thus for $m \le \mu \le M$ there exists $x_\mu \in K$ such that $f(x_\mu) = \mu$. Of course x_μ need not be unique. Since f is Riemann integrable it follows that

$$m \, \mathrm{vol}_n(K) \le \int_K f(x) \, \mathrm{d}x \le M \, \mathrm{vol}_n(K)$$

implying that

$$\int_K f(x) \, \mathrm{d}x = \mu \, \mathrm{vol}_n(K)$$

for some $m \le \mu \le M$. Thus we have proved a mean value result:

Proposition 18.17. *For a continuous function* $f : K \to \mathbb{R}$ *defined on the non-degenerate compact cell* $K \subset \mathbb{R}^n$ *there exists* $\xi \in K$ *such that*

$$\int_K f(x) \, dx = f(\xi) \mathrm{vol}_n(K). \tag{18.36}$$

Although the idea of the proof is essentially the same as in one dimensions we want to give the proof of the statement that we can approximate the integral of f by Riemann sums in detail.

Theorem 18.18. *Let* $f : K \to \mathbb{R}$ *be a Riemann integrable function on the non-degenerate compact cell* $K \subset \mathbb{R}^n$. *For every* $\epsilon > 0$ *there exists* $\delta > 0$ *such that for every partition* $Z = Z(Z_1, \ldots, Z_n) = \{x_\alpha \in K \,|\, \alpha \in A_Z\}$ *of* K *with* $\mathrm{mesh}(Z) < \delta$ *it follows for every choice of points* $\xi_\alpha \in \mathring{K}_\alpha$, $\xi = \{\xi_\alpha \in \mathring{K}_\alpha \,|\, \alpha \in A_Z\}$, *that*

$$\left| \int_K f(x) \, dx - R(f, Z, \xi) \right| \leq \epsilon. \tag{18.37}$$

Proof. Given $\epsilon > 0$ there are step functions $\varphi, \psi \in T(K)$ such that

$$\varphi \leq f \leq \psi \quad \text{and} \quad \int_K (\psi(x) - \varphi(x)) \, dx \leq \frac{\epsilon}{2}.$$

We may assume that φ and ψ are given with respect to the same partition $Z' = Z'(Z'_1, \ldots, Z'_n) = \{y_\beta \in K \,|\, \beta \in A_{Z'}\}$ with compact cells K'_β. Since f is continuous and K is compact, f is bounded and we set $M := \sup \{|f(x)| \,|\, x \in K\} \geq 0$. Clearly M is finite and we may assume $M > 0$.
Let $Z = Z(Z_1, \ldots, Z_n) = \{x_\alpha \in K \,|\, \alpha \in A_Z\}$ be a partition of K with $\mathrm{mesh}(Z) \leq (2\eta)^n$ where $\eta > 0$ will be determined later, and choose points $\xi_\alpha \in \mathring{K}_\alpha$. We define the step function $F \in T(K)$ by

$$F(x) = \begin{cases} f(\xi_\alpha), & x \in \mathring{K}_\alpha \\ 0, & x \in K \backslash \bigcup_{\alpha \in A_Z} \mathring{K}_\alpha. \end{cases}$$

It follows that

$$\int_K F(x) \, dx = \sum_{\alpha \in A_Z} f(\xi_\alpha) \mathrm{vol}_n(K_\alpha) \tag{18.38}$$

is the Riemann sum $R(f, Z, \xi)$. Moreover, F has the properties

$$\varphi(x) - 2M \leq F(x) \leq \psi(x) + 2M \tag{18.39}$$

and

$$\varphi(x) \leq F(x) \leq \psi(x) \quad \text{for} \quad x \in K_\alpha \subset K'_\beta \quad \text{for some} \quad \beta \in A_{Z'}. \tag{18.40}$$

Let $L \subset K$ be defined as

$$L := \bigcup \{ K_\alpha \, | \, \text{there exists} \ \beta \in A_{Z'} \ \text{such that} \ K_\alpha \subset K'_\beta \}$$

and define

$$s(x) := \begin{cases} 0, & x \in L \\ 2M, & x \notin L. \end{cases}$$

From (18.39) and (18.40) we deduce

$$\varphi(x) - s(x) \leq F(x) \leq \psi(x) + s(x) \quad \text{for all} \ x \in K.$$

Denote by μ the number of cells K_α on which s is not 0 and we now set $\delta = (2\eta)^n$ where

$$\eta := \frac{1}{2} \left(\frac{\epsilon}{4M\mu} \right)^{\frac{1}{n}},$$

and recall that $\text{mesh}(Z) < (2\eta)^n = \delta$ by our assumption. It follows that

$$\int_K s(x) \, \mathrm{d}x \leq 2M\mu(2\eta)^n < \frac{\epsilon}{2}$$

which implies

$$\int_K \varphi(x) \, \mathrm{d}x - \frac{\epsilon}{2} \leq \int_K F(x) \, \mathrm{d}x \leq \int_K \psi(x) \, \mathrm{d}x + \frac{\epsilon}{2}. \tag{18.41}$$

Our choice of φ and ψ further yields that

$$\int_K f \, \mathrm{d}x \leq \int_K \varphi \, \mathrm{d}x + \frac{\epsilon}{2} \quad \text{and} \quad \int_K \psi \, \mathrm{d}x \leq \int_K f \, \mathrm{d}x + \frac{\epsilon}{2} \tag{18.42}$$

and combining (18.41) and (18.42) we obtain

$$\left| \int_K f(x) \, \mathrm{d}x - \int_K F(x) \, \mathrm{d}x \right| \leq \epsilon$$

proving by (18.38) the theorem. $\qquad \square$

Now we are in a position to identify for continuous functions defined on a non-degenerate compact cell in $\mathbb{R}^n$ the Riemann integral with iterated integrals.

Theorem 18.19. *Let* $K := \bigtimes_{j=1}^n [a_j, b_j]$ *be a non-degenerate compact cell in* $\mathbb{R}^n$ *and* $f : K \to \mathbb{R}$ *a continuous function. Then*

$$\int_K f(x)\, dx = \int_{a_n}^{b_n} \left(\cdots \left(\int_{a_2}^{b_2} \left(\int_{a_1}^{b_1} f(x_1, \ldots, x_n) dx_1 \right) dx_2 \right) \cdots \right) dx_n \quad (18.43)$$

holds and we may take the integral on the right hand side in any order.

Proof. The last statement follows from Corollary 17.2. Obviously, once we understand the proof for $n = 2$ the general result follows analogously. So we want to prove for $n = 2$, $K = [a_1, b_1] \times [a_2, b_2]$, $a_j < b_j$, that

$$\int_K f(x)\, dx = \int_{[a_1,b_1] \times [a_2,b_2]} f(x)\, dx = \int_{a_2}^{b_2} \left(\int_{a_1}^{b_1} f(x_1, x_2)\, dx_1 \right) dx_2. \quad (18.44)$$

We know that the function

$$x_2 \mapsto g(x_2) := \int_{a_1}^{b_1} f(x_1, x_2)\, dx_1$$

is continuous on $[a_2, b_2]$, hence it is integrable and

$$\int_{a_2}^{b_2} g(x_2)\, dx_2 = \int_{a_2}^{b_2} \left(\int_{a_1}^{b_1} f(x_1, x_2)\, dx_1 \right) dx_2,$$

compare with Theorem 17.1. Let $Z_j = Z_j(x_{0,j}, \ldots, x_{N_j,j})$, $j = 1, 2$, be partitions of $[a_j, b_j]$, $j = 1, 2$. Then $Z = Z(Z_1, Z_2)$ is a partition of $K = [a_1, b_1] \times [a_2, b_2]$. We consider the corresponding compact, non-degenerate cells

$$K_{\mu\nu} := [x_{\mu-1,1}, x_{\mu,1}] \times [x_{\nu-1,2}, x_{\nu,2}]$$

for $\mu = 1, \ldots, N_1$ and $\nu = 1, \ldots, N_2$. It follows that

$$\mathrm{vol}_n(K_{\mu\nu}) = (x_{\mu,1} - x_{\mu-1,1})(x_{\nu,2} - \nu_{\nu-1,2}).$$

Now we choose in every $\mathring{K}_{\mu\nu}$ a point $\xi_{\mu\nu} = (\xi_{\mu\nu}^{(1)}, \xi_{\mu\nu}^{(2)})$ and consider the Riemann sum

$$R(f, Z, \xi) := \sum_{\mu=1}^{N_1} \sum_{\nu=1}^{N_2} f(\xi_{\mu\nu}) \mathrm{vol}_2(K_{\mu\nu}). \quad (18.45)$$

Since f is uniformly continuous on K, given $\epsilon > 0$ there exists $\delta > 0$ such that if $\mathrm{mesh}(Z) < \delta$ it follows that

$$|f(x_1, x_2) - f(\xi_\mu^{(1)}, \xi_\nu^{(2)})| < \epsilon \tag{18.46}$$

for all $(x_1, x_2) \in K_{\mu\nu}$ and all $1 \leq \mu \leq N_1$ and $1 \leq \nu \leq N_2$. Thus for $\mathrm{mesh}(Z) < \delta$ we find

$$\left| \int_{x_{\nu-1,2}}^{x_{\nu,2}} \left(\int_{x_{\mu-1,1}}^{x_{\mu,1}} (f(x_1, x_2) - f(\xi_\mu^{(1)}, \xi_\nu^{(2)})) \, dx_1 \right) dx_2 \right|$$

$$\leq \int_{x_{\nu-1,2}}^{x_{\nu,2}} \left(\int_{x_{\mu-1,1}}^{x_{\mu,1}} |f(x_1, x_2) - f(\xi_\mu^{(1)}, \xi_\nu^{(2)})| \, dx_1 \right) dx_2$$

$$\leq \epsilon(x_{\mu,1} - x_{\mu-1,1})(x_{\nu,2} - x_{\nu-1,2}) = \epsilon \, \mathrm{vol}_2(K_{\mu\nu}).$$

Moreover we have

$$\int_{a_2}^{b_2} \left(\int_{a_1}^{b_1} f(x_1, x_2) \, dx_1 \right) dx_2 = \sum_{\mu=1}^{N_1} \sum_{\nu=1}^{N_2} \int_{x_{\nu-1,2}}^{x_{\nu,2}} \left(\int_{x_{\mu-1,1}}^{x_{\mu,2}} f(\xi_\mu^{(1)}, \xi_\nu^{(2)}) \, dx_1 \right) dx_2 \tag{18.47}$$

and with (18.45)

$$R(f, Z, \xi) = \sum_{\mu=1}^{N_1} \sum_{\nu=1}^{N_2} \int_{x_{\nu-1,2}}^{x_{\nu,2}} \left(\int_{x_{\mu-1,1}}^{x_{\nu,1}} f(\xi_\mu^{(1)}, \xi_\nu^{(2)}) \, dx_1 \right) dx_2. \tag{18.48}$$

For $\mathrm{mesh}(Z) < \delta$ we get

$$\left| \int_K f(x) \, dx - \int_{a_2}^{b_2} \left(\int_{a_1}^{b_1} f(x_1, x_2) \, dx_1 \right) dx_2 \right|$$

$$\leq \left| \int_K f(x) \, dx - R(f, Z, \xi) \right|$$

$$+ \left| \sum_{\mu=1}^{N_1} \sum_{\nu=1}^{N_2} \int_{x_{\nu-1,2}}^{x_{\nu,2}} \left(\int_{x_{\mu-1,1}}^{x_{\mu,1}} (f(\xi_\mu^{(1)}, \xi_\nu^{(2)}) - f(x_1, x_2)) \, dx_1 \right) dx_2 \right|$$

$$\leq \left| \int_K f(x) \, dx - R(f, Z, \xi) \right| + \sum_{\mu=1}^{N_1} \sum_{\nu=1}^{N_2} \epsilon \, \mathrm{vol}_2(K_{\mu\nu})$$

$$= \left| \int_K f(x) \, dx - R(f, Z, \xi) \right| + \epsilon \, \mathrm{vol}_2(K).$$

By Theorem 18.18 we may choose δ such that (18.37) and (18.46) hold for $\text{mesh}(Z) < \delta$ implying that

$$\left| \int_K f(x) \, dx - \int_{a_2}^{b_2} \left(\int_{a_1}^{b_1} f(x_1, x_2) \, dx_1 \right) dx_2 \right| < \epsilon(1 + \text{vol}_2(K)),$$

proving the theorem (for $n = 2$). $\qquad\qquad\qquad\qquad\qquad\qquad\qquad$ $\square$

Example 18.20. A. With $K = [0, 1] \times [0, 1]$ we want to find

$$\int_K x_1 x_2 e^{x_1^2 x_2} \, dx = \int_0^1 \left(\int_0^1 x_1 x_2 e^{x_1^2 x_2} \, dx_1 \right) dx_2$$

$$= \int_0^1 \left(\int_0^1 \frac{\partial}{\partial x_1} \left(\frac{1}{2} e^{x_1^2 x_2} \right) dx_1 \right) dx_2$$

$$= \int_0^1 \frac{1}{2} e^{x_1^2 x_2} \Big|_{x_1=0}^{x_1=1} dx_2 = \frac{1}{2} \int_0^1 (e^{x_2} - 1) \, dx_2 = \frac{1}{2}(e - 2).$$

B. On $K = [0, 1] \times [0, 1]$ consider the function $g(x_1, x_2) = |2x_1 - 1|$. For the integral $\int_K g(x) \, dx$ we have

$$\int_K g(x) \, dx = \int_0^1 \left(\int_0^1 |2x_1 - 1| \, dx_1 \right) dx_2$$

$$= 2 \int_0^1 \left(\int_{\frac{1}{2}}^1 (2x_1 - 1) \, dx_1 \right) dx_2$$

$$= 2 \int_0^1 (x_1^2 - x_1) \Big|_{\frac{1}{2}}^1 \, dx_2 = \frac{1}{2}.$$

The fact that for continuous functions the Riemann integral over a non-degenerate compact cell equals the iterated integral(s) allows us to transfer some of our rules for integration. For a non-degenerate compact cell in $\mathbb{R}^n$ we introduce the space $C_0(K)$ as all continuous functions $f : K \to \mathbb{R}$ with the property that $f|_{\partial K} = 0$.

Lemma 18.21. For $f \in C^1(K) \cap C_0(K)$ the following holds

$$\int_K \frac{\partial f}{\partial x_j}(x) \, dx = 0, \quad j = 1, \ldots, n. \qquad\qquad (18.49)$$

Proof. Recall that $f \in C^1(K)$ if f has all continuous partial derivatives of first order on $\overset{\circ}{K}$ with continuous extensions to K. To prove (18.49), without loss of generality we may choose $j = 1$, and by Theorem 18.19 we find for $K = \underset{j=1}{\overset{n}{\times}}[a_j, b_j]$

$$\int_K \frac{\partial f}{\partial x_1}(x)\,\mathrm{d}x = \int_{a_n}^{b_n}\left(\cdots\left(\int_{a_2}^{b_2}\left(\int_{a_1}^{b_1}\frac{\partial}{\partial x_1}f(x_1,\ldots,x_n)\,\mathrm{d}x_1\right)\mathrm{d}x_2\right)\cdots\right)\mathrm{d}x_n$$

$$= \int_{a_n}^{b_n}\left(\cdots\left(\int_{a_2}^{b_2}\left(f(x_1,x_2,\ldots,x_n)\Big|_{x_1=a_1}^{x_1=b_1}\right)\mathrm{d}x_2\right)\cdots\right)\mathrm{d}x_n$$

$$= \int_{a_n}^{b_n}\left(\cdots\left(\int_{a_2}^{b_2}(f(b_1,x_2,\ldots,x_n)-f(a_1,x_2,\ldots,x_n))\,\mathrm{d}x_2\right)\cdots\right)\mathrm{d}x_n = 0$$

since $(a_1, x_2,\ldots,x_n), (b_1, x_2,\ldots,x_n) \in \partial K$ and $f|_{\partial K} = 0$. $\qquad\square$

Corollary 18.22. *For $f \in C^1(K) \cap C_0(K)$ and $g \in C^1(K)$ we have for $1 \le j \le n$ that*

$$\int_K\left(\frac{\partial}{\partial x_j}f(x)\right)g(x)\,\mathrm{d}x = -\int_K f(x)\left(\frac{\partial}{\partial x_j}g(x)\right)\mathrm{d}x. \qquad (18.50)$$

Proof. We note that $f \cdot g \in C^1(K) \cap C_0(K)$ and by Lemma 18.21 we find

$$0 = \int_K \frac{\partial}{\partial x_j}(f \cdot g)(x)\,\mathrm{d}x = \int_K\left(\left(\frac{\partial f}{\partial x_j}(x)\right)g(x) + f(x)\left(\frac{\partial g}{\partial x_j}(x)\right)\right)\mathrm{d}x$$

implying (18.50). $\qquad\square$

An extension of (18.50) for function not vanishing on ∂K will be discussed in Chapters 25 and 27.

Example 18.23. Since $\frac{\partial}{\partial x_1}(\sin x_1 \sin x_2) = \cos x_1 \sin x_2$ it follows that

$$\int_K(\cos x_1 \sin x_2)\,\mathrm{d}x = 0$$

for any $K = [k_1\pi, l_1\pi] \times [k_2\pi, l_2\pi]$, $k_j < l_j$, and $k_j, l_j \in \mathbb{Z}$.

Let $\alpha, \beta \in \mathbb{R}^n$ and $T : \mathbb{R}^n \to \mathbb{R}^n$ be defined by $Tx = \begin{pmatrix} \alpha_1 & & 0 \\ & \ddots & \\ 0 & & \alpha_n \end{pmatrix}x + \beta$,

i.e. $(Tx)_j = \alpha_j x_j + \beta_j$. We assume that $\alpha_j > 0$ for all $j = 1,\ldots,n$. In this case T maps the non-degenerate compact cell $K = \bigtimes_{j=1}^{n}[a_j, b_j] \subset \mathbb{R}^n$ bijectively and continuously with continuous inverse onto the non-degenerate compact cell $TK = \bigtimes_{j=1}^{n}[\alpha_j a_j + \beta_j, \alpha_j b_j + \beta_j]$. Furthermore, if $g : TK \to \mathbb{R}$ is continuous the function $g \circ T : K \to \mathbb{R}$ is continuous too and we find

$$
\int_{TK} g(y)\, \mathrm{d}y = \int_{\alpha a_n + \beta_n}^{\alpha_n b_n + \beta_n} \left(\cdots \left(\int_{\alpha_1 a_1 + \beta_1}^{\alpha_1 b_1 + \beta_1} g(y_1, \ldots, y_n)\, \mathrm{d}y_1 \right) \cdots \right) \mathrm{d}y_n
$$

$$
= \int_{\alpha a_n + \beta_n}^{\alpha_n b_n + \beta_n} \left(\cdots \left(\int_{\alpha_2 a_2 + \beta_2}^{\alpha_2 b_2 + \beta_2} \int_{a_1}^{b_1} \alpha_1 g(\alpha_1 x_1 + \beta_1, y_2, \ldots, y_n)\, \mathrm{d}x_1 \right) \mathrm{d}y_2 \cdots \right) \mathrm{d}y_n
$$

$$
= \int_{a_n}^{b_n} \left(\cdots \left(\int_{a_1}^{b_1} (g \circ T)(x_1, \ldots, x_n) \prod_{j=1}^{n} \alpha_j\, \mathrm{d}x_1 \right) \cdots \right) \mathrm{d}x_n
$$

and since the Jacobi determinant of T is just $\prod_{j=1}^{n} \alpha_j$ we have proved a first version of the transformation theorem for volume integrals.

Lemma 18.24. *Let K be a non-degenerate compact cell in $\mathbb{R}^n$ and $T : K \to \mathbb{R}^n$ be defined by $(Tx)_j = \alpha_j x_j + \beta_j$, $\alpha_j > 0$. If $g : TK \to \mathbb{R}$ is continuous then*

$$
\int_{TK} g(x)\, \mathrm{d}x = \int_K (g \circ T)(x)(\det J_T)(x)\, \mathrm{d}x. \tag{18.51}
$$

Remark 18.25. A. The argument leading to

$$
\int_{\alpha_n a_n + \beta_n}^{\alpha_n b_n + \beta_n} \left(\cdots \left(\int_{\alpha_1 a_1 + \beta_1}^{\alpha_1 b_1 + \beta_1} g(y_1, \ldots, y_n)\, \mathrm{d}y_1 \right) \cdots \right) \mathrm{d}y_n
$$

$$
= \int_{a_n}^{b_n} \left(\cdots \left(\int_{a_1}^{b_1} (g \circ T)(x_1, \ldots, x_n) \prod_{j=1}^{n} \alpha_j\, \mathrm{d}x_1 \right) \cdots \right) \mathrm{d}x_n
$$

remains correct even if some of the values α_j are zero, but in this case TK is a degenerate cell and we will return to the meaning of (18.51) in this case later.

B. If $\alpha_j = 1$ for all $j = 1, \ldots, n$, then $TK = K + \beta$ with $\beta = (\beta_1, \ldots, \beta_n)$, i.e. TK is a translation of K. Writing as before $\tau_\beta : \mathbb{R}^n \to \mathbb{R}^n$, $\tau_\beta(x) = x + \beta$, we find $T = \tau_\beta$ and since $\det J_T = \det J_{\tau_\beta} = 1$ we find

$$
\int_{\tau_\beta(K)} g(x)\, \mathrm{d}x = \int_K (g \circ T)(x)\, \mathrm{d}x, \tag{18.52}
$$

i.e. the integral over cells is translation invariant. We would like to have the same statement for rotations or more generally that the integral is invariant under $O(n)$ the orthogonal group. However, in general for $U \in O(n)$ the set $U(K)$ is not a cell and we first need an extension of the integral to other sets than cells.

Let K_j, $j = 1, \ldots, M$ be a finite collection of non-degenerate compact cells in $\mathbb{R}^n$ such that $\mathring{K}_j \cap \mathring{K}_l = \emptyset$ for $j \neq l$. If $f : \bigcup_{j=1}^{M} K_j \to \mathbb{R}$ is a bounded function and $f|_{K_j}$ is for $j = 1, \ldots, M$ Riemann integrable, we can define with $L := \bigcup_{j=1}^{M} K_j$ the integral

$$\int_L f(x)\, dx := \sum_{j=1}^{M} \int_{K_j} f(x)\, dx \tag{18.53}$$

and it is not difficult to construct Riemann sum approximations for $\int_L f(x)\, dx$. However this approach will never lead to integrals over sets having a boundary being not included in a finite number of axes-parallel hyperspaces (or hyperplanes) of $\mathbb{R}^n$, for example $\overline{B_r(0)}$ will be excluded. To make real progress we need to have a better understanding of the behaviour of boundaries of compact subsets of $\mathbb{R}^n$ when forming Riemann-type sums.

Problems

1. Consider

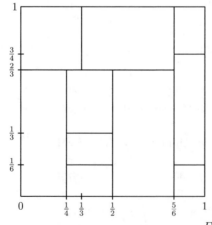

Figure 18.5

A partition of $[0,1] \times [0,1] \subset \mathbb{R}^2$ into rectangles $R_{\alpha\beta}$ is indicated in the figure above. Find minimal partitions Z_1 and Z_2 of $[0,1]$ such that every $R_{\alpha\beta}$ is the union of rectangles induced by $Z = (Z_1, Z_2)$. (A graphical solution is sufficient).

2. Prove that $T(K)$ is an algebra over $\mathbb{R}$.

3. Let V be an $\mathbb{R}$-vector space of functions $g : \Omega \to \mathbb{R}$ where $\Omega \neq \emptyset$ is a set. Further let $l : V \to \mathbb{R}$ be a linear mapping which is positivity preserving, i.e. if $g \in V$ and $g \geq 0$, i.e. $g(x) \geq 0$ for all $x \in \Omega$, then $l(g) \geq 0$. Prove that for $g, h \in V$ such that $g \leq h$, i.e. $g(x) \leq h(x)$ for all $x \in \Omega$, it follows that $l(g) \leq l(h)$. Now fix $w_0 \in \Omega$ and consider the mapping $\epsilon_{w_0} : V \to \mathbb{R}$, $\epsilon_{w_0}(g) = g(w_0)$. Prove that ϵ_{w_0} is a linear, positivity preserving mapping.

4. On the square $K := [0,1] \times [0,1]$ consider the partition $Z = (Z_1, Z_2)$ where $Z_1 = \{\frac{k}{n} | k = 0, \ldots, n\}$ and $Z_2 = \{\frac{l}{2n} | l = 0, \ldots, 2n\}$. Define the step functions $g : [0,1] \to [0,1] \to \mathbb{R}$ by

$$g(x) = \begin{cases} 1, & x \in \mathring{K}_{kl}, \ k+l \text{ is even} \\ 0, & x \in \mathring{K}_{kl}, \ k+l \text{ is odd} \end{cases}$$

where $K_{kl} = \left[\frac{k}{n}, \frac{k+l}{n}\right] \times \left[\frac{l}{2n}, \frac{l+1}{2n}\right], k = 0, \ldots, n-1, l = 0, \ldots, 2n-1$. Sketch Z for $n = 4$ and for the general case find $\int_K g(x) dx$.

5. On $K = [0,1] \times [0,1]$ take the equidistance partition $(x_k, y_l) = \left(\frac{k}{n}, \frac{l}{n}\right)$, $0 \leq k, l \leq n$. Find the Riemann sum of $f : K \to \mathbb{R}, f(x,y) = x^2 + y^2$ when the point ξ_{kl} is the midpoint of the square $[x_k, x_{k+1}] \times [y_l, y_{l+1}], k, l = 0, \ldots, n-1$.

6. Using the proof of the corresponding one-dimensional result prove (18.22).

7. Let $K \subset \mathbb{R}^n$ be a compact non-degenerate cell. Prove that if $f : K \to \mathbb{R}$ is Riemann integrable then f^+ and f^-, hence $|f|$ are Riemann integrable too.

8. Evaluate the following iterated integrals:

a) $\int_0^2 \left(\int_0^3 \frac{x-y}{(x+y)^3} dy\right) dx$ and $\int_0^3 \left(\int_0^2 \frac{x-y}{(x+y)^3} dx\right) dy$;

Hint: note that $\frac{x-y}{(x+y)^3} = \frac{2x-(x+y)}{(x+y)^3}$.

b) $\int_1^5 \left(\int_{-1}^1 \left(\int_2^4 ((x_1^2 + x_3^2) \sin x_2 + e^{x_1 + x_2 + x_3}) \, dx_3 \right) dx_2 \right) dx_1.$

9. Let $K_1 \subset \mathbb{R}^{n_1-1}, K_2 \subset \mathbb{R}^{n_2}$ be two non-degenerate cells and $I = [a, b], a \leq b$, a compact interval. Then $K := K_1 \times I \times K_2$ is a non-degenerate compact cell in $\mathbb{R}^n, n = n_1 + n_2$. Now let $f, g : K \to \mathbb{R}$ be two C^1-functions on K and assume further that
$$f(x_1, \ldots, x_{n_1-1}, a, x_{n_1+1}, \ldots, x_{n_1+n_2}) = f(x_1, \ldots, x_{n_1-1}, b, x_{n_1+1}, \ldots, x_{n_1+n_2}) = 0.$$
Prove the integration by parts formula
$$\int_K \frac{\partial f}{\partial x_{n_1}}(x)g(x)dx = -\int_K f(x)\frac{\partial g}{\partial x_{n_1}}(x)dx.$$

Now derive the estimate
$$\left(\int_K |f(x)|^2 dx \right)^{\frac{1}{2}} \leq 2\max(|a|, |b|) \left(\int_K \left| \frac{\partial f}{\partial x_{n_1}}(x) \right|^2 dx \right)^{\frac{1}{2}}. \qquad (18.54)$$

Hint: note that $|f(x)|^2 = \frac{\partial x_{n_1}}{\partial x_{n_1}} \cdot f^2(x).$

10. Define $T : \mathbb{R}^3 \to \mathbb{R}^3$ by $Tx = (3x_1, 4x_2, 2x_3) + (-1, 2, 1)$. Find $T(K)$ where $K = [-1, 1] \times [0, 2] \times [3, 4]$. Let $f : \mathbb{R}^3 \to \mathbb{R}$ be the function $f(x_1, x_2, x_3) = x_1 + x_2 + x_3$. Using Lemma 18.24 find $\int_{T(K)} f(y)dy.$

19 Boundaries in $\mathbb{R}^n$ and Jordan Measurable Sets

In this chapter we want to study boundaries ∂G of bounded subsets $G \subset \mathbb{R}^n$. The notion of a boundary is a topological one, so we fix as usual on $\mathbb{R}^n$ the Euclidean topology. By Definition 1.27 the boundary of $G \subset \mathbb{R}^n$ consists of all points $y \in \mathbb{R}^n$ such that every neighbourhood of y contains points of G and $G^\complement$, and of course we may reduce this statement to the condition that every open ball $B_\epsilon(y)$, $\epsilon > 0$, contains points of G and $G^\complement$. We have seen that for general sets $G \subset \mathbb{R}^n$ the boundary might be a very surprising set. For example the boundary of $[0,1] \cap \mathbb{Q}$ is the closed interval $[0,1]$, i.e. a set can be a proper subset of its boundary. Moreover, in the case of the set (16.17) the boundary (16.18) again does not fit to our naive idea of a boundary. Furthermore, some sets may not have a boundary. The set $\mathbb{R}^n$ has no boundary since its complement is empty.

For our purpose to define a Riemann type integral where we will use Riemann sums depending on partitions, we want to achieve that those terms in a Riemann sum depending on the boundary eventually are negligible, see Chapter 16. This we can obtain for example when we can cover ∂G with a finite number of non-degenerate compact cells having the property that the volume of their union is below a prescribed (small) bound. When considering the connection to coverings it is convenient to introduce the notion of an **open cell** in $\mathbb{R}^n$ as the interior of a non-degenerate compact cell in $\mathbb{R}^n$, i.e. $K \subset \mathbb{R}^n$ is an open cell if K is open and $\overline{K}$ is a non-degenerate compact cell. We now define

$$\operatorname{vol}_n(K) := \operatorname{vol}_n(\overline{K}) \tag{19.1}$$

for all open cells $K \subset \mathbb{R}^n$. Clearly we define for a finite collection of mutually disjoint open cells or non-degenerate compact cells $K_j \subset \mathbb{R}^n$, $j = 1, \ldots, N$,

$$\operatorname{vol}_n \left(\bigcup_{j=1}^{N} K_j \right) := \sum_{j=1}^{N} \operatorname{vol}_n(K_j). \tag{19.2}$$

Definition 19.1. *A set $F \subset \mathbb{R}^n$ is said to be a **Jordan null set** or a set of **Jordan content zero**, and for this we write $J^{(n)}(F) = 0$, if for every $\epsilon > 0$ there exists a finite covering of F by open cells $K_1, \ldots, K_N \subset \mathbb{R}^n$, $N = N(\epsilon)$, such that $\sum_{k=1}^{N} \operatorname{vol}_n(K_k) < \epsilon$.*

Example 19.2. A. Every finite set $F \subset \mathbb{R}^n$ has Jordan content zero. If F has M points, i.e. $F = \{a_1, \ldots, a_M\}$, we choose for $\epsilon > 0$ the open cube $K_j(a_j)$ with centre a_j and edges parallel to the corresponding coordinate axes with side length $\left(\frac{\epsilon}{M+1}\right)^{\frac{1}{n}}$ as an open cell to cover $\{a_j\}$. The union of these cubes, which are open cells, will cover F and since $\mathrm{vol}_n(K_j(a_j)) = \frac{\epsilon}{M+1}$ it follows that $\sum_{j=1}^{M} \mathrm{vol}_n(K_k(a_k)) = \frac{M}{M+1}\epsilon < \epsilon$.

B. Let $G := \underset{j=1}{\overset{n}{\times}}[a_j, b_j]$, $a_j < b_j$, be a non-degenerate compact cell in $\mathbb{R}^n$. We define the faces of G as

$$F_{a_j} := \left\{x \in G \,\big|\, x_j = a_j\right\} \quad \text{and} \quad F_{b_j} := \left\{x \in G \,\big|\, x_j = b_j\right\}, \quad j = 1, \ldots, n.$$

Clearly we have $\partial G = \bigcup_{j=1}^{n}(F_{a_j} \cup F_{b_j})$. We claim that every face is a set of Jordan content zero, as is ∂G. Denote by γ_j the number $\prod_{l \neq j}(b_l - a_l + 1)$ and for $0 < \epsilon < 1$ define

$$K_{a_j,\epsilon} := \left\{x \in \mathbb{R}^n \,\Big|\, a_l - \frac{\epsilon}{2} < x_l < b_l + \frac{\epsilon}{2}, \, l \neq j, \, a_j - \frac{\epsilon}{4\gamma_j} < x_j < a_j + \frac{\epsilon}{4\gamma_j}\right\},$$

see Figure 19.1.

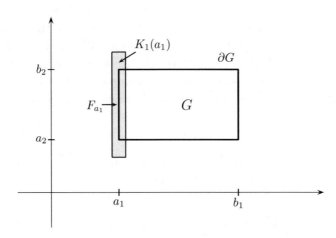

Figure 19.1

It follows that $F_{a_j} \subset K_{a_j,\epsilon}$, $K_{a_j,\epsilon}$ is an open cell, and

$$\mathrm{vol}_n(K_{a_j,\epsilon}) = \frac{\epsilon}{2\gamma_j} \prod_{l \neq j}(b_l - a_l + \epsilon) \leq \frac{\epsilon}{2\gamma_j}\gamma_j = \frac{\epsilon}{2} < \epsilon.$$

An analogous construction holds for F_{b_j}. The result is proved once we have shown that the union of a finite number of sets with Jordan content zero has Jordan content zero, and this will be done in the next lemma.

Lemma 19.3. *Let $F_1, \ldots, F_M \subset \mathbb{R}^n$ be sets of Jordan content zero. Then $\bigcup_{j=1}^{M} F_j$ has Jordan content zero too.*

Proof. Given $\epsilon > 0$, we can find open cells $K_{j,l}$, $1 \le l \le N_j$, such that $F_j \subset \bigcup_{l=1}^{N_j} K_{j,l}$ and $\sum_{l=1}^{N_j} \mathrm{vol}_n(K_{j,l}) < \frac{\epsilon}{M}$. Thus the K_{j,l_j}, $1 \le l_j \le N_j$, $j = 1, \ldots, M$, cover $\bigcup_{j=1}^{M} F_j$ and

$$\sum_{j=1}^{M} \sum_{l_j=1}^{N_j} \mathrm{vol}_n(K_{j,k_j}) < \sum_{j=1}^{M} \frac{\epsilon}{M} = \epsilon.$$

$\square$

A much larger class of sets with Jordan content zero can be constructed using

Theorem 19.4. *For a compact set $G \subset \mathbb{R}^n$, $n \ge 1$, the graph of a continuous function $f : G \to \mathbb{R}$ has Jordan content zero, i.e. $J^{(n+1)}(\Gamma(f)) = 0$.*

Proof. Since G is compact it is bounded and therefore we can find $r > 0$ such that

$$G \subset \left\{ x \in \mathbb{R}^n \mid \|x\|_\infty = \max_{1 \le j \le n} |x_j| \le r \right\} =: K_r.$$

The set K_r is a non-degenerate compact cell in $\mathbb{R}^n$ with volume $\mathrm{vol}_n(K_r) = (2r)^n$. In order to construct a covering of $\Gamma(f)$ with open cells we make use of the uniform continuity of f on the compact set G, i.e. for $\eta > 0$ we can find $\delta > 0$ such that $\|x - y\|_\infty < \delta$ implies $|f(x) - f(y)| < \frac{\eta}{2}$. (Since $\|x\|_\infty \le \|x\|_2 \le n\|x\|_\infty$ it does not matter whether we take the norm $\|.\|_\infty$ or the Euclidean norm.) We now choose $m \in \mathbb{N}$ and divide every edge of K_r into m intervals of equal length which induces a partition of K_r into $N := m^n$ compact non-degenerate cubes $K'_{r,1}, \ldots, K'_{r,N}$ with edges parallel to corresponding coordinate axes, i.e. each $K'_{r,j}$ is a cell. We assume that $\mathrm{diam}(K'_{r,j}) < \delta$, $j = 1, \ldots, N$, which we can always achieve by increasing m. Denote by $\xi_j \in K'_{r,j}$ the centre of $K'_{r,j}$ and introduce the cells

$$C'_{r,j} := K'_{r,j} \times I_j \subset \mathbb{R}^{n+1}, \quad I_j = \left(f(\xi_j) - \frac{\eta}{2}, f(\xi_j) + \frac{\eta}{2} \right).$$

It follows that $\Gamma(f) \subset \bigcup_{j=1}^{N} C'_{r,j}$ and

$$\text{vol}_{n+1}(C'_{r,j}) = \text{vol}_n(K'_{r,j})\text{vol}_1(I_j) = \eta\,\text{vol}_n(K'_{r,j}).$$

Since $\mathring{K}'_{r,j} \cap \mathring{K}'_{r,l} = \emptyset$ for $j \neq l$, we can deduce that

$$\sum_{j=1}^{N} \text{vol}_{n+1}(C'_{r,j}) = \eta \sum_{j=1}^{n} \text{vol}_n(K'_{r,j}) = \eta\,\text{vol}_n(K_r) = \eta(2r)^n.$$

But we are not done yet; the cells $C'_{r,j}$ are not open. So we replace $K'_{r,j}$ by an open cubic cell $K_{r,j}$ with the properties that $K'_{r,j} \subset K_{r,j}$ and $\text{vol}_n(K_{r,j}) \leq 2\text{vol}_n(K'_{r,j})$. The cells $C_{r,j} := K_{r,j} \times I_j$, $j = 1, \ldots, N$, are now open and they form an open covering of $\Gamma(f)$. In addition we find

$$\sum_{j=1}^{N} \text{vol}_{n+1}(C_{r,j}) \leq \eta \sum_{j=1}^{N} \text{vol}_n(K_{r,j}) \leq 2\eta \sum_{j=1}^{n} \text{vol}_n(K'_{r,j})$$

$$= 2\eta\,\text{vol}_n(K_r) = 2\eta(2r)^n.$$

Now, given $\epsilon > 0$ we choose $\eta < \frac{\epsilon}{2(2r)^n}$ to find that

$$\sum_{j=1}^{N} \text{vol}_{n+1}(C_{r,j}) < \epsilon,$$

proving the result. □

Example 19.5. Consider the set $G := \{x \in \mathbb{R}^n \mid x_1^2 + x_2^2 \leq r^2, x_2 \geq 0\}$, i.e. the closed upper half disc with radius r, see Figure 19.2.

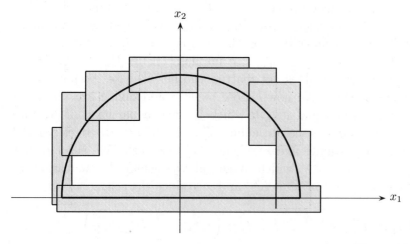

Figure 19.2

The boundary ∂G is given by $\partial G = [-r, r] \times \{0\} \cup \Gamma(f_r)$ where $f_r : [-r, r] \to \mathbb{R}$, $f_r(x_1) = \sqrt{r^2 - x_1^2}$. We know that $[-r, r] \times \{0\}$ as a face of cell in $\mathbb{R}^2$ has Jordan content zero, as does $\Gamma(f_r)$ as the graph of a continuous function. Thus $J^{(2)}(\partial G) = 0$.

In light of Example 19.5, Theorem 19.4 gives us the following idea: let $G \subset \mathbb{R}^n$ be a bounded set, hence ∂G is compact. Suppose that we can cover ∂G by open sets $U_j \subset \mathbb{R}^n$, $j \in I$, such that $\partial G \cap U_j$ has a representation as the graph of a continuous function, i.e. with $V_j \subset \mathbb{R}^{n-1}$ compact and $\varphi_j : V_j \to \mathbb{R}$ continuous it follows that $\partial G \cap U_j \subset \Gamma(\varphi_j)$, see Figure 19.3

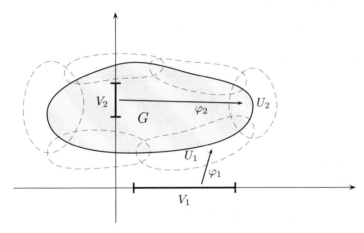

Figure 19.3

Since ∂G is compact we can cover it with finitely many sets U_j, so ∂G is the finite union of sets of Jordan content zero, hence it has Jordan content zero itself. This idea is successful, but we want to add two new considerations before we follow it up.

We want to consider sets of Jordan content zero as "small" sets. In Chapter I.32 we have already encountered small sets, i.e. null sets, and we could prove that every denumerable set is a null set, see Lemma I.32.3 or Lemma 19.8 below. However countable subsets of $\mathbb{R}^n$ need not be of Jordan content zero.

To see this consider the rational numbers in the unit interval, i.e. the set $[0, 1] \cap \mathbb{Q}$. This set is countable. From Theorem I.19.27 we know that open sets in $\mathbb{R}$ are denumerable unions of disjoint open intervals and clearly we cannot cover $[0,1] \cap \mathbb{Q}$ by a finite number of intervals with total length $\epsilon < 1$. Therefore we also need to consider null sets.

Definition 19.6. *A set $A \subset \mathbb{R}^n$ is called a **Lebesgue null set** or just a **null set** if for every $\epsilon > 0$ there exists a denumerable covering $(K_j)_{j \in \mathbb{N}}$ of A by open cells such that $\sum_{j \in \mathbb{N}} \mathrm{vol}_n(K_j) < \epsilon$, i.e.*

$$A \subset \bigcup_{j \in \mathbb{N}} K_j \quad \text{and} \quad \sum_{j \in \mathbb{N}} \mathrm{vol}_n(K_j) < \epsilon. \tag{19.3}$$

Remark 19.7. A. This definition clearly extends Definition I.32.1.
B. Since denumerable sets are either finite or countable, we may choose $\mathbb{N}$ as an index set and in the case of a finite covering $(K_j)_{j=1,\dots,N}$ we set $K_j = K_N$ for $j \geq N$.
C. If $A \subset \mathbb{R}^n$ is of Jordan content zero, then it is also a Lebesgue null set and we have already seen that the converse is in general false.

Lemma 19.8. *The countable union of Lebesgue null sets is a Lebesgue null set.*

Proof. Let $(G_j)_{j \in \mathbb{N}}$ be a countable family of Lebesgue null sets $G_j \subset \mathbb{R}^n$ and $\epsilon > 0$ be given. For $j \in \mathbb{N}$ we can find open cells $K_{j,k}$, $k \in \mathbb{N}$, such that $G_j \subset \bigcup_{k \in \mathbb{N}} K_{j,k}$ and $\sum_{k \in \mathbb{N}} \mathrm{vol}_n(K_k^j) < 2^{-j}\epsilon$. It follows that

$$\bigcup_{j \in \mathbb{N}} G_j \subset \bigcup_{j \in \mathbb{N}} \bigcup_{k \in \mathbb{N}} K_{j,k},$$

i.e. $(K_{j,k})_{j,k \in \mathbb{N}}$ is a countable covering of $\bigcup_{j \in \mathbb{N}} G_j$ by open cells, and

$$\sum_{j \in \mathbb{N}} \sum_{k \in \mathbb{N}} \mathrm{vol}_n(K_{j,k}) < \sum_{j \in \mathbb{N}} 2^{-j}\epsilon = \epsilon.$$

$\square$

Corollary 19.9. *If $G \subset \mathbb{R}^n$ is a compact Lebesgue null set then it has Jordan content zero.*

Proof. Since G is a Lebesgue null set, given $\epsilon > 0$ we can find a countable cover $(K_j)_{j \in \mathbb{N}}$ of open cells $K_j \subset \mathbb{R}^n$ with $\sum \mathrm{vol}_n(K_j) < \epsilon$. The compactness of G already implies that a finite number of cells $K_{j_1}, \dots, K_{j_N}$ will cover G and clearly $\sum_{l=1}^{N} \mathrm{vol}_n(K_{j_l}) \leq \sum_{j \in \mathbb{N}} \mathrm{vol}_n(K_j) < \epsilon$. $\square$

Our second "detour" looks at boundaries having a local representation as graph of a function. Representing locally a certain subset of $\mathbb{R}^n$ as graph of a function is not a new problem to us, we encountered this question in the context of the implicit function theorem in Chapter 10. Recall that we want to define a Riemann integral for functions defined on a compact subset of $\mathbb{R}^n$ and we have reasons to assume that this is possible if the boundary of the set is "small". Moreover, sets being locally the graph of a continuous function should be "small". If "small" means to have Jordan content zero we may proceed as follows.

Definition 19.10. *A compact set $M \subset \mathbb{R}^n$ is called of* **type** *C^k or a C^k-**type** set, $k \in \mathbb{N}_0$, if for every $x_0 \in M$ there exists a ball $B_r(x_0)$, $r = r(x_0) > 0$, and a C^k-function $u_{x_0} : Q(x_0) \to \mathbb{R}$, $Q(x_0) \subset \mathbb{R}^{n-1}$ compact, such that*

$$M \cap \overline{B_r(x_0)}$$
$$= \{(x_1, \ldots, x_{j-1}, u_{x_0}(y), x_{j+1}, \ldots, x_n) \mid (x_1, \ldots, x_{j-1}, x_{j+1}, \ldots, x_n) = y \in Q(x_0)\}, \tag{19.4}$$

where j, $1 \le j \le n$, may depend on x_0. In the case where M is of C^0-type we say simply that M is of C-type, if M is of C-type but all functions u_{x_0} are Lipschitz conditions we call M a subset of Lipschitz-type.

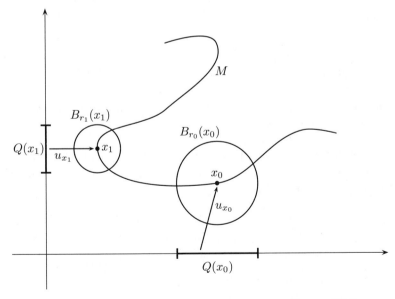

Figure 19.4

Remark 19.11. If M is of C^{k+1}-type then it is of C^k-type for $k \geq 0$.

Corollary 19.12. *If $M \subset \mathbb{R}^n$ is of C-type then it has Jordan content zero.*

Proof. For $x \in M$ choose $B_r(x)$, $r = r(x)$, and u_x as in Definition 19.10. Since M is compact we can cover M with a finite number of balls $B_{r_j}(x_j)$, $j = 1, \ldots, N$, such that $B_{r_j}(x_j)$ and u_{x_j} have these properties, i.e. $M \subset \bigcup_{j=1}^N M \cap \overline{B_{r_j}(x_j)}$, but $J^{(n)}(M \cap \overline{B_r(x_j)}) = 0$ by Theorem 19.4 implying that $J^{(n)}(M) = 0$. $\qquad \square$

Corollary 19.13. *If $G \subset \mathbb{R}^n$ is compact and ∂G of the type C^k, $k \geq 0$, then ∂G is of Jordan content zero.*

The following, almost obvious result is often very helpful as it is of theoretical interest.

Proposition 19.14. *A compact set $M \subset \mathbb{R}^n$ is of the type C^k if and only if for all $U \in O(n)$ the set $U(M)$ is of the type C^k. Further, if $\tau_a := \mathbb{R}^n \to \mathbb{R}^n$ denotes the translation $\tau_a(x) = x + a$, then M is of the type C^k if and only if $\tau_a(M)$ is of the type C^k for all $a \in \mathbb{R}^n$.*

Proof. Since U and hence U^{-1} belongs to $O(n)$ we only need to prove that if M is of the type C^k then $U(M)$ is. Let $x_0 \in M$ and $B_r(x_0)$ as well u_{x_0} be as in Definition 19.10. Let us agree to consider $Q(x_0) \subset \mathbb{R}^{n-1}$ embedded into $\mathbb{R}^n$ as subset of $\mathbb{R}_{x_1} \times \cdots \times \mathbb{R}_{x_{j-1}} \times \{0\} \times \mathbb{R}_{x_{j+1}} \times \cdots \times \mathbb{R}_{x_n}$, this is indeed as we can and should understand Figure 19.4. Now, instead of applying U only to $M \cap B_r(x_0)$ and searching for a "new" $Q(x_0)$ and u_{x_0}, we apply U to the entire configuration, i.e. Figure 19.4, and then the result becomes obvious, as the result on translation invariance. $\qquad \square$

Remark 19.15. Behind the statement of Proposition 19.14 is of course the fact that $\mathrm{vol}_n(K)$, $K \subset \mathbb{R}^n$ a cell, is invariant under translations and the operation of $O(n)$, see Problem 5.

Example 19.16. A. The set $S^{n-1} = \partial B_1(0) \subset \mathbb{R}^n$ is of the type C^∞. Indeed, if $x = (x_1, \ldots, x_n) \in S^{n-1}$ then at least for one j we have $x_j \neq 0$ and we can write $x_j = \sqrt{1 - \sum_{l \neq j} x_l^2}$ or $x_j = -\sqrt{1 - \sum_{l \neq j} x_l^2}$. Clearly this example extends to $\partial B_R(x_0) \subset \mathbb{R}^n$, $R > 0$, $x_0 \in \mathbb{R}^n$.
B. Consider the rectangle $R \subset \mathbb{R}^2$ with vertices A, B, C, D as in Figure 19.5. Clearly every edge is the graph of a C^∞-function. However ∂R is not of

the type C^k, $k \geq 1$, but it is of the type C^0, indeed of Lipschitz-type. The problems are caused by the vertices.

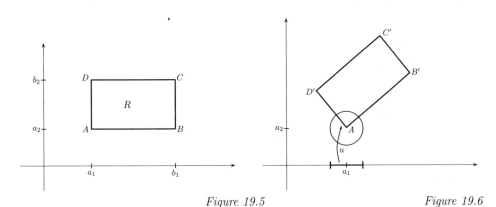

Figure 19.5   Figure 19.6

Rotating the configuration by $\frac{\pi}{4}$ with centre $A = (a_1, a_2)$, which we can achieve by a combination of translations and operations of elements of $O(2)$, we see that A is as the line segments AB' and $D'A$ a subset of the graph of the Lipschitz function $u(x) = |x - a_1| + a_2$.

Let $F : D \to \mathbb{R}$, $D \subset \mathbb{R}^n$ open, be a continuous mapping and $I := [a, b]$, $a < b$, a compact interval. Suppose that for some $\eta > 0$ the interval $[a - \eta, b + \eta]$ is contained in the range of F, i.e. $[a - \eta, b + \eta] \subset R(F)$. The pre-image $F^{-1}(I)$ is the set

$$G := \left\{ x \in D \,\middle|\, a \leq F(x) \leq b \right\} \subset \mathbb{R}^n, \tag{19.5}$$

which is a closed set as pre-image of a closed set under a continuous mapping. For the same reason we know that $F^{-1}((a, b))$, i.e. the set $\{x \in D \,|\, a < F(x) < b\}$, is open. We claim that ∂G is the set $\{x \in D \,|\, F(x) = a \text{ or } F(x) = b\}$. Clearly $F^{-1}((a, b)) \subset \mathring{G}$. Indeed, if $a < F(x_0) < b$ then there exists $0 < \epsilon < b - a$ such that $a < a + \epsilon < F(x_0) < b - \epsilon < b$ and $F^{-1}((a + \epsilon, b - \epsilon))$ is an open set in $\mathbb{R}^n$ and a subset of G. Since $\partial G = \overline{G} \setminus \mathring{G}$ it remains to prove $\partial G \subset \{x \in D \,|\, F(x) = a \text{ or } F(x) = b\}$. Let $F(x_0) = a$ (the case $F(x_0) = b$ goes analogously). By our assumption we can find $\eta > 0$ such that $(a - \eta, a + \eta) \subset R(F)$ and $F^{-1}((a - \eta, a + \eta))$ is an open set in $\mathbb{R}^n$ and a subset of D. Since $x_0 \in F^{-1}((a - \eta_0, a + \eta_0))$ we can find $\epsilon > 0$ such that $B_\epsilon(x_0) \subset F^{-1}((a - \eta, a + \eta))$ and $B_\epsilon(x_0)$ must contain points $x \in B_\epsilon(x_0)$ such that $a - \eta < F(x) < a$, i.e. $x \in G^\complement$, as well as points $y \in B_\epsilon(x_0)$ such that $a < F(y) < a + \eta$. Now we want to ask whether we can represent ∂G locally

397

as the graph of a function. But G need not be compact, take for example $\sin : \mathbb{R} \to \mathbb{R}$ and $I = [0,1]$. The periodicity of $\sin$ implies that $\sin^{-1}([0,1])$ is unbounded. To exclude this problem we assume that F is proper in the following sense:

Definition 19.17. *A continuous mapping $f; X \to Y$ between two Hausdorff spaces is called **proper** if the pre-image of every compact set in Y is compact in X.*

Corollary 19.18. *A continuous and monotone function $f : \mathbb{R} \to \mathbb{R}$ is proper.*

Proof. Let $K \subset \mathbb{R}$ be compact. It follows that $K \subset [a, b]$ for some $a < b$, $f^{-1}(K)$ is closed and $f^{-1}(K) \subset [f^{-1}(a), f^{-1}(b)]$ or $f^{-1}(K) \subset [f^{-1}(b), f^{-1}(a)]$ depending whether f is increasing or decreasing. Hence $f^{-1}(K)$ is a closed and bounded set, i.e. compact in $\mathbb{R}$. $\qquad\square$

Thus, for a continuous and proper function $F : D \to \mathbb{R}$, $D \subset \mathbb{R}^n$ open, such that $[a - \eta, b + \eta] \subset R(F)$ for some $\eta > 0$ and $a < b$ the set $G := F^{-1}(I)$, $I = [a, b]$, is a compact set in $\mathbb{R}^n$ with non-empty interior and boundary $\partial G = \{x \in D \mid F(x) = a \text{ and } F(x) = b\}$ which is of course a compact set. Suppose in addition that F is on D a C^1-function and $\operatorname{grad} F \neq 0$ in D (or in $D' \subset D$ open, $\partial G \subset D'$). For $x_0 \in \partial G$ there exists a neighbourhood $U(x_0) \subset D$ such that $\frac{\partial F}{\partial x_j}(x) \neq 0$ for all $x \in U(x_0)$ and some j, $1 \leq j \leq n$. The implicit function, Theorem 10.5, applied to $F - a$ (or $F - b$) implies now that in a neighbourhood of x_0 we can represent the solution of the equation $F(x) - a = 0$ with the help of a function g_j, i.e. with $x_j = g_j(\tilde{x})$, $\tilde{x} = (x_1, \ldots, x_{j-1}, x_{j+1}, \ldots, x_n) \in Q(x_0) \subset \mathbb{R}^{n-1} \subset \mathbb{R}^n$, we have

$$0 = F(x) - a = F(x_1, \ldots, x_{j-1}, g_j(\tilde{x}), x_{j+1}, \ldots, x_n) - a, \quad \tilde{x} \in Q(x_0).$$

Note that we need F to belong to the class C^1, we have at the moment no argument for handling the case F being just a Lipschitz continuous function or even being merely continuous. To summarise our considerations we give

Proposition 19.19. *Let $D \subset \mathbb{R}^n$ be open and $F : D \to \mathbb{R}$ a proper C^1-function. Suppose that for some $\eta > 0$ and $a < b$ the range $R(F)$ contains the interval $[a-\eta, b+\eta]$. Then the set $F^{-1}([a, b])$ is a compact set with non-empty interior the boundary of which is given by $\{x \in D \mid F(x) = a \text{ and } F(x) = b\}$. If in addition in a neighbourhood of ∂G the gradient of F does not vanish then ∂G has Jordan content zero.*

Example 19.20. For $h(x_1, x_2, x_3) = x_1^{2k} + x_2^{2l} + x_3^{2m}$, $k, l, m \in \mathbb{N}$, and $a < b$, $[a, b] \subset (0, \infty)$ the set $\{x \in \mathbb{R}^3 \mid a \le h(x_1, x_2, x_3) \le b\}$ has a boundary of Jordan content zero.

The condition $[a - \eta, b + \eta] \subset R(F)$, $\eta > 0$, is too restrictive. It excludes an application of Proposition 19.19 for example to $B_1(x_0)$ with boundary $\partial B_1(x_0) = S^{n-1}$, $x_0 \in \mathbb{R}^n$. In this case we have to take $F(x) = \left(\sum_{j=1}^n x_j^2\right)^{\frac{1}{2}}$ and find $\overline{B_1(x_0)} = \{x \in \mathbb{R}^n \mid F(x) \le 1\}$ but for no $\eta > 0$ we have $[-\eta, 1+\eta] \subset R(F)$. Thus for intervals $I = [a, b]$ where a or b is a boundary point of $R(F)$ we face a problem. However in case that $[a, b + \eta] \subset R(F)$, $\eta > 0$, and $F^{-1}(\{a\}) \subset \mathring{G}$ (or $[a - \eta, b] \subset R(F)$, $\eta > 0$, and $F^{-1}(\{b\}) \subset \mathring{G}$) we can rescue the situation.

Proposition 19.21. *Let $D \subset \mathbb{R}^n$ be open and $F : D \to \mathbb{R}$ be a proper C^1-function. Suppose that for some $\eta > 0$ we have $[a, b + \eta] \subset R(F)$, $a < b$, and $F^{-1}(\{a\}) \subset \mathring{G}$ where $G = \{x \in D \mid a \le F(x) \le b\}$, ($[a - \eta, b] \subset R(F)$ and $F^{-1}(\{b\}) \subset \mathring{G}$). Then the set $F^{-1}([a, b])$ is a compact set with non-empty interior the boundary of which is given by $\{x \in D \mid F(x) = b\}$, ($\{x \in D \mid F(x) = a\}$). Moreover, if $\operatorname{grad} F \ne 0$ in a neighbourhood of ∂G, then ∂G has Jordan content zero.*

Example 19.22. A. Consider the p-norm $\|.\|_p$, $1 \le p \le \infty$, defined on $\mathbb{R}^n$ as usual by $\|x\|_p = \left(\sum_{j=1}^n |x_j|^p\right)^{\frac{1}{p}}$. Then $\partial B_r^{(p)}(a) := \{x \in \mathbb{R}^n \mid \|x - a\|_p = r\}$ has Jordan content zero.
B. The ellipsoid $\left\{x \in \mathbb{R}^n \mid \left(\frac{x_1 - a_1}{\eta_1}\right)^2 + \cdots + \left(\frac{x_n - a_n}{\eta_n}\right)^2 \le 1\right\} \subset \mathbb{R}^n$ has a boundary of Jordan content zero.

Eventually we turn to the question how to define a volume for certain subsets $A \subset \mathbb{R}^n$. Having in mind our experience with integrals on the real line or with volume integrals for non-degenerate compact cells, we expect that when integrating the function $x \mapsto 1$ over A we obtain the volume of A. Moreover, if $A' \subset A$ we obtain the volume of A' when integrating the characteristic function $\chi_{A'}$ of A' over A, recall

$$\chi_{A'}(x) = \begin{cases} 1, & x \in A' \\ 0, & x \notin A'. \end{cases}$$

But so far A and A' must be either unions of intervals or cells. We take however the last observation as starting point for extending the notion of volume.

Definition 19.23. *A bounded set $G \subset \mathbb{R}^n$ is called **Jordan measurable** if $G \subset \mathring{K}$ for some non-degenerate compact cell K and the function $\chi_G|_K$ is Riemann integrable over K, i.e.*

$$\int_K \left(\chi_G|_K \right)(x)\,\mathrm{d}x$$

*exists. In this case we write $J^{(n)}(G) := \int_K \left(\chi_G|_K \right)(x)\,\mathrm{d}x$ and call $J^{(n)}(G)$ the **Jordan content** of G.*

Remark 19.24. A. The definition is clearly independent of the choice of K, see Problem 9.
B. If $G = K$ is a non-degenerate compact cell then $J^{(n)}(G) = J^{(n)}(K) = \mathrm{vol}_n(K)$.
C. In this chapter and in Chapter 20 we use $J^{(n)}(G)$ to denote the Jordan content of $G \subset \mathbb{R}^n$. The volume of G will later on be defined with the help of the Lebesgue measure, see Part 6 in Volume III. However, once the volume is defined it (often) coincides with the Jordan content. Therefore, in particular for making the text easily useable as a reference text, from Chapter 21 on we will also write $\mathrm{vol}_n(G)$ for $J^{(n)}(G)$.
D. From the definition of $J^{(n)}(G)$ or $\mathrm{vol}_n(G)$ we find

$$\int_* \chi_G|_K(x)dx = J^{(n)}(G) = \mathrm{vol}_n(G) = \int^* \chi_G|_K(x)dx.$$

Now let $(K_\alpha)_{\alpha \in A_Z}$ be a partition of K and $A_Z^{(1)}$ consisting of all $\alpha \in A_Z$ such that $K_\alpha \subset G$ whereas $A_Z^{(2)}$ consists of all $\beta \in A_Z$ such that $\overline{K}_\beta \cap G \neq \emptyset$. It follows that

$$\sum_{\alpha \in A_Z^{(1)}} \mathrm{vol}_n(K_\alpha) \leq \mathrm{vol}_n(G) \leq \sum_{\beta \in A_Z^{(2)}} \mathrm{vol}_n(K_\beta) \tag{19.6}$$

holds.

The central result is

Theorem 19.25. *If ∂G has Lebesgue measure zero then the bounded set $G \subset \mathbb{R}^n$ is Jordan measurable.*

The converse of this result holds too, i.e. boundaries of Lebesgue measure zero characterise Jordan measurable sets in $\mathbb{R}^n$, however we will not provide a proof here and refer to S. Hildebrandt [26].

In order to prove Theorem 19.25 we return to the Riemann integral over non-degenerate compact cells. So far we know that continuous functions are integrable, Corollary 18.15. In the case of the real line we have seen that also certain discontinuous functions, e.g. bounded monotone functions, are integrable. The notion of monotonicity cannot be extended to $\mathbb{R}^n$. However we can prove

Theorem 19.26. *Let $K \subset \mathbb{R}^n$ be a non-degenerate compact cell and $f :$ $K \to \mathbb{R}$ a bounded function. If the set of points where f is discontinuous has Jordan content zero, then f is Riemann integrable.*

Remark 19.27. We will see in Volume III that the set of points where a bounded monotone function f can be discontinuous has Lebesgue measure zero, which now yields a better understanding of the one-dimensional integrability result.

Proof of Theorem 19.25. The set of discontinuities of χ_G, $G \subset \mathring{K}$, is of course ∂G. So if ∂G is of Lebesgue measure zero, and hence by Corollary 19.9 of Jordan content zero, by Theorem 19.26 the function $\chi_G\big|_K$ is integrable and therefore G is Jordan measurable. □

Proof of Theorem 19.26. (following S. Hildebrandt [26]) For simplicity we denote $\|f\|_\infty$ by γ and the set of all discontinuities of f by $\mathrm{Dis}(f)$. Since $\mathrm{Dis}(f)$ has Jordan content zero, for $\epsilon > 0$ there exists a covering of $\mathrm{Dis}(f)$ by open cells $(K'_j)_{j=1,\ldots,N}$ such that $\overline{K'_j} \subset K$ and

$$\sum_{j=1}^{N} \mathrm{vol}_n(K'_j) < \frac{\epsilon}{4\gamma}. \tag{19.7}$$

The set $F := \bigcup_{j=1}^{N} K'_j \subset K$ is open and therefore $K\backslash F$ is compact. Since $f\big|_{K\backslash F}$ is continuous it follows that $f\big|_{K\backslash F}$ is uniformly continuous. Thus we can find $\delta > 0$ such that if $E \subset K\backslash F$ with $\mathrm{diam}(E) < \delta$ then we have for all $x, y \in E$ that $|f(x) - f(y)| < \frac{\epsilon}{2\mathrm{vol}_n(K)}$, which we can rewrite as

$$\sup_{x \in E} f(x) - \inf_{x \in E} f(x) = \sup_{x,y \in E} |f(x) - f(y)| < \frac{\epsilon}{2\mathrm{vol}_n(K)}, \tag{19.8}$$

see also Remark 19.28. Starting with the cells K'_j, $j = 1, \ldots, N$, we can construct a partition of Z into cells $(K_\alpha)_{\alpha \in I}$ such that $\mathrm{diam}(K_\alpha) < \delta$ and I admits a partition into I_1 and I_2 such that

$$K_\alpha \subset K \setminus F \text{ for } \alpha \in I_1 \text{ and } K_\alpha \subset \overline{F} \text{ for } \alpha \in I_2.$$

From (19.7) we deduce that

$$\sum_{\alpha \in I_2} \mathrm{vol}_n(K_\alpha) < \frac{\epsilon}{4\gamma}. \tag{19.9}$$

We define now the following two step functions $\varphi, \psi : K \to \mathbb{R}$ by

$$\varphi\big|_{\mathring{K}_\alpha} = \inf_{x \in K_\alpha} f(x) \text{ and } \psi\big|_{\mathring{K}_\alpha} = \sup_{x \in K_\alpha} f(x)$$

and on ∂K_α, $\alpha \in I$, we set $\varphi(x) = \psi(x) = f(x)$. We clearly have $\varphi(x) \le f(x) \le \psi(x)$ and the corresponding integrals are

$$\int_K \varphi(x)\, dx = \sum_{\alpha \in I} \left(\inf_{x \in K_\alpha} f(x) \right) \mathrm{vol}_n(K_\alpha)$$

and

$$\int_K \psi(x)\, dx = \sum_{\alpha \in I} \left(\sup_{x \in K_\alpha} f(x) \right) \mathrm{vol}_n(K_\alpha).$$

Further we find by (19.7), (19.8) and (19.9) that

$$\int_K (\psi(x) - \varphi(x))\, dx = \sum_{\alpha \in I} \left(\left(\sup_{x \in \mathring{K}_\alpha} f(x) \right) - \left(\inf_{x \in \mathring{K}_\alpha} f(x) \right) \right) \mathrm{vol}_n(K_\alpha)$$

$$= \sum_{\alpha \in I_1} \left(\left(\sup_{x \in \mathring{K}_\alpha} f(x) \right) - \left(\inf_{x \in \mathring{K}_\alpha} f(x) \right) \right) \mathrm{vol}_n(K_\alpha)$$

$$+ \sum_{\alpha \in I_2} \left(\left(\sup_{x \in \mathring{K}_\alpha} f(x) \right) - \left(\inf_{x \in \mathring{K}_\alpha} f(x) \right) \right) \mathrm{vol}_n(K_\alpha)$$

$$\le \frac{\epsilon}{2\mathrm{vol}_n(K)} \sum_{\alpha \in I_1} \mathrm{vol}_n(K_\alpha) + 2\gamma \sum_{\alpha \in I_2} \mathrm{vol}_n(K_\alpha)$$

$$< \frac{\epsilon}{2\mathrm{vol}_n(K)} \mathrm{vol}_n(K) + \frac{\epsilon}{2} = \epsilon.$$

Since φ and ψ are step functions satisfying the conditions of Theorem 18.13 we deduce that f is integrable. □

Remark 19.28. In deriving (19.8) we need to observe that we can achieve $|f(x) - f(y)| < \frac{\epsilon'}{2\mathrm{vol}_n(K)}$ for $\epsilon' < \epsilon$, and then (19.8) is obvious.

The following result enlightens the theory further, but since we do not make much use of it we do not give its proof and refer to [26].

Theorem 19.29. *A bounded function $f : K \to \mathbb{R}$ is Riemann integrable if and only if the set of its discontinuity has Lebesgue measure zero.*

Thus, while (19.25) gives a sufficient criterion for the integrability of a bounded function, Theorem 19.29 gives a characterisation, but we have to switch from Jordan content the Lebesgue measure zero for $\mathrm{Dis}(f)$.

In the chapters to come we will often encounter the following situations: $U \subset \mathbb{R}^n$ is an open (bounded) set and G is a Jordan measurable set such that $\overline{G} \subset U$. We will then start with a function $f : U \to \mathbb{R}$ or a vector field $F : U \to \mathbb{R}^n$ having on U certain smoothness properties, e.g. belonging to the class C^k, and then we want to handle (and first to define) the integral of f (or F) over G. It is therefore helpful to bring the above relation of G and U closer to our geometric setting for dealing with integrals where we prefer to cover sets by hyper-rectangles or if possible by cubes.

Lemma 19.30. *Let $U \subset \mathbb{R}^n$ be a bounded open set and G a Jordan measurable set such that $\overline{G} \subset U$. Then we can find a finite family of closed cubes $K_j \subset \mathbb{R}^n$, $j = 1, \ldots, M$, $\mathring{K}_j \cap \mathring{K}_l = \emptyset$ for $j \neq l$ such that $\overline{G} \subset \bigcup_{j=1}^M K_j \subset U$.*

Proof. Since ∂U and $\overline{G}$ are compact and disjoint, i.e. $\overline{G} \cap \partial U = \emptyset$, it follows that $\mathrm{dist}(\overline{G}, \partial U) = \tilde{\eta} > 0$. Using the fact that $\mathbb{Q} \cap \overline{G}$ is dense in $\overline{G}$ we can cover $\overline{G}$ by open cubes $C_{\frac{\eta}{4}}(x)$ with centre $x \in \overline{G} \cap \mathbb{Q}^n$ and side length $\frac{\eta}{2} \in \mathbb{Q}$, $\eta < \tilde{\eta}$, i.e. $\overline{G} = \bigcup_{x \in \mathbb{Q}^n \cap \overline{G}} C_{\frac{\eta}{4}}(x)$. We note that $C_{\frac{\eta}{4}}(x) \subset U$ and by the compactness of $\overline{G}$ we can find finitely many cubes $C_{\frac{\eta}{4}}(x_1), \ldots, C_{\frac{\eta}{4}}(x_N)$ which already cover $\overline{G}$. These open cubes may overlap, see Figure 19.7 below. However we can partition each cube $C_{\frac{\eta}{4}}(x_1), \ldots, C_{\frac{\eta}{4}}(x_N)$ into sub-cubes $K(y_l), 1 \leq l \leq m$, such that if $C_{\frac{\eta}{4}}(x_j) \cap C_{\frac{\eta}{4}}(x_l) \neq \emptyset$ this intersection is the union of sub-cubes $K(y_{l_m})$, $1 \leq l_m \leq \tilde{M}(x_j, x_l)$. Counting the sub-cubes $K(y_{l_m})$ covering such an intersection only once we obtain a covering of $\overline{G}$ by compact cubes $\overline{K(y_{l_\nu})}$, $l_\nu \in \{k \in \mathbb{N} | k \leq M\}$, with mutual disjoint interior.

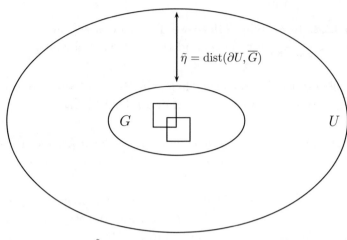

side length of cube $< \frac{\tilde{\eta}}{2}$

Figure 19.7

$\square$

Remark 19.31. The result of Lemma 19.30 gives an idea of how to approximate $J^{(n)}(G)(= \mathrm{vol}_n(G))$. We start with $\overline{G} \subset \bigcup_{j=1}^{m} \overline{K}_j$ and note that $\mathrm{vol}_n \left(\bigcup_{j=1}^{m} \overline{K}_j \right) = \sum_{j=1}^{m} \mathrm{vol}_n(\overline{K}_j)$. Denote by $2r_j$ the length of the sides of K_j. Now we divide each $\overline{K}_j$ into sub-cubes of side length $\frac{2r_j}{\nu}, j = 1, \ldots, m, \nu \in \mathbb{N}$, and we take away all sub-cubes having an empty intersection with $\overline{G}$. Thus we obtain a new covering of $\overline{G}$ by compact cubes $(\overline{K}_l^{(\nu)})_{l=1,\ldots,m_\nu}$ with mutually disjoint interior and

$$\kappa_\nu := \mathrm{vol}_n \left(\bigcup_{l=1}^{m_\nu} \overline{K}_l^{(\nu)} \right) \leq \mathrm{vol}_n \left(\bigcup_{j=1}^{m} \overline{K}_j \right).$$

The sequence $(\kappa_\nu)_{\nu \in \mathbb{N}}$ is monotone decreasing and non-negative, hence it also has a limit. This limit should be $\mathrm{vol}_n(\overline{G})$. In each step we can also consider those cubes $\overline{K}_l^{(\nu)}$ which are included in $\mathring{G}$. Denote the cubes in this collection by ${}_i\overline{K}_l^{(\nu)}$ for l from a subset $M_{i,\nu} \subset \{k \in \mathbb{N} | k \leq M_\nu\}$. It follows that

$$\lambda_\nu := \mathrm{vol}_n \left(\bigcup_{l \in M_{i,\nu}} {}_i\overline{K}_l^{(\nu)} \right) \leq \mathrm{vol}_n \left(\bigcup_{j=1}^{M} \overline{K}_j \right)$$

is an increasing sequence and hence has a limit, which again should be $\mathrm{vol}_n(\overline{G})$. Setting $\mathbb{N}_M := \{1, \ldots, M\}$ we note that with $\tilde{M}_\nu := \mathbb{N}_{M_\nu} \setminus M_{i,\nu}$ we have that $\partial G \subset \bigcup_{l \in \tilde{M}_\nu} \overline{K}_l^{(\nu)}$ and we get a further reason why Theorem 19.25 should hold.

Problems

1.   a) Prove that every plane polygon is a Jordan null set.

   b) Prove that if $M \subset \mathbb{R}^n$ is a Jordan null set then for every compact set $G \subset \mathbb{R}^m$ the set $M \times G \subset \mathbb{R}^{n+m}$ is a Jordan null set too.

2.   a) Let $R \subset [0,1]$ be a dense set, i.e. $\overline{R} = [0,1]$. Can R be a Jordan null set?

   b) Prove that $\mathbb{Q} \cap [0,1]$ is a Lebesgue null set which is dense in $[0,1]$.

3. Let I be a non-denumerable set, i.e. neither finite nor countable, and $A_j \subset \mathbb{R}$, $j \in I$, a Lebesgue null set. By giving an example show that in general $\bigcup_{j \in I} A_j$ is not a Lebesgue null set.

4. Prove that the boundary of the annulus $\overline{B_R(0)} \setminus B_r(0)$, $0 < r < R$, is of the type C^k for any $k \in \mathbb{N}$.

5. By using methods from elementary geometry prove that the volume of a non-degenerate cell $K \subset \mathbb{R}^n$ is invariant under translations and operations of $O(n)$.
   Hint: first give an elementary definition of a hyper-rectangle in a general position, i.e. the sides need not be parallel to coordinates axes, and define its volume.

6.   a) Prove that $f : (-1, 1) \to (-1, 1)$, $f(x) = 1 - x^4$ is not proper.

   Hint: consider $f^{-1}\left(\left[-\frac{3}{4}, \frac{3}{4}\right]\right)$.

   b) Prove that the composition of proper mappings is proper.

7. Let X and Y be two Hausdorff spaces. We call a mapping $f : X \to Y$ **closed** if it maps closed sets onto closed sets. Suppose that f is continuous, closed and that pre-images of points are compact. Prove that f is proper.

8. Let $B_R(0) \subset \mathbb{R}^n$ and $f : B_R(0) \to \mathbb{R}$ be a radially symmetric function, i.e. $f(x_1, \ldots, x_n) = g(||x||_2)$ where $g : [0, R) \to \mathbb{R}$. Suppose that g is a strictly monotone C^1-mapping. Prove that f is proper and for every interval $[a, b] \subset R(g)$ the set $\partial(f^{-1}([a, b]))$ has Jordan content zero.

   Hint: use Problem 7.

9. Prove that $J^{(n)}(G)$ as introduced in Definition 19.23 is independent of the choice of K.

20 Volume Integrals on Bounded Jordan Measurable Sets

In this chapter we will define and investigate the Riemann integral for certain bounded functions the domain of which are bounded Jordan measurable sets. The basic idea is to reduce the definition to that we have given for compact non-degenerate cells. At this stage we will have almost no practical way to calculate the value of an integral. However once we can identify volume integrals with iterated integrals we are in a better position.

Our starting point is

Definition 20.1. *For a bounded function $f : G \to \mathbb{R}$ defined on a Jordan measurable set $G \subset \mathbb{R}^n$ the **canonical extension** to $\mathbb{R}^n$ is the function*

$$\tilde{f}_G(x) := \begin{cases} f(x), & x \in G \\ 0, & x \notin G. \end{cases} \tag{20.1}$$

If $G \subset \mathbb{R}^n$ is bounded we can find a non-degenerate compact cell $K \subset \mathbb{R}^n$ such that $\overline{G} \subset \mathring{K}$. The function $\tilde{f}_G\big|_K$ is now a bounded function on K which may or may not be integrable.

Definition 20.2. *Let $G \subset \mathbb{R}^n$ be a bounded Jordan measurable set and $f : G \to \mathbb{R}$ a bounded function. If for some non-degenerate compact cell $K \subset \mathbb{R}^n$, $\overline{G} \subset \mathring{K}$, the function $\tilde{f}_G\big|_K$ is Riemann integrable, then we call f **Riemann integrable** on G and define its **Riemann integral** over G by*

$$\int_G f(x)\, dx := \int_K \left(\tilde{f}_G\big|_K \right)(x)\, dx. \tag{20.2}$$

Lemma 20.3. *The definition of $\int_G f(x)\, dx$ is independent of the choice of K.*

Proof. Let K_1 and K_2 be two cells with the properties stated in Definition 20.2. It follows that $\overline{G} \subset (K_1 \cap K_2)^\circ$ and $K_1 \cap K_2$ is a non-degenerate compact cell, see Figure 20.1

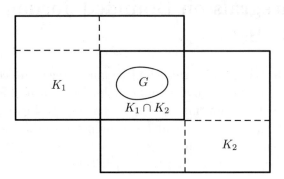

Figure 20.1

The cell $K_1 \cap K_2$ induces partitions of K_j, $j = 1, 2$, into cells $K_\alpha^{(1)}$, $\alpha = 1, \ldots, M$, and $K_\beta^{(2)}$, $\beta = 1, \ldots, N$, with $K_1^{(1)} = K_1^{(2)} = K_1 \cap K_2$, and it follows

$$\int_{K_1} \left(\tilde{f}_G \big|_{K_1} \right)(x) \, dx = \sum_{\alpha=1}^{M} \int_{K_\alpha^{(1)}} \left(\tilde{f}_G \big|_{K_\alpha^{(1)}} \right)(x) \, dx = \int_{K_1^{(1)}} \left(\tilde{f}_G \big|_{K_1^{(1)}} \right)(x) \, dx$$

$$= \int_{K_1 \cap K_2} \left(\tilde{f}_G \big|_{K_1 \cap K_2} \right)(x) \, dx = \int_{K_1^{(2)}} \left(\tilde{f}_G \big|_{K_1^{(2)}} \right)(x) \, dx$$

$$= \sum_{\beta=1}^{N} \int_{K_\beta^{(2)}} \left(\tilde{f}_G \big|_{K_\beta^{(2)}} \right)(x) \, dx = \int_{K_2} \left(\tilde{f}_G \big|_{K_2} \right))(x) \, dx.$$

$\square$

It is helpful to fix for a function $h : A \to \mathbb{R}$, $A \subset \mathbb{R}^n$, the notation

$$\text{Dis}(h) := \left\{ x \in A \, \big| \, h \text{ is discontinuous at } x \right\}. \tag{20.3}$$

Now we prove

Theorem 20.4. *Let $G \subset \mathbb{R}^n$ be a bounded Jordan measurable set and $f : G \to \mathbb{R}$ a bounded function. If $\text{Dis}(f \big|_{\mathring{G}})$ has Jordan content zero then f is Riemann integrable.*

Proof. Let $K \subset \mathbb{R}^n$ be a non-degenerate cell such that $\overline{G} \subset \mathring{K}$ and consider $g : K \to \mathbb{R}$ defined by $g := (\chi_G \tilde{f}_G) \big|_K$. We want to prove the integrability of g

by using Theorem 19.26, i.e. we aim to show that $\mathrm{Dis}(g)$ has Jordan content zero provided $\mathrm{Dis}(f|_{\mathring{G}})$ has Jordan content zero. We know that $\mathrm{Dis}(g) \subset \partial G \cup \mathrm{Dis}(f|_{\mathring{G}})$, and since G is Jordan measurable and ∂G is compact it follows that $J^{(n)}(\partial G) = 0$ as $J^{(n)}(\mathrm{Dis}(f|_{\mathring{G}})) = 0$ by assumption, so $J^{(n)}(\mathrm{Dis}(g)) = 0$ and Theorem 19.26 implies the Riemann integrability of f. $\qquad\square$

Corollary 20.5. *If for a bounded function $f : G \to \mathbb{R}$ on a Jordan measurable set G the function $f|_{\mathring{G}}$ is continuous then f is Riemann integrable over G.*

Proposition 20.6. *Let $G \subset \mathbb{R}^n$ be a bounded Jordan measurable set, $f, g : G \to \mathbb{R}$ be bounded Riemann integrable functions and $\lambda, \mu \in \mathbb{R}$. Then the following functions are Riemann integrable too:*

$$\lambda f + \mu g, \tag{20.4}$$

$$f \cdot g, \tag{20.5}$$

$$f^+ \text{ and } f^{-1}, \text{ hence } |f|^p, \ p \geq 1. \tag{20.6}$$

Moreover, the integral is linear, i.e.

$$\int_G (\lambda f + \mu g)(x)\,\mathrm{d}x = \lambda \int_G f(x)\,\mathrm{d}x + \mu \int_G g(x)\,\mathrm{d}x \tag{20.7}$$

and positivity preserving, i.e. $f \geq 0$ implies $\int_G f(x)\,\mathrm{d}x \geq 0$.

Proof. The key observation is that for the canonical extensions we have with $h = \lambda f + \mu g$ and $u = f \cdot g$

$$\tilde{h}_G = \lambda \tilde{f}_G + \mu \tilde{g}_G$$

and

$$\tilde{u}_G = \tilde{f}_G \cdot \tilde{g}_G.$$

Moreover we know that if $\tilde{f}_G|_K$ is integrable for a cell satisfying the conditions of Definition 20.2 then $\tilde{f}_G|_{K'}$ is integrable for every compact cell K' such that $K \subset K'$. Applying now Theorem 18.16 we obtain (20.4) and (20.5) and further we find with a suitable cell K

$$\int_G h(x)\,\mathrm{d}x = \int_K (\tilde{h}_G|_K)(x)\,\mathrm{d}x$$

$$= \lambda \int_K (\tilde{f}_G|_K)(x)\,\mathrm{d}x + \mu \int_K (\tilde{g}_G|_K)(x)\,\mathrm{d}x$$

$$= \lambda \int_G f(x)\,\mathrm{d}x + \mu \int_G g(x)\,\mathrm{d}x,$$

i.e. (20.7) holds.
Further we observe that

$$(f^+)\tilde{}_G(x) = \begin{cases} f(x), & f(x) \geq 0, x \in G \\ 0, & f(x) < 0, x \in G \\ 0, & x \in \mathbb{R}^n \backslash G \end{cases}$$

and

$$(\tilde{f}_G)^+ = \begin{cases} f(x), & f(x) \geq 0, x \in G \\ 0, & f(x) < 0, x \in G \\ 0, & x \in \mathbb{R}^n \backslash G, \end{cases}$$

i.e. $(f^+)\tilde{}_G = (\tilde{f}_G)^+$ and $(f^-)\tilde{}_G = (\tilde{f}_G)^-$, and now (20.6) follows when noting that $|f| = f^+ + f^-$. Since $f \geq 0$ implies $\tilde{f}_G \geq 0$ we deduce immediately that the integral is positivity preserving. $\square$

Corollary 20.7. *If $f, g : G \to \mathbb{R}$ are as in Proposition 20.6 and $f \leq g$ then the following holds*

$$\int_G f(x)\, dx \leq \int_G g(x)\, dx. \tag{20.8}$$

Proof. This corollary follows by the standard argument: if V is a vector space of real-valued functions and $l : V \to \mathbb{R}$ is linear and positivity preserving, then for $f, g \in V$ the inequality $f \leq g$ implies $l(f) \leq l(g)$:

$$0 \leq l(g - f) = l(g) - l(f).$$

$\square$

The next proposition summarises the basic inequalities for integrals. Their proofs are all along the following line: by inspection we can transfer the proof from the one-dimensional case to the case of non-degenerate cells, and now we can show the general case by using Definition 20.2. We will provide for the triangle inequality and the Cauchy-Schwarz inequality (as a special case of Hölder's inequality) the detailed proof in Problem 1, as we will give a detailed proof in Volume III when introducing the Lebesgue integral.

Proposition 20.8. *Let $G \subset \mathbb{R}^n$ be a bounded Jordan measurable set and $f, g : G \to \mathbb{R}$ bounded Riemann integrable functions. The following inequali-*

ties hold:

$$\left| \int_G f(x)\,\mathrm{d}x \right| \le \int_G |f(x)|\,\mathrm{d}x \quad (\textbf{\textit{triangle inequality}}), \qquad (20.9)$$

$$\int_G |f(x)g(x)|\,\mathrm{d}x \le \left(\int_G |f(x)|^p\,\mathrm{d}x \right)^{\frac{1}{p}} \left(\int_G |g(x)|^q\,\mathrm{d}x \right)^{\frac{1}{q}} \qquad (20.10)$$

$$\text{for } 1 < p, q < \infty \text{ and } \frac{1}{p} + \frac{1}{q} = 1 \ (\textbf{\textit{Hölder's inequality}})$$

$$\int_G |f(x)g(x)|\,\mathrm{d}x \le \left(\int_G |f(x)|^2\,\mathrm{d}x \right)^{\frac{1}{2}} \left(\int_G |g(x)|^2\,\mathrm{d}x \right)^{\frac{1}{2}} \qquad (20.11)$$

$$(\textbf{\textit{Cauchy-Schwarz inequality}})$$

$$\int_G |f(x)g(x)|\,\mathrm{d}x \le \|f\|_\infty \int_G |g(x)|\,\mathrm{d}x; \qquad (20.12)$$

$$\left(\int_G |f(x) + g(x)|^p\,\mathrm{d}x \right)^{\frac{1}{p}} \le \left(\int_G |f(x)|^p\,\mathrm{d}x \right)^{\frac{1}{p}} + \left(\int_G |g(x)|^p\,\mathrm{d}x \right)^{\frac{1}{p}} \qquad (20.13)$$

$$\text{for } 1 \le p < \infty \ (\textbf{\textit{Minkowski's inequality}}).$$

Remark 20.9. Since for a bounded Jordan measurable set $G \subset \mathbb{R}^n$ we have $J^{(n)}(G) = \int_G 1\,\mathrm{d}x$, we deduce from inequality (20.12) that

$$\int_G |f(x)|\,\mathrm{d}x \le J^{(n)}(G)\,\|f\|_\infty, \qquad (20.14)$$

whereas (20.10) implies

$$\int_G |f(x)|\,\mathrm{d}x \le (J^{(n)}(G))^{\frac{1}{q}} \left(\int_G |f(x)|^p\,\mathrm{d}x \right)^{\frac{1}{p}}, \ p \ge 1, \ \frac{1}{p} + \frac{1}{q} = 1. \qquad (20.15)$$

As in the one-dimensional case we cannot conclude that

$$\left(\int_G |f(x)|^p\,\mathrm{d}x \right)^{\frac{1}{p}} = 0, \ 1 \le p < \infty, \qquad (20.16)$$

implies that f is the zero function on G, i.e. $f(x) = 0$ for all x, see Problem 2. However we have

Proposition 20.10. *Let $G \subset \mathbb{R}^n$ be an open bounded set or a closed bounded set with non-empty interior. If (20.16) holds for a continuous function $f :$ $G \to \mathbb{R}$ then f is identically zero.*

Proof. Suppose $f(x_0) \neq 0$ for some $x_0 \in \overset{\circ}{G}$, say $f(x_0) > 0$, the case $f(x_0) < 0$ goes analogously. Then we can find a non-degenerate compact cell $K \subset \overset{\circ}{G}$ and $\epsilon > 0$ such that $f(x) > \epsilon$ for all $x \in K$ and further there exists a non-degenerate cell Z such that $G \subset \overset{\circ}{Z}$ and

$$\epsilon^p \operatorname{vol}_n(K) = \int_K \epsilon^p \, dx = \int_Z \chi_K(x) \epsilon^p \, dx$$
$$\leq \int_Z \chi_K(x) |f(x)|^p \, dx \leq \int_Z \chi_G(x) |f(x)|^p \, dx$$
$$= \int_G |f(x)|^p \, dx,$$

holds, implying

$$0 < \epsilon^p \operatorname{vol}_n(K) \leq \int |f(x)|^p \, dx$$

or

$$0 < \epsilon \leq \frac{1}{(\operatorname{vol}_n(K))^{\frac{1}{p}}} \left(\int_g |f(x)|^p \, dx \right)^{\frac{1}{p}} = 0$$

which is a contradiction. $\qquad\square$

Let us introduce the notation

$$\mathcal{N}_{p,G}(f) := \left(\int_G |f(x)|^p \, dx \right)^{\frac{1}{p}} \tag{20.17}$$

where $G \subset \mathbb{R}^n$ is a Jordan measurable set, $f : G \to \mathbb{R}$ is Riemann integrable and $p \geq 1$. The following definition we have already encountered in Example 13.10.C.

Definition 20.11. *A mapping* $\rho : X \to \mathbb{R}$ *defined on a vector space* X *is called a **semi-norm** if*

i) $\rho(x) \geq 0$ *for all* $x \in X$;

ii) $\rho(\lambda x) = |\lambda| \rho(x)$ *for all* $x \in X$ *and* $\lambda \in \mathbb{R}$;

iii) $\rho(x + y) \leq \rho(x) + \rho(y)$.

Every norm on X is of course a semi-norm, but a semi-norm need not be definite, i.e. $\rho(x) = 0$ need not imply $x = 0$.

Example 20.12. A. Let $\varphi \in C_b^k(\mathbb{R}^n)$ and for $\alpha \in \mathbb{N}_0^n$, $|\alpha| \leq k$, define

$$p_\alpha(\varphi) = \sup_{x \in \mathbb{R}^n} |\partial^\alpha \varphi(x)| = \|\partial^\alpha \varphi\|_\infty. \tag{20.18}$$

Then p_α is a semi-norm for $\alpha \neq 0 \in \mathbb{N}_0^n$ whereas p_0 is a norm. Indeed, the properties i) - iii) follow from the fact that $\|.\|_\infty$ is a norm on $C_b(\mathbb{R}^n)$. If $\alpha \neq 0$ then for every constant function $\varphi_c(x) = c$, $c \in \mathbb{R}$, we have $\partial^\alpha \varphi_c(x) = 0$ for all x, i.e. $\|\partial^\alpha \varphi_c\|_\infty = 0$, but for $c \neq 0$ the function φ_c is not the null-function or zero element in $C_b^k(\mathbb{R}^n)$.
B. By (20.17) a semi-norm is given on the vector space of all Riemann integrable functions over $G \subset \mathbb{R}^n$, which is not a norm. However $\mathcal{N}_{p,G}$ restricted to $C_b(G)$ is a norm.

As in the one-dimensional case we can prove

Theorem 20.13. *Let $G \subset \mathbb{R}^n$ a bounded Jordan measurable set and $(f_\nu)_{\nu \in \mathbb{N}}$, $f_\nu : G \to \mathbb{R}$, be a sequence of bounded continuous functions converging uniformly to $f : G \to \mathbb{R}$, i.e. $\lim_{\nu \to \infty} \|f_\nu - f\|_\infty = 0$. Then the following holds*

$$\int_G f(x)\,dx = \int_G \left(\lim_{\nu \to \infty} f_\nu(x) \right) dx = \lim_{\nu \to \infty} \int_G f_\nu(x)\,dx. \tag{20.19}$$

Proof. First we note that f as uniform limit of continuous functions must be continuous, see Theorem 2.32, and bounded with respect to the sup-norm $\|.\|_\infty$, i.e. f must be a bounded continuous functions, hence f is Riemann integrable over G. Now we deduce

$$\left| \int_G f_\nu(x)\,dx - \int_G f(x)\,dx \right| = \left| \int_G (f_\nu(x) - f(x))\,dx \right|$$

$$\leq \int_G |f_\nu(x) - f(x)|\,dx \leq \|f_\nu - f\|_{\infty,G} \int_G 1\,dx$$

$$= J^{(n)}(G)\,\|f_\nu - f\|_\infty$$

implying the convergence of $\left(\int_G f_\nu(x)\,dx \right)_{\nu \in \mathbb{N}}$ to $\int_G f(x)\,dx$. $\square$

Our next result extends the mean-value result we have proved for continuous functions defined on a non-degenerate compact cell, see Proposition 18.17.

Theorem 20.14 (Mean value theorem for volume integrals). *Let $G \subset \mathbb{R}^n$ be a Jordan measurable set and $f, g : G \to \mathbb{R}$ be two Riemann integrable* *functions, $g(x) \geq 0$ for all $x \in G$. We can find $\eta \in [\inf_{x\in G} f(x), \sup_{x\in G} f(x)]$ such that*

$$\int_G f(x)g(x)\, dx = \eta \int_G g(x)\, dx. \tag{20.20}$$

Proof. Since $g(x) \geq 0$ we have in G

$$\left(\inf_{x\in G} f(x)\right) g(x) \leq f(x)g(x) \leq \left(\sup_{x\in G} f(x)\right) g(x)$$

implying that

$$\left(\inf_{x\in G} f(x)\right) \int_G g(x)\, dx \leq \int_G f(x)g(x)\, dx \leq \left(\sup_{x\in G} f(x)\right) \int g(x)\, dx.$$

Now, if $\int g(x)\, dx = 0$ we deduce

$$\int_G f(x)g(x)\, dx = 0 = \eta \int_G g(x)\, dx$$

for every $\eta \in \mathbb{R}$. In the case $\int_G g(x)\, dx \neq 0$ we find with

$$\eta := \frac{\int_G f(x)g(x)\, dx}{\int_G g(x)\, dx}$$

equality (20.20). $\qquad\square$

Remark 20.15. A. For $g = 1$ on G it follows for some η, $\inf_{x\in G} f(x) \leq \eta \leq \sup_{x\in G} f(x)$, that

$$\int_G f(x)\, dx = \eta J^{(n)}(G).$$

If in addition G is compact and connected, as well as f continuous on G, then we know that $\inf_{x\in G} f(x)$ and $\sup_{x\in G} f(x)$ are attained and the range of f is the compact interval with these end points. Hence η is image of some $\xi \in G$, i.e.

$$\int_G f(x)\, dx = f(\xi) J^{(n)}(G), \tag{20.21}$$

which is a direct extension of (18.36).

B. In the general case of Theorem 20.14, for G compact and connected and f continuous, we find now the existence of $\zeta \in G$ such that

$$\int_G f(x)g(x)\, dx = f(\zeta) \int_G g(x)\, dx.$$

In the next results we want to study the Riemann integral as operation on the family of all Jordan measurable sets. Our first result is almost trivial.

Lemma 20.16. *Let $G \subset \mathbb{R}^n$ be a Jordan measurable set and $G' \subset G$ a Jordan measurable subset. If $f : G \to \mathbb{R}$ is a bounded function Riemann integrable over G, then $f|_{G'}$ is a bounded function Riemann integrable over G'.*

Proof. We just need to note that $\left(f|_{G'}\right)_G^{\sim} = (\chi_{G'} f)_G^{\sim}$ and by assumption $\chi_{G'}|_G$ is integrable, so we may apply (20.5) to $g = \chi_{G'}|_G$ and f. $\qquad\square$

Next we want to prove what is often called the **finite set-additivity** of the Riemann integral.

Lemma 20.17. *For two disjoint Jordan measurable sets $G_1, G_2 \subset \mathbb{R}^n$ their union $G := G_1 \cup G_2$ is again Jordan measurable and for a Riemann integrable function $f : G \to \mathbb{R}$ the following holds*

$$\int_G f(x)\,\mathrm{d}x = \int_{G_1} f(x)\,\mathrm{d}x + \int_{G_2} f(x)\,\mathrm{d}x. \qquad (20.22)$$

Proof. Since $G_1 \cap G_2 = \emptyset$ implies that $\partial G = \partial(G_1 \cup G_2) \subset \partial G_1 \cup \partial G_2$ and $J^{(n)}(\partial G_1) = J^{(n)}(\partial G_2) = 0$ by assumption, it follows that $J^{(n)}(\partial G) = 0$. Moreover we have

$$\tilde{f}_G = \chi_G \tilde{f}_G = \chi_{G_1} \tilde{f}_G + \chi_{G_2} \tilde{f}_G$$

and $\chi_{G_j} \tilde{f}_G = \left(f|_G\right)_{G_j}^{\sim}$. If $K \subset \mathbb{R}^n$ is a cell such that $\overline{G} \subset \mathring{K}$ and $\tilde{f}_G$ is integrable over K, then it follows that $\chi_{G_j} \tilde{f}_G$ is also integrable of K and (20.4) implies (20.22). $\qquad\square$

In Problem 5 we will extend (20.22) to a finite number of sets $G_j \subset \mathbb{R}^n$.

Corollary 20.18. *Let $G \subset \mathbb{R}^n$ be an open Jordan measurable set and $F \subset G$ a set of Jordan content zero, i.e. $J^{(n)}(F) = 0$. Further let $f : \overline{G} \to \mathbb{R}$ be a bounded function continuous on $G \backslash F$. Then f is integrable over $\overline{G}$ as well as over G and the following holds*

$$\int_{\overline{G}} f(x)\,\mathrm{d}x = \int_G f(x)\,\mathrm{d}x. \qquad (20.23)$$

Proof. The points of discontinuities of f on $\overline{G}$ as well as on G is a set of Jordan content zero, and therefore f is integrable over $\overline{G}$ and G. In addition we know that f is integrable over ∂G and by the previous lemma we have

$$\int_{\overline{G}} f(x)\, dx = \int_G f(x)\, dx + \int_{\partial G} f(x)\, dx.$$

Applying (20.14) we find

$$\left| \int_{\partial G} f(x)\, dx \right| \le \|f\|_\infty\, J^{(n)}(\partial G) = 0$$

which yields (20.23). $\qquad\qquad\qquad\qquad\qquad\qquad\qquad\qquad\qquad\qquad\square$

Remark 20.19. The proof of Corollary 20.18 entails: if $f : G \to \mathbb{R}$ is a bounded integrable function on a set of Jordan content zero, then $\int_G f(x)\, dx = 0$.

In Corollary 18.22 we could prove a first version of an integration by parts formula for integrable functions defined on a compact, non-degenerate cell with one of the functions vanishing on the boundary of the cell. We recall the definition of the support $\operatorname{supp} u$ of a function $u : G \to \mathbb{R}$ where $G \subset \mathbb{R}^n$ is equipped with the Euclidean metric, see (14.12),

$$\operatorname{supp} u = \overline{\{x \in G \,|\, u(x) \ne 0\}}. \qquad\qquad (20.24)$$

If $\operatorname{supp} u$ is compact we call u a function with compact support. Further we denote for a bounded, open set $G \subset \mathbb{R}^n$ the space of all continuously differentiable functions with compact support by $C_0^1(G)$.

Proposition 20.20 (Integration by parts formulae). *Let $G \subset \mathbb{R}^n$ be a bounded, open and Jordan measurable set.*
A. *For $f \in C_0^1(G)$ we have*

$$\int_G \frac{\partial}{\partial x_j} f(x)\, dx = 0 \ \text{for}\ 1 \le j \le n. \qquad\qquad (20.25)$$

B. *For $F : G \to \mathbb{R}^n$ such that $F = (f_1, \dots, f_n)$ with $f_j \in C_0^1(G)$ it follows that*

$$\int_G (\operatorname{div} F)(x)\, dx = \int_G \left(\sum_{j=1}^n \frac{\partial f_j}{\partial x_j}(x) \right)\, dx = 0. \qquad\qquad (20.26)$$

C. *If $f \in C_0^1(G)$ and $g \in C^1(G)$ then $\frac{\partial f}{\partial x_j} g$ and $f \frac{\partial g}{\partial x_j}$, $1 \leq j \leq n$, are integrable and*

$$\int_G \left(\frac{\partial f}{\partial x_j}(x) \right) g(x) \, dx = - \int_G f(x) \frac{\partial g}{\partial x_j}(x) \, dx. \tag{20.27}$$

D. *For $f \in C_0^1(G)$ and $g \in C^2(G)$ the functions $\langle \operatorname{grad} f, \operatorname{grad} g \rangle$ and $f \Delta_n g$, where Δ_n is the Laplace operator in $\mathbb{R}^n$, are integrable and we find*

$$\int_G \langle \operatorname{grad} f(x), \operatorname{grad} g(x) \rangle \, dx = - \int_G f(x) \Delta_n g(x) \, dx. \tag{20.28}$$

Proof. The function $\frac{\partial f}{\partial x_j}$ in part A, the function $\operatorname{div} F$ in part B, the functions $\frac{\partial f}{\partial x_j} g$ and $f \frac{\partial g}{\partial x_j}$ in part C and the functions $\langle \operatorname{grad} f, \operatorname{grad} g \rangle$ and $f \Delta_n g$ in part D are all continuous functions with compact support in G. Hence they are all integrable. Let $K \subset \mathbb{R}^n$ be a compact cell such that $\overline{G} \subset \mathring{K}$. Then we can extend each function under consideration to a continuous function on K which vanish on ∂K without losing regularity properties. Indeed, let $h : G \to \mathbb{R}$ be any C^k-function, $k \geq 0$, with compact support in G and K a compact set such that $\overline{G} \subset \mathring{K}$. Then $\operatorname{dist}(\operatorname{supp} h, K^{\complement}) > 0$ and $h\big|_{(\operatorname{supp} h)^{\complement}} = 0$, implying that the extension

$$\tilde{h}(x) = \begin{cases} h(x), & x \in G \\ 0, & x \in K \backslash G \end{cases}$$

is a C^k-function too. Now parts A, B and C follows from Corollary 18.22. To prove part D we note that

$$\int_G \langle \operatorname{grad} f(x), \operatorname{grad} g(x) \rangle \, dx = \sum_{j=1}^n \int_G \frac{\partial f}{\partial x_j}(x) \frac{\partial g}{\partial x_j}(x) \, dx$$

and we apply (20.27) to the functions $\frac{\partial f}{\partial x_j} \in C_0^1(G)$ and $\frac{\partial g}{\partial x_j} \in C^1(G)$ to find

$$\sum_{j=1}^n \int_G \frac{\partial f}{\partial x_j}(x) \frac{\partial g}{\partial x_j}(x) \, dx = - \sum_{j=1}^n \int_G f(x) \frac{\partial^2 g(x)}{\partial x_j^2} \, dx = - \int_G f(x) \Delta_n g(x) \, dx.$$

$$\square$$

Remark 20.21. A. Obviously (20.27) allows iteration for higher order derivatives, i.e. for $f \in C_0^{|\alpha|}(G)$ and $g \in C^{|\alpha|}(G)$ with $\alpha \in \mathbb{N}_0^n$ being a multi-index, we have

$$\int_G (D^\alpha f)(x)g(x)\, dx = (-1)^{|\alpha|} \int_G f(x) D^\alpha g(x)\, dx,$$

where $D^\alpha u = \frac{\partial^{|\alpha|} u(x)}{\partial x_1^{\alpha_1} \cdots \partial x_n^{\alpha_n}}$.

B. We still cannot handle integration by parts for both functions being involved not vanishing in a neighbourhood of ∂G. We will approach this problem in Chapter 25 after first having introduced surface integrals.

In our discussion leading to volume integrals, i.e. Chapters 16 and 17 we investigated domains $G \subset \mathbb{R}^2$ obtained by deforming the boundary of a rectangle using some well-behaved functions, see Figure 16.7 or Figure 17.1 - 17.3. This leads to a class of domains called normal domains. Some authors use the phrases elementary domains and standard domains.

Definition 20.22. *A set $G \subset \mathbb{R}^n$ is called a **normal domain** with respect to the x_j-axis or in the direction e_j if for a compact Jordan measurable set $W \subset \mathbb{R}^{n-1}$ and two continuous functions $\varphi, \psi : W \to \mathbb{R}$, $\varphi \le \psi$,*

$$G = \left\{ x \in \mathbb{R}^n \,\middle|\, \varphi(y) \le x_j \le \psi(y) \text{ and } y = (x_1, \ldots, x_{j-1}, x_{j+1}, \ldots, x_n) \in W \right\} \tag{20.29}$$

holds.

Figure 16.7 shows a normal domain with respect to the x_2-axis, as does Figure 17.1 while Figure 17.3 shows a normal domain with respect to x_1-axis. A ball $B_r(x_0) \subset \mathbb{R}^n$ is a normal domain with respect to every direction e_j, $j = 1, \ldots, n$. For $n = 2$ we have, see Figure 20.2

$$
\begin{aligned}
B_r(x_0) &= \Big\{ (x_1, x_2) \in \mathbb{R}^2 \,\big|\, x_{02} - \sqrt{r^2 - (x_1 - x_{01})^2} =: \varphi_1(x_1) \le x_2 \le \varphi_2(x_1) := \\
&\qquad x_{02} + \sqrt{r^2 - (x_1 - x_{01})^2}, -r \le x_1 \le r \Big\} \\
&= \Big\{ (x_1, x_2) \in \mathbb{R}^2 \,\big|\, x_{01} - \sqrt{r^2 - (x_2 - x_{02})^2} =: \psi_1(x_2) \le x_1 \le \psi_2(x_2) := \\
&\qquad x_{01} + \sqrt{r^2 - (x_2 - x_{02})^2}, -r \le x_2 \le r \Big\}.
\end{aligned}
$$

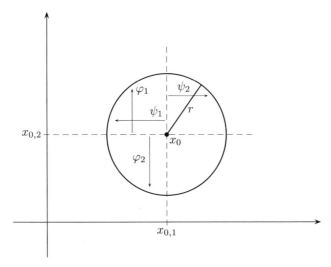

Figure 20.2

Lemma 20.23. *Every normal domain is a compact Jordan measurable set.*

Proof. Since W is compact in $\mathbb{R}^{n-1}$ we can find a non-degenerate compact cell $K := \bigtimes_{l \neq j}[a_j, b_j]$ such that $W \subset \mathring{K}$ and two continuous functions $\varphi, \psi : W \to \mathbb{R}$ such that G is given by (20.29). The functions φ and ψ attain their infimum and supremum on the compact set W and with $m := \min_{y \in W} \varphi(y)$, $M := \sup_{y \in W} \psi(y)$ we find for $\eta > 0$ that $G \subset \mathring{K}_\eta$ where

$$K_\eta := \bigtimes_{l=1}^{j-1}[a_l, b_l] \times [m - \eta, M + \eta] \times \bigtimes_{l=j+1}^{n}[a_l, b_l].$$

Thus G is bounded and by the compactness of W together with the continuity of φ and ψ it follows that G is closed, hence compact in $\mathbb{R}^n$. Moreover we have $\partial G \subset \partial W \cup \Gamma(\varphi) \cup \Gamma(\psi)$ where ∂W and the graphs of φ and ψ, $\Gamma(\varphi)$ and $\Gamma(\psi)$, have Jordan content zero. Thus G is a compact Jordan measurable set in $\mathbb{R}^n$. $\qquad\square$

The next theorem can be viewed as **Cavalieri's Principle for normal domains**.

Theorem 20.24. *Let $G \subset \mathbb{R}^n$ be a normal domain with respect to the x_j-axis given by (20.29). With $y = (x_1, \ldots, x_{j-1}, x_{j+1}, \ldots, x_n) \in W$ and $dy =*

$\mathrm{d}x_1 \cdots \mathrm{d}x_{j-1}\,\mathrm{d}x_{j+1}\cdots \mathrm{d}x_n$, *we have for* $f \in C(G)$

$$\int_G f(x)\,\mathrm{d}x = \int_W \left(\int_{\varphi(y)}^{\psi(y)} f(x)\,\mathrm{d}x_j \right) \mathrm{d}y \tag{20.30}$$

$$= \int_W \left(\int_{\varphi(x_1,\ldots,x_{j-1},x_{j+1},\ldots,x_n)}^{\psi(x_1,\ldots,x_{j-1},x_{j+1},\ldots,x_n)} \right. \tag{20.31}$$

$$\left. \times f(x_1,\ldots,x_{j-1},x_j,x_{j+1},\ldots,x_n)\,\mathrm{d}x_j \right) \mathrm{d}x_1 \cdots \mathrm{d}x_{j-1}\,\mathrm{d}x_{j+1}\cdots \mathrm{d}x_n.$$

Proof. (Following S. Hildebrandt [26]) As before let $m = \min_{y\in W} \varphi(y)$ and $M = \max_{y\in W} \psi(y)$ and define on $W \times [m, M]$ the function h by

$$h(y, z) = \begin{cases} f(x_1,\ldots,x_{j-1},z,x_{j+1},\ldots,x_n), & z \in [\varphi(y), \psi(y)] \\ h(y, \psi(y)), & z \in [\psi(y), M] \\ h(y, \varphi(y)), & z \in [m, \varphi(y)]. \end{cases}$$

This function is continuous on $W \times [m, M]$ as is the function

$$g(y, \xi, \eta) := \int_\xi^\eta h(y, z)\,\mathrm{d}z,$$

see Chapter 17. Hence

$$H(y) := g(y, \varphi(y), \psi(y)) = \int_{\varphi(y)}^{\psi(y)} h(y, z)\,\mathrm{d}z$$

is continuous on the compact, Jordan measurable set $W \subset \mathbb{R}^{n-1}$. Therefore H is integrable over W and the following holds

$$\int_W H(y)\,\mathrm{d}y = \int_W \left(\int_{\varphi(y)}^{\psi(y)} h(y, z)\,\mathrm{d}z \right) \mathrm{d}y = \int_W \left(\int_{\varphi(y)}^{\psi(y)} f(x)\,\mathrm{d}x_j \right) \mathrm{d}y.$$

Now the result follows along the lines of the proof of Theorem 18.19. $\square$

Example 20.25. The volume of a normal domain G is given by

$$\mathrm{vol}_n(G) = \int_G 1\,\mathrm{d}x = \int_W \left(\int_{\varphi(y)}^{\psi(y)} 1\,\mathrm{d}x_j \right) \mathrm{d}y = \int_W (\psi(y) - \varphi(y))\,\mathrm{d}y,$$

a formula we have already seen in Chapters 16 and 17.

Before turning to more interesting examples, we want to point out that we can iterate the above result, i.e. Theorem 20.24. In our definition of a normal domain $G \subset \mathbb{R}^n$, say with respect to the x_j-axis, W was just a compact Jordan measurable set in $\mathbb{R}^{n-1}$. Thus it could be itself a normal domain, say with respect to the x_l-axis, $l \neq j$. With $U \subset \mathbb{R}^{n-2}$, $n > 2$, and $\sigma, \tau : U \to \mathbb{R}$, $\sigma(z) \leq \tau(z)$, where U is a compact Jordan measurable set in $\mathbb{R}^{n-2}$ and σ, τ are continuous functions, $z = (x_1, \ldots, x_{k-1}, x_{k+1}, \ldots, x_{j-1}, x_{j+1}, \ldots, x_n)$ and $\mathrm{d}z = \mathrm{d}x_1 \cdots \mathrm{d}x_{k-1} \, \mathrm{d}x_{k+1} \cdots \mathrm{d}_{j-1} \, \mathrm{d}x_{j+1} \cdots \mathrm{d}x_n$ (if $k < j$) we have

$$W = \left\{ y \in \mathbb{R}^{n-1} \,\middle|\, \sigma(z) \leq y_l \leq \tau(z), \ z \in U \right\}$$

and for $f \in C(G)$ it follows that

$$\int_G f(x) \, \mathrm{d}x = \int_U \left(\int_{\sigma(z)}^{\tau(z)} \left(\int_{\varphi(y)}^{\psi(y)} f(x) \, \mathrm{d}x_j \right) \mathrm{d}x_k \right) \mathrm{d}z. \tag{20.32}$$

Thus for certain "iterated" normal domains we find after a proper renumeration of the variables

$$\int_G f(x) \, \mathrm{d}x = \int_{a_1}^{b_1} \left(\int_{\varphi_1(x_1)}^{\psi_1(x_1)} \left(\cdots \left(\int_{\varphi_{n-1}(x_1, \cdots, x_{n-1})}^{\psi_{n-1}(x_1, \cdots, x_{n-1})} f(x_1, \ldots, x_n) \, \mathrm{d}x_n \right) \cdots \right) \mathrm{d}x_2 \right) \mathrm{d}x_1. \tag{20.33}$$

Example 20.26. In $\mathbb{R}^2$ we consider the simplex $S(2) = \overline{\mathrm{conv}}(0, e_1, e_2)$ which is the closed triangle with vertices $0 \in \mathbb{R}^2$, e_1 and e_2. This is a normal domain in $\mathbb{R}^2$ since

$$S(2) = \left\{ (x_1, x_2) \in \mathbb{R}^2 \,\middle|\, 0 \leq x_2 \leq 1 - x_1, \ x_1 \in [0, 1] \right\}.$$

The volume, better the area, of this simplex is

$$\mathrm{vol}_2(S(2)) = \int_{S(2)} 1 \, \mathrm{d}x = \int_0^1 \left(\int_0^{1-x_1} 1 \, \mathrm{d}x_2 \right) \mathrm{d}x_1 = \int_0^1 (1 - x_1) \, \mathrm{d}x_1 = \frac{1}{2}.$$

The simplex $S(3) := \overline{\mathrm{conv}}(0, e_1, e_2, e_3)$ in $\mathbb{R}^3$ is an "iterated" normal domain

$$S(3) = \left\{ (x_1, x_2, x_3) \in \mathbb{R}^3 \,\middle|\, 0 \leq x_3 \leq 1 - x_1 - x_2, \ (x_1, x_2) \in S(2) \right\}$$
$$= \left\{ (x_1, x_2, x_3) \in \mathbb{R}^3 \,\middle|\, 0 \leq x_3 \leq 1 - x_1 - x_2, \ 0 \leq x_2 \leq 1 - x_1, \ x_1 \in [0, 1] \right\}$$

and we find for the volume of $S(3)$, $z = (x_1, x_2)$,

$$\text{vol}_3(S(3)) = \int_{S(2)} \left(\int_0^{1-x_1-x_2} 1 \, dx_3 \right) dz$$

$$= \int_0^1 \left(\int_0^{1-x_1} \left(\int_0^{1-x_1-x_2} 1 \, dx_3 \right) dx_2 \right) dx_1$$

$$= \int_0^1 \left(\int_0^{1-x_1} (1 - x_1 - x_2) \, dx_2 \right) dx_1$$

$$= \int_0^1 \left((1 - x_1)(1 - x_1) - \frac{(1 - x_1)^2}{2} \right) dx_1$$

$$= \frac{1}{2} \int_0^1 (1 - x_1)^2 \, dx_1 = \frac{1}{6}.$$

The next example is a "must have done" calculation for every mathematics student, namely to determine the volume of a closed ball in $\mathbb{R}^3$. This example does not only "reproduce" the well-known classical result, it is also necessary to drive this result in a consistent manner within the theory. Since volumes are translation invariant we can restrict our consideration to $\overline{B_R(0)} \subset \mathbb{R}^3$, $R > 0$, which is a further "iterated" normal domain.

Example 20.27. We want to calculate the volume of the ball $\overline{B_R(0)} \subset \mathbb{R}^3$. With

$$W = \left\{ (x, y) \in \mathbb{R}^2 \,\middle|\, x^2 + y^2 \leq R^2 \right\}$$
$$= \left\{ (x, y) \in \mathbb{R}^2 \,\middle|\, -\sqrt{R^2 - x^2} \leq y \leq \sqrt{R^2 - x^2}, x \in [-R, R] \right\}$$

we find

$$\overline{B_R(0)} = \left\{ (x, y, z) \in \mathbb{R}^3 \,\middle|\, -\sqrt{R^2 - x^2 - y^2} \leq z \leq \sqrt{R^2 - x^2 - y^2}, (x, y) \in W \right\}$$
$$= \left\{ (x, y, z) \in \mathbb{R}^3 \,\middle|\, -\sqrt{R^2 - x^2 - y^2} \leq z \leq \sqrt{R^2 - x^2 - y^2}, -\sqrt{R^2 - x^2} \right.$$
$$\left. \leq y \leq \sqrt{R^2 - x^2}, x \in [-R, R] \right\}.$$

Now we get

$$\mathrm{vol}_3(\overline{B_r(0)}) = \int_{B_R(0)} 1 \,\mathrm{d}z\,\mathrm{d}y\,\mathrm{d}x = \int_W \left(\int_{-\sqrt{R^2-x^2-y^2}}^{\sqrt{R^2-x^2-y^2}} 1\,\mathrm{d}z \right) \mathrm{d}y\,\mathrm{d}x$$

$$= 2 \int_W \sqrt{R^2 - x^2 - y^2}\,\mathrm{d}y\mathrm{d}x$$

$$= 2 \int_{-R}^{R} \left(\int_{-\sqrt{R^2-x^2}}^{\sqrt{R^2-x^2}} \sqrt{R^2 - x^2 - y^2}\,\mathrm{d}y \right) \mathrm{d}x.$$

Note that

$$\int_{-\rho}^{\rho} \sqrt{\rho^2 - y^2}\,\mathrm{d}y = \frac{1}{2}y\sqrt{\rho^2 - y^2} + \frac{\rho^2}{2}\arcsin\frac{y}{\rho}\Big|_{-\rho}^{\rho}$$

$$= \frac{\rho^2}{2}\arcsin 1 - \frac{\rho^2}{2}\arcsin(-1)$$

$$= \frac{\rho^2}{2}\left(\frac{\pi}{2} + \frac{\pi}{2}\right) = \frac{\rho^2 \pi}{2}.$$

Hence, with $\rho = \sqrt{R^2 - x^2}$ we have

$$\int_{-\sqrt{R^2-x^2}}^{\sqrt{R^2-x^2}} \sqrt{R^2 - x^2 - y^2}\,\mathrm{d}y = \frac{\pi}{2}(R^2 - x^2)$$

implying that

$$\mathrm{vol}_3(\overline{B_R(0)}) = \pi \int_{-R}^{R} (R^2 - x^2)\,\mathrm{d}x = \pi \left(R^2 x - \frac{1}{3}x^3 \right)\Big|_{-R}^{R}$$

$$= \frac{4\pi}{3}R^3,$$

i.e. as expected we obtain the classical and well-known result.

In the next example we once again calculate the volume of a body that was treated in ancient times.

Example 20.28. Let $G \subset \mathbb{R}^3$ be the compact, Jordan measurable set bounded by the (circular) cone $z^2 = \frac{h^2}{R^2}(x^2 + y^2)$, $R > 0$, $h > 0$, and the plane $z = h$, see Figure 20.3

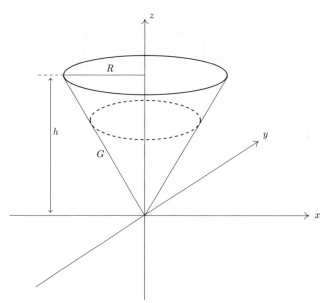

Figure 20.3

We want to find the integral of $f : G \to \mathbb{R}$ with f depending only on z, $(x, y, z) \in G$, i.e. we want to evaluate $\int_G f(z)\, dx\, dy\, dz$. The set is a normal domain with respect to the z-axis given by

$$G = \left\{ (x, y, z) \in \mathbb{R}^3 \;\middle|\; \frac{h}{R}\sqrt{x^2 - y^2} \leq z \leq h, \; (x, y) \in \overline{B_R(0)} \right\}$$

where $B_R(0) \subset \mathbb{R}^2$. Therefore we find

$$\int_G f(z)\, dx\, dy\, dz = \int_{B_R(0)} \left(\int_{\frac{h}{R}\sqrt{x^2 + y^2}}^{h} f(z)\, dz \right) dx\, dy$$

$$= \int_{B_R(0)} \left(F(h) - F\left(\frac{h}{R}\sqrt{x^2 + y^2} \right) \right) dx\, dy$$

$$= \pi R^2 F(h) - \int_{B_R(0)} F\left(\frac{h}{R}\sqrt{x^2 + y^2} \right) dx\, dy,$$

where F is a primitive of f and we have used that $\int_{B_R(0)} 1\, dz = \pi R^2$, i.e. the area of a disc with radius R. Of course it now depends on F whether we can integrate further. If we choose $f(z) = 1$ we will of course calculate the volume of G, i.e.

$$\text{vol}_3(G) = \int_G 1 \, dx \, dy \, dz = \pi R^2 h - \int_{\overline{B_R(0)}} \frac{h}{R} \sqrt{x^2 + y^2} \, dx \, dy.$$

Since $\overline{B_R(0)}$ is a two-dimensional normal domain given by

$$\overline{B_R(0)} = \left\{ (x,y) \in \mathbb{R}^2 \mid -\sqrt{R^2 - x^2} \leq y \leq \sqrt{R^2 - x^2}, \ x \in [-R, R] \right\}$$

we obtain

$$\int_{\overline{B_R(0)}} \sqrt{x^2 + y^2} \, dx \, dy = \int_{-R}^{R} \left(\int_{-\sqrt{R^2 - x^2}}^{\sqrt{R^2 - x^2}} \sqrt{x^2 + y^2} \, dy \right) dx$$

$$= 4 \int_0^R \left(\int_0^{\sqrt{R^2 - x^2}} \sqrt{x^2 + y^2} \, dy \right) dx.$$

We do not continue this calculation since in the next chapter we will learn how to transform the integral over $\overline{B_R(0)}$ to obtain (compare with Theorem 21.8)

$$\int_{\overline{B_R(0)}} \sqrt{x^2 + y^2} \, dx \, dy = \int_0^R \int_0^{2\pi} r^2 \, d\varphi \, dr = \frac{2\pi R^3}{3}$$

which yields

$$\text{vol}_3(G) = \pi R^2 h - \frac{2\pi R^3}{3} \left(\frac{h}{R} \right) = \frac{\pi R^2 h}{3}.$$

Looking at our examples two remarks are in order. First of all, our aim is to reduce a volume integral to an iterated integral (where certain initial - or end points may depend on some variables we integrate with respect to later on) and then to use the fundamental theorem of calculus and rules to evaluate one-dimensional integrals. Thus the actual integration process relies on the one-dimensional results.

Secondly, the reduction process is essentially a geometric problem. Given a compact Jordan measurable set $G \subset \mathbb{R}^n$ we try to represent it as a normal or "iterated" normal domain. To do so we depend on our knowledge in geometry.

We have seen, for example by looking at a ball $\overline{B_R(0)} \subset \mathbb{R}^n$, that a set can be a normal domain with respect to different coordinate axes. This leads to a further result on interchanging the order of integration.

Let $G \subset \mathbb{R}^n$ be a normal domain with respect to the x_j-axis and with respect to the x_k-axis, $j \neq k$. If we set $\tilde{x}_j = (x_1, \ldots, x_{j-1}, x_{j+1}, \ldots, x_n)$ then we find with suitable compact, Jordan measurable sets $V_j, V_k \subset \mathbb{R}^{n-1}$ and certain continuous functions $\psi_l : V_l \to \mathbb{R}$ and $\varphi_l : V_l \to \mathbb{R}$, $\psi_l(\tilde{x}_l) \leq \varphi_l(\tilde{x}_l)$, $l \in \{j, k\}$, that

$$
G = \left\{ x \in \mathbb{R}^n \,\middle|\, \psi_j(\tilde{x}_j) \leq x_j \leq \varphi_j(\tilde{x}_j),\ \tilde{x}_j \in V_j \right\}
$$
$$
= \left\{ x \in \mathbb{R}^n \,\middle|\, \psi_k(\tilde{x}_k) \leq x_k \leq \varphi_k(\tilde{x}_k),\ \tilde{x}_k \in V_k \right\}.
$$

For a continuous function $f : G \to \mathbb{R}$ we now find

$$
\int_G f(x)\, dx = \int_{V_j} \left(\int_{\psi_j(\tilde{x}_j)}^{\varphi_j(\tilde{x}_j)} f(x)\, dx_j \right) d\tilde{x}_j = \int_{V_k} \left(\int_{\psi_k(\tilde{x}_k)}^{\varphi_k(\tilde{x}_k)} f(x)\, dx_k \right) d\tilde{x}_k \tag{20.34}
$$

with $d\tilde{x}_j = dx_1 \cdots dx_{j-1}\, dx_{j+1} \cdots dx_n$.

In particular for $n = 2$ we have with $(x_1, x_2) = (x, y)$, $V_1 = [a, b]$ and $V_2 = [c, d]$

$$
\int_a^b \left(\int_{\psi_1(y)}^{\varphi_1(y)} f(x, y)\, dx \right) dy = \int_c^d \left(\int_{\psi_2(x)}^{\varphi_2(x)} f(x, y)\, dy \right) dx. \tag{20.35}
$$

Moreover, for $n = 3$ and $G \subset \mathbb{R}^3$ being an "iterated" normal domain with respect to both the x-axis and the y-axis we find with a suitable interval $[a, b]$

$$
G = \left\{ (x, y, z) \in \mathbb{R}^3 \,\middle|\, \psi(y, z) \leq x \leq \varphi(y, z),\ \xi(z) \leq y \leq \eta(z),\ z \in [a, b] \right\}
$$
$$
= \left\{ (x, y, z) \in \mathbb{R}^3 \,\middle|\, \alpha(x, z) \leq y \leq \beta(x, z),\ \gamma(z) \leq x \leq \delta(z),\ z \in [a, b] \right\}
$$

and for $f \in C(G)$

$$
\int_G f(x, y, z)\, dx\, dy\, dz = \int_a^b \left(\int_{\xi(z)}^{\eta(z)} \left(\int_{\psi(y,z)}^{\varphi(y,z)} f(x, y, z)\, dx \right) dy \right) dz
$$
$$
= \int_a^b \left(\int_{\gamma(z)}^{\delta(z)} \left(\int_{\alpha(x,z)}^{\beta(x,z)} f(x, y, z)\, dy \right) dx \right) dz.
$$

Of course, generalisations to the case where G is an "iterated" normal domain with respect to all three axes are possible as to the n-dimensional case.

Example 20.29. Let $G \subset \mathbb{R}^2$ be the triangle with vertices $0 \in \mathbb{R}^2$, $2e_1$ and e_2, see Figure 20.4.

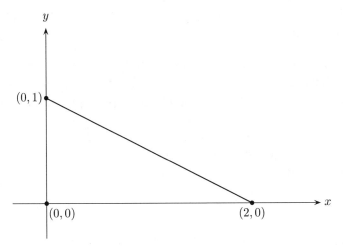

y

$(0,1)$

$(0,0)$

$(2,0)$

x

Figure 20.4

For a continuous function $f : G \to \mathbb{R}$ we want to find $\int_G f(x) \, dx$. Since

$$G = \left\{ (x_1, x_2) \in \mathbb{R}^2 \,\Big|\, 0 \le x_2 \le 1 - \frac{1}{2} x_1, \ x_1 \in [0, 2] \right\}$$
$$= \left\{ (x_1, x_2) \in \mathbb{R}^2 \,\big|\, 0 \le x_1 \le 2 - 2x_2, \ x_2 \in [0, 1] \right\}$$

we find

$$\int_0^1 \left(\int_0^{1 - \frac{1}{2} x_1} f(x_1, x_2) \, dx_2 \right) dx_1 = \int_0^1 \left(\int_0^{2 - 2x_2} f(x_1, x_2) \, dx_1 \right) dx_2.$$

We end this chapter by noting that the results extend to the situation of mappings $F : G \to \mathbb{R}^m$, $F = (f_1, \ldots, f_m)$, and each $f_j : G \to \mathbb{R}$, $j = 1, \ldots, m$, is Riemann integrable. We set

$$\int_G F(x) \, dx := \left(\int_G f_1(x) \, dx, \ldots, \int_G f_m(x) \, dx \right),$$

for the integral of F over G.

Problems

1. Let $G \subset \mathbb{R}^n$ be a bounded Jordan measurable set and $f, g : G \to \mathbb{R}$ be bounded Riemann integrable functions. Prove the triangle inequality and the Cauchy-Schwarz inequality.

2. Prove that for a Riemann integrable function $f : G \to \mathbb{R}$, where $G \subset \mathbb{R}^n$ is a Jordan measurable set, we cannot deduce in general that $\left(\int_G |f(x)|^p dx \right)^{\frac{1}{p}} = 0, p \geq 1$, implies that $f(x) = 0$ for all $x \in G$.

3. For an arbitrarily often differentiable function $u : \mathbb{R}^n \to \mathbb{R}$ and multi-indices $\alpha, \beta \in \mathbb{N}_0^n$ we define $p_{\alpha,\beta}(u) := \sup_{x \in \mathbb{R}^n} |x^\beta \partial^\alpha u(x)|$.
   The **Schwartz space** is by definition the set

   $$\mathcal{S}(\mathbb{R}^n) := \{u \in C^\infty(\mathbb{R}^n) | p_{\alpha,\beta}(u) < \infty \ \text{ for all } \ \alpha, \beta \in \mathbb{N}_0^n\}.$$

   Prove that $p_{\alpha,\beta}$ is a semi-norm on $\mathcal{S}(\mathbb{R}^n)$ and that $\mathcal{S}(\mathbb{R}^n)$ is a real vector space. Show that $u \in C_0^\infty(\mathbb{R}^n)$ belongs to $\mathcal{S}(\mathbb{R}^n)$ and that $g : \mathbb{R}^n \to \mathbb{R}$, $g(x) = e^{-\|x\|^2}$, is an element in $\mathcal{S}(\mathbb{R}^n)$ not belonging to $C_0^\infty(\mathbb{R}^n)$. (Recall that $C_0^\infty(\mathbb{R}^n) = C^\infty(\mathbb{R}^n) \cap C_0(\mathbb{R}^n)$, $C_0(\mathbb{R}^n)$ is the space of all continuous functions with compact support.)

4. Let $G \subset \mathbb{R}^n$ be an open Jordan measurable set and $f : G \to \mathbb{R}$ a continuous function. For every $x_0 \in G$ and $r > 0$ sufficiently small such that $B_r(x_0) \subset G$, use the mean value theorem in the form of (20.21) to prove that

   $$\lim_{r \to 0} \frac{1}{J^{(n)}\left(\overline{B_r(x_0)} \right)} \int_{\overline{B_r(x_0)}} f(x)dx = f(x_0).$$

5. Extend Lemma 20.17 to finitely many mutually disjoint Jordan measurable sets $G_1, \ldots, G_N \subset \mathbb{R}^n$.

6. Let $G \subset \mathbb{R}^n$ be a Jordan measurable set and $f : G \to \mathbb{R}$ a function. We call f in G piecewise continuous if we can decompose G into finitely many Jordan measurable sets $G_1, \ldots, G_N$ such that $\mathring{G}_j \cap \mathring{G}_k = \emptyset$ for $j \neq k$ and $F|_{G_j}$ is continuous for $j = 1, \ldots, N$. Prove that a piecewise continuous function on G is integrable.
   Hint: use the result of Problem 5.

7. Let $f \in C_0^1(G) \cap C^2(G)$ be a harmonic function on the bounded convex open Jordan measurable set $G \subset \mathbb{R}^n$. Use (20.28) to prove that f must be identically zero.

8. Let $G \subset \mathbb{R}^n$ be a normal domain with respect to the direction e_j. Let $a \in \mathbb{R}^n$. Is the translated set $a + G$ again a normal domain in the direction of e_j?

9. Consider the set $G \subset \mathbb{R}^2$ given in the figure below. Provide a description of G as a normal domain with respect to the direction e_1 and e_2.

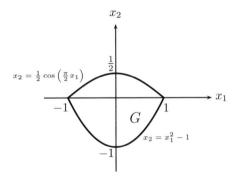

10. Let $G := \{x = (x_1, x_2) \in \mathbb{R}^2 | 0 \le x_2 \le \sqrt{x_1^2 + 2x_1}, 0 \le x_1 \le 2\}$. Use Cavalieri's principle in the form of Theorem 20.24 to find $\int_G f(x)dx$ where $f : G \to \mathbb{R}$ is the function $f(x_1, x_2) = x_1 x_2^3$.

11. Change the order of integration in the following integrals (f, g, h are assumed to be continuous in a neighbourhood of the domain of integration):

    a) $\int_0^1 \left(\int_y^1 f(x, y)dx \right) dy$;

    b) $\int_1^2 \left(\int_1^{q^2} g(p, q)dp \right) dq$;

    c) $\int_0^{\frac{\pi}{2}} \left(\int_0^{\sin r} h(t, r)dt \right) dr$.

12. By mathematical induction prove for $k, l \in \mathbb{N}_0$ that

$$\int_0^1 x^k (1 - x)^l dx = \frac{k! l!}{(k + l + 1)!}.$$

Now find the integral

$$\int_G x^n y^n dx dy$$

where $G = \{(x, y) \in \mathbb{R}^2 | x \geq 0, y \geq 0 \text{ and } x + y \leq 1\}$.

21 The Transformation Theorem: Result and Applications

One of the most powerful tools to evaluate integrals on the line is the change of variable formula or integration by substituting the variable, see Theorem I.13.12:

Let $g : [a, b] \to \mathbb{R}$ be a continuous function and let $\varphi : [\alpha, \beta] \to \mathbb{R}$ be a strictly monotone and continuously differentiable function. Suppose that $\varphi(\alpha) = a$ and $\varphi(\beta) = b$ then the following holds

$$\int_a^b g(x)\,\mathrm{d}x = \int_{\varphi(\alpha)}^{\varphi(\beta)} g(x)\,\mathrm{d}x = \int_\alpha^\beta g(\varphi(t))\varphi'(t)\,\mathrm{d}t. \qquad (21.1)$$

A few observations are helpful. The function g only needs to be continuous, but the change of variable φ is a C^1-function with C^1-inverse if $\varphi'(t) \neq 0$ for all $t \in [\alpha, \beta]$. As a consequence φ and φ^{-1} map intervals onto intervals, hence integrable sets onto integrable sets. The latter remark is important when we try to extend (21.1) to higher dimensions. If $g : G \to \mathbb{R}$ is continuous on a compact Jordan measurable set $G \subset \mathbb{R}^n$ and $\varphi : G' \to G$ is bijective, i.e. $\varphi^{-1}(G) = G'$, we need to assume that G' is Jordan measurable too. Jordan measurability of a set $G \subset \mathbb{R}^n$ is defined by a property of ∂G. Thus before we can start to extend (21.1) to higher dimensions we need to find a class of transformations mapping Jordan measurable sets in $\mathbb{R}^n$ onto Jordan measurable sets in $\mathbb{R}^n$. In particular if ∂G has Lebesgue measure zero or Jordan content zero the boundary of the image of G should have Lebesgue measure zero too.

Note that when looking at the graph of a continuous function we are dealing with a different situation. If $f : K \to \mathbb{R}$ is continuous and $K \subset \mathbb{R}^n$, say compact and $\mathring{K} \neq \emptyset$, then $f(K) \subset \mathbb{R}$ and $\Gamma(f) \subset \mathbb{R}^{n+1}$, i.e. K, $f(K)$ and $\Gamma(f)$ are for $n > 1$ subsets in spaces of different dimensions. In particular $\Gamma(f)$ can be interpreted as "lifting" $K \subset \mathbb{R}^n$ to a subset in $\mathbb{R}^{n+1}$. While we cannot make a statement at the moment about the dimensions of $\Gamma(f)$, our intuition tells us that the smallness of $\Gamma(f)$ in $\mathbb{R}^{n+1}$ should be related to the fact that $\Gamma(f)$ is of "lower" dimensions than $n+1$ and hence could be covered by very "thin" cells.

It turns out that for a start we can take diffeomorphisms as a rich class of transformations fit for our purpose.

Definition 21.1. *Let U and V be two open subsets of $\mathbb{R}^n$. We call φ : $U \to V$ a C^k-**diffeomorphism**, $k \in \mathbb{N} \cup \{\infty\}$, if φ is a bijective mapping belonging to the class C^k with inverse φ^{-1} belonging to the class C^k.*

Clearly the composition of two C^k-diffeomorphisms is a C^k-diffeomorphism. Note that a C^k-diffeomorphism is in particular a **homeomorphism**, i.e. a bijective continuous mapping with continuous inverse. Thus a C^k-diffeomorphism is in particular an open mapping, compare with Definition 3.39.

We are now interested in images (or pre-images) of subsets $G \subset U$ under a diffeomorphism $\varphi : U \to V$, $U, V \subset \mathbb{R}^n$ open. The first result deals with the image of ∂G.

Lemma 21.2. *A homeomorphism $\varphi : U \to V$ between two open sets U and V in $\mathbb{R}^n$ maps the boundary ∂G of G, $\overline{G} \subset U$, onto the boundary of $\varphi(G)$, i.e.*

$$\varphi(\partial G) = \partial \varphi(G). \tag{21.2}$$

Proof. Let $x \in \partial G$. For $n \geq k$ we have that $B_{\frac{1}{n}}(x) \subset U$ and $\varphi\left(B_{\frac{1}{n}}(x)\right)$ is an open neighbourhood of $\varphi(x)$ with points $y_n \in \varphi\left(B_{\frac{1}{n}}(x) \cap \overset{\circ}{G}\right) \subset \varphi(\overset{\circ}{G}) \subset \varphi(G)$ and $z_n \in \varphi\left(B_{\frac{1}{n}}(x) \cap G^{\complement}\right) \subset \varphi(G^{\complement}) \subset \varphi(G)^{\complement}$. Thus in every neighbourhood of $\varphi(x)$ we find points of $\varphi(G)$ and $\varphi(G)^{\complement}$, i.e. $\varphi(x) \in \partial \varphi(G)$ or $\varphi(\partial G) \subset \partial \varphi(G)$. We may apply the same argument to derive that $\varphi^{-1}(\partial(\varphi(G))) \subset \partial \varphi^{-1}(\varphi(G)) = \partial G$, which yields with the case discussed before that

$$\varphi(\partial G) \subset \partial \varphi(G) = \varphi(\varphi^{-1}(\partial \varphi(G))) \subset \varphi(\partial G)$$

or (21.2). $\qquad \square$

Usually, when working in $\mathbb{R}^n$ we use the Euclidean norm $\|.\|_2$, however in the following it is sometimes more convenient to use the sup-norm $\|.\|_\infty$ in $\mathbb{R}^n$. Since $\|x\|_\infty \leq \|x\|_2 \leq \sqrt{n}\,\|x\|_\infty$, all statements about continuity or convergence are unchanged when switching from $\|.\|_2$ to $\|.\|_\infty$ or back. The advantage we want to take is that the "balls" with respect to $\|.\|_\infty$, i.e. the sets $B_r^{(\infty)}(x) := \{y \in \mathbb{R}^n \mid \|x - y\|_\infty < r\}$ are open cells and their closures are non-degenerate compact cells. Thus working with $\|.\|_\infty$ allows us to use easily (open) coverings with cells.

Lemma 21.3. *Let $F \subset \mathbb{R}^n$ be a set of Jordan content zero and $\varphi : F \to \mathbb{R}^n$ a Lipschitz continuous mapping, i.e.*

$$\|\varphi(x) - \varphi(y)\|_\infty \leq L \|x - y\|_\infty \tag{21.3}$$

for all $x, y \in F$. Then $\varphi(F)$ is of Jordan content zero too.

Remark 21.4. Note that the estimate $\|\varphi(x) - \varphi(y)\|_2 \leq \kappa \|x - y\|_2$ implies $\|\varphi(x) - \varphi(y)\|_\infty \leq \sqrt{n}\kappa \|x - y\|_\infty$, and (21.3) gives

$$\|\varphi(x) - \varphi(y)\|_2 \leq \sqrt{n} \|\varphi(x) - \varphi(y)\|_\infty \leq L\sqrt{n} \|\varphi(x) - \varphi(y)\|_\infty \leq L\sqrt{n} \|x - y\|_2.$$

Thus it does not matter with respect to which norm the Lipschitz continuity is required.

Proof of Lemma 21.3. Since F is a set of Jordan content zero, given $\epsilon > 0$ we can find finitely many compact cells $(K_j)_{j=1,\dots,N}$ such that $F \subset \bigcup_{j=1}^{N} K_j$ and $\sum_{j=1}^{N} \mathrm{vol}_n(K_j) < \epsilon$. We may assume that the cells have vertices with rational coordinates. This allows us to cover F by a finite number of closed cubes having the same side length and $\sum_{l=1}^{M} \mathrm{vol}_n(W_l) < \epsilon$. The image of W_l under φ is by (21.3) contained in a cube V_l with $\mathrm{vol}_n(V_l) = L^n \mathrm{vol}(W_l)$. Moreover, the family $(V_l)_{l \in \mathbb{N}}$ is a cover of $\varphi(F)$ and

$$\sum_{l=1}^{M} \mathrm{vol}_n(V_l) = L^n \sum_{l=1}^{M} \mathrm{vol}_n(W_l) < L^n \epsilon \tag{21.4}$$

holds, implying that $\varphi(F)$ has Jordan content zero. $\quad\square$

Corollary 21.5. *Let $\varphi : U \to V$ be a homeomorphism between the open sets $U, V \subset \mathbb{R}^n$ and let $G \subset U$ be a bounded set such that $\overline{G} \subset U$. If G is Jordan measurable and $\varphi|_{\overline{G}}$ is Lipschitz continuous, then $\varphi(G)$ is Jordan measurable too.*

Proof. By the converse statement to Theorem 19.25, see S. Hildebrandt [26], since G is Jordan measurable ∂G is a Lebesgue null set, and since $\overline{G}$ is compact, ∂G is compact, hence by Corollary 19.9 a set of Jordan content zero. Now Lemma 21.2 applied to G yields that $\partial \varphi(G) = \varphi(\partial G)$ and by Lemma 21.3 it follows that $\partial \varphi(G)$ has Jordan content zero, hence $\varphi(G)$ is Jordan measurable. $\quad\square$

Remark 21.6. Note that Corollary 21.5 applies in particular for G being of the type C^k, $k \geq 0$.

Corollary 21.7. *Let* $\varphi : U \to V$ *be a* C^1*-diffeomorphism between the two open sets* $U, V \subset \mathbb{R}^n$. *Further let* $G \subset U$ *be a bounded set such that* $\overline{G} \subset U$. *If* G *is Jordan measurable then* $\varphi(G)$ *is Jordan measurable too.*

Proof. In the case where $\overline{G}$ is convex (or any set for which the mean value theorem holds) we find for $\varphi = (\varphi^1, \ldots, \varphi^n)$ that

$$|\varphi^j(x) - \varphi^j(y)| \leq \sqrt{n} \sup_{y \in \overline{G}} \left\| \operatorname{grad} \varphi^j(y) \right\|_2 \|x - y\|_\infty ,$$

from which we deduce the Lipschitz continuity of $\varphi|_{\overline{G}}$. For the general case we return to the proof of Lemma 21.3 and note that the crucial application of the Lipschitz continuity of φ is the estimate $\operatorname{vol}_n(V_l) \leq L^n \operatorname{vol}_n(W_l)$, where $\varphi(W_l) \subset V_l$. But $\varphi|_{W_l}$ is Lipschitz continuous with Lipschitz constant $L_l = \sqrt{n} \max_{1 \leq j \leq n} \sup_{y \in W_l} \|\operatorname{grad} \varphi^j(y)\|$ and we can derive (21.4) with L_n being replaced by $(\max_{1 \leq l \leq M} L_l)^n$. $\square$

The **transformation theorem** or **substitution formula** which we aim for is summarised in

Theorem 21.8. *Let* $\varphi : U \to V$ *be a* C^1*-diffeomorphism between the open sets* $U, V \subset \mathbb{R}^n$ *and let* $G \subset \mathbb{R}^n$ *a bounded Jordan measurable set such that* $\overline{G} \subset U$. *Further let* $f : \varphi(G) \to \mathbb{R}$ *a continuous function. Then the following holds*

$$\int_{\varphi(G)} f(y) \, \mathrm{d}y = \int_G (f \circ \varphi)(x) |\det J_\varphi(x)| \, \mathrm{d}x. \tag{21.5}$$

First let us note that in formula (21.5) every term is well defined. By Corollary 21.7 the set $\varphi(G)$ is a bounded Jordan measurable set in $\mathbb{R}^n$. Further, for $f \in C^0\left(\overline{\varphi(G)}\right)$ it follows that $f \circ \varphi \in C^0\left(\overline{G}\right)$ and since $\overline{G}$ as well as $\overline{\varphi(G)}$ are compact both f and $f \circ \varphi$ are integrable. In addition, since φ is of class C^1, $|\det J_\varphi|$ is on $\overline{G}$ a bounded and continuous function, hence both integrals in (21.5) exist.

Since f is continuous on $\overline{\varphi(G)}$ and $f \circ \varphi$ as well as $\det J_\varphi$ are continuous on $\overline{G}$, and in addition ∂G and $\partial \varphi(G)$ are of Jordan content zero, then (21.5) holds with $\varphi(G)$ and G replaced by $\overline{\varphi(G)}$ and $\overline{G}$, respectively. For practical purposes the following extensions of the transformation theorem is very useful:

Corollary 21.9. *Formula* (21.5) *still holds if for some* $F \subset \overline{G}$ *with Jordan content zero* φ *is only a* C^1-*diffeomorphism on* $U \backslash F$ *or* $\det J_\varphi \neq 0$ *only on* $\overline{G} \backslash F$.

Proof. First we note that under the assumptions of Theorem 21.8 formula (21.5) holds with G replaced by $\overline{G}$ and $\varphi(G)$ by $\varphi(\overline{G}) = \overline{\varphi(G)}$. Since F has Jordan content zero given $\epsilon > 0$, we can find finitely many open cells $K_1, \ldots, K_N \subset U$ such that $F \subset \bigcup_{j=1}^N K_j =: \tilde{K}$ and $\sum_{j=1}^N \operatorname{vol}_n(K_j) < \epsilon$. The set $H := \overline{G} \backslash \tilde{K}$ is compact and consequently we may apply (21.5) to H:

$$\int_{\varphi(H)} f(y)\, \mathrm{d}y = \int_H (f \circ \varphi)(x) |\det J_\varphi(x)|\, \mathrm{d}x.$$

Clearly $\overline{G} \backslash (U \backslash \tilde{K}) \subset \tilde{K}$ and therefore $J^{(n)}\left(\overline{G} \backslash (U \backslash \tilde{K})\right) < \epsilon$ implying that $J^{(n)}\left(\varphi(\overline{G}) \backslash \varphi(U \backslash \tilde{K})\right) < c_0 \epsilon$ where $\max_{1 \leq j \leq n} \sup_{x \in \overline{G}} \left\| \operatorname{grad} \varphi^{(j)}(x) \right\|_\infty \leq c_0$. Finally we use the set additivity of the integral to find

$$\left| \int_{\varphi(\overline{G})} f(y)\, \mathrm{d}y - \int_{\overline{G}} (f \circ \varphi)(x) |\det J_\varphi(y)|\, \mathrm{d}y \right|$$

$$= \left| \int_{\varphi(\overline{G}) \backslash \varphi(H)} f(y)\, \mathrm{d}y + \int_{\varphi(H)} f(y)\, \mathrm{d}y - \int_{\overline{G} \backslash H} (f \circ \varphi)(x) |\det J_\varphi(x)|\, \mathrm{d}x \right.$$

$$\left. - \int_H (f \circ \varphi)(x) |\det J_\varphi(x)|\, \mathrm{d}x \right|$$

$$= \left| \int_{\varphi(\overline{G}) \backslash \varphi(H)} f(y)\, \mathrm{d}y - \int_{\overline{G} \backslash H} (f \circ \varphi)(x) |\det J_\varphi(x)|\, \mathrm{d}x \right|$$

$$\leq \left| \int_{\varphi(\overline{G}) \backslash \varphi(H)} f(y)\, \mathrm{d}y \right| + \left| \int_{\overline{G} \backslash H} (f \circ \varphi)(x) |\det J_\varphi(x)|\, \mathrm{d}x \right|$$

$$\leq c_0 \|f\|_\infty \epsilon + \|f\|_\infty \|\det J_\varphi\|_{\infty, \overline{G}}\, \epsilon,$$

implying (21.5). $\qquad\square$

It is helpful to note that we can iterate (21.5).

Corollary 21.10. *Let* $U, V, W \subset \mathbb{R}^n$ *be open sets and* $\varphi : U \to V, \psi : V \to W$ *be* C^1-*diffeomorphisms. Further let* $G \subset \mathbb{R}^n$ *be an open Jordan measurable set such that* $\overline{G} \subset U$ *which implies that* $\overline{\varphi(G)} \subset V$. *For a continuous function* $f : \overline{(\psi \circ \varphi)(G)} \to \mathbb{R}$ *the following holds*

$$\int_{(\psi \circ \varphi)(G)} f(y)\mathrm{d}y = \int_G (f \circ \phi \circ \varphi)(z) |\det J_{\psi \circ \varphi}(z)|\, \mathrm{d}z. \qquad (21.6)$$

Proof. Using the chain rule we find

$$\int_{(\psi\circ\varphi)} f(y)dy = \int_{\varphi(G)} (f\circ\psi)(x)\,|\det J_\varphi(x)|\,dx$$

$$= \int_G (f\circ\psi\circ\varphi)(z)\,|\det J_\psi(\varphi(z))|\,|J_\varphi(z)|\,dz \qquad (21.7)$$

$$= \int_G (f\circ\psi\circ\varphi)(z)\,|\det J_{\psi\circ\varphi}(z)|\,dz.$$

$$(21.8)$$

$\square$

The proof of Theorem 21.8 is rather lengthy and it is better done in the context of Lebesgue's theory of integration which we will develop in Part 6 in Volume III. The main problem within a "pure" Riemann theory of integration is that the volume (or Jordan content) of a set is for sets not being a cell defined with the help of the integral. So when transforming a Jordan measurable set G to $\varphi(G)$ and both are not cells we cannot a priori separate the way the volume of G is transformed from the behaviour of the integral under the transformation. One way to overcome this difficulty is to first consider the case where the diffeomorphism φ changes only one coordinate direction, $\varphi(x_1,\ldots,x_j,\ldots,x_n) = (x_1,\ldots,x_{j-1},\varphi_j(x_1,\ldots,x_{j-1},x_{j+1},\ldots,x_n),x_{j+1},\ldots,x_n)$ and try to reduce the problem to the one-dimensional change of variable formula. The next step is to prove that (locally) every diffeomorphism can be decomposed into n such diffeomorphisms and merge these results together. This is done in great detail in H. Heuser [25], and in principle in V. A. Zorich [49]. Not least having the length of this detailed proof in mind, we have chosen not to adopt it for our presentation. Other proofs use the fact that locally we can approximate φ by its differential at a fixed point, a very detailed proof along these lines, following the ideas of J. Schwartz, is given in S. Hildebrandt [26] or J. R. L. Webb [47]. Again this is a rather lengthy argument and needs to investigate the Jordan content as a "content function" on Jordan measurable sets first. Since we will discuss in detail Lebesgue's theory of measures, we do not include these considerations here. Thus we have decided to postpone the proof of Theorem 21.8 until Part 6 in Volume III. However many advanced applications of analysis in several variables depend on the transformation theorem and therefore a discussion of the result and its applications seems to be already desirable at this stage. In particular our

436

discussion of vector analysis and applications to the classical theory of partial differential equations will depend on Theorem 21.8 or Corollary 21.9. Students should temporarily accept the result without proof in order to benefit from learning more and deeper results now - results of much importance in other parts of analysis and geometry as well as in physics and other sciences.

We now turn to applications of Theorem 21.8 and Corollary 21.9. Some of the most important applications are induced by symmetry and the use of curvilinear coordinates. However, often these transformations are not diffeomorphisms $\psi : \mathbb{R}^n \to \mathbb{R}^n$ but on certain sets, points, lines, hypersurfaces etc., $\det J_\psi$ is zero or ψ is not injective on these sets. For example in the case of polar coordinates $x = r\cos\varphi$ and $y = r\sin\varphi$ we find $(\det J_\psi)(r, \varphi) = r$ which vanishes at $r = 0$, i.e. $(x, y) = (0, 0)$, and due to the periodicity, if we allow $\varphi \in [0, 2\pi]$, we lose injectivity on the ray $x \geq 0$. If we can show that these sets have Jordan content zero, we may apply Corollary 21.9. Extending Theorem 19.4 we give

Proposition 21.11. *Let $f : \mathbb{R}^n \to \mathbb{R}$ be continuous. Then the graph $\Gamma(f) \subset \mathbb{R}^{n+1}$ has Lebesgue measure zero.*

Proof. First we cover $\mathbb{R}^n$ with cubes Q_k, $k \in \mathbb{N}$, and side length 4. Now, given $\epsilon > 0$ we can find open cells $(K_j^{(k)})_{j=1,\dots,N_k}$, $K_j^{(k)} \subset \mathbb{R}^{n+1}$, such that $\Gamma\left(f|_{\overline{Q_k}}\right) \subset \bigcup_{j=1}^{N_k} K_j^{(k)}$ and $\sum_{j=1}^{N_k} \mathrm{vol}_{n+1}(K_j^{(k)}) < 2^{-k}\epsilon$. This implies however that $\Gamma(f) \subset \bigcup_{k=1}^{\infty} \left(\bigcup_{k=1}^{N_k}(K_j^{(k)})\right)$ and

$$\sum_{k=1}^{\infty}\sum_{j=1}^{N_k} \mathrm{vol}_{n+1}(K_j^{(k)}) \leq \sum_{k=1}^{\infty} 2^{-k}\epsilon = \epsilon.$$

$\square$

Corollary 21.12. A. *Any n-dimensional subspace or hyperplane in $\mathbb{R}^{n+1}$ has $(n+1)$-dimensional Lebesgue measure zero.*
B. *Any set $F \subset \mathbb{R}^{n+1}$ contained in a compact subset of an n-dimensional subspace or hyperplane has Jordan content zero.*

Proof. **A.** A subspace or hyperplane in $\mathbb{R}^{n+1}$ is the graph of a function $g : \mathbb{R}^n \to \mathbb{R}$, $g(x) = \langle b, x \rangle + c$ with $b \in \mathbb{R}^n$ and $c \in \mathbb{R}$.
B. Since $F \subset H$, H being a subspace of hyperplane in $\mathbb{R}^{n+1}$, given

$\epsilon > 0$, we can cover F with countable many open cells $K_j \subset \mathbb{R}^{n+1}$ such that $\sum_{j=1}^{\infty} \text{vol}_n(K_j) < \epsilon$. Since F is compact, we can choose a finite subcovering. $\qquad\square$

As first concrete application we want to determine the volume (Jordan content) of the unit ball $B_1(0) \subset \mathbb{R}^n$. (Recall Remark 19.24.C, from now on we prefer to write $\text{vol}_n(G)$ for $J^{(n)}(G)$.)

Proposition 21.13. *The volume of $B_R(0) \subset \mathbb{R}^n$ is given by*

$$\text{vol}_n(B_R(0)) = \frac{\pi^{\frac{n}{2}} R^n}{\Gamma(\frac{n}{2}+1)} = \begin{cases} \frac{\pi^m R^{2m}}{m!}, & \text{if } n = 2m \\ \frac{2^{m+1} \pi^m 2^{2m+1}}{1 \cdot 3 \cdot 5 \cdots (2m+1)}, & \text{if } n = 2m+1. \end{cases} \tag{21.9}$$

Proof. We use spherical coordinate in $\mathbb{R}^n$, see Example 12.6, and introduce $\varphi : Q_R \to B_R(0)$, $Q_R := [0, R] \times [0, \pi] \times \cdots \times [0, \pi] \times [0, 2\pi]$, by (12.25), i.e.

$$(x_1, \ldots, x_n) = \varphi(r, \vartheta_1, \ldots, \vartheta_{n-2}, \vartheta_{n-1})$$
$$= (r \cos \vartheta_1, r \sin \vartheta_1 \cos \vartheta_2, \ldots, r \sin \vartheta_1 \ldots \sin \vartheta_{n-2} \sin \vartheta_{n-1})$$

which has the Jacobi determinant, see (12.26),

$$(\det J_\varphi)(r, \vartheta_1, \ldots, \vartheta_{n-1}) = r^{n-1} \sin^{n-2} \vartheta_1 \sin^{n-3} \vartheta_2 \cdots \sin \vartheta_{n-2}.$$

We note that the set where $\det J_\varphi$ is zero and the set where φ is not injective has Jordan content zero and therefore we find

$$\text{vol}_n(\overline{B_R(0)}) = \int_{\overline{B_R(0)}} 1 \, dx = \int_{\varphi(Q_1)} 1 \, dx = \int_{Q_1} |(\det J_\varphi)(r, \vartheta_1, \ldots, \vartheta_{n-1})| \, dr \, d\vartheta_1 \ldots d\vartheta_{n-1}$$

$$= \int_0^R \left(\int_0^\pi \cdots \int_0^\pi \left(\int_0^{2\pi} r^{n-1} \sin^{n-2} \vartheta_1 \sin^{n-3} \vartheta_2 \cdots \sin \vartheta_{n-2} \right. \right.$$
$$\left. \left. \times \, d\vartheta_{n-1} \, d\vartheta_{n-2} \cdots d\vartheta_1 \right) dr \right.$$
$$= \frac{2\pi R^n}{n} \left(\int_0^\pi \cdots \int_0^\pi \sin^{n-2} \vartheta_1 \sin^{n-3} \vartheta_2 \ldots \sin \vartheta_{n-2} \, d\vartheta_{n-2} \, d\vartheta_{n-3} \cdots d\vartheta_1 \right),$$

where we used that $\sin \vartheta \geq 0$ for $0 \leq \vartheta \leq \pi$. Thus the transformation theorem in form Corollary 21.9 allows us to reduce the calculation of $\text{vol}_n(B_1(0))$ to the evaluation of

$$\int_0^\pi \cdots \int_0^\pi \sin^{n-2} \vartheta_1 \sin^{n-3} \vartheta_2 \ldots \sin \vartheta_{n-2} \, d\vartheta_{n-2} \cdots d\vartheta_1 = \prod_{k=1}^{n-2} I_k$$

438

where

$$I_k = \int_0^\pi \sin^k \vartheta \, d\vartheta.$$

For $k \in \mathbb{N}$ we have that

$$I_{2k} = \frac{(2k)!\pi}{2^{2k}(k!)^2} \quad \text{and} \quad I_{2k-1} = \frac{2^{2k-1}((k-1)!)^2}{(2k-1)!}.$$

(These two integration results we can either pick from an integral table, or we may use a computer package, or we may work it out ourselves.)
For all $k \in \mathbb{N}$ we find

$$I_{k-1} I_k = \frac{2\pi}{k}$$

(see Problem 4) and therefore we obtain for $n = 2m$

$$\prod_{k=1}^{2m-2} I_k = \frac{\pi^{m-1}}{(m-1)!}$$

and for $n = 2m+1$

$$\prod_{k=1}^{2m-1} I_k = \frac{2^m \pi^{m-1}}{1 \cdot 3 \cdot 5 \cdot \ldots \cdot (2m-1)},$$

implying

$$\text{vol}_{2m}(\overline{B_R(0)}) = \frac{2\pi R^{2m}}{2m} \frac{\pi^{m-1}}{(m-1)!} = \frac{\pi^m R^{2m}}{m!}$$

and

$$\text{vol}_{2m+1}\left(\overline{B_R(0)}\right) = \frac{2\pi R^{2m+1}}{2m+1} \frac{2^m \pi^{m-1}}{1 \cdot 3 \cdot 5 \cdot \ldots \cdot (2m-1)} = \frac{2^{m+1} \pi^m R^{2m+1}}{1 \cdot 3 \cdot 5 \cdot \ldots \cdot (2m+1)}.$$

Using the fact that $\Gamma(m+1) = m!$, $\Gamma\left(\frac{1}{2}\right) = \sqrt{\pi}$ and

$$\Gamma\left(m + \frac{3}{2}\right) = \left(m + \frac{1}{2}\right) \Gamma\left(m + \frac{1}{2}\right) = \left(\prod_{k=0}^m \frac{2k+1}{2}\right) \Gamma\left(\frac{1}{2}\right)$$

$$= \prod_{k=0}^m \left(\frac{2k+1}{2}\right) \sqrt{\pi},$$

439

we eventually find

$$\mathrm{vol}_n(B_R(0)) = \mathrm{vol}_n\left(\overline{B_R(0)}\right) = \frac{\pi^{\frac{n}{2}} R^n}{\Gamma\left(\frac{n}{2} + 1\right)}. \tag{21.10}$$

□

Remark 21.14. It is convenient to introduce the notation

$$w_n := \mathrm{vol}_n(B_1(0)) = \frac{\pi^{\frac{n}{2}}}{\Gamma\left(\frac{n}{2} + 1\right)} \tag{21.11}$$

for the **volume of the unit** ball in $\mathbb{R}^n$

In Theorem I.30.14 we derived the value of the Gauss integral, i.e.

$$\int_{\mathbb{R}} e^{-x^2} \, \mathrm{d}x = \sqrt{\pi}. \tag{21.12}$$

This proof was rather involved, we needed to evaluate the Wallis product first and also certain properties of the Γ-function. Using polar coordinates in $\mathbb{R}^2$, we can derive this result much easier. In [12] K. Endl and W. Luh gave a particularly nice presentation of this standard approach:

Example 21.15. For $R > 0$ define the following sets

$$K_R := \left\{(x, y) \in \mathbb{R}^2 \,\middle|\, x \geq 0, y \geq 0, x^2 + y^2 \leq R^2\right\}$$
$$Q_R := \left\{(x, y) \in \mathbb{R}^2 \,\middle|\, 0 \leq x \leq R, 0 \leq y \leq R\right\}$$
$$K_{\sqrt{2}R} := \left\{(x, y) \in \mathbb{R}^2 \,\middle|\, x \geq 0, y \geq 0, x^2 + y^2 \leq 2R^2\right\},$$

see Figure 21.1.

On these three sets we consider the integral of the function $E(x, y) = e^{-x^2 - y^2}$. Using polar coordinates we find

$$\int_{K_R} e^{-x^2 - y^2} \, \mathrm{d}x \, \mathrm{d}y = \int_0^{\frac{\pi}{2}} \left\{\int_0^R e^{-r^2} r \, \mathrm{d}r\right\} \mathrm{d}\varphi = \frac{\pi}{4} - \frac{\pi}{4} e^{-R^2},$$

$$\int_{K_{\sqrt{2}R}} e^{-x^2 - y^2} \, \mathrm{d}x \, \mathrm{d}y = \frac{\pi}{4} - \frac{\pi}{4} e^{-2R^2},$$

440

and on the other hand we have

$$\int_{Q_R} e^{-x^2-y^2}\, dx\, dy = \int_0^R e^{-x^2}\, dx \int_0^R e^{-y^2}\, dy = \left(\int_0^R e^{-x^2}\, dx\right)^2.$$

Since

$$\int_{K_R} e^{-x^2-y^2}\, dx\, dy \le \int_{Q_R} e^{-x^2-y^2}\, dx\, dy \le \int_{K_{\sqrt{2}R}} e^{-x^2-y^2}\, dx\, dy$$

we find

$$\left(\frac{\pi}{4} - \frac{\pi}{4}e^{-R^2}\right)^{\frac{1}{2}} \le \int_0^R e^{-x^2}\, dx \le \left(\frac{\pi}{4} - \frac{\pi}{4}e^{-2R^2}\right)^{\frac{1}{2}}$$

implying for the improper integral $\int_0^\infty e^{-x^2}\, dx$ the value

$$\frac{\sqrt{\pi}}{2} = \int_0^\infty e^{-x^2}\, dx$$

and hence (21.12).

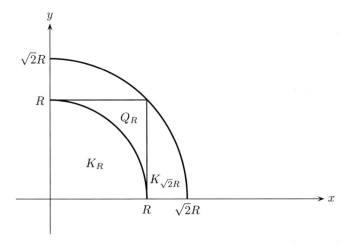

Figure 21.1

Example 21.16. Consider the set $\overline{B_R(0)}\backslash B_r(0)$ for $0 < r < R$ and $B_R(0) \subset \mathbb{R}^n$ and on $\overline{B_R(0)}\backslash B_r(0)$ a radial symmetric continuous function $f : \overline{B_R(0)}\backslash B_r(0) \to \mathbb{R}$, $f(x) = f(x_1,\ldots,x_n) = g(\rho)$, $\rho = (x_1^2 + \ldots + x_n^2)^{\frac{1}{2}} =$

$\|x\|$, with $g : [r, R] \to \mathbb{R}$ being continuous. Using spherical coordinates we find

$$\int_{\overline{B_R(0)}\backslash B_r(0)} f(x)\,dx$$

$$= \int_r^R \left(\int_0^\pi \cdots \left(\int_0^\pi \left(\int_0^{2\pi} g(\rho)\rho^{n-1} \sin^{n-1}\vartheta_1 \cdots \sin\vartheta_{n-1} \right) d\vartheta_{n-1} \right) \cdots d\vartheta_1 \right) d\rho$$

$$= nw_n \int_r^R g(\rho)\rho^{n-1}\,d\rho.$$

We are interested in two not unrelated functions defined either on $\mathbb{R}^n\backslash\{0\}$ or on the whole of $\mathbb{R}^n$, namely the functions $V_\alpha(x) = \|x\|^\alpha$, $\alpha \in \mathbb{R}$, and $W_\alpha(x) = (1 + \|x\|^2)^{\frac{\alpha}{2}}$. For $0 < r < R$ both functions are continuous on $\overline{B_R(0)}\backslash B_r(0)$, hence integrable and in particular we find for $n + \alpha \neq 0$

$$\int_{\overline{B_R(0)}\backslash B_r(0)} V_\alpha(x)\,dx = nw_n \int_r^R \rho^\alpha\rho^{n-1}\,d\rho$$

$$= nw_n \int_r^R \rho^{n+\alpha-1}\,d\rho = \frac{nw_n}{n+\alpha}\left(R^{n+\alpha} - r^{n+\alpha}\right),$$

and for $n + \alpha = 0$, i.e. $\alpha = -n$, we have

$$\int_{\overline{B_R(0)}\backslash B_r(0)} V_\alpha(x) = nw_n(\ln R - \ln r).$$

If $n + \alpha < 0$ we may consider the limit for $R \to \infty$ to find

$$\lim_{R\to\infty} \int_{\overline{B_R(0)}\backslash B_r(0)} V_\alpha(x)\,dx = \frac{nw_n}{n+\alpha}(-r^{n+\alpha}) = \frac{nw_n r^{n+\alpha}}{-n-\alpha}.$$

For $n + \alpha > 0$ however we may pass with r to 0 and get

$$\lim_{r\to 0} \int_{\overline{B_R(0)}\backslash B_r(0)} V_\alpha(x)\,dx = \frac{nw_n}{n+\alpha}R^{n+\alpha}.$$

Both limits suggest to introduce improper multidimensional integrals and write for $n + \alpha < 0$

$$\int_{\mathbb{R}^n\backslash B_r(0)} V_\alpha(x)\,dx = \int_{\mathbb{R}^n\backslash B_r(0)} \|x\|^\alpha\,dx = \lim_{R\to\infty} \int_{\overline{B_R(0)}\backslash B_r(0)} \|x\|^\alpha\,dx = \frac{nw_n r^{n+\alpha}}{-n-\alpha}$$

442

and for $n + \alpha > 0$

$$\int_{\overline{B_R(0)}} V_\alpha(x)\,\mathrm{d}x = \int_{\overline{B_R(0)}} \|x\|^\alpha \,\mathrm{d}x = \lim_{r \to 0} \int_{\overline{B_R(0)} \backslash B_r(0)} \|x\|^\alpha \,\mathrm{d}x = \frac{nw_n R^{n+\alpha}}{n+\alpha}.$$

We will discuss improper volume integrals in Chapter 22.

The function $W_\alpha(x)$ is continuous on $\overline{B_R(0)}$, thus of interest is only

$$\lim_{R \to \infty} \int_{\overline{B_R(0)}} W_\alpha(x)\,\mathrm{d}x = \lim_{R \to \infty} \int_{\overline{B_R(0)}} (1 + \|x\|^2)^{\frac{\alpha}{2}} \,\mathrm{d}x.$$

We only want to study the existence of this limit but we do not want to find its value. Of course we use spherical coordinates to find

$$\int_{\overline{B_R(0)}} W_\alpha(x)\,\mathrm{d}x = nw_n \int_0^R (1 + r^2)^{\frac{\alpha}{2}} r^{n-1}\,\mathrm{d}r.$$

If $n + \alpha < 0$ we split the integral according to

$$\int_0^R (1 + r^2)^{\frac{\alpha}{2}} r^{n-1}\,\mathrm{d}r = \int_0^1 (1 + r^2)^{\frac{\alpha}{2}} r^{n-1}\,\mathrm{d}r + \int_1^R (1 + r^2)^{\frac{\alpha}{2}} r^{n-1}\,\mathrm{d}r$$

and note that the first integral is finite and independent of R. The second integral we can estimate as follows: since $n + \alpha < 0$ implies $\alpha < 0$ and therefore $(1 + r^2)^{\frac{\alpha}{2}} \leq r^\alpha$ for $r \geq 1$ we arrive at

$$\int_1^R (1 + r^2)^{\frac{\alpha}{2}} r^{n-1}\,\mathrm{d}r \leq \int_1^R r^{\alpha+n-1}\,\mathrm{d}r,$$

and hence, by the first part of this example

$$\int_{\mathbb{R}^n} (1 + \|x\|^2)^{\frac{\alpha}{2}} \,\mathrm{d}x := \lim_{R \to \infty} nw_n \int_0^R (1 + r^2)^{\frac{\alpha}{2}} r^{n-1}\,\mathrm{d}r < \infty.$$

For $n + \alpha > 0$ we find however $(1 + r^2)^{\frac{\alpha}{2}} \geq r^\alpha$, hence $(1 + r^2)^{\frac{\alpha}{2}} r^{n-1} \geq r^{n+\alpha-1}$ and by the first part we deduce that $\lim_{R \to \infty} \int_{B_R(0)} (1 + \|x\|^2)^\alpha \,\mathrm{d}x$ diverges for $n + \alpha > 0$.

Example 21.17. For $h_j > 0$, $j = 1, 2, 3$, the set

$$\mathcal{E} := \mathcal{E}(h_1, h_2, h_3) := \left\{ x \in \mathbb{R}^3 \,\middle|\, \left(\frac{x_1}{h_1}\right)^2 + \left(\frac{x_2}{h_2}\right)^2 + \left(\frac{x_3}{h_3}\right)^2 \leq 1 \right\}$$

is an ellipsoid with half axes h_j. The mapping $\varphi : \mathbb{R}^3 \to \mathbb{R}^3$, $\varphi(y_1, y_2, y_3) = (h_1 y_1, h_2 y_2, h_3 y_3)$ maps $\overline{B_1(0)}$ onto $\mathcal{E}$ and in fact φ is a diffeomorphism with Jacobi determinant $\det J_\varphi(y_1, y_2, y_3) = h_1 h_2 h_3$. It follows for the volume of $\mathcal{E}$ that

$$\text{vol}_3(\mathcal{E}) = \int_{\varphi(\overline{B_1(0)})} 1 \, dx = \int_{\overline{B_1(0)}} |\det J_\varphi(y)| \, dy$$

$$= h_1 h_2 h_3 \int_{\overline{B_1(0)}} 1 \, dy = h_1 h_2 h_3 w_n.$$

Example 21.18. In Example 20.28 we considered a circular cone $C := \left\{ (x, y, z) \in \mathbb{R}^3 \,\middle|\, \frac{h}{R}\sqrt{x^2 + y^2} \leq z \leq h, \ (x, y) \in \overline{B_R(0)} \right\}$ where $\overline{B_R(0)} \subset \mathbb{R}^2$ and $h > 0$. Thus R is the radius of the top circle of C and h is its height, see Figure 20.3. We identified C as an "iterated" normal domain, namely

$$C = \left\{ (x, y, z) \in \mathbb{R}^3 \,\middle|\, \frac{h}{R}\sqrt{x^2 + y^2} \leq z \leq h, \ -\sqrt{R^2 - x^2} \leq y \leq \sqrt{R^2 - x^2}, \ x \in [-R, R] \right\}.$$

Using cylindrical coordinates, compare Example 12.4, we have a simpler description of C as

$$C = \left\{ (r, \varphi, z) \in [0, \infty) \times [0, 2\pi] \times \mathbb{R} \,\middle|\, 0 \leq r \leq R, \ 0 \leq \varphi \leq 2\pi, \ \frac{h}{R} r \leq z \leq h \right\}.$$

For the volume of C we find using that the Jacobi determinant for cylindrical coordinates is just r:

$$\text{vol}_3(C) = \int_C 1 \, dx \, dy \, dz = \int_{-R}^{R} \left(\int_{-\sqrt{R^2 - x^2}}^{\sqrt{R^2 - x^2}} \left(\int_{\frac{h}{R}\sqrt{x^2 + y^2}}^{h} 1 \, dz \right) dy \right) dx$$

$$= \int_0^R \left(\int_0^{2\pi} \left(\int_{\frac{h}{R} r}^{h} r \, dz \right) d\varphi \right) dr$$

$$= 2\pi \int_0^R \left(hr - \frac{h}{R} r^2 \right) dr = 2\pi \left(\frac{hr^2}{2} - \frac{hr^3}{3R} \right) \Big|_0^R$$

$$= 2\pi \left(\frac{hR^2}{2} - \frac{hR^3}{3R} \right) = \frac{\pi R^2}{3} h,$$

444

a result which we already know from Example 20.28. The careful reader will have noticed that we have in fact used a normal domain in cylindrical coordinates.

Example 21.19 (Moments of inertia). In mechanics the moments of inertia of a solid body (which we identify as a compact Jordan measurable set $G \subset \mathbb{R}^3$) with respect to the coordinate axes are defined as

$$I_x := \int_G (y^2 + z^2) \, dx \, dy \, dz$$

$$I_y := \int_G (x^2 + z^2) \, dx \, dy \, dz$$

$$I_Z := \int_G (x^2 + y^2) \, dx \, dy \, dz.$$

If G is a circular cylinder with radius R and height h, standing on the x, y-plane and having the z-axis as symmetry axis we find for the moment of inertia with respect to the symmetry axis

$$I_Z = \int_G (x^2 + y^2) \, dx \, dy \, dz.$$

Now, G is the image of $[0, R] \times [0, 2\pi] \times [0, h]$ under the transformation $\psi(r, \varphi, z) = (r \cos \varphi, r \sin \varphi, z)$ which gives of course cylindrical coordinates. Since $(\det J_\psi)(r, \varphi, z) = r$ we find

$$I_Z = \int_G (x^2 + y^2) \, dx \, dy \, dz = \int_0^R \int_0^{2\pi} \int_0^h r^3 \, dz \, d\varphi \, dr$$

$$= 2\pi h \int_0^R r^3 \, dr = \frac{\pi R^4 h}{2}.$$

Remark 21.20. In the exercises we will handle problems of the following type: a domain is described with the help of its boundary (G is bounded by a given set of "surfaces") or as intersection of some other domains and the underlying geometry allows us to introduce certain curvilinear (or other) coordinates to give a better, i.e. more suitable, description of G. On G a function is given respecting this geometry, so the integral is first transformed with the help of the transformation theorem to a more simple integral (for example due to symmetries) on some hyper-rectangle or normal domain. This new integral we try to evaluate as an iterated integral. Thus we try to employ a reduction scheme: geometry - transformation theorem - iterated one-dimensional integrals.

Problems

1.   a) Discuss the relation of the inverse mapping theorem, Theorem 10.12, with the notion of a C^k-diffeomorphism.

      b) Prove that in $\mathbb{R}^n$ an open set G is locally diffeomorphic to $B_1(0) \subset \mathbb{R}^n$, i.e. for every $x_0 \in G$ there exists an open neighbourhood $U(x_0) \subset G$ such that $U(x_0)$ is diffeomorphic to $B_1(0)$.

2.   a) Find a diffeomorphism φ mapping the open ellipsoid $\mathcal{E}_{a,b,c} := \left\{ (x, y, z) \in \mathbb{R}^3 \left| \frac{x^2}{a^2} + \frac{y^2}{b^2} + \frac{z^2}{c^2} < 1 \right| \right\}$, $a > b > c > 0$, on the open ball $B_1(0) \subset \mathbb{R}^3$.

      b) Prove that $\mathbb{R}^2 \setminus \{0\}$ and $B_2(0) \setminus \overline{B_{\frac{1}{2}}(0)}$ are C^1-diffeomorphic sets in $\mathbb{R}^2$.

3. Let $\overline{G} \subset \mathbb{R}^n$ be a compact convex set with $\mathring{G} \neq \emptyset$. Suppose that $\overline{G}$ admits only a finite number of supporting hyperplanes. Prove that $\overline{G}$ is Jordan measurable. Now give reasons why the convex hull of a finite number of points in $\mathbb{R}^n$ is Jordan measurable.

4. Justify the identity $I_{k-1}I_k = \frac{2\pi}{k}$ in the proof of Proposition 21.13.

5. Let $f : \mathbb{R}^n \to \mathbb{R}$ be a radially symmetric continuous function, i.e. $f(x) = g(||x||)$ where $g : [0, \infty) \to \mathbb{R}$. Suppose that g is a C^1-function and that $g(R) = 0$ for some $R > 0$. Prove the estimate

$$\left(\int_{B_R(0)} |f(x)|^2 dx \right)^{\frac{1}{2}} \leq \frac{2\Omega_n^{\frac{1}{2}}}{n} R^{\frac{n+1}{2}} \left(\int_0^R |g'(r)|^2 dr \right)^{\frac{1}{2}},$$

where

$$\Omega_n := \int_0^\pi \cdots \left(\int_0^\pi \left(\int_0^{2\pi} (\sin^{n-2} \vartheta_1 \sin^{n-3} \vartheta_2 \cdots \sin \vartheta_{n-2}) d\vartheta_{n-1} \right) d\vartheta_{n-2} \right) \cdots d\vartheta_1$$
$$= n w_n.$$

6.   a) Find the area of the set $G := \left\{ (x, y) \in \mathbb{R}^2 | x^2 + y^2 \leq 9, y \geq \frac{3}{2} \right\}$.

      b) Find the volume of the pipe-type domain

$$H := \left\{ (x, y, z) \in \mathbb{R}^3 | 0 < r^2 \leq x^2 + y^2 \leq R^2, 0 \leq z \leq h \right\}$$

for $r < R$ and $h > 0$.

446

7. Let $\rho : [\varphi_1, \varphi_2] \to \mathbb{R}$, $0 \le \varphi_1 \le \varphi_2 \le 2\pi$, be a strictly positive and continuous function and define in polar coordinates the set $G := \{(r, \varphi) \in [0, \infty) \times [0, 2\pi] | 0 \le r \le \rho(\varphi), \varphi_1 \le \varphi \le \varphi_2\}$. Prove that the area of G is given by

$$\text{vol}_2(G) = \frac{1}{2} \int_{\varphi_1}^{\varphi_2} \rho(\varphi)^2 d\varphi.$$

8. Using the change of variables $s = y - x$ and $t = y + x$ show that

$$\int_G e^{\frac{y-x}{x+y}} dx dy = \frac{1}{2} \sinh(1),$$

where $G = \{(x, y) \in \mathbb{R}^2 | x \ge 0, y \ge 0, x + y \le 1\}$.

9. Let $\mathcal{E} := \left\{ (x, y, z) \in \mathbb{R}^3 \big| \frac{x^2}{25} + \frac{y^2}{16} + \frac{z^2}{9} \le 1 \right\}$. Find the integral of

$h(z, y, z) = \left(1 - \frac{x^2}{25} - \frac{y^2}{16} - \frac{z^2}{9} \right)^{\frac{1}{2}}$ over the set $\mathcal{E} \cap \{(x, y, z) \in \mathbb{R}^3 | z \ge 0\}$.

10. Let $G := \{(x, y, s, t) \in \mathbb{R}^4 | x \in [2, 7], y \in [3, 6], (s, t) \in \overline{B_2(0)} \subset \mathbb{R}^2\}$. Find the integral of $f(x, y, s, t) = x(s^2 + t^2)^{\frac{1}{2}} + y^2(s^2 + t^2)^{\frac{3}{2}}$ over the set G.

11. Let $\rho : [0, \infty) \to \mathbb{R}_+$ be a continuous function and define for $x \in \overline{B}_R^{\complement}(0), B_R(0) \subset \mathbb{R}^3$,

$$N(x) := \int_{B_R(0)} \frac{\rho(||\xi||)}{||x - \xi||} d\xi.$$

Prove that $N(x) = \frac{M}{||x||}$ where $M = \int_{B_R(0)} \rho(||\xi||) d\xi$.
Hint: since we are dealing with a rotationally invariant problem we may choose x to have the coordinate $x = (0, 0, ||x||)$. Now introduce spherical coordinates (s, ϑ, φ) and justify $||x - \xi||^2 = ||x||^2 + s^2 - 2||x||s \cos \vartheta$.

22 Improper Integrals and Parameter Dependent Integrals

As in the one-dimensional case we need to extend the definition of volume integrals to either unbounded domains or unbounded functions. Already in Example 21.15 and Example 21.16 we could convince ourselves that such an extension is rather useful. Moreover, once such an extension is obtained, we may look at the case where the integrand depends on certain parameters. In one dimension we used for example parameter dependent improper integrals to define functions, such as the Γ-function.

In this section we will learn how to use improper parameter dependent integrals to solve certain partial differential equations or how to introduce integral transforms such as the Fourier transforms.

In the one-dimensional case, compare with I.28, the idea was to restrict a function we want to define an improper integral for to a sequence of subsets, say compact intervals, on which the function is integrable and then try to pass to a limit. In higher dimensions we want to do something similar and for this we need to first prove

Proposition 22.1. *For every open set $G \subset \mathbb{R}^n$ there exists a sequence $(G_k)_{k \in \mathbb{N}}$ of compact Jordan measurable subsets $G_k \subset G$ such that $G_k \subset \mathring{G}_{k+1}$ and $\bigcup_{k=1}^{\infty} G_k = G$.*

Proof. Using the maximum norm $\|.\|_\infty$ on $\mathbb{R}^n$ we define for $H \subset \mathbb{R}^n$ and $x \in \mathbb{R}^n$

$$\mathrm{dist}_\infty(x, H) := \inf \left\{ \|x - y\|_\infty \,\middle|\, y \in H \right\}. \tag{22.1}$$

The set $G^{\complement}$ is closed and therefore $\mathrm{dist}_\infty(x, G^{\complement}) > 0$ for any $x \in G$. We consider the sequence of sets

$$H_k := \left\{ x \in G \,\middle|\, \|x\|_\infty \leq k \text{ and } \mathrm{dist}_\infty(x, G^{\complement}) \geq \frac{1}{k} \right\}.$$

Since $H_k \subset \overline{B_k^{(\infty)}(0)} = \left\{ y \in \mathbb{R}^n \,\middle|\, \|y\|_\infty \leq k \right\}$, the set H_k is bounded, and since the mappings $x \mapsto \|x\|_\infty$ and $x \mapsto \mathrm{dist}_\infty(x, G^{\complement})$ are continuous, compare Example 3.28 for the latter one, the set H_k is closed, hence each H_k is a compact set $\mathbb{R}^n$. Furthermore we have $\mathrm{dist}_\infty(H_k, G^{\complement}) \geq \frac{1}{k}$, i.e. $H_k \cap G^{\complement} = \emptyset$.

The interior $\mathring{H}_k$ of H_k is the set

$$\mathring{H}_k = \left\{ x \in G \mid \|x\|_k < k \text{ and } \operatorname{dist}_\infty(x, G^\complement) > \frac{1}{k} \right\}$$

implying that $H_k \subset \mathring{H}_{k+1}$, note $k < k+1$ but $\frac{1}{k} > \frac{1}{k+1}$. By construction we have $\bigcup_{k \in \mathbb{N}} H_k \subset G$. Moreover, if $x \in G$ it follows that $\operatorname{dist}_\infty(x, G^\complement) > 0$ which yields that for some $k \in \mathbb{N}$ we find $\operatorname{dist}_\infty(x, G^\complement) \geq \frac{1}{k}$ and $\|x\|_\infty \leq k$, i.e. $G \subset \bigcup_{k \in \mathbb{N}} H_k$, implying $\bigcup_{k=1}^\infty H_k = G$. Thus the sets H_k, $k \in \mathbb{N}$, form an exhaustion of G by compact sets, however we do not know whether they are Jordan measurable. Therefore we now try to replace each H_k by a finite union of compact cells. For $x \in H_k$ we can find a compact cell $C_k(x) \subset \mathring{H}_{k+1}$, $x \in C_k(x)$, implying that $H_k \subset \bigcup_{x \in H_k} \mathring{C}_k(x)$. The compactness of H_k yields that we can cover H_k with finitely many of these cells, say $C_{k_1}(x_{k_1}), \ldots, C_{k_m}(x_{k_m})$. Now we define $G_k := \bigcup_{j=1}^m C_{k_j}(x_{k_j})$. As the union of finitely many compact cells G_k is compact and by construction we have

$$H_k \subset \mathring{G}_k \subset G_k \subset \mathring{H}_{k+1} \subset H_{k+1} \subset \mathring{G}_{k+1}$$

implying that $G_k = \overline{G}_k \subset \mathring{G}_{k+1}$, G_k is a compact Jordan measurable set and $G = \bigcup_{k+1}^\infty G_k$. □

Definition 22.2. *Given an open set $G \subset \mathbb{R}^n$. A sequence $(G_k)_{k \in \mathbb{N}}$ of compact Jordan measurable sets $G_k \subset G$ is called a **compact Jordan measurable exhaustion** of G if $G_k \subset G$ is compact, $G_k \subset \mathring{G}_{k+1}$ and $\bigcup_{k \in \mathbb{N}} G_k = G$.*

Remark 22.3. A. For simplicity we may sometimes call $(G_k)_{k \in \mathbb{N}}$ just an exhaustion.
B. In Proposition 22.1 we have proved that we can find for every open set $G \subset \mathbb{R}^n$ a compact Jordan measurable exhaustion $(G_k)_{k \in \mathbb{N}}$ of special type, namely G_k being a finite union of compact cells.
C. The proof of Proposition 22.1 also yields that $G = \bigcup_{k \in \mathbb{N}} \mathring{G}_k$.

Let $G \subset \mathbb{R}^n$ be an open, not necessarily bounded set and $(G_k)_{k \in \mathbb{N}}$ a compact Jordan measurable exhaustion of G. Further let $f : G \to \mathbb{R}$ be a continuous function (which need not be bounded). For $k \in \mathbb{N}$ the integrals $\int_{G_k} f(x)\,dx$ are well defined and finite, hence they form a sequence of real numbers. If this sequence converges, we write for a moment

$$\int_{(G_k)_{k \in \mathbb{N}}} f(x)\,dx := \lim_{k \to \infty} \int_{G_k} f(x)\,dx. \tag{22.2}$$

Lemma 22.4. *Suppose that for an open set $G \subset \mathbb{R}^n$ and a continuous function $f : G \to \mathbb{R}$ for every compact Jordan measurable exhaustion $(G_k)_{k \in \mathbb{N}}$ of G the limit $\int_{(G_k)_{k \in \mathbb{N}}} f(x) \, \mathrm{d}x$ exists. Then the limit is independent of $(G_k)_{k \in \mathbb{N}}$.*

Proof. Let $(G_k)_{k \in \mathbb{N}}$ and $(H_l)_{l \in \mathbb{N}}$ be two compact Jordan measurable exhaustions of G, in particular we have $G_k \subset \mathring{G}_{k+1}$ and $H_l \subset \mathring{H}_{l+1}$. We can find two subsequences $(G_{k_i})_{i \in \mathbb{N}}$ and $(H_{l_j})_{j \in \mathbb{N}}$ such that

$$G_{k_i} \subset \mathring{H}_{l_i} \subset H_{l_i} \subset \mathring{G}_{k_{i+1}},$$

note that G_k and H_l are compact and according to Remark 22.3.C $(\mathring{G}_k)_{k \in \mathbb{N}}$ and $(\mathring{H}_k)_{k \in \mathbb{N}}$ are open coverings of G too. It follows that the sequence $(F_\nu)_{\nu \in \mathbb{N}}$ defined by

$$(F_{2m-1})_{m \in \mathbb{N}} = (G_{k_m})_{m \in \mathbb{N}}, \quad (F_{2m})_{m \in \mathbb{N}} = (H_{l_m})_{m \in \mathbb{N}},$$

is a further compact Jordan measurable exhaustion of G and by our assumption the limits

$$\lim_{k \to \infty} \int_{G_k} f(x) \, \mathrm{d}x = \lim_{m \to \infty} \int_{G_{k_m}} f(x) \, \mathrm{d}x = \int_{(G_k)_{k \in \mathbb{N}}} f(x) \, \mathrm{d}x,$$

$$\lim_{l \to \infty} \int_{H_l} f(x) \, \mathrm{d}x = \lim_{m \to \infty} \int_{H_{l_m}} f(x) \, \mathrm{d}x = \int_{(H_l)_{l \in \mathbb{N}}} f(x) \, \mathrm{d}x$$

and

$$\lim_{\nu \to \infty} \int_{F_\nu} f(x) \, \mathrm{d}x = \lim_{m \to \infty} \int_{G_{k_m}} f(x) \, \mathrm{d}x = \lim_{m \to \infty} \int_{H_{l_m}} f(x) \, \mathrm{d}x$$

all exist implying that

$$\int_{(G_k)_{k \in \mathbb{N}}} f(x) \, \mathrm{d}x = \int_{(H_l)_{l \in \mathbb{N}}} f(x) \, \mathrm{d}x.$$

$\square$

Definition 22.5. *Let $f : G \to \mathbb{R}$ be a continuous function on the open set $G \subset \mathbb{R}^n$ and suppose that for all compact Jordan measurable exhaustions $(G_k)_{k \in \mathbb{N}}$ of G the limits $\int_{(G_k)_{k \in \mathbb{N}}} f(x) \, \mathrm{d}x$ exist, and hence by Lemma 22.4 coincide. Then we call*

$$\int_G f(x) \, \mathrm{d}x := \int_{(G_k)_{k \in \mathbb{N}}} f(x) \, \mathrm{d}x = \lim_{k \to \infty} \int_{G_k} f(x) \, \mathrm{d}x \qquad (22.3)$$

*the **improper Riemann** integral of f over G and we say that f is **improper integrable** over G.*

Corollary 22.6. *If $G \subset \mathbb{R}^n$ is an open bounded Jordan measurable set and $f \in C_b(G)$, i.e. f is a bounded continuous function on G, then the integral of f over G coincides with the improper integral of f over G, i.e. for every compact Jordan measurable exhaustion $(G_k)_{k \in \mathbb{N}}$ of G the following holds*

$$\int_G f(x)\,dx = \lim_{k \to \infty} \int_{G_k} f(x)\,dx.$$

Proof. Since G is an open bounded Jordan measurable set, given $\epsilon > 0$ we can find finitely many compact cells K_j, $j = 1, \ldots, N(\epsilon)$, such that $\mathrm{vol}_n(G \backslash \bigcup_{j=1}^{N(\epsilon)} K_j) < \epsilon$. Now let $(G_k)_{k \in \mathbb{N}}$ be a compact Jordan measurable exhaustion of G, for example constructed as in Proposition 22.1. It follows the existence of $M \in \mathbb{N}$ such that $\bigcup_{j=1}^{N(\epsilon)} K_j \subset G_k$ for $k \geq M$, hence $\mathrm{vol}_n(G \backslash G_k) < \epsilon$ for $k \geq M$ implying

$$\left| \int_G f(x)\,dx - \int_{G_k} f(x)\,dx \right| \leq \int_{G \backslash G_k} |f(x)|\,dx \leq \|f\|_\infty \mathrm{vol}_n(G \backslash G_k) \leq \epsilon \|f\|_\infty,$$

i.e.

$$\lim_{k \to \infty} \int_{G_k} f(x)\,dx = \int_G f(x)\,dx.$$

Since $(G_k)_{k \in \mathbb{N}}$ was arbitrary the corollary follows. $\qquad \square$

For practical purposes the following lemma is often quite helpful.

Lemma 22.7. *Let $f, g : G \to \mathbb{R}$ be continuous functions on the open set $G \subset \mathbb{R}^n$ and suppose $0 \leq f(x) \leq g(x)$ for all $x \in G$. If g is improper integrable over g then f is improper integrable over G and*

$$0 \leq \int_G f(x)\,dx \leq \int_G g(x)\,dx. \tag{22.4}$$

Proof. Let $(G_k)_{k \in \mathbb{N}}$ be a compact Jordan measurable exhaustion of G. The monotonicity of the integral yields that the sequence $\left(\int_{G_k} f(x)\,dx \right)_{k \in \mathbb{N}}$ of real numbers is monotone increasing and since

$$0 \leq \int_{G_k} f(x)\,dx \leq \int_{G_k} g(x)\,dx \leq \lim_{k \to \infty} \int_{G_k} g(x)\,dx = \int_G g(x)\,dx, \tag{22.5}$$

it follows that for every exhaustion the $\lim_{k \to \infty} \int_{G_k} f(x)\,dx$ exists and hence f is improper integrable and (22.5) implies (22.4). $\qquad \square$

Of course, we need some criterion to test improper integrability, so far we cannot even prove that the integrals considered in Examples 21.15 and 21.16 are improper integrals. To remedy this situation we first give

Definition 22.8. *The function $f \in C(G)$, $G \subset \mathbb{R}^n$ open, is said to be **absolutely improper integrable** if $|f|$ is improper integrable, i.e. if $\int_G |f(x)|\,\mathrm{d}x$ exists as improper integral.*

Proposition 22.9. *If $f \in C(G)$ is absolutely improper integrable then f is improper integrable.*

Proof. First we note that since f is a continuous function on G, then $|f|$, f^+ and f^- are also continuous functions on G. Moreover, $f = f^+ - f^-$ and $|f| = f^+ + f^-$ implying $f^+ \leq |f|$ as well as $f^- \leq |f|$. Thus by Lemma 22.7 we deduce that both f^+ and f^- are improper integrable implying that $f = f^+ - f^-$ is improper integrable since for every exhaustion $(G_k)_{k\in\mathbb{N}}$ we have

$$\int_{G_k} f(x)\,\mathrm{d}x = \int_{G_k} f^+(x)\,\mathrm{d}x - \int_{G_k} f^-(x)\,\mathrm{d}x$$

which yields first the existence of $\lim_{k\to\infty} \int_{G_k} f(x)\,\mathrm{d}x$ and secondly the relation $\int_G f(x)\,\mathrm{d}x = \int_G f^+(x)\,\mathrm{d}x - \int_G f^-(x)\,\mathrm{d}x$. $\qquad\square$

Proposition 22.10. *A continuous function $f : G \to \mathbb{R}$, $G \subset \mathbb{R}^n$ open, is absolutely improper integrable if and only if there is a constant $\gamma \geq 0$ such that for all Jordan measurable sets $H \subset G$ such that $\overline{H}$ is compact and $\overline{H} \subset G$ the following holds*

$$\int_H |f(x)|\,\mathrm{d}x \leq \gamma. \tag{22.6}$$

Proof. Suppose that f is absolutely improper integrable with $\int_G |f(x)|\,\mathrm{d}x = \gamma$. The definitions of the improper integral yields for $H \subset G$, $\overline{H} \subset G$, and H Jordan measurable that

$$\int_H |f(x)|\,\mathrm{d}x \leq \int_G |f(x)|\,\mathrm{d}x = \gamma.$$

Now suppose that (22.6) holds and let $(G_k)_{k\in\mathbb{N}}$ be an exhaustion of G with compact Jordan measurable sets $(G_k)_{k\in\mathbb{N}}$. It follows that

$$\int_{G_k} |f(x)|\,\mathrm{d}x \leq \gamma \quad \text{for all } k \in \mathbb{N}.$$

The sequence $\left(\int_{G_k} |f(x)|\, \mathrm{d}x \right)_{k \in \mathbb{N}}$ is a monotone increasing sequence of real numbers which is bounded from above, hence convergent. Thus f is absolutely improper integrable over G. $\qquad\square$

With Proposition 22.10 in mind we can now identify the integrals $\int_{\mathbb{R}^2} e^{-x^2 - y^2}\, \mathrm{d}x\, \mathrm{d}y$ in Example 21.15 as well as $\int_{\mathbb{R}^n \setminus B_r(0)} \|x\|^\alpha\, \mathrm{d}x$, $n + \alpha < 0$, and $\int_{\mathbb{R}^n} (1 + \|x\|^2)^{\frac{\alpha}{2}}\, \mathrm{d}x$, $n + \alpha < 0$, in Example 21.16 as improper Riemann integrals, see also Problem 4 in Chapter 22.

Example 22.11. Let $f : \mathbb{R}^n \to \mathbb{R}$, $f \geq 0$, be a continuous rotational invariant function, i.e. $f(x) = g(r)$, $r = \|x\|$, with a continuous function $g : [0, \infty) \to \mathbb{R}$, $g \geq 0$. This function f is improper integrable over $\mathbb{R}^n$ if and only if $r \mapsto g(r)r^{n-1}$ is improper integrable over $[0, \infty)$. Moreover we have

$$\int_{\mathbb{R}^n} f(x)\, \mathrm{d}x = n w_n \int_0^\infty g(\rho)\rho^{n-1}\, \mathrm{d}\rho. \tag{22.7}$$

Indeed, $\left(\overline{B_k(0)} \right)_{k \in \mathbb{N}}$, forms a compact Jordan measurable exhaustion of $\mathbb{R}^n$ and using the calculations in Example 21.16 we find

$$\int_{\overline{B_R(0)}} f(x)\, \mathrm{d}x = n w_n \int_0^R g(\rho)\rho^{n-1}\, \mathrm{d}\rho. \tag{22.8}$$

If $r \mapsto g(r)r^{n-1}$ is improper Riemann integrable, then we can pass on the right hand side to the limit to find

$$\int_{\overline{B_R(0)}} f(x)\, \mathrm{d}x \leq n w_n \int_0^\infty g(\rho)\rho^{n-1}\, \mathrm{d}\rho,$$

where we used that $g(\rho) \geq 0$. If $H \subset \mathbb{R}^n$ is any Jordan measurable set such that $\overline{H}$ is compact, then we can find $R > 0$ such that $\overline{H} \subset \overline{B_R(0)}$ and consequently

$$\int_H f(x)\, \mathrm{d}x \leq n w_n \int_0^\infty g(\rho)\rho^{n-1}\, \mathrm{d}\rho = \gamma.$$

Proposition 22.10 now implies that f is improper integrable and passing in (22.8) to the limit as R tends to ∞ we obtain (22.7). If however we know that f is improper Riemann integrable we find

$$\int_0^R g(\rho)\rho^{n-1}\, \mathrm{d}\rho \leq \frac{1}{n w_n} \int_{\mathbb{R}^n} f(x)\, \mathrm{d}x$$

and this estimate implies the improper integrability of $\rho \mapsto g(\rho)\rho^{n-1}$.

Note that if $f : \mathbb{R}^n \to \mathbb{R}$, $f(x) = g(\|x\|)$, and (22.8) holds for all $R > 0$ but we do not assume $f \geq 0$, we still can pass in (22.8) to the limit and (22.7) follows. However we cannot use our argument to deduce that f is improper integrable.

Example 22.12. First we deduce using Example 21.15 (or Theorem I.30.14) that

$$\int_{\mathbb{R}^n} e^{-\|x\|^2} \, \mathrm{d}x = \pi^{\frac{n}{2}}, \tag{22.9}$$

which follows from $e^{-\|x\|^2} = e^{-(x_1^2 + \ldots + x_n^2)} = e^{-x_1^2} \cdot \ldots \cdot e^{-x_n^2}$, and the fact that

$$\int_{\mathbb{R}^n} e^{-\|x\|^2} \, \mathrm{d}x = \lim_{R \to \infty} \int_{[-R,R]^n} e^{-\|x\|^2} \, \mathrm{d}x$$

$$= \lim_{R \to \infty} \int_{-R}^{R} \cdots \int_{-R}^{R} e^{-x_1^2} \cdot \ldots \cdot e^{-x_n^2} \, \mathrm{d}x_1 \cdots \mathrm{d}x_n$$

$$= \lim_{R \to \infty} \left(\left(\int_{-R}^{R} e^{-x_1^2} \, \mathrm{d}x_1 \right) \cdots \left(\int_{-R}^{R} e^{-x_n^2} \, \mathrm{d}x_n \right) \right)$$

$$= \left(\int_{-\infty}^{\infty} e^{-y^2} \, \mathrm{d}y \right)^n = \pi^{\frac{n}{2}}.$$

Now let $A \in M(n; \mathbb{R})$ be a positive definite matrix with eigenvalues $0 < \lambda_1 < \cdots < \lambda_n$ (counted according to multiplicity). We want to evaluate the integral

$$\int_{\mathbb{R}^n} e^{-\langle Ax, x \rangle} \, \mathrm{d}x. \tag{22.10}$$

From Problem 8 in Chapter 11 we know that $\lambda_1 \|x\|^2 \leq \langle Ax, x \rangle$ implying that $e^{-\langle Ax, x \rangle} \leq e^{-\lambda_1 \|x\|^2}$. If $e^{-\lambda_1 \|\cdot\|^2}$ is improper integrable, then the comparison result, Lemma 22.7, yields that (22.10) exists as an improper integral. However

$$\int_{-R}^{R} e^{-\lambda_1 y^2} \, \mathrm{d}y = \int_{-R}^{R} e^{-(\sqrt{\lambda_1} y)^2} \, \mathrm{d}y = \frac{1}{\sqrt{\lambda_1}} \int_{-\sqrt{\lambda_1} R}^{\sqrt{\lambda_1} R} e^{-u^2} \, \mathrm{d}u, \tag{22.11}$$

which implies that $\int_{-\infty}^{\infty} e^{-\sqrt{\lambda_1} y^2} \, \mathrm{d}y = \frac{1}{\sqrt{\lambda_1}} \pi^{\frac{1}{2}}$, hence we have

$$\int_{\mathbb{R}^n} e^{-\lambda_1 \|x\|^2} \, \mathrm{d}x = \left(\frac{\pi}{\lambda_1} \right)^{\frac{n}{2}}.$$

Thus (22.10) exists and if $y = Ux$ where $U \in O(n)$ is the matrix diagonalising A, i.e. $UAU^{-1} = \begin{pmatrix} \lambda_1 & & 0 \\ & \ddots & \\ 0 & & \lambda_n \end{pmatrix}$, we find

$$\int_{[-R,R]^n} e^{-\langle Ax,x \rangle} \, dx = \frac{1}{\sqrt{\lambda_1 \cdots \lambda_n}} \int_{-\sqrt{\lambda_1}R}^{\sqrt{\lambda_1}R} \cdots \int_{-\sqrt{\lambda_n}R}^{\sqrt{\lambda_n}R} e^{-y_1^2} \cdots e^{-y_n^2} \, dy_1 \cdots dy_n$$

which yields for $R \to \infty$ that

$$\int_{\mathbb{R}^n} e^{-\langle Ax,x \rangle} \, dx = \frac{1}{\sqrt{\lambda_1 \cdots \lambda_n}} \int_{\mathbb{R}^n} e^{-\|y\|^2} \, dy = \frac{\pi^{\frac{n}{2}}}{\sqrt{\det A}}.$$

Before looking at further examples of improper integrals we want to discuss once again parameter dependent integrals, see Chapter 14 and Chapter 17. However we now want to consider volume integrals and then improper integrals. Once more it is appropriate to note that when introducing the Lebesgue integral in Volume III certain of the following results are easier to prove and in fact even more general statements hold. For this reason we give more indications of proofs rather than very detailed arguments.

Let $G \subset \mathbb{R}^n$ be a compact Jordan measurable set and $X \subset \mathbb{R}^m$ a non-empty set. Further let $f : G \times X \to \mathbb{R}$ be a continuous function. It follows that for every $x \in X$ the function $g := f_x : G \to \mathbb{R}$, $y \mapsto g(y) = f(y, x)$ is on G continuous, hence integrable and we can define on X the new function

$$F(x) := \int_G f(y, x) \, dy. \tag{22.12}$$

We claim that F is continuous. As preparation for the proof we need

Lemma 22.13. *Let $G \subset \mathbb{R}^n$ be a compact set and $X \subset \mathbb{R}^m$ a non-empty set. Further let $f : G \times X \to \mathbb{R}$ be a continuous function. Suppose that the sequence $(x_k)_{k \in \mathbb{N}}$, $x_k \in X$, converges in $\mathbb{R}^m$ to x. Then the function $g_k : G \to \mathbb{R}$, $g_k(y) = f(y, x_k)$, converges uniformly on G to $g_x : G \to \mathbb{R}$, $g_x(y) = f(y, x)$.*

Proof. By Lemma 3.13 the set $\{x_k \mid k \in \mathbb{N}\} \cup \{x\}$ is compact in $\mathbb{R}^m$, hence bounded and closed. Since $G \subset \mathbb{R}^n$ is compact, it follows that $H := G \times (\{x_k \mid k \in \mathbb{N}\} \cup \{x\})$ is bounded and closed, hence compact in $\mathbb{R}^n \times \mathbb{R}^m$.

Therefore $f : H \to \mathbb{R}$ is uniformly continuous, i.e. for $\epsilon > 0$ there exists $\delta > 0$ such that for all $(y', x'), (\tilde{y}, \tilde{x}) \in H$ it follows that $\|(y', x') - (\tilde{y}, \tilde{x})\| < \delta$ implies $|f(y', x') - f(\tilde{y}, \tilde{x})| < \epsilon$. Since $\lim_{k \to \infty} x_k = x$ we can find $N = N(\epsilon)$ such that $k \geq N$ implies $\|x_k - x\| < \delta$ which yields for $k \geq N(\epsilon)$ that

$$|f(y, x_k) - f(y, x)| < \epsilon \quad \text{for all } y \in G,$$

i.e.

$$|g_k(y) - g_x(y)| < \epsilon \quad \text{for all } y \in G$$

or

$$\|g_k - g_x\|_\infty = \sup_{y \in G} |g_k(y) - g_x(y)| \leq \epsilon.$$

implying the uniform convergence of g_k to g_x. $\qquad \square$

Theorem 22.14 (Continuity of parameter dependent integrals). *Let $G \subset \mathbb{R}^n$ be a compact Jordan measurable set and $X \subset \mathbb{R}^m$ a non-empty set. Further let $f : G \times X \to \mathbb{R}$ be a continuous function. Then the function $F : X \to \mathbb{R}$ defined by (22.12) is continuous.*

Proof. Let $x \in X$ and $(x_k)_{k \in \mathbb{N}}$, $x_k \in X$, a sequence converging to x. With (22.12) we set

$$F(x_k) := \int_G g_k(y) \, \mathrm{d}y = \int_G f(y, x_k) \, \mathrm{d}y$$

and

$$F(x) := \int_G g_x(y) \, \mathrm{d}y = \int_G f(y, x) \, \mathrm{d}y$$

where we used the notation of Lemma 22.13. By Theorem 20.13 we have due to the uniform convergence of g_k to g_x:

$$\lim_{k \to \infty} \int_G g_k(y) \, \mathrm{d}y = \int_G g_x(y) \, \mathrm{d}y$$

or

$$\lim_{k \to \infty} f(y, x_k) = \int_G f(y, x) \, \mathrm{d}y,$$

i.e. $\lim_{k \to \infty} F(x_k) = F(x)$, implying the continuity of F. $\qquad \square$

Remark 22.15. Our proof is an adaption of O. Forster's proof in [19] for the case $G = [a, b]$.

The extension of Theorem 22.14 to improper integrals should go along the following lines: $G \subset \mathbb{R}^n$ is replaced by an open set and $(G_k)_{k \in \mathbb{N}}$ is a compact Jordan measurable exhaustion of G. As before, $X \subset \mathbb{R}^m$ is a set. The function $f : G \times X \to \mathbb{R}$ is assumed to be continuous and therefore, using Theorem 22.14, the functions

$$F_k : X \to \mathbb{R}, \quad F_k(x) = \int_{G_k} f(y, x) \, \mathrm{d}y$$

are continuous. We have to add the assumption that for $x \in X$, $y \mapsto f(y, x)$ is improper integrable, hence a function is defined on X by

$$F(x) = \int_G f(y, x) \, \mathrm{d}x.$$

In order to prove that F is continuous we need to assume that F_k converges uniformly to F on X.

Next we want to investigate the question when it is possible to differentiate a parameter dependent volume integral under the integral sign and we follow closely the proof of Lemma 14.33 and Theorem 14.34.

Lemma 22.16. *Let $G \subset \mathbb{R}^n$ be a compact Jordan measurable set and $K \subset \mathbb{R}^n$ be a compact convex set with non-empty interior. Further let $f : G \times K \to \mathbb{R}$, $(y, x) \mapsto f(y, x)$, be a continuous function. Assume that for some j, $1 \leq j \leq m$, f has the continuous partial derivative $\frac{\partial f}{\partial x_j}$, $x = (x^{(1)}, \ldots, x^{(m)})$. Let $(x_k)_{k \in \mathbb{N}}$, $x_k = (x_k^{(1)}, \ldots, x_k^{(m)}) \in K$, be a sequence in K with limit $x = (x^{(1)}, \ldots, x^{(m)}) \in K$, $x_k^{(j)} \neq x^{(j)}$, and define*

$$F_{k,j}(y) := \frac{f(y, x_k) - f(y, x)}{x_k^{(j)} - x^{(j)}}, \quad k \in \mathbb{N}$$

and

$$F_x(y) := \left(\frac{\partial f}{\partial x^{(j)}} \right)(y, x).$$

Then $(F_{k,j})_{k \in \mathbb{N}}$ converges on K uniformly to F_x.

Proof. The continuous function $\frac{\partial f}{\partial x^{(j)}}$ is uniformly continuous on the compact set $G \times K$. Therefore, for $\epsilon > 0$ there exists $\delta > 0$ such that for $x', \tilde{x} \in K$, $\|x' - \tilde{x}\| < \delta$ implies

$$\left| \frac{\partial f}{\partial x^{(j)}}(y, x') - \frac{\partial f}{\partial x^{(j)}}(y, \tilde{x}) \right| < \epsilon.$$

By the mean value theorem we can find η_k between x_k and x such that

$$F_{k,j}(y) = \left(\frac{\partial f}{\partial x^{(j)}} \right) (y, \eta_k).$$

For $\delta > 0$ as above we can find $N \in \mathbb{N}$ such that $k \geq N$ implies $\|x - x_k\| < \delta$, hence $\|x - \eta_k\| < \delta$ and therefore we have for $k \geq N$ that

$$|F_{x,j}(y) - F_{k,j}(y)| = \left| \left(\frac{\partial f}{\partial x^{(j)}} \right) (y, x) - \left(\frac{\partial f}{\partial x^{(j)}} \right) (y, \eta_k) \right| < \epsilon,$$

proving the lemma. □

Theorem 22.17 (Differentiation of parameter dependent integrals).
Let $G \subset \mathbb{R}^n$ be a compact Jordan measurable set and $K \subset \mathbb{R}^m$ a compact convex set, $\mathring{K} \neq \emptyset$. Let $f : G \times K \to \mathbb{R}$, $(y, x) \mapsto f(y, x)$, be a continuous function with continuous partial derivative $\frac{\partial f}{\partial x^{(j)}}$ for some j, $1 \leq j \leq m$. The function $F : K \to \mathbb{R}$ defined by

$$F(x) := \int_G f(y, x) \, dy \qquad (22.13)$$

has a continuous partial derivative $\frac{\partial F}{\partial x^{(j)}}$ and the following holds

$$\left(\frac{\partial F}{\partial x^{(j)}} \right) (x) = \int_G \left(\frac{\partial f}{\partial x^{(j)}} \right) (y, x) \, dy. \qquad (22.14)$$

Proof. We use the notations of Lemma 22.16. Let $(x_k)_{k \in \mathbb{N}}$ be a sequence in K converging to x, $x_k^{(j)} \neq x^{(j)}$ for all $k \in \mathbb{N}$. By Lemma 22.16 we know that the functions $F_{k,j}$ converge uniformly on K to $F_{x,j}$ and therefore we find

$$\lim_{k \to \infty} \frac{F(x_K) - F(x)}{x_k^{(j)} - x^{(j)}} = \lim_{k \to \infty} \int_G \frac{f(y, x_k) - f(y, x)}{x_k^{(j)} - x^{(j)}} \, dy$$

$$= \int_G \left(\frac{\partial f}{\partial x^{(j)}} \right) (y, x) \, dx.$$

Hence $\left(\frac{\partial f}{\partial x^{(j)}} \right) (x)$ exists and for every $x \in K$ we have

$$\left(\frac{\partial F}{\partial x^{(j)}} \right) (x) = \int_G \left(\frac{\partial f}{\partial x^{(j)}} \right) (y, x) \, dy.$$

The continuity of $\frac{\partial f}{\partial x^{(j)}}$ on the compact set $G \times K$ implies further the continuity of $\frac{\partial F}{\partial x^{(j)}}$. □

The extension of Theorem 22.17 to improper integrals follows the same idea as the extension of Theorem 22.14:

Let $G \subset \mathbb{R}^n$ be an open set with compact Jordan measurable exhaustion $(G_k)_{k \in \mathbb{N}}$. Further let $f : G \times K \to \mathbb{R}$ be a continuous function with continuous partial derivative $\frac{\partial f}{\partial x^{(j)}}$ where $K \subset \mathbb{R}^m$ is a compact convex set with non-empty interior and $(y, x) \in G \times K$. First we apply Theorem 22.17 to get

$$\frac{\partial}{\partial x^{(j)}} \int_{G_k} f(y, x) \, \mathrm{d}y = \int_{G_k} \left(\frac{\partial f}{\partial x^{(j)}} \right) (y, x) \, \mathrm{d}y,$$

and next we need to ensure that we may pass on both sides to the limit $k \to \infty$. For this we need that $\frac{\partial f}{\partial x^{(j)}}$ is improper integrable over G and uniform convergence of the partial derivatives.

Of course Theorem 22.17 has the obvious extension to higher order partial derivatives which then can be used to handle improper integrals too.

As already mentioned, using the Lebesgue integral we will obtain better results in Volume III. In "practical" cases we will start and work with concrete exhaustions and this is how we will handle the following examples.

Example 22.18. In Problem 1 of Chapter 5 we have seen that the function $h : (0, \infty) \times \mathbb{R}^n \to \mathbb{R}$, $h(t, x) = (4\pi)^{-\frac{n}{2}} e^{-\frac{\|x\|^2}{4t}}$ solves the heat equation, i.e. $\frac{\partial h}{\partial t} - \Delta_n h = 0$. In particular we have obtained

$$\frac{\partial h}{\partial t}(t, x) = (4\pi t)^{-\frac{n}{2}} \left(-\frac{n}{2t} + \frac{\|x\|^2}{4t^2} \right) e^{-\frac{\|x\|^2}{4t}}, \tag{22.15}$$

$$\frac{\partial h}{\partial x_j}(t, x) = (4\pi t)^{-\frac{n}{2}} \left(-\frac{x_j}{2t} \right) e^{-\frac{\|x\|^2}{4t}} \tag{22.16}$$

and

$$\frac{\partial^2 h}{\partial x_j^2}(t, x) = (4\pi t)^{-\frac{n}{2}} \left(-\frac{1}{2t} + \frac{x_j^2}{4t^2} \right) e^{-\frac{\|x\|^2}{4t}}. \tag{22.17}$$

Now we claim that for $f \in C_b(\mathbb{R}^n)$, i.e. a bounded continuous function on $\mathbb{R}^n$, the function

$$u(t, x) := (4\pi t)^{-\frac{n}{2}} \int_{\mathbb{R}^n} e^{-\frac{\|x-y\|^2}{4t}} f(y) \, \mathrm{d}y$$

solves the heat equation in $(0, \infty) \times \mathbb{R}^n$ too. First we note that

$$\left| e^{-\frac{\|x-y\|^2}{4t}} f(y) \right| \leq \|f\|_\infty e^{-\frac{\|x-y\|^2}{4t}}$$

implying that the integral exists as an absolutely convergent improper integral. Next we observe that if we can prove that for every $0 < T_0 < T$ and every $R > 0$ the function u satisfies the heat equation in $(T_0, T) \times B_R(0)$ or $(T_0, T) \times (-R, R)^n$ then we can conclude that u solves the heat equation in $(0, \infty) \times \mathbb{R}^n$. Let $N > 0$ and consider for $x \in B_R(0)$ and $t \in (T_0, T)$

$$u_N(t, x) := (4\pi t)^{-\frac{n}{2}} \int_{B_N(0)} e^{-\frac{\|x-y\|^2}{4t}} f(y) \, dy.$$

Using (22.15) - (22.17) we may apply Theorem 22.17 (and its extension to higher order partial derivatives) to find

$$\left(\frac{\partial}{\partial t} - \Delta_n \right) u_N(t, x) = \left(\frac{\partial}{\partial t} - \Delta_n \right) \left(\int_{B_N(0)} (4\pi t)^{-\frac{n}{2}} e^{-\frac{\|x-y\|^2}{4t}} f(y) \, dy \right)$$

$$= \int_{B_N(0)} \left(\frac{\partial}{\partial t} - \Delta_{n,x} \right) \left((4\pi t)^{-\frac{n}{2}} e^{-\frac{\|x-y\|^2}{4t}} \right) f(y) \, dy = 0.$$

Furthermore we find that for $t \in [T_0, T]$ and $x \in \overline{B_R(0)}$ the function u_N converges uniformly to u as we can prove that $\frac{\partial u_N}{\partial t}$, $\frac{\partial u_N}{\partial x_j}$ and $\frac{\partial^2 u_N}{\partial x_j^2}$ converge uniformly on $[T_0, T] \times \overline{B_R(0)}$ to $\frac{\partial u}{\partial t}$, $\frac{\partial u}{\partial x_j}$ and $\frac{\partial^2 u}{\partial x_j^2}$ respectively. This however implies $\frac{\partial u}{\partial t} - \Delta_n u = 0$ in $(T_0, T) \times B_R(0)$ and since $0 < T_0 < T$ and $0 < R$ were arbitrarily chosen we deduce that $\frac{\partial u}{\partial t} - \Delta_n u = 0$ in $(0, \infty) \times \mathbb{R}^n$.

Example 22.19. We want to show that

$$(2\pi)^{-n} \int_{\mathbb{R}^2} e^{-\frac{\|x\|^2}{2}} \cos x \cdot \xi \, dx = e^{-\frac{\|\xi\|^2}{2}}.$$

First we note that the estimate $\left| e^{-\frac{\|x\|^2}{2}} \cos x \cdot \xi \right| \leq e^{-\frac{\|x\|^2}{2}}$ implies that the integral is absolutely convergent as an improper integral. Here we used $x \cdot \xi$ as a shorter notation for the scalar product $\langle x, \xi \rangle$. With $x = (x_1, x_2)$ and $\xi = (\xi_1, \xi_2)$ we find

$$\cos x \cdot \xi = \cos(x_1 \xi_1 + x_2 \xi_2) = \cos x_1 \xi_1 \cos x_2 \xi_2 - \sin x_1 \xi_1 \sin x_2 \xi_2$$

and consequently for $R > 0$

$$\int_{[-R,R]^2} e^{-\frac{\|x\|^2}{2}} \cos x \cdot \xi \, dx = \int_{-R}^{R} \int_{-R}^{R} e^{-\frac{1}{2}(x_1^2 + x_2^2)} (\cos x_1 \xi_1 \cos x_2 \xi_2 - \sin x_1 \xi_1 \sin x_2 \xi_2) \, dx_1 \, dx_2$$

$$= \int_{-R}^{R} e^{-\frac{x_1^2}{2}} \cos x_1 \xi_1 \, dx_1 \int_{-R}^{R} e^{-\frac{x_2^2}{2}} \cos x_2 \xi_2 \, dx_2$$

$$- \int_{-R}^{R} e^{-\frac{x_1^2}{2}} \sin x_1 \xi_1 \, dx_1 \int_{-R}^{R} e^{-\frac{x_2^2}{2}} \sin x_2 \xi_2 \, dx_2$$

$$= \int_{-R}^{R} e^{-\frac{x_1^2}{2}} \cos x_1 \xi_1 \, dx_1 \int_{-R}^{R} e^{-\frac{x_2^2}{2}} \cos x_2 \xi_2 \, dx_2,$$

where we have used that $y \mapsto e^{-\frac{y^2}{2}}$ is an even function and $y \mapsto \sin y\eta$ is an odd function and therefore $\int_{-R}^{R} e^{-\frac{y^2}{2}} \sin y\eta \, dy = 0$. Since all integrals exist as absolute convergent improper integrals we find

$$(2\pi)^{-n} \int_{\mathbb{R}^2} e^{-\frac{\|x\|^2}{2}} \cos x \cdot \xi \, dx = \left((2\pi)^{-\frac{n}{2}} \int_{\mathbb{R}} e^{-\frac{x_1^2}{2}} \cos x_1 \xi_1 \, dx_1 \right) \left((2\pi)^{-\frac{n}{2}} \int_{\mathbb{R}} e^{-\frac{x_2^2}{2}} \cos x_2 \xi_2 \, dx_2 \right)$$

and our problem is reduced to prove that

$$(2\pi)^{-\frac{n}{2}} \int_{\mathbb{R}} e^{-\frac{y}{2}} \cos(\eta y) \, dy = e^{-\frac{\eta^2}{2}}.$$

Obviously we have

$$\frac{\partial}{\partial \eta} \left(\cos(y\eta) e^{-\frac{y^2}{2}} \right) = -y \sin(y\eta) e^{-\frac{y^2}{2}}$$

and therefore for $R > 0$ we find with

$$G_R(\eta) := \int_{-R}^{R} \cos(y\eta) e^{-\frac{y^2}{2}} \, dy$$

that

$$\frac{\partial}{\partial \eta} G_R(\eta) = \frac{\partial}{\partial \eta} \int_{-R}^{R} \cos(y\eta) e^{-\frac{y^2}{2}} \, dy = \int_{-R}^{R} \left(\frac{\partial}{\partial y} \cos(y\eta) \right) e^{-\frac{y^2}{2}} \, dy$$

$$= -\int_{-R}^{R} y \sin(y\eta) e^{-\frac{y^2}{2}} \, dy = \int_{-R}^{R} \sin(y\eta) \frac{\partial}{\partial y} e^{-\frac{y^2}{2}} \, dy$$

$$= -\int_{-R}^{R} \cos(y\eta) e^{-\frac{y^2}{2}} \, dy - \sin(y\eta) e^{-\frac{y^2}{2}} \Big|_{-R}^{R}$$

$$= -\eta G_R(\eta) - \sin(y\eta) e^{-\frac{y^2}{2}} \Big|_{-R}^{R}.$$

Thus, for $R \to \infty$ we obtain for $G(\eta) := \int_{-\infty}^{\infty} \cos(y\eta) \, e^{-\frac{y^2}{2}} \, dy$

$$G'(\eta) + \eta G(\eta) = 0 \ \text{ and } \ G(0) = \sqrt{2\pi}. \tag{22.18}$$

We claim that (22.18) can have only one continuous solution. Suppose that g and h are two solutions. By linearity $u = g - h$ solves

$$u'(\eta) + \eta u(\eta) = 0 \ \text{ and } \ u(0) = 0. \tag{22.19}$$

We want to prove that (22.19) for a continuous (in fact a differentiable function) implies $u = 0$. Multiplying the differential equation in (22.19) with u we obtain

$$u(\eta)u'(\eta) + \eta u^2(\eta) = 0$$

or

$$\frac{d}{d\eta}\left(\frac{u^2(\eta)}{2}\right) + \eta u^2(\eta) = 0. \tag{22.20}$$

For $a > 0$ we now integrate (22.20) over $[0, a]$ to find

$$\int_0^a \frac{d}{d\eta}\left(\frac{u^2(\eta)}{2}\right) d\eta + \int_0^a \eta u^2(\eta) \, d\eta = 0,$$

i.e.

$$\frac{u^2(a)}{2} + \int_0^a \eta u^2(\eta) \, d\eta = 0$$

which yields that

$$\int_0^a \eta u^2(\eta) \, d\eta = 0. \tag{22.21}$$

Since on $(0, a)$ the function $\eta \mapsto \eta u^2(\eta)$ is non-negative, we deduce from (22.21) that on $(0, a)$ this function must be zero, i.e. $u(\eta) = 0$ for all $\eta > 0$ and taking (22.19) into account for all $\eta \geq 0$. (For the conclusion that $\eta \mapsto \eta u^2(\eta)$ must vanish in $(0, a)$ the reader may compare with Problem 6 in Chapter I.25). The argument for $a < 0$ goes similarly. Thus we have proved that (22.18) can have at most one continuous solution. However, a direct calculation shows the $F(\eta) = \sqrt{2\pi}e^{-\frac{|\eta|^2}{2}}$ solves (22.18) and consequently

$$(2\pi)^{-\frac{1}{2}} \int_{-\infty}^{\infty} \cos(y\eta) \, e^{-\frac{|y|^2}{2}} \, dy = e^{-\frac{|\eta|^2}{2}}.$$

Problems

1.   a) For $B_R(0) \subset \mathbb{R}^n$ and $\mathbb{R}^n$ give an example of a compact Jordan measurable exhaustion.

   b) For the open disc $B_1(0) \subset \mathbb{R}^2$ construct a compact Jordan measurable exhaustion $(G_k)_{k \in \mathbb{N}}$ where each set G_k is the union of compact squares.

2. Prove that $(G_k)_{k \in \mathbb{N}}$ and $(H_k)_{k \in \mathbb{N}}$, where $G_k := \left[-k, -\frac{1}{k}\right] \cup \left[\frac{1}{k}, k\right]$ and $H_k := \left[-k, -\frac{1}{k^2}\right] \cup \left[\frac{1}{k}, k\right]$, are compact Jordan measurable exhaustions of $\mathbb{R} \setminus \{0\}$. Consider on $\mathbb{R} \setminus \{0\}$ the function function $x \mapsto \frac{1}{x^3}$. Prove that $\lim_{k \to \infty} \int_{G_k} \frac{1}{x^3} dx = 0$ whereas $\lim_{k \to \infty} \int_{H_k} \frac{1}{x^3} dx$ does not exist.

3. Let $f : \mathbb{R}^n \to \mathbb{R}$ be a continuous function and $(G_k)_{k \in \mathbb{N}}$ a sequence of compact Jordan measurable sets $G_k \subset \mathbb{R}^n$ with $\lim_{k \to \infty} \operatorname{vol}_n(G_k) = \infty$. Suppose that $f(x) \geq c > 0$ for all $x \in \bigcup_{k \in \mathbb{N}} G_k$. Prove that f is not improper Riemann integrable over $\mathbb{R}^n$.

4. We have already discussed the integrals

$$\int_{\mathbb{R}^2} e^{-x_1^2 - x_2^2} dx, \quad \int_{\mathbb{R}^n \setminus B_1(0)} ||x||^\alpha dx, \quad \text{and} \quad \int_{\mathbb{R}^n} (1 + ||x||^2)^{\frac{\alpha}{2}} dx \quad \text{for } n + \alpha < 0.$$

Now give reasons why these are improper Riemann integrals.

5. Let $f : \mathbb{R}^n \to \mathbb{R}$ be an absolutely improper integrable function and denote $\int_{\mathbb{R}^n} |f(x)| dx$ by $||f||_{L^1}$. Prove that for every $\xi \in \mathbb{R}^n$ the integral

$$\tilde{u}(\xi) := \int_{\mathbb{R}^n} \cos\langle x, \xi \rangle f(x) dx$$

exists as an absolutely improper Riemann integral and that the function $\tilde{u} : \mathbb{R}^n \to \mathbb{R}, \, \xi \mapsto \tilde{u}(\xi)$, is continuous.
Hint: show that $\tilde{u}$ is the uniform limit of $\tilde{u}_N$ where

$$\tilde{u}_N(\xi) = \int_{B_N(0)} \cos\langle x, \xi \rangle f(x) dx.$$

6. Prove that for every $\alpha > 0$ and all $\xi \in \mathbb{R}^n$ the integral

$$\int_{\mathbb{R}^n} \cos\langle \xi, x \rangle e^{-||x||^\alpha} dx$$

is absolutely convergent.

464

7. Formulate natural conditions for Theorem 22.17 to hold for higher order partial derivatives. (A proof is not required.)

8. Let $u : \mathbb{R}^n \to \mathbb{R}$ be a continuous function with compact support. Prove that $\tilde{u} : \mathbb{R}$ defined by

$$\tilde{u}(\xi) := \int_{\mathbb{R}^n} \cos\langle x, \xi \rangle u(x) dx$$

is a C^∞-function.

9. Let $u \in C(\mathbb{R}^n)$ be such that the functions $|u|$ and $x \mapsto ||x|||u(x)|$ are improper Riemann integrable over $\mathbb{R}$. Define $\tilde{u}$ as before by

$$\tilde{u}(\xi) = \int_{\mathbb{R}^n} \cos\langle x, \xi \rangle u(x) dx.$$

Prove that for all $1 \le j \le n$ the function $\tilde{u}$ has the continuous partial derivative $\frac{\partial \tilde{u}}{\partial \xi_j}$.

10. Let $u, v \in C_0(\mathbb{R}^n)$, i.e. continuous functions with compact support which we assume to be Jordan measurable. Define their **convolution** $u * v$ by

$$(u * v)(x) = \int_{\mathbb{R}^n} u(x - y) v(y) dy.$$

Prove that $u * v$ is well defined and a continuous function with support $\operatorname{supp}(u + v) = \operatorname{supp} u + \operatorname{supp} v$.

11. (**Friedrichs mollifier**)

a) With $a^{-1} := \left(\int_{||x|| \le 1} \exp((||x||^2 - 1)^{-1}) dx \right)$ define the function $j : \mathbb{R}^n \to \mathbb{R}$ by

$$j(x) = \begin{cases} a \exp((||x||^2 - 1)^{-1}), & ||x|| < 1 \\ 0, & ||x|| \ge 1. \end{cases}$$

Prove (by using established results) that $j \in C^\infty(\mathbb{R}^n)$, $\int j(x) dx = 1$, $j(x) \ge 0$, $\operatorname{supp} j \subset \overline{B_1(0)}$.

b) For $\epsilon > 0$ define $j_\epsilon(x) := \epsilon^{-n} j\left(\frac{x}{\epsilon}\right)$ and prove that $j_\epsilon \in C_0^\infty(\mathbb{R}^n)$, $\operatorname{supp} j_\epsilon \subset \overline{B_\epsilon(0)}$, $j_\epsilon \ge 0$ and $\int_{\mathbb{R}^n} j_\epsilon(x) dx = 1$.

c) For $u \in C_0(\mathbb{R}^n)$, supp u being Jordan measurable, define

$$J_\epsilon(u)(x) := (j_\epsilon * u)(x) := \int_{\mathbb{R}^n} j_\epsilon(x - y)u(y)dy.$$

Prove that

$$\int_{\mathbb{R}^n} j_\epsilon(x - y)u(y)dy = \int_{\mathbb{R}^n} u(x - y)j_\epsilon(y)dy,$$

$J_\epsilon(u) \in C^\infty(\mathbb{R}^n)$ and

$$\int_{\mathbb{R}^n} |J_\epsilon(u)(x)|^2 dx \le \int_{\mathbb{R}^n} |u(x)|^2 dx.$$

d) Show that

$$\lim_{\epsilon \to 0} \int_{\mathbb{R}^n} |J_\epsilon(u)(x) - u(x)|^2 dx = 0,$$

i.e. in the quadratic mean we can approximate a continuous function with compact support by C^∞-functions.

Part 5: Vector Calculus

23 The Scope of Vector Calculus

Those who studied Mathematics ca. 40 years ago at German universities had to study a substantial part of a "minor" subject in addition to their major subject. By tradition this was often Physics and then the theorems of Gauss, Green and Stokes, essentials for every physicist (not only for theoretical physicists), became of central importance. Accordingly these theorems played a serious part in the advanced analysis education which was often shared with physics students. To see how to **use** these theorems that everyone learnt in physics, most of all in theoretical electrodynamics, just have a look at J. D. Jackson [28]. As the students who read vector calculus, and hence those who met the theorems of Gauss, Green and Stokes, were typically in their second semester of their second year, the subject was treated in a rather concrete way relating to "nice" sets in $\mathbb{R}^2$ and $\mathbb{R}^3$ and their boundaries consisting of "well-behaved" surfaces or curves.

When theoretical mechanics was starting to be taught in the context of differentiable manifolds, see for example V. I. Arnold [5] or R. Abraham and J. E. Marsden [1], the exterior calculus of E. Cartan was introduced and the general versions of Stokes theorem for differentiable manifolds was discussed. Of course, this is not really suitable for second term second year students. Now we can also study these three theorems in the context of geometric measure theory, see H. Federer [13], M. Giaquinta, G. Modica and J. Souček. [22] and as an introduction F. Morgan [36].

There are many good reasons for the growing abstraction and generalisation of vector calculus, but at this stage of our Course we prefer to give an introduction to "classical" vector analysis in two dimensions and most of all in three dimensions. We will allow some digression since Gauss' theorem or the divergence theorem as it is sometimes called can be easily extended to a "classical" version in $\mathbb{R}^n$. In Volume VI we will discuss the extensions to differentiable manifolds, but it would be beyond the intention of our Course to put these theorems in the context of geometric measure theory.

In $\mathbb{R}^n$ the three basic first order differential operators of relevance to vector analysis are the **gradient**, the **divergence** and the **rotation** or **curl**. For $G \subset \mathbb{R}^3$ open and continuously differentiable mappings $\varphi : G \to \mathbb{R}$ as well as

$A : G \to \mathbb{R}^3$ we define

$$\operatorname{grad} \varphi := \nabla \varphi := \left(\frac{\partial \varphi}{\partial x_1}, \frac{\partial \varphi}{\partial x_2}, \frac{\partial \varphi}{\partial x_3} \right) = \sum_{j=1}^{3} \frac{\partial \varphi}{\partial x_j} e_j, \tag{23.1}$$

$$\operatorname{div} A := \nabla \cdot A := \frac{\partial A_1}{\partial x_1} + \frac{\partial A_2}{\partial x_2} + \frac{\partial A_3}{\partial x_3} = \sum_{j=1}^{3} \frac{\partial A_j}{\partial x_j}, \tag{23.2}$$

and

$$\operatorname{curl} A := \nabla \times A := \left(\frac{\partial A_2}{\partial x_3} - \frac{\partial A_3}{\partial x_2}, \frac{\partial A_1}{\partial x_3} - \frac{\partial A_3}{\partial x_1}, \frac{\partial A_2}{\partial x_1} - \frac{\partial A_1}{\partial x_2} \right) \tag{23.3}$$

where $\{e_1, e_2, e_3\} \subset \mathbb{R}^3$ (or $\{e_1, \ldots, e_n\} \subset \mathbb{R}^n$) denotes the canonical basis. The gradient maps a scalar-valued function onto a vector-valued function, i.e. a vector field, the divergence maps a vector field onto a scalar-valued function and finally the curl maps a vector field onto a vector field. As we know, the gradient and the divergence have straightforward generalisations to $\mathbb{R}^n$

$$\operatorname{grad} \varphi = \sum_{j=1}^{n} \frac{\partial \varphi}{\partial x_j} e_j, \tag{23.4}$$

and

$$\operatorname{div} A = \sum_{j=1}^{n} \frac{\partial A_j}{\partial x_j}, \tag{23.5}$$

but the curl has not such a generalisation, for extending curl we need E. Cartan's exterior calculus. However, sometimes for a vector field $A : G \to \mathbb{R}^2$, $G \subset \mathbb{R}^2$, the expression

$$\frac{\partial A_2}{\partial x_1} - \frac{\partial A_1}{\partial x_2} \tag{23.6}$$

(which is the third component of $\operatorname{curl} \tilde{A}$, $\tilde{A} = (A_1, A_2, 0)$) is treated as a projection of the curl of a vector field.

For a vector field $A : G \to \mathbb{R}^2$, $G \subset \mathbb{R}^2$, Green's theorem relates the volume integral of $\frac{\partial A_2}{\partial x_1} - \frac{\partial A_1}{\partial x_2}$ over G to a line integral of A where the curve is a certain representation of ∂G.

Gauss' theorem or divergence theorem or Gauss-Ostrogradskiï theorem relates the integral of $\operatorname{div} A$, $A : G \to \mathbb{R}^3$, $G \subset \mathbb{R}^3$, to an integral of A over

∂G, however for this ∂G must be a reasonable surface and we need to define a surface integral.

Finally, Stokes' theorem which holds for certain vector fields defined on a surface $S \subset \mathbb{R}^3$ relates the integral of the curl of the vector field over this surface to the integral of A over the "boundary" of the surface which is assumed to be a "nice" curve. The three formulae we want to give a meaning to, and then to apply to many problems are

$$\int_G \left(\frac{\partial A_2}{\partial x_1} - \frac{\partial A_1}{\partial x_2} \right) \mathrm{d}x = \int_{\partial G} A_1 \, \mathrm{d}x_1 + A_2 \, \mathrm{d}x_2 \qquad \text{(Green)}, \qquad (23.7)$$

$$\int_G \operatorname{div} A \, \mathrm{d}x = \int_{\partial G} \langle A, \vec{n} \rangle \sigma(\mathrm{d}x) \qquad \text{(Gauss)}, \qquad (23.8)$$

$$\int_S \langle \operatorname{rot} A, \vec{n} \rangle \sigma(\mathrm{d}x) = \int_{\partial S} A_1 \, \mathrm{d}x_1 + A_2 \, \mathrm{d}x_2 + A_3 \, \mathrm{d}x_3 \quad \text{(Stokes)}, \qquad (23.9)$$

where $\vec{n}$ is the normal vector to ∂G and S, respectively. In this form only Gauss' theorem has an obvious generalisation to higher dimensions which we will discuss too. But most of all we need to define and investigate these terms before we can give a precise meaning of these results.

One observation we can and shall make now is that the "dimension" of the domain of the integrals on the right hand sides (∂G or ∂S) is one less than on the left hand sides (G or S). As an open set in $\mathbb{R}^3$ ($\mathbb{R}^2$) the set G should be considered as a three-dimensional (two-dimensional) object in $\mathbb{R}^3$ ($\mathbb{R}^2$). We consider S as being two-dimensional, while ∂G and ∂S either a surface in $\mathbb{R}^3$ or a curve in $\mathbb{R}^2$ or $\mathbb{R}^3$, respectively, should be considered as a two- or one-dimensional object, respectively. Further, on the left hand sides we integrate certain derivatives while on the right hand sides we integrate the corresponding functions or vector fields. This is similar to the fundamental theorem, i.e. the one-dimensional case

$$\int_a^b f'(x) \, \mathrm{d}x = f(b) - f(a)$$

where $[a, b] \subset \mathbb{R}$ is a "one-dimensional object" while $\{a\}$ and $\{b\}$ are zero-dimensional.

To move forward, in the next chapter we will discuss the definition of the area of a surface in $\mathbb{R}^3$ and introduce surface integrals.

Problems

Although we have not developed new theory in this chapter, we would like to add some further results about the operations gradient, divergence and curl which are essentially problems on taking partial derivatives.

Before starting our problems we want to recollect the "trivial" statements. In the following λ and μ are real numbers, $\varphi, \psi : \mathbb{R}^n \to \mathbb{R}$ are at least C^1-functions and $A, B : \mathbb{R}^n \to \mathbb{R}^n$ are at least C^1-vector fields. In the case where higher order partial derivatives are involved we assume that the scalar functions and the vector fields involved have all continuous partial derivatives of the highest order appearing in the formulae. When we are working with the divergence or the gradient n can be any natural number. However, whenever terms are considered involving curl, then we assume that $n = 3$.

The operations grad, div and curl are linear:

$$\operatorname{grad}(\lambda\varphi + \mu\psi) = \lambda\operatorname{grad}\varphi + \mu\operatorname{grad}\psi;$$
$$\operatorname{div}(\lambda A + \mu B) = \lambda\operatorname{div}A + \mu\operatorname{div}B;$$
$$\operatorname{curl}(\lambda A + \mu B) = \lambda\operatorname{curl}A + \mu\operatorname{curl}B.$$

The following variants of Leibniz's rule are straightforward to prove:

$$\operatorname{grad}(\varphi\psi) = \varphi\operatorname{grad}\psi + \psi\operatorname{grad}\varphi;$$
$$\operatorname{div}(\varphi A) = \varphi\operatorname{div}A + \langle A, \operatorname{grad}\varphi\rangle;$$
$$\operatorname{div}(A \times B) = B\operatorname{curl}A - A\operatorname{curl}B;$$
$$\operatorname{curl}(\varphi A) = (\operatorname{grad}\varphi) \times A + \varphi\operatorname{grad}A.$$

1. For a C^2-function $\varphi : \mathbb{R}^3 \to \mathbb{R}$ and a C^2-vector field $A : \mathbb{R}^3 \to \mathbb{R}^3$ prove:

   a) $\operatorname{curl}(\operatorname{grad}\varphi) = 0$;

   b) $\operatorname{div}(\operatorname{curl}A) = 0$.

   Can we expect these results to hold when φ (or A) has all second order partial derivatives but they are not assumed to be continuous?

2. We know that the Laplacian Δ_n can be written as $\Delta_n\varphi = \operatorname{div}(\operatorname{grad}\varphi)$ for φ a C^2-function. It is helpful to introduce the Laplacian of a C^2-vector field $A : \mathbb{R}^n \to \mathbb{R}^n$ as $\Delta_n A = (\Delta_n A_1, \ldots, \Delta_n A_n)$. For $n = 3$ prove

$$\Delta_3 A = \operatorname{grad}(\operatorname{div}A) - \operatorname{curl}(\operatorname{curl}A).$$

3. Let $A : \mathbb{R} \times \mathbb{R}^3 \to \mathbb{R}^3, (t, x) \mapsto (A_1(t, x), A_2(t, x), A_3(t, x))$ be a C^2-vector field depending on a parameter t (interpreted as time) and denote by Δ_3, grad, div and curl the standard operations with respect to $x = (x_1, x_2, x_3)$, i.e. $\text{div} A(t, x) = \frac{\partial A_1}{\partial x_1}(t, x) + \frac{\partial a_2}{\partial x_3}(t, x) + \frac{\partial A_3}{\partial x_3}(t, x)$, etc. Further let $\varphi : \mathbb{R} \times \mathbb{R}^3 \to \mathbb{R}$ be a C^3-function. Suppose that A satisfies the vector-valued wave equation $\frac{\partial^2 A}{\partial t^2} - \Delta_3 A = 0$. Prove

a) If φ is independent of t and harmonic, then $A + \text{grad}\varphi$ is a further solution to the vector-valued wave equation.

b) If φ depends on t and satisfies the scalar wave equation, then $A + \text{grad}\varphi$ is again a solution to the vector valued wave equation.

24 The Area of a Surface in $\mathbb{R}^3$ and Surface Integrals

Our first aim is to define the area of a patch of a parametric surface $S = f(R)$, $f : Q \to \mathbb{R}^3$, $R \subset Q \subset \mathbb{R}^2$ with $\mathring{R} \neq \emptyset$. In addition we assume that $df|_{\mathring{Q}}$ is injective. We know from our discussion of the length of a parametric curve that an ad hoc definition such as (15.1) (or (7.19)) can be helpful to make progress, see Chapter 7, but needs a motivation later on, see Theorem 15.12. We now encounter a similar situation. Suppose that $R = \overline{R} \subset \mathring{Q}$ is the rectangle $R = [a, b] \times [c, d]$. Let $Z = (Z_1, Z_2)$ be a partition of R with points $z_{kl} = (u_k, v_l)$, $1 \leq k \leq N$ and $1 \leq l \leq M$. We can replace $f([u_k, u_{k+1}] \times [v_l, v_{l+1}])$ by the parallelogram

$$P_{kl} = f(u_k, v_l) + \text{span} \{f(u_{k+1}, v_l) - f(u_k, v_l), f(u_k, v_{l+1}) - f(u_k, v_l)\}.$$

The area of P_{kl} is given by

$$A(P_{kl}) = \|(f(u_{k+1}, v_l) - f(u_k, v_l)) \times (f(u_k, v_{l+1}) - f(u_k, v_l))\|$$

for which we find the approximation (by using the mean value theorem)

$$A(P_{kl}) \approx \|f_u(u_k, v_l) \times f_v(u_k, v_l)\| (u_{k+1} - u_k)(v_{l+1} - v_l).$$

Consequently an approximation of the area of $f(R)$ should be

$$\sum_{k,l} \|f_u(u_k, v_l) \times f_v(u_k, v_l)\| (u_{k+1} - u_k)(v_{l+1} - v_l)$$

which in the limit $\text{mesh}(Z) \to 0$ gives

$$\int_R \|f_u(u, v) \times f_v(u, v)\| \, du \, dv.$$

Thus we should assume that the following gives a reasonable definition of the area of a patch of a parametric surface.

Definition 24.1. Let $Q \subset \mathbb{R}^2$ be a Jordan measurable set with $\mathring{Q} \neq \emptyset$ and let $R \subset Q$ be a Jordan measurable set such that $\mathring{R} \neq \emptyset$ and $\overline{R} \subset Q$ is compact. Further let $f : Q \to \mathbb{R}^3$ be a parametric C^1-surface such that $df|_{\mathring{Q}}$ is injective. The **area** of $S := f(\overline{R})$ is defined as

$$A(S) := \int_{\overline{R}} \|(f_u \times f_v)(u, v)\| \, du \, dv. \tag{24.1}$$

Remark 24.2. From Chapter 8 we know that $f_u(u, v)$ and $f_v(u, v)$ give the directions of tangent lines to $f(Q)$ at $p = f(u, v)$, and $(f_u \times f_v)(u, v)$ is the direction of the normal line to $f(Q)$ at p.

We want to change slightly our notation. The mapping $f : Q \to \mathbb{R}^3$ has components for which we write $f(u, v) = \begin{pmatrix} x(u, v) \\ y(u, v) \\ z(u, v) \end{pmatrix}$. It follows that

$$f_u \times f_v = \begin{pmatrix} x_u \\ y_u \\ z_u \end{pmatrix} \times \begin{pmatrix} x_v \\ y_v \\ z_v \end{pmatrix} = \begin{pmatrix} y_u z_v - y_v z_u \\ z_u x_v - z_v x_u \\ x_u y_v - x_v y_u \end{pmatrix} \tag{24.2}$$

and by inspection we find

$$\begin{aligned}
\|f_u \times f_v\|^2 &= (y_u z_v - y_v z_u)^2 + (z_u x_v - z_v x_u)^2 + (x_u y_v - x_v y_u)^2 \\
&= (x_u^2 + y_u^2 + z_u^2)(x_v^2 + y_v^2 + z_v^2) \\
&\quad - (x_u x_v + y_u y_v + z_u z_v)(x_u x_v + y_u y_v + z_u z_v) \\
&= \det \begin{pmatrix} x_u^2 + y_u^2 + z_u^2 & x_u x_v + y_u y_v + z_u z_v \\ x_u x_v + y_u y_v + z_u z_v & x_v^2 + y_v^2 + z_v^2 \end{pmatrix} \\
&= \det \begin{pmatrix} \langle f_u, f_v \rangle & \langle f_u, f_v \rangle \\ \langle f_v, f_u \rangle & \langle f_v, f_v \rangle \end{pmatrix},
\end{aligned}$$

i.e.

$$\|f_u \times f_v\|^2 = \det \begin{pmatrix} \langle f_u, f_u \rangle & \langle f_u, f_v \rangle \\ \langle f_v, f_u \rangle & \langle f_v, f_v \rangle \end{pmatrix}. \tag{24.3}$$

It is common to define

$$g_{11}(u, v) := \langle f_u(u, v), f_u(u, v) \rangle, \tag{24.4}$$
$$g_{12}(u, v) := \langle f_u(u, v), f_v(u, v) \rangle, \tag{24.5}$$
$$g_{21}(u, v) := \langle f_v(u, v), f_u(u, v) \rangle, \tag{24.6}$$
$$g_{22}(u, v) := \langle f_v(u, v), f_v(u, v) \rangle, \tag{24.7}$$

and

$$g(u, v) := \det \begin{pmatrix} g_{11}(u, v) & g_{12}(u, v) \\ g_{21}(u, v) & g_{22}(u, v) \end{pmatrix} \tag{24.8}$$

which yields

$$\|f_u \times f_v\| = \sqrt{g(u, v)} \tag{24.9}$$

476

and therefore

$$A(S) = \int_{\overline{R}} \sqrt{g(u,v)} \, du \, dv. \tag{24.10}$$

This notation extends to higher dimensions, however in the classical theory of parametric surface one often uses

$$E(u,v) = g_{11}(u,v) = \langle f_u(u,v), f_u(u,v) \rangle, \tag{24.11}$$

$$F(u,v) = g_{12}(u,v) = g_{21}(u,v) = \langle f_u(u,v), f_v(u,v) \rangle, \tag{24.12}$$

$$G(u,v) = g_{22}(u,v) = \langle f_v(u,v), f_v(u,v) \rangle, \tag{24.13}$$

and consequently one finds

$$A(S) = \int_{\overline{R}} \sqrt{E(u,v)G(u,v) - F^2(u,v)} \, du \, dv. \tag{24.14}$$

Definition 24.3. *Let $f : Q \to \mathbb{R}^3$ be as in Definition 24.1.* **A.** *The function $g : Q \to \mathbb{R}$ defined by (24.8) is called the* **Gram determinant** *of the parametric surface f.*

B. *The mapping $\mathscr{I} : Q \to M(2, \mathbb{R})$, $\mathscr{I}(u,v) = \begin{pmatrix} E(u,v) & F(u,v) \\ F(u,v) & G(u,v) \end{pmatrix}$ induces for each $(u,v) \in Q$ a bilinear form $I_{(u,v)}$ on $\mathbb{R}^2$ by*

$$I_{(u,v)}(\xi, \eta) = (\xi_1, \xi_2) \begin{pmatrix} E(u,v) & F(u,v) \\ F(u,v) & G(u,v) \end{pmatrix} \begin{pmatrix} \eta_1 \\ \eta_2 \end{pmatrix}. \tag{24.15}$$

We call $I_{(u,v)}$ the **first fundamental form** *of the parametric surface f at $p = f(u,v)$.*

So far $A(S)$, $S = f(\overline{R})$, with $f : Q \to \mathbb{R}^3$ as before, depends on the parametrization of S, i.e. on $f|_{\overline{R}}$. However we expect $A(S)$ to depend only on S or equivalently the trace of $f|_{\overline{R}}$.

Theorem 24.4. *Let $f : Q \to \mathbb{R}^3$, $Q \subset \mathbb{R}^2$ open, be a C^1-surface and $R \subset Q$ an open Jordan measurable set such that $\overset{\circ}{R} \neq \emptyset$ and $\overline{R} \subset Q$ is compact. Further let $\Phi : Q' \to Q$, $Q' \subset \mathbb{R}^2$ open, be a C^1-diffeomorphism and $h := f \circ \Phi : Q' \to \mathbb{R}^3$. Then h is a parametric C^1-surface with $S = f(\overline{R}) = h(\overline{R'})$ and*

$$A(f(\overline{R})) = A(h(\overline{R'})) \tag{24.16}$$

holds, where $R' = \Phi^{-1}(R)$.

Proof. Let $R' = \Phi^{-1}(R)$. Since Φ, hence Φ^{-1}, is a C^1-diffeomorphism we know that R' is an open Jordan measurable set and $\overline{R'} = \Phi^{-1}(\overline{R})$ is a compact subset of Q'. Thus $h : Q' \to \mathbb{R}^3$ is a C^1-mapping with components $h(s,t) = (\xi(s,t), \eta(s,t), \zeta(s,t))$. With $(u,v) = \Phi(s,t)$ we find by the chain rule

$$h_s = \frac{\partial h}{\partial s} = \frac{\partial (f \circ \Phi)}{\partial s} = \frac{\partial f}{\partial u}\frac{\partial u}{\partial s} + \frac{\partial f}{\partial v}\frac{\partial v}{\partial s}$$

and

$$h_t = \frac{\partial h}{\partial t} = \frac{\partial (f \circ \Phi)}{\partial t} = \frac{\partial f}{\partial u}\frac{\partial u}{\partial t} + \frac{\partial f}{\partial v}\frac{\partial v}{\partial t}$$

which yields

$$h_s \times h_t = \left(f_u \frac{\partial u}{\partial s} + f_v \frac{\partial v}{\partial s} \right) \times \left(f_u \frac{\partial u}{\partial t} + f_v \frac{\partial v}{\partial t} \right)$$

$$= (f_u \times f_v) \left(\frac{\partial u}{\partial s}\frac{\partial v}{\partial t} - \frac{\partial u}{\partial t}\frac{\partial v}{\partial s} \right)$$

$$= (f_u \times f_v) \det J_\Phi, \tag{24.17}$$

i.e.

$$\|(h_s \times h_t)(s,t)\| = \|(f_u \times f_v)(\Phi(s,t))\| \, |\det J_\Phi(s,t)|. \tag{24.18}$$

Now the transformation theorem gives

$$\int_{R'} \|(h_s \times h_t)(s,t)\| \, ds \, dt = \int_{R'} \|(f_u \times f_v)(\Phi(s,t))\| \, |\det J_\Phi(s,t)| \, ds \, dt$$

$$= \int_{R'} \|(f_u \times f_v)(u,v)\| \, du \, dv,$$

or $A(h(\overline{R'})) = A(f(\overline{R}))$. $\qquad\square$

We now want to turn to examples. First we have a look at the (surface) area of a graph of a function $h : Q \to \mathbb{R}$ inducing the parametric surface

$$S = \Gamma(h) = \left\{ \begin{pmatrix} u \\ v \\ h(u,v) \end{pmatrix} \middle| (u,v) \in Q \right\}.$$

Example 24.5 (Area of a graph). We want to find the area of a surface obtained as the graph of a function. For this let $Q \subset \mathbb{R}^2$ be an open Jordan measurable set and $h : Q \to \mathbb{R}$ a C^1-function. The mapping defined by

$f : Q \to \mathbb{R}^3$, $f(u, v) = \begin{pmatrix} u \\ v \\ h(u, v) \end{pmatrix}$, gives a parametric surface with $f_u(u, v) =$

$\begin{pmatrix} 1 \\ 0 \\ h_u(u, v) \end{pmatrix}$ and $f_v(u, v) = \begin{pmatrix} 0 \\ 1 \\ h_v(u, v) \end{pmatrix}$, hence $\mathrm{d}f_{(u,v)}$ is injective. Further it
follows that

$$
\begin{aligned}
g(u, v) &= \langle f_u, f_v \rangle \langle f_v, f_v \rangle - \langle f_u, f_v \rangle^2 \\
&= (1 + h_u^2(u, v))(1 + h_v^2(u, v)) - h_u^2(u, v)h_v^2(u, v) \\
&= 1 + h_u^2(u, v) + h_v^2(u, v),
\end{aligned}
$$

which yields that for every compact Jordan measurable set $R \subset Q$ with
$S := f(R) \subset \mathbb{R}^3$ we have

$$
A(S) = A(\Gamma(h|_R)) = \int_R \sqrt{1 + h_u^2(u, v) + h_v^2(u, v)} \, \mathrm{d}u \, \mathrm{d}v. \qquad (24.19)
$$

Of course for R not compact we may still try to use (24.19) but the integral
might be understood as improper integral. In concrete cases a specific investigation is often the better way forward than developing a "general theory"
now.

Example 24.6 (Surface area of a sphere). The sphere $S^2 \subset \mathbb{R}^3$ can be
looked at as a parametric surface with $S^2 = f([0, \pi], [0, 2\pi))$, $f : [0, \pi] \times$
$[0, 2\pi) \to \mathbb{R}^3$ given by $f(\vartheta, \varphi) = (\sin \vartheta \cos \varphi, \sin \vartheta \sin \varphi, \cos \vartheta)$, compare with
Example 8.6.C. Thus we can find $A(S^2)$ by using (24.1) or (24.10). However
we can find $A(S^2)$ as $2A(\Gamma(h))$ where $h : \overline{B_1(0)} \to \mathbb{R}$, $h(u, v) = \sqrt{1 - u^2 - v^2}$,
$\overline{B_1(0)} \subset \mathbb{R}^2$. By (24.19) this leads to

$$
\begin{aligned}
A(\Gamma(h)) &= \int_{B_1(0)} \sqrt{1 + h_u^2(u, v) + h_v^2(u, v)} \, \mathrm{d}u \, \mathrm{d}v \\
&= \int_{B_1(0)} \left(1 + \frac{u^2}{1 - u^2 - v^2} + \frac{v^2}{1 - u^2 - v^2} \right)^{\frac{1}{2}} \mathrm{d}u \, \mathrm{d}v \\
&= \int_{B_1(0)} \frac{1}{\sqrt{1 - u^2 - v^2}} \, \mathrm{d}u \, \mathrm{d}v \\
&= \int_0^1 \int_0^{2\pi} \frac{r}{\sqrt{1 - r^2}} \, \mathrm{d}\varphi \, \mathrm{d}r = 2\pi \int_0^1 \frac{r}{\sqrt{1 - r^2}} \, \mathrm{d}r \\
&= 2\pi,
\end{aligned}
$$

and therefore we find

$$A(S^2) = 4\pi. \tag{24.20}$$

The area of a sphere in $\mathbb{R}^3$ with radius r is of course given by $4\pi r^2$ using (24.10).

Example 24.7 (Surface area of a torus). In Example 8.6.C we discussed the torus $T_{a,b} \subset \mathbb{R}^3$, $a > b > 0$ as a parametric surface given by the mapping $f : [0, 2\pi] \times [0, 2\pi] \to \mathbb{R}^3$, $f(u, v) = ((a + b\cos u)\cos v, (a + b\cos u)\sin v, b\sin u)$. This leads to

$$f_u(u, v) = (-b\sin u \cos v, -b\sin u \sin v, b\cos u)$$

and

$$f_v(u, v) = (-(a + b\cos u)\sin v, (a + b\cos u)\cos v, 0)$$

implying

$$\langle f_u, f_v\rangle(u, v) = b^2,$$
$$\langle f_v, f_v\rangle(u, v) = (a + b\cos u)^2,$$
$$\langle f_u, f_v\rangle(u, v) = 0$$

which yields

$$\sqrt{g(u, v)} = b(a + b\cos u)$$

and therefore by (24.10)

$$A(T_{a,b}) = \int_0^{2\pi} \int_0^{2\pi} b(a + b\cos u)\, du\, dv = 4\pi^2 ab.$$

Example 24.8 (Surface area of a surface of revolution). Let $h, k : [a, b] \to \mathbb{R}$ be C^1-functions, $h > 0$, and $f : [a, b] \times [0, 2\pi] \to \mathbb{R}^3$ be the corresponding surface of revolution, i.e. $f(u, v) = (h(u)\cos v, h(u)\sin v, k(u))$ see Example 8.7. Since

$$f_u(u, v) = \begin{pmatrix} h'(u)\cos v \\ h'(u)\sin v \\ k'(u) \end{pmatrix} \quad \text{and} \quad f_v(u, v) = \begin{pmatrix} -h(u)\sin v \\ h(u)\cos v \\ 0 \end{pmatrix}$$

480

we find accordingly

$$\|f_u(u,v) \times f_v(u,v)\| = \left\| \begin{pmatrix} -k'(u)h(u)\cos v \\ -k'(u)h(u)\sin v \\ h'(u)h(u) \end{pmatrix} \right\|$$
$$= \sqrt{(k'(u)^2 + h'(u)^2)\,h(u)^2}$$
$$= |h(u)|\sqrt{k'(u)^2 + h'(u)^2}.$$

Thus we have for $S = f([a,b] \times [0, 2\pi])$, see (24.1),

$$A(S) = \int_a^b \int_0^{2\pi} |h(u)|\sqrt{k'(u)^2 + h'(u)^2}\,\mathrm{d}v\,\mathrm{d}u$$
$$= 2\pi \int_a^b |h(u)|\sqrt{k'(u)^2 + h'(u)^2}\,\mathrm{d}u.$$

For example if C_l is a circular cylinder with symmetry axis of length l as well radius of the ground circle of length l we find $h(u) = k(u) = u$, $u \in (0, l)$, and consequently

$$A(C_l) = 2\pi \int_0^l \sqrt{2}u\,\mathrm{d}u = \sqrt{2}\pi l^2.$$

These examples suggest, as did our motivation before, that our ad hoc definition of the area of a parametric surface is reasonable. One may ask, as in the case of curves, whether we can derive (24.1) or (24.10) by first introducing a notion of rectifiable surfaces. Such an approach is possible, however it falls into the realm of geometric measure theory and we will not touch this now, also see H. Federer [13] as classical reference and F. Morgan [36] who gives a nice introduction.

In the following, if not stated otherwise, we will always assume that $Q \subset \mathbb{R}^2$ is an open set and $\emptyset \neq \mathring{R} \subset \overline{R} \subset Q$ where R is a Jordan measurable set such that $\overline{R}$ is compact. Further $f : Q \to \mathbb{R}^3$ is a C^1-function and $\mathrm{d}f_{(u,v)}$ is assumed to be injective for all $(u,v) \in Q$. The trace S of $f_{\overline{R}}$ is the surface we want to use as domain of mappings. As $S \subset \mathbb{R}^3$ it carries the relative topology of $\mathbb{R}^3$, see Proposition 1.24, and we can speak about continuous functions $\psi : S \to \mathbb{R}$ or continuous vector fields $F : S \to \mathbb{R}^3$. A point $p \in S$ can be represented as $p = f(u,v)$ and therefore we can write $\psi(p) = \psi(f(u,v))$ where $\psi \circ f : \overline{R} \to \mathbb{R}$ is now a function. It makes sense to call $\psi|_{\mathring{R}}$ a C^k-function if $(\psi \circ f) : \overline{R} \to \mathbb{R}$ is a C^k-function.

Definition 24.9. *Let* $\psi : S \to \mathbb{R}$ *be a continuous function. The integral of* ψ *over* S *is defined as*

$$\int_S \psi(p)\sigma(dp) := \int_S \psi \, d\sigma := \int_{\overline{R}} \psi(f(u,v))\sqrt{g(u,v)} \, du \, dv. \qquad (24.21)$$

Remark 24.10. A. Since $1 : S \to \mathbb{R}$, $1(p) = 1$ for all $p \in S$, is continuous we find

$$A(S) = \int_S 1 \, d\sigma = \int_{\overline{R}} \sqrt{g(u,v)} \, du \, dv \qquad (24.22)$$

which is analogous to $J^{(n)}(H) = \int_H 1 \, dx$ for a bounded Jordan measurable set $H \subset \mathbb{R}^n$.

B. Note that (24.21) is analogous to (15.43), i.e. the formula for the integral of a continuous, scalar-valued function $f : \text{tr}(\gamma) \to \mathbb{R}$ defined on a C^1-curve $\gamma : [a, b] \to \mathbb{R}^n$.

Let $R = \overline{R}_1 \cup \overline{R}_2$ be a partition of $\overline{R}$ into two compact Jordan measurable sets with non-empty interior $\mathring{R}_j$ such that $\mathring{R}_1 \cap \mathring{R}_2 = \emptyset$ and suppose that $\overline{R}_1 \cap \overline{R}_2$ has Jordan content zero. By Corollary 20.18 it follows that

$$\int_{\overline{R}} \psi(f(u,v))\sqrt{g(u,v)} \, du \, dv = \int_{\overline{R}_1} \psi(f(u,v))\sqrt{g(u,v)} \, du \, dv + \int_{\overline{R}_2} \psi(f(u,v))\sqrt{g(u,v)} \, du \, dv$$

and therefore, with $S_j = f(\overline{R}_j)$, $S = S_1 \cup S_2$ and

$$A(S) = A(S_1) + A(S_2) \qquad (24.23)$$

which of course extends under appropriate conditions to finitely many sets $\overline{R}_j$ or S_j respectively.

Since the trace of a Lipschitz curve $\gamma : [a, b] \to \mathbb{R}^2$ has in $\mathbb{R}^2$ Jordan content zero, see Lemma 21.3, it follows that if $\overline{R}_1 \cap \overline{R}_2 = \text{tr}(\gamma)$ for some Lipschitz curve γ then (24.23) holds. This also implies that if $\partial R = \partial \overline{R}$ is the trace of a Lipschitz curve then we have with $\mathring{S} = f(\mathring{R})$

$$\int_{\mathring{S}} \psi \, d\sigma = \int_S \psi \, d\sigma.$$

So far the definition of $\int_S \psi \, d\sigma$ depends on the parametrization f of S. However the next result shows that the definition is independent of the parametrization.

Theorem 24.11. *Let* $f : Q \to \mathbb{R}^3$, $R \subset Q$, S *and* ψ *be as stated before in Definition 24.9. Moreover let* $\Phi : Q' \to Q$, $Q' \subset \mathbb{R}^2$ *open, be a* C^1-*diffeomorphism. Define* $h : f \circ \Phi : Q' \subset \mathbb{R}^3$. *Then the following holds with* $R' = \Phi^{-1}(R)$

$$\int_{\overline{R'}} (\psi \circ h)(s,t) \, \|(h_s \times h_t)(s,t)\| \, ds \, dt = \int_{\overline{R}} (\psi \circ f)(u,v) \, \|(f_u \times f_v)(u,v)\| \, du \, dv.$$

Proof. Using the proof of Theorem 24.4 we note that

$$(h_s \times h_t)(s,t) = (f_u \times f_v) \circ \Phi(s,t) \det J_\Phi(s,t)$$

and the transformation theorem yields once again

$$\int_{\overline{R'}} (\psi \circ h)(s,t) \, \|(h_s \times h_t)(s,t)\| \, ds \, dt$$

$$= \int_{\overline{R'}} ((\psi \circ f) \circ \Phi)\,(s,t) \, \|(h_s \times h_t)(s,t)\| \, ds \, dt$$

$$= \int_{\overline{R'}} ((\psi \circ f)(\Phi(s,t))) \, \|(f_u \times f_v)(\Phi(s,t))\| \, |\det J_\Phi(s,t)| \, ds \, dt$$

$$= \int_{\overline{R}} (\psi \circ f)(u,v) \, \|(f_u \times f_v)(u,v)\| \, du \, dv$$

$$= \int_S \psi(p)\sigma(dp).$$

$\square$

The next example shows how to evaluate a surface integral.

Example 24.12. We want to find the integral

$$\int_S \psi(p)\sigma(dp)$$

where $S = \{(x,y,z) \in \mathbb{R}^3 \mid z = 4 - x^2 - y^2, x^2 + y^2 \le 4\}$ and $\psi(p) = \psi(x,y,z) = x^2 + y^2 - z + 4$. As we can see from Figure 24.1 the surface S is part of the paraboloid $z = 4 - x^2 - y^2$.

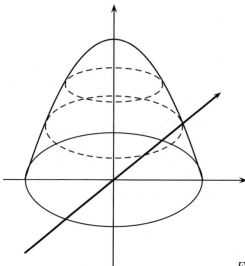

Figure 24.1

For S we find a representation as the graph of $h : \overline{B_2(0)} \to \mathbb{R}$, $h(u, v) = 4 - u^2 - v^2$, so $S = f\left(\overline{B_2(0)}\right)$ with $f(u, v) = \begin{pmatrix} u \\ v \\ 4 - u^2 - v^2 \end{pmatrix}$, which yields

$$f_u(u, v) = \begin{pmatrix} 1 \\ 0 \\ -2u \end{pmatrix}, \ f_v(u, v) = \begin{pmatrix} 0 \\ 1 \\ -2v \end{pmatrix} \text{ and therefore } g(u, v) = 1 + 4u^2 + 4v^2,$$

and further

$$\psi(p) = \psi(u, v, h(u, v)) = 2(u^2 + v^2).$$

Consequently we have

$$\int_S \psi(p) \, dp = \int_{B_2(0)} \psi(f(u, v)) \sqrt{g(u, v)} \, du \, dv$$

$$= \int_{B_2(0)} 2(u^2 + v^2) \sqrt{1 + 4u^2 + 4v^2} \, du \, dv$$

$$= \int_0^2 \int_0^{2\pi} 2r^2 \sqrt{1 + 4r^2} \, r \, d\varphi \, dr$$

484

$$= 4\pi \int_0^2 r^3 \sqrt{1 + 4r^2} \, dr$$

$$= \frac{\pi}{4} \int_1^{\sqrt{17}} \rho^2 (\rho^2 - 1) \, d\rho = \frac{2}{15} + \frac{1}{5} 17^{\frac{5}{2}} - \frac{1}{3} 17^{\frac{3}{2}},$$

where we used the substitution $\rho = \sqrt{1 + 4r^2}$.

We now want to define a surface integral for a vector field $F : S \to \mathbb{R}^3$ where S is as before a (patch of a) parametric surface $f : Q \to \mathbb{R}^3$, $Q \subset \mathbb{R}^2$ open, $S = f(R)$ where R is a Jordan measurable set with $\emptyset \neq \mathring{R}$, $\overline{R} \subset Q$ and $\overline{R}$ is compact. As usual we assume $df_{(u,v)}$ to be injective. Points on S are of the type

$$p = f(u, v) = (x(u, v), y(u, v), z(u, v)),$$

and F is given by components

$$F(p) = F(x, y, z) = F(f(u, v)) = (U(x, y, z), V(x, y, z), W(x, y, z))$$

(and the notations such as $U(p)$ or $U(f(u, v))$ are obvious extensions). Our definition of a surface integral for vector fields follows the line of that of a line integral along a curve, see Definition 15.15.

Definition 24.13. *Let $S \subset \mathbb{R}^3$ be as above with $R = \overline{R} \subset Q$ compact and Jordan measurable and let $F : S \to \mathbb{R}^3$ be a continuous vector field on S. The* **surface integral** *of F over S is defined as*

$$\int_S F \cdot d\sigma := \int_S F(p) \cdot \sigma(dp) := \int_R \langle F(f(u, v)), f_u(u, v) \times f_v(u, v) \rangle \, du \, dv.$$
$$(24.24)$$

Note that $\int_S F \cdot d\sigma$ is a scalar, not a vector. There are different and helpful ways to express (24.24). First we may introduce as in Definition 8.4.B the vector $N(u, v) := f_u(u, v) \times f_v(u, v)$ to write the right hand side of (24.24) as

$$\int_R \langle F(f(u, v)), N(u, v) \rangle \, du \, dv, \qquad (24.25)$$

and with $\vec{n} := \frac{N(u,v)}{\|N(u,v)\|}$, see (8.9), we further find

$$\int_R \langle F(f(u,v)), N(u,v) \rangle \, du \, dv$$

$$= \int_R \langle F(f(u,v)), \vec{n}(u,v) \rangle \, \|N(u,v)\| \, du \, dv$$

$$= \int_R \langle F(f(u,v)), \vec{n}(u,v) \rangle \sqrt{g(u,v)} \, du \, dv. \qquad (24.26)$$

With (24.26) in mind we will also use the notation

$$\int_S \langle F(p), \vec{n}(p) \rangle \sigma(dp) := \int_S F \cdot d\sigma \qquad (24.27)$$

where we write with a small abuse of notation $\vec{n}(p)$ also for the normal vector to S as function of $p \in S$. Next we may work out the scalar product:

$$\langle F(f(u,v)), N(u,v) \rangle = U(f(u,v))N_1(u,v) + V(f(u,v))N_2(u,v) + W(f(u,v))N_3(u,v)$$

where $N = (N_1, N_2, N_3)$ was used. Thus it follows that

$$\int_R \langle F(f(u,v)), N(u,v) \rangle \, du \, dv = \int_R U(f(u,v))N_1(u,v) \, du \, dv$$

$$+ \int_R V(f(u,v))N_2(u,v) \, du \, dv$$

$$+ \int_R W(f(u,v))N_3(u,v) \, du \, dv.$$

Using our previous notation $f(u,v) = (x(u,v), y(u,v), z(u,v))$, recall (24.2), we observe

$$N_1 = y_u z_v - y_v z_u,$$
$$N_2 = z_u x_v - z_v x_u,$$
$$N_3 = x_u y_v - x_v y_u.$$

The projection of $f(Q)$ onto the (y, z)-plane can be identified with the set $\{(\eta, \zeta) \in \mathbb{R}^2 \,|\, \eta = y(u,v), \zeta = z(u,v), \, u,v \in Q\}$ and then N_1 is the Jacobi determinant of $\varphi_{y,z}, \, (u,v) \mapsto (y(u,v), z(u,v))$, for which we may write

$$\frac{\partial(y, z)}{\partial(u, v)} := \left(\det J_{\varphi_{y,z}} \right)(u,v).$$

486

Analogously $\frac{\partial(z,x)}{\partial(u,v)}$ and $\frac{\partial(x,y)}{\partial(u,v)}$ are defined. This yields with (24.24)

$$\int_S F(p) \cdot \sigma(\mathrm{d}p)$$
$$= \int_R U(f(u,v)) \frac{\partial(y,z)}{\partial(u,v)} \, \mathrm{d}u \, \mathrm{d}v + \int_R V(f(u,v)) \frac{\partial(z,x)}{\partial(u,v)} \, \mathrm{d}u \, \mathrm{d}v + \int_R W(f(u,v)) \frac{\partial(x,y)}{\partial(u,v)} \, \mathrm{d}u \, \mathrm{d}v \qquad (24.28)$$

This formula makes sense and may help to evaluate the integral. By tradition, in particular in physics books, authors also write

$$\int_S F \cdot \mathrm{d}\sigma = \int_S U(x,y,z) \, \mathrm{d}y \, \mathrm{d}z + \int_S V(x,y,z) \, \mathrm{d}z \, \mathrm{d}x + \int_S W(x,y,z) \, \mathrm{d}x \, \mathrm{d}y, \quad (24.29)$$

or noting that $\frac{\partial(z,x)}{\partial(u,v)} = -\frac{\partial(x,z)}{\partial(u,v)}$

$$\int_S F \cdot \mathrm{d}\sigma = \int_S U(x,y,z) \, \mathrm{d}y \, \mathrm{d}z - \int_S V(x,y,z) \, \mathrm{d}x \, \mathrm{d}z + \int_S W(x,y,z) \, \mathrm{d}x \, \mathrm{d}y. \quad (24.30)$$

While both formulae are quite "suggestive" and give (physicists) some intuition, we think that for beginners they are more confusing than helpful. Note that switching from (24.29) to (24.30) is a change of sign in the second integral on the right hand side. This we will understand much better when eventually in Volume VI we investigate differential forms (or even k-forms). A further often useful geometric consideration is the following. The normal vector $\vec{n} = \frac{N}{\|N\|}$ is a unit vector in $\mathbb{R}^3$ and its components are obtained by projecting $\vec{n}$ onto the e_1-, e_2- and e_3-axis, respectively. These projections are given by $\langle \vec{n}, e_j \rangle$, but note $\vec{n} = \vec{n}(u,v)$. The quantities

$$\cos \gamma_j := \langle \vec{n}, e_j \rangle$$

are called the **direction cosines** of S at $p = f(u,v)$ (with respect to the normal vector $\vec{n}$). Thus we find

$$\int_R U(f(u,v)) N_1(u,v) \, \mathrm{d}u \, \mathrm{d}v = \int_R U(f(u,v)) \cos(\gamma_1(u,v)) \sqrt{g(u,v)} \, \mathrm{d}u \, \mathrm{d}v,$$

and hence

$$\int_S F \cdot \mathrm{d}\sigma = \int_R \left(U(f(u,v)) \cos \gamma_1(u,v) + V(f(u,v)) \cos \gamma_2(u,v) \right.$$
$$\left. + W(f(u,v)) \cos \gamma_3(u,v) \right) \sqrt{g(u,v)} \, \mathrm{d}u \, \mathrm{d}v. \qquad (24.31)$$

Again, (24.31) is a well defined and understood equality. The often used formula

$$\int_S F \cdot d\sigma = \int_S (U \cos \gamma_1 + V \cos \gamma_2 + W \cos \gamma_3) \, d\sigma$$

is for a beginner often misleading as short hand for (24.31).

Returning to our original definition (24.24) it is obvious how to obtain under suitable condition results such as

$$\int_S F \cdot d\sigma = \int_{S_1} F \cdot d\sigma + \int_{S_2} F \cdot d\sigma, \quad S = S_1 \cup S_2, \quad \mathring{S}_1 \cap \mathring{S}_2 = \emptyset,$$

or

$$\int_S F \cdot d\sigma = \int_{\mathring{S}} F \cdot d\sigma.$$

A non-trivial problem now is the independence of the definition of the parametrization. Consider $\Phi : \mathbb{R}^2_{(\xi,\eta)} \to \mathbb{R}^2_{(u,v)}$ given by $\Phi(\xi, \eta) = (u(\xi, \eta), v(\xi, \eta)) := (\eta, \xi)$ which just changes the order of the variables. For $h : \mathbb{R}^2_{(\xi,\eta)} \to \mathbb{R}$ and $f : \mathbb{R}^2_{(u,v)} \to \mathbb{R}$ two C^1-functions related by $h = f \circ \Phi$ we find

$$h_\xi = f_v \circ \Phi \quad \text{and} \quad h_\eta = f_u \circ \Phi,$$

implying that

$$(h_\xi \times h_\eta)(\xi, \eta) = ((f_v \times f_u) \circ \Phi)(\xi, \eta) = -((f_u \times f_v) \circ \Phi)(\xi, \eta),$$

which yields

$$\langle F(h(\xi, \eta)), (h_\xi \times h_\eta)(\xi, \eta) \rangle = -\langle F(f \circ \Phi)(\xi, \eta), (f_u \times f_v) \circ \Phi(\xi, \eta) \rangle.$$

Note that the transformations Φ has the Jacobi determinant

$$\det \begin{pmatrix} \frac{\partial \eta}{\partial \xi} & \frac{\partial \eta}{\partial \eta} \\ \frac{\partial \xi}{\partial \xi} & \frac{\partial \xi}{\partial \eta} \end{pmatrix} = \det \begin{pmatrix} 0 & 1 \\ 1 & 0 \end{pmatrix} = -1,$$

hence

$$\langle F(h(\xi, \eta)), (h_\xi \times h_\eta)(\xi \times \eta) \rangle = \langle F(f \circ \Phi)(\xi, \eta), (f_u \times f_v) \circ \Phi(\xi, \eta) \rangle \det J_\Phi(\xi, \eta).$$

This formula holds in general as we can see by using (24.17). So we are led to make a distinction whether a change of coordinate has a positive or a negative Jacobi determinant.

Definition 24.14. *Let* Q_1, $Q_2 \subset \mathbb{R}^n$ *be two open sets and* $\Phi : Q_1 \to Q_2$ *a* C^1-*diffeomorphism. We call* Φ ***orientation preserving*** *or* ***orientation faithful*** *if* $\det J_\Phi > 0$ *in* Q_1.

Since $\det J_{\Phi^{-1}} = \frac{1}{\det J_\Phi}$ we conclude that Φ is orientation preserving if and only if Φ^{-1} is. From our previous considerations we conclude

Proposition 24.15. *For a* C^1-*coordinate change* $\Phi : Q' \to Q$ *the following holds with* $R' = \Phi^{-1}(R)$

$$\int_{\overline{R'}} \langle F(h(\xi,\eta)), h_\xi \times h_\eta)(\xi,\eta)\rangle \, \mathrm{d}\xi \, \mathrm{d}\eta$$

$$= \mathrm{sgn}(\det J_\Phi) \int_{\overline{R}} \langle F(f(u,v)), (f_u \times f_v)(u,v)\rangle \, \mathrm{d}u \, \mathrm{d}v. \quad (24.32)$$

Corollary 24.16. *The surface integral* (24.24) *is invariant under an orientation preserving change of coordinates.*

Let S be a parametric surface and $p \in S$. At p we want to consider the normal line to S. While the tangent plane and the normal line as a function of the parameters (u,v) is always well defined, they may not be well defined as a function of $p = f(u,v)$. For this we must assume that f is injective, i.e. S has no double points. More precisely, whenever f is injective in a neighbourhood of $f^{-1}(p)$ we can speak of a tangent plane and normal line to S at p.

Suppose that S is a parametric surface as before however with an injective parametrization f. Using (8.8) and (8.9) we find that $\det(\vec{t}_1, \vec{t}_2, \vec{n}) > 0$. We now call $\vec{n}$ the positivity orientated unit normal vector to S at p and $-\vec{n}$ the negatively orientated unit normal vector to S at p. In this sense for every injectively parametrized surface S the field of positively orientated unit normal vectors induces a (positive) **orientation** on S. The orientation of S is unchanged under an orientation preserving change of coordinates and will be reversed if the Jacobi determinant of the change of coordinates is negative. With this new parametrization the orientation is now given by $-\vec{n}$ when $\vec{n}$ is the unit normal vector associated with the original parametrization.

Eventually we want to evaluate some surface integrals of vector fields.

Example 24.17. For a patch $S = f(R)$ of a parametric C^1-surface $f : Q \to \mathbb{R}^3$, $R \subset Q \subset \mathbb{R}^2$, R Jordan measurable, compact and Q open, and the unit normal vector field $\vec{n} : S \to \mathbb{R}^3$, $\vec{n} = \vec{t}_1 \times \vec{t}_2$, we find

$$\int_S \vec{n} \cdot d\sigma = \int_R \langle \vec{n}(u, v), \vec{n}(u, v) \rangle \sqrt{g(u, v)} \, du \, dv$$

$$= \int_R 1 \sqrt{g(u, v)} \, du \, dv = A(S).$$

Example 24.18. The equation $3x + 2y + 6z = 12$ defines a plane in $\mathbb{R}^3$. On the patch S of this plane defined by the conditions $x \geq 0$, $y \geq 0$ and $z \geq 0$ the vector field $F(x, y, z) = (12z, -9x, 6y)$ is given. We want to calculate the surface integral $\int_S F \cdot d\sigma$.

With $h(u, v) = 2 - \frac{u}{2} - \frac{v}{3}$ a parametrization of S is given by $f : R \to \mathbb{R}^3$, $f(u, v) = (u, v, h(u, v))$ where R is the triangle $R = \{(u, v) \in \mathbb{R}^2 \,|\, u \geq 0, v \geq 0, 3u + 2v \leq 12\}$. It follows that $(f_u \times f_v)(u, v) = \begin{pmatrix} 1 \\ 0 \\ -\frac{1}{2} \end{pmatrix} \times \begin{pmatrix} 0 \\ 1 \\ \frac{1}{3} \end{pmatrix} = \begin{pmatrix} \frac{1}{2} \\ \frac{1}{3} \\ 1 \end{pmatrix}$

and therefore we find

$$F(f(u, v)) = F(x, y, z) = (24 - 6u - 4v, -9u, 6v)$$

and

$$\langle F(f(u, v)), (f_u \times f_v)(u, v) \rangle = \left\langle \begin{pmatrix} 24 - 6u - 4v \\ -9u \\ 6v \end{pmatrix}, \begin{pmatrix} \frac{1}{2} \\ \frac{1}{3} \\ 1 \end{pmatrix} \right\rangle$$

$$= 12 - 6u + 4v$$

which yields

$$\int_S F \cdot d\sigma = \int_R \langle F(f(u, v)), (f_u \times f_v)(u, v) \rangle \, du \, dv$$

$$= \int_0^4 \int_0^{\frac{12-3u}{2}} (12 - 6u - 4v) \, dv \, du$$

$$= \int_0^4 \left(-18u + \frac{9}{2} u^2 \right) du = -48.$$

For the following two examples a remark is needed due to the partial periodicity of parametrization which leaves us either with a non-compact domain of integration or with double points. We will discuss this problem in a wider context in the beginning of the next chapter and take for the moment for granted that our calculations are justified. The key observation will be that a surface integral of a continuous function or vector field shall not change when a set of surface area zero is added or removed.

Example 24.19. Let $S := \big\{(x, y, z) \in \mathbb{R}^3 \,\big|\, x^2 + y^2 = 9,\, 0 \le z \le z_0\big\}$ be part of the cylinder $C = \big\{(x, y, z) \in \mathbb{R}^3 \,\big|\, x^2 + y^2 = 9,\, z \in \mathbb{R}\big\}$ and on S consider the vector field $F = \begin{pmatrix} x \\ y \\ h(z) \end{pmatrix}$ with some continuous function $h : [0, z_0] \to \mathbb{R}$. We want to find $\int_S F \cdot d\sigma$.

A parametrization of S is given by $f : [0, 2\pi] \times [0, z_0] \to \mathbb{R}^3$ $f(u, v) = \begin{pmatrix} \cos u \\ \sin u \\ v \end{pmatrix} = \begin{pmatrix} x \\ y \\ z \end{pmatrix}$ and with $f_u(u, v) = \begin{pmatrix} -\sin u \\ \cos u \\ 0 \end{pmatrix}$, $f_v(u, v) = \begin{pmatrix} 0 \\ 0 \\ 1 \end{pmatrix}$ we find

$$(f_u \times f_v)(u, v) = \begin{pmatrix} \cos u \\ \sin u \\ 0 \end{pmatrix} \text{ which implies}$$

$$\langle F(f(u, v)), (f_u \times f_v)(u, v)\rangle = \left\langle \begin{pmatrix} \cos u \\ \sin u \\ h(v) \end{pmatrix}, \begin{pmatrix} \cos u \\ \sin u \\ 0 \end{pmatrix}\right\rangle = \cos^2 u + \sin^2 u = 1$$

and consequently

$$\int_S F \cdot d\sigma = \int_0^{2\pi} \int_0^{z_0} \langle F(f(u, v)), (f_u \times f_v)(u, v)\rangle \, dv \, du$$

$$= \int_0^{2\pi} \int_0^{z_0} 1 \, dv \, du = 2\pi z_0.$$

Example 24.20. Consider on $S = \partial B_r(0) = \big\{(x, y, z) \in \mathbb{R}^3 \,\big|\, x^2 + y^2 + z^2 = r^2\big\}$ a radial vector field $F(x, y, z) = h(r) \begin{pmatrix} x \\ y \\ z \end{pmatrix}$. We want to investigate $\int_S F \cdot d\sigma$. The surface S admits the parametrization $f : [0, \pi] \times [0, 2\pi] \to \mathbb{R}^3$ where

$$f(\vartheta, \varphi) = (r \sin \vartheta \cos \varphi, r \sin \vartheta \sin \varphi, r \cos \vartheta) = (x, y, z).$$

491

Since

$$f_\vartheta(\vartheta, \varphi) = \begin{pmatrix} r \cos \vartheta \cos \varphi \\ r \cos \vartheta \sin \varphi \\ -r \sin \vartheta \end{pmatrix} \quad \text{and} \quad f_\varphi(\vartheta, \varphi) = \begin{pmatrix} -r \sin \vartheta \sin \varphi \\ r \sin \vartheta \cos \varphi \\ 0 \end{pmatrix}$$

we have

$$(f_u \times f_v)(\vartheta, \varphi) = \begin{pmatrix} r^2 \sin^2 \vartheta \cos \varphi \\ r^2 \sin^2 \vartheta \sin \varphi \\ r^2 \sin \vartheta \cos \varphi \end{pmatrix} = (r \sin \vartheta) f(\vartheta, \varphi).$$

Now, with $F(x, y, z) = F(f(\vartheta, \varphi)) = h(r) \begin{pmatrix} x \\ y \\ z \end{pmatrix} = h(r) f(\vartheta, \varphi)$ it follows that

$$\langle F(f(\vartheta, \varphi)), (f_\vartheta \times f_\varphi)(\vartheta, \varphi) \rangle = (r \, h(r) \sin \vartheta) \langle f(\vartheta, \varphi), f(\vartheta, \varphi) \rangle$$
$$= r^3 h(r) \sin \vartheta,$$

which implies

$$\int_S F \cdot d\sigma = \int_0^\pi \int_0^{2\pi} r^3 h(r) \sin \vartheta \, d\varphi \, d\vartheta$$
$$= 2\pi r^3 h(r) \int_0^\pi \sin \vartheta \, d\vartheta = 4\pi r^3 h(r).$$

If for example F is the Newton force of attraction of a mass m_1 at the origin and a mass m_2 at the point $p = (x, y, z)$, i.e.

$$F(p) = -\gamma \frac{m_1 m_2}{r^3} p$$

where r is the distance from the origin to p, then we find

$$\int_{\partial B_r(0)} F(p) \cdot \sigma(\mathrm{d}p) = -4\pi \gamma m_1 m_2.$$

Problems

1. Find the coefficients of the first fundamental form of the following surfaces:

a) S is the graph of a C^1-function $h : Q \to \mathbb{R}$ where $Q \subset \mathbb{R}^2$ is an open Jordan measurable set;

b) S is the boundary of the ball $B_r(0) \subset \mathbb{R}^3$;

c) S is a surface of revolution.

2. Let $h : \mathbb{R}^2 \to \mathbb{R}$ depend only on the first variable, i.e. $h(u, v) = g(u)$ where $g : \mathbb{R} \to \mathbb{R}$ is a C^1-function. Give a geometric interpretation of $\Gamma(h) \subset \mathbb{R}^3$ as a surface where $g(u) = \begin{pmatrix} u \\ 0 \\ g(u) \end{pmatrix}$ is considered as a curve in $\mathbb{R}^3$. For $R := [a, b] \times [c, d]$, $a < b$ and $c < d$, find the area of $S = \Gamma(h|_R) \subset \mathbb{R}^3$. Now choose $g(u) = \frac{u^2}{2}$ and $R = [2, 4] \times [-1, 1]$ and find the area of the corresponding surface $\tilde{S} = \Gamma(\tilde{h})$, $\tilde{h}(u, v) = \frac{u^2}{2}$, $(u, v) \in [2, 4] \times [-1, 1]$.

3. Find the Gram determinant of the parametric surface $f : \mathbb{R}^2 \to \mathbb{R}^3$,
$$f(u, v) = \begin{pmatrix} u + v \\ u - v \\ uv \end{pmatrix}.$$
Given the function $\psi : \mathbb{R}^3 \to \mathbb{R}$, $\psi(x, y, z) = 5\sqrt{5}(x^2 + y^2 + 4)^{\frac{3}{2}}$, find the integral $\int_S \psi \cdot d\sigma$ where $S = f([2, 5] \times [3, 4])$.

4. Let $S := f([a, b] \times [0, 2\pi])$, $a < b$, be a surface of revolution where
$$f : [a, b] \times [0, 2\pi] \to \mathbb{R}^3, \quad f(u, v) = \begin{pmatrix} h(u) \cos v \\ h(u) \sin v \\ k(u) \end{pmatrix}$$
with C^1-functions $h, k : [a, b] \to \mathbb{R}$, $h > 0$. Further let $F : \mathbb{R}^3 \to \mathbb{R}^3$, $F(x, y, z) = \begin{pmatrix} F_1(x, y, z) \\ F_2(x, y, z) \\ F_3(x, y, z) \end{pmatrix}$ be a continuous vector field. Show that

$$\int_S F \cdot d\sigma = \int_a^b \left(\int_0^{2\pi} \Big(h'(u)h(u)F_3(f(u, v)) - k'(u)h(u) \sin v F_2(f(u, v)) \right.$$
$$\left. - k'(u)h(u) \cos v F_1(f(u, v)) \Big) dv \right) du.$$

Now find an expression for $\int_S F \cdot d\sigma$ when S is as before and $F(x, y, z) = \begin{pmatrix} x \\ y \\ z \end{pmatrix}$.

5. Use the result of Problem 4 to evaluate the integral $\int_S F \cdot d\sigma$ where

$$F(x, y, z) = \begin{pmatrix} x \\ y \\ z \end{pmatrix} \text{ and } S \text{ is the surface of revolution given by } f :$$

$$[-1, 1] \times [0, 2\pi] \to \mathbb{R}^3, \ f(u, v) = \begin{pmatrix} \sqrt{1 - u^2} \cos v \\ \sqrt{1 - u^2} \sin v \\ u \end{pmatrix}.$$

6. Consider the surface $S = S_1 \cup S_2 \subset \mathbb{R}^3$ where $S_1 = \{(x, y, z) \in \mathbb{R}^3 | x^2 + y^2 \leq 1, z = 0\}$ and $S_2 = \{(x, y, z) \in \mathbb{R}^3 | x^2 + y^2 + z^2 \leq 1, z > 0\}$. Further

let $F : S \to \mathbb{R}^3$ be the vector field defined by $F(x, y, z) = \begin{pmatrix} 0 \\ 0 \\ 1 \end{pmatrix}$ for

$(x, y, z) \in S_1$ and $F(x, y, z) = \begin{pmatrix} x \\ y \\ z \end{pmatrix}$ for $(x, y, z) \in S_2$. Find the value

of $\int_S F \cdot d\sigma$.

7. Find $\int_S F \cdot d\sigma$ where $S = \partial B_R(0) \cap \{(x, y, z) \in \mathbb{R}^3 | z \geq 0\}$ and $F :$

$\mathbb{R}^3 \to \mathbb{R}^3$ is given by $F(x, y, z) = \begin{pmatrix} g_1(\rho) \\ g_2(\rho) \\ g_3(\rho) \end{pmatrix}$, $\rho^2 = x^2 + y^2 + z^2$, and

$g_j : [0, \infty) \to \mathbb{R}, j = 1, 2, 3$, is a continuous function.

25 Gauss' Theorem in $\mathbb{R}^3$

For suitable domains $G \subset \mathbb{R}^3$ with boundaries ∂G, interpreted as a surface, and C^1-vector fields $F : G \to \mathbb{R}^3$ we want to prove

$$\int_G \operatorname{div} F \, dx = \int_{\partial G} F \cdot d\sigma. \tag{25.1}$$

If F belongs to the class C^1 and G is a compact Jordan measurable set in $\mathbb{R}^3$, then $\operatorname{div} F$ is a continuous function which is integrable over G. Thus the volume integral on the left hand side of (25.1) is well defined for a large class of domains. The situation is more complicated with the surface integral on the right hand side. The boundary ∂G is by definition the set $\overline{G} \backslash \mathring{G}$ and for $G = \overline{G}$ being compact, ∂G is itself compact. However for the definition of $\int_{\partial G} F \cdot d\sigma$ we need ∂G to be the trace of a parametric surface of class C^1. This does not even hold for a cube.

A way forward is to look at the definition of a surface integral, i.e. (24.24), and recall that for every set $N \subset R$, $J^{(2)}(N) = 0$ we have

$$\int_R \langle F(f(u, v)), (f_u \times f_v)(u, v) \rangle \, du \, dv = \int_{R \backslash N} \langle F(f(u, v)), (f_u \times f_v)(u, v) \rangle \, du \, dv$$

where $R \subset Q$ is compact $f : Q \to \mathbb{R}^3$ is as in Definition 24.13 and $F : R \to \mathbb{R}^3$ is a continuous vector field. Now let $\gamma : I \to R$ be a Lipschitz curve. Then $J^{(2)}(\gamma(I)) = 0$ and therefore

$$\int_R \langle F(f(u, v)), (f_u \times f_v)(u, v) \rangle \, du \, dv = \int_{R \backslash \gamma(I)} \langle F(f(u, v)), (f_u \times f_v)(u, v) \rangle \, du \, dv.$$

For $1 \leq l \leq k$ let $S_l = f_l(R_l)$, $R_l \subset Q_l \subset \mathbb{R}^2$, Q_l open and R_l compact, be a C^1-parametric surface. In addition assume that either $S_j \cap S_l = \emptyset$ or $S_j \cap S_l = \operatorname{tr}(\gamma_{jl})$ where $\gamma_{jl} : I_{jl} \to \mathbb{R}^3$ is a Lipschitz curve. Then we define for $S := S_1 \cup \cdots \cup S_k$ and a continuous vector field $F : S \to \mathbb{R}^3$ the surface integral

$$\int_S F \cdot d\sigma := \sum_{j=1}^k \int_{S_j} F \cdot d\sigma, \tag{25.2}$$

an idea which we can also use for defining a surface integral for scalar valued functions. Note that instead of $\gamma_{jl} : I_{jl} \to \mathbb{R}^3$ we may work with

$\eta_{jl} : \tilde{I}_{jl} \to R_j \cap R_l$ and then define $\gamma_{jl} := f_j \circ \eta_{jl}$ or $\tilde{\gamma}_{jl} = f_l \circ \eta_{jl}$ and assuming $\operatorname{tr}(\gamma_{jl}) = \operatorname{tr}(\tilde{\gamma}_{jl})$.

We now want to discuss a lengthy example which eventually already gives us Gauss' theorem for the unit cube. We follow closely H. Fischer and H. Kaul [18]. Consider the unit cube $W = [0,1] \times [0,1] \times [0,1] \subset \mathbb{R}^3$, i.e. $W = \{(x_1, x_2, x_3) \in \mathbb{R}^3 \mid 0 \le x_j \le 1, j = 1, 2, 3\}$. The boundary ∂W is the union of the following six surfaces

$$S_{10} := \{(0, x_2, x_3) \mid x_2, x_3 \in [0,1]\}, \quad S_{11} := \{(1, x_2, x_3) \mid x_2, x_3 \in [0,1]\},$$
$$S_{20} := \{(x_1, 0, x_3) \mid x_1, x_3 \in [0,1]\}, \quad S_{21} := \{(x_1, 1, x_3) \mid x_1, x_3 \in [0,1]\},$$
$$S_{30} := \{(x_1, x_2, 0) \mid x_1, x_2 \in [0,1]\}, \quad S_{31} := \{(x_1, x_2, 1) \mid x_1, x_2 \in [0,1]\}.$$

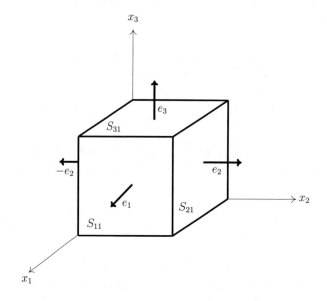

Figure 25.1

Clearly $S_{j_0} \cap S_{j_1} = \emptyset$ whereas $S_{j\alpha} \cap S_{l\alpha}$ for $j \ne l$, $\alpha = 1, 2$, and $S_{j_0} \cap S_{l_1}$ for $j = l$ are edges of the cube, i.e. line segments of length 1 parallel to a coordinate axis with end points being vertices of W. Each surface $S_{j\alpha}$, $j = 1, 2, 3$,

$\alpha = 0, 1$, is a C^1-parametric surface with parametrization $f_{j\alpha} : Q \to \mathbb{R}^3$, $Q = [0, 1] \times [0, 1]$, given by

$$
\begin{aligned}
f_{10} &: Q \to \mathbb{R}^3, & f_{10}(u, v) &= (0, v, u), \\
f_{20} &: Q \to \mathbb{R}^3, & f_{20}(u, v) &= (u, 0, v), \\
f_{30} &: Q \to \mathbb{R}^3, & f_{30}(u, v) &= (v, u, 0), \\
f_{11} &: Q \to \mathbb{R}^3, & f_{11}(u, v) &= (1, u, v), \\
f_{21} &: Q \to \mathbb{R}^3, & f_{21}(u, v) &= (v, 1, u), \\
f_{31} &: Q \to \mathbb{R}^3, & f_{31}(u, v) &= (u, v, 1).
\end{aligned}
$$

The corresponding tangent vectors (which coincide with the unit tangent vectors) and normal vectors are

$$
\begin{aligned}
f_{10,u} &= e_3, & f_{10,v} &= e_2, & \vec{n}_{10} &= f_{10,u} \times f_{10,v} = -e_1, \\
f_{20,u} &= e_1, & f_{20,v} &= e_3, & \vec{n}_{20} &= f_{20,u} \times f_{20,v} = -e_2, \\
f_{30,u} &= e_2, & f_{30,v} &= e_1, & \vec{n}_{30} &= f_{30,u} \times f_{30,v} = -e_3, \\
f_{11,u} &= e_2, & f_{11,v} &= e_3, & \vec{n}_{11} &= f_{11,u} \times f_{11,v} = e_1, \\
f_{21,u} &= e_3, & f_{21,v} &= e_1, & \vec{n}_{21} &= f_{21,u} \times f_{21,v} = e_2, \\
f_{31,u} &= e_1, & f_{31,v} &= e_2, & \vec{n}_{31} &= f_{31,u} \times f_{31,v} = e_3.
\end{aligned}
$$

We have chosen the parametrizations such that all normal vectors point outside the cube W and we will discuss the reasons for this soon.

Now let $F : H \to \mathbb{R}^3$ be a C^1-vector field where $W \subset H$ and $H \subset \mathbb{R}^3$ is open. With $Q = [0, 1] \times [0, 1]$ as before and $F = (F_1, F_2, F_3)$ we find using (25.2)

$$
\int_{\partial W} F \cdot d\sigma = \sum_{j=1}^{3} \sum_{\alpha=0}^{1} \int_{S_{j\alpha}} F \cdot d\sigma
$$

$$
= \sum_{j=1}^{3} \sum_{\alpha=0}^{1} \int_{Q} \langle F(f_{j\alpha}(u, v)), \vec{n}_{j\alpha} \rangle \, du \, dv
$$

$$= \int_Q \langle F(0, v, u), \vec{n}_{10} \rangle \, du \, dv + \int_Q \langle F(1, u, v), \vec{n}_{11} \rangle \, du \, dv$$

$$+ \int_Q \langle F(u, 0, v), \vec{n}_{20} \rangle \, du \, dv + \int_Q \langle F(v, 1, u), \vec{n}_{21} \rangle \, du \, dv$$

$$+ \int_Q \langle F(v, u, 0), \vec{n}_{30} \rangle \, du \, dv + \int_Q \langle F(u, v, 1), \vec{n}_{31} \rangle \, du \, dv$$

$$= - \int_Q F_1(0, v, u) \, du \, dv + \int_Q F_1(1, u, v) \, du \, dv$$

$$- \int_Q F_2(u, 0, v) \, du \, dv + \int_Q F_2(v, 1, u) \, du \, dv$$

$$- \int_Q F_3(v, u, 0) \, du \, dv + \int_Q F_3(u, v, 1) \, du \, dv.$$

Since $\int_Q F_1(0, v, u) \, du \, dv = \int_Q F_1(0, u, v) \, du \, dv$ and similar formulae hold for other integrals we arrive at

$$\int_{\partial W} F \cdot d\sigma = \int_Q (F_1(1, u, v) - F_1(0, u, v)) \, du \, dv$$

$$+ \int_Q (F_2(u, 1, v) - F_2(u, 0, v)) \, du \, dv$$

$$+ \int_Q (F_3(u, 1, v) - F_3(u, 0, v)) \, du \, dv.$$

Our conditions on F_j allows us to apply the fundamental theorem to find

$$\int_{\partial W} F \cdot d\sigma = \int_0^1 \left(\int_0^1 \left(\int_0^1 \frac{\partial F_1}{\partial x_1}(x_1, u, v) \, dx_1 \right) du \right) dv$$

$$+ \int_0^1 \left(\int_0^1 \left(\int_0^1 \frac{\partial F_2}{\partial x_2}(u, x_2, v) \, dx_2 \right) du \right) dv$$

$$+ \int_0^1 \left(\int_0^1 \left(\int_0^1 \frac{\partial F_3}{\partial x_3}(u, v, x_3) \, dx_3 \right) du \right) dv$$

$$= \int_W (\operatorname{div} F)(x) \, dx,$$

i.e. we have proved Gauss' theorem for a cube.

498

Proposition 25.1. *Let $H \subset \mathbb{R}^3$ be an open set and $W \subset H$. Further let $F : H \to \mathbb{R}^3$ be a C^1-vector field. Then the following holds*

$$\int_W (\operatorname{div} F)(x) \, \mathrm{d}x = \int_{\partial W} F(p) \cdot \sigma(\mathrm{d}p). \qquad (25.3)$$

Remark 25.2. Obviously our derivation of (25.3) still works when $W = [0,1] \times [0,1] \times [0,1]$ is replaced by any hyper-rectangle $R = [a_1 \times b_1] \times [a_2, b_2] \times [a_3, b_3]$ with $b_j < a_j$.

The strategy to now prove (25.1) for more general domains is to look at domains obtained by "deforming" the cube. However first we want to get a better understanding of the notion of an outward normal vector field and its relation to the orientation of a parametric surface.

In our calculation leading to (25.3) it was important that the normal vectors to S_{j0} and S_{j1} pointed in opposite directions, both pointing outward from W. For an injective parametric C^1-surface $f : Q \to \mathbb{R}^3$, $Q \subset \mathbb{R}^2$ open and connected, we can at each point $p = f(u, v) \in S = \operatorname{tr}(f)$ form the normal vector $f_u \times f_v$, and with $\vec{t}_1 = \frac{f_u}{\|f_u\|}$, $\vec{t}_2 = \frac{f_v}{\|f_v\|}$, $\vec{n} = \frac{f_u \times f_v}{\|f_u \times f_v\|}$ we always have $\det(\vec{t}_1, \vec{t}_2, \vec{n}) = 1$, i.e. these three unit vectors determine an orientation in $\mathbb{R}^3$, and for every $p = f(u, v)$ the corresponding vectors $\vec{t}_1(u, v)$, $\vec{t}_2(u, v)$ and $\vec{n}(u, v)$ determine the same orientation in $\mathbb{R}^3$ which we can call the canonical orientation of S given the parametrization f.

With $\vec{n}$ the vector $-\vec{n}$ is also a unit vector orthogonal to $\operatorname{span}(\vec{t}_1, \vec{t}_2)$. Thus at every point $p = f(u, v)$ we have two distinct normal vectors $\pm \vec{n}(u, v)$. Since f is a C^1-mapping $(u, v) \mapsto \vec{n}(u, v)$ and $(u, v) \mapsto -\vec{n}(u, v)$ are continuous, recall $\vec{n}(u, v) = f_u(u, v) \times f_v(u, v)$. Moreover $\det(\vec{t}_1, \vec{t}_2, \vec{n}) = 1$ and $\det(\vec{t}_1, \vec{t}_2, -\vec{n}) = -1$. These two determinants are continuous mappings in (u, v) and since Q is by assumption connected, we cannot find a non-constant continuous vector field ν on S with $\nu(f(u, v)) \in \{\vec{n}(u, v), -\vec{n}(u, v)\}$. In this sense the choice of $\vec{n} = \vec{t}_1 \times \vec{t}_2$ or $-\vec{n} = \vec{t}_2 \times \vec{t}_1$ determine a fixed orientation on S as long as the parametrization is injective and Q is connected. A change of parameter Φ will not change the orientation as long as $\det J_\Phi > 0$, if $\det J_\Phi < 0$ then the orientation will be changed under this parameter transformation.

However we have to exclude parametrizations allowing double points since in such a case tangent and normal vectors are not necessarily determined by

$p \in S$ but by $(u, v) \in Q$ and we may have for $p = f(u_1, v_1) = f(u_2, v_2)$, $(u_1, v_1) \neq (u_2, v_2)$ different vectors $f_u(u_1, v_1)$, $f_u(u_2, v_2)$ or $f_v(u_1, v_1)$ and $f_v(u_2, v_2)$. This already causes problems with a sphere if we want to have as parameter domain a compact set $Q \subset \mathbb{R}^2$ which we prefer for integration. Moreover, the surfaces we are interested in now are the boundaries of open bounded sets in $\mathbb{R}^3$, thus we do not expect a parametrization defined on a compact set and which is injective. We may however succeed to cover such a boundary with traces of finitely many parametric C^1-surfaces with injective C^1-parametrizations, see Example 25.5 where we will discuss this for the sphere.

So let $G \subset \mathbb{R}^3$ be a connected open bounded set with $\partial G \subset \mathbb{R}^3$ having the property that $\partial G = \bigcup_{k=1}^{M} S_k$ where $S_k = f_k(Q_k)$, $Q_k \subset \mathbb{R}^2$ open and connected, and $f_k : Q \to \mathbb{R}^3$ is an injective C^1-parametrization of a surface $S_k \subset \partial G$, i.e. S_k is a patch of ∂G. We assume in addition that whenever $S_k \cap S_l \neq \emptyset$ then the mappings $f_l^{-1} \circ f_k$ and $f_k^{-1} \circ f_l$ defined on $f_k^{-1}(S_k \cap S_l)$ and $f_l^{-1}(S_k \cap S_l)$ are continuously differentiable. We call these mappings **local parameter changes** of ∂G.

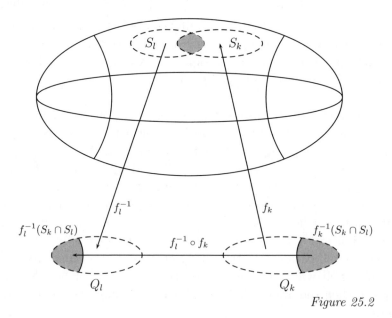

Figure 25.2

On each $S_k \subset \partial G$ we can define the normal unit vectors $\vec{n}$ and $-\vec{n}$, one vector will point outward of G, the other inward to G.

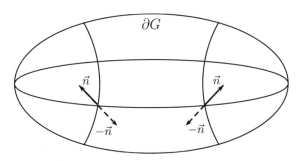

Figure 25.3

But this formulation and Figure 25.3 is not quite correct. When dealing with tangent or normal vectors to ∂G at p so far we worked with vectors in $\mathbb{R}^3$ with components with respect to the canonical basis $\{e_1, e_2, e_3\}$, i.e. graphically, the normal vectors to S_k at p_1 and p_2 must both be drawn starting at the origin! The normal line at $p = f(u, v)$ is obtained by translating the line $\{\lambda \vec{n} \mid \lambda \in \mathbb{R}\}$ to the parallel line passing through p. Thus when speaking about the inward or outward normal vector we first have to translate $\{\lambda \vec{n}_p \mid \lambda \in \mathbb{R}\}$ to $p \in \partial G$ and then we have to decide which of the two translated vectors $\pm \vec{n}_p$ is pointing inward and which is pointing outward, respectively. With this clarification in mind we will now address $\vec{n}$ as either inward or outward pointing normal vector to ∂G.

Let $G \subset \mathbb{R}^3$ be a bounded, connected open set and suppose that we can cover ∂G with the traces of finitely many, say M, parametric C^1-surfaces $S_k = f_k(Q_k)$, $Q_k \subset \mathbb{R}^2$, having the properties stated above. Every $p \in \partial G$ belongs to some S_k, $1 \leq k \leq M$, $p = f_k(u^{(k)}, v^{(k)})$ with $(u^{(k)}, v^{(k)}) \in Q_k$. Hence we can define a p the two unit normal vectors $\vec{n}_{f_k}(u^{(k)}, v^{(k)})$ and $-\vec{n}_{f_k}(u^{(k)}, v^{(k)})$ and the conditions posed on ∂G and the covering $(S_k)_{k=1,\dots,M}$ imply that if $p = f_k(u^{(k)}, v^{(k)}) = f_l(u^{(l)}, v^{(l)})$, $k \neq l$, then we must have $\vec{n}_{f_k}(u^{(k)}, v^{(k)}) = \pm \vec{n}_{f_l}(u^{(l)}, v^{(l)})$.

Definition 25.3. *Let $G \subset \mathbb{R}^3$ be a bounded, connected open set and suppose that ∂G can be covered by finitely many injective parametric C^1-surfaces S_k, $1 \leq k \leq M$, with the properties mentioned above.*
A. *A **normal unit vector field** on ∂G is by definition a vector field $N : \partial G \to \mathbb{R}^3$ such that $N(p)$ is a unit normal vector to ∂G at p, i.e. $N(p) \in \{n_{f_k}(u^{(k)}, v^{(k)}), -n_{f_k}(u^{(k)}, v^{(k)})\}$ for some k, $1 \leq k \leq M$.*
B. *We call ∂G a **closed orientated surface** if we can find on ∂G a continuous normal unit vector field. We say that such a vector field determines*

an orientation on ∂G.

C. *If ∂G is a closed orientated surface we call a normal unit vector field the **outward normal vector field** if one and hence all of its vectors point outward of G. The orientation determined by the outward normal vector field is called the **positive orientation** of ∂G.*

Remark 25.4. On ∂G we have two orientations, the positive orientation as defined and if all vectors of a normal unit vector field point inward to G we get the **negative orientation**. Under local coordinate changes with positive Jacobi determinant the orientation remains unchanged.

Example 25.5. Consider the sphere $S^2 := \{(x, y, z) \in \mathbb{R}^3 \mid x^2 + y^2 + z^2 = 1\}$. We can cover S^2 by the traces of the following six C^1-parametric surfaces $\varphi_j : B_1(0) \to \mathbb{R}^3$, $B_1(0) \subset \mathbb{R}^2$, $j = 1, \ldots, 6$, where

$$\varphi_1(u, v) = (u, v, \sqrt{1 - u^2 - v^2}), \quad \varphi_2(v, u) = (v, u, -\sqrt{1 - u^2 - v^2})$$
$$\varphi_3(u, v) = (v, \sqrt{1 - u^2 - v^2}, u), \quad \varphi_4(u, v) = (u, -\sqrt{1 - u^2 - v^2}, v)$$
$$\varphi_5(u, v) = (\sqrt{1 - u^2 - v^2}, u, v), \quad \varphi_6(u, v) = (-\sqrt{1 - u^2 - v^2}, v, u),$$

see Figure 25.4, also compare with [11].

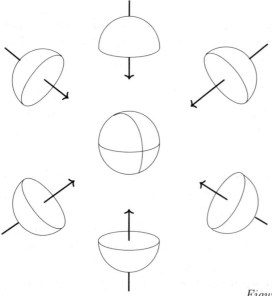

Figure 25.4

Each of these mappings is injective, so if we set $S_j := \varphi_j(B_1(0)) \subset S^2 \subset \mathbb{R}^3$, the inverse $\varphi_j^{-1} : S_j \to B_1(0)$ is well defined. For example we have

$$\varphi_1^{-1} : S_1 \to B_1(0), \quad (x, y, z) \mapsto (x, y)$$

or

$$\varphi_5^{-1} : S_5 \to B_1(0), \quad (x, y, z) \mapsto (y, z).$$

Let us consider more closely $S_1 \cap S_5$. Note that $S_1 = \{(x, y, z) \in S^2 \,|\, z > 0\}$ and $S_5 = \{(x, y, z) \in S^2 \,|\, x > 0\}$. Thus we have

$$S_1 \cap S_5 = \{(x, y, z) \in S^1 \,|\, x > 0, \, z > 0\} = \varphi_5(\{(u, v) \in B_1(0) \,|\, v > 0\}).$$

It now follows on $\{(u, v) \in B_1(0) \,|\, v > 0\}$ that

$$\left(\varphi_1^{-1} \circ \varphi_5\right) = \varphi_1^{-1}\left(\sqrt{1 - u^2 - v^2}, u, v\right) = \left(\sqrt{1 - u^2 - v^2}, u\right),$$

which yields for the Jacobi matrix

$$\left(J_{\varphi_1^{-1} \circ \varphi_5}\right)(u, v) = \begin{pmatrix} \frac{-u}{\sqrt{1-u^2-v^2}} & \frac{-v}{\sqrt{1-u^2-v^2}} \\ 1 & 0 \end{pmatrix}$$

and consequently

$$\left(\det J_{\varphi_1^{-1} \circ \varphi_5}\right)(u, v) = \frac{v}{\sqrt{1 - u^2 - v^2}} > 0$$

since $v > 0$. Thus the change of the parametrization of $S_1 \cap S_5$ from $\varphi_1 : \{(u, v) \in B_1(0) \,|\, u > 0\} \to \mathbb{R}^3$ to $\varphi_5 : \{(u, v) \in B_1(0) \,|\, v > 0\} \to \mathbb{R}^3$ is orientation preserving. Similarly we can prove that the other changes of parametrizations are orientation preserving. Next we use these parametrizations to calculate the corresponding normal vectors to S^2. With $A = \sqrt{1 - u^2 - v^2}$ we have

$$(\varphi_{1,u} \times \varphi_{1,v})(u, v) = \begin{pmatrix} 1 \\ 0 \\ -\frac{u}{A} \end{pmatrix} \times \begin{pmatrix} 0 \\ 1 \\ -\frac{v}{A} \end{pmatrix} = \frac{1}{A}\begin{pmatrix} u \\ v \\ A \end{pmatrix} = \frac{1}{A}\varphi_1(u, v),$$

$$(\varphi_{2,u} \times \varphi_{2,v})(u, v) = \begin{pmatrix} 0 \\ 1 \\ \frac{u}{A} \end{pmatrix} \times \begin{pmatrix} 1 \\ 0 \\ \frac{v}{A} \end{pmatrix} = \frac{1}{A}\begin{pmatrix} u \\ v \\ -A \end{pmatrix} = \frac{1}{A}\varphi_2(u, v),$$

$$(\varphi_{3,u} \times \varphi_{3,v})(u,v) = \begin{pmatrix} 0 \\ -\frac{u}{A} \\ 1 \end{pmatrix} \times \begin{pmatrix} 1 \\ -\frac{v}{A} \\ 0 \end{pmatrix} = \frac{1}{A} \begin{pmatrix} u \\ A \\ v \end{pmatrix} = \frac{1}{A}\varphi_3(u,v),$$

$$(\varphi_{4,u} \times \varphi_{4,v})(u,v) = \begin{pmatrix} 1 \\ \frac{u}{A} \\ 0 \end{pmatrix} \times \begin{pmatrix} 0 \\ \frac{v}{A} \\ 1 \end{pmatrix} = \frac{1}{A} \begin{pmatrix} u \\ -A \\ v \end{pmatrix} = \frac{1}{A}\varphi_4(u,v),$$

$$(\varphi_{5,u} \times \varphi_{5,v})(u,v) = \begin{pmatrix} -\frac{u}{A} \\ 1 \\ 0 \end{pmatrix} \times \begin{pmatrix} -\frac{v}{A} \\ 0 \\ 1 \end{pmatrix} = \frac{1}{A} \begin{pmatrix} A \\ u \\ v \end{pmatrix} = \frac{1}{A}\varphi_5(u,v),$$

$$(\varphi_{6,u} \times \varphi_{6,v})(u,v) = \begin{pmatrix} \frac{u}{A} \\ 0 \\ 1 \end{pmatrix} \times \begin{pmatrix} \frac{v}{A} \\ 1 \\ 0 \end{pmatrix} = \frac{1}{A} \begin{pmatrix} -A \\ u \\ v \end{pmatrix} = \frac{1}{A}\varphi_6(u,v).$$

Thus in each case $\varphi_{j,u} \times \varphi_{j,v}$ is a positive multiple of the vector φ_j, and φ_j points from the origin $0 \in \mathbb{R}^3$ to S^2, hence $\varphi_{j,u} \times \varphi_{j,v}$ is the outward normal vector of length $\|\varphi_{j,u} \times \varphi_{j,v}\| = \frac{1}{A}$. So the outward unit normal vector field on S^2 is as we should expect given by the vector field $(x,y,z) \mapsto (x,y,z)$, $(x,y,z) \in S^2$, and for $(x,y,z) \in S_j$ we find $(x,y,z) = A(\varphi_{j,u} \times \varphi_{j,v})$.

Remark 25.6. Our purpose in this chapter is to prove Gauss' theorem and the approach to try to cover ∂G by finitely many well-behaved traces of parametric surfaces fits our purpose. However, it turns out that for many other considerations a better approach is to introduce differentiable manifolds M as we do in Volume VI. Here the basic idea is to map patches of M onto open subsets of $\mathbb{R}^3$ (or $\mathbb{R}^n$) such that coordinate changes become differentiable. In our example we only need to switch from φ_j to φ_j^{-1}. We now leave this problem with this remark but will come back to it in Volume VI.

With these preparations we now want to extend Proposition 25.1 to more general domains. In order to have a simple façon de parler we give

Definition 25.7. *A bounded open and connected Jordan measurable set $G \subset \mathbb{R}^3$ is called a **Gauss domain** if for some bounded open set $G_0 \subset \mathbb{R}^3$ we have that $\overline{G} \subset G_0$, ∂G is a closed orientated C^1-surface (which however may have several connectivity components) and for every C^1-vector field $F : G_0 \to \mathbb{R}^3$ the following holds*

$$\int_G \operatorname{div} F \, dx = \int_{\partial G} F \cdot d\sigma. \tag{25.4}$$

Of course, in (25.4) we can replace the volume integral over G by that over $\overline{G}$.

First we note that if we can decompose a Gauss domain G into two Gauss domains such that $G_1 \cap G_2 = \emptyset$ and $\overline{G}_1 \cap \overline{G}_2$ is a C^1-parametric surface S, then it follows

$$\int_{\partial G} F \cdot d\sigma = \int_G \operatorname{div} F \, dx = \int_{G_1} \operatorname{div} F \, dx + \int_{G_2} \operatorname{div} F \, dx = \int_{\partial G_1} F \cdot d\sigma + \int_{\partial G_2} F \cdot d\sigma. \quad (25.5)$$

The two boundaries $\partial G_1 = (\partial G_1 \cap \partial G) \cup S$ and $\partial G_2 = (\partial G_2 \cap \partial G) \cup S$ are by assumption orientated surfaces and in both cases we have to take the positive orientation which implies that the orientation of S as part of ∂G_1 is opposite to the orientation of S as part of ∂G_2, see Figure 25.5.

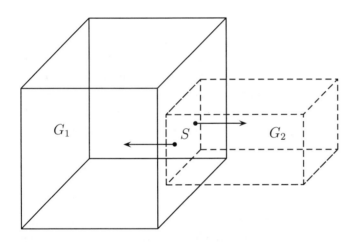

Figure 25.5

In order to distinguish the different orientation on S we write S_{G_1} and S_{G_2} where S_{G_j} has the orientation induced by ∂G_j. Using (25.5) we deduce

$$\int_{\partial G} F \cdot d\sigma = \int_{\partial G_1} F \cdot d\sigma + \int_{\partial G_2} F \cdot d\sigma$$

$$= \int_{\partial G_1 \cap \partial G} F \cdot d\sigma + \int_{S_{G_1}} F \cdot d\sigma + \int_{\partial G_2 \cap \partial G} F \cdot d\sigma + \int_{S_{G_2}} F \cdot d\sigma$$

$$= \int_{\partial G} F \cdot d\sigma + \int_{S_{G_1}} F \cdot d\sigma + \int_{S_{G_2}} F \cdot d\sigma$$

and must conclude

$$\int_{S_{G_1}} F \cdot \mathrm{d}\sigma = -\int_{S_{G_2}} F \cdot \mathrm{d}\sigma. \tag{25.6}$$

This of course is no surprise. If we leave the point set S and the vector field F unchanged but change the sign of the normal vector, the surface integral $\int_S F \cdot \mathrm{d}\sigma$ must change the sign. Changing our point of view, if G_1 and G_2 are two Gauss domains in $\mathbb{R}^3$ such that $G_1 \cap G_2 = \emptyset$ but $\overline{G}_1 \cap \overline{G}_2$ is a parametric C^1-surface without double points, then $\left(\overline{G}_1 \cup \overline{G}_2\right)^{\circ}$ is a Gauss domain with the relative interior of $S = \overline{G}_1 \cap \overline{G}_2$ being a subset of $\left(\overline{G}_1 \cup \overline{G}_2\right)^{\circ}$. Certainly these considerations extend to finitely many Gauss domains.

Let $G_1, \ldots, G_M \subset \mathbb{R}^3$ be Gauss domains such that $G_k \cap G_l = \emptyset$ and $G := \left(\bigcup_{k=0}^{M} \overline{G}_k\right)^{\circ}$ is connected. If $\overline{G}_k \cap \overline{G}_l \neq \emptyset$ then we assume that $S_{k,l} := \overline{G}_k \cap \overline{G}_l$ is a C^1-parametric surface without double points and the relative interior $\mathring{S}_{kl}$ of S_{kl} is a subset of G. By $S_{kl,k}$ we denote S_{kl} equipped with the orientation induced by ∂G_k, and by $S_{kl,l}$ we denote S_{kl} equipped with the orientation induced by ∂G_l, i.e. $S_{kl,k}$ and $S_{kl,l}$ have opposite orientations. Further we assume that if $S_{kl} \cap S_{ij} \neq \emptyset$ then it is the trace of a Lipschitz curve (which may include a constant curve, i.e. a point). These assumptions imply for every continuous vector field that

$$\int_{S_{kl,k}} F \cdot \mathrm{d}\sigma = -\int_{S_{kl,l}} F \cdot \mathrm{d}\sigma,$$

and further that for every surface S_{kl} the integral of F over S_{kl} (with any chosen orientation) equals the sum of the surface integrals over the surfaces obtained by removing from S_{kl} all intersections with surfaces S_{ij}. In Figure 25.5 we have given an indication for such a situation using hyper-rectangles. We claim that G is a Gauss domain. To see this we start with G and a C^1-vector field defined in a neighbourhood of G, for example in $\bigcup_{k=1}^{M} G_{0k}$, $\overline{G}_k \subset G_{0k}$. We decompose G into the Gauss domains G_k, $k = 1, \ldots, M$, and hence we have

$$\int_G \operatorname{div} F \, \mathrm{d}x = \sum_{k=1}^{M} \int_{G_k} \operatorname{div} F \, \mathrm{d}x.$$

Since G_k is a Gauss domain we find further

$$\int_{G_k} \operatorname{div} F \, \mathrm{d}x = \int_{\partial G_k} F \cdot \mathrm{d}\sigma$$

and

$$\int_{\partial G_k} F \cdot d\sigma = \int_{\partial G_k \cap \partial G} F \cdot d\sigma + \sum_j{}' \int_{S_{kj,k}} F \cdot d\sigma$$

where the sum $\sum_j{}'$ is taken over all surfaces $S_{kj,k}$ (but only those which are non-empty will contribute). Each integral $\int_{S_{kj,k}} F \cdot d\sigma$ has a counterpart $\int_{S_{kj,j}} F \cdot d\sigma$ in the decomposition of $\int_{\partial G_j} F \cdot d\sigma$ and these two integrals due to the opposite orientation of $S_{kj,k}$ and $S_{kj,j}$ add up to zero. Thus we obtain

$$\sum_{k=1}^{M} \int_{\partial G_k} F \cdot d\sigma = \sum_{k=1}^{M} \int_{\partial G_k \cap \partial G} F \cdot d\sigma = \int_{\partial G} F \cdot d\sigma,$$

and therefore

$$\int_G \operatorname{div} F \, dx = \int_{\partial G} F \cdot d\sigma,$$

i.e. G is a Gauss domain.

Our next step is to prove Gauss' theorem for certain normal domains. We need some preparations.

Let $W \subset \mathbb{R}^2$ be a compact Jordan measurable set such that $\partial W = \bigcup_{k=1}^{M} S_k$, $S_k = \gamma_k([0,1])$, where $\gamma_k : [0,1] \to \mathbb{R}^2$ is a regular C^1-curve. We assume for $k \neq l$ that either $S_k \cap S_l = \emptyset$ or $S_k \cap S_l$ consists of common end points of γ_k and γ_l. Further let $\varphi, \psi : \tilde{W} \to \mathbb{R}$, $\varphi \leq \psi$, be C^1-functions where $W \subset \tilde{W}$ and $\tilde{W} \subset \mathbb{R}^2$ is an open set. Consider the domain

$$G = \left\{ x = (y,z) \,\middle|\, y = (y_1, y_2) \in \mathring{W}, \; \varphi(y) < z < \psi(y) \right\}. \tag{25.7}$$

The boundary of ∂G is given by $\partial G = \partial_\psi G \cup \partial_\varphi G \cup \partial_W G$ where

$$\partial_\psi G = \left\{ (y,z) \in \mathbb{R}^3 \,\middle|\, y \in W, \; z = \psi(y) \right\}$$
$$\partial_\varphi G = \left\{ (y,z) \in \mathbb{R}^3 \,\middle|\, y \in W, \; z = \varphi(y) \right\}$$

and

$$\partial_W G = \left\{ (y,z) \in \mathbb{R}^3 \,\middle|\, y \in \partial W, \; \varphi(y) < z < \psi(y) \right\}.$$

For points in $\{(y,z) \in \mathbb{R}^3 \,|\, y \in \partial W, \; z = \psi(y)\} \cup \{(y,z) \in \mathbb{R}^3 \,|\, y \in \partial W \; z = \varphi(y)\}$ the (outer) normal to ∂G may not exist as it may not exist for points in $\{(y,z) \in \mathbb{R}^3 \,|\, y \in S_k \cap S_l, \; \varphi(y) < z < \psi(y)\}$. However these are sets of measure zero, in fact Jordan content zero and we can decompose ∂G into a finite number of sets on each of which the (outer) normal vector is defined and a continuous function which is bounded. In such a case we will just speak about the (outer) normal or normal vector to ∂G, not mentioning anymore that it is defined only up to an exceptional set. Given the existence of the normal vector as above we can integrate continuous functions over ∂G, see the remark following (25.2).

Lemma 25.8. *Let $G \subset \mathbb{R}^3$ be as in (25.7) and $F_3 \in C^1(\tilde{G})$, $G \subset \tilde{G}$, $\tilde{G} \subset \mathbb{R}^3$ open. With $n(x) = (\vec{n}_1(x), \vec{n}_2(x), \vec{n}_3(x))$ denoting the outer normal to ∂G we have*

$$\int_G \frac{\partial F_3}{\partial x_3}(x)\,\mathrm{d}x = \int_G \frac{\partial F_3(y_1, y_2, z)}{\partial z}\,\mathrm{d}z\,\mathrm{d}y_1\,\mathrm{d}y_2 = \int_{\partial G} F_3(x)\vec{n}_3(x)\sigma(\mathrm{d}x). \quad (25.8)$$

Proof. For F_3 as above we find

$$\int_G \frac{\partial F_3}{\partial x_3}(x)\,\mathrm{d}x = \int_W \left(\int_{\varphi(y)}^{\psi(y)} \frac{\partial F_3}{\partial z}(y, z)\,\mathrm{d}z \right) \mathrm{d}y$$

$$= \int_W (F_3(y, \psi(y)) - F_3(y, \varphi(y)))\,\mathrm{d}y.$$

The outer unit normal to $\partial_W G$ is

$$\vec{n}(y, z) = (\nu(y), 0)$$

where $\nu(y)$ is the outer unit normal to ∂W. Further, since $\partial_\varphi G$ and $\partial_\psi G$ are graphs of functions, according to Example 8.6.A we find for the outer unit normal on $\partial_\psi G$ and $\partial_\varphi G$

$$\vec{n}(y, z) = \frac{1}{(1 + \|\operatorname{grad}\psi(y)\|^2)^{\frac{1}{2}}}(-\operatorname{grad}\psi(y), 1)$$

and

$$\vec{n}(y, z) = \frac{1}{(1 + \|\operatorname{grad}\varphi(y)\|^2)^{\frac{1}{2}}}(\operatorname{grad}\varphi(y), -1),$$

respectively, where $\operatorname{grad}\psi$ and $\operatorname{grad}\varphi$ are two-dimensional gradients. Thus we find

$$\vec{n}_3\big|_{\partial_W G} = 0, \quad \vec{n}_3\big|_{\partial_\psi G} = \frac{1}{(1 + \|\operatorname{grad}\psi(\cdot)\|^2)^{\frac{1}{2}}}, \quad \vec{n}_3\big|_{\partial_\varphi G} = \frac{-1}{(1 + \|\operatorname{grad}\varphi(\cdot)\|^2)^{\frac{1}{2}}},$$

which yields using Definition 24.9

$$\int_W F_3(y, \psi(y)) \, \mathrm{d}y = \int_W F_3(y, \psi(y)) \vec{n}_3(y, \psi(y))(1 + \|\psi(y)\|^2)^{\frac{1}{2}} \, \mathrm{d}y$$

$$= \int_{\partial_\psi G} F_3(x) n_3(x) \sigma(\mathrm{d}x),$$

$$-\int_W F_3(y, \varphi(y)) \, \mathrm{d}y = \int_W F_3(y, \psi(y)) \vec{n}_3(y, \psi(y))(1 + \|\varphi(y)\|^2)^{\frac{1}{2}} \, \mathrm{d}y$$

$$= \int_{\partial_\varphi G} F_3(x) \vec{n}_3(x) \sigma(\mathrm{d}x),$$

and

$$\int_{\partial_W G} F_3(x) \vec{n}_3(x) \sigma(\mathrm{d}x) = 0.$$

Adding up we arrive at

$$\int_W \big(F_3(y, \psi(y)) - F_3(y, \varphi(y)) \big) \mathrm{d}y$$

$$= \int_{\partial_W G} F_3(x) \vec{n}_3(x) \sigma(\mathrm{d}x) + \int_{\partial_\psi G} F_3(x) \vec{n}_3(x) \sigma(\mathrm{d}x) + \int_{\partial_\varphi G} F_3(x) \vec{n}_3(x) \sigma(\mathrm{d}x)$$

$$= \int_{\partial G} F_3(x) \vec{n}_3(x) \sigma(\mathrm{d}x),$$

proving the lemma. □

In order to have a short way to refer to domains as G given in (25.7) we call such a domain for a moment a **pre-Gauss domain with respect to e_3**. Of course, pre-Gauss domains are special normal domains.

Theorem 25.9. *Let $G \subset \mathbb{R}^3$ be a pre-Gauss domain with respect to all three coordinate directions then G is a Gauss domain, i.e. for every C^1-vector field $F : \tilde{G} \to \mathbb{R}^3$, $\overline{G} \subset \tilde{G}$ and $\tilde{G}$ open, Gauss' theorem holds:*

$$\int_G \operatorname{div} F(x) \, \mathrm{d}x = \int_{\partial G} F(x) \cdot \sigma(\mathrm{d}x).$$

Proof. Since G satisfies for all three directions the assumptions of Lemma 25.8 we have for $j = 1, 2, 3$,

$$\int_G \frac{\partial F}{\partial x_j}(x) \, \mathrm{d}x = \int_{\partial G} F_j(x) \vec{n}_j(x) \sigma(\mathrm{d}x)$$

and adding up yields

$$\int_G \operatorname{div} F(x)\,dx = \int_G \left(\frac{\partial F_1}{\partial x_1}(x) + \frac{\partial F_2}{\partial x_2}(x) + \frac{\partial F_3}{\partial x_3}(x)\right) dx$$

$$= \int_G \left(F_1(x)\vec{n}_1(x) + F_2(x)\vec{n}_2(x) + F_3(x)\vec{n}_3(x)\right) \sigma(dx)$$

$$= \int_{\partial G} \langle F(x), \vec{n}(x)\rangle \sigma(dx) = \int_{\partial G} F \cdot d\sigma.$$

$\square$

Corollary 25.10. *An open Jordan measurable set $G \subset \mathbb{R}^3$ which can be decomposed into a finite number of sets G_j with each G_j being a pre-Gauss domain with respect to all three directions e_1, e_2 and e_3 is a Gauss domain.*

As mentioned above, Gauss' theorem holds for more general domains than those in Corollary 25.10, but for a first go and many concrete problems this class is sufficiently large.

As a first application of Gauss' theorem we prove an **integration by parts** formula.

Proposition 25.11. *Let $G \subset \mathbb{R}^3$ be a Gauss domain and $\tilde{G} \subset \mathbb{R}^3$ an open set such that $\overline{G} \subset \tilde{G}$. For $u, v \in C^1(\tilde{G})$ and $1 \le j \le 3$ the following holds*

$$\int_G \frac{\partial u}{\partial x_j}(x)v(x)\,dx = \int_{\partial G} u(x)v(x)\vec{n}_j(x)\sigma(d(x)) - \int_G u(x)\frac{\partial v}{\partial x_j}(x)\,dx. \quad (25.9)$$

Proof. We introduce on $\tilde{G}$ the C^1-vector field $F = (F_1, F_2, F_3)$ with $F_j = u \cdot v$ and $F_l = 0$ for $l \ne j$. From Gauss' theorem we deduce

$$\int_G \operatorname{div} F\,dx = \int_{\partial G} \langle F, \vec{n}\rangle(x)\sigma(dx)$$

and note that

$$\operatorname{div} F = \frac{\partial}{\partial x_j}(uv) = \left(\frac{\partial}{\partial x_j}u\right)v + u\frac{\partial}{\partial x_j}v$$

and that

$$\langle F, \vec{n}\rangle(x) = uv\vec{n}_j,$$

implying that

$$\int_G \left(\frac{\partial u}{\partial x_j}(x)v(x) + u(x)\frac{\partial v}{\partial x_j}(x)\right) \mathrm{d}x = \int_{\partial G} u(x)v(x)\vec{n}_j(x)\sigma(\mathrm{d}x)$$

which yields (25.9). $\qquad\qquad\qquad\qquad\qquad\qquad\qquad\qquad\qquad$ $\square$

In the case where u or v vanishes on ∂G we obtain the formula

$$\int_{\partial G} \frac{\partial u}{\partial x_j}(x)v(x)\,\mathrm{d}x = -\int_{\partial G} u(x)\frac{\partial v}{\partial x_j}(x)\,\mathrm{d}x \qquad (25.10)$$

which we could already derive under more restrictive conditions in Chapter 20, Proposition 20.20.C, however for the n-dimensional case. In Chapter 27, we will extend Gauss' theorem to n-dimensions and (25.9) and (25.10) will follow for the general case. A first consequence of the integration by parts formula is a simple version of **Poincaré's inequality**, also compare with Problem 9 in Chapter 18.

Lemma 25.12. *Let $G \subset \mathbb{R}^3$ be a Gauss domain and $u \in C^1(\tilde{G})$, where $\tilde{G} \subset \mathbb{R}^3$ is an open set such that $\overline{G} \subset \tilde{G}$. If $u|_{\partial G} = 0$ then*

$$\|u\|_{L^2} \le 2\max_{x\in\overline{G}}|x_1|\left\|\frac{\partial u}{\partial x_1}\right\|_{L^2} \qquad (25.11)$$

where

$$\|v\|_{L^2} = \left(\int_G |v(x)|^2\,\mathrm{d}x\right)^{\frac{1}{2}} = \left(\int_G v^2(x)\,\mathrm{d}x\right)^{\frac{1}{2}}.$$

Proof. The following holds

$$\int_G u^2(x)\,\mathrm{d}x = \int_G \left(\frac{\partial x_1}{\partial x_1}\right)u^2(x)\,\mathrm{d}x = -\int_G x_1\frac{\partial}{\partial x_1}u^2(x)\,\mathrm{d}x$$

$$= -2\int_G x_1 u(x)\frac{\partial u}{\partial x_1}(x)\,\mathrm{d}x$$

$$\le 2\max_{x\in\overline{G}}|x_1|\int_G |u(x)|\left|\frac{\partial u}{\partial x_1}(x)\right|\,\mathrm{d}x$$

$$\le 2\max_{x\in\overline{G}}|x_1|\left(\int_G u^2(x)\,\mathrm{d}x\right)^{\frac{1}{2}}\left(\int_G \left(\frac{\partial u}{\partial x_1}(x)\right)^2\,\mathrm{d}x\right)^{\frac{1}{2}},$$

where in the last step we used the Cauchy-Schwarz inequality. For $\|u\|_{L^2} = 0$ the estimate (25.11) is trivial, the other case follows from our inequality

$$\|u\|_{L^2}^2 \leq \left(2 \max_{x \in \overline{G}} |x_1| \right) \|u\|_{L^2} \left\| \frac{\partial u}{\partial x_1} \right\|_{L^2}.$$

$\square$

We want to find some more concrete examples of Gauss domains in $\mathbb{R}^3$. First we note that if G is a Gauss domain and $a \in \mathbb{R}^3$ then $a + G$ is a Gauss domain too which follows from the translation invariance of the integral. But we have not (yet) studied the translation invariance of arbitrary surface integrals. However, it is obvious by looking at (25.7) that if G is a pre-Gauss domain with respect to e_j then $a + G$ is also a pre-Gauss domain with respect to e_j. Hence, if G is a pre-Gauss domain with respect to all three coordinate axes, then $a + G$ is such a domain too. Moreover, if G is a pre-Gauss domain with respect to e_j and if $U(\varphi)$ is a rotation with fixed axis $\{\lambda e_j \,|\, \lambda \in \mathbb{R}\}$, then we see by inspection that $U(G)$ is again a pre-Gauss domain with respect to the same axis. Using results on the structure of rotations in $\mathbb{R}^3$ we can deduce that the image under a rotation of a pre-Gauss domain with respect to all three axes is again of this type. With these remarks in mind we can easily extend the following classes of examples by translating and rotating.

Example 25.13. The ball $B_r(0) \subset \mathbb{R}^3$, $r > 0$, is a Gauss domain as is the cylinder $Z_{r,h} := \{(x,y,z) \in \mathbb{R}^3 \,|\, x^2 + y^2 < r, \, 0 < z < h\}$. Indeed, for $\overline{B_r(0)}$ we have the representation

$$\overline{B_r(0)} = \left\{ (x_1, x_2, x_3) \,\middle|\, (x_j, x_k) \in \overline{B_r^{(2)}}(0) \subset \mathbb{R}^2 - \sqrt{r^2 - x_j^2 - x_k^2} \leq x_l \leq \sqrt{r^2 - x_j^2 - x_k^2}, \right.$$
$$\left. j \neq k, \, j \neq l, \, l \neq k \right\}.$$

For $\overline{Z}_{r,h}$ we have the representation

$$\overline{Z}_{r,h} = \left\{ (x,y,z) \in \mathbb{R}^3 \,\middle|\, (x,y) \in \overline{B_r^{(2)}} \subset \mathbb{R}^2 \, 0 \leq z \leq h \right\}$$
$$= \left\{ (x,y,z) \in \mathbb{R}^3 \,\middle|\, (x,z) \in [-r,r] \times [0,h], \, -\sqrt{r^2 - x^2} \leq y \leq \sqrt{r^2 - x^2} \right\}$$
$$= \left\{ (x,y,z) \in \mathbb{R}^3 \,\middle|\, (y,z) \in [-r,r] \times [0,h], \, -\sqrt{r^2 - y^2} \leq x \leq \sqrt{r^2 - y^2} \right\}.$$

Example 25.14. The set $B_R(0)\backslash\overline{B_r(0)} \subset \mathbb{R}^3, 0 < r < R$, is a Gauss domain. For a vector field $F : B_{R+\epsilon}(0) \to \mathbb{R}^3$ of class C^1 we note

$$\int_{B_R(0)\backslash\overline{B_r(0)}} \text{div}\, F\, dx = \int_{B_R(0)} \text{div}\, F\, dx - \int_{B_r(0)} \text{div}\, F\, dx$$

$$= \int_{\partial B_R(0)} \langle F, \vec{n}\rangle\, d\sigma - \int_{\partial B_r(0)} \langle F, \vec{n}\rangle\, d\sigma,$$

note that $\partial\left(B_R(0)\backslash\overline{B_r(0)}\right) = \partial B_R(0) \cup \partial B_r(0)$, but the outer normal to $\partial B_r(0) \subset \partial\left(B_R(0)\backslash\overline{B_r(0)}\right)$ has opposite direction to the outer normal to $\partial B_r(0)$ when $\partial B_r(0)$ is considered as boundary of $B_r(0)$ alone. Thus we find

$$\int_{B_R(0)\backslash\overline{B_r(0)}} \text{div}\, F\, dx = \int_{\partial(B_R(0)\backslash\overline{B_r(0)})} \langle F, \vec{n}\rangle\, d\sigma,$$

implying that $B_R(0)\backslash\overline{B_r(0)}$ is a Gauss domain.

Obviously Example 25.14 has the following generalisation: Let $K \subset G \subset \mathbb{R}^3$ be two Gauss domains and suppose that $\partial G \cap \partial K = \emptyset$. Then $G\backslash\overline{K}$ is a Gauss domain.

We will now study two further examples of applications of Gauss' theorem, however much more will be provided in Chapter 27.

Example 25.15. On the sphere $\partial B_R(0) = \{(x_1, x_2, x_3) \in \mathbb{R}^3 \mid x_1^2 + x_2^2 + x_3^2 = R^2\}$ we want to find the surface integral

$$\int_{\partial B_R(0)} F \cdot d\sigma$$

where $F(x_1, x_2, x_3) = \left(\dfrac{x_1}{(x_1^2+x_2^2+x_3^2)^{\frac{1}{2}}}, \dfrac{x_2}{(x_1^2+x_2^2+x_3^2)^{\frac{1}{2}}}, \dfrac{x_3}{(x_1^2+x_2^2+x_3^2)^{\frac{1}{2}}}\right)$. We want to apply Gauss' theorem, i.e.

$$\int_{\partial B_R(0)} F \cdot d\sigma = \int_{B_R(0)} \text{div}\, F\, dx$$

and for this we need div F. For $1 \leq j \leq 3$ we find

$$\frac{\partial}{\partial x_j}\left(\frac{x_j}{(x_1^2 + x_2^2 + x_3^2)^{\frac{1}{2}}}\right) = \frac{1}{(x_1^2 + x_2^2 + x_3^2)^{\frac{1}{2}}} + x_j\frac{\partial}{\partial x_j}\left(\frac{1}{(x_1^2 + x_2^2 + x_3^2)^{\frac{1}{2}}}\right)$$

$$= \frac{1}{(x_1^2 + x_2^2 + x_3^2)^{\frac{1}{2}}} - \frac{x_j^2}{(x_1^2 + x_2^2 + x_3^2)^{\frac{3}{2}}} = \frac{x_1^2 + x_2^2 + x_3^2 - x_j^2}{(x_1^2 + x_2^2 + x_3^2)^{\frac{3}{2}}}$$

which yields

$$\operatorname{div} F(x_1, x_2, x_3) = \frac{2(x_1^2 + x_2^2 + x_3^2)}{(x_1^2 + x_2^2 + x_3^2)^{\frac{3}{2}}} = 2(x_1^2 + x_2^2 + x_3^2)^{\frac{1}{2}}.$$

Using spherical coordinates we find further

$$\int_{B_R(0)} \operatorname{div} F \, \mathrm{d}x = 2\int_{B_R(0)} (x_1^2 + x_2^2 + x_3^2)^{\frac{1}{2}} \, \mathrm{d}x$$

$$= 2\int_0^R \int_0^\pi \int_0^{2\pi} (\rho^2)^{\frac{1}{2}} \rho^2 \sin\vartheta \, \mathrm{d}\varphi \, \mathrm{d}\vartheta \, \mathrm{d}\rho$$

$$= 8\pi \int_0^R \rho^3 \, \mathrm{d}\rho = 2\pi R^4.$$

Example 25.16. In this example we will see that we can calculate the volume of a Gauss domain using a surface integral. For a Gauss domain $G \subset \mathbb{R}^3$ the volume is given by

$$\operatorname{vol}_3(G) = \int_G 1 \, \mathrm{d}x.$$

If $F : \tilde{G} \to \mathbb{R}^3$, $\overline{G} \subset \tilde{G}$ and $\tilde{G} \subset \mathbb{R}^3$ open, is a C^1-vector field with div $F = 1$ then we obtain

$$\operatorname{vol}_3(G) = \int_G 1 \, \mathrm{d}x = \int_G \operatorname{div} F \, \mathrm{d}x = \int_{\partial G} F \cdot \mathrm{d}\sigma. \qquad (25.12)$$

A possible choice for F could be the vector field $F(x_1, x_2, x_3) = \frac{1}{3}(x_1, x_2, x_3)$ for which we have indeed div $F = 1$. When G is a pre-Gauss domain with respect to each axis e_j we find

$$\operatorname{vol}_3(G) = \frac{1}{3}\int_{\partial G} (x_1\vec{n}_1(x) + x_2\vec{n}_2(x) + x_3\vec{n}_3(x))\sigma(\mathrm{d}x).$$

In fact using Lemma 25.8 we find for a pre-Gauss domain with respect to e_j that

$$\text{vol}_3(G) = \int_{\partial G} x_j \vec{n}_j(x)\sigma(\mathrm{d}x).$$

Combining (25.12) with the mean value theorem in the form of (20.21) we can obtain more insight in the meaning of the divergence. For a continuous function $f : \tilde{G} \to \mathbb{R}$ and every connected open Jordan measurable set $G \subset \overline{G} \subset \tilde{G} \subset \mathbb{R}^n$ such that $\overline{G}$ is compact we have by (20.21) for some $\xi \in G$

$$\frac{1}{\text{vol}_n(G)} \int_G f(x)\,\mathrm{d}x = \frac{1}{\text{vol}_n(\overline{G})} \int_{\overline{G}} f(x)\,\mathrm{d}x = f(\xi). \tag{25.13}$$

Now if $(G_k)_{k\in\mathbb{N}}$ is a sequence of open connected Jordan measurable sets such that $\overline{G}_k$ is a compact subset of G, $x_0 \in \bigcap_{k\in\mathbb{N}} \overline{G}_k$ and $\lim_{k\to\infty}(\text{diam}(\overline{G}_k)) = 0$, then there exists a sequence of points $(\xi_k)_{k\in\mathbb{N}}$, $\xi_k \in \overline{G} \subset G$, such that

$$f(\xi_k) = \frac{1}{\text{vol}_n(G_k)} \int_{G_k} f(x)\,\mathrm{d}x. \tag{25.14}$$

Since for $k \to \infty$ it follows that $\lim_{k\to\infty} \xi_k = x_0$ we find

$$f(x_0) = \lim_{k\to\infty} \frac{1}{\text{vol}_n(G_k)} \int_{G_k} f(x)\,\mathrm{d}x. \tag{25.15}$$

Now we add the assumption that each set G_k is in addition a Gauss domain and we replace f by $\text{div}\, F$ where $F : \tilde{G} \to \mathbb{R}^3$ is a C^1-vector field, hence $\text{div}\, F$ is a continuous function on $\tilde{G}$. Now (25.15) reads as

$$\text{div}\, F(x_0) = \lim_{k\to\infty} \left(\frac{1}{\text{vol}_n(G_k)} \int_{G_k} \text{div}\, F\,\mathrm{d}x \right) = \lim_{k\to\infty} \left(\frac{1}{\text{vol}_n(G_k)} \int_{\partial G_k} \langle F, \vec{n} \rangle\,\mathrm{d}\sigma \right).$$

Since $\langle F, \vec{n} \rangle$ is the projection of F in the direction of $\vec{n}$, physicists interpret $\frac{1}{\text{vol}(G)} \int_{\partial G} \langle F, \vec{n} \rangle\,\mathrm{d}\sigma$ as flow of F through ∂G per unit volume. Hence $\text{div}\, F(x_0)$ is the flow of F at x_0 in the direction of the outer normal per unit volume.

Problems

1. Let $W = [-2, 2] \times [-3, 3] \times [-4, 4]$ and $F : \mathbb{R}^3 \to \mathbb{R}^3$ the vector field

$$F(x, y, z) = \begin{pmatrix} 3xz^2 \\ 4xy^2 \\ 5yz \end{pmatrix}.$$ Use Gauss' theorem for hyper-rectangles to find

$\int_{\partial W} F \cdot \mathrm{d}\sigma$.

2. Use Gauss' theorem to prove that

$$\int_{B_R(0)} \operatorname{div} F\, dxdydz = 4\pi R^2$$

where $B_R(0) \subset \mathbb{R}^3$ is the closed ball with centre 0 and radius R and $F : \mathbb{R}^3 \to \mathbb{R}^3$ is the vector field $F(x, y, z) = \frac{1}{R} \begin{pmatrix} x \\ y \\ z \end{pmatrix}$.

3. Consider the cylinder $C = \{(x, y, z) \in \mathbb{R}^3 | x^2 + y^2 \le R^2, 0 \le z \le a\}$ for $R > 0$ and $a > 0$. Further let $G \subset \mathbb{R}^3$ be an open set such that $C \subset G$. For $u \in C^1(G)$ and $v \in C(G)$ such that $v(x, y, z) = w(x, y)$, i.e. v is independent of z, prove

$$\int_C \frac{\partial u}{\partial z} v\, dxdydz = \int_K u(x, y, a) w(x, y)\, dxdy - \int_K u(x, y, 0) w(x, y)\, dxdy$$

where $K = \{(x, y) \in \mathbb{R}^2 | x^2 + y^2 \le R^2\}$.

4. Let $\tilde{G} \subset \mathbb{R}^3$ be an open bounded set and $\emptyset \ne G \subset \tilde{G}$ a Gauss domain. Define $\tilde{C}_0^1(G) := \{u|_G | u \in C^1(\tilde{G}), u|_{\partial G} = 0\}$. Now give a proof that $(\tilde{C}_0^1(G), ||| \cdot |||_H)$ is a normed space where

$$|||u|||_G^2 = \int_G ||\operatorname{grad} u||^2 dx.$$

Hint: use the Poincaré inequality from Lemma 25.12 to show that by

$$< u, v >_G := \int_G \langle \operatorname{grad} u, \operatorname{grad} v \rangle dx$$

a scalar product is given on $\tilde{C}_0^1(G)$ and use the fact that if $\langle \cdot, \cdot \rangle$ is a scalar product on a $\mathbb{R}$-vector space H, then a norm is given on H by $\langle u, u \rangle^{\frac{1}{2}}$, see Problem 9 in Chapter 1.

5. Let $G \subset \mathbb{R}^3$ be a Gauss domain $\overline{G} \subset \tilde{G} \subset \mathbb{R}^3$ where $\tilde{G}$ is an open set. For $u, v \in C^2(\tilde{G})$ prove **Green's first identity**:

$$\int_G (\Delta_3 u) v\, dx + \int_G \langle \operatorname{grad} u, \operatorname{grad} v \rangle dx = \int_{\partial G} v \frac{\partial u}{\partial \vec{n}} d\sigma$$

where $\frac{\partial v}{\partial \vec{n}} = \langle \vec{n}, \operatorname{grad} v \rangle$ is the directional derivative of v in the direction of the outward pointing normal to ∂G. For v being harmonic in G deduce

$$\int_G \|\operatorname{grad} v\|^2 dx = \frac{1}{2} \int_{\partial G} \langle \vec{n}, \operatorname{grad} v^2 \rangle d\sigma = \frac{1}{2} \int_{\partial G} \frac{\partial v^2}{\partial \vec{n}} d\sigma.$$

6. Let $G \subset \mathbb{R}^3$ be a Gauss domain $\overline{G} \subset \tilde{G} \subset \mathbb{R}^3$ where $\tilde{G}$ is an open set. For $u, v \in C^2(\tilde{G})$ prove **Green's second identity**:

$$\int_G (v\Delta_3 u - u\Delta_3 v)dx = \int_{\partial G} \left(v\frac{\partial u}{\partial \vec{n}} - u\frac{\partial v}{\partial \vec{n}} \right) d\sigma.$$

Deduce that for a harmonic function v we have

$$\int_G v\Delta_3 u\, dx = \int_{\partial G} \left(v\frac{\partial u}{\partial \vec{n}} - u\frac{\partial v}{\partial \vec{n}} \right) d\sigma$$

as well as

$$\int_{\partial G} \frac{\partial v}{\partial \vec{n}} d\sigma = 0.$$

7. Use results of Problem 6, in particular Green's second identity, to show for a harmonic function $v : B_{1+\epsilon}(0) \to \mathbb{R}$, $B_{1+\epsilon}(0) \subset \mathbb{R}^3$, $\epsilon > 0$, that

$$2\int_{S^2} v\, d\sigma - \int_{S^2} \frac{\partial v}{\partial \vec{n}} d\sigma = \sigma \int_{\overline{B_1(0)}} v\, dx$$

where $S^2 = \partial B_1(0)$ is the unit sphere in $\mathbb{R}^3$.

Hint: take in Green's second identity u such that $\Delta_3 u = 6$.

8. Let $G \subset \mathbb{R}^3$ be an open set and suppose that for all balls $\overline{B_r(a)} \subset G$ the following holds

$$(*) \qquad \int_{\partial B_r(a)} \langle \vec{n}, E \rangle d\sigma = 4\pi \int_{B_r(a)} \rho(x)dx,$$

where $E : G \to \mathbb{R}^3$ is a C^1-vector field and $\rho : G \to \mathbb{R}$ a continuous function. The identity $(*)$ can be seen as an integral formulation of Gauss' law of electro-statics with E being the electric field and ρ the charge density, see J. D. Jackson [28]. Prove that $(*)$ implies $\operatorname{div} E = 4\pi\rho$.

517

26 Stokes' Theorem in $\mathbb{R}^2$ and $\mathbb{R}^3$

We start with a lengthy example. Let $Q \subset \mathbb{R}^2$ be an open set and $R = [a, b] \times [c, d]$, $a < b$, $c < d$, a subset of Q, i.e. $R \subset Q$. Further let $F : Q \to \mathbb{R}^2$ be a C^1-vector field, $F = \begin{pmatrix} F_1 \\ F_2 \end{pmatrix}$. It follows that $\frac{\partial F_k}{\partial x_j}$, $k, j = 1, 2$, is integrable over R and

$$\int_R \left(\frac{\partial F_2}{\partial x_1}(x) - \frac{\partial F_1}{\partial x_2}(x) \right) dx \tag{26.1}$$

is well defined. The boundary of R is the set

$$\partial R = [a, b] \times \{c\} \cup \{b\} \times [c, d] \cup [a, b] \times \{d\} \cup \{a\} \times [c, d],$$

see Figure 26.1.

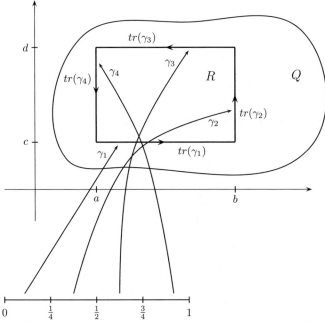

Figure 26.1

519

We can represent ∂R as trace of $\gamma : [0, 1] \to \mathbb{R}^2$ where

$$\gamma_1(t) := \gamma\big|_{[0,\frac{1}{4}]}(t) = \begin{pmatrix} 4(b - a)t + a \\ c \end{pmatrix},$$

$$\gamma_2(t) := \gamma\big|_{[\frac{1}{4},\frac{1}{2}]}(t) = \begin{pmatrix} b \\ 4(d - c)t + (2c - d) \end{pmatrix},$$

$$\gamma_3(t) := \gamma\big|_{[\frac{1}{2},\frac{3}{4}]}(t) = \begin{pmatrix} 4(a - b)t + (3b - 2a) \\ d \end{pmatrix},$$

$$\gamma_4(t) := \gamma\big|_{[\frac{3}{4},1]}(t) = \begin{pmatrix} a \\ 4(c - d)t + (4d - 3c) \end{pmatrix},$$

hence γ is a piecewise C^1-curve and for suitable functions the line integral over γ is defined as $\int_\gamma = \int_{\gamma_1} + \int_{\gamma_2} + \int_{\gamma_3} + \int_{\gamma_4}$. The regularity of F allows us to calculate the volume integral of $\frac{\partial F_2}{\partial x_1}$ over R as iterated integral and we get

$$\int_R \frac{\partial F_2}{\partial x_1}(x)\,\mathrm{d}x = \int_c^d \left(\int_a^b \frac{\partial F_2}{\partial x_1}(x_1, x_2)\,\mathrm{d}x_1 \right)\mathrm{d}x_2$$

$$= \int_c^d (F_2(b, x_2) - F_2(a, x_2))\,\mathrm{d}x_2 \tag{26.2}$$

as well as

$$\int_R \frac{\partial F_1}{\partial x_2}(x)\,\mathrm{d}x = \int_a^b (F_1(x_1, d) - F_1(x_1, c))\,\mathrm{d}x_1. \tag{26.3}$$

Using the parametrization of ∂R with the help of γ we first note

$$\int_\gamma F(x) \cdot \mathrm{d}x = \int_0^1 \langle F(\gamma(t)), \dot{\gamma}(t) \rangle\,\mathrm{d}t$$

$$= \int_0^{\frac{1}{4}} \left\langle \begin{pmatrix} F_1(\gamma_1(t)) \\ F_2(\gamma_1(t)) \end{pmatrix}, \begin{pmatrix} 4(b - a) \\ 0 \end{pmatrix} \right\rangle\,\mathrm{d}t$$

$$+ \int_{\frac{1}{4}}^{\frac{1}{2}} \left\langle \begin{pmatrix} F_1(\gamma_2(t)) \\ F_2(\gamma_2(t)) \end{pmatrix}, \begin{pmatrix} 0 \\ 4(b - c) \end{pmatrix} \right\rangle\,\mathrm{d}t$$

$$+ \int_{\frac{1}{2}}^{\frac{3}{4}} \left\langle \begin{pmatrix} F_1(\gamma_3(t)) \\ F_2(\gamma_3(t)) \end{pmatrix}, \begin{pmatrix} 4(a - b) \\ 0 \end{pmatrix} \right\rangle\,\mathrm{d}t$$

$$+ \int_{\frac{3}{4}}^1 \left\langle \begin{pmatrix} F_1(\gamma_4(t)) \\ F_2(\gamma_4(t)) \end{pmatrix}, \begin{pmatrix} 0 \\ 4(c - d) \end{pmatrix} \right\rangle\,\mathrm{d}t$$

$$= \int_0^{\frac{1}{4}} 4(b-a)F_1(\gamma_1(t))\,\mathrm{d}t + \int_{\frac{1}{2}}^{\frac{3}{4}} 4(a-b)F_1(\gamma_3(t))\,\mathrm{d}t$$

$$+ \int_{\frac{1}{4}}^{\frac{1}{2}} 4(d-c)F_2(\gamma_2(t))\,\mathrm{d}t + \int_{\frac{3}{4}}^{1} 4(c-d)F_2(\gamma_4(t))\,\mathrm{d}t.$$

Furthermore we find

$$\int_0^{\frac{1}{4}} \left\langle \begin{pmatrix} F_1(\gamma_1(t)) \\ 0 \end{pmatrix}, \dot{\gamma}_1(t) \right\rangle \mathrm{d}t = \int_0^{\frac{1}{4}} 4(b-a)F_1(\gamma_1(t))\,\mathrm{d}t$$

$$= \int_0^{\frac{1}{4}} 4(b-a)F_1(4(b-a)t+a, c)\,\mathrm{d}t = \int_a^b F_1(x_1, c)\,\mathrm{d}x_1,$$

$$\int_{\frac{1}{4}}^{\frac{1}{2}} \left\langle \begin{pmatrix} F_1(\gamma_2(t)) \\ 0 \end{pmatrix}, \dot{\gamma}_2(t) \right\rangle \mathrm{d}t = \int_{\frac{1}{4}}^{\frac{1}{2}} \left\langle \begin{pmatrix} F_1(\gamma_2(t)) \\ 0 \end{pmatrix}, \begin{pmatrix} 0 \\ 4(d-c) \end{pmatrix} \right\rangle \mathrm{d}t = 0,$$

$$\int_{\frac{1}{2}}^{\frac{3}{4}} \left\langle \begin{pmatrix} F_1(\gamma_3(t)) \\ 0 \end{pmatrix}, \dot{\gamma}_3(t) \right\rangle \mathrm{d}t = \int_{\frac{1}{2}}^{\frac{3}{4}} 4(a-b)F_1(\gamma_3(t))\,\mathrm{d}t$$

$$= \int_{\frac{1}{2}}^{\frac{3}{4}} 4(a-b)F_1(4(a-b)t+(3b-2a), d)\,\mathrm{d}t$$

$$= \int_b^a F_1(x_1, d)\,\mathrm{d}x_1 = -\int_a^b F_1(x_1, d)\,\mathrm{d}x_1,$$

and

$$\int_{\frac{3}{4}}^{1} \left\langle \begin{pmatrix} F_1(\gamma_4(t)) \\ 0 \end{pmatrix}, \dot{\gamma}_4(t) \right\rangle \mathrm{d}t = \int_{\frac{3}{4}}^{1} \left\langle \begin{pmatrix} F_1(\gamma_4(t)) \\ 0 \end{pmatrix}, \begin{pmatrix} 0 \\ 4(c-d) \end{pmatrix} \right\rangle \mathrm{d}t = 0,$$

which leads to

$$\int_0^1 \left\langle \begin{pmatrix} F_1(\gamma(t)) \\ 0 \end{pmatrix}, \dot{\gamma}(t) \right\rangle \mathrm{d}t = -\left(\int_a^b (F_1(x_1, d) - F_1(x_1, c))\,\mathrm{d}x_1 \right)$$

$$= -\int_R \frac{\partial F_1}{\partial x_2}(x)\,\mathrm{d}x,$$

or

$$\int_\gamma \begin{pmatrix} F_1 \\ 0 \end{pmatrix} \cdot \mathrm{d}x = -\int_R \frac{\partial F_1}{\partial x_2}(x)\,\mathrm{d}x \qquad (26.4)$$

and analogously we find

$$\int_\gamma \begin{pmatrix} 0 \\ F_2 \end{pmatrix} \cdot \mathrm{d}x = \int_R \frac{\partial F_2}{\partial x_1}(x)\,\mathrm{d}x \tag{26.5}$$

which yields

$$\int_R \left(\frac{\partial F_2}{\partial x_1}(x) - \frac{\partial F_1}{\partial x_2}(x) \right) \mathrm{d}x = \int_\gamma F \cdot \mathrm{d}x$$
$$= \int_0^1 \langle F(\gamma(t)), \dot{\gamma}(t) \rangle\,\mathrm{d}t.$$

The equality

$$\int_R \left(\frac{\partial F_2}{\partial x_1}(x) - \frac{\partial F_1}{\partial x_2}(x) \right) \mathrm{d}x = \int_0^1 \langle F(\gamma(t)), \dot{\gamma}(t) \rangle\,\mathrm{d}t \tag{26.6}$$

is a formula in which all terms are well defined and we can easily understand. By tradition, the right hand side is often written as

$$\int_{\partial R} F_1\,\mathrm{d}x_2 + \int_{\partial R} F_2\,\mathrm{d}x_2$$

which allows the following interpretation: If G_1 and G_2 are two continuous functions on ∂R then $\int_{\partial R} G_1\,\mathrm{d}x_1$ is the line integral of the vector field $\begin{pmatrix} G_1 \\ 0 \end{pmatrix}$ and $\int_{\partial R} G_2\,\mathrm{d}x_2$ is the line integral of the vector field $\begin{pmatrix} 0 \\ G_2 \end{pmatrix}$, and this notation or interpretation holds for more general curves. Formula (26.6) is a first version of Stokes' theorem.

Note that in order to derive (26.6) we have used a special parametrization γ of ∂R. However any other parametrization η of ∂R which parametrizes ∂R as simply closed piecewise C^1-curve having the same orientation will lead to tangent vectors $\dot{\eta}$ pointing at each $(x_1, x_2) \in \partial R \backslash \{(a, c), (a, d), (b, c), (b, d)\}$ in the same direction as the corresponding tangent vectors $\dot{\gamma}$ and hence will lead again to (26.6). The reason behind this is that in this situation the unit tangent vectors are determined by ∂R and not by the parametrization.

We want to prove (26.6) for more general domains. The calculation leading to (26.4) and (26.5) suggests that a normal domain with respect to one axis will give one of these results, hence a normal domain with respect to both axes leads to (26.6).

For $a < b$ let $\varphi : [a, b] \to \mathbb{R}$, $\varphi_1(x_1) \leq \varphi_2(x_1)$ for $x_1 \in [a, b]$, be continuously differentiable functions. We consider the normal domain with respect to the x_2-axis given by

$$G := \left\{ (x_1, x_2) \in \mathbb{R}^2 \,\middle|\, a \leq x_1 \leq b,\ \varphi_1(x_1) \leq x_2 \leq \varphi_2(x_1) \right\}. \qquad (26.7)$$

The boundary of ∂G has four parts

$$\partial G = \Gamma(\varphi_1) \cup \{b\} \times [\varphi_1(b), \varphi_2(b)] \cup \Gamma(\varphi_2) \cup \{a\} \times [\varphi_1(a), \varphi_2(a)],$$

see Figure 26.2.

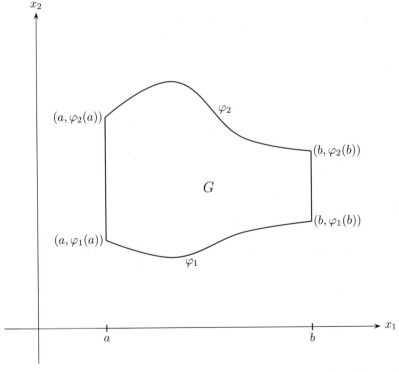

Figure 26.2

Again we can consider ∂G as trace of a piecewise C^1-parametric curve, for example $\gamma : [0,1] \to \mathbb{R}^2$ with

$$\gamma_1(t) = \gamma\big|_{[0,\frac{1}{4}]}(t) := \begin{pmatrix} 4(b-a)t + a \\ \varphi_1(4(b-a)t + a) \end{pmatrix},$$

$$\gamma_2(t) = \gamma\big|_{[\frac{1}{4},\frac{1}{2}]}(t) := \begin{pmatrix} b \\ 4(\varphi_2(b) - \varphi_1(b))t + (2\varphi_1(b) - \varphi_2(b)) \end{pmatrix},$$

$$\gamma_3(t) = \gamma\big|_{[\frac{1}{2},\frac{3}{4}]}(t) := \begin{pmatrix} 4(a-b)t + (3b - 2a) \\ \varphi_2(4(a-b)t + (3b - 2a)) \end{pmatrix},$$

$$\gamma_4(t) = \gamma\big|_{[\frac{3}{4},1]}(t) := \begin{pmatrix} a \\ 4(\varphi_1(a) - \varphi_2(a))t + (4\varphi_2(a) - 3\varphi_1(a)) \end{pmatrix}$$

and a line integral over γ is again understood as the sum of the line integrals over γ_j.

Now let $Q \subset \mathbb{R}^2$ be an open set such that for the compact set G we have $G \subset Q$. Moreover let $F : Q \to \mathbb{R}^2$, $F = \begin{pmatrix} F_1 \\ F_2 \end{pmatrix}$, be a C^1-vector field and consider the integral

$$-\int_G \frac{\partial F_1}{\partial x_2}(x)\,\mathrm{d}x = -\int_a^b \left(\int_{\varphi_1(x_1)}^{\varphi_2(x_1)} \frac{\partial F_1}{\partial x_2}(x_1, x_2)\,\mathrm{d}x_2 \right) \mathrm{d}x_1$$

$$= -\int_a^b (F_1(x_1, \varphi_2(x_1)) - F_1(x_1, \varphi_2(x_1)))\,\mathrm{d}x_1.$$

For the vector field $\begin{pmatrix} F_1 \\ 0 \end{pmatrix}$ we find using $\left\langle \begin{pmatrix} F_1 \\ 0 \end{pmatrix}(\gamma_2(t)), \dot{\gamma}_2(t) \right\rangle = \left\langle \begin{pmatrix} F_1 \\ 0 \end{pmatrix}(\gamma_4(t)), \dot{\gamma}_4(t) \right\rangle = 0$ that

$$\int_0^1 \left\langle \begin{pmatrix} F_1 \\ 0 \end{pmatrix}(\gamma(t)), \dot{\gamma}(t) \right\rangle \mathrm{d}t = \int_0^{\frac{1}{4}} \left\langle \begin{pmatrix} F_1 \\ 0 \end{pmatrix}(\gamma_1(t)), \dot{\gamma}_1(t) \right\rangle \mathrm{d}t + \int_{\frac{1}{2}}^{\frac{3}{4}} \left\langle \begin{pmatrix} F_1 \\ 0 \end{pmatrix}(\gamma_3(t)), \dot{\gamma}_3(t) \right\rangle \mathrm{d}t$$

$$= \int_0^{\frac{1}{4}} 4(b-a)F_1(4(b-a)t + a, \varphi_1(4(b-a)t + a))\,\mathrm{d}t$$

$$+ \int_{\frac{1}{2}}^{\frac{3}{4}} 4(a-b)F_1(4(a-b)t + (3b - 2a), \varphi_2(4(a-b)t + (3b - 2a)))\,\mathrm{d}t$$

$$= \int_a^b F_1(x_1, \varphi_1(x_1))\,\mathrm{d}x_1 - \int_a^b F_1(x_1, \varphi_2(x_1))\,\mathrm{d}x_1.$$

Thus we obtain with $\gamma = (\gamma^{(1)}, \gamma^{(2)})$

$$-\int_G \frac{\partial F_1}{\partial x_2}(x)\,\mathrm{d}x = \int_0^1 \langle \begin{pmatrix} F_1 \\ 0 \end{pmatrix}(\gamma(t)), \dot{\gamma}(t)\rangle \mathrm{d}t = \int_0^1 F_1(\gamma(t))\dot{\gamma}^{(1)}(t)\,\mathrm{d}t. \quad (26.8)$$

When G is a normal domain with respect to the x_1-axis, i.e. if with C^1-functions $\psi_1, \psi_2; [c,d] \to \mathbb{R}$, $\psi_1(x_2) \le \psi_2(x_2)$, we have

$$G = \{(x_1, x_2) \in \mathbb{R}^2 \mid c \le x_2 \le d, \psi_1(x_2) \le x_1 \le \psi_2(x_2)\} \subset Q, \quad (26.9)$$

we obtain with a similar calculation

$$\int_G \frac{\partial F_2}{\partial x_1}(x)\,\mathrm{d}x = \int_0^1 \langle \begin{pmatrix} 0 \\ F_2 \end{pmatrix}(\eta(t)), \dot{\eta}(t)\rangle \mathrm{d}t = \int_0^1 F_2(\eta(t))\dot{\eta}^{(2)}(t)\,\mathrm{d}t \quad (26.10)$$

where $\eta = (\eta^{(1)}, \eta^{(2)})$ is the parametrization of ∂G constructed along the lines we have constructed γ, but now using the functions ψ_1 and ψ_2. Hence we have proved for these two special parametrizations of ∂G

$$\int_G \left(\frac{\partial F_2}{\partial x_1}(x) - \frac{\partial F_1}{\partial x_2}(x) \right)\mathrm{d}x = \int_0^1 F_1(\gamma(t))\dot{\gamma}^{(1)}(t)\,\mathrm{d}t + \int_0^1 F_2(\eta(t))\dot{\eta}^{(2)}(t)\,\mathrm{d}t. \quad (26.11)$$

First we note that if we replace γ and η by a new parametrization of ∂G, again each parametrizing a simply closed piecewise C^1-curve with the same orientation as γ and η, respectively, formula (26.11) remains unchanged. So let $\gamma = (\gamma^{(1)}, \gamma^{(2)})$, $\gamma : I_\gamma \to \mathbb{R}^2$, and $\eta = (\eta^{(1)}, \eta^{(2)})$, $\eta : I_\eta \to \mathbb{R}^2$, be two anticlockwise orientated simply closed piecewise C^1-parametrizations of ∂G, γ referring to G as a normal domain with respect to the x_2-axis and η referring to G as a normal domain with respect to the x_1-axis. Then **Green's theorem** holds.

Theorem 26.1. *For G and ∂G, as well as $\gamma : I_\gamma \to \mathbb{R}^2$ and $\eta : I_\eta \to \mathbb{R}^2$ as above and for a C^1-vector field $F = \begin{pmatrix} F_1 \\ F_2 \end{pmatrix} : Q \to \mathbb{R}^2$ where $Q \subset \mathbb{R}^2$ is an open set such that $G \subset Q$ the following holds*

$$\int_G \left(\frac{\partial F_2}{\partial x_1}(x) - \frac{\partial F_1}{\partial x_2}(x) \right)\mathrm{d}x = \int_{I_\gamma} F_1(\gamma(t))\dot{\gamma}^{(1)}(t)\,\mathrm{d}t + \int_{I_\eta} F_2(\eta(t))\dot{\eta}^{(2)}(t)\,\mathrm{d}t. \quad (26.12)$$

Remark 26.2. A. The common way to write (26.12) is

$$\int_G \left(\frac{\partial F_2}{\partial x_1}(x) - \frac{\partial F_1}{\partial x_2}(x) \right) \mathrm{d}x = \int_{\partial G} F_1 \, \mathrm{d}x_1 + \int_{\partial G} F_2 \, \mathrm{d}x_2. \tag{26.13}$$

B. Some authors call Theorem 26.1 Gauss' theorem or the divergence theorem in the plane, others call it Stokes' theorem in the plane.

A few words to justify the notation in (26.12) are in order. The parametrization of ∂G as a simply closed curve implies that for all anticlockwise orientated piecewise C^1-parametrizations the corresponding unit tangent vectors to ∂G coincide (at all points they are defined). So whenever ∂G contains an open line segment S_2 parallel to the x_2-axis, then $\dot{\gamma}^{(2)}\big|_{S_2} = 0$ and whenever $S_1 \subset \partial G$ is an open line segment parallel to the x_1-axis then $\dot{\eta}^{(1)}\big|_{S_1} = 0$. Thus the integrals over I_γ and I_η can be decomposed as

$$\int_{I_\gamma} F_1(\gamma(t))\dot{\gamma}^{(1)}(t) \, \mathrm{d}t = \int_{I_\gamma^{(1)}} F_1(\gamma(t))\dot{\gamma}^{(1)}(t) \, \mathrm{d}t + \int_{I_\gamma^{(2)}} F_1(\gamma(t))\dot{\gamma}^{(1)}(t) \, \mathrm{d}t \tag{26.14}$$

where $I_\gamma^{(1)}$ and $I_\gamma^{(2)}$ are disjoint closed sub-intervals of I_γ (one of which might even be empty), and

$$\int_{I_\eta} F_2(\eta(t))\dot{\eta}^{(2)}(t) \, \mathrm{d}t = \int_{I_\eta^{(1)}} F_2(\eta(t))\dot{\eta}^{(2)}(t) \, \mathrm{d}t + \int_{I_\eta^{(2)}} F_2(\eta(t))\dot{\eta}^{(2)}(t) \, \mathrm{d}t \tag{26.15}$$

where again $I_\eta^{(1)}$ and $I_\eta^{(2)}$ are disjoint closed sub-intervals of I_η (one of which might even be empty) such that on $I_\gamma \backslash (I_\gamma^{(1)} \cup I_\gamma^{(2)})$ we have $\dot{\gamma}^{(1)}(t) = 0$ as we have on $I_\eta \backslash (I_\eta^{(1)} \cup I_\eta^{(2)})$ that $\dot{\eta}^{(2)}(t) = 0$. As a normal domain with respect to the x_j-axis ∂G can only contain at most two open line segments parallel to this axis. On $I_\gamma^{(j)}$ and $I_\eta^{(j)}$ we may use the domain defining functions to parametrize ∂G, which however need not always be the best (easiest) parametrization of ∂G.

Before we try to extend Theorem 26.1 to more general domains, we want to discuss an interesting consequence and some applications.

Corollary 26.3. *In the situation of Theorem 26.1 the area of G is given by*

$$\mathrm{vol}_2(G) = \frac{1}{2} \left(\int_{\partial G} x_1 \, \mathrm{d}x_2 - \int_{\partial G} x_2 \, \mathrm{d}x_1 \right) = \frac{1}{2} \int_{I_\eta} \eta^{(1)}(t)\dot{\eta}^{(2)}(t) \, \mathrm{d}t - \frac{1}{2} \int_{I_\gamma} \gamma^{(2)}(t)\dot{\gamma}^{(1)}(t) \, \mathrm{d}t. \tag{26.16}$$

Proof. We take in (26.12) the vector field $F(x) = \begin{pmatrix} -x_2 \\ x_1 \end{pmatrix}$ to find

$$\int_G \left(\frac{\partial F_2}{\partial x_1}(x) - \frac{\partial F_1}{\partial x_2}(x) \right) \mathrm{d}x = \int_G (1+1)\,\mathrm{d}x = 2 \int_G 1\,\mathrm{d}x = 2\mathrm{vol}_2(G).$$

$\square$

Example 26.4. The disc $B_r(0) = \{(x_1, x_2) \in \mathbb{R}^2 \mid x_1^2 + x_2^2 \le r^2\}$ is a normal domain with respect to the x_2-axis as well as with respect to the x_1-axis. Considering $B_r(0)$ as a normal domain with respect to the x_2-axis we can use the parametrization $\gamma : [-r, r] \to \mathbb{R}^2$, $\gamma(x_1) = \begin{pmatrix} r\cos\left(\pi\frac{x_1}{r}\right) \\ r\sin\left(\pi\frac{x_1}{r}\right) \end{pmatrix}$, and we find

$$-\frac{1}{2}\int_{I_\gamma} \gamma^{(2)}(t)\dot{\gamma}^{(1)}(t)\,\mathrm{d}t = -\frac{1}{2}\int_{-r}^{r} \left(r\sin\left(\pi\frac{x_1}{r}\right) \right) \frac{\mathrm{d}}{\mathrm{d}x_1}\left(r\cos\left(\pi\frac{x_1}{r}\right) \right)) \mathrm{d}x_1$$

$$= \frac{\pi r}{2}\int_{-r}^{r} \left(\sin\left(\pi\frac{x_1}{r}\right) \right)^2 \mathrm{d}x_1 = \frac{\pi r^2}{2}.$$

Considering $B_r(0)$ as a normal domain with respect to the x_1-axis we can use $\eta : [-r, r] \to \mathbb{R}^2$, $\eta(x_2) = \begin{pmatrix} -r\sin\left(\pi\frac{x_2}{r}\right) \\ r\cos\left(\pi\frac{x_2}{r}\right) \end{pmatrix}$, as parametrization of $\partial B_r(0)$ and we get

$$\frac{1}{2}\int_{I_\eta} \eta^{(1)}(t)\dot{\eta}^{(2)}(t)\,\mathrm{d}t = \frac{1}{2}\int_{-r}^{r} \left(-r\sin\left(\pi\frac{x_2}{r}\right) \right) \frac{\mathrm{d}}{\mathrm{d}x_2}\left(r\cos\left(\pi\frac{x_2}{r}\right) \right) \mathrm{d}x_2$$

$$= \frac{\pi r}{2}\int_{-r}^{r} \left(\sin\left(\pi\frac{x_2}{r}\right) \right)^2 \mathrm{d}x_2 = \frac{\pi r^2}{2}$$

which yields the known result

$$\mathrm{vol}_2(B_r(0)) = \pi r^2.$$

We would like to note that we have chosen the two parametrizations such that for γ we start with $(-1, 0)$ so that x_1 can be used as an independent variable and for η we start with $(0, 1)$ where we used x_2 as a variable. Of course, we may also use the parametrizations $\gamma(t) = \eta(t) = \begin{pmatrix} r\cos 2\pi t \\ r\sin 2\pi t \end{pmatrix}$, $t \in [0, 1]$. Then we find $F_1(\gamma(t)) = -r \sin 2\pi t$, $F_2(\eta(t)) = r \cos 2\pi t$, $\dot{\gamma}^{(1)}(t) = -2\pi r \sin 2\pi t$, $\eta^{(2)}(t) = 2\pi r \cos 2\pi t$ which leads to

$$\frac{1}{2}\int_{I_\gamma} F_1(\gamma(t))\dot{\gamma}^{(1)}\,\mathrm{d}t + \int_{I_\eta} F_2(\eta(t))\dot{\eta}^{(2)}\,\mathrm{d}t$$

$$= \frac{1}{2}\int_0^1 (-r\sin 2\pi t)(-2\pi r \sin 2\pi t)\,\mathrm{d}t + \frac{1}{2}\int_0^1 (r\cos 2\pi t)(2\pi r \cos 2\pi t)\,\mathrm{d}t$$

$$= \pi r^2 \int_0^1 (\sin^2 2\pi t + \cos^2 2\pi t)\,\mathrm{d}t = \pi r^2.$$

Example 26.5. Consider the ellipse $\mathcal{E} := \left\{ (x_1, x_2) \in \mathbb{R}^2 \,\middle|\, \frac{x_1^2}{a^2} + \frac{x_2^2}{b^2} \le 1 \right\}$ the boundary of which we can parametrize for example by $\gamma : [0, 2\pi] \to \mathbb{R}^2$, $\gamma(t) = \begin{pmatrix} a\cos t \\ b\sin t \end{pmatrix}$. Let $F(x_1, x_2) = \begin{pmatrix} x_1^3 + 2x_2 - \sin(1 + x_1^2) \\ e^{-x_2^2} - 3x_1 + \ln(1 + x_2^2) \end{pmatrix}$. We want to evaluate

$$\int_{\partial\mathcal{E}} F_1\,\mathrm{d}x_1 + \int_{\partial\mathcal{E}} F_2\,\mathrm{d}x_2.$$

Using Green's theorem we find

$$\int_{\partial\mathcal{E}} F_1\,\mathrm{d}x_1 + \int_{\partial\mathcal{E}} F_2\,\mathrm{d}x_2 = \int_{\mathcal{E}} \left(\frac{\partial F_2}{\partial x_1} - \frac{\partial F_1}{\partial x_2} \right)\mathrm{d}x$$

$$= -\int_{\mathcal{E}} (2 + 3)\,\mathrm{d}x = -5\mathrm{vol}_2(\mathcal{E}) = -5\pi ab.$$

Example 26.6. Let $G \subset \mathbb{R}^2$ be as in Theorem 26.1 and $Q \subset \mathbb{R}^2$ be an open set such that $G \subset Q$. Consider for a scalar-valued C^2-function $u : Q \to \mathbb{R}^2$ the vector field $F = \begin{pmatrix} -\frac{\partial u}{\partial x_2} \\ \frac{\partial u}{\partial x_1} \end{pmatrix}$. Applying Green's theorem we find

$$\int_G \left(\frac{\partial^2 u}{\partial x_1^2} + \frac{\partial^2 u}{\partial x_2^2} \right)\mathrm{d}x = \int_G \left(\frac{\partial F_2}{\partial x_1} - \frac{\partial F_1}{\partial x_2} \right)\mathrm{d}x$$

$$= -\int_{\partial G} \frac{\partial u}{\partial x_2}\,\mathrm{d}x_1 + \int_{\partial G} \frac{\partial u}{\partial x_1}\,\mathrm{d}x_2.$$

In particular for u harmonic, i.e $\Delta_2 u = 0$, we find

$$\int_{\partial G} \frac{\partial u}{\partial x_1}\,\mathrm{d}x_2 - \int_{\partial G} \frac{\partial u}{\partial x_2}\,\mathrm{d}x_1 = 0.$$

Before extending the class of domains for which Green's theorem holds, we remind the reader that line integrals will change their sign when we change the orientation of the parametrization, compare with Remark 15.19 (which of course extends to scalar-valued functions).

Now let G_1 and G_2 be two compact sets in $\mathbb{R}^2$ with boundaries ∂G_1 and ∂G_2 each allowing an anticlockwise piecewise C^1-parametrization $\gamma_j : I_j \to \mathbb{R}^2$ as a simply closed curve. In addition we assume that $\mathring{G}_1 \cap \mathring{G}_2 = \emptyset$ but $G_1 \cap G_2 \subset \partial G_1 \cap \partial G_2$ is part of $\mathrm{tr}(\gamma_1) \cap \mathrm{tr}(\gamma_2)$, in fact let us assume that for some open sub-intervals I_j^0 we have $\partial G_1 \cap \partial G_2 = \mathrm{tr}\left(\gamma_1\big|_{I_1^0}\right) = \mathrm{tr}\left(\gamma_2\big|_{I_2^0}\right)$, see Figure 26.3.

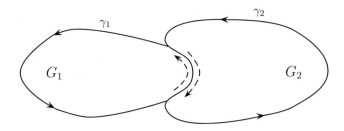

Figure 26.3

The boundary of $G_1 \cap G_2$ can now be parametrized by $\gamma := \gamma_1\big|_{I_1 \setminus I_1^0} \oplus \gamma_2\big|_{I_2 \setminus I_2^0}$ which is again a piecewise C^1-curve, anticlockwise orientated and parametrizes $\partial(G_1 \cup G_2)$ as a simply closed curve. Since for every continuous vector field or scalar-valued function the line integrals over $\gamma_1\big|_{I_1^0}$ and $\gamma_2\big|_{I_2^0}$ have different orientations they add up to 0. Thus we find

$$\int_\gamma F(p) \cdot \mathrm{d}p = \int_{\gamma_1} F(p) \cdot \mathrm{d}p + \int_{\gamma_2} F(p) \cdot \mathrm{d}p \qquad (26.17)$$

which extends to finitely many domains and to scalar-valued functions. This observation leads to

Theorem 26.7. *Let $Q \subset \mathbb{R}^2$ be an open set and $G \subset Q$ a compact set the boundary ∂G of which is the trace of a simply closed, piecewise C^1-curve*

which is anticlockwise orientated. Suppose that we can decompose G into a finite number $G_1, \ldots, G_M$ of normal domains with respect to both axes such that the domain defining functions are of class C^1. Further suppose that $\mathring{G}_j \cap \mathring{G}_k = \emptyset$ for $j \neq k$. Denote by $\gamma_j : I_j \to \mathbb{R}^2$ a piecewise C^1-parametrization of ∂G_j as a simply closed, anticlockwise orientated curve. We assume further that $\partial G_j \cap \partial G_k$, $j \neq k$, is either empty or is the union of finitely many points and finitely many, say $N_{j,k}$, portions of the type $\operatorname{tr}\left(\gamma_j\big|_{I^{0,l}_{\gamma_j, \gamma_k}}\right) = \operatorname{tr}\left(\gamma_k\big|_{I^{0,l}_{\gamma_k \cdot \gamma_j}}\right)$ *where $I^{0,l}_{\gamma_j, \gamma_k}$ is an open sub-interval of I_{γ_j}, $l = 1, \ldots, N_{j,k}$. For every C^1-vector field $F = \begin{pmatrix} F_1 \\ F_2 \end{pmatrix} : Q \to \mathbb{R}^2$ Green's theorem holds on G, i.e.*

$$\int_G \left(\frac{\partial F_2}{\partial x_1} - \frac{\partial F_1}{\partial x_2} \right) \mathrm{d}x = \int_{\partial G} F_1 \, \mathrm{d}x_1 + \int_{\partial G} F_2 \, \mathrm{d}x_2. \qquad (26.18)$$

(In short: when we can decompose G into finitely many domains being normal to both axes, then Green's theorem holds for G.)

Proof. For every domain G_j from the decomposition we find

$$\int_{G_j} \left(\frac{\partial F_2}{\partial x_1} - \frac{\partial F_1}{\partial x_2} \right) \mathrm{d}x = \int_{\partial G_j} F_1 \, \mathrm{d}x + \int_{\partial G_j} F_2 \, \mathrm{d}x_2. \qquad (26.19)$$

Clearly we have

$$\int_G \left(\frac{\partial F_2}{\partial x_1} - \frac{\partial F_1}{\partial x_2} \right) \mathrm{d}x = \sum_{j=1}^{M} \int_{G_j} \left(\frac{\partial F_2}{\partial x_1} - \frac{\partial F_1}{\partial x_2} \right) \mathrm{d}x. \qquad (26.20)$$

Using (26.12) we find

$$\int_{\partial G_j} F_1 \, \mathrm{d}x_1 + \int_{\partial G_j} F_2 \, \mathrm{d}x_2 = \int_{I_{\gamma_j}} F_1(\gamma_j(t)) \dot{\gamma}_j(t) \, \mathrm{d}t + \int_{I_{\eta_j}} F_2(\eta_j) \dot{\eta}_j(t) \, \mathrm{d}t$$

with the obvious meaning of $\gamma_j : I_{\gamma_j} \to \mathbb{R}^2$ and $\eta_j : I_{\eta_j} \to \mathbb{R}^2$. By (26.19) and (26.20) we have

$$\int_G \left(\frac{\partial F_2}{\partial x_1} - \frac{\partial F_1}{\partial x_2} \right) \mathrm{d}x = \sum_{j=1}^{M} \left(\int_{I_{\gamma_j}} F_1(\gamma_j(t)) \dot{\gamma}_j(t) \, \mathrm{d}t + \int_{I_{\eta_j}} F_2(\eta_j(t)) \dot{\eta}_j(t) \, \mathrm{d}t \right). \, (26.21)$$

For each of the two sums on the right hand side of (26.21) the observation made above yields that if $\mathrm{tr}(\gamma_j) \cap \mathrm{tr}(\gamma_k)$ is not empty or a point, then $\mathrm{tr}(\gamma_j) \cap \mathrm{tr}(\gamma_k)$ is of the type $\mathrm{tr}\left(\gamma_j\big|_{I^{0,l}_{\gamma_j,\gamma_k}}\right) = \mathrm{tr}\left(\gamma_k\big|_{I^{0,l}_{\gamma_k,\gamma_j}}\right)$ and corresponding line integrals add up to zero, the same argument applies for the η_j's. Moreover $\partial G = \overline{\bigcup_{k \neq j} \gamma_j \left(I_{\gamma_j} \setminus \left(\bigcup_l I^{0,l}_{\gamma_j,\gamma_k} \right) \right)}$ and we find

$$\sum_{j=1}^{M} \int_{I_{\gamma_j}} F_1(\gamma_j(t)) \dot{\gamma}_j(t)\,\mathrm{d}t = \sum_{j=1}^{M} \int_{I_{\gamma_j} \setminus \bigcup_l I^{0,l}_{\gamma_j,\gamma_k}} F_1(\gamma_j(t)) \dot{\gamma}_j(t)\,\mathrm{d}t = \int_{\partial G} F_1\,\mathrm{d}x_1$$

as well as

$$\sum_{j=1}^{M} \int_{I_{\eta_j}} F_2(\eta_j(t)) \dot{\eta}_j(t)\,\mathrm{d}t = \sum_{j=1}^{M} \int_{I_{\eta_j} \setminus \bigcup_l I^{0,l}_{\eta_j,\eta_k}} F_1(\eta_j(t)) \dot{\eta}_j(t)\,\mathrm{d}t = \int_{\partial G} F_2\,\mathrm{d}x_2.$$

$\square$

Remark 26.8. Note that the main content of Theorem 26.7 once we have proved Theorem 26.1 is hidden in Figure 26.3 and the observation that line integrals with opposite orientations cancel. The formulations and the proof of Theorem 26.7 is an attempt of "formally correct bookkeeping".

Corollary 26.9. *In the situation of Theorem 26.7 we have*

$$\mathrm{vol}_2(G) = \frac{1}{2} \int_{\partial G} x_1\,\mathrm{d}x_2 - \frac{1}{2} \int_{\partial G} x_2\,\mathrm{d}x_1. \tag{26.22}$$

Remark 26.10. We have preferred to work with piecewise C^1-parametrizations, in most considerations rectifiable curves will do.

A simply closed C^1-curve in $\mathbb{R}^2$ can be considered as a bijective image of S^1 and as topological space S^1 is pathwise connected. Hence the boundary of a domain considered in Theorem 26.7 is connected. To generalise Green's theorem further to domains which allow holes and hence boundaries having several connectivity components, see Figure 26.4, we need more topology and we will deal with these issues in Volume III Part 7.

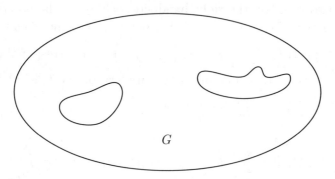

G

Figure 26.4

Example 26.11. Consider the set

$$G := \left\{ (x_1, x_2) \in \mathbb{R}^2 \,\middle|\, 4 \leq x_1^2 + x_2^2 \leq 9 \right\} \setminus \left\{ (x_1, x_2) \in \mathbb{R}^2 \,\middle|\, x_1 > 0, \, x_2 < 0 \right\}$$

$$= \left\{ (r, \varphi) \in [0, \infty) \times [0, 2\pi] \,\middle|\, 2 \leq r \leq 3, \, 0 \leq \varphi \leq \frac{3\pi}{2} \right\}$$

which is not a normal domain with respect to both axes but the three domains G_1, G_2, G_3 in Figure 26.5 are:

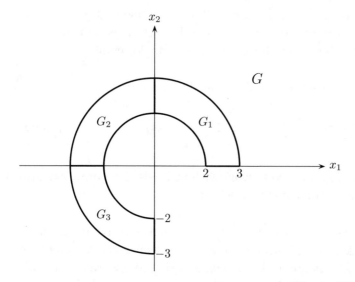

Figure 26.5

Given the vector field $F(x_1, x_2) = \begin{pmatrix} x_1 x_2 \\ x_1^2 \end{pmatrix}$. By Theorem 26.7 we have

$$\int_{\partial G} F_1 \, dx_1 + \int_{\partial G} F_2 \, dx_2 = \int_G \left(\frac{\partial F_2}{\partial x_1} - \frac{\partial F_1}{\partial x_2} \right) dx$$

$$= \int_G (2x_1 - x_1) \, dx = \int_G x_1 \, dx$$

$$= \int_2^3 \int_0^{\frac{3\pi}{2}} (r \cos \varphi) r \, d\varphi \, dr = \left(\frac{r^3}{3} \Big|_2^3 \right) \left(\sin \varphi \Big|_0^{\frac{3\pi}{2}} \right) = -\frac{19}{3}.$$

The next example picks up our discussion of the integrability conditions (15.31).

Example 26.12. Let $F : Q \to \mathbb{R}^2$, $F = \begin{pmatrix} F_1 \\ F_2 \end{pmatrix}$, be a C^1-vector field and $\overline{B_r(0)} \subset Q$. Suppose that for all circles $\partial B_\rho(\xi)$, $\xi = (\xi_1, \xi_2)$, with $\overline{B_\rho(\xi)} \subset B_r(0)$ the following holds

$$\int_{\partial B_\rho(\xi)} F_1 \, dx_1 + \int_{\partial B_\rho(\xi)} F_2 \, dx_2 = 0. \tag{26.23}$$

Then in $B_r(0)$ the integrability conditions $\frac{\partial F_2}{\partial x_1} - \frac{\partial F_1}{\partial x_2} = 0$ is fulfilled. Suppose that for some point $\xi = (\xi_1, \xi_2) \in B_r(0)$ the following holds

$$\frac{\partial F_2}{\partial x_1}(\xi) - \frac{\partial F_1}{\partial x_2}(\xi) > 0, \tag{26.24}$$

the case $\frac{\partial F_2}{\partial x_1}(\xi) - \frac{\partial F_1}{\partial x_2}(\xi) < 0$ is handled analogously. By continuity of $\frac{\partial F_2}{\partial x_1} - \frac{\partial F_1}{\partial x_2}$ it follows that (26.24) holds in some ball $B_\rho(\xi) \subset B_r(0)$ and therefore

$$0 = \int_{\partial B_\rho(\xi)} F_1 \, dx_1 + \int_{\partial B_\rho(\xi)} F_2 \, dx_2 = \int_{B_\rho(\xi)} \left(\frac{\partial F_2}{\partial x_1} - \frac{\partial F_1}{\partial x_2} \right)(x) \, dx > 0$$

which is a contradiction. Conversely, Green's theorem immediately implies that if the integrability condition holds, then for all circle $|x - \xi| = \rho$ in $B_r(0)$ relation (26.23) is satisfied.

Our next aim is to prove a first version of Stokes' theorem in $\mathbb{R}^3$. The general Stokes' theorem deals with the integrals of k-forms on m-dimensional sub-manifolds of n-dimensional manifolds and will be dealt with in Volume VI,

Part 16. Here we restrict ourselves to a topological simple situation and dealing with "smooth" vector fields in $\mathbb{R}^3$ defined in an open neighbourhood of a C^2-parametric surface with "nice" parameter domain admitting a "smooth" boundary curve. We start by introducing the geometric context.

In the following $G \subset \mathbb{R}^2$ is a (compact) normal domain with respect to both coordinate axes and the domain defining functions are assumed to be of class C^1. We assume that ∂G is parametrized as a simply closed (regular) curve $\gamma : [a, b] \to \mathbb{R}^2$, $a < b$, $\gamma(t) = (\gamma_1(t), \gamma_2(t))$, with piecewise C^1 components γ_1, γ_2 and we assume that γ is anticlockwise oriented and regularity refers to the C^1 parts of γ. The domain of G is supposed to be a subset of the open set $Q \subset \mathbb{R}^2$ and points in Q, hence G, have coordinates (u, v). On Q we consider a C^2-parametric surface $g : Q \to \mathbb{R}^3$ with injective mapping g having the components x, y, z, i.e. $g(u, v) = \begin{pmatrix} x(u, v) \\ y(u, v) \\ z(u, v) \end{pmatrix}$ with C^2-functions $x, y, z : Q \to \mathbb{R}$. The surface we are interested in is $S := g(G)$. Since G is compact S is compact, hence bounded in $\mathbb{R}^3$. Therefore we can find an open (bounded) neighbourhood $U \subset \mathbb{R}^3$ of S and on U we can consider a C^1-vector field $F : U \to \mathbb{R}^3$ with components F_1, F_2, F_3, i.e. $F = \begin{pmatrix} F_1 \\ F_2 \\ F_3 \end{pmatrix}$. Thus on S we can consider F (and F_j) as a function of $(x, y, z) = (x(u, v), y(u, v), z(u, v))$, i.e. for $(u, v) \in G$ we have

$$F(g(u, v)) = F(x(u, v), y(u, v), z(u, v)).$$

Hence we can discuss F by looking at the functions $F_j \circ g : G \to \mathbb{R}$. Moreover we can lift ∂G with the help of g and γ to a parametric curve in $\mathbb{R}^3$ by considering $g \circ \gamma : [a, b] \to S$ which is a simply closed piecewise C^1-curve in $\mathbb{R}^3$ with components $(g \circ \gamma)(t) = \begin{pmatrix} (x \circ \gamma)(t) \\ (y \circ \gamma)(t) \\ (z \circ \gamma)(t) \end{pmatrix}$. In fact we can consider $(g \circ \gamma)([a, b])$ as "boundary curve" of S which needs however more topological considerations (using relative topologies) and this interpretation is not necessary for our formulation of Stokes' theorem.

We want to investigate the integral

$$\int_{g \circ \gamma} F_1 \, dx,$$

recall that this is the line integral of the vector field $\begin{pmatrix} F_1 \\ 0 \\ 0 \end{pmatrix}$ along $g \circ \gamma$ and therefore

$$\int_{g \circ \gamma} F_1 \, dx = \int_a^b F_1((x \circ \gamma)(t), (y \circ \gamma)(t), (z \circ \gamma)(t)) \frac{d(x \circ \gamma)(t)}{dt} \, dt. \quad (26.25)$$

The chain rule yields

$$\frac{d(x \circ \gamma)(t)}{dt} = \frac{\partial x(\gamma(t))}{\partial u} \dot{\gamma}_1(t) + \frac{\partial x(\gamma(t))}{\partial v} \dot{\gamma}_2(t)$$

which implies for (26.25)

$$\int_{g \circ \gamma} F_1 \, dx = \int_a^b F_1((x \circ \gamma)(t), (y \circ \gamma)(t), (z \circ \gamma)(t)) \left(\frac{\partial x(\gamma(t))}{\partial u} \dot{\gamma}_1(t) + \frac{\partial x(\gamma(t))}{\partial v} \dot{\gamma}_2(t) \right) dt$$

$$= \int_\gamma F_1(u, v) \frac{\partial x}{\partial u}(u, v) \, du + \int_\gamma F_1(u, v) \frac{\partial x}{\partial v}(u, v) \, dv. \quad (26.26)$$

Now we may apply Green's theorem to (26.26) to obtain

$$\int_\gamma F_1(u, v) \frac{\partial x}{\partial u}(u, v) \, du + \int_\gamma F_1(u, v) \frac{\partial x}{\partial v}(u, v) \, dv$$

$$= \int_G \left(\frac{\partial}{\partial u} \left(F_1 \frac{\partial x}{\partial v} \right) - \frac{\partial}{\partial v} \left(F_1 \frac{\partial x}{\partial u} \right) \right) du \, dv.$$

Since F_1 is a C^1-function and x is a C^2-function on Q we can work out $\frac{\partial}{\partial u} \left(F_1 \frac{\partial x}{\partial v} \right) - \frac{\partial}{\partial v} \left(F_1 \frac{\partial x}{\partial u} \right)$ using Leibniz's rule and the chain rule. For this we note

$$\frac{\partial}{\partial u} \left(F_1 \frac{\partial x}{\partial v} \right) - \frac{\partial}{\partial v} \left(F_1 \frac{\partial x}{\partial u} \right) = \frac{\partial F_1}{\partial u} \frac{\partial x}{\partial v} + F_1 \frac{\partial^2 x}{\partial u \partial v} - \frac{\partial F_1}{\partial v} \frac{\partial x}{\partial u} - F_1 \frac{\partial^2 x}{\partial v \partial u}$$

$$= \frac{\partial F_1}{\partial u} \frac{\partial x}{\partial v} - \frac{\partial F_1}{\partial v} \frac{\partial x}{\partial u}, \quad (26.27)$$

as well as

$$\frac{\partial F_1}{\partial u} = \frac{\partial F_1((x(u,v), y(u,v), z(u,v))}{\partial u} \tag{26.28}$$
$$= \frac{\partial F_1}{\partial x}\frac{\partial x}{\partial u} + \frac{\partial F_1}{\partial y}\frac{\partial y}{\partial u} + \frac{\partial F_1}{\partial z}\frac{\partial z}{\partial u}$$

and

$$\frac{\partial F_1}{\partial v} = \frac{\partial F_1(x(u,v), y(u,v), z(u,v))}{\partial v} \tag{26.29}$$
$$= \frac{\partial F_1}{\partial x}\frac{\partial x}{\partial v} + \frac{\partial F_1}{\partial y}\frac{\partial y}{\partial v} + \frac{\partial F_1}{\partial z}\frac{\partial z}{\partial v}.$$

Combining (26.27) - (26.29) it follows that

$$\frac{\partial F_1}{\partial u}\frac{\partial x}{\partial v} - \frac{\partial F_1}{\partial v}\frac{\partial x}{\partial u}$$
$$= \left(\frac{\partial F_1}{\partial x}\frac{\partial x}{\partial u} + \frac{\partial F_1}{\partial y}\frac{\partial y}{\partial u} + \frac{\partial F_1}{\partial z}\frac{\partial z}{\partial u} \right)\frac{\partial x}{\partial v}$$
$$- \left(\frac{\partial F_1}{\partial x}\frac{\partial x}{\partial v} + \frac{\partial F_1}{\partial y}\frac{\partial y}{\partial v} + \frac{\partial F_1}{\partial z}\frac{\partial z}{\partial v} \right)\frac{\partial x}{\partial u}$$
$$= \frac{\partial F_1}{\partial y}\left(\frac{\partial y}{\partial u}\frac{\partial x}{\partial v} - \frac{\partial y}{\partial v}\frac{\partial x}{\partial u} \right) + \frac{\partial F_1}{\partial z}\left(\frac{\partial z}{\partial u}\frac{\partial x}{\partial v} - \frac{\partial z}{\partial v}\frac{\partial x}{\partial u} \right),$$

and using as in Chapter 24 the notations

$$\frac{\partial(x,y)}{\partial(u,v)} = \frac{\partial x}{\partial u}\frac{\partial y}{\partial v} - \frac{\partial x}{\partial v}\frac{\partial y}{\partial u},$$
$$\frac{\partial(z,x)}{\partial(u,v)} = \frac{\partial z}{\partial u}\frac{\partial x}{\partial v} - \frac{\partial z}{\partial v}\frac{\partial x}{\partial u},$$

we find

$$\frac{\partial}{\partial u}\left(F_1\frac{\partial x}{\partial v} \right) - \frac{\partial}{\partial v}\left(F_1\frac{\partial x}{\partial v} \right) = -\frac{\partial F_1}{\partial y}\frac{\partial(x,y)}{\partial(u,v)} + \frac{\partial F_1}{\partial z}\frac{\partial(z,x)}{\partial(u,v)}. \tag{26.30}$$

Hence it follows again with the notation of Chapter 24

$$\int_{g\circ\gamma} F_1\, dx = \int_G \left(-\frac{\partial F_1}{\partial y}\frac{\partial(x,y)}{\partial(u,v)} + \frac{\partial F_1}{\partial z}\frac{\partial(z,x)}{\partial(u,v)} \right) du\, dv$$
$$= -\int_S \frac{\partial F_1}{\partial y}\, dx\, dy + \int_S \frac{\partial F_1}{\partial z}\, dz\, dx. \tag{26.31}$$

With a completely analogous argument one can derive

$$\int_{g\circ\gamma} F_2\,\mathrm{d}y = -\int_S \frac{\partial F_2}{\partial z}\,\mathrm{d}y\,\mathrm{d}z + \int_S \frac{\partial F_2}{\partial x}\,\mathrm{d}x\,\mathrm{d}y \qquad (26.32)$$

and

$$\int_{g\circ\gamma} F_3\,\mathrm{d}z = -\int_S \frac{\partial F_3}{\partial x}\,\mathrm{d}z\,\mathrm{d}x + \int_S \frac{\partial F_3}{\partial y}\,\mathrm{d}y\,\mathrm{d}z. \qquad (26.33)$$

Adding (26.31) - (26.33) we obtain

$$\int_{g\circ\gamma} \langle F(\gamma(t)), \dot{\gamma}(t)\rangle \mathrm{d}t = \int_{g\circ\gamma} F_1\,\mathrm{d}x + \int_{g\circ\gamma} F_2\,\mathrm{d}y + \int_{g\circ\gamma} F_3\,\mathrm{d}z$$

$$= \int_S \left(\frac{\partial F_3}{\partial y} - \frac{\partial F_2}{\partial z}\right)\mathrm{d}y\,\mathrm{d}z + \int_S \left(\frac{\partial F_1}{\partial z} - \frac{\partial F_3}{\partial x}\right)\mathrm{d}z\,\mathrm{d}x + \int_S \left(\frac{\partial F_2}{\partial x} - \frac{\partial F_1}{\partial y}\right)\mathrm{d}x\,\mathrm{d}y$$

$$= \int_s \left\{ \left(\frac{\partial F_3}{\partial y} - \frac{\partial F_2}{\partial z}\right)\frac{\partial(y,z)}{\partial(u,v)} + \left(\frac{\partial F_1}{\partial z} - \frac{\partial F_3}{\partial x}\right)\frac{\partial(z,x)}{\partial(u,v)} + \right.$$

$$\left. + \left(\frac{\partial F_2}{\partial x} - \frac{\partial F_1}{\partial y}\right)\frac{\partial(x,y)}{\partial(u,v)}\right\}\mathrm{d}\sigma$$

$$= \int_S \operatorname{curl} F \cdot \mathrm{d}\sigma$$

where as in Chapter 24

$$N(N_1, N_2, N_3) = \left(\frac{\partial(y,z)}{\partial(u,v)}, \frac{\partial(z,x)}{\partial(u,v)}, \frac{\partial(x,y)}{\partial(u,v)}\right)$$

and

$$\operatorname{curl} F = \begin{pmatrix} \frac{\partial F_3}{\partial y} - \frac{\partial F_2}{\partial z} \\ \frac{\partial F_1}{\partial z} - \frac{\partial F_3}{\partial x} \\ \frac{\partial F_2}{\partial x} - \frac{\partial F_1}{\partial y} \end{pmatrix}.$$

Thus, eventually we have proved a version of **Stokes' theorem** in $\mathbb{R}^3$:

Theorem 26.13. *Let S, in particular g and γ, be as above. For every C^1-vector field defined in an open neighbourhood of S we have*

$$\int_{g\circ\gamma} F(p)\cdot \mathrm{d}p = \int_S \operatorname{curl} F\cdot \mathrm{d}\sigma = \int_G \langle \operatorname{curl} F(g(u,v)), g_u(u)\times g_v(v)\rangle \mathrm{d}u\,\mathrm{d}v, \quad (26.34)$$

where p denotes the points in $(g\circ\gamma)([a,b])$.

Remark 26.14. Once Theorem 26.1 is established, the proof of Theorem 26.13 is straightforward, however we still want to mention that in our presentation we followed closely H. Heuser [25].

In Example 26.5 we have made use of Green's theorem to evaluate line integrals noting the fact that volume integrals are in general more easy to handle than line integrals. In general surface integrals are more difficult to treat than line integrals and in the next example we will see how Stokes' theorem can be used to calculate surface integrals by reducing them to line integrals. However the main point in the following example is to go step by step through all definitions and results in light of a concrete example.

Example 26.15. Let $h : B_r(0) \to \mathbb{R}^3$ be given by $h(u, v) = \begin{pmatrix} u \\ v \\ \sqrt{r^2 - u^2 - v^2} \end{pmatrix}$
and for $0 < \epsilon < r$ consider $\overline{B_{(r^2-\epsilon^2)^{\frac{1}{2}}}(0)} \subset B_r(0) \subset \mathbb{R}^2$. We are interested in the surface

$$S_\epsilon := h\left(\overline{B_{(r^2-\epsilon^2)^{\frac{1}{2}}}(0)}\right) = \{(x, y, z) \in \mathbb{R}^3 \mid x^2 + y^2 + z^2 = r^2, z \geq \epsilon\} \quad (26.35)$$

which has as its border the circle $C_\epsilon := \{(x, y, z) \in \mathbb{R}^3 \mid x^2 + y^2 + z^2 = r^2, z = \epsilon\}$. We can parametrize C_ϵ with $\gamma_\epsilon : [0, 2\pi] \to B_r(0)$, $\gamma_\epsilon(t) = \begin{pmatrix} (r^2 - \epsilon^2)^{\frac{1}{2}} \cos t \\ (r^2 - \epsilon^2)^{\frac{1}{2}} \sin t \end{pmatrix}$, and we find

$$(h \circ \gamma_\epsilon)(t) = \begin{pmatrix} (r^2 - \epsilon^2)^{\frac{1}{2}} \cos t \\ (r^2 - \epsilon^2)^{\frac{1}{2}} \sin t \\ \epsilon \end{pmatrix}, \quad (h \circ \gamma_\epsilon)^{\cdot}(t) = \begin{pmatrix} -(r^2 - \epsilon^2)^{\frac{1}{2}} \sin t \\ (r^2 - \epsilon^2)^{\frac{1}{2}} \cos t \\ 0 \end{pmatrix}. \quad (26.36)$$

Denoting coordinates in $B_r(0)$ with (u, v) we further get

$$h_u(u, v) = \begin{pmatrix} 1 \\ 0 \\ \frac{-u}{\sqrt{r^2 - u^2 - v^2}} \end{pmatrix}, \quad h_v(u, v) = \begin{pmatrix} 0 \\ 1 \\ \frac{-v}{\sqrt{r^2 - u^2 - v^2}} \end{pmatrix}$$

which yields

$$(h_u \times h_v)(u, v) = \begin{pmatrix} \frac{u}{\sqrt{r^2 - u^2 - v^2}} \\ \frac{v}{\sqrt{r^2 - u^2 - v^2}} \\ \end{pmatrix}. \quad (26.37)$$

Since $B_r(0)$ and $B_{(r^2-\epsilon^2)^{\frac{1}{2}}}(0)$ are best described in polar coordinates (ρ, φ), $u = \rho \cos \varphi$, $v = \rho \sin \varphi$, we note for later purposes that

$$(h_u \times h_v)(\rho \cos \varphi, \rho \sin \varphi) = \frac{1}{\sqrt{r^2 - \rho^2}} \begin{pmatrix} \rho \cos \varphi \\ \rho \sin \varphi \\ \sqrt{r^2 - \rho^2} \end{pmatrix}. \tag{26.38}$$

For the vector field $F : h(B_r(0)) \to \mathbb{R}^3$, $F(x, y, z) = \begin{pmatrix} 3x - 4y \\ y \\ -y^2 z^2 \end{pmatrix}$ we find

$$(\operatorname{curl} F)(x, y, z) = \begin{pmatrix} \frac{\partial F_3}{\partial y} - \frac{\partial F_2}{\partial z} \\ \frac{\partial F_1}{\partial z} - \frac{\partial F_3}{\partial x} \\ \frac{\partial F_2}{\partial x} - \frac{\partial F_1}{\partial y} \end{pmatrix} = \begin{pmatrix} -2yz^2 \\ 0 \\ 4 \end{pmatrix} = \begin{pmatrix} -2\rho \sin \varphi (r^2 - \rho^2) \\ 0 \\ 4 \end{pmatrix} \tag{26.39}$$

and therefore

$$\begin{aligned}
&\langle \operatorname{curl} F(h(u, v)), (h_u \times h_v)(u, v) \rangle \\
&= \left(\left(\frac{\partial F_3}{\partial y} \right)(h(u, v)) - \left(\frac{\partial F_2}{\partial z} \right)(h(u, v)) \right) \frac{u}{\sqrt{r^2 - u^2 - v^2}} \\
&+ \left(\left(\frac{\partial F_1}{\partial z} \right)(h(u, v)) - \left(\frac{\partial F_3}{\partial x} \right)(h(u, v)) \right) \frac{v}{\sqrt{r^2 - u^2 - v^2}} \\
&+ \left(\left(\frac{\partial F_2}{\partial x} \right)(h(u, v)) - \left(\frac{\partial F_1}{\partial y} \right)(h(u, v)) \right) \\
&= -2\rho \sin \varphi (r^2 - \rho^2)\rho \cos \varphi + 0 + 4 \cdot 1 = -2\rho^2 (r^2 - \rho^2) \sin \varphi \cos \varphi + 4.
\end{aligned}$$

Now we get

$$\begin{aligned}
\int_{S_\epsilon} \operatorname{curl} F \cdot d\sigma &= \int_{B_{(r^2-\epsilon^2)^{\frac{1}{2}}}(0)} \langle \operatorname{curl} F(h(u, v)), (h_u \times h_v)(u, v) \rangle du\, dv \\
&= \int_0^{(r^2-\epsilon^2)^{\frac{1}{2}}} \int_0^{2\pi} (-2\rho^2 (r^2 - \rho^2) \sin \varphi \cos \varphi + 4)\rho\, d\varphi\, d\rho \\
&= -\int_0^{(r^2-\epsilon^2)^{\frac{1}{2}}} 2\rho^3 (r^2 - \rho^2)\, d\rho \int_0^{2\pi} \sin \varphi \cos \varphi\, d\varphi + 2\pi \int_0^{(r^2-\epsilon^2)^{\frac{1}{2}}} 4\rho\, d\rho \\
&= 4\pi (r^2 - \epsilon^2),
\end{aligned}$$

where we used the orthogonality of sin and cos on $[0, 2\pi]$, i.e. the fact that

$$\int_0^{2\pi} \sin\varphi \cos\varphi \, d\varphi = 0.$$

Thus we have arrived at

$$\int_{S_\epsilon} \operatorname{curl} F \cdot d\sigma = 4\pi(r^2 - \epsilon^2). \tag{26.40}$$

Now we calculate the corresponding line integral:

$$\int_{h \circ \gamma_\epsilon} F \cdot dp = \int_0^{2\pi} \langle F(h(\gamma_\epsilon(t))), (h \circ \gamma_\epsilon)^{\cdot}(t) \rangle dt$$

$$= \int_0^{2\pi} \left(2(r^2 - \epsilon^2)^{\frac{1}{2}} \cos t - 4(r^2 - \epsilon^2)^{\frac{1}{2}} \sin t \right)$$

$$\times \left(-(r^2 - \epsilon^2)^{\frac{1}{2}} \sin t + (r^2 - \epsilon^2)^{\frac{1}{2}} \sin t (r^2 - \epsilon^2)^{\frac{1}{2}} \cos t \right) dt$$

$$= 4 \int_0^{2\pi} (r^2 - \epsilon^2) \sin^2 t \, dt,$$

where we used again the orthogonality of sin and cos on $[0, 2\pi]$. Since

$$\int_0^{2\pi} \sin^2 t \, dt = \frac{t}{2} - \frac{\sin 2t}{4} \Big|_0^{2\pi} = \pi$$

we eventually find as expected

$$\int_{h \circ \gamma_\epsilon} F \cdot dp = 4\pi(r^2 - \epsilon^2). \tag{26.41}$$

Note that in this example we may pass to the limit $\epsilon \to 0$ although for $\epsilon = 0$ the conditions stated in Theorem 26.13 are not fulfilled.

We want to use Stokes' theorem to continue our discussion of the integrability condition, Proposition 15.27. For a vector field $F : \Omega \to \mathbb{R}^3$, $\Omega \subset \mathbb{R}^3$ open, this condition reads as

$$\frac{\partial F_1}{\partial y} = \frac{\partial F_2}{\partial x}, \quad \frac{\partial F_2}{\partial z} = \frac{\partial F_3}{\partial y}, \quad \frac{\partial F_3}{\partial x} = \frac{\partial F_1}{\partial z}, \tag{26.42}$$

the other relations are trivial. But (26.42) is of course equivalent to

$$\text{curl } F = \begin{pmatrix} \frac{\partial F_3}{\partial y} - \frac{\partial F_2}{\partial z} \\ \frac{\partial F_1}{\partial z} - \frac{\partial F_3}{\partial x} \\ \frac{\partial F_2}{\partial x} - \frac{\partial F_1}{\partial y} \end{pmatrix} = 0. \tag{26.43}$$

Now let $Q \subset \mathbb{R}^2$, $G \subset Q$, $h : Q \subset \mathbb{R}^3$, $S := h(G) \subset \Omega$, $\gamma : [a, b] \to G$, $\gamma([a, b]) = \partial G$ all be as in Stokes' theorem, Theorem 26.13, and denote by Γ the curve $h \circ \gamma$. For every such surface S with border curve Γ Stokes' theorem holds and with (26.43) will give

$$\int_\Gamma F \cdot dp = 0.$$

We try a partial converse of this result. Let $\Omega \subset \mathbb{R}^3$ be an open set and $F : \Omega \to \mathbb{R}^3$ a C^1-vector field. Further suppose that for each of the surfaces

$$B_{\rho_j}^{(j)}(x_0, y_0, z_0) \subset \Omega, \quad \rho_j > 0,$$

where

$$B_{\rho_1}^{(1)}(x_0, y_0, z_0) = \left\{ (x, y, z) \in \mathbb{R}^3 \mid (x - x_0)^2 + (y - y_0)^2 \leq \rho_1^2, z = z_0 \right\},$$
$$B_{\rho_2}^{(2)}(x_0, y_0, z_0) = \left\{ (x, y, z) \in \mathbb{R}^3 \mid (y - y_0)^2 + (z - z_0)^2 \leq \rho_2^2, x = x_0 \right\},$$
$$B_{\rho_3}^{(3)}(x_0, y_0, z_0) = \left\{ (x, y, z) \in \mathbb{R}^3 \mid (x - x_0)^2 + (z - z_0)^2 \leq \rho_3^2, y = y_0 \right\}$$

the following holds

$$\int_{\partial B_{\rho_j}^{(j)}(x_0, y_0, z_0)} F \cdot dp = 0 \tag{26.44}$$

(where we interpret $\partial B_{\rho_j}^{(j)}(x_0, y_0, z_0)$ as anticlockwise parametrized simply closed curve). We claim that under these conditions curl $F = 0$ in Ω. From Stokes' theorem we deduce

$$0 = \int_{\partial B_{\rho_1}^{(1)}(x_0, y_0, z_0)} F \cdot dp = \int_{B_{\rho_1}^{(1)}(x_0, y_0, z_0)} \text{curl } F \cdot d\sigma = \int_{B_{\rho_1}^{(1)}(x_0, y_0, z_0)} \left(\frac{\partial F_2}{\partial x} - \frac{\partial F_1}{\partial y} \right) d\sigma, \tag{26.45}$$

$$0 = \int_{\partial B_{\rho_2}^{(2)}(x_0, y_0, z_0)} F \cdot dp = \int_{B_{\rho_2}^{(2)}(x_0, y_0, z_0)} \text{curl } F \cdot d\sigma = \int_{B_{\rho_2}^{(2)}(x_0, y_0, z_0)} \left(\frac{\partial F_3}{\partial y} - \frac{\partial F_2}{\partial z} \right) d\sigma, \tag{26.46}$$

$$0 = \int_{\partial B_{\rho_3}^{(3)}(x_0, y_0, z_0)} F \cdot dp = \int_{B_{\rho_3}^{(3)}(x_0, y_0, z_0)} \text{curl } F \cdot d\sigma = \int_{B_{\rho_3}^{(3)}(x_0, y_0, z_0)} \left(\frac{\partial F_1}{\partial z} - \frac{\partial F_3}{\partial x} \right) d\sigma, \tag{26.47}$$

for suitable radii $\rho_1, \rho_2, \rho_3 > 0$. Now the standard argument applies. Suppose that curl $F(x_0, y_0, z_0) \neq 0$. Each component which is not zero at (x_0, y_0, z_0) must be not zero in a ball $B_\rho(x_0, y_0, z_0) \subset \Omega$ which however will contradict the corresponding equality of the equalities (26.45) - (26.27). Thus we can deduce that the vanishing of certain line integrals in $\mathbb{R}^3$ implies the integrability condition.

Problems

1. Consider $G = \{(x_1, x_2) \in \mathbb{R}^2 \mid -\sqrt{1 - x_1^2} \leq x_2 \leq \cos \frac{\pi}{2} x_1 - 1 \leq x_1 \leq 1\}$ which is a normal domain with respect to the x_2-axis and is easy to see that G is also a normal domain with respect to the x_1-axis. (Just sketch G and reflect in the line $x_1 = x_2$.) Prove that $\gamma : [-1, 3] \to \mathbb{R}^2$,

$$\gamma(t) := \begin{cases} \begin{pmatrix} -t \\ \cos \frac{\pi}{2} t \end{pmatrix}, & t \in [-1, 1] \\ \begin{pmatrix} t - 2 \\ -\sqrt{1 - (t - 2)^2} \end{pmatrix}, & t \in [1, 3] \end{cases}$$

is a parametrization of ∂G by a piecewise C^1-curve. Now use Green's theorem to show that

$$\int_G (x_1 - 1) dx = \frac{4}{\pi} - \frac{\pi}{2}.$$

Hint: consider the vector field $F(x_1, x_2) = \begin{pmatrix} x_2 \\ \frac{1}{2} x_1^2 \end{pmatrix}$.

2. Consider the set

$$G := \{(x_1, x_2) \in \mathbb{R}^2 \mid -2\sqrt{1 - x_1^2} \leq x_2 \leq \sqrt{1 - x_1^2}, -1 \leq x_1 \leq 1\}.$$

Sketch G, convince yourself that G is a normal domain with respect to the x_2- as well as the x_1-axis, and show that a parametrization of ∂G

is given by $\gamma : [0, 2\pi] \to \mathbb{R}^2$,

$$\gamma(t) = \begin{cases} \begin{pmatrix} \cos t \\ \sin t \end{pmatrix}, & t \in [0, \pi] \\ \begin{pmatrix} \cos t \\ 2 \sin t \end{pmatrix}, & t \in [\pi, 2\pi] \end{cases}$$

which is a piecewise C^∞-mapping. Use Green's theorem to calculate the area $A(G)$ of G. (Before starting any calculation derive by elementary geometric considerations that $A(G) = \frac{3\pi}{2}$.)

3. Consider the set

$$G = \left\{ (x_1, x_2) \in \mathbb{R}^2 \left| \left(\frac{x_1}{9}\right)^2 + \left(\frac{x_2}{4}\right)^2 \leq 1 \right| x_1 < 0 \right\} \setminus$$
$$\left\{ (x_1, x_2) \in \mathbb{R}^2 \left| \left(\frac{x_1}{3}\right)^2 + \left(\frac{x_2}{2}\right)^2 \leq 1 \right| x_1 < 0 \right\}.$$

Sketch the set G and decompose it into two domains being normal domains with respect to both axes. Use Green's theorem to find

$$\int_{\partial G} F_1 dx_1 + \int_{\partial G} F_2 dx_2 = 30$$

where $F(x_1, x_2) = \begin{pmatrix} -2x_2 \\ 3x_1 x_2 \end{pmatrix}$.

4. Let $\tilde{G} \subset \mathbb{R}^3$ be an open set and $G \subset \tilde{G}$ a (non-empty) Gauss domain. Suppose further that $F : \tilde{G} \to \mathbb{R}^3$ is a C^2-vector field. Prove that $\int_{\partial G} \mathrm{curl} F \cdot d\sigma = 0$.

5. For $a < b$ consider the surface $S := \{(x, y, z) \in \mathbb{R}^3 | x^2 + y^2 = 1, a \leq z \leq b\}$ which is a patch of the cylinder over the unit circle which is bounded by two curves, the circle $\partial_1 S := \{(x, y, b) \in \mathbb{R}^3 | x^2 + y^2 = 1\}$ and $\partial_2 S := \{(x, y, a) \in \mathbb{R}^3 | x^2 + y^2 = 1\}$. Thus we cannot apply our version of Stokes' theorem immediately. Nonetheless prove that

$$\int_S \mathrm{curl} F \cdot d\sigma = \int_{\gamma_1} F \cdot d\rho + \int_{\gamma_2} F \cdot d\rho$$

holds where γ_j parametrizes $\partial_j S$ with γ_1 being anticlockwise and γ_2 being clockwise oriented, and F is a C^1-vector field defined in an open neighbourhood of S.

Hint: slice S along a curve connecting $\partial_1 S$ and $\partial_2 S$ to obtain a surface bounded by one curve only with four parts, two being identical but with opposite orientation, and then apply Stokes' theorem, see the figure below.

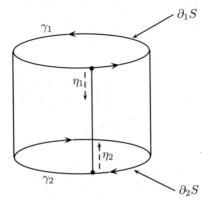

Figure 26.6

27 Gauss' Theorem for $\mathbb{R}^n$

In Chapter 25 we discussed Gauss' theorem for certain domains $G \subset \mathbb{R}^3$, i.e. the formula

$$\int_G \operatorname{div} F \, \mathrm{d}x = \int_{\partial G} F \cdot \mathrm{d}\sigma \qquad (27.1)$$

where $F : G_0 \to \mathbb{R}^3$ is a C^1-vector field defined on an open set $G_0 \subset \mathbb{R}^3$ such that the compact set $\overline{G}$ is contained in G_0, i.e. $\overline{G} \subset G_0$. There is no problem to extend the meaning of the integral on the left hand side of (27.1) to a setting in $\mathbb{R}^n$. We need to start with an open set $G_0 \subset \mathbb{R}^n$ and a C^1-vector field $F : G_0 \to \mathbb{R}^n$ and then we consider a Jordan measurable set $G \subset \mathbb{R}^n$ such that $\overline{G}$ is compact and $\overline{G} \subset G_0$. For $F = (F_1, \cdots, F_n)$ we have already defined $\operatorname{div} F = \sum_{j=1}^{n} \frac{\partial F_j}{\partial x_j}$ which is under our assumptions a continuous function on $\overline{G}$ hence integrable over G and $\int_G \operatorname{div} F \, \mathrm{d}x$ is well defined. However the term on the right hand side of (27.1) causes serious problems. In $\mathbb{R}^3$ we have considered ∂G to be decomposed into a finite number of surfaces S_j such that $S_j \cap S_l$ is either empty or a set we can neglect for integration. In addition S_j was assumed to satisfy conditions which allow us to define a surface integral over S_j for continuous functions or vector fields. But even in $\mathbb{R}^3$ the definition of surface integrals was a problem: we could give ad hoc definitions (Definition 24.9, Definition 24.13) but these have been definitions based on analytic considerations with some heuristic geometrical justification - a proper notion of a surface area and hence of a rectifiable surface was and still is beyond the scope of our current state of knowledge, see however Volume III, Part 6. Thus our first goal must be to find a proper understanding of the meaning of $\int_{\partial G} F \cdot \mathrm{d}\sigma$. We will now give a rather lengthy motivation for an appropriate definition, a motivation based on an approach to prove (27.1) for compact cells. The reader who is willing to accept a corresponding ad hoc definition of "surface" integrals in $\mathbb{R}^n$ may immediately continue with Definition 27.1.

Let $F : G_0 \to \mathbb{R}^n$ be a C^1-vector field defined on the open set $G_0 \subset \mathbb{R}^n$ and let $K := \bigtimes_{j=1}^{n} [a_j, b_j]$ be a non-degenerate compact cell in $\mathbb{R}^n$. We need some notations. Fix $1 \le k \le n$ and set

$$\tilde{K}_{(k-1)} := \begin{cases} \emptyset, & k = 1 \\ \bigtimes_{j=1}^{k-1}[a_j, b_j], & 1 < k < n-1 \end{cases}$$

$$K'_{(k+1)} := \begin{cases} \bigtimes_{j=k+1}^{n}[a_j, b_j], & k+1 < n \\ \emptyset, & k = n \end{cases}$$

and put

$$\tilde{K}^{(k)} := \tilde{K}_{(k-1)} \times K'_{(k+1)}.$$

Clearly we have $K = \tilde{K}_{(k-1)} \times [a_k, b_k] \times K'_{(k+1)}$. For $x \in K$ we write $x = (\tilde{x}_{(k-1)}, x_k, x'_{(k+1)})$ with $\tilde{x}_{(k-1)} \in \tilde{K}_{(k-1)}$, $x'_{(k+1)} \in K'_{(k+1)}$ and $x_k \in [a_k, b_k]$. With this notation we can write

$$\int_K \operatorname{div} F \, \mathrm{d}x = \sum_{k=1}^{n} \int_K \frac{\partial F_k}{\partial x_k} \, \mathrm{d}x$$

$$= \sum_{k=1}^{n} \int_{\tilde{K}^{(k)}} \left(\int_{a_k}^{b_k} \frac{\partial F_k}{\partial x_k} \, \mathrm{d}x_k \right) \mathrm{d}\tilde{x}_{k-1} \, \mathrm{d}x'_{k+1}$$

where

$$\int_{\tilde{K}^{(k)}} h \, \mathrm{d}\tilde{x}_{k-1} \, \mathrm{d}x'_{k+1} = \int_{K'_{(k+1)}} \left(\int_{\tilde{K}_{(k-1)}} h(\tilde{x}_{k-1}, x'_{k+1}) \, \mathrm{d}\tilde{x}_{k-1} \right) \mathrm{d}x'_{k+1}.$$

Now we consider

$$\int_{\tilde{K}^{(k)}} \left(\int_{a_k}^{b_k} \frac{\partial F_k}{\partial x_k} \, \mathrm{d}x_k \right) \mathrm{d}\tilde{x}_{k-1} \mathrm{d}x'_{k+1} = \int_{\tilde{K}^{(k)}} (F_k(x_1, \ldots, x_{k-1}, b_k, x_{k+1}, \ldots, x_n)$$
$$- F_k(x_1, \ldots, x_{k-1}, a_k, x_{k+1}, \ldots, x_n)) \mathrm{d}x'_{k+1} \, \mathrm{d}\tilde{x}_{k-1}.$$

The sets $\tilde{K}_{(k-1)} \times \{b_k\} \times K'_{(k+1)}$ and $\tilde{K}_{(k-1)} \times \{a_k\} \times K'_{(k+1)}$ are subsets of ∂K and each of them are subsets of a hyperplane orthogonal to the coordinate axis e_k. We want to use this observation to understand the geometric situation further. But we note here that with $c_k \in \{a_k, b_k\}$ we can interpret the integral

$$\int_{\tilde{K}^{(k)}} F_k(x_1, \ldots, x_{k-1}, c_k, x_{k+1}, \ldots, x_n) \mathrm{d}\tilde{x}_{k+1} \, \mathrm{d}\tilde{x}_{k-1}$$

as a surface integral in $\mathbb{R}^n$.

We can write

$$\tilde{K}_{(k-1)} \times \{c_k\} \times K'_{(k+1)} = c_k e_k + \tilde{K}_{(k-1)} \times \{0\} \times K'_{(k+1)}$$

and for $y \in \tilde{K}_{(k-1)} \times \{0\} \times K'_{(k+1)}$ it follows that $\langle y, e_k \rangle = 0$. In this sense $\tilde{K}_{(k-1)} \times \{c_k\} \times K'_{(k+1)}$ can be interpreted to be orthogonal to e_k or e_k to give a normal vector to $\tilde{K}_{(k-1)} \times \{c_k\} \times K'_{(k+1)}$. In addition we note that in the case $c_k = b_k$ the direction of e_k is outward of K, while for $c_k = a_k$ the direction $-e_k$ is outward of K. Thus when we introduce

$$\partial_k K := \tilde{K}_{(k-1)} \times \{b_k\} \times K'_{(k+1)} \cup \tilde{K}_{(k-1)} \times \{a_k\} \times K'_{(k+1)} = \partial_{k,+}K \cup \partial_{k,-}K,$$

and

$$\vec{n}_{(k)} : \partial_k K \to \mathbb{R}^n, \quad \vec{n}_{(k)}(x) = \begin{cases} e_k, & x \in \partial_{k,+}K \\ -e_k, & x \in \partial_{k,-}K \end{cases}$$

we can write

$$\int_{\tilde{K}^{(k)}} \left(F_k(x_1, \ldots, x_{k-1}, b_k, x_{k+1}, \ldots, x_n) - F_k(x_1, \ldots, x_{k-1}, a_k, x_{k+1}, \ldots, x_n) \right) dx'_{k+1} \, d\tilde{x}_{k-1}$$

$$= \int_{\partial_k K} \langle F(x), \vec{n}_{(k)}(x) \rangle \, dx'_{k+1} \, d\tilde{x}_{k-1} \tag{27.2}$$

$$= \int_{\partial_{k,+}K} \langle F(x), \vec{n}_{(k)}(x) \rangle \, dx'_{k+1} \, d\tilde{x}_{k-1} + \int_{\partial_{k,-}K} \langle F(x), \vec{n}_{(k)}(x) \rangle \, dx'_{k+1} \, d\tilde{x}_{k-1}.$$

Next we want to look at $\partial_{k,\pm}K$ as $(n-1)$-dimensional parametric "surface" in $\mathbb{R}^n$, in fact as graph of a function. For this we consider the function φ_{c_k}, $c_k \in \{a_k, b_k\}$ defined by

$$\varphi_{c_k} : \tilde{K}_{(k-1)} \times K'_{(k+1)} \to \mathbb{R}, \quad \varphi_{c_k}(\tilde{x}_{k-1}, x'_{k+1}) = c_k, \tag{27.3}$$

and the corresponding parametric surface

$$f_{c_k} : \tilde{K}_{(k-1)} \times K'_{(k+1)} \to \mathbb{R}^n \tag{27.4}$$
$$f_{c_k}(\tilde{x}_{k-1}, x'_{k+1}) = (\tilde{x}_{k-1}, \varphi_{c_k}(\tilde{x}_{k-1}, x'_{k+1}), x'_{k+1})$$

which has trace $\operatorname{tr}(f_{c_k}) = \tilde{K}_{(k-1)} \times \{c_k\} \times K'_{(k+1)}$. It follows that $\langle \frac{\partial}{\partial x_l} f_{c_k}, e_k \rangle = 0$ for every $l \neq k$ and the Jacobi matrix $J_{f_{c_k}}$ of f is given by the $n \times (n-1)$

matrix

$$
J_{f_{c_k}}(\tilde{x}) = \left(\frac{\partial f_{c_k,j}}{\partial x_l}(\tilde{x})\right)_{\substack{j=1,\ldots,n \\ l=1,\ldots,k-1,k+1,\ldots,n}} = \begin{pmatrix} 1 & & & & & & 0 \\ & \ddots & & & & & \\ & & 1 & & & & \\ & & & 0 & & & \\ & & & & 1 & & \\ & & & & & \ddots & \\ 0 & & & & & & 1 \end{pmatrix}, \quad (27.5)
$$

with 0 in the kth row. Here we have denoted the components of f_{c_k} by $f_{c_k,j}$ and used the fact that $\frac{\partial f_{c_k,k}}{\partial x_l} = \delta_{jl}$ and (in symbolic notation)

$$
J_{f_{c_k}}(\tilde{x}) = \begin{pmatrix} \operatorname{grad}_{(k-1)} f_{c_k,1}(\tilde{x}) \\ \vdots \\ \operatorname{grad}_{(k-1)} f_{c_k,n}(\tilde{x}) \end{pmatrix}, \quad (27.6)
$$

where grad_m stands for the gradient in $\mathbb{R}^m$.

Let us introduce the **Gram** or **Gauss matrix** associated with f_{c_k} as

$$
\operatorname{Gr}(f_{c_k})(\tilde{x}) = J_{f_{c_k}}^t(\tilde{x}) J_{f_{c_k}}(\tilde{x}) \quad (27.7)
$$

which is an $(n-1) \times (n-1)$ matrix, hence admits a determinant called the **Gram determinant** of f_{c_k}:

$$
\operatorname{g}(f_{c_k}(\tilde{x})) := \det \operatorname{Gr}(f_{c_k})(\tilde{x}). \quad (27.8)
$$

In our special case we find (with 0 now in the kth column)

$$
J_{f_{c_k}}^t = \begin{pmatrix} 1 & & & & & & 0 \\ & \ddots & & & & & \\ & & 1 & & & & \\ & & & 0 & & & \\ & & & & 1 & & \\ & & & & & \ddots & \\ 0 & & & & & & 1 \end{pmatrix}
$$

and therefore

$$J_{f_{c_k}}^t J_{f_{c_k}}(\tilde{x}) = \begin{pmatrix} 1 & & & & & 0 \\ & \ddots & & & & \\ & & 1 & & & \\ & & & 0 & & \\ & & & & 1 & \\ & & & & & \ddots \\ 0 & & & & & 1 \end{pmatrix} \begin{pmatrix} 1 & & & & & 0 \\ & \ddots & & & & \\ & & 1 & & & \\ & & & 0 & & \\ & & & & 1 & \\ & & & & & \ddots \\ 0 & & & & & 1 \end{pmatrix}$$

$$= \begin{pmatrix} 1 & & 0 \\ & \ddots & \\ 0 & & 1 \end{pmatrix} = \mathrm{id}_{\mathbb{R}^{n-1}}$$

with $g(f_{c_k}(\tilde{x})) = 1$. In order to understand this calculation and to put this digression into context, let us return to a parametric surface $h : Q \to \mathbb{R}^3$, $Q \subset \mathbb{R}^2$ open. The Jacobi matrix of h is given by

$$J_h(u, v) = (h_u, h_v)(u, v) = \begin{pmatrix} \frac{\partial h_1}{\partial u} & \frac{\partial h_1}{\partial v} \\ \frac{\partial h_2}{\partial u} & \frac{\partial h_2}{\partial v} \\ \frac{\partial h_3}{\partial u} & \frac{\partial h_3}{\partial v} \end{pmatrix} (u, v) = \begin{pmatrix} \mathrm{grad}_2\, h_1(u, v) \\ \mathrm{grad}_2\, h_2(u, v) \\ \mathrm{grad}_2\, h_3(u, v) \end{pmatrix}$$

and consequently we have

$$\mathrm{Gr}(h)(u, v) = \begin{pmatrix} \frac{\partial h_1}{\partial u} & \frac{\partial h_2}{\partial u} & \frac{\partial h_3}{\partial u} \\ \frac{\partial h_1}{\partial v} & \frac{\partial h_2}{\partial v} & \frac{\partial h_3}{\partial v} \end{pmatrix} \begin{pmatrix} \frac{\partial h_1}{\partial u} & \frac{\partial h_1}{\partial v} \\ \frac{\partial h_2}{\partial u} & \frac{\partial h_2}{\partial v} \\ \frac{\partial h_3}{\partial u} & \frac{\partial h_3}{\partial v} \end{pmatrix} (u, v)$$

$$= \begin{pmatrix} \langle \mathrm{grad}\, h_1, \mathrm{grad}\, h_1 \rangle & \langle \mathrm{grad}\, h_1, \mathrm{grad}\, h_2 \rangle \\ \langle \mathrm{grad}\, h_1, \mathrm{grad}\, h_2 \rangle & \langle \mathrm{grad}\, h_2, \mathrm{grad}\, h_2 \rangle \end{pmatrix}$$

and

$$g(h(u, v)) = \left(\langle \mathrm{grad}\, h_1, \mathrm{grad}\, h_1 \rangle \langle \mathrm{grad}\, h_2, \mathrm{grad}\, h_2 \rangle - \langle \mathrm{grad}\, h_1, \mathrm{grad}\, h_2 \rangle^2 \right) (u, v)$$

which is however (24.8). Thus, the Gram determinant (27.8) can be seen as extensions of (24.8). Nothing can prevent us from introducing these notations for more general $(n-1)$-dimensional parametric "surfaces". But first back to (27.2). We find with $\psi_{c_k}(f_{c_k}(\tilde{x})) = \langle F, \vec{n} \rangle|_{\partial_{k,\pm}K}$ that

$$\int_{\partial_{k,\pm}K} \langle F(x), \vec{n}(x) \rangle \, \mathrm{d}\tilde{x}_{k-1} \, \mathrm{d}x'_{k+1} = \int_{\partial_{k,\pm}K} \psi_{c_k}(f_{c_k}(\tilde{x})) \sqrt{g(f_{c_k}(\tilde{x}))} \, \mathrm{d}\tilde{x}$$

with the obvious interpretation that $\partial_{k,+}K$ corresponds to f_{b_k} and that $\partial_{k,-}K$ corresponds to f_{a_k}. So let us introduce some helpful notations. We decompose ∂K according to

$$\partial K = \bigcup_{k=1}^{n} \left(\tilde{K}_{(k-1)} \times \{b_k\} \times K'_{(k+1)} \right) \cup \bigcup_{k=1}^{n} \left(\tilde{K}_{(k-1)} \times \{a_k\} \times K'_{(k+1)} \right)$$

and define

$$\int_{\partial K} F \cdot d\sigma := \sum_{k=1}^{n} \int_{\partial_{k,+}K} \langle F, \vec{n} \rangle \big|_{\partial_{k,+}K} \, d\tilde{x} + \sum_{k=1}^{n} \int_{\partial_{k,-}K} \langle F, \vec{n} \rangle \big|_{\partial_{k,-}K} \, d\tilde{x}.$$

With these definitions (27.1) follows for K. Of course, a next step could be to switch from K to a normal domain of the type $G = \{x \in \mathbb{R}^n \,|\, \varphi(\tilde{x}) \leq x_k \leq \psi(\tilde{x}), \tilde{x} \in \tilde{K}_{(k-1)} \times K'_{(k+1)}\}$ with suitable functions φ and ψ, $\varphi \leq \psi$. We do not want to give all proofs of generalising (27.1) for a reasonable class of domains, but we hope that the considerations made above give some motivations and indications that the following definitions are meaningful and the results shall hold. We refer to Volume VI for a more rigorous treatment.

Definition 27.1. A. *Let $Q_0 \subset \mathbb{R}^{n-1}$ be an open set and $\overline{Q} \subset Q_0$ a compact Jordan measurable subset, $\overset{\circ}{Q} \neq \emptyset$. Further let $f : Q_0 \to \mathbb{R}^n$ be a C^k-mapping. We call $f\big|_{\overline{Q}}$ a **regular parametric $(n-1)$-dimensional surface** in $\mathbb{R}^n$ if $J_f(u)$ has rank $n-1$ for all $u \in \overline{Q}$. The **trace** $\mathrm{tr}(f\big|_{\overline{Q}})$ of f is by definition $f(\overline{Q})$. If in addition $f\big|_{\overline{Q}}$ is injective, we call f an **immersed** $(n-1)$-dimensional **surface** in $\mathbb{R}^n$.*
B. *For a regular parametric $(n-1)$-dimensional surface belonging to the class C^1 we define its **Gram determinant** by*

$$g(f(u)) := \det\left(J_f^t(u) J_f(u) \right), \quad u = (u_1, \ldots, u_{n-1}) \in \overline{Q}. \tag{27.9}$$

Remark 27.2. From

$$\begin{pmatrix} \frac{\partial f_1}{\partial u_1} & \cdots & \frac{\partial f_n}{\partial u_1} \\ \vdots & & \vdots \\ \frac{\partial f_1}{\partial u_{n-1}} & \cdots & \frac{\partial f_n}{\partial u_{n-1}} \end{pmatrix} \begin{pmatrix} \frac{\partial f_1}{\partial u_1} & \cdots & \frac{\partial f_1}{\partial u_{n-1}} \\ \vdots & & \vdots \\ \frac{\partial f_n}{\partial u_1} & \cdots & \frac{\partial f_n}{\partial u_{n-1}} \end{pmatrix}$$
$$= \left(\sum_{j=1}^{n} \frac{\partial f_j}{\partial u_l} \frac{\partial f_j}{\partial u_k} \right)_{l,k=1,\ldots,n-1}$$

it follows that $J_f^t(u)J_f(u)$ is a symmetric matrix and further we find for $\xi \in \mathbb{R}^{n-1}$

$$\langle J_f^t(u)J_f(u)\xi, \xi\rangle = \langle J_f(x)\xi, J_f(x)\xi\rangle \geq 0,$$

implying that $J_f^t(u)J_f(u)$ is a positive semi-definite matrix, hence $g(f(u)) \geq 0$ for all $u \in \overline{Q}$. It also follows that $J_f^t(u)J_f(u)$ is positive definite in Q if and only if $g(f(u)) > 0$ for all $u \in \overline{Q}$ which is equivalent to the fact that $J_f^t(u)J_f(u)$ has rank $n-1$ for all $u \in \overline{Q}$.

Now we are in position to define the "area" of $\mathrm{tr}(f)$ and surface integrals. We restrict ourselves first to injective, regular parametric surfaces $S = \mathrm{tr}(f)$.

Definition 27.3. *Let $Q_0 \subset \mathbb{R}^{n-1}$ be an open set and $\overline{Q} \subset Q_0$ a compact Jordan measurable subset, $\mathring{Q} \neq \emptyset$. Further let $f : Q_0 \to \mathbb{R}^n$ be a C^1-mapping with $J_f|_{\overline{Q}}$ having rank $n-1$. In addition assume that $f|_{\mathring{Q}}$ is injective. Denote the trace of $f|_{\overline{Q}}$ by S.*
A. *The **area** of S is defined as*

$$\mathrm{area}_{n-1}(S) := \int_{\overline{Q}} \sqrt{g(f(u))}\, du. \tag{27.10}$$

B. *For a continuous function $\psi : Q_0 \to \mathbb{R}$ we define its integral over S as*

$$\int_S \psi \, d\sigma := \int_{\overline{Q}} \psi(f(u))\sqrt{g(f(u))}\, du. \tag{27.11}$$

Since Q is Jordan measurable we note that

$$\int_{\overline{Q}} \psi(f(u))\sqrt{g(f(u))}\, du = \int_{\mathring{Q}} \psi(f(u))\sqrt{g(f(u))}\, du. \tag{27.12}$$

Now let $G \subset \mathbb{R}^n$ be a bounded Jordan measurable set and assume that ∂G has a decomposition $\partial G = \bigcup_{j=1}^N S_j$ where $S_j = \mathrm{tr}(f_j|_{\overline{Q}_j})$ and f_j is a regular injective parametric $(n-1)$-dimensional C^1-surface, $f_j : Q_{0,j} \to \mathbb{R}^n$, $Q_{0,j} \in \mathbb{R}^{n-1}$ open, and $\overline{Q}_j \subset Q_{0,j}$. In addition we assume that if $S_j \cap S_l$ is not empty then $S_j \cap S_l = f_j(R_j) = f_l(R_l)$ where $R_j \subset \overline{Q}_j$ and $R_l \subset \overline{Q}_l$ are sets of $(n-1)$-dimensional Jordan content zero. In this situation we say that G has a **piecewise C^1-regular boundary**. We now define the area of ∂G

$$\mathrm{area}_{n-1}(\partial G) = \sum_{j=1}^N \mathrm{area}_{n-1}(S_j) = \sum_{j=1}^N \int_{\overline{Q}_j} \sqrt{g(f_j(u))}\, du. \tag{27.13}$$

Further, if $U = U(\partial G)$ is an open neighbourhood of ∂G and $\psi : U \to \mathbb{R}$ is a continuous function we define

$$\int_{\partial G} \psi \, d\sigma := \sum_{j=1}^{N} \int_{S_j} \psi(f_j(u)) \sqrt{g(f_j(u))} \, du. \qquad (27.14)$$

Finally, in order to extend Gauss' theorem we need the notion of outward normal to ∂G. Let $f\big|_{\overline{Q}}$ be a regular, injective parametric $(n-1)$-dimensional surface in $\mathbb{R}^n$, $f : Q_0 \to \mathbb{R}^n$, $\overline{Q} \subset Q_0$. Since $J_f(u)$ has rank $n-1$ for all $u \in \overline{Q}$ it follows that $\{f_{u_1}(u), \ldots, f_{u_{n-1}}(u)\}$ is an independent set in $\mathbb{R}^n$. Hence its span is an $(n-1)$-dimensional subspace $T_{f(u)}S \subset \mathbb{R}^n$ which we can interpret as (pre-)tangent space to $S = \mathrm{tr}(f)$ at $f(u)$. Thus we can find a vector $N(u) \in \mathbb{R}^n \backslash \{0\}$ which is orthogonal to $T_{f(u)}S$. We may choose $N(u)$ such that $\det(f_{u_1}(u), \ldots, f_{u_{n-1}}(u), N(u)) > 0$. Since we obtain $N(u)$ from $\{f_{u_1}(u), \ldots, f_{u_{n-1}}(u)\}$ by solving a system of linear equations on every open connectivity component of $\overline{Q}$ the mapping $u \mapsto N(u)$ is continuous. This choice of $N(u)$, $u \in \overline{Q}$, induces an orientation on S which we call the positive orientation. Of course, switching from $N(u)$ to $-N(u)$ induces the other possible orientation.

As in the case of 3-dimensional parametric surfaces we can now introduce unit tangent vectors $\vec{t}_1(u), \ldots, \vec{t}_{n-1}(u)$ where $\vec{t}_j(u) = \dfrac{f_{u_j}(u)}{\|f_{u_j}(u)\|}$ and the **unit normal vector** $\vec{n}(u) := \dfrac{N(u)}{\|N(u)\|}$. Note that the assumption that f is injective implies that $\vec{t}_j$ and $\vec{n}$ depend on $f(u) = p \in S$ and not $u \in \overline{Q}$. For the positive orientation we now obtain in the case where $\{\vec{t}_j(u) \,|\, j = 1, \ldots, n-1\}$ forms an orthogonal system $\det(\vec{t}_1(u), \ldots, \vec{t}_{n-1}(u), \vec{n}(u)) = 1$.

Now if $G \subset \mathbb{R}^n$ is a bounded Jordan measurable set and $\partial G = \bigcup_{j=1}^{N} S_j$ is a piecewise C^1-regular boundary we may choose on each S_j the unit normal vector pointing outward of G, again we write $\vec{n}(u)$ (or $\vec{n}(x)$, $x = f(u)$) for this vector. For a C^1-vector field $F : G_0 \to \mathbb{R}^n$, $\overline{G} \subset G_0$ and $G_0 \subset \mathbb{R}^n$ open the integral

$$\int_{\partial G} F \cdot d\sigma := \int_{\partial G} \langle F, \vec{n} \rangle \, d\sigma := \sum_{j=1}^{N} \int_{S_j} \langle F(f_j(u)), \vec{n}(f_j(u)) \rangle \sqrt{g(f_j(u))} \, du$$

$$(27.15)$$

is well defined. Now we may return to Gauss' theorem.

Definition 27.4. *We call a bounded Jordan measurable set $G \subset \mathbb{R}^n$, $\mathring{G} \neq \emptyset$, a **Gauss domain** in $\mathbb{R}^n$ if for some open set $G_0 \subset \mathbb{R}^n$ we have $\overline{G} \subset G_0$, ∂G is a piecewise regular boundary and*

$$\int_G \operatorname{div} F \, \mathrm{d}x = \int_{\partial G} \langle F, \vec{n} \rangle \, \mathrm{d}\sigma \qquad (27.16)$$

holds for all C^1-vector fields $F : G_0 \to \mathbb{R}^n$.

Now the arguments which lead to Theorem 25.9 carry over:

First we can prove that on certain normal domains W_k with respect to the axis e_k we have

$$\int_{W_k} \frac{\partial F_k}{\partial x_k}(x) \, \mathrm{d}x = \int_{\partial W_k} F_k(x) \, \vec{n}_k(x) \sigma(\mathrm{d}x)$$

where $\vec{n}_k$ is the kth component of the normal vector $\vec{n}$. Then we can show that if G is such a normal domain with respect to each axis $e_1, \ldots, e_n$, then G is a Gauss domain, and eventually we can once more prove that if we can decompose G into a finite number of Gauss domains then G is a Gauss domain too.

A different question is that for a classification of all Gauss domains. Clearly, hyper-rectangles or the n-dimensional ball $B_R(x)$ are Gauss domains as are their diffeomorphic images. However a satisfactory answer to this question needs more (differential) geometry in $\mathbb{R}^n$ and results from geometric measure theory. Our main purpose so far in this Chapter has been to give the student who needs Gauss' theorem in $\mathbb{R}^n$ either in physics (or engineering or mechanics) or in some course on partial differential equations an idea how to understand all terms in formula (27.16) and some indication how one can prove this formula. In the following we will make use of (27.16) to discuss some applications. Our first results are related to integration by parts formulae.

Theorem 27.5. *Let $G \subset \mathbb{R}^n$ be a Gauss domain and $u, v : G_0 \to \mathbb{R}$, $G \subset G_0$, be two C^1-functions. Then we have for $1 \leq j \leq n$*

$$\int_G \frac{\partial u}{\partial x_j} v \, \mathrm{d}x = \int_{\partial G} (u \, v) \vec{n}_j \, \mathrm{d}\sigma - \int_G u \frac{\partial v}{\partial x_j} \, \mathrm{d}x. \qquad (27.17)$$

Proof. We introduce on G_0 the vector field $F : G_0 \to \mathbb{R}^n$ by $F = (F_1, \ldots, F_n)$ $= (0, \ldots, 0, uv, 0, \ldots, 0)$ where $F_j = uv$. From (27.16) we deduce

$$\int_G \text{div } F \, dx = \int_{\partial G} \langle F, \vec{n} \rangle \, d\sigma,$$

but $\text{div} F = \frac{\partial u}{\partial x_j} v + u \frac{\partial v}{\partial x_j}$ and $\langle F, \vec{n} \rangle = F_j \vec{n}_j$ and the result follows. $\qquad \square$

Of course we may iterate the result of Theorem 27.5 for higher order partial derivatives, for example we may find for C^2-functions

$$\int_G \frac{\partial^2 u}{\partial x_j \partial x_l} v \, dx = \int_{\partial G} \left(\frac{\partial u}{\partial x_l} v \right) \vec{n}_j \, d\sigma - \int_G \frac{\partial u}{\partial x_l} \frac{\partial v}{\partial x_j} \, dx$$

$$= \int_{\partial G} \left(\frac{\partial u}{\partial x_l} v \right) \vec{n}_j \, d\sigma - \int_{\partial G} \left(u \frac{\partial v}{\partial x_j} \right) \vec{n}_l \, d\sigma$$

$$+ \int_G u \frac{\partial^2 v}{\partial x_l \partial x_j} \, dx.$$

In particular for $j = l$ we obtain

$$\int_G \frac{\partial^2 u}{\partial x_j^2} v \, dx = \int_{\partial G} \left(\frac{\partial u}{\partial x_j} \vec{n}_j \right) v \, d\sigma - \int_{\partial G} u \left(\frac{\partial v}{\partial x_j} \vec{n}_j \right) d\sigma \qquad (27.18)$$

$$+ \int_G u \frac{\partial^2 v}{\partial x_j^2} \, dx,$$

and summing up from $j = 1$ to $j = n$ we find

$$\int_G \left(\sum_{j=1}^n \frac{\partial^2 u}{\partial x_j^2} \right) v \, dx = \int_{\partial G} \sum_{j=1}^n \left(\vec{n}_j \frac{\partial u}{\partial x_j} \right) v \, d\sigma - \int_G u \sum_{j=1}^n \left(\vec{n}_j \frac{\partial v}{\partial x_j} \right) d\sigma \quad (27.19)$$

$$+ \int_G u \sum_{j=1}^n \frac{\partial^2 v}{\partial x_j^2} \, dx.$$

With $\Delta_n = \sum_{j=1}^n \frac{\partial^2}{\partial x_j^2}$ being the Laplace operator and using the notation of directional derivatives, see Definition 6.14 and Theorem 6.17, we may rewrite the last identity as

$$\int_G (\Delta_n u) v \, dx = \int_{\partial G} (D_{\vec{n}} u) v \, d\sigma - \int_{\partial G} u (D_{\vec{n}} v) \, d\sigma + \int_G u \Delta_n v \, dx$$

which is usually called **Green's second identity** and often written as

$$\int_G (v\Delta_n u - u\Delta_n v)\,\mathrm{d}x = \int_{\partial G}\left(v\frac{\partial u}{\partial \vec{n}} - u\frac{\partial v}{\partial \vec{n}}\right)\mathrm{d}\sigma \qquad (27.20)$$

and $\frac{\partial u}{\partial \vec{n}} = \langle \operatorname{grad} u, \vec{n}\rangle$ is called the **normal derivative** of u in the direction of the outward normal of ∂G. When treating the Laplace operator in more detail in Volume IV we will use Green's second identity extensively. We want to derive some further interesting formulae involving the Laplacian. We may choose in (27.17) u to be $\frac{\partial u}{\partial x_j}$ and then we obtain

$$\int_G \frac{\partial^2 u}{\partial x_j^2}v\,\mathrm{d}x = \int_G \left(\frac{\partial u}{\partial x_j}n_j\right)v\,\mathrm{d}\sigma - \int_G \frac{\partial u}{\partial x_j}\frac{\partial v}{\partial x_j}\,\mathrm{d}x, \qquad (27.21)$$

implying

$$\int_G (\Delta_n u)v\,\mathrm{d}x = \int_{\partial G}\frac{\partial u}{\partial \vec{n}}v\,\mathrm{d}\sigma - \int_G \left(\sum_{j=1}^n \frac{\partial u}{\partial x_j}\frac{\partial v}{\partial x_j}\right)\mathrm{d}x. \qquad (27.22)$$

If we now take $v = 1$ we arrive at

$$\int_G \Delta_n u\,\mathrm{d}x = \int_{\partial G}\frac{\partial u}{\partial \vec{n}}\,\mathrm{d}\sigma. \qquad (27.23)$$

Thus we have proved, compare with Problem 6 in Chapter 25,

Corollary 27.6. *For a harmonic function u defined in a neighbourhood of a Gauss domain $G \subset \mathbb{R}^n$ the following holds*

$$\int_{\partial G}\frac{\partial u}{\partial \vec{n}}\,\mathrm{d}\sigma = 0. \qquad (27.24)$$

Next let u be a C^2-function in the open set $G_0 \subset \mathbb{R}^n$ and harmonic in the Gauss domain $G \subset G_0$. In addition suppose that $u\big|_{\partial G} = 0$. From (27.22) we derive with $u = v$ that in this case

$$\int_G \|\operatorname{grad} u\|^2\,\mathrm{d}x = 0. \qquad (27.25)$$

As seen before, this implies for $1 \le j \le n$ that $\frac{\partial u}{\partial x_j} = 0$ in $\overset{\circ}{G}$. If we assume that $\overline{G}$ is pathwise connected this implies that u must be identical to 0. We sketch the proof for the case where G is convex. In this case we first deduce by using the mean value theorem that u must be constant in $\overset{\circ}{G}$ implying that u has a unique continuous, namely constant continuation to $\overline{G}$. But $u\big|_{\partial G} = 0$ and we have proved

Corollary 27.7. *If G is a convex (pathwise connected) Gauss domain, $G \subset G_0$ and $u : G \to \mathbb{R}$ is a harmonic function vanishing on ∂G, then u is identically zero.*

We introduce the **Dirichlet problem** for the Laplace operator for an open set $G \subset \mathbb{R}^n$ as the problem:
Find a C^2-function $u : G_0 \to \mathbb{R}$, $G \subset G_0$, such that $u\big|_{\overset{\circ}{G}}$ is harmonic and $u\big|_{\partial G} = g$ where $g = h\big|_{\partial G}$ and h is a continuous function defined in a neighbourhood of ∂G. We claim

Corollary 27.8. *If G is a convex (pathwise connected) Gauss domain, $G \subset G_0$, then the Dirichlet problem has at most one solution.*

Proof. Suppose that u_1 and u_2 solve the Dirichlet problem in G with boundary data g. It follows that $u_1 - u_2$ is a harmonic function in G vanishing on ∂G. Hence by Corollary 27.7 it must vanish in $\overline{G}$. $\qquad\square$

Finally we want to consider a problem for the heat equation. Let $\overline{G} \subset Q \subset \mathbb{R}^n$ be a compact (Jordan measurable) Gauss domain and Q an open set. In the cylinder $\overline{G}_{T_0} := \overline{G} \times [0, T_0]$ we suppose that we can solve the problem

$$\frac{\partial}{\partial t} u(x, t) - \Delta_n u(x, t) = 0, \quad (x, t) \in \overset{\circ}{G} \times (0, T_0], \tag{27.26}$$

$$u(x, t) = f(x), \quad x \in \overline{G}, \tag{27.27}$$

$$u(x, t) = 0, \quad (x, t) \in \partial G \times (0, T_0] \tag{27.28}$$

by a continuous function $u : \overline{G}_{T_0} \to \mathbb{R}$ which has a continuous partial derivative with respect to t and all second order partial derivatives with respect to $x \in \mathbb{R}^n$ which we assume to be continuous on $\overline{G}$.

If we multiply (27.26) by u and integrate over $\overline{G}$ we find

$$\int_{\overline{G}} u_t u \, dx = \int_{\overline{G}} u \Delta_n u \, dx$$

which yields

$$\int_{\overline{G}} \frac{1}{2} \frac{\partial}{\partial t} (u^2(x, t)) \, dx = - \int_{\overline{G}} \|\operatorname{grad} u(x, t)\|^2 \, dx \leq 0 \tag{27.29}$$

where we have used that $u(x, t) = 0$ for $x \in \partial G$ and grad (as Δ_n) refers to the variable $x \in \mathbb{R}^n$. Thus it follows

$$\int_{\overline{G}} \frac{1}{2} \frac{\partial}{\partial t}(u^2(x, t))\, \mathrm{d}x \leq 0$$

and integrating with respect to t we get for every $T \leq T_0$

$$\int_0^T \frac{1}{2} \int_{\overline{G}} \frac{\partial}{\partial t}(u^2(x, t))\, \mathrm{d}x\, \mathrm{d}t = \frac{1}{2} \int_{\overline{G}} \left(\int_0^T \frac{\partial}{\partial t}(u^2(x, t))\, \mathrm{d}t \right)\, \mathrm{d}x$$

$$= \frac{1}{2} \int_{\overline{G}} \left(u^2(x, T) - u^2(x, 0) \right)\, \mathrm{d}x$$

$$= \frac{1}{2} \int_{\overline{G}} (u^2(x, T) - f^2(x))\, \mathrm{d}x \leq 0,$$

or

$$\int_{\overline{G}} u^2(x, T)\, \mathrm{d}x \leq \int_{\overline{G}} f^2(x)\, \mathrm{d}x. \tag{27.30}$$

Estimate (27.30) is not only a nice bound for the solution in terms of the initial data, from (27.30) we can also derive a uniqueness result for the problem (27.26) - (27.29). Indeed suppose that we have two solutions u_1 and u_2. The difference $u_1 - u_2$ must be a solution of (27.26) - (27.29) with f being replaced by the constant function 0. Now (27.30) implies for all $0 \leq T \leq T_0$

$$\int_{\overline{G}} (u_1 - u_2)^2(x, T)\, \mathrm{d}x \leq 0$$

implying $u_1 = u_2$ on $\overline{G} \times [0, T_0]$.

In Volume IV we will see many more applications of Gauss' theorem in the study of harmonic functions and more generally in the study of partial differential equations.

Problems

1. Let $Q \subset \mathbb{R}^3$ be an open set and $G \neq \emptyset$ where $\overline{G}$ is a compact subset of Q. Further let $h : Q \to \mathbb{R}$ be a continuously differentiable function. The mapping $f : \overline{G} \to \mathbb{R}^4$ given by

$$f(u_1, u_2, u_3) = \begin{pmatrix} u_1 \\ u_2 \\ u_3 \\ h(u_1, u_2, u_3) \end{pmatrix}$$

defines a C^1-parametric "surface" in $\mathbb{R}^4$ which is the graph of the function h. Find the Gram determinant of f.

2. Give a parametrization of the upper unit sphere in $\mathbb{R}^4$ and find the corresponding Gram determinant.

   Hint: represent the upper unit sphere in $\mathbb{R}^4$ as the graph of a function and use Problem 1.

3. Let $\tilde{G} \subset \mathbb{R}^4$ be an open set and $G \subset \tilde{G}$ a non-empty bounded and pathwise connected Gauss domain. Further let $u : \tilde{G} \to \mathbb{R}$ be a C^2-function satisfying in $\overset{\circ}{G}$ the partial differential equation $\sum_{k=1}^{n} a_k \frac{\partial^2 u}{\partial x_k^2} = 0$, $a_k > 0$, and the boundary condition $u|_{\partial G} = 0$. Prove that u is identically zero in $\overline{G}$.

Appendices

Appendix I: Vector Spaces and Linear Mappings

By its very definition the derivative at $x_0, Df(x_0)$, of a mapping $f : G \to \mathbb{R}^m, G \subset \mathbb{R}^n$, is a linear approximation of the increment $f(x) - f(x_0)$, i.e. $f(x) - f(x_0) = Df(x_0)(x - x_0) + \varphi(x, x_0)$, where $Df(x_0) : \mathbb{R}^n \to \mathbb{R}^m$ is linear and $\lim\limits_{x \to x_0} \dfrac{\varphi(x, x_0)}{\|x - x_0\|} = 0$. Hence linear mappings are necessarily objects we need to understand for Analysis. Further, we use the eigenvalues of the Hesse matrix at x_0 of $f : G \to \mathbb{R}, G \subset \mathbb{R}^n$, in order to decide whether f at a critical value x_0, i.e. grad $f(x_0) = 0$, has a local extreme value. Thus we need to understand eigenvalues (and eigenvectors) in order to handle basic questions in Analysis.

Moreover, we consider $\mathbb{R}^n$ also as a geometric object, i.e. as an Euclidean space in which we can study in particular lines, (hyper-) surfaces, more generally subspaces and their translations. In fact, approximating the increment $f(x) - f(x_0)$ by a linear mapping has as its geometric counterpart the replacement of the graph of $f : G \to \mathbb{R}, G \subset \mathbb{R}^n$, in a neighbourhood of $(x_0, f(x_0))$ by a line $(n = 1)$, a plane $(n = 2)$ or more generally by a hyperplane.

In addition we are used to looking at symmetries such as invariance under rotations or translations, etc. All these considerations and many more require a good knowledge of vector spaces and linear mappings as taught in standard courses like "Vector Spaces" or "Linear Algebra". In this Appendix we summarise and discuss (most of) those results needed in our course. This Appendix should be seen as a reference text only and not as a textbook. It is expected that this material has already been studied. In addition, while so far we have not used complex numbers in our course, this will change in Volume III, we of course assume the reader (at least a second year student) to have already studied the algebraic properties of the complex field $\mathbb{C}$, and therefore we will formulate many results for the $\mathbb{K}$-vector space with $\mathbb{K} \in \{\mathbb{Q}, \mathbb{R}, \mathbb{C}\}$. The geometric considerations will be done either in the Euclidean setting ($\mathbb{K} = \mathbb{R}$) or the unitary setting ($\mathbb{K} = \mathbb{C}$). Algebraic notions such as groups (subgroups), rings, fields or algebras are supposed to be known. However, when speaking about $\mathbb{K}$-vector spaces, here $\mathbb{K}$ is limited to $\mathbb{Q}, \mathbb{R}$ and $\mathbb{C}$, i.e. we do not handle vector spaces over general fields. In writing this appendix we made use of G. Fischer [17] and [16], R. Lingenberg [32] and F. Lorenz [34]-[35]. A good reference text is also S. Lang [31].

Definition A.I.1. *A $\mathbb{K}$-vector space V or $(V, \mathbb{K})$ is an Abelian group $(V, +)$ such that a scalar multiplication $\sigma : \mathbb{K} \times V \to V$, $\sigma(\lambda, x) = \lambda x$, is defined satisfying the following axioms:*

$$(\lambda + \mu)x = \lambda x + \mu x; \tag{A.I.1}$$

$$\lambda(x + y) = \lambda x + \lambda \mu; \tag{A.I.2}$$

$$\lambda(\mu x) = (\lambda \mu)x; \tag{A.I.3}$$

$$1x = x; \tag{A.I.4}$$

for all $\lambda, \mu \in \mathbb{K}, x, y \in V$ and $1 \in \mathbb{K}$ being the neutral element for multiplication.

Remark A.I.2. In the following we will denote by 0 the zero vector, the zero element in $\mathbb{K}$, the zero mapping, say from a set G to $\mathbb{K}$ or to $(V, \mathbb{K})$, the zero matrix, etc. Only in a few cases where a distinction is really necessary we will use a notation such as O_V or $O_{\mathbb{K}}$ etc.

Elements in a vector space V are called **vectors**, elements in $\mathbb{K}$ are called **scalars**. Clearly, $\mathbb{K}$ is itself a vector space as is $\mathbb{K}^n, n \in \mathbb{N}$, with the standard operations. A further example of a vector space is the set of all functions $f : G \to \mathbb{K}$ where $G \neq \emptyset$ is a set and $(f + g)(x) := f(x) + g(x), (\lambda f)(x) = \lambda f(x)$. We have already identified $C(G), C^k(G), G \subset \mathbb{R}^n$ open, as $\mathbb{R}$-vector spaces. A set $W \subset V$ is called a **subspace** of $(V, \mathbb{K})$ if it is a $\mathbb{K}$-vector space itself. For example $\mathbb{R}^n \times \{0\} \subset \mathbb{R}^{n+1}$ is a subspace or $C^1(G) \subset C(G)$. Moreover, if $W_j \subset V, j \in J$, are subspaces then $\bigcap_{j \in J} W_j$ is a subspace of V too.

Definition A.I.3. *For finitely many vectors $x_k \in V, k = 1, \ldots, m$, in a $\mathbb{K}$-vector space V and scalars $\lambda_k \in \mathbb{K}$ we call*

$$\sum_{k=1}^{m} \lambda_k x_k = \lambda_1 x_1 + \cdots + \lambda_m x_m \tag{A.I.5}$$

*the **linear combination** of $x_1, \ldots, x_m$ and the scalars $\lambda_1, \ldots, \lambda_m$.*

The reader should always have in mind that a linear combination is a finite sum, a formal or a convergent series $\sum_{k=1}^{\infty} \lambda_k x_k$ is not a linear combination.

We call $\sum_{k=1}^{m} \lambda_k x_k$ a **proper linear combination** if $x_k \neq x_l$ for $k \neq l$. If $\sum_{k=1}^{m} \lambda_k x_k = 0$, we call $\sum_{k=1}^{m} \lambda_k x_k$ a **trivial linear combination**.

Definition A.I.4. *For $M \subset V$ we define its **span** by*

$$\mathrm{span}(M) = \{x \in V \mid x = \sum_{k=1}^{m} \lambda_k x_k, x_k \in M, \lambda_k \in \mathbb{K}, m \in \mathbb{N}\}. \qquad (A.I.6)$$

*Thus $\mathrm{span}(M)$, sometimes called the **linear hull** of M, is the set of all linear combinations formed from elements in M. Obviously $\mathrm{span}(M)$ is a subspace of V.*

The following definition is by far the most important one in the whole theory.

Definition A.I.5. *Let V be a $\mathbb{K}$-vector space and $\emptyset \neq M \subset V$ a non-empty set. We call M **linearly independent** or a set of **linearly independent vectors** if for every proper, trivial linear combination of elements in M, i.e.*
$\sum_{j=1}^{m} \lambda_j x_j = 0$, *$x_j \in M$, $x_j \neq x_k$ for $j \neq k$ and $\lambda_j \in \mathbb{K}, m \in \mathbb{N}$, it follows that $\lambda_j = 0, j = 1, \ldots, m$. If M is not linearly independent we call it a **linearly dependent** set and its elements **linearly dependent vectors**.*

This definition says: if M is linearly independent we cannot form a proper, trivial linear combination out of its elements with non-vanishing coefficients.

Definition A.I.6. *Let V be a $\mathbb{K}$-vector space. We call V **finite dimensional** if there exists $N \in \mathbb{N}$ such that every linearly independent set $M \subset V$ has at most N elements. If V is not finite dimensional we call it an **infinite dimensional vector space**.*

Theorem A.I.7. *If V is finite dimensional then every maximal linearly independent set M has the same number of elements called the **dimension**, $\dim V$, of V. Here a finite linearly independent set M is said to be maximal linearly independent if for every $x \in V \backslash M$ it follows that $M \cup \{x\}$ is linearly dependent.*

The space $\mathbb{K}^n$ is finite dimensional with dim $\mathbb{K}^n = n$. The space of all polynomials $p : \mathbb{R} \to \mathbb{R}$ of degree less or equal to m is a finite dimensional $\mathbb{R}$-vector space of dimension $m+1$. The space of all mappings $f : E \to \mathbb{R}$ is a vector space (over $\mathbb{R}$) and it is finite dimensional if and only if E is finite. In this case the number of elements of E is the dimension of this vector space. In particular $C^k(G), G \subset \mathbb{R}^n$ open, is infinite dimensional.

Important to note: span(M) of a finite set M is a finite dimensional space. However, although span(M) consists only of linear combinations, i.e. finite sums, in general span(M) will be not finite dimensional. For example the span of $\{p_\nu : \mathbb{R} \to \mathbb{R} \mid p_\nu(x) = x^\nu, \nu \in \mathbb{N}_0\}$ is the set of all polynomials $p : \mathbb{R} \to \mathbb{R}$ which is an infinite dimensional vector space.

Definition A.I.8. *Let V be a finite dimensional $\mathbb{K}$-vector space of dimension dim $V = n$. Any linearly independent subset $B \subset V$ of n elements is called a **basis** of V.*

Thus a basis of a (finite) dimensional vector space is a maximal linearly independent subset. If $B = \{b_1, \ldots, b_n\}$ is a basis of V then we have

$$V = \text{span}(B) = \left\{ \sum_{j=1}^{n} \lambda_j b_j \mid \lambda_j \in \mathbb{K} \right\}.$$

Moreover, the coefficients λ_j in the basis representation of x, $x = \sum_{j=1}^{n} \lambda_j b_j$, are uniquely determined and the mapping $\beta : V \to \mathbb{K}^n, x \mapsto \beta(x) := (\lambda_1, \ldots, \lambda_n)$ is bijective.

If $M \subset V$ is a linearly independent set of the n-dimensional $\mathbb{K}$-vector space V with $m < n$ elements we can find vectors $b_{m+1}, \ldots, b_n \in V$ such that $M \cup \{b_{m+1}, \ldots, b_n\}$ is a basis of V.

If $B = \{b_1, \ldots, b_n\}$ is a basis of V, dim $V = n$, then span $\{b_1, \ldots, b_k\}$ is a k-dimensional subspace of V.

Now we turn to mappings respecting the vector space structure.

Definition A.I.9. *Let V and W be two $\mathbb{K}$-vector spaces. We call a mapping $A : V \to W$ a **linear mapping** or a **linear operator** if for all $x, y \in V$ and $\lambda, \mu \in \mathbb{K}$ we have*

$$A(\lambda x + \mu y) = \lambda A x + \mu A y. \tag{A.I.7}$$

The identity $\mathrm{id}_V : V \to V, \mathrm{id}(x) = x$ is a linear mapping as is the zero mapping $0 : V \to W, 0x = 0$ for all $x \in V$. (Note 0 in $0x$ is a mapping, 0 on the right hand side of $0x = 0$ is the zero vector in W.) For every linear mapping we have $A0 = 0$. We define the **kernel** of A by

$$\ker(A) = \{x \in V \mid Ax = 0\} \tag{A.I.8}$$

and the **range** of A by

$$\mathrm{ran}(A) = \{y \in W \mid y = Ax \text{ for some } x \in V\}. \tag{A.I.9}$$

Lemma A.I.10. *For a linear mapping $A : V \to W$ the kernel is a subspace of V and the range is a subspace of W.*

We define for $A_1, A_2 : V \to W$ and $\lambda, \mu \in \mathbb{K}$ the mapping $\lambda A_1 + \mu A_2 : V \to W$ by $(\lambda A_1 + \mu A_2)x = \lambda A_1 x + \mu A_2 x$ which is again a linear mapping and it follows that $\mathrm{Hom}(V, W)$ defined by $\mathrm{Hom}(V, W) := \{A : V \to W \mid A \text{ is linear}\}$ is itself a vector space. Note that in the context of analysis and in particular functional analysis $\mathrm{Hom}(V, W)$ is denoted by $L(V, W)$. Of particular interest is the (algebraic) **dual space** $V^* := Hom(V, \mathbb{K})$.

When $A : V_1 \to V_2$ and $B : V_2 \to V_3$ are linear mappings between the $\mathbb{K}$-vector spaces $V_j, j = 1, 2, 3$, then $B \circ A : V_1 \to V_3$ is linear too. In particular $\mathrm{Hom}(V) := \mathrm{Hom}(V, V)$ is an algebra with identity id_V.

Let V and W be finite dimensional $\mathbb{K}$-vector spaces with basis $B_V = \{v_1, \ldots, v_n\}$ and $B_W := \{w_1, \ldots, w_m\}$, respectively.

Given a linear mapping $A : V \to W$ we can expand $x \in V$ with respect to B_V, i.e. $x = \sum_{j=1}^{n} \lambda_j v_j$, and then we find

$$Ax = A\left(\sum_{j=1}^{n} \lambda_j v_j\right) = \sum_{j=1}^{n} \lambda_j A v_j$$

with $Av_j \in W$. Now we can expand Av_j with respect to B_W:

$$Av_j = \sum_{k=1}^{m} (Av_j)_k w_k$$

to find for $x = \sum_{j=1}^{n} \lambda_j v_j$

$$Ax = \sum_{j=1}^{n} \lambda_j \sum_{k=1}^{m} (Av_j)_k w_k = \sum_{k=1}^{m} \mu_k w_k$$

where

$$\mu_k = \sum_{j=1}^{n} \lambda_j (Av_j)_k.$$

Introducing

$$a_{kj} = (Av_j)_k$$

it follows that

$$\mu_k = \sum_{j=1}^{n} a_{kj}\lambda_j.$$

Now we identify V with $\mathbb{K}^n$ by $\beta_{B_V} : V \to \mathbb{K}^n, \beta_{B_V}(x) = \beta_{B_V}(\Sigma\lambda_j v_j) = \begin{pmatrix} \lambda_1 \\ \vdots \\ \lambda_n \end{pmatrix}$ and we identify W with $\mathbb{K}^m$ by $\beta_{B_W} : W \to \mathbb{K}^m, \beta_{B_W}(\Sigma\mu_k w_k) = \begin{pmatrix} \mu_1 \\ \vdots \\ \mu_m \end{pmatrix}$ and we arrive at the matrix representation of A with respect to the bases B_V and B_W:

$$\begin{pmatrix} \mu_1 \\ \vdots \\ \mu_m \end{pmatrix} = \begin{pmatrix} a_{11} & \cdots & a_{1n} \\ \vdots & & \vdots \\ a_{m1} & \cdots & a_{mn} \end{pmatrix} \begin{pmatrix} \lambda_1 \\ \vdots \\ \lambda_n \end{pmatrix}. \tag{A.I.10}$$

Of course, a change of B_V and/or B_W will change the matrix representation of A.

The vectors $\begin{pmatrix} a_{11} \\ \vdots \\ a_{m1} \end{pmatrix}, \ldots, \begin{pmatrix} a_{1n} \\ \vdots \\ a_{mn} \end{pmatrix}$ are called the **column vectors** of the matrix and the vectors $(a_{11}, \ldots, a_{1n}), \ldots, (a_{n1}, \ldots, a_{mn})$ are the **row vectors** of the matrix. The **rank** of a matrix is the maximal number of independent column vectors which equals the maximal number of independent row vectors. If B and D are two bases of $V, \dim V = n$, and S is the transition matrix form B to D, then for every linear mapping $A : V \to V$ we have

$$A_D = S^{-1} A_B S, \tag{A.I.11}$$

where A_B denotes the matrix representation of A with respect to the basis B. It is now easy to see that $\beta_{B_V} : V \to \mathbb{K}^n$, $\dim V = n$, is linear and bijective and for obvious reasons when doing Analysis we prefer to work in $\mathbb{R}^n$ (or $\mathbb{C}^n$) than to work in a general n-dimensional $\mathbb{K}$-vector space V.

In $\mathbb{K}^n$ the n **unit vectors** $e_1, \ldots, e_n$, are $e_j = (0, \ldots, 0, 1, 0, \ldots, 0) \in \mathbb{K}^n$ with 1, the multiplicative unit in $\mathbb{K}$, in the jth position and they form a basis of $\mathbb{K}^n$. When working in $\mathbb{R}^n$ or with linear mappings $A : \mathbb{R}^n \to \mathbb{R}^m$ we always will use (if not explicitly stated otherwise) the **canonical basis** $\{e_1, \ldots, e_n\}$ and/or $\{e_1, \ldots, e_m\}$. Usually we write e_j for the jth unit vector for any $\mathbb{R}^n, n \in \mathbb{N}, j \leq n$. Only occasionally we write $\{e_1, \ldots, e_n\} \subset \mathbb{R}^n$ and $\{\tilde{e}_1, \ldots, \tilde{e}_m\} \subset \mathbb{R}^m$ for the canonical bases in $\mathbb{R}^n$ and $\mathbb{R}^m$, respectively.

Fixing in $\mathbb{R}^n$ and $\mathbb{R}^m$ the canonical basis then we can identify $\text{Hom}(\mathbb{R}^m, \mathbb{R}^n)$ by the vector space $M(n, m; \mathbb{R})$ of all $m \times n$ matrices. So we may use

$$\text{Hom}(\mathbb{R}^m, \mathbb{R}^n) = L(\mathbb{R}^m, \mathbb{R}^n) \cong M(n, m; \mathbb{R}) \cong \mathbb{R}^{nm}.$$

If $A : \mathbb{K}^n \to \mathbb{K}^m$ and $B : \mathbb{K}^m \to \mathbb{K}^l$ are two linear mappings with matrix representations $(a_{\alpha\beta})_{\substack{\alpha = 1, \ldots, m \\ \beta = 1, \ldots, n}}$ and $(\beta_{\gamma\delta})_{\substack{\gamma = 1, \ldots, l \\ \delta = 1, \ldots, m}}$ with respect to the canonical bases in $\mathbb{K}^n, \mathbb{K}^m$ and $\mathbb{K}^l$, then $C = BA : \mathbb{K}^n \to \mathbb{K}^l$ has the matrix representation $(c_{\mu\nu})_{\substack{\mu = 1, \ldots, l \\ \nu = 1, \ldots, n}}$ given by

$$\begin{pmatrix} c_{11} & \cdots & c_{1n} \\ \vdots & & \vdots \\ c_{l1} & \cdots & c_{ln} \end{pmatrix} = \begin{pmatrix} b_{11} & \cdots & b_{1m} \\ \vdots & & \vdots \\ b_{l1} & \cdots & b_{lm} \end{pmatrix} \begin{pmatrix} a_{11} & \cdots & a_{1n} \\ \vdots & & \vdots \\ a_{m1} & \cdots & a_{mn} \end{pmatrix} \qquad \text{(A.I.12)}$$

where

$$c_{\mu\nu} = \sum_{\rho=1}^{m} b_{\mu\rho} a_{\rho\nu}. \qquad \text{(A.I.13)}$$

We take it for granted that the reader knows the basic theory of how to solve linear systems of equations and its relations to linear mappings, i.e. how the equation $Ax = b$ can be transformed for a linear mapping $A : V \to W$, $\dim V = n$, $\dim W = m$, into the system

$$\begin{aligned} a_{11}x_1 + \cdots + a_{1n}x_n &= b_1 \\ \vdots \qquad\qquad &\quad \vdots \\ a_{m1}x_1 + \cdots + a_{mn}x_n &= b_m \end{aligned} \qquad \text{(A.I.14)}$$

and vice versa. Further we take it for granted that the reader knows the basic rules on manipulating matrices. Since the "best" we may expect is that $A : V \to W$ maps linearly independent vectors in V onto linearly independent vectors in W, we know that if $\dim V < \dim W$ then A cannot be surjective, hence the equation $Ax = b$ or equivalently the system (A.I.14) is not always solvable. If on the other hand $\dim W < \dim V$ then A must have a non-trivial kernel. Furthermore, if $y \in V$ is one solution to $Ax = b$ then all solutions to $Ax = b$ are given by $y + \ker A$, where as usual $a + M$ stands for the set $\{z = a + m \mid m \in M\}$. To solve (A.I.14) we may use the Gauss algorithm, which is however more a "theoretical" tool than a practical one.

In the case $\dim V = \dim W$ the question of unique solvability of $Ax = b$ is equivalent to find the inverse of the mapping A and we can discuss this equivalently as the problem to find the corresponding inverse matrix. We restrict ourselves (for the moment) to $\mathbb{R}^n$ with canonical basis. The tool we want to use is the theory of determinants and for this we need to recollect some facts on permutations.

For a set $X \neq \emptyset$ the set of all bijective mappings $\sigma : X \to X$ forms a group with respect to composition, this group is denoted by $S(X)$ and its elements are called **permutations**. Often we write $\sigma_1 \sigma_2$ instead of $\sigma_1 \circ \sigma_2$ with $\sigma_1, \sigma_2 \in S(X)$ for the composition of these mappings. If X is a finite set of N elements we may identify X with $\{1, \ldots, N\} =: \mathbb{N}_N$ and we call $S(\mathbb{N}_N)$ the **symmetric group** of order N and denote it by S_N. The group S_N has $N!$ elements and for $k \in \mathbb{N}_N$ fixed $S_N(k) := \{\sigma \in S_N \mid \sigma(k) = k\}$ is a subgroup of S_N isomorphic to S_{N-1}. For $1 \leq i < j \leq N$ we call $\tau = \tau_{ij} \in S_N$ a transposition if $\tau_{ij}(i) = j$, $\tau_{ij}(j) = i$ and $\tau(k) = k$ for $k \neq i, j$. Every permutation $\sigma \in S_N$ is a product of transpositions, i.e. $\sigma = \tau_1 \cdot \ldots \cdot \tau_r$, $r \leq N$, but the representation is not necessarily unique. However all representations of σ as a product of transpositions consist either of an even or an odd number of transpositions. We define the **sign of a permutation** $\sigma, \mathrm{sgn}\sigma$, to be 1 if we need an even number of transpositions for its decomposition, and $\mathrm{sgn}(\sigma) = -1$ if an odd number is needed. Accordingly we call σ an even or an odd permutation. For $\sigma, \tau \in S_N$ we have $\mathrm{sgn}(\sigma\tau) = \mathrm{sgn}(\sigma)\mathrm{sgn}(\tau)$.

Definition A.I.11. *We call $D : \mathbb{R}^n \times \cdots \times \mathbb{R}^n \to \mathbb{R}$, n copies of $\mathbb{R}^n$, a* ***determinant form*** *if D satisfies the following conditions:*

(D.1) $\quad D(a_1, \ldots, a_{j-1}, \lambda x + \mu y, a_{j+1}, \ldots, a_n)$
$\qquad = \lambda D(a_1, \ldots, a_{j-1}, x, a_{j+1}, \ldots, a_n) + \mu D(a_1, \ldots, a_{j-1}, y, a_{j+1}, \ldots, a_n)$

for all $1 \leq j \leq n$ with $a_j, x, y \in \mathbb{R}^n$ and $\lambda, \mu \in \mathbb{R}$;

(D.2) $\qquad\qquad D(a_1, \ldots, a_n) = 0$ if $a_j = a_k$ for some $j \neq k$;

(D.3) $\qquad\qquad D(e_1, \ldots, e_n) = 1.$

Remark A.I.12. A. Clearly, (D.1) is a linearity condition, i.e. D is linear in each of its components, hence it is a multilinear form; in this case an n-linear form. The condition (D.2) is called the **alternating** condition, hence D is an **alternating multilinear form** which according to (D.3) is normalised. We may give a slightly different interpretation to D. Let $A = (a_{j_k})_{j,k=1,\ldots,n}$ be an $n \times n$ matrix and let us introduce the **column vectors** $a_k = \begin{pmatrix} a_{1k} \\ \vdots \\ a_{nk} \end{pmatrix}, k = 1, \ldots, n$. The vector $(a_{j1}, \ldots, a_{jn})$ is called the jth **row vector**. We can now define

$$\det(A) = D(a_1, \ldots, a_n). \qquad (A.I.15)$$

Conversely, if $a_1, \ldots, a_n \in \mathbb{R}^n$ are given we can form the corresponding matrix A with column vectors $a_1, \ldots, a_n$.

B. An equivalent characterisation of a determinant form is

(D.1') $\qquad\qquad D(a_1, \ldots, a_k + a_l, \ldots, a_n) = D(a_1, \ldots, a_n), \ k \neq l$;

(D.2') $\qquad\qquad D(a_1, \ldots, \lambda a_k, \ldots, a_n) = \lambda D(a_1, \ldots, a_n), \ \lambda \in \mathbb{R}$;

(D.3') $\qquad\qquad D(e_1, \ldots, e_n) = 1.$

Theorem A.I.13. *There exists a unique determinant form* det *defined on* $\mathbb{R}^n \times \cdots \times \mathbb{R}^n$ *or equivalently on all* $n \times n$ *matrices, and we call* $\det A$ *the* **determinant** *of* A, A *being an* $n \times n$ *matrix. If* $A = (a_{jk})_{j,k=1,\ldots,n}$ *then the following holds*

$$\det A = \sum_{\sigma \in S_n} (\text{sgn}\,\sigma) a_{1\sigma(1)} \cdot \ldots \cdot a_{n\sigma(n)}. \qquad (A.I.16)$$

Here are some rules for determinants. In the following $\lambda \in \mathbb{R}$, A and B are $n \times n$ matrices and $A = (a_{kl})_{k,l=1,\ldots,n}$ has the column vectors $a_j = \begin{pmatrix} a_{1j} \\ \vdots \\ a_{nj} \end{pmatrix}, j = 1, \ldots, n.$

$\det(\lambda A) = \det(\lambda a_1, \ldots, \lambda a_n) = \lambda^n \det(a_1, \ldots, a_n) = \lambda^n \det A;$

$$(A.I.17)$$

if $a_j = 0$ for some $1 \leq j \leq n$ then $\det A = \det(a_1, \ldots, a_n) = 0$;

$$\text{(A.I.18)}$$

$$\det(a_1, \ldots, a_k, \ldots, a_l, \ldots, a_n) = -\det(a_1, \ldots, a_l, \ldots, a_k, \ldots, a_n), k \neq l,$$

$$\text{(A.I.19)}$$

interchanging two column vectors changes the sign of $\det A$;

$$\det(A+B) = \det A + \det B;$$

$$\text{(A.I.20)}$$

$$\det(AB) = (\det A)(\det B);$$

$$\text{(A.I.21)}$$

if A^{-1} exists then $\det(A^{-1}) = (\det A)^{-1}$;

$$\text{(A.I.22)}$$

$\det A \neq 0$ if and only if $\{a_1, \ldots, a_n\}$ is a linearly independent set.

$$\text{(A.I.23)}$$

If $A = (a_{kl})_{k,l=1,\ldots,n}$ is an $n \times n$ matrix we define the **transposed matrix** A^t by

$$A^t = \begin{pmatrix} a_{11} & \cdots & a_{n1} \\ \vdots & & \vdots \\ a_{1n} & \cdots & a_{nn} \end{pmatrix}, \qquad \text{(A.I.24)}$$

and we find

$$\det A^t = \det A. \qquad \text{(A.I.25)}$$

While formula (A.I.16) allows us to calculate $\det A$, it is worth deriving a more practical rule reducing the calculation of the determinant of an $n \times n$ matrix to the evaluation of the determinants of $(n-1) \times (n-1)$ matrices. We need some definitions first. Let

$$A = \begin{pmatrix} a_{11} & \cdots & a_{1l} & \cdots & a_{1n} \\ \vdots & & \vdots & & \\ a_{k1} & \cdots & a_{kl} & \cdots & a_{kn} \\ \vdots & & \vdots & & \\ a_{n1} & \cdots & a_{nl} & \cdots & a_{nn} \end{pmatrix}$$

be a given $n \times n$ matrix. By deleting the lth column and the kth row vector we obtain a new $(n-1) \times (n-1)$ matrix M_{kl}

$$M_{kl} := \begin{pmatrix} a_{11} & \cdots & a_{1l-1}a_{1l+1} & \cdots & a_{1n} \\ \vdots & & & & \\ a_{k-11} & \cdots & a_{k-1l-1}a_{k-1l+1} & \cdots & a_{k-1n} \\ a_{k+11} & \cdots & a_{k+1l-1}a_{k+1l+1} & \cdots & a_{k+1n} \\ \vdots & & \vdots & & \\ a_{n1} & & a_{nl-1}a_{nl+1} & \cdots & a_{nn} \end{pmatrix} \quad \text{(A.I.26)}$$

and the determinant of M_{kl}, is called the **minor of the element** a_{kl} whereas $A_{kl} := (-1)^{k+l} \det M_{kl}$ is called the **co-factor** of a_{kl}. By definition the kth **principal minor** $\delta_k(A)$ of an $n \times n$ matrix $A, k \le n$, is the determinant of the matrix obtained by deleting in A the last $n-k$ rows and columns:

$$\begin{pmatrix} a_{11} & \cdots & a_{1k-1} & a_{1k} & \cdots & a_{1n} \\ \vdots & & \vdots & & & \\ a_{k-11} & \cdots & a_{k-1k-1} & a_{k-1k} & \cdots & a_{k-1n} \\ a_{k1} & \cdots & a_{kk-1} & a_{kk} & \cdots & a_{kn} \\ \vdots & & \vdots & & & \\ a_{n1} & \cdots & a_{nk-1} & a_{nk} & \cdots & a_{nn} \end{pmatrix}$$

hence

$$\delta_k(A) = \det \begin{pmatrix} a_{11} & \cdots & a_{1k-1} \\ \vdots & & \\ a_{k-11} & \cdots & a_{k-1k-1} \end{pmatrix}.$$

Theorem A.I.14. *For $1 \le k \le l \le n$ the following holds*

$$\det A = \sum_{k=1}^{n} a_{kl} A_{kl} = \sum_{l=1}^{n} a_{kl} A_{kl}. \quad \text{(A.I.27)}$$

Iterating (A.I.27) we will end up with (A.I.16). A well known result is

Theorem A.I.15. *If $n = m$ then the system(A.I.14) has a unique solution for all $b \in \mathbb{R}^n$ if and only if $\det A \ne 0$.*

For $n = m$ **Cramer's rule** to solve (A.I.14) reads as follows: Denote by C_k the matrix obtained from A by replacing $a_k = \begin{pmatrix} a_{1k} \\ \vdots \\ a_{nk} \end{pmatrix}$ by the vector $b = \begin{pmatrix} b_1 \\ \vdots \\ b_n \end{pmatrix}$. If $\det A \neq 0$ then the system (A.I.14) has a unique solution which is given by

$$x_k = \frac{\det C_k}{\det A}, k = 1, \ldots, n. \tag{A.I.28}$$

Cramer's rule has some theoretical merits, for n large however it is not a practical tool to solve a linear $n \times n$ system of equations. However we can use Cramer's rule to derive a formula for A^{-1} if $\det A \neq 0$.

Theorem A.I.16. *An $n \times n$ matrix is invertible if and only if $\det A \neq 0$ and then its inverse is given by*

$$A^{-1} = \left(\left(\frac{A_{kl}}{\det} \right)_{k,l=1,\ldots,n} \right)^t = \left(\left(\frac{(-1)^{k+l} \det M_{kl}}{\det A} \right)_{k,l=1,\ldots,n} \right)^t. \tag{A.I.29}$$

So far, when handling determinants we have dealt with matrices, but we should recall that a matrix is a representation of a linear mapping, and the representation depends on the choice of the basis. It turns out that the determinant is invariant under a change of basis, so $\det A$ is in fact a "number" attached to the linear mapping independent of its matrix representation. A further such number is the **trace** of a matrix $A = (a_{kl})_{k,l=1,\ldots,n}$. For $A = (a_{kl})_{k,l=1,\ldots,n}$ its trace is defined by

$$\mathrm{tr}(A) = \mathrm{tr} \begin{pmatrix} a_{11} & \cdots & a_{1n} \\ \vdots & & \\ a_{n1} & \cdots & a_{nn} \end{pmatrix} := \sum_{k=1}^{n} a_{kk} \tag{A.I.30}$$

and it turns out that $\mathrm{tr}(A)$ is invariant under a change of basis, hence $\mathrm{tr}(A)$ is a "number" attached to the linear mapping represented by the matrix $(a_{kl})_{k,l=1,\ldots,n}$.

Let V be an n-dimensional $\mathbb{K}$-vector space and $W_1, W_2 \subset V$ subspaces. The **sum** of W_1 and W_2 is by definition the span of their union

$$W_1 + W_2 := \mathrm{span}(W_1 \cup W_2).$$

If $W_1 \cap W_2 = \{0\}$ then we call $\mathrm{span}(W_1 \cup W_2)$ the **direct sum** of W_1 and W_2 and for this we write

$$W_1 \oplus W_2 \quad (\text{direct sum of } W_1, W_2, W_1 \cap W_2 = \{0\}). \qquad (\text{A.I.31})$$

If $W_1 \subset V$ is a subspace and $\dim V = n < \infty$, then there exists $W_2 \subset V$, a further subspace, such that $W_1 \oplus W_2 = V$.

Let V be a finite dimensional $\mathbb{K}$-vector space and let $W_j \subset V, 1 \leq j \leq m \leq n$, be subspaces. If $W_k \cap W_l = \{0\}$ for $k \neq l$ and span $\{W_1 \cup \cdots \cup W_m\} = V$ we call

$$V = W_1 \oplus \cdots \oplus W_m \qquad (\text{A.I.32})$$

a decomposition of V into a direct sum.

Example A.I.17. Let $B = \{b_1, \ldots, b_n\}$ be a basis of V and $\{B_1, \ldots, B_m\}$ a partition of B. If $W_j = \mathrm{span}(B_j), j = 1, \ldots, m$, then $V = W_1 \oplus \cdots \oplus W_m$.

Lemma A.I.18. *The subspaces $W_j \subset V, j = 1, \ldots, m, \dim V = n$, form a decomposition of V into a direct sum, $V = W_1 \oplus \cdots \oplus W_m$, if and only if every $x \in V$ admits a unique decomposition $x = w_1 + \cdots + w_m$ with $w_j \in W_j$.*

Let V be a $\mathbb{K}$-vector space and $A : V \to V$ be a linear mapping. We call $W \subset V$ an **invariant subspace** with respect to A or an invariant subspace of A if $AW \subset W$. Of course V and $\{0\}$ are always invariant subspaces and we are interested in non-trivial invariant subspaces of A. Let $W \subset V, m = \dim W < \dim V = n$, be invariant under A and $\{c_1, \ldots, c_m\}$ be a basis of W. We may ask whether we can extend $\{c_1, \ldots, c_m\}$ to a basis $\{c_1, \ldots, c_m, d_{m+1}, \ldots, d_n\}$ of V such that $U := \mathrm{span}\{d_{m+1}, \ldots, d_n\}$ is an invariant subspace of A. When this is possible we have a decomposition of V into a direct sum of two invariant subspaces of A and A would be completely determined by its properties on W and U, i.e. if we know $A|_W$ and $A|_U$ we know A. The search for invariant subspaces leads to eigenvalues, eigenvectors and eigenspaces.

Definition A.I.19. *Let V be a $\mathbb{K}$-vector space and $A : V \to V$ be a linear mapping. We call $\lambda \in \mathbb{K}$ an **eigenvalue** of A and $v \in V, v \neq 0$, **eigenvector** of A with respect to λ if*

$$Av = \lambda v \text{ or } (A - \mathrm{id}_V)v = 0. \qquad (\text{A.I.33})$$

Let $\lambda \in \mathbb{K}$ be an eigenvalue of A. For $\mu_1, \mu_2 \in \mathbb{K}$ and eigenvectors v_1, v_2 it follows that

$$A(\mu_1 v_1 + \mu_2 v_2) = \mu_1 A v_1 + \mu_2 A v_2 = \lambda(\mu_1 v_1 + \mu_2 v_2),$$

hence the set of all eigenvectors to the eigenvalue λ together with $0 \in V$ form a subspace of V, the **eigenspace** W_λ of λ. If V is finite dimensional as we will always assume in the following and W_λ is the eigenspace to $A : V \to V$ with respect to the eigenvalue λ then we call $\dim W_\lambda$ the **geometric multiplicity** of λ.

An example of an eigenspace is of course $\ker(A)$ provided it is not trivial which is the eigenspace of A to the eigenvalue $0 \in \mathbb{K}$. If $\lambda_1 \neq \lambda_2$ are two eigenvalues of A then $W_{\lambda_1} \cap W_{\lambda_2} = \{0\}$. Hence, if A has k distinct eigenvalues $\lambda_1, \ldots, \lambda_k$ with geometric multiplicities $m_1, \ldots, m_k$ such that $m_1 + \cdots + m_k = n = \dim V$, then V admits a decomposition into a direct sum of the eigenspaces of A, i.e. $V = W_{\lambda_1} \oplus \cdots \oplus W_{\lambda_k}$. In this case we may choose in each eigenspace W_{λ_l} a basis of eigenvectors $D_l = \{d_1^{(l)}, \ldots, d_{m_l}^{(l)}\}$ and $D = D_1 \cup \cdots \cup D_{m_k}$ is a basis of V consisting of eigenvectors of A. With respect to a basis of eigenvectors A has a very simple matrix representation: if $x \in V$ and $D = \{d_1^{(1)}, \ldots, d_{m_k}^{(k)}\}$ is a basis of eigenvectors of A as above then we find

$$x = \sum_{l=1}^{k} \sum_{j=1}^{m_l} \xi_j^{(l)} d_j^{(l)} \tag{A.I.34}$$

and

$$Ax = \sum_{l=1}^{k} \sum_{j=1}^{m_l} \xi_j^{(l)} A d_j^{(l)} = \sum_{l=1}^{k} \sum_{j=1}^{m_l} \xi_j^{(l)} \lambda_l d_j^{(l)}, \tag{A.I.35}$$

and therefore the matrix representation $\tilde{A}$ of A when we choose the basis D

in V has the form

$$
\tilde{A} = \begin{pmatrix}
\lambda_1 & 0 & & & & & & & \\
 & \ddots & & & & & 0 & & \\
0 & & \lambda_1 & & & & & & \\
 & & 0 & \lambda_2 & 0 & & & & \\
 & & & & \ddots & & & & \\
 & & & & 0 & \lambda_2 & 0 & & \\
 & & & & & & \ddots & & \\
 & & & & & & & \lambda_k & 0 \\
 & 0 & & & & & & \ddots & \\
 & & & & & & & 0 & \lambda_k
\end{pmatrix}
\tag{A.I.36}
$$

i.e. $\tilde{A}$ is a diagonal matrix with its eigenvalues in the diagonal. Of course, $\tilde{A}$ has a particular simple form and we may ask: when does A have a basis of eigenvectors? In such a case we call A **diagonalisable**. From (A.I.33) we can derive an easy criterion to find eigenvalues of A: A number $\lambda \in \mathbb{K}$ is an eigenvalue of $A : V \to V$, V being a $\mathbb{K}$-vector space of finite dimension $\dim V = n$, if and only if

$$
\chi(\lambda) := \det(A - \lambda id_V) = 0.
\tag{A.I.37}
$$

This is a polynomial χ in λ, called the **characteristic polynomial** of A, which has the leading term $(-1)^n \lambda^n$, hence it is of degree n. Over $\mathbb{C}$ this polynomial has n roots, however some might have a certain multiplicity, i.e. over $\mathbb{C}$ we have

$$
\chi(\lambda) = \prod_{j=1}^{r} (\lambda_j - \lambda)^{\kappa_j}, \lambda_j \in \mathbb{C},
\tag{A.I.38}
$$

and $\kappa_1 + \cdots + \kappa_r = n$. We call κ_j the **algebraic multiplicity** of λ_j. We can further prove that if

$$
\chi(\lambda) = \sum_{k=0}^{n} \alpha_k \lambda^k,
$$

then $\alpha_n = (-1)^n, \alpha_{n-1} = (-1)^{n-1} \operatorname{tr}(A)$ and $\alpha_0 = \det A$, which holds for any field $\mathbb{K}$.

If $\mathbb{K} = \mathbb{R}$ the characteristic polynomial may not factorise into linear factors since complex roots may occur. If λ_j is such a complex root, then $\bar{\lambda}_j$ is a further one and both have the same algebraic multiplicity. This implies that over $\mathbb{R}$ the characteristic polynomial factorises into linear and quadratic factors.

It is convenient when dealing with eigenvalues to work with complex-valued roots first and do, if necessary, some adjustments for $\mathbb{R}$-vector spaces.

In general $\dim W_{\lambda_j} \leq \kappa_j$, i.e. the geometric multiplicity of an eigenvalue is dominated by its algebraic multiplicity. The following result gives a criterion for A being diagonalisable:

Theorem A.I.20. *A linear mapping $A : V \to V$ over a finite dimensional $\mathbb{C}$-vector space V admits a basis of eigenvectors, i.e. is diagonalisable, if and only if for every eigenvalue the geometric multiplicities is equal to the arithmetic multiplicity.*

For $\mathbb{R}$-vector spaces we have to add the condition that the characteristic polynomial has a factorisation into linear factors. In both cases the diagonalisability of A is equivalent to

$$V = W_{\lambda_1} \oplus \cdots \oplus W_{\lambda_k} \tag{A.I.39}$$

where $\lambda_1, \ldots, \lambda_k$ are the eigenvalues of A and $W_{\lambda_j}, j = 1, \ldots, k$ the corresponding eigenspaces.

The most important class of real matrices (or linear mappings $A : V \to V$ in a real vector space) admitting a basis of eigenvectors are symmetric matrices. We call A symmetric if $A = A^t$ and we have

Theorem A.I.21. *If A is a symmetric $n \times n$ matrix over $\mathbb{R}$ then all eigenvalues are real and A is diagonalisable.*

This theorem has an analogue in $\mathbb{C}$, however both results are best understood when "adding" geometry to our consideration, i.e. when introducing scalar products and when starting to study Euclidean and unitary vector spaces. Let V be an $\mathbb{R}$-vector space. A **scalar product** on V is a mapping $\langle \cdot, \cdot \rangle : V \times V \to \mathbb{R}$ such that for all $x, y, z \in V$ and $\lambda, \mu \in \mathbb{R}$ the following hold

$$\langle \lambda x + \mu y, z \rangle = \lambda \langle x, z \rangle + \mu \langle y, z \rangle; \tag{A.I.40}$$

$$\langle x, y \rangle = \langle y, x \rangle; \tag{A.I.41}$$

$$\langle x, x \rangle \geq 0 \text{ and } \langle x, x \rangle = 0 \text{ if and only if } x = 0. \tag{A.I.42}$$

Property (A.I.40) is a linearity condition in the first argument which however with the symmetry condition (A.I.41) implies that a scalar product in an $\mathbb{R}$-vector space is also linear in the second argument, i.e.

$$\langle x, \lambda y + \mu z \rangle = \lambda \langle x, y \rangle + \mu \langle x, z \rangle, \qquad \text{(A.I.43)}$$

while (A.I.42) is the requirement that $\langle \cdot, \cdot \rangle$ is positive definite.

If W is a $\mathbb{C}$-vector space we call $\langle \cdot, \cdot \rangle : W \times W \to \mathbb{C}$ a **(complex) scalar product** on W if for all $x, y, z \in W$ and $\lambda, \mu \in \mathbb{C}$ (A.I.40) and (A.I.42) hold, while (A.I.41) is replaced by

$$\langle x, y \rangle = \overline{\langle x, y \rangle}, \qquad \text{(A.I.44)}$$

where $\bar{a}$ is the complex conjugate of $a \in \mathbb{C}$. Note that (A.I.44) implies that $< x, x > \in \mathbb{R}$, and in combination with (A.I.41) this yields for a (complex) scalar product in a $\mathbb{C}$-vector space

$$\langle x, \lambda y + \mu z \rangle = \overline{\lambda} \langle x, y \rangle + \bar{\mu} \langle x, z \rangle. \qquad \text{(A.I.45)}$$

Authors are split on this topic; some request a complex scalar product to be linear with respect to the second argument, i.e. (A.I.43) and then they obtain for the first argument

$$\langle \lambda x + \mu y, z \rangle = \overline{\lambda} \langle x, z \rangle + \overline{\mu} \langle y, z \rangle. \qquad \text{(A.I.46)}$$

For every scalar product we have the **Cauchy-Schwarz inequality**

$$| \langle x, y \rangle | \leq \|x\| \|y\| \qquad \text{(A.I.47)}$$

where $\|x\| := \langle x, x \rangle^{1/2}$ is the norm associated with the scalar product. The **Euclidean scalar product** on $\mathbb{R}^n$ is defined by

$$\langle x, y \rangle := \sum_{k=1}^{n} x_k y_k, \ x = (x_1, \ldots, x_n), \ y = (y_1, \ldots, y_n) \in \mathbb{R}^n, \qquad \text{(A.I.48)}$$

and the **unitary scalar product** on $\mathbb{C}^n$ is

$$\langle w, z \rangle = \sum_{k=1}^{n} w_k \overline{z}_n, w = (w_1, \ldots, w_n), \ z = (z_1, \ldots, z_n) \in \mathbb{C}^n. \qquad \text{(A.I.49)}$$

Note that for $z \in \mathbb{C}^n$ we have the representation

$$z = (z_1, \ldots, z_n) = (x_1 + iy_1, \ldots, x_n + iy_n) = x + iy$$

and

$$\overline{z} = (\overline{z}_1, \ldots, \overline{z}_n) = (x_1 - iy_1, \ldots, x_n - iy_n) = x - iy$$

where $(x_1, \ldots, x_n), (y_1, \ldots, y_n) \in \mathbb{R}^n$.

Let $B = \{b_1, \ldots, b_n\}$ be a basis of the $\mathbb{R}$-vector space V and $\langle \cdot, \cdot \rangle$ a scalar product on V. For $x = \sum_{k=1}^{n} x_k b_k$ and $y = \sum_{k=1}^{n} y_k b_k$ we find

$$\langle x, y \rangle = \left\langle \sum_{k=1}^{n} x_k b_k, \sum_{l=1}^{n} y_l b_l \right\rangle = \sum_{k,l=1}^{n} x_k y_l \langle b_k, b_l \rangle. \tag{A.I.50}$$

Hence we can associate with $\langle \cdot, \cdot \rangle$ a matrix $S_B = (\langle b_k, b_l \rangle)_{k,l=1,\ldots,n}$ (depending of course on B) which determines by

$$\sum_{k,l=1}^{n} (S_B)_{kl} x_k y_l = \sum_{k,l=1}^{n} \langle b_k, b_l \rangle x_k y_k$$

the scalar product and a similar result holds in the complex case. In fact, such a representation holds for more general objects.

Definition A.I.22. *Let $V, \dim V = n$, be an $\mathbb{R}$-vector space. We call $\beta : V \times V \to \mathbb{R}$ a **bilinear form** if for all $x, y, z \in V$ and $\lambda, \mu \in \mathbb{R}$ we have*

$$\beta(\lambda x + \mu y, z) = \lambda \beta(x, z) + \mu \beta(y, z) \tag{A.I.51}$$

and

$$\beta(x, \lambda y + \mu z) = \lambda \beta(x, y) + \mu \beta(x, z). \tag{A.I.52}$$

If

$$\beta(x, y) = \beta(y, x) \text{ for all } x, y \in V \tag{A.I.53}$$

*we call β **symmetric**. Further, if β is symmetric and*

$$\beta(x, x) > 0 \quad (\geq 0, < 0, \leq 0) \tag{A.I.54}$$

*we call β **positive definite** (**positive semi-definite, negative definite, negative semi-definite**).*

If β is a bilinear form on V we call $x \mapsto \beta(x,x)$ the associated **quadratic form**.

Thus a scalar product is a positive definite, hence symmetric, bilinear form, and vice versa every positive definite bilinear form gives a scalar product on V.

As discussed previously for a scalar product we can associate with a bilinear form β a matrix $A_{\beta,B}$ with respect to a fixed basis B in V. Conversely, if $A = (a_{kl})_{k,l=1,\ldots,n}$ is an $n \times n$ matrix on $\mathbb{R}^n$ with respect to the canonical basis (for simplicity) we can define on $\mathbb{R}^n$ a bilinear form by

$$\beta_A(x,y) := \sum_{k,l=1}^n a_{kl} x_k y_l, \quad x = (x_1, \ldots, x_n), \ y = (y_1, \ldots, y_n) \in \mathbb{R}^n.$$

Clearly, A is symmetric, i.e. $A = A^t$, if and only if β_A is symmetric. Further we call A a **positive definite** (**positive semi-definite, negative definite, negative semi-definite**) **matrix** if the corresponding bilinear form has this property.

A natural question arises whether we can classify those matrices giving rise to a positive definite (etc.) bilinear form:

Theorem A.I.23. *A matrix $A = (a_{kl})_{k,l=1,\ldots,n}, a_{kl} \in \mathbb{R}$, is the matrix of a positive definite (positive semi-definite, negative definite, negative semi-definite) symmetric bilinear form on an $\mathbb{R}$-vector space $V, \dim V = n$, if and only if all its eigenvalues $\lambda_1, \ldots, \lambda_n$ (counted according to their multiplicity) are strictly positive (non-negative, strictly negative, non-positive).*

We also note that in $\mathbb{R}^n$ equipped with the Euclidean scalar product for every $n \times n$ matrix A we have

$$\langle Ax, y \rangle = \langle x, A^t y \rangle \tag{A.I.55}$$

whereas for a symmetric matrix A we have

$$\langle Ax, y \rangle = \langle x, Ay \rangle. \tag{A.I.56}$$

In Theorem 9.12 we discussed further characterisations of positive definite matrices and some of these will now be discussed in more detail. For this we will change once more our point of view, geometry will become more important and in this connection also properties invariant under certain changes of bases or under certain groups of transformations.

We know that in $\mathbb{R}^2$ and $\mathbb{R}^3$ the Euclidean scalar product $\langle \cdot, \cdot \rangle$ has the representation

$$\langle x, y \rangle = \|x\|\|y\| \cos \varphi \tag{A.I.57}$$

where φ is the angle between x and y (in mathematical positive orientation). In particular $\langle x, y \rangle = 0, x, y \neq 0$ if $|\cos \varphi| = 0$, i.e. x and y are orthogonal or perpendicular. Furthermore $|\langle x, y \rangle| = \|x\|\|y\|$ if and only if x and y are linearly dependent, i.e. $x = \lambda y$. The latter can be proved in $\mathbb{R}^n$ and using the Euclidean scalar product $\langle \cdot, \cdot \rangle$ in $\mathbb{R}^n$ we can now define the cosine of the angle between x and y, $x, y \in \mathbb{R}^n \backslash \{0\}$, as

$$\cos \varphi = \frac{\langle x, y \rangle}{\|x\|\|y\|}. \tag{A.I.58}$$

In particular we call $x, y \in \mathbb{R}^n \backslash \{0\}$ **orthogonal** if $\langle x, y \rangle = 0$. Since $\|x\| := \langle x, x \rangle^{1/2}$ is a norm on $\mathbb{R}^n$, in the Euclidean space, i.e. $\mathbb{R}^n$ equipped with the Euclidean scalar product, we can handle the length $\|x\|$ of vectors, hence the distance $\|x - y\|$ of two vectors, and the angle of two vectors $x, y \in \mathbb{R}^n \backslash \{0\}$. This can be extended to any **scalar product space**, i.e. a finite dimensional vector space V over $\mathbb{R}$ with scalar product $\langle \cdot, \cdot \rangle_V$ and corresponding norm $\|.\|_V$. Again $\|x\|_V$ and $\|x - y\|_V$ will determine length and distance whereas $\cos \varphi = \frac{\langle x, y \rangle_V}{\|x\|_V \|y\|_V}$ allows us to introduce an angle between x and y. In particular in a scalar product space the notion of **orthogonality** is defined for $x, y \in V \backslash \{0\}$ by

$$x \perp_V y \text{ if and only if } \langle x, y \rangle_V = 0. \tag{A.I.59}$$

(If the scalar product is fixed or clear from the context we write just $x \perp y$.) We may ask for those transformations of V, i.e. linear mappings $T : V \to V$, which leave these objects, i.e. length, distance and angle, invariant. This reduces to the question of which linear mappings leave the scalar product invariant, i.e. when does

$$\langle Tx, Ty \rangle_V = \langle x, y \rangle_V \text{ for all } x, y \in V, \tag{A.I.60}$$

where $T : V \to V$ is linear? Clearly, id_V has this property. If T satisfies (A.I.60) then

$$\|Tx\|_V^2 = \langle Tx, Tx \rangle_V = \langle x, x \rangle_V = \|x\|_V^2,$$

i.e. T preserves length and distance. In particular since $\|x\|_V = 0$ if and only if $x = 0$ it follows that $Tx = 0$ if and only if $x = 0$, i.e. T is bijective.

Moreover if both T and S satisfy (A.I.60) then

$$\langle T \circ Sx, T \circ Sy \rangle_V \; = \; \langle Tx, Ty \rangle_V \; = \; \langle x, y \rangle_V$$

as well as

$$< T^{-1}x, T^{-1}y >_V \; = \; < TT^{-1}x, TT^{-1}y >_V \; = \; < x, y >_V \; .$$

Thus the invariance of the scalar product $\langle \cdot, \cdot \rangle_V$ leads to a family of linear transformations $T : V \to V$ which are invertible, includes the identity and are closed under composition, hence they form a group with respect to composition. We prefer to now study this group in $\mathbb{R}^n$ with the Euclidean scalar product. Denote by $M(n, \mathbb{R})$ the algebra of all $n \times n$ matrices over $\mathbb{R}$ and by $GL(n; \mathbb{R})$ the **general linear group**, i.e. all invertible matrices $A \in M(n, \mathbb{R})$.

Definition A.I.24. *The subgroup $O(n) \subset GL(n; \mathbb{R})$ consisting of all matrices leaving the Euclidean scalar product in $\mathbb{R}^n$ invariant is called the **orthogonal group** in n dimensions.*

A matrix $T \in GL(n; \mathbb{R})$ belongs to $O(n)$ if and only if $T^{-1} = T^t$ which follows from

$$\langle x, y \rangle \; = \; \langle Tx, Ty \rangle \; = \; \langle T^t T x, y \rangle,$$

and therefore we find

$$(\det T)^2 = \det(TT^t) = \det(\mathrm{id}_n) = 1,$$

i.e. $|\det T| = 1$ for $T \in O(n)$. We call $T \in O(n)$ with $\det T = 1$ a rotation. The set $SO(n)$ of all rotations form a subgroup of $O(n)$ called the **special orthogonal group** in n dimension. If $H \subset \mathbb{R}^n$ is a hyperplane passing through the origin $O \in \mathbb{R}^n$, i.e. H is an $(n-1)$-dimensional subspace, we can define the reflexion S_H at H which belongs to $O(n)$ and has the properties $\det S_H = -1$ and $S_H^2 = \mathrm{id}$. Moreover, if $T \in O(n), \det T = -1$ then there exists a reflexion S_H and a rotation $R \in O(n)$ such that $T = S_H \circ R$.
As $T \in O(n)$ is bijective it maps a basis $B = \{b_1, \ldots, b_n\}$ of $\mathbb{R}^n$ into a basis $D = \{d_1, \ldots, d_n\}, d_j = Tb_j$, and we have

$$\|d_j\| = \|b_j\| \text{ as well as } \langle d_k, d_l \rangle \; = \; \langle b_k, b_l \rangle,$$

i.e. basis vectors in D have the same length as their pre-images in B and the cosine of the angle between d_k and d_l is the same as that between b_k and b_l.

Definition A.I.25. A. *A basis $B = \{b_1, \ldots, b_n\}$ in an n-dimensional scalar product space $(V, \langle \cdot, \cdot \rangle_V)$ is called an **orthogonal basis** if*

$$\langle b_k, b_l \rangle_V = \delta_{kl} = \begin{cases} 1, k = l \\ 0, k \neq l. \end{cases} \quad \text{In this case we have } \|b_k\|_V = 1, \text{ i.e. } b_k \text{ has}$$

length 1.

B. *A basis $B = \{b_1, \ldots, b_n\}$ of $\mathbb{R}^n$ is called **positively oriented** or to have **positive orientation** if the determinant of the matrix $(b_1, \ldots, b_n), b_j$ being the jth column vector of this matrix, is positive. Otherwise we call B **negatively oriented**.*

Theorem A.I.26. *If $T \in O(n)$ then it maps an orthogonal basis of $\mathbb{R}^n$ onto an orthonormal basis, conversely if a matrix maps orthonormal bases onto orthonormal bases it must be an element of $O(n)$. Furthermore, T preserves the orientation if and only if $T \in SO(n)$.*

This result implies in particular that the column vectors of $T \in O(n)$ form an orthogonal basis of $\mathbb{R}^n$ and every orthonormal basis in $\mathbb{R}^n$ gives rise, in this way, to an element of $O(n)$.

The following result is of central importance and has far reaching generalisations such as the "spectral theorem" in functional analysis and operator theory, see Volume V.

Theorem A.I.27. *Let A be a symmetric $n \times n$ matrix over $\mathbb{R}$. Then there exists an orthonormal matrix $U \in O(n)$ such that*

$$U^t A U = U^{-1} A U = \begin{pmatrix} \lambda_1 & & 0 \\ & \ddots & \\ 0 & & \lambda_n \end{pmatrix} \quad \text{(A.I.61)}$$

is a diagonal matrix with eigenvalues $\lambda_1, \ldots, \lambda_n \in \mathbb{R}$ of A each counted according to its multiplicity. Moreover $\mathbb{R}^n$ has an orthogonal basis consisting of eigenvectors of A. If β_A is the bilinear form associated with A, with respect to the basis of eigenvectors of A it is a diagonal form, i.e.

$$\beta_{U^t A U}(x, y) = \sum_{k=1}^{n} \lambda_k x_k y_k.$$

From Theorem A.I.27 it is straightforward to Theorem 9.12.

A very interesting and useful result is the following **polar decomposition** of non-singular matrices A, i.e. $A \in GL(n; \mathbb{R})$.

Theorem A.I.28. *For $A \in GL(n; \mathbb{R})$ there exists $U \in O(n)$ and a symmetric positive definite matrix S such that $A = US$.*

Certain of these considerations do hold in analogous form in the complex case. We have to replace symmetric matrices by **Hermitian matrices**. If A is the $n \times n$-complex matrix

$$A = \begin{pmatrix} a_{11} & \cdots & a_{1n} \\ \vdots & & \\ a_{n1} & \cdots & a_{nn} \end{pmatrix}$$

we define

$$A^\dagger = \begin{pmatrix} \overline{a}_{11} & \cdots & \overline{a}_{n1} \\ \vdots & & \\ \overline{a}_{1n} & \cdots & \overline{a}_{nn} \end{pmatrix}.$$

We call A **Hermitian** if $A = A^\dagger$, and we can also speak of a **Hermitian bilinear form** on $\mathbb{C}^n$:

$$\beta_A(x, y) := \langle Ax, y \rangle, \quad x, y \in \mathbb{C}^n$$

where $\langle \cdot, \cdot \rangle$ is now the scalar product in $\mathbb{C}^n$ and $A = A^\dagger$. Note: the name Hermitian bilinear form, although commonly used should be replaced by the **Hermitian sesquilinear form** since $\langle \cdot, \cdot \rangle$ is not bilinear in $\mathbb{C}^n$ but sesquilinear, i.e. linear in the first argument, additive in the second but "conjugate homogeneous" in the second argument $\langle z, \lambda w \rangle = \overline{\lambda} \langle z, w \rangle$.

The analogue to $O(n)$ and $SO(n)$ are the **unitary group** $U(n)$ and the **special unitary group** $SU(n)$. A complex $n \times n$ matrix U belongs to $U(n)$ if it leaves the unitary scalar product $< \cdot, \cdot >$ in $\mathbb{C}^n$ invariant i.e.

$$\langle Ux, Uy \rangle = \langle x, y \rangle \text{ for all } x, y \in \mathbb{C}^n,$$

which is equivalent to $U^\dagger = U^{-1}$, and $U \in U(n)$ belongs to $SU(n)$ if $\det U = 1$, whereas for $U \in U(n)$ we have $\det U \in S^1 = \{z \in \mathbb{C} \mid |z| = 1\}$.

A more thorough discussion about Hermitian forms, $U(n)$ and related topics will be given in Volume III.

Finally we turn to the cross product in $\mathbb{R}^3$. We only want to list some of its basic properties, we refer also to Chapter 16 where we discuss its relation to volume.

The **cross product** for $a = (a_1, a_2, a_3)$, $b = (b_1, b_2, b_3) \in \mathbb{R}^3$ is defined by

$$a \times b = (a_2 b_3 - a_3 b_2, a_3 b_1 - a_1 b_3, a_1 b_2 - a_2 b_1) \tag{A.I.62}$$

and it has the following properties for $a, b, c \in \mathbb{R}^3$ and $\lambda, \mu \in \mathbb{R}$

$$a, b \in \mathbb{R}^3 \text{ are independent if and only if } a \times b \neq 0; \tag{A.I.63}$$

$$\|a \times b\| = \|a\| \|b\| \sin \varphi,$$

$\varphi \in [0, \pi]$ is the angle between a and b we assume $\|a\| \neq 0 \neq \|b\|$; (A.I.64)

$$(a \times b) \perp \lambda a \text{ and } (a \times b) \perp \mu b$$

for all $\lambda, \mu \in \mathbb{R}$ and independent vectors $a, b \in \mathbb{R}^3$; (A.I.65)

$$(\lambda a + \mu b) \times c = \lambda (a \times c) + \mu (b \times c) \text{ and } a \times (\lambda b + \mu c) = \lambda (a \times b) + \mu (a \times c) \tag{A.I.66}$$

which is a linearity statement;

$$(a \times b) \times c = \langle a, c \rangle b - \langle b, c \rangle a; \tag{A.I.67}$$

$$a \times b = -b \times a; \tag{A.I.68}$$

$$(a \times b) \times c + (b \times c) \times a + (c \times a) \times b = 0. \tag{A.I.69}$$

In particular, the cross product is neither commutative nor associative. A simple way to remember how to calculate $a \times b$ is to look at the very formal determinant

$$a \times b = \text{``det''} \begin{pmatrix} e_1 & e_2 & e_3 \\ a_1 & a_2 & a_3 \\ b_1 & b_2 & b_3 \end{pmatrix}$$

$$= e_1 \det \begin{pmatrix} a_2 & a_3 \\ b_2 & b_3 \end{pmatrix} - e_2 \det \begin{pmatrix} a_1 & a_3 \\ b_1 & b_3 \end{pmatrix} + e_3 \det \begin{pmatrix} a_1 & a_2 \\ b_1 & b_2 \end{pmatrix}$$

$$= (a_2 b_3 - a_3 b_2) e_1 + (a_3 b_1 - a_1 b_3) e_2 + (a_1 b_2 - a_2 b_1) e_3$$

where $\{e_1, e_2, e_3\}$ is the canonical basis in $\mathbb{R}^3$.

The cross product, as useful as it is in $\mathbb{R}^3$, looks at first glance strange as it does not have an obvious analogue in other dimensions. We will see in Volume VI that the proper context to understand the cross product is the theory of alternating multilinear forms.

Appendix II: Two Postponed Proofs of Part 3

In this appendix we will provide the complete proofs of Theorem 3.18 and Proposition 14.10.

As a preparation of the proof of Theorem 3.18 and reformulating Problem 2 in Chapter 3:

Lemma A.II.1. *Let* $(x_n)_{n\in\mathbb{N}}$ *be a Cauchy sequence in the metric space* (X, d). *If* x_0 *is an accumulation point of* $(x_n)_{n\in\mathbb{N}}$ *then* $(x_n)_{n\in\mathbb{N}}$ *converges to* x_0.

Proof of Theorem 3.18

We start with proving that i) implies ii). So we assume that X is compact and $(x_k)_{n\in\mathbb{N}}$ is a sequence in X. Denote by A_n the closure of $\{x_k \mid k \geq n\}$. We claim that $\bigcap_{n\in\mathbb{N}} A_n \neq \emptyset$. Suppose that $\bigcap_{n\in\mathbb{N}} A_n = \emptyset$. In this case the open sets $U_n := A_n^{\complement}, n \in \mathbb{N}$, would form an open covering of X,

$$X = \bigcup_{n\in\mathbb{N}} U_n = \bigcup_{n\in\mathbb{N}} A_n^{\complement} = (\bigcap_{n\in\mathbb{N}} A_n)^{\complement}.$$

Since X is compact there would exist a finite subcovering $U_{n_0}, \ldots, U_{n_M}$ of X implying that $\bigcap_{j=0}^{M} A_{n_j} \neq \emptyset$. However $\emptyset \neq A_\kappa \subset A_j$ for all $j = 0, \ldots, M$ where $\kappa = \max\{n_0, \ldots, n_M\}$. This is however a contradiction. Hence there exists $x_0 \in \bigcap_{n\in\mathbb{N}} A_n$, but by construction x_0 is an accumulation point of $(x_n)_{n\in\mathbb{N}}$.

Now we prove that ii) implies iii) or that if every infinite sequence in X has an accumulation point then X is complete and totally bounded. If $(x_n)_{n\in\mathbb{N}}$ is a Cauchy sequence, by assumption it must have an accumulation point x_0. Hence every Cauchy sequence in X has a limit, i.e. X is complete. We now prove that X is totally bounded. If not, for some $\varepsilon > 0$ we could not cover X with a finite number of open balls with radius ε. We define a sequence $(x_n)_{n\in\mathbb{N}}$ inductively as follows: assuming $d(x_i, x_j) \geq \varepsilon$ for $i \neq j$ and $1 \leq i \leq n - 1, 1 \leq j \leq n - 1$, the union of open balls with centre $x_i, 1 \leq i \leq n - 1$, and radius ε does not cover X, hence there exists $x_n \in X$ such that $d(x_i, x_n) \geq \varepsilon$ for $i < n$. We claim that $(x_n)_{n\in\mathbb{N}}$ does not have an accumulation point. Suppose that x_0 is an accumulation point of $(x_n)_{n\in\mathbb{N}}$ and that the sequence $(x_{n_k})_{k\in\mathbb{N}}$ converges to x_0. Then for some k_0 it will follow that $d(x_0, x_{n_k}) \leq \frac{\varepsilon}{2}$ and therefore $d(x_{n_j}, x_{n_k}) < \varepsilon$ for $j \geq k_0$ and $k \geq k_0, j \neq k$, which contradicts the definition of $(x_n)_{n\in\mathbb{N}}$. Finally we show that iii) implies i), i.e. if X is complete and totally bounded then X is compact. Suppose that $(U_j)_{j \in J}$ is an open covering of X such that no finite subcovering of $(U_j)_{j \in J}$ will cover X too. We construct the following family of open balls $(K_n)_{n \in \mathbb{N}} : K_{n-1}$ has radius $\frac{1}{2^{n-1}}$ and no finite number of elements of $(U_j)_{j \in J}$ will cover $K_{n-1}, K_n \subset K_{n-1}$ with the analogous properties. Since X is by assumption totally bounded we can find a finite cover of X with open balls $V_k, 1 \leq k \leq n$, having radius $\frac{1}{2^n}$. Among the balls V_k having a non-empty intersection with K_{n-1} we can choose one to be K_n, i.e. one which cannot be covered by a finite subfamily of $(U_j)_{j \in J}$. Indeed, if each of the V_k's can be covered by a finite subfamily of $(U_j)_{j \in J}$, then K_{n-1} could be covered by a finite subfamily of $(U_j)_{j \in J}$. Let x_n be the centre of K_n. Since $K_n \cap K_{n-1} \neq \emptyset$ the triangle inequality yields $d(x_{n-1}, x_n) \leq \frac{1}{2^{n-1}} + \frac{1}{2^n} \leq \frac{1}{2^{n-2}}$. For $n \leq \nu < \mu$ it now follows that

$$d(x_\nu, x_\mu) \leq d(x_\nu, x_{\nu+1}) + \cdots + d(x_{\mu-1}, x_\mu) \leq \frac{1}{2^{\nu-1}} + \cdots + \frac{1}{2^{\mu-1}} \leq \frac{1}{2^{n-2}},$$

which means that $(x_n)_{n \in \mathbb{N}}$ is a Cauchy sequence in X and hence by the completeness of X it has a limit x_0. Let $j_0 \in J$ such that $x_0 \in U_{j_0}$. Then there exists $\varepsilon > 0$ such that $B_\varepsilon(x_0) \subset U_{j_0}$. Moreover, the definition of x_0 implies the existence of $n_0 \in \mathbb{N}$ such that $d(x_0, x_n) < \frac{\varepsilon}{2}$ and $\frac{1}{2^n} < \frac{\varepsilon}{2}$. Again, by the triangle inequality we have to deduce that $K_n \subset K_\varepsilon(x_0) \subset U_{j_0}$ which however is a contradiction to the fact that K_n cannot be covered by a finite number of sets belonging to $(U_j)_{j \in \mathbb{N}}$. $\qquad \square$

(This proof is taken from J. Dieudonne [10].)

In order to fill the gap in the proof of Proposition 14.10 we need to show

Lemma A.II.2. *There exists a sequence of polynomials $(p_k)_{k \in \mathbb{N}_0}$ such that $p_k|_{[-1,1]}$ converges uniformly on $[-1,1]$ to $|.|$, i.e.*

$$\lim_{k \to \infty} \left(\sup_{x \in [-1,1]} |p_k(x) - |x|| \right) = 0.$$

The proof of this Lemma uses the Theorem of Dini which we have discussed in Problem 15 in Chapter 3.

Proof of Lemma A.II.2

(Using H. Bauer [6]). Consider the sequence of polynomials defined by

$p_0(x) = 0$ and

$$p_{k+1}(x) := p_k(x) + \frac{1}{2}(x^2 - p_k^2(x)), k = 0, 1, \dots \qquad \text{(A.II.1)}$$

We want to prove that this sequence converges on [-1,1] uniformly to $|.|$. The strategy is to prove that on $[-1, 1]$ the sequence $(p_k)_{k \in \mathbb{N}_0}$ is pointwise increasing and bounded from above, i.e. $p_k(x) \le p_{k+1}(x)$ for $x \in [-1, 1]$ and $p_k(x) \le M$ for all $k \in \mathbb{N}$. This implies that $(p_k)_{k \in \mathbb{N}_0}$ has a pointwise limit $q(x), x \in [-1, 1]$, which by Dini's theorem must also be a uniform limit on $[-1, 1]$. Passing in (A.II.1) to the limit as $k \to \infty$ we obtain

$$q(x) = q(x) + \frac{1}{2}(x^2 - q^2(x))$$

or $x^2 = q(x)$ implying $q(x) = |x|$.

First we use mathematical induction to derive $-|x| \le p_k(x) \le |x|$. From (A.II.1) we deduce

$$|x| - p_{k+1}(x) = |x| - p_k(x) - \frac{1}{2}(|x^2| - p_k^2(x)) = (|x| - p_k(x))(1 - \frac{1}{2}(|x| + p_k(x))) \qquad \text{(A.II.2)}$$

and

$$|x| + p_{k+1}(x) = |x| + p_k(x) + \frac{1}{2}(|x|^2 - p_k^2(x)) = (|x| + p_k(x))(1 + \frac{1}{2}(|x| + p_k(x))). \qquad \text{(A.II.3)}$$

Clearly $-|x| \le p_0 \le |x|$ for $x \in [-1, 1]$, and now assume that $-|x| \le p_k(x) \le |x|$ for all $x \in [-1, 1]$. This implies $0 \le |x| - p_k(x) \le 2$ and $0 \le |x| + p_k(x) \le 2$ and it follows that $|x| - p_{k+1}(x) \ge 0$ whereas (A.II.3) implies $|x| + p_{k+1}(x) \ge 0$ for $x \in [-1, 1]$. Hence $-|x| \le p_{k+1}(x) \le |x|$. We find in particular that $|x|^2 - p_k^2(x) \ge 0$ for all $x \in [-1, 1]$ and all $k \in \mathbb{N}_0$, and therefore by the definition of p_k we arrive at $p_{k+1}(x) \ge p_k(x)$. In addition we have that $|p_k(x)| \le 1$ for all $x \in [-1, 1]$. Thus $(p_k)_{k \in \mathbb{N}_0}$ satisfies all properties outlined in the beginning of the proof and the result follows. □

Solutions to Problems of Part 3

Chapter 1

1. For $\epsilon > 0$ and $t, s \geq 0$ we find

$$f(s + t + \epsilon) - f(t + \epsilon) = \int_{t+\epsilon}^{s+t+\epsilon} f'(r) \, dr$$

$$= \int_{\epsilon}^{s+\epsilon} f'(t + r) \, dr \leq \int_{\epsilon}^{s+\epsilon} f'(r) \, dr = f(s + \epsilon) - f(\epsilon),$$

implying

$$f(s + t + \epsilon) \leq f(s + \epsilon) + f(t + \epsilon) - f(\epsilon)$$

and for $\epsilon \to 0$ we obtain $f(s + t) \leq f(s) + f(t)$.

Now we prove that $d_f(x, y) := f(\|x - y\|)$ is a metric on $\mathbb{R}^n$. Clearly $d_f(x, y) \geq 0$ and $d_f(x, y) = 0$ implies $f(\|x - y\|) = 0$, i.e. $\|x - y\| = 0$, or $x = y$. Furthermore we have

$$d_f(x, y) = f(\|x - y\|) = f(\|y - x\|) = d_f(y, x).$$

Finally, the triangle inequality follows from

$$d_f(x, y) = f(\|x - y\|) = f(\|x - z + z - y\|)$$
$$\leq f(\|x - z\| + \|z - y\|)$$
$$\leq f(\|x - z\|) + f(\|z - y\|) = d_f(x, z) + d_f(z, y).$$

2. First we investigate $d^{(\alpha)}$. For $\alpha \leq 1$ we can apply the result of Problem 1 when using $f_\alpha : [0, \infty] \to \mathbb{R}$, $f_\alpha(t) = t^\alpha$, which satisfies all assumptions, hence $f_\alpha(\|x - y\|) = \|x - y\|^\alpha$ is for every norm $\|.\|$ on $\mathbb{R}^n$ a metric on $\mathbb{R}^n$.

For $\alpha > 1$ we claim that $d^{(\alpha)}(x, y) = \|x - y\|^\alpha$ is not a metric on $\mathbb{R}^n$. Clearly, only the triangle inequality may fail to hold. Suppose $\alpha > 1$ and that for all $x, yz \in \mathbb{R}^n$

$$\|x - y\|^\alpha \leq \|x - z\|^\alpha = \|y - z\|^\alpha$$

holds for a norm $\|.\|$ on $\mathbb{R}^n$. For $z = 0$ this implies

$$\|x - y\|^\alpha \leq \|x\|^\alpha + \|y\|^\alpha. \tag{*}$$

Now we choose $x = -y$ and observe

$$\left\| \frac{x}{\|x\|} - \frac{y}{\|y\|} \right\|^\alpha = \left\| \frac{2x}{\|x\|} \right\|^\alpha = 2^\alpha$$

as well as

$$\left\| \frac{x}{\|x\|} \right\|^\alpha = \left\| \frac{y}{\|y\|} \right\|^\alpha = \frac{1}{\|y\|^\alpha} \|y\|^\alpha = 1.$$

Thus (*) would imply $2^\alpha \leq 2$ which for $\alpha > 1$ is a contradiction. Therefore $d^{(\alpha)}$ is for $\alpha > 1$ not a metric in $\mathbb{R}^n$.

Next we ask when $\|.\|_{(\alpha)}$ is a norm. For $\alpha \geq 1$ we know by Corollary I.23.18 that

$$\|x\|_{(\alpha)} := \left(\sum_{j=1}^{n} |x_j|^\alpha \right)^{\frac{1}{\alpha}}$$

is a norm. We claim that for $\alpha < 1$ the term $\|.\|$ does not define a norm on $\mathbb{R}^n$. Again, only the triangle inequality may fail. Suppose for $\alpha < 1$ and all $x, y \in \mathbb{R}^n$ the following holds

$$\|x + y\|_{(\alpha)} \leq \|x\|_{(\alpha)} + \|y\|_{(\alpha)} , \tag{**}$$

or

$$\left(\sum_{j=1}^{n} |x_j + y_j|^\alpha \right)^{\frac{1}{\alpha}} \leq \left(\sum_{j=1}^{n} |x_j|^\alpha \right)^{\frac{1}{\alpha}} + \left(\sum_{j=1}^{n} |y_j|^\alpha \right)^{\frac{1}{\alpha}} .$$

Choosing $x_j = y_j = 0$ for $j > 2$ we consider the case $n = 2$, i.e.

$$(|x_1 + y_1|^\alpha + |x_2 + y_2|^\alpha)^{\frac{1}{\alpha}} \leq (|x_1|^\alpha + |x_2|^\alpha)^{\frac{1}{\alpha}} + (|y_1|^\alpha + |y_2|^\alpha)^{\frac{1}{\alpha}} .$$

With $x_1 = y_2 = 1$ and $x_2 = y_1 = 0$ we find

$$(1^\alpha + 1^\alpha)^{\frac{1}{\alpha}} = 2^{\frac{1}{\alpha}} \leq 1 + 1 = 2,$$

but $\alpha < 1$, i.e. $\frac{1}{\alpha} > 1$ and again we have a contradiction. Consequently, for $\alpha < 1$ the term $\|x\|_{(\alpha)}$ does not define a norm on $\mathbb{R}^n$.

3. Clearly we have $d((a_k)_{k\in\mathbb{N}}, (b_k)_{k\in\mathbb{N}}) \geq 0$ and $d((a_k)_{k\in\mathbb{N}}, (b_k)_{k\in\mathbb{N}}) = 0$ implies $\sum_{k=1}^{\infty} 2^{-k} |a_k - b_k| = 0$ or $|a_k - b_k| = 0$ for all $k \in \mathbb{N}$, which gives $a_k = b_k$ for all $k \in \mathbb{N}$, i.e. $(a_k)_{k\in} = (b_k)_{k\in\mathbb{N}}$. Furthermore $d((a_k)_{k\in\mathbb{N}}, (b_k)_{k\in\mathbb{N}}) = \sum_{k=1}^{\infty} 2^{-k} |a_k - b_k| = \sum_{k=1}^{\infty} 2^{-k} |b_k - a_k| = d((b_k)_{k\in\mathbb{N}}, (a_k)_{k\in\mathbb{N}})$, i.e. we have proved the symmetry. Finally, for $(a_k)_{k\in\mathbb{N}}, (b_k)_{k\in\mathbb{N}}$ and $(c_k)_{k\in\mathbb{N}}$ we find

$$d((a_k)_{k\in\mathbb{N}}, (b_k)_{k\in\mathbb{N}}) = \sum_{k=1}^{\infty} 2^{-k} |a_k - b_k|$$

$$= \sum_{k=1}^{\infty} 2^{-k} |a_k - c_k + c_k - b_k| \leq \sum_{k=1}^{\infty} 2^{-k} \left(|a_k - c_k| + c_k - b_k| \right)$$

$$= \sum_{j=1}^{\infty} 2^{-k} |a_k - c_k| + \sum_{j=1}^{\infty} 2^{-k} |c_k - b_k|$$

$$= d((a_k)_{k\in\mathbb{N}}, (c_k)_{k\in\mathbb{N}}) + d((c_k)_{k\in\mathbb{N}}, (b_k)_{k\in\mathbb{N}}).$$

4. Obviously $d_\gamma(f, g) \geq 0$ and if $d_\gamma(f, g) = 0$ it follows that

$$\gamma(x)|f(x) - g(x)| \leq \sup_{x\in[a,b]} \gamma(x)|f(x) - g(x)| = 0.$$

590

Since $\gamma(x) > 0$ we deduce $f(x) = g(x)$ for all $x \in [a, b]$, i.e. $f = g$. Moreover we have

$$d_\gamma(f, g) = \sup_{x \in [a,b]} \gamma(x)|f(x) - g(x)| = \sup_{x \in [a,b]} \gamma(x)|g(x) - f(x)| = d_\gamma(g, f),$$

hence d_γ is symmetric. Finally, for $f, g, h \in C([a, b])$ we find

$$\begin{aligned}
d_\gamma(f, g) &= \sup_{x \in [a,b]} \gamma(x)|f(x) - g(x)| = \sup_{x \in [a,b]} \gamma(x)|f(x) - h(x) + h(x) - g(x)| \\
&\leq \sup_{x \in [a,b]} \gamma(x)\left(|f(x) - h(x)| + |h(x) - g(x)|\right) \\
&\leq \sup_{x \in [a,b]} \gamma(x)|f(x) - h(x)| + \sup_{x \in [a,b]} \gamma(x)|h(x) - g(x)| = d_\gamma(f, h) + d_\gamma(h, g).
\end{aligned}$$

Thus d_γ is a metric on $C([a, b])$.

5. For $(v_1, \ldots, v_n) \in V$ it follows that

$$\|(v_1, \ldots, v_n)\|_V = \sum_{j=1}^n \|v_j\|_j \geq 0$$

and if $\|(v_1, \ldots, v_n)\|_V = 0$ then $\|v_j\|_j = 0$ for all $j = 1, \ldots, n$, i.e. $v_j = 0 \in V_j$ implying that $(v_1, \ldots, v_n) = 0 \in V$. Moreover, for $\lambda \in \mathbb{R}$ we find

$$\begin{aligned}
\|\lambda(v_1, \ldots, v_n)\|_v &= \sum_{j=1}^n \|\lambda v_j\|_j = \sum_{j=1}^n |\lambda| \|v_j\|_j \\
&= |\lambda| \sum_{j=1}^n \|v_j\|_j = |\lambda| \|(v_1, \ldots, v_n)\|_V.
\end{aligned}$$

Eventually we observe for $(v_1, \ldots, v_n), (w_1, \ldots, w_n) \in V$

$$\begin{aligned}
\|(v_1, \ldots, v_n) + (w_1, \ldots, w_n)\|_V &= \|(v_1 + w_1, \ldots, v_n + w_n)\| \\
&= \sum_{j=1}^n \|v_j + w_j\|_j \leq \sum_{j=1}^n \left(\|v_j\|_j + \|w_j\|_j\right) \\
&= \|(v_1, \ldots, v_n)\|_V + \|(w_1, \ldots, w_n)\|_V
\end{aligned}$$

which yields that $\|.\|_V$ is indeed a norm.

6. First we note that it is sufficient to prove with $\delta_q > 0$

$$\gamma_q \|x\|_q \leq \|x\|_1 \leq \delta_q \|x\|_q \tag{$*$}$$

for all $q \geq 1$. Instead $(*)$ implies

$$\|x\|_q \leq \frac{1}{\gamma_q} \|x\|_1 \leq \frac{\delta_p}{\gamma_q} \|x\|_p$$

591

and

$$\|x\|_p \le \frac{1}{\gamma_p} \|x\|_1 \le \frac{\delta_q}{\gamma_p} \|x\|_q$$

i.e.

$$\|x\|_q \le \frac{\delta_p}{\gamma_q} \|x\|_p \le \frac{\delta_p \delta_q}{\gamma_p} \gamma_q \|x\|_q$$

or

$$\frac{\gamma_q}{\delta_p} \|x\|_q \le \|x\|_p \le \frac{\delta_q}{\gamma_p} \|x\|_q .$$

Now, for $q > 1$ and $q' = \frac{q}{q-1}$, i.e. $\frac{1}{q} + \frac{1}{q'} = 1$ we find using Hölder's inequality

$$\|x\|_1 = \sum_{j=1}^n |x_j| \cdot 1 \le \left(\sum_{j=1}^n |x_j|^q \right)^{\frac{1}{q}} \left(\sum_{j=1}^n 1^{q'} \right)^{\frac{1}{q'}} = n^{\frac{q-1}{q}} \|x\|_q ,$$

i.e. we find that $\delta_q = n^{\frac{q-1}{q}}$ in (*).

In order to prove the lower bound we first note that for $a, b \ge 0$ and $p > 1$ we have that

$$(a^p + b^p) \le (a + b)^p. \tag{**}$$

This is equivalent to $\left(1 + \left(\frac{b}{a} \right)^p \right) \le \left(1 + \frac{b}{a} \right)^p$ for $p > 1$ and $0 < \frac{b}{a} \le 1$. The function $f_p(\lambda) = (1 + \lambda)^p - (1 + \lambda^p)$ is on $[0, 1]$ monotone increasing since $f_p(0) = 0$ and for $\lambda > 0$ we have $f'(\lambda) = p(1 + \lambda)^{p-1} - p\lambda^{p-1}$. Hence $f_p(\lambda) \ge 0$ and (**) is proved. By induction we now find

$$\sum_{j=1}^n |x_j|^p = \left(\sum_{j=1}^{n-1} |x_j|^p + |x_n|^p \right)$$

$$\le \left(\sum_{j=1}^{n-1} |x_j|^p \right)^p + |x_n|^p$$

$$\le \left(\sum_{j=1}^{n-1} |x_j| + |x_n| \right)^p$$

$$= \left(\sum_{j=1}^n |x_j| \right)^p ,$$

which yields the first inequality in (*) with $\gamma_p = 1$.

7. We know that $\varphi : M(n; \mathbb{R}) \to \mathbb{R}^{n^2}$ defined by

$$\varphi(A) := (a_{11}, \dots, a_{1n}, a_{21}, \dots, a_{2n}, \dots, a_{n1}, \dots, a_{nn})$$

592

is a vector space isomorphism and $\|A\| = \|\varphi(A)\|_{\mathbb{R}^{n^2}}$, where for the moment $\|.\|_{\mathbb{R}^{n^2}}$ denotes the Euclidean norm in $\mathbb{R}^{n^2}$. Therefore it follows

$$\|A\| \geq 0 \quad \text{and} \quad \|A\| = 0 \quad \text{implies} \quad \varphi(A) = 0, \text{ i.e. } A = 0;$$

$$\|\lambda A\| = \|\varphi(\lambda A)\|_{\mathbb{R}^{n^2}} = \|\lambda \varphi(A)\|_{\mathbb{R}^{n^2}} = |\lambda| \, \|\varphi(A)\|_{\mathbb{R}^{n^2}} \leq |\lambda| \, \|A\|;$$

and

$$\|A + B\| = \|\varphi(A + B)\|_{\mathbb{R}^{n^2}} = \|\varphi(A) + \varphi(B)\|_{\mathbb{R}^{n^2}}$$
$$\leq \|\varphi(A)\|_{\mathbb{R}^{n^2}} + \|\varphi(B)\|_{\mathbb{R}^{n^2}} = \|A\| + \|B\|.$$

Moreover, for $x \in \mathbb{R}^n$ we find using the Cauchy-Schwarz inequality

$$\|Ax\|_2 = \left\| \begin{pmatrix} \sum_{l=1}^n a_{1l}x_l \\ \vdots \\ \sum_{l=1}^n a_{nl}x_l \end{pmatrix} \right\| = \left(\sum_{k=1}^n \left(\sum_{l=1}^n a_{kl}x_l \right)^2 \right)^{\frac{1}{2}}$$

$$\leq \left(\sum_{k=1}^n \left(\left(\sum_{l=1}^n a_{kl}^2 \right)^{\frac{1}{2}} \left(\sum_{l=1}^n x_l^2 \right)^{\frac{1}{2}} \right)^2 \right)^{\frac{1}{2}}$$

$$= \left(\sum_{k=1}^n \left(\sum_{l=1}^n a_{kl}^2 \right) \left(\sum_{l=1}^n x_l^2 \right) \right)^{\frac{1}{2}}$$

$$= \left(\sum_{k=1}^n \sum_{l=1}^n a_{kl}^2 \right)^{\frac{1}{2}} \left(\sum_{l=1}^n x_l^2 \right)^{\frac{1}{2}} = \|A\| \, \|x\|_2.$$

8.　a)　We have to prove:

(i) $\|p\| = 0$ if and only if p is the polynomial $p(x) = 0$ for all $x \in \mathbb{R}$, i.e. p is the null-polynomial;

(ii) $\|\lambda p\| = |\lambda| \, \|p\|$ for all $\lambda \in \mathbb{R}$ and $p \in P$;

(iii) $\|p + q\| \leq \|p\| + \|q\|$ for all $p, q \in P$.

(i) If $p(x) = \sum_{k=0}^m a_k x^k$ in the null polynomial then all coefficients of p are zero, i.e. $\sum_{k=0}^m |a_k| = 0$, or $\|p\| = 0$. Conversely if $\sum_{k=0}^m |a_k| = 0$ then all a_k must be 0, i.e. p is the null-polynomial.

(ii) For $p(x) = \sum_{k=0}^m a_k x^k$ we have $(\lambda p)(x) = \sum_{k=0}^m (\lambda a_k) x^k$ and therefore

$$\|\lambda p\| = \sum_{k=0}^m |\lambda a_k| = |\lambda| \sum_{k=0}^m |a_k| = |\lambda| \, \|p\|.$$

(iii) Let $p(x) = \sum_{k=0}^m a_K x^k$ and $q(x) = \sum_{l=0}^n b_l x^l$. Assume that $m \leq n$ and set $a_k = 0$ for $k > m$, to find

$$p(x) = q(x) = \sum_{j=0}^n (a_j + b_j) x^j.$$

Now it follows that

$$\|p+q\| = \sum_{j=0}^{n} |a_j + b_j| \le \sum_{j=0}^{n} (|a_j| + |b_j|) = \sum_{j=0}^{n} |a_j| + \sum_{j=0}^{n} |b_j|$$

$$= \sum_{j=0}^{m} |a_j| + \sum_{j=0}^{n} |b_j| = \|p\| + \|q\| \,.$$

b) For $f, g \in C([a,b])$ we already know the triangle inequality (Corollary I.25.20) and we also have

$$\int_a^b |\lambda f(t)| \, \mathrm{d}t = \int_a^b |\lambda||f(t)| \, \mathrm{d}t = |\lambda| \int_a^b |f(t)| \, \mathrm{d}t.$$

Thus it remains to prove that $\|f\|_1 = 0$ if and only if f is identically zero on $[a,b]$. Clearly, if f is identically zero in $[a,b]$ then $\int_a^b |f(t)| \, \mathrm{d}t = 0$. Now suppose that $f \in C([a,b])$ and $\int_a^b |f(t)| \, \mathrm{d}t = 0$. Suppose $f(t_0) \ne 0$ for some $t_0 \in (a,b)$. Then $f(t) \ne 0$ for all $t \in (t_0 - \eta, t_0 + \eta) \subset (a,b)$ with a suitably small $\eta > 0$. Consequently for all $t \in \left[t_0 - \frac{eta}{2}, t_0 + \frac{\eta}{2}\right]$ we have $0 < \min\left\{|f(t)| \,\big|\, t \in [t_0 - \frac{\eta}{2}, t_0 + \frac{\eta}{2}]\right\} \le |f(t)|$, implying that

$$0 < \int_{t_0 - \eta/2}^{t_0 + \eta/2} |f(t)| \, \mathrm{d}t \le \int_a^b |f(t)| \, \mathrm{d}t$$

which is a contradiction to the assumption $\int_a^b |f(t)| \, \mathrm{d}t = 0$.

9. For $v = 0$ the result is trivial as it is trivial that $B(u,v) = 0$. The positive definiteness of B yields for all $\lambda \in \mathbb{R}$ and $u, v \in H$, $v \ne 0$,

$$0 \le B(u - \lambda v, u - \lambda v) = B(u,u) - 2\lambda B(u,v) + \lambda^2 B(v,v)$$

which gives for $\lambda = \frac{B^2(u,v)}{B(v,v)} \ne 0$ that

$$0 \le B(u,u) - \frac{2B^2(u,v)}{B(v,v)} + \frac{B^2(u,v)}{B^2(v,v)} B(v,v)$$

or

$$0 \le B(u,u)B(v,v) - 2B^2(u,v) + B^2(u,v),$$

i.e.

$$B^2(u,v) \le B(u,u)B(v,v)$$

which gives $|B(u,v)| \le B^{\frac{1}{2}}(u,u)B^{\frac{1}{2}}(v,v)$.

Next we show that $\|u\|_H := B^{\frac{1}{2}}(u,u)$ is a norm. Clearly, $\|u\|_H \ge 0$ and since B is positive definite it follows that $\|u\|_H = 0$ if and only if $u = 0$. Further, for $\lambda \in \mathbb{R}$ we have

$$\|\lambda u\|_H^2 = B^{\frac{1}{2}}(\lambda u, \lambda u) = (\lambda^2 B(u,u))^{\frac{1}{2}} = |\lambda| B^{\frac{1}{2}}(u,u) = |\lambda| \|u\|_H.$$

The triangle inequality follows from the Cauchy-Schwarz inequality:

$$\|u+v\|_H^2 = B(u+v, u+v) = B(u,u) + 2B(u,v) + B(v,v)$$
$$\leq B(u,u) + 2B^{\frac{1}{2}}(u,u)B^{\frac{1}{2}}(v,v) + B(v,v)$$
$$= \|u\|_H^2 + 2\|u\|_H\|v\|_H + \|v\|_H = (\|u\|_H + \|v\|_H)^2.$$

10. First we determine the boundary ∂A. Recall that $x \in \partial A$ if every neighbourhood of x contains points of A and $A^{\complement}$. For every $0 < \epsilon < 1$ we have

$$-\frac{\epsilon}{2} \notin A \quad \text{but} \quad -\frac{\epsilon}{2} \in (-\epsilon, \epsilon), \quad \text{and} \quad A \cap (-\epsilon, \epsilon)$$

$$\frac{\epsilon}{2} + 1 \notin A \quad \text{but} \quad \frac{\epsilon}{2} + 1 \in (-\epsilon + 1, \epsilon + 1) \quad \text{and} \quad 1 \in (-\epsilon + 1, \epsilon + 1) \cap A$$

$$-\frac{\epsilon}{2} + 5 \notin A \quad \text{but} \quad -\frac{\epsilon}{2} + 5 \in (-\epsilon + 5, \epsilon + 5) \quad \text{and} \quad 5 \in (-\epsilon + 5, \epsilon + 5) \cap A.$$

Thus $\{0, 1, 5\} \subset \partial A$. We know that $[2,3] \cap \mathbb{Q}$ has a boundary $[2,3]$ i.e. for every $x \in [2,3]$ there exists a neighbourhood $U(x)$ such that $U(x) \cap A \neq \emptyset$ but $U(x)$ contains points not belonging to $[2,3] \cap \mathbb{Q}$.
If we consider $V(x) := U(x) \cap \left(\frac{3}{2}, \frac{7}{2}\right)$, it follows that for every $x \in [2,3]$ the neighbourhood $V(x)$ contains points of A and $A^{\complement}$. Consequently $\partial A \subset \{0, 1, 5\} \cup [2,3]$. Now, if $x \in \{0, 1, 5\} \cup [2,3]$ and $x \in A^{\complement}$ then with $\eta := \min\{|x|, |x-1|, |x-2|, |x-3|, |x-5|\}$ it follows that $\left(x - \frac{\eta}{2}, x + \frac{\eta}{2}\right) \cap A = \emptyset$. In the case that $x \notin \{0, 1, 5\} \cup [2,3]$ and $x \in A$ then $x \in (0,1)$ and for $\eta = \min\{x, 1-x\}$ it follows that $\left(x - \frac{\eta}{2}, x + \frac{\eta}{2}\right) \cap A^{\complement} = \emptyset$. Thus we have

$$\partial A = \{0, 1, 5\} \cup [2,3].$$

Since $\bar{A} = A \cup \partial A$ we find

$$\bar{A} = [0,1] \cup [2,3] \cup \{5\}$$

and since $\mathring{A} = A \setminus \partial A$ we get
$$\mathring{A} = (0,1).$$

11. If $A \subset X$ is open then A is a neighbourhood of each of its points, i.e. for every $x \in A$ there exists $B_\epsilon(x) \subset A$. Conversely, if for every $x \in A$ there exists $B_\epsilon(x) \subset A$ then A is a neighbourhood of all of its points, hence A is open.
Let $y \in X$ and N_1, N_2 be the neighbourhoods of y. There exists $\epsilon_1 > 0$ and $\epsilon_2 > 0$ such that $B_{\epsilon_1}(y) \subset N_1$ and $B_{\epsilon_2}(y) \subset N_2$. Since with $\eta := \min(\epsilon_1, \epsilon_2)$ we have $B_\eta(y) = B_{\epsilon_1}(y) \cap B_{\epsilon_2}(y) \subset N_1 \cap N_2$ it follows that $N_1 \cap N_2$ is a neighbourhood of y. Finally if N_1 is a neighbourhood of y then $B_\epsilon(y) \subset N_1 \subset N$ for some $\epsilon > 0$, hence $B_\epsilon(y) \subset N$ and therefore N is a neighbourhood of y too.

12. Let $B \subset X$ be open and $B \subset A$. For every $x \in B$ we can find $\epsilon > 0$ such that $B_\epsilon(x) \subset B \subset A$, i.e. $x \in B \subset A$ and $\{y \in X \mid d(x,y) < \epsilon\} \subset B \subset A$, which yields $\{y \in A \mid d_A(x,y) < \epsilon\} \subset B$, and therefore B is open in (A, d_A).

13. By definition a set B in $(A, d_A) = ((a, b], d|_{(a,b]})$ is open if for every $x \in (a, b]$ there exists $\epsilon > 0$ such that

$$\{y \in (a, b] \,|\, d|_{(a,b]} < \epsilon\} \subset (a, b].$$

Now let $a < c$ and $x \in (c, b]$. We take $\epsilon := \frac{1}{2}\min(c - x, b - x)$ to find $\{y \in (a, b] \,|\, d|_{(a,b]}(x, y) < \epsilon\} \subset (c, b]$, implying that for $a < c$ the set $(c, b]$ is open in $((a, b], d|_{(a,b]})$. Note that $(c, b]$ is of course not open in $\mathbb{R}$.

14. Let $A \subset X$ be open and $x_0 \in A$. Further let N be a neighbourhood of x_0. Thus we can find $\eta > 0$ such that $B_\eta(x_0) \subset A \cap N$. Therefore N contains points of $A \setminus \{x_0\}$, namely the set $B_\eta(x_0) \setminus \{x_0\}$ and N contains points of $(A \setminus \{x_0\})^{\complement}$, namely $\{x_0\}$. Therefore $x_0 \in \partial A$.

Now consider in the metric space $(\mathbb{R}, |.|)$ the set $[0, 1] \cup \{2\}$. Since $[0, 1]$ and $\{2\}$ are closed it follows that $B := [0, 1] \cup \{2\}$ is a closed set. We claim that $\{2\}$ is not a boundary point of $[0, 1]$, i.e. $2 \in B$ but $2 \notin \partial(B \setminus \{2\})$. Since $B \setminus \{2\} = [0, 1]$ by Example 1.21.A. we have $\partial(B \setminus \{2\}) = \{0, 1\}$.

15. Clearly $\mathcal{P}(X)$ satisfies (i) $-$ (iii):

$\emptyset, X \in \mathcal{P}(X)$, and if $U, V \in \mathcal{P}(X)$, i.e. $U \subset X$ and $V \subset X$, then $U \cap V \subset X$, i.e. $U \cap V \in \mathcal{P}(X)$. Moreover, for any collection $U_j \subset X$, $j \in J$, i.e. $U_j \in \mathcal{P}(X)$, we find $\bigcup_{j \in J} U_j \subset X$ or $\bigcup_{j \in J} U_j \in \mathcal{P}(X)$.

Next let $\mathcal{O}_j \subset \mathcal{P}(X)$, $j \in J$, be a collection of topologies in X and define $\mathcal{O} := \bigcap_{j \in J} \mathcal{O}_j$. Since $\emptyset, X \in \mathcal{O}$, i.e. $U, V \in \bigcap_{j \in J} \mathcal{O}_j$ it follows that $U, V \in \mathcal{O}_j$ for all $j \in J$, hence $U \cap V \in \mathcal{O}_j$ for all $j \in J$ since $\mathcal{O}_j$ is a topology. This implies of course $U \cap V \in \bigcap_{j \in J} \mathcal{O}_j$. Finally, if $U_l \in \mathcal{O}$, $l \in L$, then $U_l \in \bigcap_{j \in J} \mathcal{O}_j$, i.e. $U_l \in \mathcal{O}_j$ for all $j \in J$ and $l \in L$. Hence $\bigcup_{l \in L} U_l \in \mathcal{O}_j$ for all $j \in J$ implying that $\bigcup_{l \in L} U_l \in \bigcap_{j \in J} \mathcal{O}_j$, i.e. $\mathcal{O} = \bigcap_{j \in J} \mathcal{O}_j$ is indeed a topology.

Now we can conclude that $\mathcal{O}_\mathcal{A}$ is a topology since $\mathcal{P}(X)$ is a topology such that $\mathcal{A} \subset \mathcal{P}(X)$ and therefore the intersection is not taken over the empty set. Clearly $\mathcal{O}_\mathcal{A}$ must be the smallest topology containing $\mathcal{A}$ since for every topology $\tilde{\mathcal{O}}$ containing $\mathcal{A}$ we have $\mathcal{O}_\mathcal{A} \subset \tilde{\mathcal{O}}$.

Chapter 2

1. Let $(x_k)_{k \in \mathbb{N}}$ be a sequence in X converging with respect to the metric d_1 to $x \in X$. This means that for $\epsilon > 0$ there exists $N_1 = N_1(\epsilon) \in \mathbb{N}$ such that $k \geq N_1$ implies $d_1(x, x_k) < \epsilon$. Thus for $k \geq N_1$ we also find $d_2(x, x_k) < \kappa_2 \epsilon$ implying the convergence of $(x_k)_{k \in \mathbb{N}}$ to x with respect to the metric d_2. Conversely if $(x_k)_{k \in \mathbb{N}}$ converges with respect to the metric d_2 to x, then for $\epsilon > 0$ there exists $N_2 = N_2(\epsilon)$ such that $k \geq N_2$ implies $d_2(x, x_k) < \epsilon$ which yields for $k \geq N_2$ also $d_1(x, x_k) < \frac{\epsilon}{\kappa_1}$, i.e. the convergence of $(x_k)_{k \in \mathbb{N}}$ to x with respect to d_1.

2. For every metric space (X, d) we know that $(x_k)_{k \in \mathbb{N}}$, $x_k \in X$, converges to $x \in X$ if and only if $\lim_{k \to \infty} d(x, x_k) = 0$, and this limit is a limit in $\mathbb{R}$. Suppose $(x_k)_{k \in \mathbb{N}}$ converges in $(\mathbb{R}^n, \|.\|_2)$ to x, hence $\lim_{k \to \infty} \|x_k - x\|_2 = 0$ and the continuity of ψ

implies

$$\lim_{k\to\infty} d_\psi(x, x_k) = \lim_{k\to\infty} \psi(\|x_k - x\|_2)$$
$$= \psi(\lim_{k\to\infty} \|x_k - x\|_2) = \psi(0) = 0.$$

Conversely, if $\lim_{k\to\infty} d_\psi(x, x_k) = 0$ then $\lim_{k\to\infty} \psi(\|x_k - x\|_2) = 0$ which is equivalent to $\psi(\lim_{k\to\infty} \|x_k - x\|_2) = 0$. However, $0 \in \mathbb{R}^n$ is by assumption the only zero of ψ, so, again using the continuity of ψ we deduce that $\lim_{k\to\infty} \|x_k - x\|_2 = 0$.

3. First we show that $\delta(x, y) := \min(1, d(x, y))$ is a metric on X if d is a metric on X. Clearly, $\delta(x, y) \geq 0$ and $\delta(x, y) = 0$ means $d(x, y) = 0$ or, since d is a metric, $x = y$. Moreover, since $d(x, y) = d(y, x)$ we find that $\delta(x, y) = \delta(y, x)$. The triangle inequality we see as follows. Now let $x, y, z \in X$. If $\delta(x, z) = 1$ and $\delta(z, y) = 1$, then $\delta(x, y) \leq \delta(x, z) + \delta(z, y)$ since $\delta(x, y) \leq 1$. If $\delta(x, z) < 1$ and $\delta(z, y) < 1$ then $\delta(x, z) = d(x, z)$ and $\delta(z, y) = d(z, y)$ and therefore $\delta(x, y) = \min(1, d(x, y)) \leq d(x, y) \leq d(x, z) + d(z, y) = \delta(x, z) + \delta(z, y)$. If $\delta(x, z) < 1$ and $\delta(z, y) = 1$ then $\delta(x, y) = \min(1, d(x, y)) \leq 1 \leq \delta(x, z) + \delta(z, y)$, and the case $\delta(x, z) = 1$ and $\delta(z, y) < 1$ goes analogously.

In order to see that convergence with respect to δ is equivalent to convergence with respect to d we only need to note that for every $0 < \epsilon < 1$ the two estimates $\delta(x, y) < \epsilon$ and $d(x, y) < \epsilon$ are equivalent.

4. We observe that for every pair of sequences $(a_k)_{k\in\mathbb{N}}$ and $(b_k)_{k\in\mathbb{N}}$ of real numbers $\frac{|a_k - b_k|}{1 + |a_k - b_k|} \leq 1$ and therefore

$$\sum_{k=1}^{\infty} \frac{1}{2^k} \frac{|a_k - b_k|}{1 + |a_k - b_k|} \leq \sum_{k=1}^{\infty} \frac{1}{2^k} = 1,$$

i.e. the series always converges and d_S is well defined. Clearly $d_S(a, b) \geq 0$ and $d_S(a, b) = 0$ if and only if $a_k = b_k$ for all $k \in \mathbb{N}$, i.e. $(a_k)_{k\in\mathbb{N}} = (b_k)_{k\in\mathbb{N}}$. Since $|a_k - b_k| = |b_k - a_k|$ the symmetry relation $d_S(a, b) = d_S(b, a)$ is trivial. The triangle inequality is seen as follows: the function $x \mapsto \frac{x}{1+x}$ is on $\mathbb{R}_+$ monotone increasing since $\frac{d}{dx}\left(\frac{x}{1+x}\right) = \frac{1}{1+x^2}$. Since $|a_k - b_k| \leq |a_k - c_k| + |c_k - b_k|$ we deduce

$$\frac{|a_k - b_k|}{1 + |a_k - b_k|} \leq \frac{|a_k - c_k| + |c_k - b_k|}{1 + |a_k - c_k| + |c_k - b_k|}$$
$$\leq \frac{|a_k - c_k|}{1 + |a_k - c_k|} + \frac{|c_k - b_k|}{1 + |c_k - b_k|},$$

hence

$$\sum_{k=1}^{\infty} \frac{1}{2^k} \frac{|a_k - b_k|}{1 + |a_k - b_k|} \leq \sum_{k=1}^{\infty} \frac{1}{2^k} \frac{|a_k - c_k|}{1 + |a_k - c_k|} + \sum_{k=1}^{\infty} \frac{1}{2^k} \frac{|c_k - b_k|}{1 + |c_k - b_k|}$$

or

$$d_S(a, b) \leq d_S(a, c) + d_S(c, b).$$

597

Now let $(a^l)_{l \in \mathbb{N}}$, $a^{(l)} \in S$, be a sequence converging in d_S to $a \in S$. This implies that for $\epsilon > 0$ there exists $N = N(\epsilon) \in \mathbb{N}$ such that for $k_0 \in \mathbb{N}$ and $l \geq N(\epsilon)$ it follows that

$$\sum_{k=1}^{\infty} \frac{1}{2^k} \frac{|a_k^l - a_k|}{1 + |a_k^l - a_k|} < \frac{\epsilon}{2^{k_0}},$$

thus for $k_0 \in \mathbb{N}$ fixed we deduce for $l \geq N$ that

$$\frac{1}{2^{k_0}} \frac{|a_{k_0}^l - a_{k_0}|}{1 + |a_{k_0}^l - a_{k_0}|} < \frac{\epsilon}{2^{k_0}},$$

or

$$\frac{|a_{k_0}^l - a_{k_0}|}{1 + |a_{k_0}^l - a_{k_0}|} < \epsilon,$$

i.e. $\lim_{l \to \infty} \frac{|a_{k_0}^l - a_{k_0}|}{1 + |a_{k_0}^l - a_{k_0}|} = 0$ which implies $\lim_{l \to \infty} |a_{k_0}^l - a_{k_0}| = 0$. Hence we have proved that $\lim_{l \to \infty} d_S(a^l, a) = 0$ implies for all $k \in \mathbb{N}$ that $\lim_{l \to \infty} a_k^{(l)} = a_k$. Now suppose that for all $k \in \mathbb{N}$ we have $\lim_{l \to \infty} a_k^l = a_k$. Given $\epsilon > 0$ we first note the existence of $N_0 = N_0(\epsilon)$ such that $\sum_{k=N_0}^{\infty} \frac{1}{2^k} < \frac{\epsilon}{2}$. Since $\frac{|a_k^{(l)} - a_k|}{1 + |a_k^{(l)} - a_k|} < 1$ we deduce that

$$\sum_{k=N_0}^{\infty} \frac{1}{2^k} \frac{|a_k^{(l)} - a_k|}{1 + |a_k^{(l)} - a_k|} < \frac{\epsilon}{2}.$$

Moreover, by $\lim_{l \to \infty} a_k^{(l)} = a_k$ we can find $N_1 = N_1(\epsilon)$ such that for $l \geq N_1$ it follows that

$$\sum_{k=1}^{N_0 - 1} \frac{1}{2^k} \frac{|a_k^{(l)} - a_k|}{1 + |a_k^{(l)} - a_k|} < \frac{\epsilon}{2}.$$

Both estimates together yield for $l \geq \max(N_0, N_1)$ that

$$d_S(a^{(l)}, a) = \sum_{k=1}^{\infty} \frac{1}{2^k} \frac{|a_k^{(l)} - a_k|}{1 + |a_k^{(l)} - a_k|} < \epsilon.$$

5. In order to prove that $(g_k)_{k \in \mathbb{N}}$ converges in $(C([-\pi, \pi]), \|\cdot\|_1)$ to g_0, $g_0(x) = 0$ for all $x \in [-\pi, \pi]$, we need to show that

$$\lim_{k \to \infty} \int_{-\pi}^{\pi} |g_k(x) - g_0(x)| dx = \lim_{k \to \infty} \int_{-\pi}^{\pi} |g_k(x)| dx$$

$$= \lim_{k \to \infty} \int_{-\pi}^{\pi} \frac{|x|^k}{\pi^k} dx = 0.$$

We note that

$$\int_{-\pi}^{\pi} \frac{|x|^k}{\pi^k} dx = 2 \int_{0}^{\pi} \frac{x^k}{\pi^k} dx = \frac{2}{\pi^k} \frac{x^{k+1}}{k+1} \Big|_{0}^{\pi} = \frac{2\pi}{k+1}.$$

and hence we have indeed

$$\lim_{k \to \infty} \int_{-\pi}^{\pi} |g_k(x) - g_0(x)| dx = \lim_{k \to \infty} \frac{2\pi}{k+1} = 0.$$

For $x = \pi$ we find $g_k(\pi) = \frac{\pi^k}{\pi^k} = 1$, i.e. $\lim_{k \to \infty} g_k(\pi) = 1$, and for $x = -\pi$ we have $g_k(-\pi) = \frac{(-\pi)^k}{\pi^k} = (-1)^k$ and hence $(g_k(-\pi))_{k \in \mathbb{N}}$ has no limit at all. Thus on $[-\pi, \pi]$ the sequence $(g_k)_{k \in \mathbb{N}}$ does not converge pointwise and therefore it cannot converge uniformly to g_0.

6. If $(x_k)_{k \in \mathbb{N}}$ converges to $x \in X$, for every $\epsilon > 0$ there exists $N \in \mathbb{N}$ such that $k \geq N$ implies $d(x, x_k) < \epsilon$, i.e. $x_k \in B_\epsilon(x)$. Thus, if U is a neighbourhood of x, we choose $\epsilon > 0$ such that $B_\epsilon(x) \subset U$ and it follows that the existence of $N(U) \in \mathbb{N}$ such that $k \geq N(U)$ implies $x_k \in U$. Conversely, if for every neighbourhood U of x there exists $N(U) \in \mathbb{N}$ such that $k \geq N(U)$ implies $x_k \in U$ we can choose as U the open ball $B_\epsilon(x)$, $\epsilon > 0$, to find that for $\epsilon > 0$ there exists $N(\epsilon) = N(B_\epsilon(x))$ such that $k \geq N(\epsilon)$ implies $x_k \in B_\epsilon(x)$, i.e. $d(x, x_k) < \epsilon$.

7. The injectivity of j is almost trivial: if $x_1 \neq x_2$ then $d_Y(j(x_1), j(x_2)) = d_X(x_1, x_2) > 0$, implying that $j(x_1) \neq j(x_2)$. Moreover, if $(x_k)_{k \in \mathbb{N}}$ converges in (X, d_X) we find for every $\epsilon > 0$ some $N(\epsilon) \in \mathbb{N}$ such that $k \geq N(\epsilon)$ yields

$$d_Y(j(x_k), j(x)) = d_X(x_k, x) < \epsilon,$$

i.e. $(j(x_k))_{k \in \mathbb{N}}$ converges in (Y, d_Y) to $j(x)$.

Finally, consider $Y = \mathbb{R}$ with the Euclidean metric, i.e. the metric induced by the absolute value, and $X = \mathbb{Q}$ with the same metric. Then the identity on $\mathbb{R}$ induces an isometry $j : \mathbb{Q} \to \mathbb{R}$, $j(x) = x$, however, while $\mathbb{R}$ is complete, $\mathbb{Q}$ is not complete with respect to the Euclidean metric.

8. Consider the function $\chi_{[\frac{1}{2}, 1]} : [0, 1] \to \mathbb{R}$. The functions $h_k : [0, 1] \to \mathbb{R}$, $h \geq 2$, defined by

$$h_k(x) = \begin{cases} 0, & 0 \leq x \leq \frac{1}{2} - \frac{1}{k} \\ \frac{k}{2}x + \frac{1}{2} - \frac{k}{4}, & \frac{1}{2} - \frac{1}{k} \leq x \leq \frac{1}{2} + \frac{1}{k} \\ 1, & \frac{1}{2} + \frac{1}{k} \leq x \leq 1 \end{cases}$$

are continuous and moreover we find

$$\int_0^1 |h_k(x) - \chi_{[\frac{1}{2}, 1]}(x)| dx = \frac{1}{2k},$$

see the following figure.

599

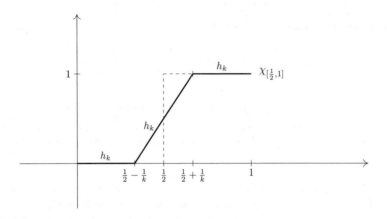

Note that the integral is just twice the area of the triangle with vertices $(\frac{1}{2} - \frac{1}{k}, 0)$, $(\frac{1}{2}, 0)$, $(\frac{1}{2}, \frac{1}{2})$ which is of course $\frac{1}{4k}$. Thus $(h_k)_{k \in \mathbb{N}}$ is with respect to $\|\cdot\|_1$ a Cauchy sequence in $(C([0,1]), \|\cdot\|_1)$ but its limit is not in $C([0,1])$ since it is discontinuous, hence the space $(C([0,1]), \|\cdot\|_1)$ is not a Banach space.

9. Since

$$\int_a^b |u(t) + v(t)| dt \leq \int_a^b |u(t)| dt + \int_a^b |v(t)| dt$$

and

$$\int_a^b |\lambda u(t)| dt = |\lambda| \int_a^b |u(t)| dt,$$

it is clear that $C\mathcal{L}^1((a,b))$ is a vector space. Further we note that $\|\cdot\|_1$ is non-negative, homogeneous and the triangle inequality holds (as we have already used above). To prove that $\|\cdot\|_1$ is a norm on $C\mathcal{L}^1((a,b))$ we need to show that for $u \in C\mathcal{L}^1((a,b))$ the fact that $\|u\|_1 = 0$ implies $u = 0$, i.e. $u(x) = 0$ for all $x \in (a,b)$. Suppose that $u(x_0) \neq 0$, $x_0 \in (a,b)$. We may assume $u(x_0) > 0$, the case $u(x_0) < 0$ goes analogously. By continuity we find an open interval $(x_0 - \epsilon, x_0 + \epsilon) \subset (a,b)$ such that $u(x) \geq \gamma > 0$ in $(x_0 - \epsilon, x_0 + \epsilon)$. Now it follows

$$0 < \gamma \cdot \epsilon \leq \int_{x_0 - \frac{\epsilon}{2}}^{x_0 + \frac{\epsilon}{2}} u(x) dx \leq \int_a^b |u(x)| dx = 0,$$

which is a contradiction. Hence $\|u\|_1 = 0$ and u continuous implies $u = 0$ in (a,b). Consider the function $f_n : (a,b) \to \mathbb{R}$ defined by

$$f_n(x) = \frac{1}{n(b-a)^{\frac{1}{n}}} (x - a)^{\frac{1}{n} - 1}, \quad n \geq 2.$$

These are all unbounded functions on (a,b), however

$$\int_a^b |f_n(x)| dx = \frac{1}{n(b-a)^{\frac{1}{n}}} \int_a^b (x-a)^{\frac{1}{n} - 1} dx = 1,$$

600

hence the set $\{f_n|n \geq 2\}$ is bounded in $C\mathcal{L}^1((a,b))$.

10. For $k \in \mathbb{N}_0$ we find $\|u\|_{k,\infty} \geq 0$ and $\|u\|_{k,\infty} = 0$ implies $\|u^{(n)}\|_\infty$ for all $0 \leq n \leq k$. In particular we have $\|u\|_\infty = 0$ which yields $u = 0$. Moreover, for $\lambda \in \mathbb{R}$ we find

$$\|\lambda u\|_{k,\infty} = \sum_{n=0}^{k} \left\| \lambda u^{(n)} \right\|_\infty = |\lambda| \sum_{n=0}^{k} \left\| u^{(n)} \right\|_\infty = |\lambda| \|u\|_{k,\infty}$$

and for $u, v \in C^k([a,b])$ it follows

$$\|u + v\|_{k,\infty} = \sum_{n=0}^{k} \left\| u^{(n)} + v^{(n)} \right\|_\infty \leq \sum_{n=0}^{k} \left\| u^{(n)} \right\|_\infty + \sum_{n=0}^{k} \left\| v^{(n)} \right\|_\infty$$
$$= \|u\|_{k,\infty} + \|v\|_{k,\infty}.$$

Thus $\|\cdot\|_{k,\infty}$ is a norm. Similarly, for $\|u\|_{k,\infty}$ we have $\|u\|_{k,\infty} \geq 0$ and $\|u\|_{k,\infty} = 0$ implies $\|u\|_\infty = 0$, i.e. $u = 0$. In addition, for $\lambda \in \mathbb{R}$ we have

$$\|\lambda u\|_{k,\infty} = \max_{0 \leq n \leq k} \left\| \lambda u^{(n)} \right\|_\infty = |\lambda| \max_{0 \leq n \leq k} \left\| u^{(n)} \right\|_\infty = |\lambda| \|u\|_{k,\infty}$$

as well as for $u, v \in C^k([a,b])$

$$\|u + v\|_{k,\infty} = \max_{0 \leq n \leq k} \left\| u^{(n)} + v^{(n)} \right\|_\infty \leq \max_{0 \leq n \leq k} \left(\left\| u^{(n)} \right\|_\infty + \left\| v^{(n)} \right\|_\infty \right)$$
$$= \|u\|_{k,\infty} + \|v\|_{k,\infty}.$$

Only $\left\| u^{(0)} \right\|_\infty = \|u\|_\infty$ is a norm, since $\left\| u^{(n)} \right\|_\infty = 0$, $n \geq 1$, implies $u(x) = \sum_{l=0}^{n-1} a_l x^l$ with arbitrary coefficients a_l.

11. The mapping $f : (\mathbb{R}, |\cdot|) \to (\mathbb{R}, |\cdot|)$ is unbounded since

$$\text{diam} \left\{ y \in \mathbb{R} | y = f(x) | x \in \mathbb{R} \right\} = \text{diam} \left\{ x \in \mathbb{R} | x \in \mathbb{R} \right\} = \sup \left\{ |x - y| | x, y \in \mathbb{R} \right\} = \infty$$

and the same holds for $f : (\mathbb{R}, \arctan |\cdot|) \to (\mathbb{R}, |\cdot|)$ since only the metric in the target space counts for our consideration. However, for $f : (\mathbb{R}, |\cdot|) \to (\mathbb{R}, \arctan |\cdot|)$ we find

$$\text{diam} \left\{ y \in \mathbb{R} | y = f(x), x \in \mathbb{R} \right\} = \sup \left\{ \arctan |x - y| | x, y \in \mathbb{R} \right\} = \frac{\pi}{2}.$$

We know that $\arctan : \mathbb{R} \to \mathbb{R}$ is continuous, thus $x_k \to x$ implies $\arctan x_k \to \arctan x$ and this already implies the continuity of ii), whereas the continuity of i) is trivial. To handle the continuity of iii) we note that $\arctan |x_k - x| \to 0$ implies $x_k \to x$ and hence iii) is continuous too.

12. It is the triangle inequality which does not hold for $\|\cdot\|_\alpha$. We first prove

$$\|x + y\|_\alpha \leq 2^{1 + \frac{1}{\alpha}} (\|x\|_\alpha + \|y\|_\alpha).$$

Note for $x_j, y_j \in \mathbb{R}$ that

$$|x_j + y_j|^\alpha \leq (|x_j + y_j|)^\alpha \leq 2^\alpha \max(|x_j|^\alpha, |y_j|^\alpha) \leq 2^\alpha (|x_j|^\alpha + |y_j|^\alpha),$$

implying

$$\sum_{j=1}^{n} |x_j + y_j|^\alpha \leq 2^\alpha \left(\sum_{j=1}^{n} |x_j|^\alpha + \sum_{j=1}^{n} |y_j|^\alpha \right),$$

or

$$\left(\sum_{j=1}^{n} |x_j + y_j|^\alpha \right)^{\frac{1}{\alpha}} \leq 2 \left(\sum_{j=1}^{n} |x_j|^\alpha + \sum_{j=1}^{n} |y_j|^\alpha \right)^{\frac{1}{\alpha}}$$

$$\leq 2 \cdot 2^{\frac{1}{\alpha}} \max \left(\left(\sum_{j=1}^{n} |x_j|^\alpha \right)^{\frac{1}{\alpha}}, \left(\sum_{j=1}^{n} |y_j|^\alpha \right)^{\frac{1}{\alpha}} \right)$$

$$= 2^{1+\frac{1}{\alpha}} \left(\left(\sum_{j=1}^{n} |x_j|^\alpha \right)^{\frac{1}{\alpha}} + \left(\sum_{j=1}^{n} |y_j|^\alpha \right)^{\frac{1}{\alpha}} \right),$$

i.e.

$$\|x + y\|_\alpha \leq 2^{1+\frac{1}{\alpha}} (\|x\|_j + \|y\|_j).$$

To show that $\|\cdot\|_\alpha$ is not a norm for $0 < \alpha < 1$ we consider $x = \frac{e_1}{2}$ and $y = \frac{e_2}{2}$. First we note that

$$\left\| \frac{e_1}{2} \right\|_\alpha = \left(\left(\frac{1}{2} \right)^\alpha \right)^{\frac{1}{\alpha}} = \frac{1}{2} = \left\| \frac{e_2}{2} \right\|_\alpha,$$

so $\left\| \frac{e_1}{2} \right\|_\alpha + \left\| \frac{e_2}{2} \right\|_\alpha = 1$. However

$$\left\| \frac{e_1 + e_2}{2} \right\|_\alpha = \left(\left(\frac{1}{2} \right)^\alpha + \left(\frac{1}{2} \right)^\alpha \right)^{\frac{1}{\alpha}} = 2^{\frac{1}{\alpha}} \cdot \frac{1}{2},$$

and since $0 < \alpha < 1$ it follows that $\frac{1}{\alpha} > 1$, hence $2^{\frac{1}{\alpha}} > 2$ which yields that

$$\left\| \frac{e_1 + e_2}{2} \right\|_\alpha > \left\| \frac{e_1}{2} \right\|_\alpha + \left\| \frac{e_2}{2} \right\|_\alpha.$$

(Also compare with Problem 2 in Chapter 1.)

13. With $z = \lambda x + (1 - \lambda)y$, $0 \leq \lambda \leq 1$, we find

$$\|x - z\| + \|z - y\| = \|x - \lambda x - (1 - \lambda)y\| + \|\lambda x + (1 - \lambda)y - y\|$$
$$= \|(1 - \lambda)x - (1 - \lambda)y\| + \|\lambda x - \lambda y\|$$
$$= (1 - \lambda) \|x - y\| + \lambda \|x - y\| = \|x - y\|.$$

In order to prove that for $0 < \alpha < 1$ there is no point $z \in \mathbb{R}^n$ such that for $x, y \in \mathbb{R}^n$, $x \neq y$, the following holds

$$\|x - y\|^\alpha = \|x - z\|^\alpha + \|z - y\|^\alpha,$$

we observe that the metric $\|x - y\|^\alpha$ is invariant under translation and therefore we may choose $y = 0$, so we consider the problem to find $z \in \mathbb{R}^n$, $z \neq 0$ and $z \neq x$ for $x \in \mathbb{R}^n$, $x \neq 0$, such that

$$\|x\|^\alpha = \|x - z\|^\alpha + \|z\|^\alpha.$$

We note that we cannot have $z = \lambda x$ for $\lambda \in \mathbb{R} \setminus \{0, 1\}$ since this implies

$$\|x\|^\alpha = \|x - \lambda x\|^\alpha + \|\lambda x\|^\alpha = |1 - \lambda|^\alpha \|x\|^\alpha + |\lambda|^\alpha \|x\|^\alpha,$$

i.e. $1 = |1 - \lambda|^\alpha + |\lambda|^\alpha$ which cannot hold for $\lambda \in \mathbb{R} \setminus \{0, 1\}$ and for $0 < \alpha < 1$.

14. For two real numbers a, b we know, compare with Lemma I.2.7, that

$$\max\{a, b\} = a \vee b = \frac{1}{2}(a + b + |a - b|)$$

and

$$\max\{a, b\} = a \wedge b = \frac{1}{2}(a + b - |a - b|).$$

Since $x \mapsto |x|$ is a continuous function the two functions

$$x \mapsto \frac{1}{2}(f(x) + g(x) + |f(x) - g(x)|) = (f \vee g)(x)$$

and

$$x \mapsto \frac{1}{2}(f(x) + g(x) - |f(x) - g(x)|) = (f \wedge g)(x)$$

are continuous from X to $\mathbb{R}$ as compositions of continuous functions. This implies of course that f^+, f^- and $|f|$ are continuous.

15. First we note that $x \in \mathbb{R}^m$, $x_j \in [a, b]$, is fixed and the variable is $u \in C([a, b])$. Let $(u_\nu)_{\nu \in \mathbb{N}}$ be a sequence in $C([a, b])$ converging uniformly to $u \in C([a, b])$. Then all the sequences $(u_\nu(x_j))_{\nu \in \mathbb{N}}$, $1 \leq j \leq m$, converge and $\lim_{\nu \to \infty} u_\nu(x_j) = u(x_j)$. Consequently we find that $\lim_{\nu \to \infty}(u_\nu(x_1), ..., u_\nu(x_m)) = (u(x_1), ..., u(x_m))$, i.e pr_x maps convergent sequences in $C([a, b])$ to convergent sequences in $\mathbb{R}^m$, and hence $\text{pr}_x : C([a, b]) \to \mathbb{R}^m$ is continuous.

16. Clearly $\|u\|_\infty \geq 0$ and if $\|u\|_\infty = 0$ then $\|u_j\|_\infty = 0$ for all $j = 1, ..., n$, hence $u = (u_1, ..., u_n) = 0$, the null function in $C([a, b], \mathbb{R}^n)$. For $\lambda \in \mathbb{R}$ we have

$$\|\lambda u\|_\infty = \max_{1 \leq k \leq n} \|\lambda u_j\|_\infty = |\lambda| \max_{1 \leq k \leq n} \|u_j\|_\infty = |\lambda| \|u\|_\infty,$$

and for $u, v \in C([a, b], \mathbb{R}^n)$ it follows that

$$\|u + v\|_\infty = \max_{1 \leq k \leq n} \|u_j + v_j\|_\infty \leq \max_{1 \leq k \leq n} (\|u_j\|_\infty + \|v_j\|_\infty)$$
$$\leq \max_{1 \leq k \leq n} \|u_j\|_\infty + \max_{1 \leq k \leq n} \|v_j\|_\infty = \|u\|_\infty + \|v\|_\infty.$$

603

Thus $\|\cdot\|_\infty$ is a norm on $C([a,b], \mathbb{R}^n)$. In order to prove the continuity of h we observe that for $u, v \in C([a,b], \mathbb{R}^n)$ we have

$$\|h(u) - h(v)\|_\infty = \left\| \sum_{j=1}^n A_j u_j - \sum_{j=1}^n A_j v_j \right\|_\infty$$

$$= \left\| \sum_{j=1}^n A_j (u_j - v_j) \right\|_\infty \leq \sum_{j=1}^n |A_j| \|u_j - v_j\|_\infty$$

$$\leq \left(\sum_{j=1}^n |A_j| \right) \max_{1 \leq j \leq n} \|u_j - v_j\|_\infty = \left(\sum_{j=1}^n |A_j| \right) \|u - v\|_\infty.$$

Therefore, given $\epsilon > 0$ we choose $\delta = \frac{\epsilon}{\sum_{j=1}^n |A_j|}$, assuming $\sum_{j=1}^n |A_j| \neq 0$, to find for $\|u - v\|_\infty < \delta$ that

$$\|h(u) - h(v)\|_\infty \leq \left(\sum_{j=1}^n |A_j| \right) \|u - v\|_\infty < \left(\sum_{j=1}^n |A_j| \right) \frac{\epsilon}{\sum_{j=1}^n |A_j|} = \epsilon.$$

In the case $\sum_{j=1}^n |A_j| = 0$ we have $A = 0$, and hence $h(u) = 0$ for all $u \in C([a,b], \mathbb{R}^n)$, which implies of course the continuity of h.

17. For $u, v \in C^1([a,b])$ we have

$$\left\| \frac{d}{dx}(u - v) \right\|_\infty \leq \left\| \frac{d}{dx}(u - v) \right\|_\infty + \|u - v\|_\infty = \|u - v\|_{\infty,1},$$

implying the continuity of $\frac{d}{dx} : (C^1([a,b]), \|\cdot\|_{\infty,1}) \to (C([a,b]), \|\cdot\|_\infty)$. Now we consider for $k \in \mathbb{N}$ the functions $f_k : [a,b] \to \mathbb{R}$, $f_k(x) = \frac{1}{k} \sin \frac{4\pi k x}{a+b}$. First we note that

$$\|f_k\|_\infty = \sup_{x \in [a,b]} \left| \frac{1}{k} \sin \frac{4\pi k x}{a+b} \right| \leq \frac{1}{k}$$

implying that $(f_k)_{k \in \mathbb{N}}$ converges in $(C([a,b]), \|\cdot\|_\infty)$ to the zero function f, $f(x) = 0$ for all $x \in [a,b]$. Next we observe that $\frac{d}{dx} f_k(x) = \frac{4\pi}{a+b} \cos \frac{4\pi k x}{a+b}$ which yields

$$\left\| \frac{df_k}{dx} \right\|_\infty = \sup_{x \in [a,b]} \left| \frac{4\pi}{a+b} \cos \frac{4\pi k x}{a+b} \right| = \frac{4\pi}{|a+b|}$$

which follows from $\cos(\frac{4k\pi}{a+b}(\frac{a+b}{2})) = \cos 2\pi k = 1$. Thus $\frac{df_k}{dx}$ does not converge in the norm $\|\cdot\|_\infty$ to $\frac{df}{dx} = 0$, and therefore $\frac{d}{dx} : (C^1([a,b]), \|\cdot\|_\infty) \to (C([a,b]), \|\cdot\|_\infty)$ is not continuous.

18. Since $(x - y)^2 = x^2 - 2xy + y^2$ we find that

$$\int_0^1 (x - y)^2 u(y) dy = x^2 \int_0^1 u(y) dx - 2x \int_0^1 y u(y) dy + \int_0^1 y^2 u(y) dy,$$

and therefore $x \mapsto \int_0^1 (x-y)^2 u(y) dy$ is a polynomial, hence continuous. For $u, v \in C([0,1])$, we find

$$Tu(x) - Tv(x) = \int_0^1 (x-y)^2 u(y) dy - \int_0^1 (x-y)^2 v(y) dy$$

or

$$|Tu(x) - Tv(x)| = |\int_0^1 (x-y)^2 (u(y) - v(y)) dy|$$

$$\leq \int_0^1 (x-y)^2 dy \|u-v\|_\infty$$

$$= \left(-\frac{(x-y)^3}{3} \Big|_0^1 \right) \|u-v\|_\infty$$

$$= \left(\frac{x^3}{3} - \frac{(x-1)^3}{3} \right) \|u-v\|_\infty$$

$$\leq \left(\frac{1}{3} + \frac{1}{3} \right) \|u-v\|_\infty = \frac{2}{3} \|u-v\|_\infty$$

which yields

$$\|Tu - Tv\|_\infty \leq \frac{2}{3} \|u-v\|_\infty,$$

i.e. the Lipschitz continuity of T. Note that our estimate $\frac{x^3}{3} - \frac{(x-1)^3}{3} \leq \frac{2}{3}$ for $x \in [0,1]$ is not the sharpest, indeed the function $\frac{x^3}{3} - \frac{(x-3)^3}{3} = x^2 - x + \frac{1}{3}$ has on $[0,1]$ a relative minimum for $x = \frac{1}{2}$ and an absolute maximum for $x = 0$ and $x = 1$ with value $\frac{1}{3}$, so we can improve our estimate to $\|Tu - Tv\|_\infty \leq \frac{1}{3} \|u-v\|_\infty$.

19. For a Lipschitz continuous mapping we have with some $K > 0$

$$d_Y(f(x_1), f(x_2)) \leq K d_X(x_1, x_2).$$

If $A \subset X$ is a bounded set, then we find

$$\mathrm{diam}_X(A) = \sup \{d_X(x_1, x_2) | x_1, x_2 \in A\} = M < \infty$$

which implies that

$$\sup \{d_Y(f(x_1), f(x_2)) | x_1, x_2 \in A\} \leq K \sup \{d_X(x_1, x_2) | x_1, x_2 \in A\} = K \, \mathrm{diam}_X(A),$$

and hence

$$\mathrm{diam}_Y(f(A)) \leq K \, \mathrm{diam}_X(A)$$

and $f(A) \subset Y$ is bounded.

20. Since the convergence of $\sum_{k=1}^\infty x_k$ is defined as the convergence of the sequence of partial sums $S_N = \sum_{k=1}^N x_k$, a Cauchy criterion for the convergence of $\sum_{k=1}^\infty x_k$ is derived from a Cauchy criterion for the sequence $(S_N)_{N \in \mathbb{N}}$: for every $\epsilon > 0$ there exists $\nu = \nu(\epsilon) \in \mathbb{N}$ such that $N, M \geq \nu(\epsilon)$ implies $\|S_N - S_M\| < \epsilon$. This we can reformulate as: for every $\epsilon > 0$ there exists $\nu(\epsilon) \in \mathbb{N}$ such that $N > M > \nu(\epsilon)$ implies $\left\| \sum_{k=M}^N x_k \right\| < \epsilon$.

21. We prove that the sequence of partial sums is a Cauchy sequence. Since in a Banach space a Cauchy sequence converges the result will follow. For $N > M$ we find

$$\left\| \sum_{k=M}^{N} a_k x_k \right\| \leq \sum_{k=M}^{N} |a_k| \, \|x_k\| \leq \sum_{k=M}^{N} |a_k| \|x_0\|^k \leq \sum_{k=M}^{N} |a_k| R^k.$$

Since the radius of convergence of the power series $\sum_{k=0}^{\infty} a_k t^k$ is ρ and $R < \rho$, we know that $\sum_{k=0}^{\infty} a_k R^k$ converges absolutely (see Theorem I.29.4), so $(\sum_{k=0}^{N} |a_k| R^k)_{N \in \mathbb{N}}$ is a Cauchy sequence and therefore, given $\epsilon > 0$ we can find $\nu = \nu(\epsilon) \in \mathbb{N}$ such that $N > M > \nu(\epsilon)$ implies

$$\left\| \sum_{k=M}^{N} a_k x_k \right\| \leq \sum_{k=M}^{N} |a_k| R^k < \epsilon.$$

Chapter 3

1. If $(z_n)_{n \in \mathbb{N}}$ converges to z, then each of its subsequences must converge to z too. Now suppose that the sequences $(x_n)_{n \in \mathbb{N}}$ and $(y_n)_{n \in \mathbb{N}}$ converge to z. Given $\epsilon > 0$ there exist $N_1(\epsilon), N_2(\epsilon) \in \mathbb{N}$ such that $n \geq N_1(\epsilon)$ implies $d(x_n, z) < \epsilon$ and $n \geq N_2(\epsilon)$ implies $d(y_n, z) < \epsilon$, hence for $n \geq \max(N_1(\epsilon), N_2(\epsilon))$ we have $d(z_n, z) < \epsilon$ implying the result.

2. Since $(x_n)_{n \in \mathbb{N}}$ is a Cacuhy sequence, for $\epsilon > 0$ there exists $N_1(\epsilon) \in \mathbb{N}$ such that $n, m \geq N_1(\epsilon)$ implies $d(x_n, x_m) < \frac{\epsilon}{2}$. The convergence of the subsequence $(x_{n_k})_{k \in \mathbb{N}}$ to x implies the existence of $N_2(\epsilon) \in \mathbb{N}$ such that $n_k \geq N_2(\epsilon)$ yields $d(x_{n_k}, x) < \frac{\epsilon}{2}$. Thus for $N := \max(N_1(\epsilon), N_2(\epsilon))$ we find for $n_k \geq n \geq N$ that

$$d(x_n, x) \leq d(x_{n_k}, x_n) + d(x_{n_k}, x) < \frac{\epsilon}{2} + \frac{\epsilon}{2} = \epsilon,$$

implying the convergence of $(x_n)_{n \in \mathbb{N}}$ to x.

3. Note that $x \in \bar{Y}$ if and only if $B_{\frac{1}{n}}(x) \cap Y \neq \emptyset$ for all $n \in \mathbb{N}$. Hence, for every $n \in \mathbb{N}$ there exists $y \in Y$ such that $d(x, y) < \frac{1}{n}$, but $\text{dist}(x, y) = \inf \{d(x, y)|y \in Y\}$ which implies the equivalence of

   i) $x \in \bar{Y}$;

   ii) $\text{dist}(x, y) < \frac{1}{n}$ for all $n \in \mathbb{N}$;

   iii) $\text{dist}(x, y) = 0$.

4. Let $(U_l)_{l \in I}$ be an open covering of $\cup_{j=1}^{N} K_j$. Then $(U_l)_{l \in I}$ is also an open covering for each K_j, $j = 1, ..., N$. Hence there exists sets $U_1^{(j)}, ..., U_{m_j}^{(j)}$ such that $K_j \subset \cup_{k=1}^{m_j} U_k^{(j)}$ implying that $\{U_{k_j}^{(j)} | j = 1, ..., N, k_j = 1, ...m_j\}$ is an open covering of $\cup_{j=1}^{N} K_j$, i.e. $\cup_{j=1}^{N} K_j$ is compact. In $\mathbb{R}$ the intervals $[-n, n]$, $n \in \mathbb{N}$, are compact while $\cup_{n=1}^{\infty} [-n, n] = \mathbb{R}$ is not.

5. Let $x \in U$. The density of Y in X implies that $Y \cap B_{\frac{1}{n}}(x) \neq \emptyset$ for every $n \in \mathbb{N}$. Thus we can form a sequence $(x_n)_{n \in \mathbb{N}}$, $x_n \in Y \cap B_{\frac{1}{n}}(x)$. Since U is open and $x \in U$, for $n \geq N$ it follows that $B_{\frac{1}{n}}(x) \subset U$, i.e. $x_n \in Y \cap U$ for these n. Further the sequence $(x_n)_{n \in \mathbb{N}}$ converges to x. Thus $x \in \overline{Y \cap U}$ implying $U \subset \overline{Y \cap U}$.

6. Consider on $\mathbb{R}$ the metric $\delta(x, y) = \min(1, |x-y|)$. With respect to this metric every set, in particular $\mathbb{R}$, is bounded, however we cannot cover $\mathbb{R}$ by a finite number of balls $B_{\epsilon}^{\delta}(x_j)$, $j = 1, ..., N$, if $\epsilon < 1$, since then $\delta(x, y) = |x - y|$.

7. Just take $X = Y = \mathbb{R}$ and $f : X \to Y$, $f(x) = c$ for all $x \in \mathbb{R}$, $c \in \mathbb{R}$ fixed.

8. Let $(x_k)_{k \in \mathbb{N}}$ be a sequence in K_1 converging to $x_0 \in K_1$, and let $(y_k)_{k \in \mathbb{N}}$ be a sequence in K_2 such that $f(x_k, y_k) = g(x_k)$. Since K_2 is compact we may choose a subsequence $(y_{k_j})_{j \in \mathbb{N}}$ converging to some $y_0 \in K_2$. The continuity of f implies that $(f(x_{k_j}, y_{k_j}))_{j \in \mathbb{N}}$ converges to $f(x_0, y_0)$. We claim that $f(x_0, y_0) = g(x_0)$. Suppose this does not hold. Then there exists $\tilde{y}_0 \in K_2$ such that $f(x_0, \tilde{y}_0) > f(x_0, y_0)$. This yields the existence of some $\epsilon > 0$ such that $B_{\epsilon}((x_0, y_0)) \cap B_{\epsilon}((x_0, \tilde{y}_0)) = \emptyset$ and $f|_{B_{\epsilon}((x_0, \tilde{y}_0))} > f|_{B_{\epsilon}((x_0, y_0))}$ implying $f(x_{k_j}, \tilde{y}_0) > f(x_{k_j}, y_{k_j})$ which is a contradiction. Thus for any sequence $(x_k)_{k \in \mathbb{N}}$ which converges to x_0 there exists a subsequence $(x_{k_j})_{j \in \mathbb{N}}$ such that $(g(x_{k_j}))_{j \in \mathbb{N}}$ converges to $g(x_0)$. This however implies that for any sequence $(x_k)_{k \in \mathbb{N}}$ which converges to x_0 it follows that $(g(x_k))_{k \in \mathbb{N}}$ converges to $g(x_0)$.

9. If $f : X \to Y$ is continuous then for every compact set $K \subset X$ the mapping $f|_K$ is also continuous since for every open set $U \subset Y$ the pre-image of U under f is the set $K \cap f^{-1}(U)$ and since $f^{-1}(U)$ is open in X it follows that $K \cap f^{-1}(U)$ is open in K (with the relative topology), hence $f|_K$ is continuous. To see the converse let $(x_k)_{k \in \mathbb{N}}$, $x_k \in X$, be a sequence converging in X to x. The set $\{x_k | k \in \mathbb{N}\} \cup \{x\}$ is compact and the continuity of f on this set implies that $(f(x_k))_{k \in \mathbb{N}}$ converges to $f(x)$, i.e. f is continuous on X.

10. Without loss of generality we may assume $n = 2$ and $j = 1$. First we note that the projection of an open ball $B_{\rho}(x) \subset \mathbb{R}^2$ onto the x_1-axis is the open interval $(x_1 - \rho, x_1 + \rho)$, see the following figure

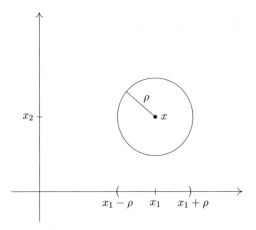

If $U \subset \mathbb{R}^2$ is open then for $x \in U$ there exists $B_\rho(x) \subset U$ and hence for every $x_1 \in \mathrm{pr}_1(U)$ there exists $\rho > 0$ such that $(x_1 - \rho, x_1 + \rho) \in \mathrm{pr}_1(U)$ implying that $\mathrm{pr}_1(U)$ is open.

11. Recall that $C \subset \mathbb{R}^n$ is convex if for every pair of points $x, y \in C$ the line segment connecting x with y, i.e. $\{z = \lambda x + (1 - \lambda)y | 0 \leq \lambda \leq 1\}$, belongs to C. Since the line segment connecting x with y is a path $\gamma(\lambda) = \lambda x + (1 - \lambda)y$, C is pathwise connected. Convex sets are studied in more detail in Chapter 13, some applications are discussed in Chapter 6.

12.   a) First note that a line segment in $\mathbb{R}^n$ is connected, in fact it is pathwise connected. Since for a chain $S = S_{a_0,a_1} \cup S_{a_1,a_2} \cup .... \cup S_{a_{N-1},a_N}$ we find that $S_{a_j,a_{j+1}} \cap S_{a_{j+1},a_{j+2}}$ contains at least the point a_{j+1} we can apply Corollary 3.33.

    b) This result follows immediately from Corollary 3.35: S is the union of the connected sets S_{x,x_0} and they all contain x_0.

13. For $U \subset X$ open we have $\partial U = \bar{U} \setminus \mathring{U} = \bar{U} \setminus U$. Thus $\partial U = \emptyset$ if and only if U is closed. The connectivity of X implies however that X and $\emptyset$ are the only sets being both closed and open. Since by assumption X is connected, $U \neq \emptyset$ and $U \neq X$, we conclude $\partial U \neq \emptyset$.

14. This follows immediately from Proposition 3.37.

15. Define $h_n(x) := f(x) - f_n(x) \geq 0$. We have to prove that $(h_n)_{n \in \mathbb{N}}$ converges uniformly on K to the zero function 0. The pointwise convergence of $(h_n)_{n \in \mathbb{N}}$ to 0 implies that for $\epsilon > 0$ and $x \in K$ there exists $N_{\epsilon,x}$ such that $0 \leq h_{N_{\epsilon,x}}(x) < \frac{\epsilon}{2}$. The continuity of h_n and the monotonicity of the sequence $(h_n)_{n \in \mathbb{N}}$ implies that for every $x \in K$ exists an open neighbourhood $U(x)$ such that $0 \leq h_n(y) < \epsilon$ for $n \geq N_{\epsilon,x}$ and all $y \in U(x)$. The family $(U(x))_{x \in K}$ is an open covering of K. Hence by compactness we can cover K by finitely many sets $U(x_1), ..., U(x_m)$. For $M_0 := \max\{N_{\epsilon,x_1}, ..., N_{\epsilon,x_M}\}$ it follows that $n \geq M_0$ implies for all $x \in K$

that $0 \leq h_n(x) < \epsilon$, i.e. $\sup_{x \in K} |h_n(x)| \leq \epsilon$, which means that $(h_n)_{n \in \mathbb{N}}$ converges uniformly to 0.

16. (This problem is taken from A. W. Knapp [30]. Also compare this problem with our construction of the square root in Example 17 in Chapter I.17.)

Given the recursion $p_{k+1}(x) = p_k(x) + \frac{1}{2}(x - p_k^2(x))$, $p_0(x) = 0$, we find that $p_{k+1}(x) - p_k(x) = \frac{1}{2}(x - p_k^2(x))$ and by induction we see that for $x \in [0, 1]$ it follows that $p_{k+1}(x) \geq p_k(x)$. Moreover we find

$$p_{k+1}(x) = p_k(x) + \frac{1}{2}(\sqrt{x} + p_k(x))(\sqrt{x} - p_k(x))$$
$$\leq p_k(x) + (\sqrt{x} - p_k(x)) = \sqrt{x},$$

where we used that $\sqrt{x} + p_k(x) \leq 2$ for $x \in [0, 1]$. Thus for every $x \in [0, 1]$ the sequence $(p_k(x))_{k \in \mathbb{N}}$ is increasing and bounded, hence convergent. Passing in

$$p_{k+1}(x) = p_k(x) + \frac{1}{2}(x - p_k^2(x))$$

to the limit we arrive at

$$p(x) = p(x) + \frac{1}{2}(x - p^2(x)),$$

or $p(x) = \sqrt{x}$, but now we can apply Dini's theorem to find that the convergence is uniform on $[0, 1]$.

Chapter 4

1. a) On the line $x = 1$ we find for $y \neq -1$ that $f(1, y) = 0$ and for $y = -1$ we have $f(1, -1) = 0$ by definition, so $\lim_{\substack{y \to -1 \\ x=1}} f(x, y) = 0$.

   b) On the line $y = -1$ we have for $x \neq 1$ that $f(x, -1) = 0$ and for $x = 1$ it follows that $f(1, -1) = 0$, i.e. $\lim_{\substack{x \to 1 \\ y=-1}} f(x, y) = 0$.

   c) Along the line $(y + 1) = a(x - 1)$, $a \neq 0$, and $(x, y) \neq (1, -1)$ we note that $y = a(x - 1) - 1$ and hence

   $$f(x, y) = f(x, a(x - 1) - 1)$$
   $$= \frac{(x - 1)(a(x - 1) - 1 + 1)^2}{(x - 1)^2 + (a(x - 1) - 1 + 1)^4}$$
   $$= \frac{a^2(x - 1)^3}{(x - 1)^2 + a^4(x - 1)^4}$$
   $$= \frac{a^2(x - 1)}{1 + a^4(x - 1)^2},$$

   implying

   $$\lim_{\substack{(x,y) \to (1,-1) \\ (y+1)=a(x-1)}} f(x, y) = \lim_{x \to 1} \frac{a^2(x - 1)}{1 + a^4(x - 1)^2} = 0.$$

d) For $(y+1)^2 = (x-1)$ but $(x,y) \neq (1,-1)$ we find

$$f(x,y) = \frac{(x-1)(x-1)}{(x-1)^2 + (x-1)^2} = \frac{1}{2},$$

which yields

$$\lim_{\substack{(x,y)\to(1,-1) \\ (y+1)^2=(x-1)}} f(x,y) = \frac{1}{2}(\neq f(1,-1)).$$

2. The straight lines passing through $(2,2)$ are given by $(y-2) = a(x-2)$ for $a \in \mathbb{R}$. This implies for $f(x,y)$, $(x,y) \neq (2,2)$, that

$$f(x,y) = \frac{(x-2)^2 a(x-2)}{(x-2)^6 + a^2(x-2)^2} = \frac{a(x-2)}{(x-2)^4 + a^2}.$$

If $a \neq 0$ it follows that

$$\lim_{\substack{(x,y)\to(2,2) \\ (y-2)=a(x-2)}} f(x,y) = 0,$$

while for $a = 0$ we have $y = 2$ and therefore $f(x,2) = 0$ for all $x \neq 2$, and the limit for $x \to 2$ is again 0. However for $(x,y) \neq (2,2)$ and $(y-2) = (x-2)^3$ we have

$$f(x,y) = \frac{(x-2)^2(x-2)^3}{(x-2)^6 + (x-2)^6} = \frac{1}{x-2}$$

and $f(x,y)$ becomes unbounded for $x \to 2$ and $(y-2) = (x-2)^3$, in particular the limit for $(x,y) \to (2,2)$ along the parabola $(y-2) = (x-2)^3$ does not exist.

Remark: Let $f : G \to \mathbb{R}$, $G \subset \mathbb{R}^n$ open, be a function and $x_0 \in G$. If f has the property that along all straight line segments belonging to G and passing through x_0 the limit $f(x)$ for $x \to x_0$ exists and equals $f(x_0)$, then we call f **radially continuous** at x_0. Problems 1 and 2 have shown that radial continuity does not imply continuity.

3. Along any straight line parallel to the line $x = y$ the function χ_H is constant, either with value 1 or with value 0, hence it is continuous on all these lines, but of course it is not continuous on $\mathbb{R}^2$. Interesting is the line $x = y$ itself: On this line χ_H is constant with value 0, hence continuous along this line, but along any other line passing through a point $(x,y) = (x,x)$ the function χ_H is not continuous.

4. In order to be continuous in G, f must be continuous at every point $x \in G$. For $x \in G$ we find $\rho > 0$ such that $B_\rho(x) \subset G$ and for all $y \in B_\rho(x)$ holds (with $0 < \alpha \leq 1$ being the Hölder exponent of f).

$$\|f(x) - f(y)\| \leq c_x \|x - y\|^\alpha.$$

Given $\epsilon > 0$ we choose $\delta = \min\left(\rho, \frac{\epsilon^{\frac{1}{\alpha}}}{c_x^{\frac{1}{\alpha}}}\right)$ to find for $\|x - y\| < \delta$ that

$$\|f(x) - f(y)\| \leq c_x \|x - y\|^\alpha < c_x \left(\frac{\epsilon^{\frac{1}{\alpha}}}{c_x^{\frac{1}{\alpha}}}\right)^\alpha = \epsilon,$$

i.e. f is continuous at x.

The mapping $g : \mathbb{R} \to \mathbb{R}$, $g(x) = x^2$, is locally but not globally Hölder continuous. The local Hölder continuity holds by the following argument:

$$|g(x) - g(y)| = |x^2 - y^2| = |x + y||x - y|,$$

so for $\rho = 1$ and $0 < \alpha \leq 1$ we find for $y \in B_1(x)$ that

$$|g(x) - g(y)| = (|x + y||x - y|^{1-\alpha})|x - y|^{\alpha}$$

and further, since $y \in B_1(x)$, i.e. $|y| \leq |x| + 1$,

$$|g(x) - g(y)| \leq (2|x| + 1)^{2-\alpha}|x - y|^{\alpha}$$

proving the local Hölder continuity for g with any exponent $0 < \alpha \leq 1$. However g is not globally Hölder continuous for any $0 < \alpha \leq 1$. Suppose g were globally Hölder continuous with exponent $0 < \alpha \leq 1$. Then we find

$$\frac{|g(x) - g(y)|}{|x - y|^{\alpha}} \leq c \text{ for all } x \neq y, x, y \in \mathbb{R}.$$

This means

$$c \geq \frac{|g(x) - g(y)|}{|x - y|^{\alpha}} = \frac{|x - y||x + y|}{|x - y|^{\alpha}} = |x + y||x - y|^{1-\alpha}$$

for all $x, y \in \mathbb{R}$. Taking $y = 2x$ we deduce $c \geq 3|x|^{2-\alpha}$ which is impossible.

5. Since f is locally Hölder continuous with exponent $0 < \alpha \leq 1$ on $[-2, 2]$ we find some $\rho \in (0, \frac{1}{2})$ such that with some $K > 0$

$$|f(y)| = |f(y) - f(0)| \leq K|y - 0|^{\alpha} = K|y|^{\alpha}$$

for all $y \in B_{\rho}(0) = (-\rho, \rho)$. Now

$$\int_0^1 \frac{f(y)}{y} dy = \int_0^{\rho} \frac{f(y)}{y} dy + \int_{\rho}^1 \frac{f(y)}{y} dy,$$

and the second integral is finite as integral of a continuous function on a compact interval. For first integral we note for $0 < \epsilon < \rho$

$$\left| \int_{\epsilon}^{\rho} \frac{f(y)}{y} dy \right| \leq \int_{\epsilon}^{\rho} \frac{|f(y)|}{y} dy \leq K \int_{\epsilon}^{\rho} \frac{|y|^{\alpha}}{y} dy$$

$$= K \int_{\epsilon}^{\rho} y^{1-\alpha} dy = \frac{K}{2 - \alpha}(\rho^{2-\alpha} - \epsilon^{2-\alpha}),$$

and in the limit we have

$$\int_0^{\rho} \frac{f(y)}{y} dy = \lim_{\epsilon \to 0} \int_{\epsilon}^{\rho} \frac{f(y)}{y} dy \text{ exists,}$$

611

hence $\int_0^1 \frac{f(y)}{y} dy$ exists.

Remark: For a non-zero constant function the result does not hold, the singularity of $\frac{1}{y}$ is too "strong" and $\int_0^1 \frac{c}{y} dy$ does not exist. We see here that the better regularity of f at $y = 0$ compensates the singularity of $y \mapsto \frac{1}{y}$ at 0. When handling potentials, say the Newton potential in Part 5, we see higher dimensional analogues of this type of compensation of singularities.

6. Let $x \in G$ be any point. Since G is open we can find $\rho > 0$ such that $\overline{B_\rho(x)} \subset B_{2\rho}(x) \subset G$, but $\overline{B_\rho(x)}$ is compact, hence $f(\overline{B_\rho(x)})$ is compact as the image of a compact set under a continuous function. Since compact sets are bounded it follows that f is locally bounded.

The identity $\text{id} : \mathbb{R}^n \to \mathbb{R}^n$ is of course continuous, hence locally bounded, but since $\text{id}(\mathbb{R}^n) = \mathbb{R}^n$ it is not globally bounded.

7. a) We just need to run through $[a, b]$ in the reverse order, so $\gamma : [a, b] \to \mathbb{R}^2$, $\gamma(t) = (-t + a + b, f(-t + a + b))$, is a curve with the desired property.

b) The curve $\gamma : [0, 2\pi] \to \mathbb{R}^2$, $\gamma(t) = (a \cos t, b \sin t)$ is continuous and simply closed and its trace is $\mathcal{E}$.

c) The curve $\gamma : [0, \frac{1}{2}] \to \mathbb{R}^2$, $\gamma(t) = (\cos 4\pi t, \sin 4\pi t)$ has the demanded properties.

8. The Lipschitz condition implies

$$\|\gamma(t_j) - \gamma(t_{j-1})\| \le K(t_j - t_{j-1})$$

and therefore

$$V(\gamma, Z) = \sum_{j=1}^{k+1} \|\gamma(t_j) - \gamma(t_{j-1})\| \le \sum_{j=1}^{k+1} \kappa(t_j - t_{j-1}) = \kappa(b - a).$$

Since the bound is independent of Z, we deduce

$$V(\gamma) \le \kappa(b - a).$$

Curves for which $V(\gamma)$ is finite are called **rectifiable** curves and we will study them in greater detail soon in Chapter 15. The argument fails when the Lipschitz condition is replaced by a Hölder condition with exponent $0 < \alpha < 1$. We may take an equidistance partition $Z(t_1, ..., t_{k+1})$, $t_j - t_{j-1} = \frac{b-a}{k+1}$, $j = 1, ..., k + 1$, to find

$$\sum_{j=1}^{k+1} \|\gamma(t_j) - \gamma(t_{j-1})\| \le \kappa(b - a)^\alpha \sum_{j=1}^{k+1} \frac{1}{(k+1)^\alpha}$$

$$= \kappa(b - a)^\alpha (k+1) \frac{1}{(k+1)^\alpha} = K(b - a)^\alpha (k+1)^{1-\alpha}.$$

Since for $k \to \infty$ the right hand side diverges for $\alpha < 1$, we cannot deduce that $V(\gamma)$ is in this case finite.

Note: this is not (yet) a proof that γ is not rectifiable, this only shows, as required, that the argument used for Lipschitz curves does not work in this case.

9. First we look at the following figure:

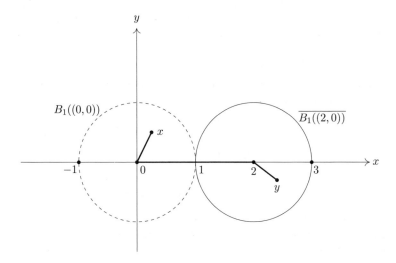

We note that $[0,2] \times \{0\} \subset B_1((0,0)) \cap \overline{B_1((2,0))}$: For $0 \leq x \leq 1$ it follows that $(x,0) \in B_1((0,0))$ and for $1 \leq x \leq 2$ it follows that $(x,0) \in \overline{B_1((2,0))}$. Now, if x_1, x_2 are two points belonging either both to $B_1((0,0))$ or to $\overline{B_1((2,0))}$ we can connect them with an arc within these sets since both sets are convex, hence the line segment connecting x_1 with x_2 will do. If however $x_1 \in B_1((0,0))$ and $x_2 \in \overline{B_1((2,0))}$ we first connect x_1 with $(0,0)$ by a line segment belonging entirely to $B_1((0,0))$, then we connect $(0,0)$ to $(2,0)$ by the line segment $[0,2] \times \{0\}$ which belongs to $B_1((0,0)) \cup \overline{B_1((2,0))}$, and eventually we can connect in $\overline{B_1((2,0))}$ the point $(2,0)$ with x_2 by a line segment. Thus $B_1((0,0)) \cup \overline{B_1((2,0))}$ is arcwise connected.

10. Since $M(n,\mathbb{R})$ can be identified with $\mathbb{R}^{n^2}$ we can identify $GL(n,\mathbb{R})$ as subset of $\mathbb{R}^{n^2}$. On $M(n,\mathbb{R})$, hence on $\mathbb{R}^{n^2}$, we have

$$\begin{pmatrix} a_{11} \\ \vdots \\ a_{1n} \\ \vdots \\ a_{nn} \end{pmatrix} \mapsto \det \begin{pmatrix} a_{11} & \cdots & a_{1,n} \\ \vdots & & \vdots \\ a_{n1} & \cdots & a_{nn} \end{pmatrix} = \sum_{\sigma \in S_n} (\operatorname{sgn}\sigma) a_{1\sigma(1)} \cdots a_{n\sigma(n)},$$

implying the continuity of det. Now

$$GL(n,\mathbb{R}) = \{A \in M(n,\mathbb{R})| \det A \neq 0\} = \det^{-1}(\mathbb{R} \setminus \{0\})$$

and the openness of $\mathbb{R} \setminus \{0\}$ implies the openness of $GL(n,\mathbb{R})$ in $M(n,\mathbb{R})$. Furthermore we have

$$\{A \in GL(n,\mathbb{R})| \det A > 0\} = \{A \in GL(n,\mathbb{R})| \det A < 0\}^c.$$

613

Both sets are open, the first is the set $\det^{-1}((0,\infty))$, the second is the set $\det^{-1}((-\infty,0))$. Hence their complements are closed, but this implies that these sets are both open and closed, and therefore $GL(n,\mathbb{R})$ is not connected but has two connectivity components. Since $\det(\mathrm{id}_n) = 1$ the identity component in $GL(n,\mathbb{R})$ is the set

$$\det^{-1}((0,\infty)) = \{A \in GL(n,\mathbb{R})|\det A > 0\}.$$

11. Elements in $SO(2)$ are proper rotations about the origin, thus $U \in SO(2)$ if there exists $\varphi \in [0, 2\pi]$ such that

$$U = \begin{pmatrix} \cos\varphi & -\sin\varphi \\ \sin\varphi & \cos\varphi \end{pmatrix}.$$

We now consider $\gamma : [0, 2\pi] \to GL(n,\mathbb{R}) \subset M(n,\mathbb{R})$ defined by

$$\gamma(\varphi) = U(\varphi) = \begin{pmatrix} \cos\varphi & -\sin\varphi \\ \sin\varphi & \cos\varphi \end{pmatrix}$$

which is continuous and $\gamma(\varphi) = \gamma(\varphi + 2\pi)$. Moreover, $\gamma|_{[0,2\pi)}$ is injective which follows from basic properties of the trigonometrical functions $\cos$ and $\sin$, i.e. the relations of their values.

Chapter 5

1. We have

$$\frac{\partial h(t,x)}{\partial t} = \frac{\partial}{\partial t}\left((4\pi t)^{-\frac{n}{2}} e^{\frac{-\|x\|^2}{4t}}\right)$$

$$= \left(\frac{\partial}{\partial t}(4\pi t)^{-\frac{n}{2}}\right) e^{\frac{-\|x\|^2}{4t}} + (4\pi t)^{-\frac{n}{2}}\frac{\partial}{\partial t}\left(e^{\frac{-\|x\|^2}{4t}}\right)$$

$$= (4\pi t)^{-\frac{n}{2}}\left(\frac{-n}{2t}\right) e^{\frac{-\|x\|^2}{4t}} + (4\pi t)^{-\frac{n}{2}}\frac{\partial}{\partial t}\left(\frac{-\|x\|^2}{4t}\right) e^{\frac{-\|x\|^2}{4t}}$$

$$= (4\pi t)^{-\frac{n}{2}}\left(-\frac{n}{2t} + \frac{\|x\|^2}{4t^2}\right) e^{\frac{-\|x\|^2}{4t}},$$

as well as

$$\frac{\partial h(t,x)}{\partial x_j} = \frac{\partial}{\partial x_j}\left((4\pi t)^{-\frac{n}{2}} e^{\frac{-\|x\|^2}{4t}}\right)$$

$$= (4\pi t)^{-\frac{n}{2}}\left(\frac{-2x_j}{4t}\right) e^{\frac{-\|x\|^2}{4t}} = (4\pi t)^{-\frac{n}{2}}\left(\frac{-x_j}{2t}\right) e^{\frac{-\|x\|^2}{4t}},$$

and

$$\frac{\partial^2 h(t,x)}{\partial x_j^2} = (4\pi t)^{-\frac{n}{2}}\frac{\partial}{\partial x_j}\left(\frac{-x_j}{2t} e^{\frac{-\|x\|^2}{4t}}\right)$$

$$= (4\pi t)^{-\frac{n}{2}}\left(\frac{-1}{2t} e^{\frac{-\|x\|^2}{4t}} - \frac{x_j}{2t}\left(\frac{-x_j}{2t}\right) e^{\frac{-\|x\|^2}{4t}}\right)$$

$$= (4\pi t)^{-\frac{n}{2}}\left(\frac{-1}{2t} + \frac{x_j^2}{4t^2}\right) e^{\frac{-\|x\|^2}{4t}},$$

which implies

$$\Delta_n h(t,x) = \sum_{j=1}^{n} \frac{\partial^2 h(t,x)}{\partial x_j^2} = (4\pi t)^{-\frac{n}{2}} \sum_{j=1}^{n} \left(\frac{-1}{2t} + \frac{x_j^2}{4t^2} \right) e^{-\frac{\|x\|^2}{4t}}$$

$$= (4\pi t)^{-\frac{n}{2}} \left(\frac{-n}{2t} + \frac{\|x\|^2}{4t^2} \right) e^{-\frac{\|x\|^2}{4t}},$$

and consequently $(\frac{\partial}{\partial t} - \Delta_u) h(t,x) = 0$.

2. First we note that

$$\left(\frac{1}{r} \frac{\partial}{\partial r} \left(r \frac{\partial}{\partial r} \right) + \frac{1}{r^2} \frac{\partial^2}{\partial \varphi^2} \right) \left(\sum_{k=0}^{N} c_k r^k \cos k\varphi \right)$$

$$= \sum_{k=0}^{N} c_k \left(\frac{1}{r} \frac{\partial}{\partial r} \left(r \frac{\partial}{\partial r} \right) + \frac{1}{r^2} \frac{\partial^2}{\partial \varphi^2} \right) (r^k \cos k\varphi).$$

Now we consider

$$\left(\frac{1}{r} \frac{\partial}{\partial r} \left(r \frac{\partial}{\partial r} \right) + \frac{1}{r^2} \frac{\partial^2}{\partial \varphi^2} \right) (r^k \cos k\varphi)$$

$$= \frac{1}{r} \frac{\partial}{\partial r} \left(r \frac{\partial}{\partial r} \right) (r^k \cos k\varphi) + \frac{1}{r^2} \frac{\partial^2}{\partial \varphi^2} (r^k \cos k\varphi)$$

$$= \frac{1}{r} \frac{\partial}{\partial r} (r k r^{k-1} \cos k\varphi) + \frac{1}{r^2} \frac{\partial}{\partial \varphi} (-k r^k \sin k\varphi)$$

$$= \frac{1}{r} k \cos k\varphi \frac{\partial}{\partial r} r^k - k r^{k-2} \frac{\partial}{\partial \varphi} \sin k\varphi$$

$$= k^2 r^{k-2} \cos k\varphi - k^2 r^{k-2} \cos k\varphi = 0,$$

which yields

$$\left(\frac{1}{r} \frac{\partial}{\partial r} \left(r \frac{\partial}{\partial r} \right) + \frac{1}{r^2} \frac{\partial^2}{\partial \varphi^2} \right) \left(\sum_{k=0}^{N} c_k r^k \cos^k \varphi \right) = 0.$$

3. Since

$$L = \frac{1}{\sin \vartheta} \frac{\partial}{\partial \vartheta} \left(\sin \vartheta \frac{\partial}{\partial \vartheta} \right) + \frac{1}{\sin^2 \vartheta} \frac{\partial^2}{\partial \varphi^2}$$

$$= \frac{\partial^2}{\partial \vartheta^2} + \frac{\cos \vartheta}{\sin \vartheta} \frac{\partial}{\partial \vartheta} + \frac{1}{\sin^2 \vartheta} \frac{\partial^2}{\partial \varphi^2}$$

we find

$$LY_{11}(\vartheta, \varphi) = \left(\frac{\partial^2}{\partial \vartheta^2} + \frac{\cos \vartheta}{\sin \vartheta} \frac{\partial}{\partial \vartheta} + \frac{1}{\sin^2 \vartheta} \frac{\partial^2}{\partial \varphi^2} \right) Y_{11}(\vartheta, \varphi)$$

$$= \left(\frac{\partial^2}{\partial \vartheta^2} + \frac{\cos \vartheta}{\sin \vartheta} \frac{\partial}{\partial \vartheta} + \frac{1}{\sin^2 \vartheta} \frac{\partial^2}{\partial \varphi^2} \right) \sqrt{\frac{3}{8\pi}} \sin \vartheta \cos \varphi$$

$$= \sqrt{\frac{3}{8\pi}} \left(- \sin \vartheta \cos \varphi + \frac{\cos^2 \vartheta \cos \varphi}{\sin \vartheta} - \frac{\cos \varphi}{\sin \vartheta} \right)$$

$$= \sqrt{\frac{3}{8\pi}} \left(\sin \vartheta \cos \varphi - 2 \sin \vartheta \cos \varphi + \frac{\cos^2 \vartheta \cos \varphi}{\sin \vartheta} - \frac{\cos \varphi}{\sin \vartheta} \right)$$

$$= -2 \sqrt{\frac{3}{8\pi}} \sin \vartheta \cos \varphi + \sqrt{\frac{3}{8\pi}} \left(\frac{\sin^2 \vartheta \cos \varphi}{\sin \vartheta} + \frac{\cos^2 \vartheta \cos \varphi}{\sin \vartheta} - \frac{\cos \varphi}{\sin \vartheta} \right)$$

$$= -2 Y_{11}(\vartheta, \varphi) + \sqrt{\frac{3}{8\pi}} \left(\frac{(\sin^2 \vartheta + \cos^2 \vartheta) \cos \varphi}{\sin \vartheta} - \frac{\cos \varphi}{\sin \vartheta} \right)$$

$$= -2 Y_{11}(\vartheta, \varphi).$$

4. First note that for $T \in O(n)$

$$T = \begin{pmatrix} t_{11} & \cdots & t_{1n} \\ \vdots & & \vdots \\ t_{n1} & \cdots & t_{nn} \end{pmatrix}$$

we have with $\vec{t_j} = \begin{pmatrix} t_{1j} \\ \vdots \\ t_{nj} \end{pmatrix}$ the two relations

$$\langle \vec{t_j}, \vec{t_j} \rangle = 1 \text{ and } \langle \vec{t_j}, \vec{t_l} \rangle = 0 \text{ for } j \neq l.$$

Now we observe

$$(u \circ T)(x) = u(Tx) = u \left(\sum_{l_1=1}^{n} t_{1l_1} x_{l_1}, ..., \sum_{l_n=1}^{n} t_{nl_n} x_{l_n} \right)$$

which yields

$$\frac{\partial}{\partial x_j}(u \circ T)(x) = \frac{\partial}{\partial x_j} \left(\sum_{l_1=1}^{n} t_{1l_1} x_{l_1} \right) u_{x_1}(Tx) + ... + \frac{\partial}{\partial x_j} \left(\sum_{l_n=1}^{n} t_{nl_n} x_{l_n} \right) u_{x_n}(Tx)$$

$$= t_{1j} u_{x_1}(Tx) + ... + t_{nj} u_{x_n}(Tx)$$

and further

$$\frac{\partial^2}{\partial x_j^2}(u \circ T)(x) = \frac{\partial}{\partial x_j}\left(\sum_{k=1}^{n} t_{kj} u_{x_k}(Tx)\right)$$

$$= \sum_{k=1}^{n} t_{kj} \frac{\partial}{\partial x_j} u_{x_k}(Tx)$$

$$= \sum_{k=1}^{n} t_{kj} \sum_{m=1}^{n} t_{mj} u_{x_k x_m}(Tx),$$

which implies

$$(\Delta_n(u \circ T))(x) = \sum_{j=1}^{n}\sum_{k=1}^{n}\sum_{m=1}^{n} t_{kj} t_{mj} u_{x_k} u_{x_m}(Tx)$$

$$= \sum_{k=1}^{n}\sum_{m=1}^{n} \delta_{km} u_{x_k} u_{x_m}(Tx)$$

$$= \sum_{k=1}^{n} u_{x_k} u_{x_k}(Tx) = (\Delta_n u)(Tx).$$

5. For $(x, y) \neq (0, 2)$ we may apply the standard rules of differentation to find

$$\frac{\partial}{\partial x}\left(\frac{x^2(y-2)^2}{x^6 + (y-2)^6}\right) = \frac{\frac{\partial}{\partial x}(x^2(y-2)^2)(x^6 + (y-2)^6) - x^2(y-2)^2\frac{\partial}{\partial x}(x^6 + (y-2)^6)}{(x^6 + (y-1)^6)^2}$$

$$= \frac{2x(y-2)^2(x^6 + (y-2)^6) - 6x^7(y-2)^2}{(x^6 + (y-2)^6)^2}$$

$$= 2x(y-2)^2 \frac{(y-2)^6 - 2x^6}{(x^6 + (y-2)^6)^2}$$

and

$$\frac{\partial}{\partial y}\left(\frac{x^2(y-2)^2}{x^6 + (y-2)^6}\right) = \frac{\frac{\partial}{\partial y}(x^2(y-2)^2)(x^6 + (y-2)^6) - x^2(y-2)^2\frac{\partial}{\partial y}(x^6 + (y-2)^6)}{(x^6 + (y-1)^6)^2}$$

$$= \frac{2x^2(y-2)(x^6 + (y-2)^6) - x^2(y-2)^2 6(y-2)^5}{(x^6 + (y-2)^6)^2}$$

$$= 2x^2(y-2)\frac{x^6 - 2(y-2)^6}{(x^6 + (y-2)^6)^2}.$$

For $(x, y) = (0, 2)$ we have to investigate

$$\lim_{x \to 0} \frac{f(x, 2) - f(0, 2)}{x} = \lim_{x \to 0} \frac{x^2(2-2)^2}{x^7} = 0$$

617

and

$$\lim_{y \to 2} \frac{f(0, y) - f(0, 2)}{y - 2} = \lim_{y \to 2} \frac{0(y - 2)^2}{(y - 2)^7} = 0.$$

Thus we have

$$f_x(x, y) = \begin{cases} 2x(y - 2)^2 \frac{(y-2)^6 - 2x^6}{(x^6 + (y-2)^6)^2}, & (x, y) \neq (0, 2) \\ 0, & (x, y) = (0, 2) \end{cases}$$

and

$$f_y(x, y) = \begin{cases} 2x^2(y - 2) \frac{x^6 - 2(y-2)^6}{(x^6 + (y-2)^6)^2}, & (x, y) \neq (0, 2) \\ 0, & (x, y) = (0, 2) \end{cases}.$$

For the line $x = y - 2$ passing through $(0, 2)$ we find

$$f(y - 2, y) = \frac{(y - 2)^4}{2(y - 2)^6} = \frac{1}{2(y - 2)^2} \text{ for } y \neq 2$$

implying that

$$\lim_{\substack{(x,y) \to (0,2) \\ x = y - 2}} f(x, y)$$

does not exist, i.e. f is not continuous at $(0, 2)$.

6. Again, except for the critical point $(0, 0)$ we can apply standard rules to find partial derivatives. For the calculation to follow we note that

$$\frac{\partial}{\partial x} \frac{1}{\sqrt{x^2 + y^2}} = \frac{-x}{(x^2 + y^2)^{\frac{3}{2}}} \text{ and } \frac{\partial}{\partial y} \frac{1}{\sqrt{x^2 + y^2}} = \frac{-y}{(x^2 + y^2)^{\frac{3}{2}}}$$

and for a suitable function $h(x, y)$:

$$\frac{\partial}{\partial x} \ln(\ln h(x, y)) = \frac{\frac{\partial h}{\partial x}(x, y)}{h(x, y) \ln h(x, y)} \text{ and } \frac{\partial}{\partial y} \ln(\ln h(x, y)) = \frac{\frac{\partial h}{\partial y}(x, y)}{h(x, y) \ln h(x, y)}.$$

Now, for $(x, y) \neq (0, 0)$ we get

$$\frac{\partial}{\partial x} \left(xy \ln \left(\ln \frac{1}{\sqrt{x^2 + y^2}} \right) \right) = \frac{\partial}{\partial x} (xy) \ln \left(\ln \frac{1}{\sqrt{x^2 + y^2}} \right) + xy \frac{\partial}{\partial x} \ln \left(\ln \frac{1}{\sqrt{x^2 + y^2}} \right)$$

$$= y \ln \left(\ln \frac{1}{\sqrt{x^2 + y^2}} \right) + xy \frac{\frac{\partial}{\partial x} \frac{1}{\sqrt{x^2 + y^2}}}{\frac{1}{\sqrt{x^2 + y^2}} \ln \frac{1}{\sqrt{x^2 + y^2}}}$$

$$= y \ln \left(\ln \frac{1}{\sqrt{x^2 + y^2}} \right) - x^2 y \frac{\frac{1}{(x^2 + y^2)^{\frac{3}{2}}}}{\frac{1}{(x^2 + y^2)^{\frac{1}{2}}} \ln \frac{1}{\sqrt{x^2 + y^2}}}$$

$$= y \ln \left(\ln \frac{1}{\sqrt{x^2 + y^2}} \right) - \frac{x^2 y}{(x^2 + y^2) \ln \frac{1}{\sqrt{x^2 + y^2}}}$$

$$= y \ln \left(\ln \frac{1}{\sqrt{x^2 + y^2}} \right) + \frac{x^2 y}{(x^2 + y^2) \ln \frac{1}{\sqrt{x^2 + y^2}}},$$

and analogously

$$\frac{\partial}{\partial y} g(x,y) = x \ln\left(\ln \frac{1}{\sqrt{x^2+y^2}} \right) + \frac{y^2 x}{(x^2+y^2)\ln\sqrt{x^2+y^2}}.$$

For $(x,y) = (0,0)$ we find, since $g(0,y) = g(x,0) = 0$ that $g_x(0,0)$ and $g_y(0,0) = 0$. Hence we arrive at

$$\frac{\partial}{\partial x}\left(xy \ln\left(\ln \frac{1}{\sqrt{x^2+y^2}} \right) \right) = \begin{cases} y \ln\left(\ln \frac{1}{\sqrt{x^2+y^2}} \right) + \frac{x^2 y}{(x^2+y^2)\ln\sqrt{x^2+y^2}} & (x,y) \neq (0,0) \\ 0 & (x,y) = (0,0) \end{cases}$$

and

$$\frac{\partial}{\partial y}\left(xy \ln\left(\ln \frac{1}{\sqrt{x^2+y^2}} \right) \right) = \begin{cases} x \ln\left(\ln \frac{1}{\sqrt{x^2+y^2}} \right) + \frac{x^2 y}{(x^2+y^2)\ln\sqrt{x^2+y^2}} & (x,y) \neq (0,0) \\ 0 & (x,y) = (0,0) \end{cases}$$

Since $g_x(x,0) = 0$ and $g_y(0,y) = 0$ we find that g_{xx} and g_{yy} exist for all $(x,y) \in \mathbb{R}^2$, and in particular we have $g_{xx}(0,0) = g_{yy}(0,0) = 0$. However, for $(x,y) = (0,0)$ the mixed second order derivatives $\frac{\partial}{\partial y \partial x} g(x,y)$ and $\frac{\partial}{\partial x \partial y} g(x,y)$ do not exist:

For $(0,y)$ we find

$$g_x(0,y) - g_x(0,0) = y \ln\left(\ln \frac{1}{\sqrt{y^2}} \right)$$

implying that

$$\frac{g_x(0,y) - g_x(0,0)}{y} = \ln\left(\ln \frac{1}{|y|} \right)$$

and for $y \to 0$ the limit $\ln\left(\ln \frac{1}{|y|} \right)$ does not exist as a real number since $\lim_{y \to 0} \ln\left(\ln \frac{1}{|y|} \right) = \infty$. Analogously we find

$$\frac{g_y(x,0) - g_x(0,0)}{x} = \ln\left(\ln \frac{1}{|x|} \right),$$

and it follows that $\frac{\partial^2}{\partial x \partial y} g$ and $\frac{\partial^2}{\partial y \partial x} g$ do not exist at $(0,0)$.

7. We claim that

$$\frac{\partial}{\partial x_j} \prod_{k=1}^{N} v_k(x) = \sum_{l=1}^{N} \prod_{k=1}^{N} w_{kl}(x)$$

where

$$w_{kl}(x) := \begin{cases} \frac{\partial v_l}{\partial x_j}, & k = l \\ v_k, & k \neq l \end{cases}.$$

In other words

$$\frac{\partial}{\partial x_j} \prod_{k=1}^{N} v_k(x) = \frac{\partial v_1}{\partial x_j}(x)v_2(x) \cdot \ldots \cdot v_N(x) + v_1(x)\frac{\partial v_2}{\partial x_j}(x)v_3(x) \cdot' ldots \cdot v_N(x) + \ldots$$

$$\ldots + v_1(x) \cdot \ldots \cdot v_{N-1}(x)\frac{\partial v_N(x)}{\partial x_j},$$

and the proof uses of course mathematical induction. For $N = 1$ nothing is to prove and for $N = 2$ we just apply Leibniz's rule

$$\frac{\partial}{\partial x_j} v_1(x)v_2(x) = \frac{\partial v_1}{\partial x_j}(x)v_2(x) + v_1(x)\frac{\partial v_2}{\partial x_j}(x).$$

Now if

$$\frac{\partial}{\partial x_j} \prod_{k=1}^{N} v_k(x) = \frac{\partial v_1}{\partial x_j}(x)v_2(x) \cdot \ldots \cdot v_N(x) + \ldots + v_1(x) \cdot \ldots \cdot v_{N-1}(x)\frac{\partial v_N}{\partial x_j}(x)$$

we find

$$\frac{\partial}{\partial x_j} \prod_{k=1}^{N+1} v_k(x) = \frac{\partial}{\partial x_j} \prod_{k=1}^{N} \tilde{v}_k(x)$$

with $\tilde{v}_k = v_k$ for $1 \le k \le N-1$ and $\tilde{v}_N = v_N v_{N+1}$.

It follows that

$$\frac{\partial}{\partial x_j} \prod_{k=1}^{N+1} v_k(x) = \frac{\partial}{\partial x_j} \prod_{k=1}^{N} \tilde{v}_k(x)$$

$$= \frac{\partial v_1}{\partial x_j}(x)v_2(x) \cdot \ldots \cdot \tilde{v}_N(x) + \ldots$$

$$+ v_1(x) \cdot \ldots \cdot v_{N-2}(x)\frac{\partial v_{N-1}}{\partial x_j}(x)\tilde{v}_N(x)$$

$$+ v_1(x) \cdot \ldots \cdot v_{N-1}(x)\frac{\partial \tilde{v}_N}{\partial x_j}(x)$$

$$= \frac{\partial v_1}{\partial x_j}(x)v_2(x) \cdot \ldots \cdot v_N(x)v_{N+1}(x) + \ldots$$

$$+ v_1(x) \cdot \ldots \cdot v_{N-2}(x)\frac{\partial v_{N-1}}{\partial x_j}(x)v_N(x)v_{N+1}(x)$$

$$+ v_1(x) \cdot \ldots \cdot v_{N-1}(x)\left(\frac{\partial v_N}{\partial x_j}(x)v_{N+1}(x) + v_N(x)\frac{\partial v_{N+1}}{\partial x_j}\right)(x),$$

and the result follows.

8. We find by the chain rule

$$\frac{\partial}{\partial \rho} H(\rho, \varphi) = \left(\frac{\partial f}{\partial x}\right)(\rho \cos \varphi, \rho \sin \varphi)\frac{\partial(\rho \cos \varphi)}{\partial \rho} + \left(\frac{\partial f}{\partial y}\right)(\rho \cos \varphi, \rho \sin \varphi)\frac{\partial(\rho \sin \varphi)}{\partial \rho}$$

and

$$\frac{\partial}{\partial\varphi}H(\rho,\varphi) = \left(\frac{\partial f}{\partial x}\right)(\rho\cos\varphi,\rho\sin\varphi)\frac{\partial(\rho\cos\varphi)}{\partial\rho} + \left(\frac{\partial f}{\partial y}\right)(\rho\cos\varphi,\rho\sin\varphi)\frac{\partial(\rho\sin\varphi)}{\partial\rho}$$

or

$$\frac{\partial H(\rho,\varphi)}{\partial\rho} = \left(\frac{\partial f}{\partial x}\right)(\rho\cos\varphi,\rho\sin\varphi)\cos\varphi + \left(\frac{\partial f}{\partial y}\right)(\rho\cos\varphi,\rho\sin\varphi)\sin\varphi$$

$$\frac{\partial H(\rho,\varphi)}{\partial\varphi} = \left(\frac{\partial f}{\partial x}\right)(\rho\cos\varphi,\rho\sin\varphi)(-\rho\sin\varphi) + (\frac{\partial f}{\partial y})(\rho\cos\varphi,\rho\sin\varphi)(\rho\cos\varphi)$$

implying (while suppressing the arguments for H and f)

$$\left(\frac{\partial H}{\partial\rho}\right)^2 = \left(\frac{\partial f}{\partial x}\right)^2\cos^2\varphi + 2\frac{\partial f}{\partial x}\frac{\partial f}{\partial y}\cos\varphi\sin\varphi + \left(\frac{\partial f}{\partial y}\right)^2\sin^2\varphi$$

$$\left(\frac{1}{\rho}\frac{\partial H}{\partial\varphi}\right)^2 = \left(\frac{\partial f}{\partial x}\right)^2\sin^2\varphi - 2\frac{\partial f}{\partial x}\frac{\partial f}{\partial y}\cos\varphi\sin\varphi + \left(\frac{\partial f}{\partial y}\right)^2\cos^2\varphi$$

which by using $\sin^2\varphi + \cos^2\varphi = 1$ eventually yields that

$$\left(\frac{\partial H}{\partial\rho}\right)^2 + \left(\frac{1}{\rho}\frac{\partial H}{\partial\varphi}\right)^2 = \left(\frac{\partial f}{\partial x}\right)^2 + \left(\frac{\partial f}{\partial y}\right)^2.$$

9. Since for $0 < x < \pi$ we have $0 < \sin x < 1$, we conclude that $g(x) = \sqrt{\sin x}$ is defined and $(\sqrt{\sin x})^4 = \sin^2 x$. Therefore it follows that

$$f(x,g(x)) = \cos^2 x + (\sqrt{\sin x})^4 - 1 = \cos^2 x + \sin^2 x - 1 = 0,$$

which implies

$$0 = \frac{d}{dx}f(x,g(x)).$$

On the other hand we have

$$\frac{d}{dx}f(x,g(x)) = \frac{\partial f}{\partial x}(x,g(x)) + \frac{\partial f}{\partial y}(x,g(x))g'(x)$$

and for $\frac{\partial f}{\partial y}(x,g(x)) \neq 0$ we arrive at

$$g'(x) = -\frac{\frac{\partial f}{\partial x}(x,g(x))}{\frac{\partial f}{\partial y}(x,g(x))} = -\frac{-2\cos x\sin x}{4\sqrt{\sin x}^{-3}} = \frac{\cos x}{2\sqrt{\sin x}},$$

which we get of course by a direct calculation using the chain rule.

621

10. We first claim that

$$(*) \qquad \frac{\partial^m}{\partial x_j^m}(1 + x_2^2 + \ldots + x_n^2)^{\frac{-k}{2}} = \frac{P_{m,j}(x)}{(1 + x_1^2 + \ldots + x_n^2)^{\frac{k+2m}{2}}}$$

where $P_{m,j}$ is a polynomial of degree less or equal to m. For $m = 0$ this is clear, we just take $P_{0,j}(x) = 1$ for all $x \in \mathbb{R}^n$. If $(*)$ holds for m, then we find

$$\frac{\partial^{m+1}}{\partial x_j^{m+1}}(1 + x_1^2 + \ldots + x_n^2)^{\frac{-k}{2}} = \frac{\partial}{\partial x_j}\frac{P_{m,j}(x)}{(1 + x_1^2 + \ldots + x_n^2)^{\frac{k+2m}{2}}}$$

$$= \frac{(\frac{\partial}{\partial x_j}P_{m,j}(x))(1 + x_1^2 + \ldots x_n^2) - 2(\frac{k+2m}{2})x_j P_{m,j}(x)}{(1 + x_1^2 + \ldots + x_n^2)^{\frac{k+2m+2}{2}}}$$

$$= \frac{P_{m+1,j}(x)}{(1 + x_1^2 + \ldots + x_n^2)^{\frac{k+2m+2}{2}}}$$

where

$$P_{m+1,j}(x) := \left(\frac{\partial}{\partial x_j}P_{m,j}(x)\right)(x_1^2 + \ldots x_n^2) - (k + 2m)x_j P_{m,j}(x)$$

is a polynomial of degree less or equal to $m + 1$. If $Q(x)$ is a polynomial of degree l then we have for some $K > 0$

$$(**) \qquad |Q(x)| \leq K(1 + x_1^2 + \ldots x_n^2)^{\frac{l}{2}}$$

which we will prove immediately, but first we note that $(*)$ combined with $(**)$ yields

$$\left|\frac{\partial^m}{\partial x_j^m}(1 + x_1^2 + \ldots x_n^2)^{\frac{-k}{2}}\right| \leq \frac{|P_{m,j}(x)|}{(1 + x_1^2 + \ldots + x_n^2)^{\frac{k+2m}{2}}}$$

$$\leq \frac{K_{m,j}(1 + x_1^2 + \ldots + x_n^2)^{\frac{m}{2}}}{(1 + x_n^2 + \ldots + x_n^2)^{\frac{k+2m}{2}}}$$

$$= K_{m,j}(1 + x_1^2 + \ldots = x_n^2)^{\frac{k+m}{2}}.$$

To see $(**)$ it is sufficient to note that for $\alpha \in \mathbb{N}_0^n$, $|\alpha| = \alpha_1 + \ldots + \alpha_n \leq l$ it follows that

$$|x^\alpha| \leq c_\alpha(1 + x_1^2 + \ldots + x_n^2)^{\frac{l}{2}}$$

which follows from

$$|x^\alpha| = |x_1^{\alpha_1} \ldots x_n^{\alpha_n}|$$

and for $x_j \in \mathbb{R}$ we have

$$|x_j^{\alpha_j}| \leq c(1 + x_j^2)^{\frac{\alpha_j}{2}} \leq c(1 + x_1^2 + \ldots + x_n^2)^{\frac{\alpha_j}{2}}$$

implying the result.

11.   a) We have

$$\partial^{(2,1,2)}\left(e^{-\|x\|^2}\right) = \partial^{(2,1,2)}\left(e^{-x_1^2-x_2^2-x_3^2}\right)$$

$$= \partial^{(2,1,1)}\left((-2x_3)e^{-x_1^2-x_2^2-x_3^2}\right)$$

$$= \partial^{(2,1,0)}\left((4x_3^2-2)e^{-x_1^2-x_2^2-x_3^2}\right)$$

$$= \partial^{(2,0,0)}\left((4x_3^2-2)(-2x_2)e^{-x_1^2-x_2^2-x_3^2}\right)$$

$$= \partial^{(1,0,0)}\left((4x_3^2-2)(-2x_2)(-2x_1)e^{-x_1^2-x_2^2-x_3^2}\right)$$

$$= (4x_3^2-2)(-2x_2)(4x_1^2-2)e^{-x_1^2-x_2^2-x_3^2}$$

$$= -2(4x_3^2-2)x_2(4x_1^2-2)e^{-\|x\|^2}.$$

b) Since $x^\alpha = x_1^{\alpha_1}....x_n^{\alpha_n}$, if $\alpha_l = 0$ or $\alpha_k = 0$, then e^{x^α} is independent of x_l or x_k and we have of course $\frac{\partial^2}{\partial x_k \partial x_l}e^{x^\alpha} = 0$. For $\alpha_l \neq 0$ and $\alpha_k \neq 0$ we obtain

$$\frac{\partial^2}{\partial x_k \partial x_l}e^{x^\alpha} = \frac{\partial^2}{\partial x_k \partial x_l}e^{x_1^{\alpha_1}....x_n^{\alpha_n}}$$

$$= \frac{\partial}{\partial x_k}\left(\frac{\partial}{\partial x_l}(x_1^{\alpha_1}....x_n^{\alpha_n})\right)e^{x_1^{\alpha_1}....x_n^{\alpha_n}}$$

$$= \frac{\partial}{\partial x_k}(x_1^{\alpha_1}....x_{l-1}^{\alpha_{l-1}}(\alpha_l x_l^{\alpha_l-1})x_{l+1}^{\alpha_{l+1}}....x_n^{\alpha_n})e^{x_1^{\alpha_1}....x_n^{\alpha_n}}$$

$$= \frac{\partial}{\partial x_k}(\alpha_l x^{\alpha-\epsilon_l}e^{x^\alpha})$$

$$= \alpha_l \alpha_k x^{\alpha-\epsilon_l-\epsilon_k}e^{x^\alpha}.$$

c) First we note that

$$\partial^\beta p(x) = \partial^\beta \sum_{|\alpha|\leq m} c_\alpha x^\alpha = \sum_{|\alpha|\leq m} c_\alpha \partial^\beta x^\alpha,$$

so we need to prove that if $|\beta| > m$ then for $|\alpha| \leq m$ it follows that $\partial^\beta x^\alpha = 0$. Note that

$$\partial^\beta x^\alpha = \partial_1^{\beta_1}....\partial_n^{\beta_n}(x_1^{\alpha_1}....x_n^{\alpha_n}) = \prod_{j=1}^n \partial_j^{\beta_j} x_j^{\alpha_j}$$

and $\partial_j^{\beta_j} x_j^{\alpha_j} = 0$ if $\beta_j > \alpha_j$. Since $\alpha_1 + ... + \alpha_n \leq m < \beta_1 + ... + \beta_n$ there must be at least one j such that $\alpha_j < \beta_j$ implying

$$\prod_{j=1}^n \partial_j^{\beta_j} x_j^{\alpha_j} = 0.$$

d) Consider

$$\partial^\gamma \cos\langle a, x\rangle = \partial^\gamma \cos\left(\sum_{j=1}^n a_j x_j\right) = \partial_1^{\gamma_1}....\partial_n^{\gamma_n}\cos\left(\sum_{j=1}^n a_j x_j\right).$$

We note that

$$\partial_k^{\gamma_k} \cos\left(\sum_{j=1}^n a_j x_j\right) = \sigma(\gamma_k) a_k^{\gamma_k} \operatorname{trig}_{\gamma_k}\left(\sum_{j=1}^n a_j x_j\right)$$

where $\sigma(\gamma_k) \in \{-1, 1\}$ and $\operatorname{trig}_{\gamma_k} \in \{\sin, \cos\}$. Therefore

$$\partial^\gamma \cos\langle a, x\rangle = \sigma(\gamma_1)....\sigma(\gamma_n) a_1^{\gamma_1}....a_n^{\gamma_n} \operatorname{trig}_{\gamma_1}\langle a, x\rangle....\operatorname{trig}_{\gamma_n}\langle a, x\rangle$$

which implies

$$|\partial^\gamma \cos\langle a, x\rangle| \le |a_1^{\gamma_1}....a_n^{\gamma_n}| = |a^\gamma|.$$

Chapter 6

1. With $F_1(\rho, \varphi, z) = \rho\cos\varphi$, $F_2(\rho, \varphi, z) = \rho\sin\varphi$ and $F_3(\rho, \varphi, z) = z$ we find

$$J_F(\rho, \varphi, z) = \begin{pmatrix} \frac{\partial F_1}{\partial \rho}(\rho, \varphi, z) & \frac{\partial F_1}{\partial \varphi}(\rho, \varphi, z) & \frac{\partial F_1}{\partial z}(\rho, \varphi, z) \\ \frac{\partial F_2}{\partial \rho}(\rho, \varphi, z) & \frac{\partial F_2}{\partial \varphi}(\rho, \varphi, z) & \frac{\partial F_2}{\partial z}(\rho, \varphi, z) \\ \frac{\partial F_3}{\partial \rho}(\rho, \varphi, z) & \frac{\partial F_3}{\partial \varphi}(\rho, \varphi, z) & \frac{\partial F_3}{\partial z}(\rho, \varphi, z) \end{pmatrix}$$

$$= \begin{pmatrix} \cos\varphi & -\rho\sin\varphi & 0 \\ \sin\varphi & \rho\cos\varphi & 0 \\ 0 & 0 & 1 \end{pmatrix}$$

implying

$$\det J_F(\rho, \varphi, z) = \det\begin{pmatrix} \cos\varphi & -\rho\sin\varphi \\ \sin\varphi & \rho\cos\varphi \end{pmatrix} = \rho.$$

2. Using the components $G_1(u, v, z) = \frac{1}{2}(u^2 - v^2)$, $G_2(u, v, z) = uv$ and $G_3(u, v, z) = z$ we get

$$J_G(u, v, , z) = \begin{pmatrix} \frac{\partial G_1}{\partial u}(u, v, z) & \frac{\partial G_1}{\partial v}(u, v, z) & \frac{\partial G_1}{\partial z}(u, v, z) \\ \frac{\partial G_2}{\partial u}(u, v, z) & \frac{\partial G_2}{\partial v}(u, v, z) & \frac{\partial G_2}{\partial z}(u, v, z) \\ \frac{\partial G_3}{\partial u}(u, v, z) & \frac{\partial G_3}{\partial v}(u, v, z) & \frac{\partial G_3}{\partial z}(u, v, z) \end{pmatrix}$$

$$= \begin{pmatrix} u & -v & 0 \\ v & u & 0 \\ 0 & 0 & 1 \end{pmatrix}$$

which yields

$$\det J_G(u, v, z) = \det\begin{pmatrix} u & -v \\ v & u \end{pmatrix} = u^2 + v^2.$$

3. The following holds

$$J_H(u, v, \varphi) = \begin{pmatrix} v\cos\varphi & u\cos\varphi & -uv\sin\varphi \\ v\sin\varphi & u\sin\varphi & uv\cos\varphi \\ u & -v & 0 \end{pmatrix}$$

and therefore

$$\det J_H(u,v,\varphi) = u \det \begin{pmatrix} u\cos\varphi & -uv\sin\varphi \\ u\sin\varphi & uv\cos\varphi \end{pmatrix}$$

$$+ v \det \begin{pmatrix} v\cos\varphi & -uv\sin\varphi \\ v\sin\varphi & uv\cos\varphi \end{pmatrix}$$

$$= u^3 v + v^3 u.$$

4. First we need $J_K(\xi,\eta,\varphi)$ which is given by

$$J_K(\xi,\eta,\varphi) = \begin{pmatrix} \cosh\xi\sin\eta\cos\varphi & \sinh\xi\cos\eta\cos\varphi & -\sinh\xi\sin\eta\sin\varphi \\ \cosh\xi\sin\eta\sin\varphi & \sinh\xi\cos\eta\sin\varphi & \sinh\xi\sin\eta\cos\varphi \\ \sinh\xi\cos\eta & -\cosh\xi\sin\eta & 0 \end{pmatrix}$$

implying

$$\det J_K(\xi,\eta,\varphi) = \sinh\xi\cos\eta \det \begin{pmatrix} \sinh\xi\cos\eta\cos\varphi & -\sinh\xi\sin\eta\sin\varphi \\ \sinh\xi\cos\eta\sin\varphi & \sinh\xi\sin\eta\cos\varphi \end{pmatrix}$$

$$+ \cosh\xi\sin\eta \det \begin{pmatrix} \cosh\xi\sin\eta\cos\varphi & -\sinh\xi\sin\eta\sin\varphi \\ \cosh\xi\sin\eta\sin\varphi & \sinh\xi\sin\eta\cos\varphi \end{pmatrix}$$

$$= \sinh\xi\cos\eta(\sinh^2\xi(\cos\eta\sin\eta\cos^2\varphi + \cos\eta\sin\eta\sin^2\varphi))$$

$$+ \cosh\xi\sin\eta(\cosh\xi\sinh\xi(\sin^2\eta\cos^2\varphi + \sin^2\eta\sin^2\varphi))$$

$$= \sinh^3\xi\cos^2\eta\sin\eta + \cosh^2\xi\sinh\xi\sin^2\eta$$

$$= \sinh\xi\sin\eta(\sinh^2\xi\cos^2\eta + \cosh^2\xi\sin^2\eta)$$

$$= \sinh\xi\sin\eta(\sinh^2\xi\cos^2\eta + (1 + \sinh^2\xi\sin^2\eta))$$

$$= \sinh\xi\sin\eta(\sinh^2\xi + \sin^2\eta).$$

5. Note that we only need to handle the function $\vartheta \mapsto S(2,\vartheta,\frac{\pi}{4}) = (\sqrt{2}\sin\vartheta, \sqrt{2}\sin\vartheta, 2\cos\vartheta)$ in a neighbourhood of $\vartheta = \frac{\pi}{2}$, and it follows that

$$S(2,\vartheta,\frac{\pi}{4}) - S(2,\frac{\pi}{2},\frac{\pi}{4}) = \frac{d}{d\vartheta}(\sqrt{2}\sin\vartheta, \sqrt{2}\sin\vartheta, 2\cos\vartheta)\big|_{\vartheta=\frac{\pi}{2}}(\vartheta - \frac{\pi}{2}) + \Phi(\vartheta)$$

$$= (\sqrt{2}\cos\frac{\pi}{2}, \sqrt{2}\cos\frac{\pi}{2}, -2\sin\frac{\pi}{2})(\vartheta - \frac{\pi}{2}) + \Phi(\vartheta)$$

$$= (0,0,-2)(\vartheta - \frac{\pi}{2}) + \Phi(\vartheta)$$

$$= \pi - 2\vartheta + \Phi(\vartheta)$$

where $\lim_{\vartheta\to\frac{\pi}{2}}\left(\frac{\Phi(\vartheta)}{\vartheta-\frac{\pi}{2}}\right) = 0$. Hence the linear approximation of $S(2,\vartheta,\frac{\pi}{4})$ is given by $\vartheta \mapsto \pi - 2\vartheta$.

6. For the calculation we first have to "fix" our identification of $M(m,n,\mathbb{R})$ with $\mathbb{R}^{mn}$ which we do in the following way

$$\begin{pmatrix} a_{11} & \cdots & a_{1n} \\ \vdots & & \vdots \\ a_{m1} & \cdots & a_{mn} \end{pmatrix} \mapsto (a_{11}, \ldots a_{1n}, a_{21}, \ldots, a_{2n}, \ldots a_{m1}, \ldots, a_{mn})^t.$$

625

Thus $A(x) = J_f(x) = \left(\frac{\partial f_j}{\partial x_k}(x)\right)_{\substack{j=1,\ldots,m \\ k=1,\ldots,n}}$ we identify with

$$x \mapsto \left(\frac{\partial f_1}{\partial x_1}(x), \ldots, \frac{\partial f_1}{\partial x_n}(x), \frac{\partial f_2}{\partial x_1}(x), \ldots, \frac{\partial f_2}{\partial x_n}(x), \ldots, \frac{\partial f_m}{\partial x_1}(x), \ldots, \frac{\partial f_m}{\partial x_m}(x)\right)^t$$

and this yields for $J_A(x) = J_{J_f}(x)$ (with respect to our fixed identification of $M(n, m, \mathbb{R})$ with $\mathbb{R}^{mn}$:

$$J_A(x) = \begin{pmatrix} \frac{\partial^2 f_1}{\partial x_1^2}(x) & \cdots & \frac{\partial^2 f_1}{\partial x_n \partial x_1}(x) \\ \vdots & & \vdots \\ \frac{\partial^2 f_1}{\partial x_1 \partial x_n}(x) & \cdots & \frac{\partial^2 f_1}{\partial x_n^2}(x) \\ \frac{\partial^2 f_2}{\partial x_1^2}(x) & \cdots & \frac{\partial^2 f_2}{\partial x_n \partial x_1}(x) \\ \vdots & & \vdots \\ \frac{\partial^2 f_2}{\partial x_1 \partial x_n}(x) & \cdots & \frac{\partial^2 f_2}{\partial x_n^2}(x) \\ \vdots & & \vdots \\ \frac{\partial^2 f_m}{\partial x_1 \partial x_n}(x) & \cdots & \frac{\partial^2 f_m}{\partial x_n^2}(x) \end{pmatrix}.$$

Let us introduce the matrices $J_{f_j}^2(x)$, $j = 1, \ldots, m$,

$$J_{f_j}^2(x) := \left(\frac{\partial^2 f_j}{\partial x_k \partial x_l}(x)\right)_{k,l=1,\ldots,n}$$

which allows us to write $J_A(x)$ as a block matrix

$$J_A(x) = \begin{pmatrix} J_{f_1}^2(x) \\ \vdots \\ J_{f_m}^2(x) \end{pmatrix}$$

and we find $(x - x_0)$ interpreted as the block matrix

$$J_f(x) - J_f(x_0) = \begin{pmatrix} J_{f_1}^2(x) \\ \vdots \\ J_{f_m}^2(x) \end{pmatrix} (x - x_0)^t + \psi(x - x_0)$$

$$= \begin{pmatrix} J_{f_1}^2(x)(x - x_0)^t \\ \vdots \\ J_{f_m}^2(x)(x - x_0)^t \end{pmatrix} + \psi(x - x_0)$$

where $\psi : B_{\rho(x_0)} \to M(m, n, \mathbb{R})$ and $\lim_{x \to x_0} \frac{\psi(x-x_0)}{\|x-x_0\|} = 0$.

626

7. Again we first identify $M(2, \mathbb{R})$ with $\mathbb{R}^4$ by

$$\begin{pmatrix} a_{11} & a_{12} \\ a_{21} & a_{22} \end{pmatrix} \mapsto (a_{11}, a_{12}, a_{21}, a_{22})^t,$$

and we now rewrite

$$U(r, \varphi) = \begin{pmatrix} r\cos\varphi & -r\sin\varphi \\ r\sin\varphi & r\cos\varphi \end{pmatrix} \mapsto \begin{pmatrix} r\cos\varphi \\ -r\sin\varphi \\ r\sin\varphi \\ r\cos\varphi \end{pmatrix},$$

which gives (after the identification)

$$J_U(r, \varphi) \begin{pmatrix} \frac{\partial(r\cos\varphi)}{\partial r} & \frac{\partial(r\cos\varphi)}{\partial\varphi} \\ \frac{\partial(-r\sin\varphi)}{\partial r} & \frac{\partial(-r\sin\varphi)}{\partial\varphi} \\ \frac{\partial(r\sin\varphi)}{\partial r} & \frac{\partial(r\sin\varphi)}{\partial\varphi} \\ \frac{\partial(r\cos\varphi)}{\partial r} & \frac{\partial(r\cos\varphi)}{\partial\varphi} \end{pmatrix} = \begin{pmatrix} \cos\varphi & -r\sin\varphi \\ -\sin\varphi & -r\cos\varphi \\ \sin\varphi & r\cos\varphi \\ \cos\varphi & -r\sin\varphi \end{pmatrix}.$$

We may turn again to the block matrix interpretation, but it is obvious that we need a more clever way to handle higher order differentials, and for this we will introduce tensor analysis later on in this Course.

8. The Jacobi matrix of $f : \mathbb{R}^3 \to \mathbb{R}^2$, $f(x, y, z) = \begin{pmatrix} (x^2 + y^2 + z^2 + 1)^{\frac{1}{2}} \\ x + y + z \end{pmatrix}$ is

$$J_f(x, y, z) = \begin{pmatrix} \frac{x}{(x^2+y^2+z^2+1)^{\frac{1}{2}}} & \frac{y}{(x^2+y^2+z^2+1)^{\frac{1}{2}}} & \frac{z}{(x^2+y^2+z^2+1)^{\frac{1}{2}}} \\ 1 & 1 & 1 \end{pmatrix}$$

and that of $g : \mathbb{R}^2 \to \mathbb{R}^3$, $g(p, g) = \begin{pmatrix} p + q \\ p - q + 2p \\ 0 \end{pmatrix}$ is

$$J_g(p, q) = \begin{pmatrix} 1 & 1 \\ q + 2 & p \\ 0 & 0 \end{pmatrix}.$$

Now we get for $J_{g \circ f}$ with $Q(x, y, z) = (x^2 + y^2 + z^2 + 1)^{\frac{1}{2}}$.

$J_{g \circ f}(x, y, z)$

$$= \begin{pmatrix} 1 & 1 \\ x + y + z + 2 & (x^2 + y^2 + z^2 + 1)^{\frac{1}{2}} \\ 0 & 0 \end{pmatrix} \begin{pmatrix} \frac{x}{Q(x,y,z)} & \frac{y}{Q(x,y,z)} & \frac{z}{Q(x,y,z)} \\ 1 & 1 & 1 \end{pmatrix}$$

$$= \begin{pmatrix} \frac{x+Q(x,y,z)}{Q(x,y,x)} & \frac{y+Q(x,y,z)}{Q(x,y,x)} & \frac{z+Q(x,y,z)}{Q(x,y,x)} \\ \frac{x(x+y+z+2)}{Q(x,y,z)} + Q(x,y,z) & \frac{y(x+y+z+2)}{Q(x,y,z)} + Q(x,y,z) & \frac{z(x+y+z+2)}{Q(x,y,z)} + Q(x,y,z) \\ 0 & 0 & 0 \end{pmatrix}.$$

For the Jacobi matrix of $f \circ g$ we find

$$J_{f \circ g}(p,q) = \begin{pmatrix} \frac{\partial f_1}{\partial x}\frac{\partial g_1}{\partial p} + \frac{\partial f_1}{\partial y}\frac{\partial g_2}{\partial p} + \frac{\partial f_1}{\partial z}\frac{\partial g_3}{\partial p} & \frac{\partial f_1}{\partial x}\frac{\partial g_1}{\partial q} + \frac{\partial f_1}{\partial y}\frac{\partial g_2}{\partial q} + \frac{\partial f_1}{\partial z}\frac{\partial g_2}{\partial q} \\ \frac{\partial f_2}{\partial x}\frac{\partial g_1}{\partial p} + \frac{\partial f_2}{\partial y}\frac{\partial g_2}{\partial p} + \frac{\partial f_2}{\partial z}\frac{\partial g_3}{\partial p} & \frac{\partial f_2}{\partial x}\frac{\partial g_1}{\partial q} + \frac{\partial f_2}{\partial y}\frac{\partial g_2}{\partial q} + \frac{\partial f_2}{\partial z}\frac{\partial g_3}{\partial q} \end{pmatrix}$$

which gives with

$$R(p,q) = (g_1^2(p,q) + g_2^2(p,q) + g_3^2(p,q) + 1)^{\frac{1}{2}}$$
$$= ((p+q)^2 + (pq+2p)^2 + 1)^{\frac{1}{2}}$$

$$J_{f \circ g}(p,q) = \begin{pmatrix} \frac{p+q}{R(p,q)} & \frac{(pq+2p)}{R(p,q)} & 0 \\ 1 & 1 & 1 \end{pmatrix} \begin{pmatrix} 1 & 1 \\ q+2 & p \\ 0 & 0 \end{pmatrix}$$
$$= \begin{pmatrix} \frac{p+q+(q+2)(pq+2p)}{R(p,q)} & \frac{p+q+p(pq+2p)}{R(p,q)} \\ q+3 & p+1 \end{pmatrix}.$$

9. By the chain rule we have

$$d(\mathrm{pr}_j \circ f)(x) = (d\,\mathrm{pr}_j)_{f(x)} \circ df(x),$$

or in terms of the Jacobi matrices

$$J_{\mathrm{pr}_j \circ f}(x) = J_{\mathrm{pr}_j}(f(x))J_f(x).$$

Now we have for $x \in G$

$$J_f(x) = \begin{pmatrix} \frac{\partial f_1}{\partial x_1}(x) & \cdots & \frac{\partial f_1}{\partial x_n}(x) \\ \vdots & & \vdots \\ \frac{\partial f_k}{\partial x_1}(x) & \cdots & \frac{\partial f_k}{\partial x_n}(x) \end{pmatrix}$$

and for $\mathrm{pr}_j : \mathbb{R}^k \to \mathbb{R}$, $(y_1, ..., y_k) \mapsto y$, $1 \le j \le n$, we find

$$J_{\mathrm{pr}_j}(y) = (0, ..., 0, 1, 0, ..., 0)$$

with 1 in the jth position. This yields

$$J_{\mathrm{pr}_j \circ f}(x) = (0, ..., 0, 1, 0, ..., 0) \begin{pmatrix} \frac{\partial f_1}{\partial x_1}(x) & \cdots & \frac{\partial f_1}{\partial x_n}(x) \\ \vdots & & \vdots \\ \frac{\partial f_k}{\partial x_1}(x) & \cdots & \frac{\partial f_k}{\partial x_n}(x) \end{pmatrix}$$
$$= \left(\frac{\partial f_j}{\partial x_1}(x), ..., \frac{\partial f_j}{\partial x_n}(x) \right) = J_{f_j}(x).$$

Of course, the result is not surprising since

$$(\mathrm{pr}_j \circ f)(x) = f_j(x).$$

10. For $x \in G$ we find

$$
\mathrm{grad}\left(\frac{1}{g(x)}\right) = \left(\frac{\partial}{\partial x_1}\left(\frac{1}{g(x)}\right), \dots, \frac{\partial}{\partial x_n}\left(\frac{1}{g(x)}\right)\right)
$$

$$
= \left(\frac{-\frac{\partial g}{\partial x_1}(x)}{g^2(x)}, \dots, \frac{-\frac{\partial g}{\partial x_n}(x)}{g^2(x)}\right)
$$

$$
-\frac{1}{g^2(x)}\left(\frac{\partial g}{\partial x_1}(x), \dots, \frac{\partial g}{\partial x_n}(x)\right) = -\frac{1}{g^2(x)}\,\mathrm{grad}\,g(x).
$$

11. Let $g : [0, \infty) \to \mathbb{R}$ be homogeneous of degree α, i.e. $g(\lambda t) = \lambda^\alpha g(t)$ for $\lambda > 0$. For the derivative of g we find

$$
\lambda\left(\frac{dg}{dt}\right)(\lambda t) = \frac{d}{dt}\left(g(\lambda t)\right) = \frac{d\lambda^\alpha g(t)}{dt} = \lambda^\alpha \frac{dg(t)}{dt},
$$

hence for g' we find

$$
g'(\lambda t) = \lambda^{\alpha-1} g'(t),
$$

i.e. g' is homogeneous of degree $\alpha - 1$. By iterating the argument we find that $g^{(k)}$ is homogeneous of degree $\alpha - k$. Now we find

$$
\frac{\partial f}{\partial x_k}(x) = \frac{\partial}{\partial x_k} g((x_1^2 + \dots + x_n^2)^{\frac{1}{2}})
$$

$$
= \frac{x_k}{(x_1^2 + \dots + x_n^2)^{\frac{1}{2}}} g'((x_1^2 + \dots + x_n^2)^{\frac{1}{2}})
$$

and hence with

$$
\delta_{kj} = \begin{cases} 1, & k = j \\ 0, & k \neq j \end{cases}
$$

$$
\frac{\partial^2 f}{\partial x_j \partial x_k}(x) = \frac{\partial}{\partial x_j}\left(\frac{x_k}{(x_1^2 + \dots + x_n^2)^{\frac{1}{2}}} g'((x_1^2 + \dots + x_n^2)^{\frac{1}{2}})\right)
$$

$$
= \frac{(x_1^2 + \dots x_n^2)\delta_{kj} - x_k x_j}{(x_1^2 + \dots + x_n^2)^{\frac{3}{2}}} g'((x_1^2 + \dots + x_n^2)^{\frac{1}{2}})
$$

$$
+ \frac{x_k x_j}{(x_1^2 + \dots + x_n^2)} g''((x_1^2 + \dots + x_n^2)^{\frac{1}{2}})
$$

$$
= \frac{r^2 \delta_{kj} - x_k x_j}{r^3} g'(r) + \frac{x_k x_j}{r^2} g''(r).
$$

For $\lambda > 0$ we now find

$$
\left(\frac{\partial^2 f}{\partial x_j \partial x_k}\right)(\lambda x) = \frac{\lambda^2 r^2 \delta_{kj} - \lambda x_k x_j}{\lambda^3 r^3} g'(\lambda r) + \frac{\lambda^2 x_k x_j}{\lambda^2 r^2} g''(\lambda r)
$$

$$
= \frac{r^2 \delta_{kj} - x_k x_j}{r^3} \frac{1}{\lambda} \lambda^{\alpha-1} g'(r) + \frac{x_k x_j}{r^2} \lambda^{\alpha-2} g(r)
$$

$$
= \lambda^{\alpha-2}\left(\frac{\partial^2 f}{\partial x_j \partial x_k}\right)(x),
$$

so $\frac{\partial^2 f}{\partial x_j \partial x_k}$ is homogeneous of degree $\alpha - 2$.

629

12. With $\vec{t_x}(x_0, y_0) = -\frac{y_0}{r^2}$ and $\vec{t_y}(x_0, y_0) = -\frac{x_0}{r^2}$ we find

$$\left(\frac{\partial}{\partial \vec{t}} f\right)(x_0, y_0) = \vec{t_x}(x_0, y_0)\left(\frac{\partial f}{\partial x}\right)(x_0, y_0) + \vec{t_y}(x_0, y_0)\left(\frac{\partial f}{\partial y}\right)(x_0, y_0)$$

$$= -\frac{y_0}{r^2}\frac{\partial h(x^2 + y^2)}{\partial x}\Big|_{(x,y)=(x_0,y_0)} + \frac{x_0}{r^2}\frac{\partial h(x^2 + y^2)}{\partial y}\Big|_{(x,y)=(x_0,y_0)}$$

$$= -\frac{y_0}{r^2}(2x_0 h'(r^2)) + \frac{x_0}{r^2}(2y_0 h'(r^2))$$

$$= -\frac{2x_0 y_0}{r^2}h'(r^2) + \frac{2x_0 y_0}{r^2}h'(r^2) = 0.$$

13. Since

$$\frac{\partial f}{\partial v}(x_0) = \sum_{j=1}^{n} v_j \frac{\partial f}{\partial x_j}(x_0)$$

the Cauchy-Schwarz inequality yields

$$\left|\frac{\partial f}{\partial v}(x_0)\right| \le \|\mathrm{grad}\, f(x_0)\|\|v\| = \|\mathrm{grad}\, f(x_0)\|,$$

hence

$$-\|\mathrm{grad}\, f(x_0)\| \le \frac{\partial f}{\partial v}(x_0) \le \|\mathrm{grad}\, f(x_0)\|.$$

With $w := \frac{\mathrm{grad}\, f(x_0)}{\|\mathrm{grad}\, f(x_0)\|}$ we have

$$\frac{\partial f}{\partial w}(x_0) = \frac{1}{\|\mathrm{grad}\, f(x_0)\|}\sum_{j=1}^{n}\frac{\partial f}{\partial x_j}(x_0)\frac{\partial f}{\partial x_j}(x_0) = \|\mathrm{grad}\, f(x_0)\| > 0,$$

and the result follows. Note that

$$\inf_{\|v\|=1}\frac{\partial f}{\partial v}(x_0) = \frac{\partial f}{\partial u}(x_0) = -\|\mathrm{grad}\, f(x_0)\|$$

where $u := -\frac{\mathrm{grad}\, f(x_0)}{\|\mathrm{grad}\, f(x_0)\|}$.

14.   a) Since G is convex, with x and y the line segment $x + \vartheta(y - x)$, $0 \le \vartheta \le 1$, belongs to G and by the mean value theorem we find

$$g(x) - g(y) = \langle \mathrm{grad}\, g(x + \vartheta(y - x)), x - y\rangle = 0,$$

i.e. $g(x) = g(y)$ for all $x, y \in G$ meaning that g is constant in G.

Note that we can replace the convexity of G by the assumption that we can join any two points x and y by a finite chain of line segments.

  b) As in part a) we know that by convexity of G, with x and y the line segment $x + \vartheta(y - x)$, $0 \le \vartheta \le 1$, belongs to G. Applying the Cauchy-Schwarz inequality to

$$f(x) - f(y) = \langle \mathrm{grad}\, f(x + \vartheta(y - x)), x - y\rangle$$

we get

$$|f(x) - f(y)| = |\langle \operatorname{grad} f(x + \vartheta(y - x)), x - y\rangle|$$
$$\leq \|\operatorname{grad} f(x + \vartheta(y - x))\| \, \|x - y\|$$
$$\leq M \, \|x - y\|\,.$$

Chapter 7

1. The limit

$$(*) \qquad f'(t_0) := \lim_{h \to 0} \frac{f(t_0 + h) - f(t_0)}{h}$$

exists if and only if for $1 \leq j \leq k$ the limits

$$f_j'(t_0) = \lim_{h \to 0} \frac{f_j(t_0 + h) - f_j(t_0)}{h}$$

exist. Thus $(*)$ exists if and only if with a function $\varphi_{t_0,j}$ defined in a neighbourhood of 0 and satisfying

$$\lim_{h \to 0} \frac{\varphi_{t_0,j}(h)}{h} = 0$$

we have

$$f_j(t_0 + h) - f_j(t_0) = f_j'(t_0)h + \varphi_{t_0,j}(h),$$

which is of course equivalent to

$$f(t_0 + h) - f(t_0) = f'(t_0)h + \varphi_{t_0}(h)$$

with $\varphi_{t_0}(h) = (\varphi_{t_0,1}(h), ..., \varphi_{t_0,k}(h))$ satisfying $\lim_{h \to 0} \frac{\varphi_{t_0}(h)}{h} = 0$.

2. a) We claim that the following type of Leibniz's rule holds

$$\frac{d}{dt}(A(t)u(t)) = A'(t)u(t) + A(t)u'(t).$$

To see this we note that

$$A(t)u(t) = \begin{pmatrix} \sum_{j=1}^{n} a_{1j}(t)u_j(t) \\ \vdots \\ \sum_{j=1}^{n} a_{mj}(t)u_j(t) \end{pmatrix}$$

and therefore we find

631

$$\frac{d}{dt}(A(t)u(t)) = \frac{d}{dt}\begin{pmatrix} \sum_{j=1}^n a_{1j}(t)u_j(t) \\ \vdots \\ \sum_{j=1}^n a_{mj}(t)u_j(t) \end{pmatrix}$$

$$= \begin{pmatrix} \frac{d}{dt}\sum_{j=1}^n a_{1j}(t)u_j(t) \\ \vdots \\ \frac{d}{dt}\sum_{j=1}^n a_{mj}u_j(t) \end{pmatrix} = \begin{pmatrix} \sum_{j=1}^n \frac{d}{dt}(a_{1j}(t)u_j(t)) \\ \vdots \\ \sum_{j=1}^n \frac{d}{dt}(a_{mj}(t)u_j(t)) \end{pmatrix}$$

$$= \begin{pmatrix} \sum_{j=1}^n (a_{1j}'(t)u_j(t) + a_{1j}(t)u_j'(t)) \\ \vdots \\ \sum_{j=1}^n (a_{mj}'(t)u_j(t) + a_{mj}(t)u_j'(t)) \end{pmatrix}$$

$$= \begin{pmatrix} \sum_{j=1}^n a_{1j}'(t)u_j(t) \\ \vdots \\ \sum_{j=1}^n a_{mj}'(t)u_j(t) \end{pmatrix} + \begin{pmatrix} \sum_{j=1}^n a_{1j}(t)u_j'(t) \\ \vdots \\ \sum_{j=1}^n a_{mj}(t)u_j'(t) \end{pmatrix}$$

$$= A'(t)u(t) + A(t)u'(t).$$

b) First we note that for $0 < a < b < 1$ the function $t \mapsto A(t)u(t)$ is integrable as continuous function (with values in $\mathbb{R}^n$) on a compact interval $[a, b]$. We have

$$A(t)u(t) = \begin{pmatrix} \sum_{l=1}^n a_{1l}(t)u_l(t) \\ \vdots \\ \sum_{l=1}^n a_{ml}(t)u_l(t) \end{pmatrix}$$

and therefore

$$\left\| \int_a^b A(t)u(t)dt \right\| = \left\| \begin{pmatrix} \sum_{l=1}^n \int_a^b a_{1l}(t)u_l(t)dt \\ \vdots \\ \sum_{l=1}^n \int_a^b a_{ml}(t)u_l(t)dt \end{pmatrix} \right\|$$

$$= \left(\sum_{k=1}^m \left(\sum_{l=1}^n \int_a^b a_{kl}(t)u_l(t)dt \right)^2 \right)^{\frac{1}{2}}.$$

By the Cauchy-Schwarz inequality for integrals we find

$$\left| \int_a^b a_{kl}(t)u_l(t)dt \right| \le \left(\int_a^b a_{kl}^2(t)dt \right)^{\frac{1}{2}} \left(\int_a^b u_l^2(t)dt \right)^{\frac{1}{2}}$$

and the Cauchy-Schwarz inequality for sums now yields

$$\left(\sum_{l=1}^n \int_a^b a_{kl}(t)u_l(t)dt \right)^2 \le \left(\sum_{l=1}^n \left(\int_a^b a_{kl}^2(t)dt \right)^{\frac{1}{2}} \left(\int_a^b u_l^2(t)dt \right)^{\frac{1}{2}} \right)^2$$

$$\le \left(\sum_{l=1}^n \int_a^b a_{kl}^2(t)dt \right) \left(\sum_{l=1}^n \int_a^b u_l^2(t)dt \right)$$

and therefore we get

$$\left(\sum_{k=1}^{m}\left(\sum_{l=1}^{n}\int_{a}^{b}a_{kl}(t)u_{l}(t)dt\right)^{2}\right)^{\frac{1}{2}}$$

$$\leq\left(\sum_{k=1}^{m}\left(\sum_{l=1}^{n}\int_{a}^{b}a_{kl}^{2}(t)dt\right)\left(\sum_{l=1}^{n}\int_{a}^{b}u_{l}^{2}(t)dt\right)\right)^{\frac{1}{2}}$$

$$=\left(\int_{a}^{b}\left(\sum_{k=1}^{m}\sum_{l=1}^{n}a_{kl}^{2}(t)\right)dt\right)^{\frac{1}{2}}\left(\int_{a}^{b}\sum_{l=1}^{n}u_{l}^{2}(t)\right)^{\frac{1}{2}}$$

$$=\left(\int_{a}^{b}\left(\sum_{k=1}^{m}\sum_{l=1}^{n}a_{kl}^{2}(t)\right)dt\right)^{\frac{1}{2}}\left(\int_{a}^{b}\|u(t)\|^{2}dt\right)^{\frac{1}{2}}.$$

Since

$$\left(\int_{a}^{b}\left(\sum_{k=1}^{m}\sum_{l=1}^{n}a_{kl}^{2}(t)\right)dt\right)^{\frac{1}{2}}\leq\left(\max_{\substack{1\leq k\leq m\\1\leq l\leq n}}\|a_{kl}\|_{\infty,[a,b]}\right)\left(\int_{a}^{b}\left(\sum_{k=1}^{m}\sum_{l=1}^{n}1\right)dt\right)^{\frac{1}{2}}$$

$$=(mn(b-a))^{\frac{1}{2}}\max_{\substack{1\leq k\leq m\\1\leq l\leq n}}\|a_{kl}\|_{\infty,[a,b]},$$

the second estimate follows.

3.   a) We note that

$$\frac{d}{dt}\langle u(t),v(t)\rangle=\frac{d}{dt}\sum_{j=1}^{n}u_{j}(t)v_{j}(t)$$

$$=\sum_{j=1}^{n}(u_{j}^{'}v_{j}(t)+u_{j}(t)v_{j}^{'}(t))$$

$$=\sum_{j=1}^{n}u_{j}^{'}(t)v_{j}(t)+\sum_{j=1}^{n}u_{j}(t)v_{j}^{'}(t)$$

$$=\left\langle\frac{du}{dt}(t),v(t)\right\rangle+\left\langle u(t),\frac{dv}{dt}(t)\right\rangle.$$

633

b) The following holds

$$\frac{d}{dt}(\alpha(t) \times \beta(t)) = \frac{d}{dt}\begin{pmatrix} \alpha_2(t)\beta_2(t) - \alpha_3(t)\beta_2(t) \\ \alpha_3(t)\beta_1(t) - \alpha_1(t)\beta_3(t) \\ \alpha_1(t)\beta_2(t) - \alpha_2(t)\beta_1(t) \end{pmatrix}$$

$$= \begin{pmatrix} \alpha_2'(t)\beta_3(t) + \alpha_2(t)\beta_3'(t) - \alpha_3'\beta_2(t) - \alpha_3(t)\beta_2'(t) \\ \alpha_3'(t)\beta_1(t) + \alpha_3(t)\beta_1'(t) - \alpha_1'(t)\beta_3(t) - \alpha_1(t)\beta_3'(t) \\ \alpha_1'(t)\beta_2(t) + \alpha_1(t)\beta_2'(t) - \alpha_2'(t)\beta_1(t) - \alpha_2(t)\beta_1'(t) \end{pmatrix}$$

$$= \begin{pmatrix} \alpha_2'\beta_3(t) - \alpha_3'(t)\beta_2(t) \\ \alpha_3'(t)\beta_1(t) - \alpha_1'(t)\beta_3(t) \\ \alpha_1'(t)\beta_2(t) - \alpha_2'(t)\beta_1(t) \end{pmatrix} + \begin{pmatrix} \alpha_2(t)\beta_3'(t) - \alpha_3(t)\beta_2'(t) \\ \alpha_3(t)\beta_1'(t) - \alpha_1(t)\beta_3'(t) \\ \alpha_1(t)\beta_2'(t) - \alpha_2(t)\beta_1'(t) \end{pmatrix}$$

$$= \frac{d\alpha}{dt}(t) \times \beta(t) + \alpha(t) \times \frac{d\beta}{dt}(t).$$

4.   a) We already know that $\text{tr}(\gamma) \subset \mathcal{E}$. Now let $(x, y) \in \mathcal{E}$. We have to find $t \in [0, 2\pi]$ such that $(\alpha \cos t, \beta \sin t) = (x, y)$. For this we connect the point (x, y) by a line segment l to the origin $(0, 0)$ and take t to be the angle between the positive x-axis and the line segment l, see the figure below

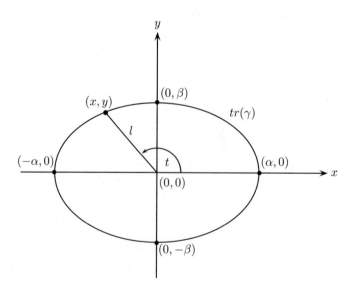

b) For $t \in \mathbb{R}$ we have

$$\left\| \left(\frac{1-t^2}{1+t^2}, \frac{2t}{1+t^2} \right) \right\|^2 = \frac{(1-t^2)^2}{(1+t^2)^2} + \frac{(2t)^2}{(1+t^2)^2}$$
$$= \frac{1+2t^2+t^4}{(1+t^2)^2} = \frac{(1+t^2)^2}{(1+t^2)^2} = 1,$$

which shows that $\mathrm{tr}(\gamma) \subset S^1$.

c) With $x = 3t\cos t$, $y = 3t\sin t$, $z = 5t$ we have

$$\frac{x^2}{9} + \frac{y^2}{9} - \frac{z^2}{25} = \frac{9t^2\cos^2 t}{9} + \frac{9t^2\sin^2 t}{9} - \frac{25t^2}{25}$$
$$= t^2\cos^2 t + t^2\sin^2 t - t^2 = 0.$$

Note that $\left\{ (x,y,x) \in \mathbb{R}^3 \,\middle|\, \frac{x^2}{9} + \frac{y^2}{9} - \frac{z^2}{25} = 0 \right\}$ is a circular cone with vertex $0 \in \mathbb{R}^3$ and symmetry axis being the z-axis.

5. a) For the velocity vector we get

$$\dot{\varphi}(t) = (a - b\cos t, b\sin t).$$

A point is singular if $\dot{\varphi}(t_0) = 0$, i.e. $a - b\cos t_0 = 0$ and $b\sin t_0 = 0$. The latter equation implies $t_0 = k\pi$, $k \in \mathbb{Z}$. For $t_0 = 2l\pi$ we have $\cos t_0 = 1$, hence $a - b\cos t_0 = a - b$, and for $b < a$ it follows $a - b\cos t_0 \neq 0$. For $t_0 = (2l+1)\pi$ we have $\cos t_0 = -1$ and therefore $a - b\cos t_0 = a + b > 0$, thus for $0 < b < a$ all points are regular. However for $a = b$ all points with parameter $t_0 = 2l\pi$ are singular.

The trace of φ is a **cycloid**.

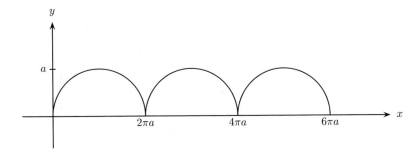

b) Since for $\gamma_1(t) = (t^5, t^4)$ we have $\dot{\gamma}_1(t) = (5t^4, 4t^3)$ it follows that $\gamma(0)$ is a singular point, but all other points are regular. Further, since $t \mapsto t^5$ is injective γ_1 has no multiple points.

635

Now we turn to γ_2. First we calculate $\dot{\gamma}_2(t) = (2t+1, 3t^2 + 2t - 2)$. The only singular point could occur for $t = -\frac{1}{2}$, but $3(-\frac{1}{2})^2 + 2(-\frac{1}{2}) - 2 = -\frac{9}{4}$, hence $\dot{\gamma}_2(t) \neq 0$ for all t, i.e. all points are regular.

For a double point, and hence for any multiples point of γ we must find $t_0, s_0 \in \mathbb{R}$ such that $\gamma_2(t_0) = \gamma_2(s_0)$, i.e. $\gamma_2^{(1)}(t_0) = \gamma_2^{(1)}(s_0)$ and $\gamma_2^{(2)}(t_0) = \gamma_2^{(2)}(s_0)$

In our case this implies

$$(*) \qquad (t_0 - 1)(t_0 + 2) = (s_0 - 1)(s_0 + 2)$$

and

$$(**) \qquad t_0(t_0 - 1)(t_0 + 2) = s_0(s_0 - 1)(s_0 + 2)$$

where we made use of the factorisations

$$t^2 + t - 2 = (t - 1)(t + 2)$$

and

$$t^3 + t^2 - 2t = t(t - 1)(t + 2).$$

Now if $(t_0 - 1)(t_0 + 2) = (s_0 - 1)(s_0 + 2) \neq 0$ then $(**)$ implies $t_0 = s_0$, i.e. there will be no multiple point. However for $t_0 = 1$ and $s_0 = -2$ we have $t_0 \neq s_0$, but

$$\gamma_2(1) = (0,0) = \gamma_2(-2),$$

hence the point $(0,0)$ is a double point for γ.

6. a) The velocity vector at t is given by

$$\dot{\gamma}(t) = (ae^{at}\cos t - e^{at}\sin t, ae^{at}\sin t + e^{at}\cos t)$$

and for $t_1 = \frac{\pi}{2}$ we have $\gamma(\frac{\pi}{2}) = (-e^{\frac{a\pi}{2}}, ae^{\frac{a\pi}{2}})$ and therefore the tangent line is given by

$$g_{\frac{\pi}{2}}(s) = \begin{pmatrix} -e^{\frac{a\pi}{2}} \\ ae^{\frac{a\pi}{2}} \end{pmatrix} s + \begin{pmatrix} 0 \\ e^{\frac{a\pi}{2}} \end{pmatrix}.$$

b) In general a tangent line to $\gamma : I \to \mathbb{R}^n$ at t_0 is given by $g_{\gamma(t_0)}(s) = \dot{\gamma}(t_0)s + \gamma(t_0)$. Thus in the case of the circular helix where $\gamma(t) = (r\cos t, r\sin t, bt)$, and hence $\dot{\gamma}(t_0) = (-r\sin t_0, r\cos t_0, b)$, we find

$$g_{\gamma(t_0)}(s) = \begin{pmatrix} -r\sin t_0 \\ r\cos t_0 \\ b \end{pmatrix} s + \begin{pmatrix} r\cos t_0 \\ r\sin t_0 \\ bt_0 \end{pmatrix}.$$

The periodicity of sin and cos implies

$$g_{\gamma(t_0 + 2\pi k)}(s) = \begin{pmatrix} -r\sin(t_0 + 2\pi k) \\ r\cos(t_0 + 2\pi k) \\ b \end{pmatrix} s + \begin{pmatrix} r\cos(t_0 + 2\pi k) \\ r\sin(t_0 + 2\pi k) \\ b(t_0 + 2\pi k) \end{pmatrix}$$

$$= \begin{pmatrix} -r\sin t_0 \\ r\cos t_0 \\ b \end{pmatrix} s + \begin{pmatrix} r\cos t_0 \\ r\sin t_0 \\ b(t_0 + 2\pi k) \end{pmatrix}$$

which yields

$$g_{\gamma(t_0)}(s) - g_{\gamma(t_0+2\pi k)}(s)$$

$$= \begin{pmatrix} -r\sin t_0 \\ r\cos t_0 \\ b \end{pmatrix} s + \begin{pmatrix} r\cos t_0 \\ r\sin t_0 \\ bt_0 \end{pmatrix} - \begin{pmatrix} -r\sin t_0 \\ r\cos t_0 \\ b \end{pmatrix} s - \begin{pmatrix} r\cos t_0 \\ r\sin t_0 \\ b(t_0+2\pi k) \end{pmatrix}$$

$$= \begin{pmatrix} 0 \\ 0 \\ -2b\pi k \end{pmatrix}.$$

Hence the trace $g_{\gamma(t_0+2\pi k)}$ is obtained by translating the trace of $g_{\gamma(t_0)}$ by the

vector $\eta = \begin{pmatrix} 0 \\ 0 \\ 2bk\pi \end{pmatrix}$ and therefore these two lines are parallel.

7. a) From elementary geometry the result is clear: the normal line to a point on a circle has to pass through the centre. Now we use our "differential geometric" approach. If γ is the parametrization $\gamma(t) = (r\cos t, r\sin t)$ we find for the tangent line passing through $\gamma(t_0)$ the equation

$$g_{t_0}(s) = \begin{pmatrix} -r\sin t_0 \\ r\cos t_0 \end{pmatrix} s + \begin{pmatrix} r\cos t_0 \\ r\sin t_0 \end{pmatrix}$$

and consequently the normal line is given by

$$n_{t_0}(s) = \begin{pmatrix} -r\cos t_0 \\ -r\sin t_0 \end{pmatrix} s + \begin{pmatrix} r\cos t_0 \\ r\sin t_0 \end{pmatrix}$$

which implies for all t_0 that $n_{t_0}(1) = 0 \in \mathbb{R}^2$.

b) The tangent line to γ at φ_0 is given by

$$g_{\varphi_0}(s) = \dot\gamma(\varphi_0)s + \gamma(\varphi_0)$$

with $\dot\gamma(\varphi_0) = \dot\rho(\varphi_0)\begin{pmatrix} \cos\varphi_0 \\ \sin\varphi_0 \end{pmatrix} + \rho(\varphi_0)\begin{pmatrix} -\sin\varphi_0 \\ \cos\varphi_0 \end{pmatrix}$. This yields for the normal line at φ_0

$$n_{\varphi_0}(s) = \begin{pmatrix} -\dot\rho(\varphi_0)\sin\varphi_0 - \rho(\varphi_0)\cos\varphi_0 \\ \dot\rho(\varphi_0)\cos\varphi_0 - \rho(\varphi_0)\sin\varphi_0 \end{pmatrix} s + \rho(\varphi_0)\begin{pmatrix} \cos\varphi_0 \\ \sin\varphi_0 \end{pmatrix}$$

$$= \begin{pmatrix} -\dot\rho(\varphi_0)\sin\varphi_0 + \rho(\varphi_0)(1-s)\cos\varphi_0 \\ \dot\rho(\varphi_0)\cos\varphi_0 + \rho(\varphi_0)(1-s)\sin\varphi_0 \end{pmatrix}.$$

If for example $\gamma(\varphi) = \alpha\varphi\begin{pmatrix} \cos\varphi \\ \sin\varphi \end{pmatrix}$ is the Archimedean spiral we find $\dot\rho(\varphi_0) = \alpha$ for all φ_0 and

$$n_{\varphi_0}(1) = \alpha\begin{pmatrix} -\sin\varphi_0 \\ \cos\varphi_0 \end{pmatrix},$$

implying that $\varphi_0 \mapsto n_{\varphi_0}(1)$, $\varphi_0 \in [0, 2\pi]$ is the circle with centre 0 and radius α_0. A similar result holds for the logarithmic spiral.

637

8. a) It makes sense to introduce the function $h(s) = s + \sqrt{s^2 + 1}$ and write $\gamma(s) = (\frac{1}{2}h(s), \frac{1}{2}\frac{1}{h(s)}, \frac{\sqrt{2}}{2}\log h(s))$ and now we find

$$\dot{\gamma}(s) = \left(\frac{1}{2}\dot{h}(s), -\frac{1}{2}\frac{1}{h(s)^2}\dot{h}(s), \frac{\sqrt{2}}{2}\frac{1}{h(s)}\dot{h}(s)\right) = \dot{h}(s)\left(\frac{1}{2}, -\frac{1}{2}\frac{1}{h(s)^2}, \frac{\sqrt{2}}{2}\frac{1}{h(s)}\right).$$

Since $\dot{h}(s) = 1 + \frac{s}{\sqrt{s^2+1}} = \frac{s+\sqrt{s^2+1}}{\sqrt{s^2+1}}$ we find

$$|\dot{\gamma}(s)| = \frac{1}{2}\frac{s + \sqrt{s^2+1}}{2\sqrt{s^2+1}}\left(1 + \frac{1}{h(s)^4} + 2\frac{1}{h(s)^2}\right)^{\frac{1}{2}}$$

$$= \frac{1}{2}\frac{h(s)}{\sqrt{s^2+1}}\left(\frac{h(s)^4 + 2h(s) + 1}{h(s)^4}\right)^{\frac{1}{2}}$$

$$= \frac{1}{2}\frac{h(s)}{\sqrt{s^2+1}}\left(\frac{(h(s)^2+1)^2}{(h(s))^4}\right)^{\frac{1}{2}} = \frac{1}{2}\frac{h(s)^2+1}{h(s)(\sqrt{s^2+1})}$$

$$= \frac{s^2 + s\sqrt{s^2+1} + 1}{(s + \sqrt{s^2+1})(\sqrt{s^2+1})} = 1.$$

b) This line segment connects the points $\begin{pmatrix} 0 \\ -4 \end{pmatrix}$ and $\begin{pmatrix} 4 \\ 4 \end{pmatrix}$, hence it has length $\left\|\begin{pmatrix} 4 \\ 8 \end{pmatrix}\right\| = (16+64)^{\frac{1}{2}} = \sqrt{80}$. Thus we take instead of $\begin{pmatrix} 2 \\ 4 \end{pmatrix}$ the vector $\frac{1}{\sqrt{20}}\begin{pmatrix} 2 \\ 4 \end{pmatrix}$, and it follows for

$$\dot{\gamma} = \frac{1}{\sqrt{20}}\begin{pmatrix} 2 \\ 4 \end{pmatrix}t + \begin{pmatrix} 2 \\ 0 \end{pmatrix}$$

that

$$\dot{\tilde{\gamma}}(t) = \frac{1}{\sqrt{20}}\begin{pmatrix} 2 \\ 4 \end{pmatrix}, \text{ i.e. } |\dot{\tilde{\gamma}}(t)| = \left(\frac{4}{20} + \frac{16}{20}\right)^{\frac{1}{2}} = 1,$$

i.e. $\tilde{\gamma} : \mathbb{R} \to \mathbb{R}$ is parametrized with respect to the arc length. Now we seek for t_0 and t_1 such that

$$\tilde{\gamma}(t_0) = \begin{pmatrix} 0 \\ -4 \end{pmatrix} \text{ and } \tilde{\gamma}(t_1) = \begin{pmatrix} 4 \\ 4 \end{pmatrix}$$

or

$$\frac{1}{\sqrt{20}}\begin{pmatrix} 2 \\ 4 \end{pmatrix}t_0 + \begin{pmatrix} 2 \\ 0 \end{pmatrix} = \begin{pmatrix} 0 \\ -4 \end{pmatrix}$$

and

$$\frac{1}{\sqrt{20}}\begin{pmatrix} 2 \\ 4 \end{pmatrix}t_1 + \begin{pmatrix} 2 \\ 0 \end{pmatrix} = \begin{pmatrix} 4 \\ 4 \end{pmatrix},$$

which yields $t_0 = -\sqrt{20}$ and $t_1 = \sqrt{20}$. Indeed

$$\tilde{\gamma} : [-\sqrt{20}, \sqrt{20}] \to \mathbb{R}^2, t \mapsto \frac{1}{\sqrt{20}}\begin{pmatrix} 2 \\ 4 \end{pmatrix}t + \begin{pmatrix} 2 \\ 0 \end{pmatrix},$$

is a parametric curve which is parametrized with respect to its arc length since $|\dot{\tilde{\gamma}}(t)| = 1$ for all t. Moreover, it is the line segment connecting $\begin{pmatrix} 0 \\ -4 \end{pmatrix}$ and $\begin{pmatrix} 4 \\ -4 \end{pmatrix}$. Note that the interval $[-\sqrt{20}, \sqrt{20}]$ has the length $\sqrt{20} + \sqrt{20} = 2\sqrt{20} = \sqrt{80}$ as it should.

Remark: In this solution we made much use of our prior knowledge that the trace of the curve under consideration is a line segment.

9. a) Recall that the arc length parameter is given by $s = s(t) = \int_{t_0}^{t} \|\dot{\gamma}(r)\| \, dr$ if $\gamma : I \to \mathbb{R}^3$ and $t_0 \in I$. Thus we have $\gamma(t) = \tilde{\gamma}(s(t))$. In the following the derivatives with respect to t are denoted by $\dot{}$ and those with respect to s we denote by $'$. We find

$$\dot{\gamma} = \tilde{\gamma}' \dot{s}, \quad \ddot{\gamma} = \tilde{\gamma}' \ddot{s} + \tilde{\gamma}'' \dot{s}^2$$

therefore

$$\dot{\gamma} \times \ddot{\gamma} = \tilde{\gamma}' \dot{s} \times (\tilde{\gamma}' \ddot{s} + \tilde{\gamma}'' \dot{s}^2)$$

and since $\dot{s} = \|\dot{\gamma}\|$ and $\tilde{\gamma}' \times \tilde{\gamma}' = 0$ we get

$$\dot{\gamma} \times \ddot{\gamma} = \dot{s}^3 (\tilde{\gamma}' \times \tilde{\gamma}'') = \|\dot{\gamma}\|^3 (\tilde{\gamma}' \times \tilde{\gamma}'').$$

Now we recall that $\tilde{\gamma}' = \vec{t}$, $\tilde{\gamma}''$ and $\tilde{\gamma}'$ are orthogonal and $\|\tilde{\gamma}'\| = 1$, $\|\tilde{\gamma}''\| = \left\| \vec{t}' \right\| = \kappa$ to conclude that

$$\|\dot{\gamma} \times \ddot{\gamma}\| = \|\dot{\gamma}\|^3 \|\tilde{\gamma}'\| \|\tilde{\gamma}''\| = \|\dot{\gamma}\|^3 \kappa,$$

or $\kappa = \frac{\|\dot{\gamma} \times \ddot{\gamma}\|}{\|\dot{\gamma}\|^3}$.

b) We know

$$\dot{\gamma}(s) = \vec{t}(s)$$

$$\ddot{\gamma}(s) = \kappa(s) \vec{n}(s)$$

$$\dddot{\gamma}(s) = \kappa(s) \vec{n}' + \dot{\kappa}(s) \vec{n}(s) = -\kappa^2(s) \vec{t}(s) + \dot{\kappa}(s) \vec{n}(s) - \kappa(s) \tau(s) \vec{b}(s),$$

and therefore

$$\dot{\gamma}(s) \times \ddot{\gamma}(s) = \vec{t}(s) \times \kappa(s) \vec{n}(s) = \kappa(s) \vec{t}(s) \times \vec{n}(s) = \kappa(s) \vec{b}(s),$$

hence

$$\langle \dot{\gamma}(s) \times \ddot{\gamma}(s), \dddot{\gamma}(s) \rangle$$

$$= \langle \kappa(s) \vec{b}(s), -\kappa^2 \vec{t}(s) + \dot{\kappa}(s) \vec{n}(s) - \kappa(s) \tau(s) \vec{b}(s) \rangle$$

$$= -\kappa^3(s) \langle \vec{b}(s), \vec{t}(s) \rangle + \kappa(s) \dot{\kappa}(s) \langle \vec{b}(s), \vec{n}(s) \rangle$$

$$- \kappa^2(s) \tau(s) \langle \vec{b}(s), \vec{b}(s) \rangle$$

$$= -\kappa^2(s) \tau(s),$$

implying

$$\frac{-\langle \dot{\gamma}(s) \times \ddot{\gamma}(s), \dddot{\gamma}(s) \rangle}{\kappa^2(s)} = \tau(s)$$

10.   a) First we note

$$\dot{\gamma}(t) = (-r\sin t, r\cos t, h)$$

hence

$$\|\dot{\gamma}(t)\| = (r^2 \sin^2 t + r^2 \cos^2 t + h^2)^{\frac{1}{2}} = (r^2 + h^2)^{\frac{1}{2}}.$$

Thus with $t = (r^2 + h^2)^{-\frac{1}{2}} s$ we find for

$$\tilde{\gamma}(s) = \gamma(\varphi(s)), \quad \varphi(s) = (r^2 + h^2)^{-\frac{1}{2}} s,$$

$$\tilde{\gamma}(s) = (r\cos(r^2 + h^2)^{-\frac{1}{2}} s), r\sin((r^2 + h^2)^{-\frac{1}{2}} s), h(r^2 + h^2)^{-\frac{1}{2}} s),$$

and consequently

$$\|\dot{\tilde{\gamma}}(s)\|$$

$$= \left(\frac{r^2}{r^2 + h^2} \sin^2((r^2 + h^2)^{-\frac{1}{2}} s) + \frac{r^2}{r^2 + h^2} \cos^2((r^2 + h^2)^{-\frac{1}{2}} s) + \frac{h^2}{r^2 + h^2} \right)^{\frac{1}{2}}$$

$$= \left(\frac{r^2 + h^2}{r^2 + h^2} \right)^{\frac{1}{2}} = 1.$$

b)   Since $\dot{\gamma}(t) = (-3\sin t, 3\cos t, 4)$ we have

$$\|\dot{\gamma}(t)\| = (9\sin^2 t + 9\cos^2 t + 16)^{\frac{1}{2}} = 5,$$

and we find, compare also with Problem 10a), that the parametrization with respect to the arc length is given by

$$\tilde{\gamma}(s) = \left(3\cos\frac{s}{5}, 3\sin\frac{s}{5}, \frac{4}{5} s \right), s \in \tilde{I} = [0, 20\pi].$$

Therefore we obtain

$$\dot{\tilde{\gamma}}(s) = \left(-\frac{3}{5}\sin\frac{s}{5}, \frac{3}{5}\cos\frac{s}{5}, \frac{4}{5} \right) = \vec{t}(s)$$

$$\ddot{\tilde{\gamma}}(s) = \left(-\frac{3}{25}\cos\frac{s}{5}, -\frac{3}{25}\sin\frac{s}{5}, 0 \right) = \kappa(s)\vec{n}(s)$$

$$\dddot{\tilde{\gamma}}(s) = \left(\frac{3}{125}\sin\frac{s}{5}, -\frac{3}{125}\cos\frac{s}{5}, 0 \right),$$

implying

$$\kappa(s) = \|\ddot{\tilde{\gamma}}(s)\| = \left(\left(\frac{3}{25}\right)^2 \cos^2\frac{s}{5} + \left(\frac{3}{25}\right)^2 \sin^2\frac{s}{5} + 0 \right)^{\frac{1}{2}} = \frac{3}{25}.$$

Thus $\tilde{\gamma}$ has constant curvature. Moreover we have

$$\tau(s) = -\frac{\langle \dot{\tilde{\gamma}}(s) \times \ddot{\tilde{\gamma}}(s), \dddot{\tilde{\gamma}}(s) \rangle}{(\frac{3}{25})^2}.$$

Now

$$\dot{\tilde{\gamma}}(s) \times \ddot{\tilde{\gamma}} = \begin{pmatrix} (\frac{3}{5}\cos\frac{s}{5}) \cdot 0 - \frac{4}{5}(-\frac{3}{25}\sin\frac{s}{5}) \\ \frac{4}{5}(-\frac{3}{25}\cos\frac{s}{5}) - (-\frac{3}{5}\sin\frac{s}{5}) \cdot 0 \\ (-\frac{3}{5}\sin\frac{s}{5})(-\frac{3}{25}\sin\frac{s}{5}) - (\frac{3}{5}\cos\frac{s}{5})(-\frac{3}{25}\cos\frac{s}{5}) \end{pmatrix} = \begin{pmatrix} \frac{12}{125}\sin\frac{s}{5} \\ -\frac{12}{125}\cos\frac{s}{5} \\ \frac{9}{125} \end{pmatrix}$$

$$\langle \dot{\tilde{\gamma}}(s) \times \ddot{\tilde{\gamma}}, \dddot{\tilde{\gamma}}(s) \rangle = \left\langle \begin{pmatrix} \frac{12}{125}\sin\frac{s}{5} \\ -\frac{12}{125}\cos\frac{s}{5} \\ \frac{9}{125} \end{pmatrix}, \begin{pmatrix} \frac{3}{125}\sin\frac{s}{5} \\ -\frac{3}{125}\cos\frac{s}{5} \\ 0 \end{pmatrix} \right\rangle = \frac{36}{125^2},$$

which gives

$$\tau(s) = -\frac{\frac{36}{125^2}}{\frac{9}{25^2}} = -\frac{4}{25}.$$

Thus the torsion is constant too.

(Note, when τ is defined with the other sign, in this example we would get, of course, $\tau(s) = \frac{4}{25}$.)

The tangent vector at s is

$$\vec{t}(s) = \dot{\tilde{\gamma}}(s) = \left(-\frac{3}{5}\sin\frac{s}{5}, \frac{3}{5}\cos\frac{s}{5}, \frac{4}{5} \right),$$

and the normal vector is

$$\vec{n}(s) = \frac{\ddot{\tilde{\gamma}}}{\kappa(s)} = \frac{1}{\frac{3}{25}}\left(-\frac{3}{25}\cos\frac{s}{5}, -\frac{3}{25}\sin\frac{s}{5}, 0 \right) = \left(-\cos\frac{s}{5}, -\sin\frac{s}{5}, 0 \right)$$

which gives the osculating plane at s_0 as

$$E_{\gamma(s_0)} = \left\{ (3\cos\frac{s_0}{5}, 3\sin\frac{s_0}{5}, \frac{4}{5}s_0) + \lambda\vec{t}(s_0) + \mu\vec{n}(s_0) | \lambda, \mu \in \mathbb{R} \right\}$$

$$= \left\{ (3\cos\frac{s_0}{5} - \frac{3\lambda}{5}\sin\frac{s_0}{5} - \mu\cos\frac{s_0}{5}, 3\sin\frac{s_0}{5} + \frac{3\lambda}{5}\cos\frac{s_0}{5} - \mu\sin\frac{s_0}{5}, \right.$$

$$\left. \frac{4}{5}s_0 + \frac{4\lambda}{5}) | \lambda, \mu \in \mathbb{R} \right\}.$$

11. A vector (a, b) normal to $\dot{\gamma}(t)$, $\|\dot{\gamma}(t)\| = 1$, must satisfy the equation $\dot{\gamma}_1(t)a + \dot{\gamma}_2(t)b = 0$ which implies that

$$\begin{pmatrix} a \\ b \end{pmatrix} = \lambda \begin{pmatrix} -\dot{\gamma}_2(t) \\ \dot{\gamma}_1(t) \end{pmatrix}$$

for some $\lambda \in \mathbb{R}$. Since $\frac{1}{2}\frac{d}{dt}\|\dot\gamma(t)\|^2 = \dot\gamma_1(t)\ddot\gamma_1(t) + \dot\gamma_2(t)\ddot\gamma_2(t) = 0$ we can already deduce that

$$\begin{pmatrix} \ddot\gamma_1(t) \\ \ddot\gamma_2(t) \end{pmatrix} = \lambda_0 \begin{pmatrix} -\dot\gamma_2(t) \\ \dot\gamma_1(t) \end{pmatrix}$$

and since by assumption $\|\dot\gamma(t)\| = 1$, hence $\left\| \begin{pmatrix} -\dot\gamma_2(t) \\ \dot\gamma_1(t) \end{pmatrix} \right\| = 1$, we deduce that $|\lambda_0| = \|\ddot\gamma(t)\|$.

12. The point of intersection is of course given by $R_{\varphi_0}(\rho_0) = C_{\rho_0}(\varphi_0) = \rho_0 \begin{pmatrix} \cos\varphi_0 \\ \sin\varphi_0 \end{pmatrix}$, see the figure below.

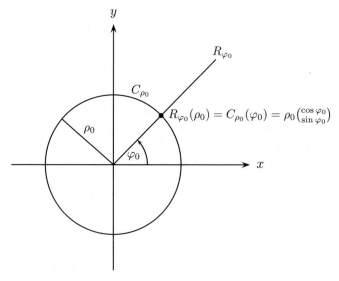

Since R_{φ_0} is a line segment it coincides with parts of its tangent line, so $g_{R_{\varphi_0}}(s) = \begin{pmatrix} \cos\varphi_0 \\ \sin\varphi_0 \end{pmatrix} s$, whereas the tangent line to C_{ρ_0} at φ_0 is given by

$$g_{C_{\rho_0}}(t) = \rho_0 \begin{pmatrix} -\sin\varphi_0 \\ \cos\varphi_0 \end{pmatrix} + \rho_0 \begin{pmatrix} \cos\varphi_0 \\ \sin\varphi_0 \end{pmatrix}$$

and we find at the intersection point

$$\left\langle \begin{pmatrix} \cos\varphi_0 \\ \sin\varphi_0 \end{pmatrix}, \rho_0 \begin{pmatrix} -\sin\varphi_0 \\ \cos\varphi_0 \end{pmatrix} \right\rangle = \rho_0(-\cos\varphi_0 \sin\varphi_0 + \sin\varphi_0 \cos\varphi_0) = 0,$$

i.e. these two coordinate curves are orthogonal at their intersection point.

13. According to Example 7.34 we have

$$\cos \alpha = \frac{1}{\sqrt{1 + g'(\frac{\pi}{4})^2}} \frac{1}{\sqrt{1 + h'(\frac{\pi}{4})^2}} \left| 1 + g'(\frac{\pi}{4}) h'(\frac{\pi}{4}) \right|$$

$$= \frac{1}{\sqrt{1 + \frac{1}{2}}} \frac{1}{\sqrt{1 + \frac{1}{2}}} \left| 1 - \sin \frac{\pi}{4} \cos \frac{\pi}{4} \right|$$

$$= \frac{2}{3} \left| 1 - \frac{1}{2} \right| = \frac{1}{3}.$$

This refers approximately to $\alpha \approx 19.5°$, also see the following

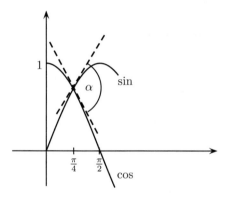

Note that this diagram has not been drawn to the correct scale in order to clearly illustrate the solution.

Chapter 8

1. We are asked to translate the curve with trace in $\mathbb{R}^3$ given by $\left\{ \begin{pmatrix} x \\ 0 \\ h(x) \end{pmatrix} \middle| x \in (0, 1) \right\}$ along the y-axis, so we arrive at

$$S = \left\{ \begin{pmatrix} x \\ y \\ h(x) \end{pmatrix} \middle| x \in (0, 1), y \in \mathbb{R} \right\}.$$

Since S is the graph of $g : (0, 1) \times \mathbb{R} \to \mathbb{R}$, given by $g(x, y) = h(x)$, it is clear that S is the trace of a parametric surface, see as illustrated in the following figure:

643

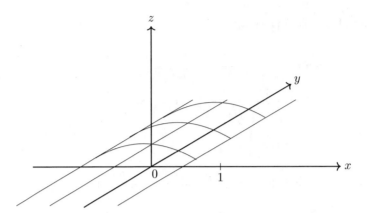

2. We can follow Example 8.3. Since $0 \leq f(x,y) \leq r$ we find first that

$$f^{-1}(c) = \emptyset \text{ for } c \in (-\infty, 0) \cup (r, \infty).$$

and further

$$f^{-1}(r) = \{(0,0)\}.$$

Finally suppose $0 < c = f(x,y) < r$. Then it follows that

$$r^2 - \frac{x^2}{a^2} - \frac{y^2}{b^2} = c^2$$

or

$$(*) \qquad \frac{x^2}{a^2} + \frac{y^2}{b^2} = r^2 - c^2$$

implying that $f^{-1}(c)$ is in this case the ellipse defined by equation $(*)$.

3.   a) We first write P as (trace of a) parametric surface, $f : \mathbb{R}^2 \to \mathbb{R}^3$, $(u,v) \mapsto$

$$f(u,v) = \begin{pmatrix} u \\ v \\ u^2 + v^2 + 1 \end{pmatrix}. \text{ The tangent vectors are } \begin{pmatrix} 1 \\ 0 \\ 2u \end{pmatrix} \text{ and } \begin{pmatrix} 0 \\ 1 \\ 2v \end{pmatrix}, \text{ so for}$$

the point $p = (2,2,9)$ we find the tangent plane

$$\left\{ \begin{pmatrix} 2 \\ 2 \\ 9 \end{pmatrix} + \lambda \begin{pmatrix} 1 \\ 0 \\ 4 \end{pmatrix} + \mu \begin{pmatrix} 0 \\ 1 \\ 4 \end{pmatrix} \Big| \lambda, \mu \in \mathbb{R} \right\}.$$

The normal vector to P at $(2,2,9)$ is given by

$$\vec{n} = \begin{pmatrix} 1 \\ 0 \\ 4 \end{pmatrix} \times \begin{pmatrix} 0 \\ 1 \\ 4 \end{pmatrix} = \begin{pmatrix} -4 \\ -4 \\ 1 \end{pmatrix}$$

which implies that the normal line passing through $p = (2,2,9)$ is the line

$$\left\{ \begin{pmatrix} 2 \\ 2 \\ 9 \end{pmatrix} + \sigma \begin{pmatrix} -4 \\ -4 \\ 1 \end{pmatrix} \Big| \sigma \in \mathbb{R} \right\}.$$

b) Again we first represent H as parametric surface with the help of $h : \mathbb{R}^2 \to \mathbb{R}^3$,

$h(u, v) = \begin{pmatrix} u \\ v \\ u^2 - v^2 \end{pmatrix}$. This gives the tangent vectors $\begin{pmatrix} 1 \\ 0 \\ 2u \end{pmatrix}$ and $\begin{pmatrix} 0 \\ 1 \\ -2v \end{pmatrix}$, so

for $u = v$ we have $\begin{pmatrix} 1 \\ 0 \\ 2u \end{pmatrix}$ and $\begin{pmatrix} 0 \\ 1 \\ -2u \end{pmatrix}$, whereas for $u = -v$ we find $\begin{pmatrix} 1 \\ 0 \\ 2u \end{pmatrix}$,

$\begin{pmatrix} 0 \\ 1 \\ 2u \end{pmatrix}$. Thus the corresponding tangent planes and normal lines are

$$\left\{ \begin{pmatrix} x \\ x \\ 0 \end{pmatrix} + \lambda \begin{pmatrix} 1 \\ 0 \\ 2x \end{pmatrix} + \mu \begin{pmatrix} 0 \\ 1 \\ -2x \end{pmatrix} \bigg| \lambda, \mu \in \mathbb{R} \right\}$$

$$\left\{ \begin{pmatrix} x \\ x \\ 0 \end{pmatrix} + \sigma \begin{pmatrix} -2x \\ 2x \\ 1 \end{pmatrix} \bigg| \sigma \in \mathbb{R} \right\}$$

and

$$\left\{ \begin{pmatrix} x \\ -x \\ 0 \end{pmatrix} + \lambda \begin{pmatrix} 1 \\ 0 \\ 2x \end{pmatrix} + \mu \begin{pmatrix} 0 \\ 1 \\ 2x \end{pmatrix} \bigg| \lambda \mu \in \mathbb{R} \right\}$$

$$\left\{ \begin{pmatrix} x \\ -x \\ 0 \end{pmatrix} + \sigma \begin{pmatrix} -2x \\ -2x \\ 1 \end{pmatrix} \bigg| \sigma \in \mathbb{R} \right\}$$

4. a) Since $f_u(u_0, v_0) = \begin{pmatrix} h'(u_0) \cos v_0 \\ h'(u_0) \sin v_0 \\ k'(u_0) \end{pmatrix}$ and $f_v(u_0, v_0) = \begin{pmatrix} -h(u_0) \sin v_0 \\ h(u_0) \cos v_0 \\ 0 \end{pmatrix}$ we

find

$$f_u(u_0, v_0) \times f_v(u_0, v_0) = \begin{pmatrix} -k'(u_0)h(u_0) \cos v_0 \\ -k'(u_0)h(u_0) \sin v_0 \\ h'(u_0)h(u_0) \end{pmatrix}$$

and therefore the tangent plane to M at p is given by

$$\left\{ \begin{pmatrix} h(u_0) \cos v_0 \\ h(u_0) \sin v_0 \\ k(u_0) \end{pmatrix} + \lambda \begin{pmatrix} h'(u_0) \cos v_0 \\ h'(u_0) \sin v_0 \\ k'(u_0) \end{pmatrix} + \mu \begin{pmatrix} -h(u_0) \sin v_0 \\ h(u_0) \cos v_0 \\ 0 \end{pmatrix} \bigg| \lambda, \mu \in \mathbb{R} \right\}$$

and the normal line to M at p is

$$\left\{ \begin{pmatrix} h(u_0) \cos v_0 \\ h(u_0) \sin v_0 \\ k(u_0) \end{pmatrix} + \lambda \begin{pmatrix} -k'(u_0)h(u_0) \cos v_0 \\ -k'(u_0)h(u_0) \sin v_0 \\ h'(u_0)h(u_0) \end{pmatrix} \bigg| \lambda \in \mathbb{R} \right\}.$$

645

b) With $h(u) = k(u) = u$ we find for the tangent plane

$$\left\{ \begin{pmatrix} u_0 \cos v_0 \\ u_0 \sin v_0 \\ u_0 \end{pmatrix} + \lambda \begin{pmatrix} \cos v_0 \\ \sin v_0 \\ 1 \end{pmatrix} + \mu \begin{pmatrix} -u_0 \sin v_0 \\ u_0 \cos v_0 \\ 0 \end{pmatrix} \middle| \lambda, \mu \in \mathbb{R} \right\}.$$

We observe that the ray $\left\{ \lambda \begin{pmatrix} \cos v_0 \\ \sin v_0 \\ 1 \end{pmatrix} \middle| \lambda > 0 \right\}$ belongs to the tangent plane as well as to the cone. For the normal line we obtain

$$\left\{ \begin{pmatrix} u_0 \cos v_0 \\ u_0 \sin v_0 \\ u_0 \end{pmatrix} + \lambda \begin{pmatrix} -u_0 \cos v_0 \\ -u_0 \sin v_0 \\ u_0 \end{pmatrix} \middle| \lambda \in \mathbb{R} \right\}$$

which means that the normal vectors are not only as expected orthogonal to the tangent plane at $p = \begin{pmatrix} u_0 \cos v_0 \\ u_0 \sin v_0 \\ u_0 \end{pmatrix}$ but also to the point p itself.

5. The catenoid is given by the parametrization $f(u, v) = (\cosh u \cos v, \cosh u \sin v, u)$ and its trace is sketched in the figure below

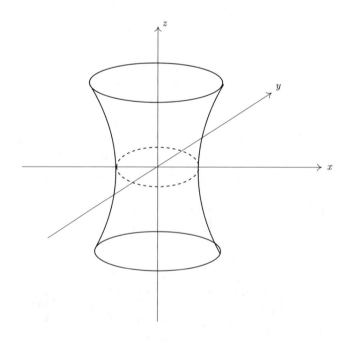

The parametric curve $u \mapsto f(u, v_0) = \begin{pmatrix} \cosh u \cos v_0 \\ \cosh u \sin v_0 \\ u \end{pmatrix}$ has the tangent vector

$$\frac{d}{du} f(u, v_0) = \begin{pmatrix} \sinh u \cos v_0 \\ \sinh u \sin v_0 \\ 1 \end{pmatrix}$$

and the parametric curve $v \mapsto f(u_0, v)$ has the tangent vector

$$\frac{d}{dv} f(u_0, v) = \begin{pmatrix} -\cosh u_0 \sin v \\ \cosh u_0 \cos v \\ 0 \end{pmatrix}.$$

At the intersection $p = f(u_0, v_0)$ we now find

$$\langle \frac{df}{du}(u_0, v_0), \frac{df}{dv}(u_0, v_0) \rangle = \left\langle \begin{pmatrix} \sinh u_0 \cos v_0 \\ \sinh u_0 \sin v_0 \\ 1 \end{pmatrix}, \begin{pmatrix} -\cosh u_0 \sin v_0 \\ \cosh u_0 \cos v_0 \\ 0 \end{pmatrix} \right\rangle$$
$$= -\sinh u_0 \cos v_0 \cosh u_0 \sin v_0$$
$$+ \sinh u_0 \sin v_0 \cosh u_0 \cos v_0 = 0,$$

implying that the coordinate lines are orthogonal.

6. A straightforward calculation yields

$$(dS)_{(u,v)} = \frac{4}{(u^2 + v^2 + 4)^2} \begin{pmatrix} -u^2 + v^2 + 4 & -2uv \\ -2uv & u^2 - v^2 + 4 \\ 4u & 4v \end{pmatrix}.$$

We claim that for all $(u, v) \in \mathbb{R}^2$ this Jacobi matrix is injective. To prove this we assume that the two column vectors are dependent, i.e. for some $\lambda \neq 0$

$$\lambda \begin{pmatrix} -u^2 + v^2 + 4 \\ -2uv \\ 4u \end{pmatrix} = \begin{pmatrix} -2uv \\ u^2 - v^2 + 4 \\ 4v \end{pmatrix}.$$

Now, if $u = 0$ and $\lambda \neq 0$ then it follows that $v = 0$ and the vectors

$$\begin{pmatrix} 4 \\ 0 \\ 0 \end{pmatrix}, \begin{pmatrix} 0 \\ 4 \\ 0 \end{pmatrix}$$

are independent. If $u \neq 0$, then $\lambda = \frac{v}{u}$ and

$$\lambda(-2uv) = \frac{v}{u}(-2uv) = -2v^2 = u^2 - v^2 + 4$$

647

implies $0 = u^2 + v^2 + 4$, which can never happen, so $(dS)_{(u,v)}$ is injective for all $(u,v) \in \mathbb{R}^2$, and therefore S is a parametric surface. Furthermore we find for $(x,y,z) = \left(\frac{4u}{u^2+v^2+4}, \frac{4v}{u^2+v^2+4}, \frac{2u^2+2v^2}{u^2+v^2+4} \right)$ that

$$
\begin{aligned}
x^2 + y^2 + (z-1)^2 &= \left(\frac{4u}{u^2+v^2+4} \right)^2 + \left(\frac{4v}{u^2+v^2+4} \right)^2 + \left(\frac{2u^2+v^2}{u^2+v^2+4} - 1 \right)^2 \\
&= \frac{1}{(u^2+v^2+4)^2} (16u^2 + 16v^2 + (2u^2 + 2v^2 - u^2 - v^2 - 4)^2) \\
&= \frac{1}{(u^2+v^2+4)^2} (16u^2 + 16v^2 + (u^2 + v^2 - 4)^2) \\
&= \frac{1}{(u^2+v^2+4)^2} (u^2 + v^2 + 4)^2 = 1,
\end{aligned}
$$

i.e. $\operatorname{tr}(S) \subset \partial B_1((0,0,1))$.

Chapter 9

1. For $x = (x_1, ..., x_n) \in \mathbb{R}^n \backslash \{0\}$ there exists $R > 0$ such that $y := \left(\frac{x_1}{R}, ..., \frac{x_n}{R} \right) \in B_\epsilon(0) \backslash \{0\}$, for example we may take $R = \frac{2\|x\|}{\epsilon}$ implying

$$
\|y\|^2 = \frac{\epsilon^2 x_1^2}{4\|x\|^2} + ... + \frac{\epsilon^2 x_n^2}{4\|x\|^2} = \frac{\epsilon^2}{4},
$$

i.e. $\|y\| < \frac{\epsilon}{2}$ and hence $y \in B_\epsilon(0) \backslash \{0\}$. If $\beta_A(y,y) \geq 0$ it follows for $x \in \mathbb{R}^n \backslash \{0\}$ that

$$
\beta_A(x,x) = R^2 \beta_A(y,y) \geq 0
$$

and for $x = 0$ the inequality is trivial. In the case where $\beta_A(y,y) > 0$ we deduce further $\beta_A(x,x) > 0$ for $x \in \mathbb{R}^n \backslash \{0\}$.

2. Since A is symmetric we can find an orthonormal basis $b_1, ..., b_n$ of eigenvectors and every $\xi \in \mathbb{R}^n$ admits a representation $\xi = \sum_{j=1}^n \eta_j b_j$. For $\beta_A(\xi,\xi)$ this implies

$$
\beta_A(\xi,\xi) = \sum_{j=1}^n \lambda_j \eta_j^2.
$$

This representation yields immediately that if $\lambda_1 > 0$ then A is positive definite and if $\lambda_1 \geq 0$ then A is positive semi-definite, recall we have the order $\lambda_1 \leq \lambda_2 \leq ... \leq \lambda_n$. Furthermore, if $\lambda_n < 0$ ($\lambda_n \leq 0$), then A is negative (semi-)definite. If $\lambda_1 < 0$ and $\lambda_n > 0$ we choose $\xi \in \operatorname{span}\{b_1, b_n\}$ to obtain $\beta_A(\xi,\xi) = \lambda_1 \gamma_1^2 + \lambda_n \gamma_n^2$ with $\lambda_1 \cdot \lambda_n < 0$ which implies that A is indefinite.

3. a) Clearly $|||A||| \geq 0$ and if $|||A||| = 0$ then we find a sequence $(c_n)_{n \in \mathbb{N}}$, $\lim_{n \to \infty} c_n = 0$, such that for all $x \in \mathbb{R}^n$ it follows $\|Ax\| \leq c_n \|x\|$ implying

$\|Ax\| = 0$ for all $x \in \mathbb{R}^n$, i.e. $A = 0$. Moreover we find for $\lambda \in \mathbb{R}$, $\lambda \neq 0$,

$$
\begin{aligned}
|||\lambda A||| &= \inf \left\{ c > 0 \,\middle|\, |||\lambda Ax||| \leq c\,\|x\| \right\} \\
&= \inf \left\{ c > 0 \,\middle|\, |\lambda| \|Ax\| \leq c\,\|x\| \right\} \\
&= \inf \left\{ c > 0 \,\middle|\, \|Ax\| \leq \frac{c}{|\lambda|}\,\|x\| \right\} \\
&= \inf \left\{ |\lambda| d > 0 \,\middle|\, \|Ax\| \leq d\,\|x\| \right\} = |\lambda| \,|||A|||,
\end{aligned}
$$

and the case $\lambda = 0$ is trivial. Further we observe that

$$
\begin{aligned}
|||A + B||| &= \inf \left\{ c > 0 \,\middle|\, \|Ax + Bx\| \leq c\,\|x\| \right\} \\
&\leq \inf \left\{ c_1 > 0 \,\middle|\, \|Ax\| \leq c_1\,\|x\| \right\} + \inf \left\{ c_2 > 0 \,\middle|\, \|Bx\| \leq c_2\,\|x\| \right\} \\
&= |||A||| + |||B|||,
\end{aligned}
$$

i.e. $|||\cdot|||$ is indeed a norm and in addition the following holds

$$
\|ABx\| \leq |||A|||\,\|Bx\| \leq |||A|||\,|||B|||\,\|x\|
$$

implying

$$
|||AB||| \leq |||A|||\,|||B|||.
$$

Note that we cannot expect that in general equality holds in $|||AB||| \leq |||A|||\,|||B|||$. Just take $A = \begin{pmatrix} 1 & 0 \\ 0 & 0 \end{pmatrix}$ and $B = \begin{pmatrix} 0 & 0 \\ 1 & 0 \end{pmatrix}$ and note that $AB = \begin{pmatrix} 0 & 0 \\ 0 & 0 \end{pmatrix}$. Finally we prove $|||A||| = \sup_{x \neq 0} \frac{\|Ax\|}{\|x\|}$. Since $\|Ax\| \leq |||A|||\,\|x\|$ we have

$$
\sup_{x \neq 0} \frac{\|Ax\|}{\|x\|} \leq |||A|||.
$$

On the other hand we find for $x \neq 0$

$$
\frac{\|Ax\|}{\|x\|} \leq \sup_{x \neq 0} \frac{\|Ax\|}{\|x\|},
$$

or

$$
|||Ax||| \leq \left(\sup_{x \neq 0} \frac{\|Ax\|}{\|x\|} \right) \|x\|
$$

which implies that $|||A||| \leq \sup_{x \neq 0} \frac{\|Ax\|}{\|x\|}$, i.e. $|||A||| = \sup_{x \neq 0} \frac{\|Ax\|}{\|x\|}$.

b) We consider

$$
\sup_{x \neq 0} \frac{\|Ax\|_\infty}{\|x\|_\infty} = \sup_{x \neq 0} \frac{\max_{1 \leq j \leq n} \left| \sum_{k=1}^n a_{jk} x_k \right|}{\max_{1 \leq l \leq n} |x_l|} \leq \max_{1 \leq j \leq n} \sum_{k=1}^n |a_{jk}|,
$$

thus

$$
|||A||| \leq \max_{1 \leq j \leq n} \sum_{k=1}^n |a_{jk}|.
$$

We claim that $|||A||| = \max_{1\leq j\leq n}\sum_{k=1}^{n}|a_{jk}|$. For this let j_0 such that

$$\sum_{k=1}^{n}|a_{j_0 k}| = \max_{1\leq j\leq n}\sum_{k=1}^{n}|a_{jk}|.$$

Choose $z\in\mathbb{R}^n$ with $z_k = 1$ if $a_{j_0 k}\geq 0$ and $z_k = -1$ if $a_{j_0 k} < 0$. It follows that $\|z\|_\infty = \max_{1\leq k\leq n}|z_k| = 1$ and

$$\|Az\|_\infty = \max_{1\leq j\leq n}\left|\sum_{k=1}^{n}a_{jk}z_k\right| \geq \left|\sum_{k=1}^{n}a_{j_0 k}z_k\right|$$

$$= \sum_{k=1}^{n}|a_{j_0 k}| = \max_{1\leq j\leq n}\sum_{k=1}^{n}|a_{jk}|,$$

implying that $|||A||| = \max_{1\leq j\leq n}\sum_{k=1}^{n}|a_{jk}|$. This norm is called the **column sum norm** of A.

4. Since $A(x_0)$ is positive definite we know that $\beta_{A(x_0)}(\xi,\xi)\geq\gamma_0\|\xi\|^2$ for all $\xi\in\mathbb{R}^n$, $\gamma_0 > 0$. This implies

$$\beta_{A(x)}(\xi,\xi) = \beta_{A(x_0)}(\xi,\xi) + \beta_{A(x)-A(x_0)}(\xi,\xi)$$
$$= \langle A(x_0)\xi,\xi\rangle + \langle(A(x)-A(x_0))\xi,\xi\rangle$$
$$\geq \gamma_0\|\xi\|^2 - \|(A(x)-A(x_0))\xi\|\,\|\xi\|$$
$$\geq \gamma_0(-\|A(x)-A(x_0)\|)\|\xi\|^2,$$

where we used the matrix norm with respect to the Euclidean norm $\|\cdot\|$ in $\mathbb{R}^n$. Since A is continuous we can find for $\epsilon = \frac{\gamma_0}{2}$ a number $r > 0$ such that $B_r(x_0)\subset G$ and $\|A(x)-A(x_0)\| < \epsilon = \frac{\gamma_0}{2}$ for all $x\in B_r(x_0)$, which implies for all $x\in B_r(x_0)$ and $\xi\in\mathbb{R}^n$

$$\beta_{A(x)}(\xi,\xi)\geq\frac{\gamma_0}{2}\|\xi\|^2.$$

5. First we note that

$$\frac{\partial f}{\partial x}(x,y) = \frac{\partial}{\partial x}\left(\arctan\frac{y}{x} + \arctan\frac{x}{y}\right)$$

$$= \frac{-y}{x^2(1+\frac{y^2}{x^2})} + \frac{1}{y(1+\frac{x^2}{y^2})}$$

$$= \frac{-y}{x^2+y^2} + \frac{y}{x^2+y^2} = 0,$$

and analogously we find $\frac{\partial f}{\partial y}(x,y) = 0$. Thus, in the convex set $\{(x,y)\in\mathbb{R}^2 | x > 0, y > 0\}$ the grad of f is identically zero. This implies by the mean value theorem that f must be constant:

$$f(x_1,y_1)-f(x_2,y_2) = \langle\mathrm{grad}\,f((x_1,y_1)+\vartheta((x_1-x_2,y_1-y_2))),(x_1-x_2,y_1-y_2)\rangle = 0.$$

Since $f(1,1) = \arctan 1 + \arctan 1 = \frac{\pi}{2}$ we find for all $x,y > 0$ that $f(x,y) = \frac{\pi}{2}$.

6. For $y \in G$ fixed we get for $x_1, x_2 \in G$

$$f(x_1, y) - f(x_2, y) = \frac{\partial f}{\partial x}(x_1 + \vartheta(x_1 - x_2), y)(x_1 - x_2) + \frac{\partial f}{\partial y}(x_1 + \vartheta(x_1 - x_2), y)(y - y) = 0$$

implying for all $y \in G$ that $f(x_1, y) = f(x_2, y)$, i.e. $f(x_1, y) = f(x_2, y) = h(y)$ for some function h, for example $h(y) := f(x_1, y)$.

7.   a) We note that

$$\frac{\partial f}{\partial x}(x, y) = (y \cos(x + y) - \sin(x + y))e^{xy}$$

$$\frac{\partial f}{\partial y}(x, y) = (x \cos(x + y) - \sin(x + y))e^{xy}$$

$$\frac{\partial^2 f}{\partial x^2}(x, y) = (y^2 - 2y \sin(x + y) - \cos(x + y))e^{xy}$$

$$\frac{\partial^2 f}{\partial x \partial y}(x, y) = \frac{\partial^2 f}{\partial y \partial x}(x, y) = (\cos(x + y) + xy \cos(x + y)$$
$$- y \sin(x + y) - x \sin(x + y))e^{xy}$$

$$\frac{\partial^2 f}{\partial y^2}(x, y) = (x^2 - 2x \sin(x + y) - \cos(x + y))e^{xy}$$

which yields

$$f(0,0) = 1, \frac{\partial f}{\partial x}(0,0) = \frac{\partial f}{\partial y}(0,0) = 0, \frac{\partial^2 f}{\partial x^2}(0,0) = \frac{\partial^2 f}{\partial y^2}(0,0) = -1,$$

$$\frac{\partial^2 f}{\partial x \partial y}(0,0) = \frac{\partial^2 f}{\partial y \partial x}(0,0) = 0,$$

and therefore the Taylor polynomial of order 2 of f about $(0,0)$ is

$$T_f^{(2)}(x, y) = \sum_{|\alpha| \leq 2} \frac{D^\alpha f}{\alpha!}(0,0)x^{\alpha_1}y^{\alpha_2} = 1 - \frac{x^2}{2} - \frac{y^2}{2}.$$

b) We find

$$\frac{\partial g}{\partial x}(x, y) = \frac{1}{x}, \quad \frac{\partial g}{\partial y}(x, y) = \frac{1}{y},$$

$$\frac{\partial^2 g}{\partial x^2}(x, y) = -\frac{1}{x^2}, \frac{\partial^2 g}{\partial y^2}(x, y) = -\frac{1}{y^2}, \frac{\partial^2 g}{\partial x \partial y}(x, y) = \frac{\partial^2 g}{\partial y \partial x}(x, y) = 0,$$

$$\frac{\partial^3 g}{\partial x^3}(x, y) = \frac{2}{x^3}, \frac{\partial^3 g}{\partial y^3}(x, y) = \frac{2}{y^3}, \frac{\partial^3 g}{\partial y^2 \partial x}(x, y) = \frac{\partial^3 g}{\partial x^2 \partial y}(x, y) = 0,$$

and therefore we find for the Taylor polynomial of order 3 of g about $(1, 1)$

using the fact that all mixed partial derivatives are identically zero

$$T_g^{(3)}(x,y) = g(1,1) + \frac{\partial g}{\partial x}(1,1)(x-1) + \frac{\partial g}{\partial y}(1,1)(y-1)$$

$$+ \frac{1}{2!}\frac{\partial^2 g}{\partial x^2}(1,1)(x-1)^2 + \frac{1}{2!}\frac{\partial^2 g}{\partial y^2}(1,1)(y-1)^2$$

$$+ \frac{1}{3!}\frac{\partial^3 g}{\partial x^3}(1,1)(x-1)^3 + \frac{1}{3!}\frac{\partial^3}{\partial y^3}(1,1)(y-1)^3$$

$$= (x-1) + (y-1) - \frac{(x-1)^2}{2} - \frac{(y-1)^2}{2} + \frac{(x-1)^3}{3} + \frac{(y-1)^3}{3}.$$

8. We use the Taylor formula, Theorem 9.4, which we can read as

$$f(x) - \sum_{|\alpha| \le k} \frac{D^\alpha f(y)}{\alpha!}(x-y)^\alpha = \sum_{|\alpha|=k+1} \frac{D^\alpha f(y + \vartheta(x-y))}{\alpha!}(x-y)^\alpha.$$

For $k = 2$ and $n = 2$ we have

$$f(x) - f(y) - \frac{\partial f}{\partial x_1}(y)(x_1 - y_1) - \frac{\partial f}{\partial x_2}(y)(x_2 - y_2) - \frac{1}{2}\frac{\partial^2 f}{\partial x_1^2}(y)(x_1 - y_1)^2$$

$$- 2\frac{\partial^2 f}{\partial x_1 \partial x_2}(y)(x_1 - y_1)(x_2 - y_2) - \frac{1}{2}\frac{\partial^2 f}{\partial x_2^2}(y)(x_2 - y_2)^2$$

$$= \sum_{|\alpha|=3} \frac{D^2 f(y + \vartheta(x-y))}{\alpha!}(x-y)^\alpha.$$

For $f(x) = \cos x_1 \cos x_2$ we have with $y = (y_1, y_2) = (0,0)$

$$f(y) = 1,$$

$$\frac{\partial f}{\partial x_1}(y) = -\sin y_1 \cos y_2 = 0, \frac{\partial f}{\partial x_2}(y) = -\cos y_1 \sin y_2 = 0,$$

$$\frac{\partial^2 f}{\partial x_1^2}(y) = -\cos y_1 \cos y_2 = -1, \frac{\partial^2 f}{\partial x_2^2}(y) = -\cos y_1 \cos y_2 = -1,$$

$$\frac{\partial^2 f}{\partial x_1 \partial x_2}(y) = \frac{\partial^2 f}{\partial x_2 \partial x_1}(y) = \sin y_1 \sin y_2 = 0,$$

which yields

$$\left| \cos x_1 \cos x_2 - 1 + \frac{x_1^2}{2} + \frac{x_2^2}{2} \right| \le \sum_{|\alpha| \le 3} \frac{|D^\alpha f(y + \vartheta(x-y))|}{\alpha!}|(x-y)^\alpha|.$$

Now we note that $|D^\alpha f(z)| \le 1$ for all α, $|\alpha| = 3$, and moreover we have

$$|(x-y)^\alpha| = |x_1 - y_1|^{\alpha_1}|x_2 - y_2|^{\alpha_2}|x_3 - y_3|^\alpha,$$

implying that $|(x - y)^\alpha| \leq \epsilon^3$ if $\|x - y\| < \epsilon$. Therefore we find for $x \in B_\epsilon(0)$

$$\sum_{|\alpha|=3} \frac{|D^\alpha f(\vartheta x)|}{\alpha!} |x^\alpha| \leq \left(\frac{1}{3!} + \frac{1}{2!1!} + \frac{1}{1!2!} + \frac{1}{3!}\right) \epsilon^3 = \frac{4}{3}\epsilon^3,$$

and we eventually get for $x \in B_\epsilon(0)$

$$\left| \cos x_1 \cos x_2 - 1 + \frac{x_1^2}{2} + \frac{x_2^2}{2} \right| \leq \frac{4}{3}\epsilon^3.$$

9. Consider $h : \mathbb{R}^n \to \mathbb{R}$, $h(x) = (x_1 + \dots + x_n)^k$. For the derivatives of h we find

$$\frac{\partial}{\partial x_j} h(x) = k(x_1 + \dots + x_n)^{k-1}$$

$$\frac{\partial^2}{\partial x_{j_1} \partial x_{j_2}} h(x) = k(k - 1)(x_1 + \dots + x_n)^{k-2}$$

$$\vdots$$

$$\frac{\partial^m}{\partial x_{j_1} \dots \partial x_{j_m}} h(x) = k(k - 1) \dots (k + 1 - m)(x_1 + \dots + x_n)^{k-m}$$

implying $D^\alpha f(0) = 0$ for $|\alpha| < k$, $D^\alpha f(0) = k!$ for $|\alpha| = k$, and $D^\alpha f(0) = 0$ for $|\alpha| > k$. Applying Taylor's formula to h we note for the remainder term

$$\sum_{|\alpha|=k+1} \frac{(D^\alpha f)(x + \vartheta(z - x))}{\alpha!} (z - x)^\alpha = 0.$$

Moreover, all Taylor polynomials of order less than k about 0 are identically zero, and we arrive at

$$(x_1 + \dots + x_n)^k = h(x) - h(0) = \sum_{|\alpha|=k} \frac{k!}{\alpha!} x^\alpha.$$

10. On $\mathbb{R}^n \backslash \{0\}$ the function $g(x) = \frac{\beta_A(x,x)}{\|x\|^2}$ is arbitrarily often differentiable. Now let $x_0 \neq 0$ be a critical point, i.e. $\operatorname{grad} f(x_0) = 0$. Note that

$$\frac{\partial}{\partial x_j} g(x) = \frac{\partial}{\partial x_j} \frac{\langle Ax, x \rangle}{\|x\|^2} = \frac{\partial}{\partial x_j} \frac{\sum_{k,l=1}^n a_{kl} x_k x_l}{\|x\|^2}$$

$$= \frac{\|x\|^2 \frac{\partial}{\partial x_j} (\sum_{k,l=1}^n a_{kl} x_k x_l - 2x_j \sum_{k,l=1}^n a_{kl} x_k x_l)}{\|x\|^4}$$

$$= \frac{2\|x\|^2 (Ax)j - 2x_j \langle Ax, x \rangle}{\|x\|^4},$$

which yields for a critical point $x_0 \neq 0$ that

$$0 = 2\|x_0\|^2 (Ax_0)_j - 2(x_0)_j \langle Ax_0, x_0 \rangle$$

or

$$Ax_0 = \frac{\langle Ax_0, x_0 \rangle}{\|x_0\|^2} x_0,$$

i.e. x_0 is eigenvector to A with eigenvalue $\frac{\langle Ax_0, x_0 \rangle}{\|x_0\|^2} = \frac{\beta_A(x_0, x_0)}{\|x_0\|^2}$.

11.　a) With $x = (x_1, x_2, x_3)$ and $f(x_1, x_2, x_3) = (x_3 + \sqrt{1 + x_1^2}) - x_2 x_3$ we have

$$\frac{\partial f}{\partial x_1}(x) = \frac{x_1}{\sqrt{1 + x_1^2}}, \quad \frac{\partial f}{\partial x_2}(x) = -x_3, \quad \frac{\partial f}{\partial x_3}(x) = 1 - x_2,$$

$$\frac{\partial^2 f}{\partial x_1^2}(x) = \frac{1}{(1 + x_1^2)^{\frac{3}{2}}}, \frac{\partial^2 f}{\partial x_1 \partial x_2}(x) = \frac{\partial^2 f}{\partial x_2 \partial x_1}(x) = 0, \frac{\partial^2 f}{\partial x_1 \partial x_3}(x) = \frac{\partial^2 f}{\partial x_3 \partial x_1}(x) = 0,$$

$$\frac{\partial^2 f}{\partial x_2^2}(x) = 0, \frac{\partial^2 f}{\partial x_2 \partial x_3}(x) = \frac{\partial^2 f}{\partial x_3 \partial x_2}(x) = -1, \frac{\partial^2 f}{\partial x_3^2} = 0,$$

which leads to

$$(\mathrm{Hess} f)(x) = \begin{pmatrix} \frac{1}{(1+x_1^2)^{\frac{3}{2}}} & 0 & 0 \\ 0 & 0 & -1 \\ 0 & -1 & 0 \end{pmatrix}$$

b) We note that $g(x, y) = g_{11}(x)g_{22}(x) - g_{12}^2(y)$ implying

$$\frac{\partial g}{\partial x}(x, y) = g_{11}'(x)g_{22}(x) + g_{11}(x)g_{22}'(x)$$

$$\frac{\partial g}{\partial y}(x, y) = -2g_{12}'(y)g_{12}(y)$$

$$\frac{\partial^2 g}{\partial x^2}(x, y) = g_{11}''(x)g_{22}(x) + 2g_{11}'(x)g_{22}'(x) + g_{11}(x)g_{22}''(x)$$

$$\frac{\partial^2 g}{\partial y^2}(x, y) = -2g_{12}''(y)g_{12}(y) - 2(g_{12}'(y))^2$$

$$\frac{\partial^2 g}{\partial x \partial y}(x, y) = \frac{\partial^2 g(x, y)}{\partial y \partial x} = 0$$

which yields

$(\mathrm{Hess}\ g)(x, y) =$
$$\begin{pmatrix} g_{11}''(x)g_{22}(x) + 2g_{11}'(x)g_{22}'(x) + g_{11}(x)g_{22}''(x) & 0 \\ 0 & -2g_{12}''(y)g_{12}(y) - 2(g_{12}'(y))^2 \end{pmatrix}.$$

12.　a) The critical points of g are given by

$$\frac{\partial g}{\partial x}(x_0, y_0) = \frac{\partial g}{\partial y}(x_0, y_0) = 0,$$

654

i.e.

$$y_0 e^{-(x_0+y_0)} - x_0 y_0 e^{-(x_0+y_0)} = 0$$

and

$$x_0 e^{-(x_0+y_0)} - x_0 y_0 e^{-(x_0+y_0)} = 0,$$

or, since $e^{-(x_0+y_0)} \neq 0$,

$$y_0 - x_0 y_0 = 0 \text{ and } x_0 - x_0 y_0 = 0.$$

If $x_0 \neq 0$ this implies $y_0 = 1$ and $x_0 = 1$, but $x_0 = 0$ implies $y_0 = 0$, so we have the critical points $(0,0)$ and $(1,1)$. For the Hesse matrix of g we find

$$\frac{\partial^2 g}{\partial x^2}(x,y) = (-2y + xy)e^{-(x+y)}$$

$$\frac{\partial^2 g}{\partial y^2}(x,y) = (-2x + xy)e^{-(x+y)}$$

$$\frac{\partial^2 g}{\partial x \partial y}(x,y) = \frac{\partial^2 g}{\partial y \partial x}(x,y) = (1 - x - y + xy)e^{-(x+y)}$$

leading to

$$(\text{Hess } g)(0,0) = \begin{pmatrix} 0 & 1 \\ 1 & 0 \end{pmatrix}, \quad (\text{Hess } g)(1,1) = \begin{pmatrix} -\frac{1}{e^2} & 0 \\ 0 & -\frac{1}{e^2} \end{pmatrix}.$$

The eigenvalue of $\begin{pmatrix} 0 & 1 \\ 1 & 0 \end{pmatrix}$ are determined by $\det \begin{pmatrix} -\lambda & 1 \\ 1 & -\lambda \end{pmatrix} = 0$, i.e. $\lambda^2 - 1 = 0$, giving $\lambda_1 = 1$, $\lambda_1 = -1$. Thus $\begin{pmatrix} 0 & 1 \\ 1 & 0 \end{pmatrix}$ is indefinite and g has no isolated local extreme value at $(0,0)$. However, $\begin{pmatrix} -\frac{1}{e^2} & 0 \\ 0 & -\frac{1}{e^2} \end{pmatrix}$ is negative definite implying that g has an isolated local maximum at $(1,1)$ with value $g(1,1) = \frac{1}{e^2}$.

b) In order to determine the critical points we look at

$$0 = \frac{\partial h}{\partial u}(u_0, v_0) = 2u_0 - 4u_0(v_0^2 + u_0^2)$$

and

$$0 = \frac{\partial h}{\partial v}(u_0, v_0) = -2v_0 - 4v_0(v_0^2 + u_0^2),$$

which implies

$$0 = u_0(1 - 2(v_0^2 + u_0^2))$$

and

$$0 = v_0(1 + 2(v_0^2 + u_0^2)),$$

i.e. $(u_1, v_1) = (0,0)$, $(u_2, v_2) = (\frac{1}{2}\sqrt{2}, 0)$, $(u_3, v_3) = (-\frac{1}{2}\sqrt{2}, 0)$.

The Hesse matrix of h at (u, v) is

$$(\text{Hess } h)(u, v) = \begin{pmatrix} -12u^2 - 4v^2 + 2 & -8uv \\ -8uv & -4u^2 - 12v^2 - 2 \end{pmatrix}$$

which gives

$$(\text{Hess } h)(0, 0) = \begin{pmatrix} 2 & 0 \\ 0 & -2 \end{pmatrix}$$

which is an indefinite matrix and therefore we have no isolated local extreme value at $(0, 0)$. However

$$(\text{Hess } h)(\frac{1}{2}\sqrt{2}, 0) = (\text{Hess } h)(-\frac{1}{2}\sqrt{2}, 0) = \begin{pmatrix} -4 & 0 \\ 0 & -4 \end{pmatrix}$$

is negative definite and hence we have local isolated maxima and $(\frac{1}{2}\sqrt{2}, 0)$ and $(-\frac{1}{2}\sqrt{2}, 0)$.

13. First we note that $f(x) \geq 0$ and for the gradient of f we find

$$\text{grad } f(x) = \left(2\sum_{l=1}^{k}(x_1 - \xi_1^l), ..., 2\sum_{l=1}^{k}(x_n - \xi_n^l) \right)$$

which implies for a critical value x^0

$$\sum_{l=1}^{k}(x_j^0 - \xi_j^l) = 0, \ j = 1, ..., n$$

or $kx_j^0 = \sum_{l=1}^{k} \xi_j^l$, i.e. $x_j^0 = \frac{1}{k}\sum_{l=1}^{m} \xi_j^l$, which yields

$$x^0 = \frac{1}{k}\sum_{l=1}^{m} \xi^l.$$

For the mixed second order derivatives of f the following holds

$$\frac{\partial^2 f(x)}{\partial x_j \partial x_i} = \sum_{l=1}^{k} \|x - \xi^l\|^2 = 0, j \neq i,$$

while

$$\frac{\partial^2 f}{\partial x_j^2}(x) = 2k.$$

Hence $(\text{Hess } f)(x) = 2k \, \text{id}_n$ which is positive definite and therefore we have indeed an isolated minimum of f at x^0.

14. For $r = \|x\| = (x_1^2 + \ldots + x_n^2)^{\frac{1}{2}}$ we find if $x \neq 0$ that $\frac{\partial}{\partial x_k} \|x\| = \frac{x_k}{\|x\|} = \frac{x_k}{r}$ implying that

$$\frac{\partial^2}{\partial x_l \partial x_k} \|x\| = \frac{\partial}{\partial x_l} \frac{x_k}{\|x\|} = \frac{\delta_{kl}}{\|x\|} + x_k \frac{\partial}{\partial x_l} \frac{1}{\|x\|}$$
$$= \frac{\delta_{kl}}{\|x\|} - \frac{x_k x_l}{\|x\|^3} = \frac{r^2 \delta_{kl} - x_k x_l}{r^3}.$$

Now we find for $x \neq 0$

$$\frac{\partial f}{\partial x_k} = \frac{\partial}{\partial x_k} g(\|x\|) = g'(\|x\|) \frac{\partial}{\partial x_k} \|x\|$$

and

$$\frac{\partial^2 f}{\partial x_l \partial x_k}(x) = \frac{\partial}{\partial x_l} \left(g'(\|x\|) \frac{\partial}{\partial x_k} \|x\| \right)$$
$$= \left(\frac{\partial}{\partial x_l} g'(\|x\|) \right) \frac{\partial}{\partial x_k} \|x\| + g'(\|x\|) \frac{\partial^2}{\partial x_l \partial x_k} \|x\|$$
$$= g''(\|x\|) \frac{\partial}{\partial x_l} \|x\| \frac{\partial}{\partial x_k} \|x\| + g'(\|x\|) \frac{\partial^2}{\partial x_l \partial x_k} \|x\|$$
$$= g''(r) \frac{x_k x_l}{r^2} + g'(r) \left(\frac{r^2 \delta_{kl} - x_k x_l}{r^3} \right),$$

thus

$$(\text{Hess } f)(x) \left(g''(r) \frac{x_k x_l}{r^2} + g'(r) \left(\frac{r^2 \delta_{kl} - x_k x_l}{r^3} \right) \right)_{k,l=1,\ldots,n}.$$

Since $\Delta_n f = \text{tr}(\text{Hess } f)$ it follows that

$$\Delta_n f(x) = \sum_{k=1}^{n} g''(r) \frac{x_k^2}{r^2} + \sum_{k=1}^{n} \frac{r^2 g'(r) \delta_{kk}}{r^2} - \sum_{k=1}^{n} g'(r) \frac{x_k^2}{r^3}$$
$$= g''(r) + \frac{n g'(r)}{r} - \frac{g'(r)}{r} = g''(r) + \frac{n-1}{r} g'(r).$$

Now we can deduce that in order that $f(x) = g(\|x\|)$ is harmonic in $\mathbb{R}^n \setminus \{0\}$ the function g must satisfy for $r > 0$ the differential equation $g''(r) + \frac{n-1}{r} g'(r) = 0$.

15. Assume that x_1 and x_2, $x_1 \neq x_2$, are two critical points of f. Since by assumption $(\text{Hess } f)(x_1)$ and $(\text{Hess} f)(x_2)$ are positive definite, f has an isolated minimum at x_1 and at x_2. Consider the function h defined on the line segment connecting x_1 and x_2, i.e. on $\{y = \lambda x_1 + (1 - \lambda)x_2 | 0 \leq \lambda \leq 1\}$, by $h(\lambda) = f(\lambda x_1 + (1 - \lambda)x_2)$. This is a C^2-function $h : [0, 1] \to \mathbb{R}$ having a local minimum at $\lambda = 0$ and at $\lambda = 1$. If h is not constant, which is excluded by our assumption, then h must have some

657

local maximum at some $\lambda_0 \in (0,1)$. At λ_0 we must have $h''(\lambda_0) \leq 0$. Hence

$$
\begin{aligned}
0 \geq h''(\lambda_0) &= \frac{d^2}{d\lambda^2} f(\lambda x_1 + (1-\lambda)x_2) \\
&= \frac{d^2}{d\lambda^2} f(x_0 + \lambda_0(x_1 - x_2)) \\
&= \langle (\text{Hess} f)(x_0 + \lambda_0(x_1 - x_2))(x_1 - x_2), (x_1 - x_2) \rangle,
\end{aligned}
$$

but by our assumption we have $\langle \text{Hess} f(y)z, z \rangle > 0$ for all $y \in \mathbb{R}^n$ and all $z \in \mathbb{R}^n$. Thus we arrive at a contradiction and conclude that f can have at most one critical point.

Chapter 10

1. We have to solve the equation $f(x,y) = \frac{x^4}{16} + \frac{y^2}{9} - 25 = 0$ for y, i.e. $\frac{y^2}{9} = 25 - \frac{x^4}{16}$ or $y^2 = 225 - \frac{9x^4}{16}$. Since we are seeking g with non-negative values, we take

$$
g(y) = \sqrt{225 - \frac{9x^4}{16}}
$$

with domain $D(y) = \{x \in \mathbb{R} | |x| \leq 2\sqrt{5}\}$.

2. First we note that

$$
z'(t) = 2x'(t)x(t)y^3(t) + 3x^2(t)y^2(t)y'(t)
$$

as well as

$$
4x^3(t)x'(t) + y'(t) = 1
$$

and

$$
2x'(t)x(t) + 2y'(t)y(t) = 2t.
$$

This yields the system

$$
\begin{pmatrix} 4x^3(t) & 1 \\ x(t) & y(t) \end{pmatrix} \begin{pmatrix} x'(t) \\ y'(t) \end{pmatrix} = \begin{pmatrix} 1 \\ t \end{pmatrix}
$$

with solution

$$
\begin{pmatrix} x'(t) \\ y'(t) \end{pmatrix} = \frac{1}{x(t)(4x^2(t)y(t) - 1)} \begin{pmatrix} y(t) & -1 \\ -x(t) & 4x^3(t) \end{pmatrix} \begin{pmatrix} 1 \\ t \end{pmatrix},
$$

or

$$
x'(t) = \frac{y(t) - t}{x(t)(4x^2(t)y(t) - 1)}, \quad y'(t) = \frac{-x(t) + 4x^3(t)t}{x(t)(4x^2(t)y(t) - 1)}
$$

from which we deduce for $x(t) \neq 0$ and $4x^2(t)y(t) \neq 1$

$$
\begin{aligned}
z'(t) &= \frac{2x(t)y^2(t) - 2x(t)y(t)t - 3x^3(t)y^2(t) + 12x^5(t)y^2(t)t}{x(t)(4x^2(t)y(t) - 1)} \\
&= \frac{y(t)(2y(t) - 2t - 3x^2(t)y(t) + 12x^4(t)y(t)t)}{4x^2(t)y(t) - 1}.
\end{aligned}
$$

3. First we observe that for $y = 0$ the system

$$x^2 + y^2 - 2z^2 = 0 \text{ and } 2x^2 + y^2 + z^2 = 4$$

reduces to $x^2 = 2z^2$ and $2x^2 + z^2 = 4$ which has the particular solution $x = \sqrt{\frac{8}{5}} > 0$ and $z = \sqrt{\frac{4}{5}} > 0$. Next we introduce the mapping $h = (h_1, h_2)$ by

$$h_1(y, x, z) = x^2 + y^2 - 2z^2, \quad h_2(y, x, z) = 2x^2 + y^2 + z^2 - 4,$$

and we find

$$\frac{\partial h}{\partial(x, z)}(y, x, z) = \begin{pmatrix} 2x & -4z \\ 4x & 2z \end{pmatrix}$$

and for $(x, z) = \left(\sqrt{\frac{8}{5}}, \sqrt{\frac{4}{5}} \right)$ we find

$$\det \frac{\partial h}{\partial(x, z)}\left(0, \sqrt{\frac{8}{5}}, \sqrt{\frac{4}{5}} \right) = \det \begin{pmatrix} 2\sqrt{\frac{8}{5}} & -4\sqrt{\frac{4}{5}} \\ 4\sqrt{\frac{8}{5}} & 2\sqrt{\frac{4}{5}} \end{pmatrix} = 4\sqrt{32} \neq 0.$$

Hence $\frac{\partial f}{\partial(x,z)}\left(0, \sqrt{\frac{8}{5}}, \sqrt{\frac{4}{5}} \right)$ is invertible and we may apply the implicit function theorem, Theorem 10.5, which yields the existence of a neighbourhood $U_1(0), 0 \in \mathbb{R}$, and a neighbourhood $U_2\left(\sqrt{\frac{8}{5}}, \sqrt{\frac{4}{5}} \right) \subset \mathbb{R}^2$ as well as a mapping $H : U_1(0) \to U_2\left(\sqrt{\frac{8}{5}}, \sqrt{\frac{4}{5}} \right)$, $H(y) = (f(y), g(y))$ such that $h(y, H(y)) = h(y, f(y), g(y)) = 0$ and $f(0) = \sqrt{\frac{8}{5}}$, $g(0) = \sqrt{\frac{4}{5}}$. Of course we may take $U_1(0)$ to be of the form $(-\epsilon, \epsilon)$. The continuity of f and g allows us to choose $\epsilon > 0$ such that both f and g are positive for $|y| < \epsilon$.

4. For $\frac{\partial f}{\partial u}$ we find

$$\frac{\partial f}{\partial u}(x, u) = \begin{pmatrix} 2u_1 & -1 \\ 1 & -4u_2 \end{pmatrix}$$

and since $\det \frac{\partial f}{\partial u}(x, u) = 1 - 8u_1 u_2$, it follows that for $u_1 u_2 \neq \frac{1}{8}$ the matrix $\frac{\partial f}{\partial u}(x, u)$ is invertible. Thus once more the implicit function theorem gives in a neighbourhood $U_1(x_0) \times U_2(u_0)$ of $(x_0, u_0) \in \mathbb{R}^4$, $u_{01} u_{02} \neq \frac{1}{8}$, the existence of $g : U_1(x_0) \to U_2(u_0)$ such that $(u_1, u_2) = g(x_1, x_2)$, $(x_1, x_2) \in U_1(x_0)$. Of course we can now take as neighbourhoods open balls $B_{\rho_1}(x_0) \subset U_1(x_0)$ and $B_{\rho_2}(u_0) \subset U_2(u_0)$. Further we have

$$\left(\frac{\partial f}{\partial u} \right)^{-1}(x, u) = \frac{1}{1 - 8u_1 u_2} \begin{pmatrix} -4u_2 & 1 \\ -1 & 2u_1 \end{pmatrix}$$

and further

$$\frac{\partial f}{\partial x}(x, u) = \begin{pmatrix} -3 & -1 \\ -1 & 2 \end{pmatrix}$$

which implies

$$(dg) = -\left(\frac{\partial f}{\partial u}\right)^{-1}\left(\frac{\partial f}{\partial x}\right),$$

i.e.

$$
\begin{aligned}
(dg)(x) &= \frac{-1}{1 - 8u_1 u_2}\begin{pmatrix} -4u_2 & 1 \\ -1 & 2u_1 \end{pmatrix}\begin{pmatrix} -3 & -1 \\ -1 & 2 \end{pmatrix} \\
&= \frac{1}{1 - 8u_1 u_2}\begin{pmatrix} 1 - 12u_2 & -4u_2 - 2 \\ 2u_1 - 3 & -4u_1 - 1 \end{pmatrix} \\
&= \frac{1}{1 - 8g_1(x_1, x_2)g_2(x_1, x_2)}\begin{pmatrix} 1 - 12g_2(x_1, x_2) & -4g_2(x_1, x_2) - 2 \\ 2g_1(x_1, x_2) - 3 & -4g_1(x_1, x_2) - 1 \end{pmatrix}.
\end{aligned}
$$

5. The explicit, relatively simple form of f allows us to make explicit calculations. We write

$$f(x_1, x_2, x_3) = \begin{pmatrix} y_1 \\ y_2 \\ y_3 \end{pmatrix} = \begin{pmatrix} \frac{x_1}{1 + x_1 + x_2 + x_3} \\ \frac{x_2}{1 + x_1 + x_2 + x_3} \\ \frac{x_3}{1 + x_2 + x_2 + x_3} \end{pmatrix}$$

to find the system

$$\begin{pmatrix} (1 - y_1) & -y_1 & -y_1 \\ -y_2 & (1 - y_2) & -y_2 \\ -y_3 & -y_3 & (1 - y_3) \end{pmatrix}\begin{pmatrix} x_1 \\ x_2 \\ x_3 \end{pmatrix} = \begin{pmatrix} y_1 \\ y_2 \\ y_3 \end{pmatrix}.$$

It follows that

$$\det\begin{pmatrix} (1 - y_1) & -y_1 & -y_1 \\ -y_2 & (1 - y_2) & -y_2 \\ -y_3 & -y_3 & (1 - y_3) \end{pmatrix} = 1 - y_1 - y_2 - y_3$$

which does not vanish for $1 \neq y_1 + y_2 + y_3$ and we find further

$$
\begin{aligned}
&\begin{pmatrix} (1 - y_1) & -y_1 & -y_1 \\ -y_2 & (1 - y_2) & -y_2 \\ -y_3 & -y_3 & (1 - y_3) \end{pmatrix}^{-1} \\
&= \frac{1}{1 - y_1 - y_2 - y_3}\begin{pmatrix} 1 - y_2 - y_3 & y_1 & y_1 \\ y_2 & 1 - y_1 - y_3 & y_2 \\ y_3 & y_3 & 1 - y_1 - y_2 \end{pmatrix}.
\end{aligned}
$$

This implies

$$
\begin{aligned}
\begin{pmatrix} x_1 \\ x_2 \\ x_3 \end{pmatrix} = f^{-1}(y) &= \frac{1}{1 - y_1 - y_2 - y_3}\begin{pmatrix} 1 - y_2 - y_3 & y_1 & y_1 \\ y_2 & 1 - y_1 - y_3 & y_2 \\ y_3 & y_3 & 1 - y_1 - y_2 \end{pmatrix}\begin{pmatrix} y_1 \\ y_2 \\ y_3 \end{pmatrix} \\
&= \frac{1}{1 - y_1 - y_2 - y_3}\begin{pmatrix} y_1 \\ y_2 \\ y_3 \end{pmatrix}.
\end{aligned}
$$

This implies that f maps $G := \mathbb{R}^3 \backslash \{x \in \mathbb{R}^3 | x_1 + x_2 + x_3 = -1\}$ bijectively onto $f(G) = \mathbb{R}^3 \backslash \{y \in \mathbb{R}^3 | y_1 + y_2 + y_3 = 1\}$ with inverse f^{-1} as above. This allows a direct calculation of $d(f^{-1})$. For this note that for $k \neq l$

$$\frac{\partial}{\partial y_k} \left(\frac{y_l}{1 - y_1 - y_2 - y_3} \right) = \frac{y_l}{(1 - y_1 - y_2 - y_3)^2}$$

whereas for $k \neq l$, $k \neq j$ and $l \neq j$

$$\frac{\partial}{\partial y_k} \left(\frac{y_k}{1 - y_1 - y_2 - y_3} \right) = \frac{1 - y_l - y_j}{1 - y_1 - y_2 - y_3}.$$

Thus we get

$$(df^{-1})(y_1, y_2, y_3) = \frac{1}{(1 - y_1 - y_2 - y_3)^2} \begin{pmatrix} 1 - y_2 - y_3 & y_1 & y_1 \\ y_2 & 1 - y_1 - y_3 & y_2 \\ y_3 & y_3 & 1 - y_1 - y_2 \end{pmatrix}.$$

(This problem is taken from H. Heuser [25].)

6. Let $\psi : J \to G$ be a level line of f, i.e. we have $f(\psi_1(s), \psi_2(s)) = c$ for all $s \in J$, and let $\gamma : I \to G$ be a gradient line. Suppose that $\gamma(t_0) = \psi(s_0)$ for some $t_0 \in I$ and $s_0 \in J$. It follows that

$$0 = \frac{d}{ds} f(\psi(s)) = \langle (\operatorname{grad} f)(\psi(s)), \dot{\psi}(s) \rangle$$

and for the point $\gamma(t_0) = \psi(s_0)$ we find

$$\langle (\operatorname{grad} f)(\gamma(t_0)), \dot{\gamma}(s_0) \rangle = 0.$$

Since $(\operatorname{grad} f)(\gamma(t_0)) = \frac{\dot{\gamma}(t_0)}{\kappa(t_0)}$ we deduce that

$$\langle \dot{\gamma}(t_0), \dot{\psi}(s_0) \rangle = 0,$$

hence the level line and the gradient line are orthogonal at their intersection point.

7. Write $G = (G_1, ..., G_m)$. Since $G : U_1 \times U_2 \to \mathbb{R}^m$ is continuous and $G(0, 0) = 0$ it follows for the components G_j that they are continuous and $G_j(0, 0) = 0$ for $l \leq j \leq m$. The continuity implies that for $\epsilon > 0$ we can find $\delta_j > 0$ such that $|g_j(x, 0)| < \epsilon$ if $\|x\| < \delta_j$. Thus for $\delta = \min_{1 \leq j \leq m} \delta_j$ we have that $\|x\| < \delta$ implies that $\|G(x, 0)\| < \sqrt{n}\epsilon$ which yields $\sup_{\|x\| < \delta} \|G(x, 0)\| \leq \sqrt{n}\epsilon$ implying (10.39) for δ sufficiently small.

Chapter 11

1. Since $x > 0$ and $y > 0$ we can solve the equation $xy = 1$, for example by $y = \frac{1}{x}$, and then our constraint problem becomes the unconstraint problem: find on $(0, \infty)$ the

extreme values of $x \mapsto f(x, \frac{1}{x}) = \frac{x^p}{p} + \frac{1}{qx^q}$, $\frac{1}{p} + \frac{1}{q} = 1$. Differentiating with respect to x yields

$$\frac{d}{dx}\left(\frac{x^p}{p} + \frac{1}{qx^q}\right) = \frac{px^{p-1}}{p} + \frac{1}{q}(-q)\frac{1}{x^{q+1}} = x^{p-1} - x^{-q-1}.$$

The equation $x^{p-1} = x^{-q-1}$ has on $(0, \infty)$ the only solution $x_0 = 1$ which yields $y_0 = \frac{1}{x_0} = 1$. For $x \to \infty$ the term $\frac{x^p}{p} + \frac{1}{qx^q}$ tends to ∞ and for $x \to 0$ this term tends to ∞ too. Since $x_0 = 1$ is the only critical value $f(x, \frac{1}{x})$ must have a minimum at $x_0 = 1$ implying that $f(x, y)$ must have under the condition $xy = 1$ a minimum at $(1, 1)$ which is $f(1, 1) = \frac{1}{p} + \frac{1}{q} = 1$. If $u, v \in \mathbb{R}\backslash\{0\}$ and $x = \frac{|u|}{|uv|^{\frac{1}{p}}}$, $y = \frac{|v|}{|uv|^{\frac{1}{q}}}$ it follows that $x > 0$ and $y > 0$ as well as $xy = 1$. Consequently we find

$$1 = f(1,1) \le f(x, y) = \frac{|u|^p}{p(|uv|^{\frac{1}{p}})^p} + \frac{|v|^q}{q(|uv|^{\frac{1}{q}})^q}$$

or

$$|uv| \le \frac{|u|^p}{p} + \frac{|v|^q}{q}$$

for all $u, v \in \mathbb{R}\backslash\{0\}$ but for $u = 0$ or $v = 0$ the inequality is trivial.

2. a) We may choose $K = \overline{B_1((-2,0))} \cup \overline{B_1((2,0))}$ and $(\xi, \eta) = (0,0) \in \mathbb{R}^2$. The two points where $\text{dist}(K, (\xi, \eta))$ is attained are $(-1, 0)$ and $(1, 0)$, see the following figure

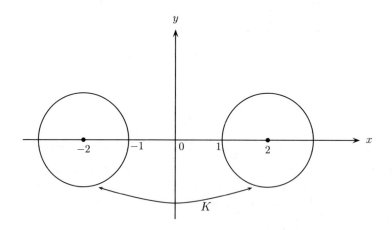

Clearly K is compact since closed and bounded but it is not convex, for example we cannot join $(-2, 0)$ and $(2, 0)$ by a line segment belonging to K.

b) We may take $G = B_1^{\complement}(0)$ and $(\xi, \eta) = (0, 0)$. The set G is unbounded, hence cannot be compact and the points $(-1, 0)$ and $(1, 0)$ we cannot join by a line segment belonging to G, so G is also not convex. Since $\partial G = S^1$ we know that $\|(x,y) - (0,0)\| = 1$ for all $(x,y) \in S^1 = \partial G$, see the figure below:

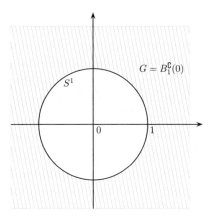

3. First we note that γ is well defined: the cube Q is a compact convex set in $\mathbb{R}^3$ and $G \cap Q = \emptyset$. Therefore, for every $g(x) \in G$ there exists a unique $\gamma(x) \in Q$ such that $\|\gamma(x) - g(x)\| = \operatorname{dist}(Q, g(x))$. In order to find $\operatorname{tr}\gamma \subset Q$ we make use of the symmetry, i.e. we note that $\gamma(x)$ must belong to the x-z-plane. In this plane we have the following graphical representation

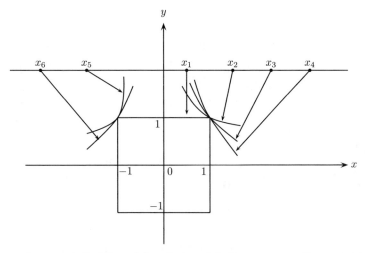

This graph suggest that for $g(x)$, $-1 \leq x \leq 1$, the corresponding point is $\gamma(x) = (x, 0, 1)$, $-1 \leq x \leq 1$, whereas for $x > 1$ all points $g(x) \in G$ are mapped onto $(1, 0, 1)$ and all points $g(x) \in G$, $x < -1$ are mapped onto $(-1, 0, 1)$. If this were true, the

trace of γ would be the set $\{(x,0,1)| -1 \leq x \leq 1\}$. We discuss the situation for $x > 1$, the case $-1 \leq x \leq 1$ is trivial, and the case $x < -1$ goes analogously. Let $x > 1$ and $\tilde{g}(x) \in \tilde{G}$ where $\tilde{G}$ is the projection of G to the x-z-plane. Denote by $r(x) = \|\tilde{g}(x) - (1,1)\|$. The circle with centre $\tilde{g}(x)$ and radius $r(x)$ intersects $\tilde{Q}$, the projection of Q into the x-z-plane, only in the point $(1,1)$. If $0 < \rho < r(x)$, then the circle with radius ρ and centre $\tilde{g}(x)$ does not hit $\tilde{Q}$ nor does any circle with centre $\tilde{g}(x)$ and radius $\rho > r(x) + 2\sqrt{2}$. For $r(x) < \rho < r(x) + 2\sqrt{2}$ the circle with centre $\tilde{g}(x)$ will intersect $\partial\tilde{Q}$ at two points, but the point with smallest distant is unique, so indeed $g(x)$ is mapped for $x > 1$ onto $\gamma(x) = (1,0,1)$.

Note that we have given a solution almost entirely based on elementary geometry, however the uniqueness result we used is based (at least in our Course) on analysis.

4. For $a > 0$, $b > 0$ and $c > 0$ we have to maximise the function $f(a,b,c) = \sqrt{l(l-a)(l-b)(l-c)}$ under the condition that $2l = a + b + c$, or $h(a,b,c) = 2l - a - b - c = 0$. With the Lagrange multiplier λ we look at

$$F(a,b,c,\lambda) = \sqrt{l(l-a)(l-b)(l-c)} + \lambda(2l - a - b - c)$$

and we find

$$\frac{\partial F}{\partial a}(a,b,c,\lambda) = -\frac{l(l-b)(l-c)}{2\sqrt{l(l-a)(l-b)(l-c)}} - \lambda$$

$$\frac{\partial F}{\partial b}(a,b,c,\lambda) = -\frac{l(l-a)(l-c)}{2\sqrt{l(l-a)(l-b)(l-c)}} - \lambda$$

$$\frac{\partial F}{\partial c}(a,b,c,\lambda) = -\frac{l(l-a)(l-b)}{2\sqrt{l(l-a)(l-b)(l-c)}} - \lambda$$

$$\frac{\partial F}{\partial \lambda}(a,b,c,\lambda) = -2l - a - b - c,$$

which yields for a critical point $(a_0, b_0, c_0, \lambda_0)$

$$\lambda_0 = -\frac{l(l-b_0)(l-c_0)}{2\sqrt{l(l-a_0)(l-b_0)(l-c_0)}} = -\frac{l(l-a_0)(l-c_0)}{2\sqrt{l(l-a_0)(l-b_0)(l-c_0)}}$$

$$= -\frac{l(l-a_0)(l-b_0)}{2\sqrt{l(l-a_0)(l-b_0)(l-c_0)}}$$

and

$$2l - a_0 - b_0 - c_0 = 0.$$

Since $0 < a_0, b_0, c_0 < 2l$ we conclude first

$$(*) \quad \begin{cases} (l-b_0)(l-c_0) = (l-a_0)(l-c_0) \\ (l-a_0)(l-c_0) = (l-a_0)(l-b_0) \\ (l-b_0)(l-c_0) = (l-a_0)(l-b_0). \end{cases}$$

First suppose $l = b_0$ implying that either a_0 or c_0 must be equal to l implying that one of the quantity must be 0, and the same type of argument exclude a_0 and c_0

to be equal to l. Now we deduce from $(*)$ that $a_0 = b_0 = c_0$, i.e. ABC must be an equivalent triangle with side length $a = \frac{2l}{3}$. Note that for the result we do not need to calculate the explicit value of λ_0.

5. We have to minimise the function $f(x) = x_1^k + \dots + x_n^k$, $x_j > 0$, $n \in \mathbb{N}$, $k > 1$, under the constraint $g(x) = x_1 + \dots + x_n - 1 = 0$. With $F(x_1, \dots, x_n, \lambda) := f(x) - \lambda g(x) = \sum_{j=1}^n x_j^k + \lambda \left(\sum_{j=1}^n x_j - 1 \right)$ (where we have chosen for convenience $-\lambda$ instead of λ as multiplier) we find for critical points of F

$$\frac{\partial F}{\partial x_j}(x_1, \dots, x_n, \lambda) = kx_j^{k-1} - \lambda = 0, \ 1 \le j \le n,$$

and

$$\frac{\partial F}{\partial \lambda}(x_1, \dots, x_n, \lambda) = x_1 + \dots + x_n - 1 = 0.$$

This gives $x_j = \left(\frac{\lambda}{k} \right)^{\frac{1}{k-1}}$, hence $n \left(\frac{\lambda}{k} \right)^{\frac{1}{k-1}} = 1$ or $\lambda = k \left(\frac{1}{n} \right)^{k-1}$ and therefore $x_j = \frac{1}{n}$, which yields $f(x) = (\frac{1}{n}, \dots, \frac{1}{n}) = n \left(\frac{1}{n} \right)^k = \left(\frac{1}{n} \right)^{k-1}$. On the set $\{ x \in \mathbb{R}^n | x_j > 0 \} \cap \{ x \in \mathbb{R}^n | x_1 + \dots + x_n = 1 \}$ we have only one critical point and for $x = e_j$ we have $f(e_j) = 1 > \left(\frac{1}{n} \right)^{k-1}$, hence at this point we must have a minimum and we deduce for $x_j > 0$ and $x_1 + \dots + x_n = 1$ that

$$\left(\frac{1}{n} \right)^{k-1} \le x_1^k + \dots + x_n^k.$$

For $x_j > 0$ we define $y_j := \frac{x_j}{x_1 + \dots + x_n}$ to find that $y_1 + \dots + y_n = 1$ and therefore with $S = \sum_{j=1}^n x_j$

$$\frac{1}{n^k} \le \frac{y_1^k + \dots + y_n^k}{n} = \frac{\left(\frac{x_1}{S} \right)^k + \dots + \left(\frac{x_n}{S} \right)^k}{n}$$

or

$$\frac{S^k}{n^k} = \frac{\left(\sum_{j=1}^n x_j \right)^k}{n^k} = \left(\frac{1}{n} \sum_{j=1}^n x_j \right)^k \le \frac{1}{n} \sum_{j=1}^n x_j^k.$$

6. It looks as if we now have to employ two Lagrange multipliers, one for the condition $g(x, y, z) = x + y - z = 0$ and one for the condition $h(x, y, z) = x^2 + y^2 + z^2 - 8 = 0$. However, the first condition is solvable, say for $z = x + y$, and we consider the problem: find the extreme values of

$$f_1(x, y) = x + 4y + 2x + 2y = 3x + 6y$$

665

under the condition

$$\tilde{h}_1(x, y) = x^2 + y^2 + (x + y)^2 - 8 = 0,$$

or after dividing by 2

$$h_1(x, y) = x^2 + y^2 + xy - 4 = 0,$$

so we now study

$$F(x, y, \lambda) = 3x + 6y + \lambda(x^2 + y^2 + xy - 4),$$

and obtain the conditions

$$\frac{\partial F}{\partial x}(x, y, \lambda) = 3 - 2\lambda x + \lambda y = 0$$

$$\frac{\partial F}{\partial y}(x, y, \lambda) = 6 - 2\lambda y + \lambda x = 0$$

$$\frac{\partial F}{\partial \lambda}(x, y, \lambda) = x^2 + y^2 + xy - 4 = 0.$$

The first two equations yield $x = 0$ and $y = \frac{3}{-\lambda}$ and the third equation gives $\lambda = \mp\frac{3}{2}$, so the points we have to investigate are $x_1 = 0$, $y_1 = 2$, $z_1 = 2$ and $x_2 = 0$, $y_2 = -2$, $z_2 = -2$ with values

$$f(0, 2, 2) = 8 - 4 = 4$$

and

$$f(0, -2, -2) = -8 + 4 = -4.$$

The points $(0, 2, 2)$ and $(0, -2, -2)$ are inner points of $B_{\sqrt{8}}(0) \cap \{(x, y, z) \in \mathbb{R}^3 | x + y - z = 0\}$ and therefore we have a local maximum at $(0, 2, 2)$ and a local minimum at $(0, -2, -2)$.

7. We now have two constraints and need two Lagrange multipliers λ and μ. Thus we consider

$$F(x_1, ..., x_n, y_1, ..., y_n, \lambda, \mu) = \sum_{j=1}^{n} x_j y_j - \lambda \left(\sum_{j=1}^{n} x_j^2 - 1 \right) - \mu \left(\sum_{j=1}^{n} y_j^2 - 1 \right),$$

where we have chosen again for convenience as Lagrange multipliers $-\lambda$ and $-\mu$ instead of λ and μ. The necessary conditions for an extreme value becomes λ and μ.

$$\frac{\partial F}{\partial x_j}(x_1, ..., x_n, y_1, ..., y_n, \lambda, \mu) = y_j - 2\lambda x_j = 0,$$

$$\frac{\partial F}{\partial y_j}(x_1, ..., x_n, y_1, ..., y_n, \lambda, \mu) = x_j - 2\mu y_j = 0,$$

$$\frac{\partial F}{\partial \lambda}(x_1, ..., x_n, y_1, ..., y_n, \lambda, \mu) = \sum_{j=1}^{n} x_j^2 - 1 = 0,$$

$$\frac{\partial F}{\partial \mu}(x_1, ..., x_n, y_1, ..., y_n, \lambda, \mu) = \sum_{j=1}^{n} y_j^2 - 1 = 0$$

which gives $\lambda = \frac{y_j^0}{2x_j^0} > 0$, $\mu = \frac{x_j^0}{2y_j^0} > 0$, note $x_j^0 > 0$, $y_j^0 > 0$ by assumption. This implies

$$\sum_{j=1}^{n} y_j^{0^2} = 4\lambda^2 \sum_{j=1}^{n} x_j^{0^2} \text{ and } \sum_{j=1}^{n} x_j^{0^2} = 4\mu^2 \sum_{j=1}^{n} y_j^{0^2},$$

or

$$1 = 4\lambda^2 \text{ and } 1 = 4\mu^2,$$

i.e. $\lambda = \mu = \frac{1}{2}$ which now gives $x_j^0 = y_j^0$ and therefore for these values

$$\sum_{j=1}^{k} x_j^0 y_j^0 = \sum_{j=1}^{n} x_j^{0^2} = \sum_{j=1}^{n} y_j^{0^2} = 1.$$

If we allow ourselves for a moment to use the Cauchy-Schwarz inequality we can conclude that this extreme value must be a maximum since

$$\sum_{j=1}^{k} x_j^0 y_j^0 \leq \left(\sum_{j=1}^{n} x_j^{0^2} \right)^{\frac{1}{2}} \left(\sum_{j=1}^{n} y_j^{0^2} \right)^{\frac{1}{2}}.$$

However, we would like to have an argument independent of the Cauchy-Schwarz inequality. If we are successful we may give the following new proof of the Cauchy-Schwarz inequality: for $\xi_j, \eta_j \in \mathbb{R}$, $\xi_j \neq 0$, $\eta_j \neq 0$, $1 \leq j \leq n$, define

$$x_j = \frac{|\xi_j|}{\|\xi\|}, \quad y_j = \frac{|\eta_j|}{\|\eta\|}$$

where $\|\xi\| = \left(\sum_{j=1}^{n} \xi_j^2 \right)^{\frac{1}{2}}$ is the Euclidean norm of $\xi = (\xi_1, ..., \xi_n)$. It follows that $x_j > 0$ and $y_j > 0$ and further $\sum_{j=1}^{n} x_j^2 = 1$ and $\sum_{j=1}^{n} y_j^2 = 1$. Therefore, we obtain (under the assumption we are dealing with a maximum of $\sum_{j=1}^{n} x_j y_j$ under the constraints $\sum_{j=1}^{n} x_j^2 = \sum_{j=1}^{2} y_j^2 = 1$ at $x_j = y_j$)

$$\left| \sum_{j=1}^{n} \frac{\xi_j}{\|\xi\|} \frac{\eta_j}{\|\eta\|} \right| \leq \sum_{j=1}^{n} \frac{\xi_j}{\|\xi\|} \frac{\eta_j}{\|\eta\|} \leq 1$$

or

$$\left| \sum_{j=1}^{n} \xi_j \eta_j \right| \leq \sum_{j=1}^{n} |\xi_j \eta_j| \leq \|\xi\| \|\eta\|$$

667

where the case of ξ_j or η_j being 0 is again trivial. So we want to have a new argument for having a maximum for $\sum_{j=1}^{n} x_j y_j$ at $x_j^0 = y_j^0$, $\sum_{j=1}^{n} x_j^2 = \sum_{j=1}^{n} y_j^2 = 1$. This is however quite easy: for $x_j, y_j > 0$ we always have $x_j y_j \leq \frac{x_j^2 + y_j^2}{2}$ implying

$$\sum_{j=1}^{n} x_j y_j \leq \left(\sum_{j=1}^{n} x_j^2 + \sum_{j=1}^{n} y_j^2 \right) \text{ and for } \sum_{j=1}^{n} x_j^2 = \sum_{j=1}^{n} y_j^2 = 1 \text{ it follows}$$

that $\sum_{j=1}^{n} x_j y_j \leq 1$.

8. Now we want to use the Lagrange multipliers rule, Theorem 11.3, not to find extreme values, but we want to use the existence of extreme values to deduce a characterisation of eigenvalues. In order to apply Theorem 11.3 we consider f as a function on $\mathbb{R}^n \backslash \{0\}$ excluding the critical value $x = 0$. On the compact set S^{n-1} there exists a minimiser ξ^1 of $f(x) = \sum_{k,l=1}^{n} a_{kl} x_k x_l$, i.e. f has a local minimum under the constraint $g(x) = \|x\|^2 - 1 = 0$. By Theorem 11.3 follows the existence $\tilde{\lambda}_1 \in \mathbb{R}$ such that

$$(\operatorname{grad} f)(\xi^1) + \tilde{\lambda}_1 \operatorname{grad} g(\xi^1) = 0.$$

Since $(\operatorname{grad} f)(\xi^1) = 2A\xi^1$ and $\operatorname{grad} g(\xi^1) = 2\xi^1$ we conclude

$$2A\xi^1 + 2\tilde{\lambda}_1 \xi^1 = 0,$$

implying that $\lambda_1 := -\tilde{\lambda}_1$ is an eigenvalue of A. Now we note that $S^{n-1} \cap H(\xi^1)$ is compact as intersection of a compact and a closed set. Therefore there exists a minimiser ξ^2 of $f\big|_{S^{n-1} \cap H(\xi^1)}$ or in other words, f has a local extreme value at ξ^2 under the condition $g(\xi) = \|\xi\|^2 - 1 = 0$ and $h_1(x) := \langle x, \xi^1 \rangle = 0$. Applying Theorem 11.3 yields the existence of $\mu_1, \mu_2 \in \mathbb{R}$ such that

$$(\operatorname{grad} f)(\xi^2) + \mu_1 (\operatorname{grad} g)(\xi^2) + \mu_2 (\operatorname{grad} h_1)(\xi^2) = 0,$$

or

$$2A\xi^2 + 2\mu_1 \xi^2 + \mu_2 \xi^1 = 0,$$

where we used that $\operatorname{grad} h_1(x) = \xi^1$ for all x. Taking in this equation the scalar product with ξ^1, noting that $\langle \xi^2, \xi^1 \rangle = 0$ by the definition of $H(\xi^1)$ and $\langle \xi^1, \xi^1 \rangle = 1$, since $\xi^1 \in S^{n-1}$, we obtain

$$2\langle A\xi^2, \xi^1 \rangle + 2\mu_1 \langle \xi^2, \xi^1 \rangle + \mu_2 \langle \xi^1, \xi^1 \rangle = 0$$

but the symmetry of A implies

$$\langle A\xi^2, \xi^1 \rangle = \langle \xi^2, A\xi^1 \rangle = \lambda_1 \langle \xi^2, \xi^1 \rangle$$

which yields $\mu_2 = 0$ and consequently $A\xi^2 = \lambda_2 \xi^2$ with $\lambda_2 = -\mu_1$.

Obviously we may continue this process by minimising f on

$$\{x \in \mathbb{R}^n \mid \|x\|^2 - 1 = 0, \langle x, \xi^1 \rangle = 0, \langle x, \xi^2 \rangle = 0\},$$

and more generally we may find the jth eigenvalue λ_j of A given the eigenvalues $\lambda_1, \ldots, \lambda_{j-1}$ with corresponding eigenvectors $\xi^1, \ldots, \xi^{j-1} \in S^{n-1}$, by minimising f on the set

$$\{x \in \mathbb{R}^n \mid \|x\|^2 - 1 = 0, \langle x, \xi^1 \rangle = \ldots = \langle x, \xi^{j-1} \rangle = 0\}.$$

Maybe the most important message of this problem is: eigenvalue problems are often related to extremal problems.

9.   a) The equation $\frac{\partial \psi}{\partial c}(x, y, c) = 0$ yields $5(x - c)^4 = 0$, or $x = c$ which gives with $\psi(x, y, c) = 0$ the condition $y = 0$, i.e. the envelope of this family of parabolas of order 5 is the x-axis, see the figure below

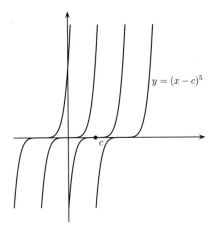

$y = (x - c)^5$

b)   We find $\frac{\partial \varphi}{\partial \alpha}(x, y, \alpha) = x \cos \alpha - y \sin \alpha = 0$ which yields with $x \sin \alpha + y \cos \alpha = 1$

$$\begin{pmatrix} \cos \alpha & -\sin \alpha \\ \sin \alpha & \cos \alpha \end{pmatrix} \begin{pmatrix} x \\ y \end{pmatrix} = \begin{pmatrix} 0 \\ 1 \end{pmatrix}$$

and since $\begin{pmatrix} \cos \alpha & -\sin \alpha \\ \sin \alpha & \cos \alpha \end{pmatrix}^{-1} = \begin{pmatrix} \cos \alpha & \sin \alpha \\ -\sin \alpha & \cos \alpha \end{pmatrix}$ we find $x = \sin \alpha$ and $y = \cos \alpha$. Thus we find $x^2 + y^2 = 1$ and the envelope is the unit circle S^1:

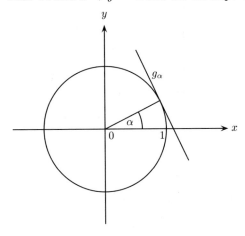

10. With $\varphi(x, y, z, c) = (2+c)x + (3+c)y + (4+c)z - c^2 = 0$ we find

$$\frac{\partial \varphi}{\partial c}(x, y, z, c) = x + y + z - 2c = 0$$

or

$$c = \frac{x + y + z}{2}.$$

This now gives the equation

$$0 = 2x + 3y + 4z + c(x + y + z) - c^2$$

$$= 2x + 3y + 4z + \frac{x + y + z}{2}(x + y + z) - \frac{(x + y + z)^2}{4}$$

$$= 2x + 3y + 4z + \frac{(x + y + z)^2}{y}$$

or

$$(x + y + z)^2 + 4(2x + 3y + 4z) = 0.$$

The corresponding surface is called an elliptic paraboloid.

Chapter 12

1. a) The set A is given in polar coordinates as

$$A = \left\{ (r, \varphi) \in [0, \infty) \times [0, 2\pi) \mid 1 \le r \le 2, 0 \le \varphi \le \frac{\pi}{2} \right\}$$

also see the figure below

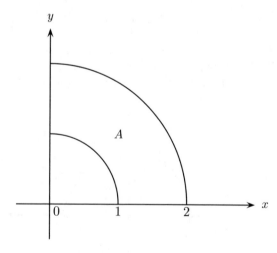

   b) For the set A we find

$$A = \left\{ (r, \varphi, z) \in [0, \infty) \times [0, 2\pi) \times \mathbb{R} \mid 0 \le r \le 1, 0 \le \varphi < 2\pi, 0 \le z \le 1 \right\}$$
$$\cup \left\{ (r, \varphi, z) \in [0, \infty) \times [0, 2\pi) \times \mathbb{R} \mid 0 \le r \le 1, 0 \le \varphi < \pi, 1 \le z \le 2 \right\}.$$

2. The coordinate lines are the curves

$$u \mapsto \gamma_{\varphi_0}(u) = \begin{pmatrix} \cosh u \cos \varphi_0 \\ \sinh u \sin \varphi_0 \end{pmatrix}, u \geq 0, \text{ and } \varphi \mapsto \gamma_{u_0}(\varphi) = \begin{pmatrix} \cosh u_0 \cos \varphi \\ \sinh u_0 \sin \varphi \end{pmatrix},$$

$0 \leq \varphi < 2\pi$, which yields

$$\dot{\gamma}_{\varphi_0}(u) = \begin{pmatrix} \sinh u \cos \varphi_0 \\ \cosh u \sin \varphi_0 \end{pmatrix} \text{ and } \dot{\gamma}_{u_0}(\varphi) = \begin{pmatrix} -\cosh u_0 \sin \varphi \\ \sinh u_0 \cos \varphi \end{pmatrix}$$

and for the intersection point $\gamma_{\varphi_0}(u_0) = \gamma_{u_0}(\varphi_0)$ we find

$$\langle \begin{pmatrix} \sinh u_0 \cos \varphi_0 \\ \cosh u_0 \sin \varphi_0 \end{pmatrix}, \begin{pmatrix} -\cosh u_0 \sin \varphi_0 \\ \sinh u_0 \cos \varphi_0 \end{pmatrix} \rangle$$

$$= -\sinh u_0 \cosh u_0 \cos \varphi_0 \sin \varphi_0 + \sinh u_0 \cosh u_0 \cos \varphi_0 \sin \varphi_0 = 0,$$

i.e. these are orthogonal coordinates. The coordinate lines look as in the following figure, compare with M. R. Spiegel [44].

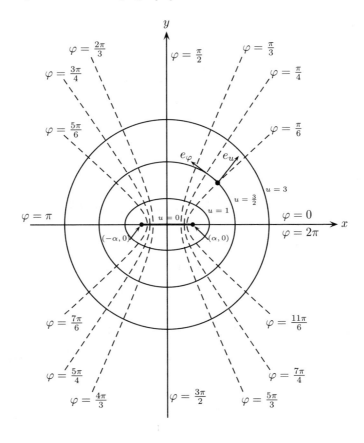

671

3. In the new coordinates $\xi = \frac{1}{2}(x+y)$ and $\eta = \frac{1}{2}(x-y)$ we find

$$\frac{\partial u}{\partial x} = \frac{\partial \xi}{\partial x}\frac{\partial v}{\partial \xi} + \frac{\partial \eta}{\partial x}\frac{\partial v}{\partial \eta} \quad \text{and} \quad \frac{\partial u}{\partial y} = \frac{\partial \xi}{\partial y}\frac{\partial v}{\partial \xi} + \frac{\partial \eta}{\partial y}\frac{\partial v}{\partial \eta}$$

implying that $\frac{\partial u}{\partial x} + \frac{\partial u}{\partial y} = \frac{\partial v}{\partial \xi}$, i.e. in the (ξ, η)-coordinates we now have the simple equation $v_{\xi} = 0$ instead of $u_x + u_y = 0$. This equation means that v is independent of ξ, thus with some function $h : \mathbb{R} \to \mathbb{R}$ we must have $v(\xi, \eta) = h(\eta)$ or $u(x,y) = h\left(\frac{x-y}{2}\right) = \tilde{h}(x-y)$. A straightforward calculation yields

$$u_x(x,y) + u_y(x,y) = \tilde{h}_x(x-y) - \tilde{h}_y(x-y) = 0.$$

Since $u(x,0) = \tilde{h}(x)$ the initial condition $u(x,0) = f(x)$ now implies that $u(x,y) = f(x-y)$ is a solution to the initial value problem. Hence introducing the appropriate coordinates helps to obtain a solution.

4. We know that

$$\frac{\partial g}{\partial r}(r, \varphi) = \cos\varphi\left(\frac{\partial f}{\partial x}\right)(r\cos\varphi, r\sin\varphi) + \sin\varphi\left(\frac{\partial f}{\partial y}\right)(r\cos\varphi, r\sin\varphi)$$

and therefore it follows for the C^2-functions g and f

$$\begin{aligned}
\frac{\partial^2 g}{\partial r \partial \varphi}(r, \varphi) &= \frac{\partial^2 g}{\partial \varphi \partial r}(r, \varphi) \\
&= \frac{\partial}{\partial \varphi}\left(\cos\varphi\left(\frac{\partial f}{\partial x}\right)(r\cos\varphi, r\sin\varphi) + \sin\varphi\left(\frac{\partial f}{\partial y}\right)(r\cos\varphi, r\sin\varphi)\right) \\
&= -\sin\varphi\left(\frac{\partial f}{\partial x}\right)(r\cos\varphi, r\sin\varphi) + \cos\varphi\frac{\partial}{\partial \varphi}\left(\left(\frac{\partial f}{\partial x}\right)(r\cos\varphi, r\sin\varphi)\right) \\
&\quad + \cos\varphi\left(\frac{\partial f}{\partial y}\right)(r\cos\varphi, r\sin\varphi) + \sin\varphi\frac{\partial}{\partial \varphi}\left(\left(\frac{\partial f}{\partial y}\right)(r\cos\varphi, r\sin\varphi)\right)
\end{aligned}$$

$$= -\sin\varphi\left(\frac{\partial f}{\partial x}\right)(r\cos\varphi, r\sin\varphi) + \cos\varphi\left(-r\sin\varphi\left(\frac{\partial^2 f}{\partial x^2}\right)(r\cos\varphi, r\sin\varphi)\right.$$

$$\left. + r\cos\varphi\left(\frac{\partial^2 f}{\partial y\partial x}\right)(r\cos\varphi, r\sin\varphi)\right)$$

$$+ \cos\varphi\left(\frac{\partial f}{\partial y}\right)(r\cos\varphi, r\sin\varphi) + \sin\varphi\left(-r\sin\varphi\left(\frac{\partial^2 f}{\partial x\partial y}\right)(r\cos\varphi, r\sin\varphi)\right.$$

$$\left. + r\cos\varphi\left(\frac{\partial^2 f}{\partial y^2}\right)(r\cos\varphi, r\sin\varphi)\right)$$

$$= -\sin\varphi\left(\frac{\partial f}{\partial x}\right)(r\cos\varphi, r\sin\varphi) + \cos\varphi\left(\frac{\partial f}{\partial y}\right)(r\cos\varphi, r\sin\varphi)$$

$$- r\sin\varphi\cos\varphi\left(\frac{\partial^2 f}{\partial x^2}\right)(r\cos\varphi, r\sin\varphi)$$

$$+ r(\cos^2\varphi - \sin^2\varphi)\left(\frac{\partial^2 f}{\partial x\partial y}\right)(r\cos\varphi, r\sin\varphi)$$

$$+ r\cos\varphi\sin\varphi\left(\frac{\partial^2 f}{\partial y^2}\right)(r\cos\varphi, r\sin\varphi).$$

5. We know that

$$\left(\frac{d}{dr}x\right)(r_0, \vartheta_0, \varphi_0) = (\sin\vartheta_0\cos\varphi_0, \sin\vartheta_0\sin\varphi_0, \cos\vartheta_0)$$

$$\left(\frac{d}{d\vartheta}x\right)(r_0, \vartheta_0, \varphi_0) = (r_0\cos\vartheta_0\cos\varphi_0, r_0\cos\vartheta_0\sin\varphi_0, -r_0\sin\vartheta_0)$$

$$\left(\frac{d}{d\varphi}x\right)(r_0, \vartheta_0, \varphi_0) = (-r_0\sin\vartheta_0\sin\varphi_0, r_0\sin\vartheta_0\cos\varphi_0, 0),$$

which gives

$$\langle\frac{d}{dr}x(r_0, \vartheta_0, \varphi_0), \frac{d}{d\vartheta}x(r_0, \vartheta_0, \varphi_0)\rangle$$

$$= \langle(\sin\vartheta_0\cos\varphi_0, \sin\vartheta_0\sin\varphi_0, \cos\vartheta_0), (r_0\cos\vartheta_0\cos\varphi_0, r_0\cos\vartheta_0\sin\varphi_0, -r_0\sin\vartheta_0)\rangle$$

$$= r_0(\sin\vartheta_0\cos\vartheta_0\cos^2\varphi_0 + \cos\vartheta_0\sin\vartheta_0\sin^2\varphi_0 - \cos\vartheta_0\sin\vartheta_0)$$

$$= r_0(\sin\vartheta_0\cos\vartheta_0(\cos^2\varphi_0 + \sin^2\varphi_0) - \cos\vartheta_0\sin\vartheta_0) = 0,$$

$$\langle\frac{\partial}{\partial\vartheta}x(r_0, \vartheta_0, \varphi_0), \frac{d}{d\varphi}x(r_0, \vartheta_0, \varphi_0)\rangle$$

$$= \langle(r_0\cos\vartheta_0\cos\varphi_0, r_0\cos\vartheta_0\sin\varphi_0, -r\sin\vartheta_0), (-r_0\sin\vartheta_0\sin\varphi_0, r_0\sin\vartheta_0\cos\varphi_0, 0)\rangle$$

$$= r_0^2(-\cos\vartheta_0\sin\vartheta_0\cos\varphi_0\sin\varphi_0 - \cos\vartheta_0\sin\vartheta_0\sin\varphi_0\cos\varphi_0) = 0,$$

and

$$\langle\frac{d}{d\varphi}x(r_0, \vartheta_0, \varphi_0), \frac{d}{dr}x(r_0, \vartheta_0, \varphi_0)\rangle$$

$$= \langle(-r_0\sin\vartheta_0\sin\varphi_0, r\sin\vartheta_0\cos\varphi_0, 0), (\sin\vartheta_0\cos\varphi_0, \sin\vartheta_0\sin\varphi_0, \cos\vartheta_0)\rangle$$

$$= r_0(-\sin^2\vartheta_0\cos\varphi_0\sin\varphi_0 + \sin^2\vartheta_0\cos\varphi_0\sin\varphi_0) = 0,$$

proving the orthogonality of spherical coordinates in $\mathbb{R}^3$.

6. We find easily

$$x_1^2 + x_2^2 + x_3^2 = r^2 \sin^2 \vartheta \cos^2 \varphi + r^2 \sin^2 \vartheta \sin^2 \varphi + r^2 \cos^2 \vartheta$$
$$= r^2 (\sin^2 \vartheta (\cos^2 \varphi + \sin^2 \varphi) + \cos^2 \vartheta) = r^2,$$

so we have

$$r = \sqrt{x_1^2 + x_2^2 + x_3^2}.$$

This leads to

$$(*) \qquad \cos \vartheta = \frac{x_3}{r} = \frac{x_3}{\sqrt{x_1^2 + x_2^2 + x_3^2}}$$

and for $0 \leq \vartheta \leq \pi$ this gives a unique value for ϑ, which we may write with some care as

$$\vartheta = \arccos \frac{x_3}{\sqrt{x_1^2 + x_2^2 + x_3^2}},$$

but in some sense $(*)$ is a more appropriate formula. As in the case of polar coordinates, depending on whether $\cos \varphi$ or $\sin \varphi$ is not zero we find

$$\cot \varphi = \frac{x_2}{x_1} \text{ and } \tan \varphi = \frac{x_1}{x_2},$$

and we need to be a bit careful when applying to these equations the inverse functions arccot and arctan, we need to choose the correct "branch" which we understand better after having studied complex-valued functions of a complex variable in Volume III.

7. The case $n = 3$ is the easiest

$$x_3 = r \sin \vartheta_1 \sin \vartheta_2, x_2 = r \sin \vartheta_1 \cos \vartheta_2, x_1 = r \cos \vartheta_1$$

with $\vartheta_1 \in [0, 2\pi)$, $\vartheta_2 \in [0, \pi]$. Hence by setting $\varphi := \vartheta_1$ and $\vartheta := \vartheta_2$ we obtain

$$x_1 = r \cos \vartheta, x_2 = r \sin \vartheta \cos \varphi, x_3 = \sin \vartheta \sin \varphi$$

which are up to the enumeration of variables the spherical coordinates in $\mathbb{R}^3$. Now for $n = 2$ we find that only $\vartheta_{2-1} = \vartheta_1$ enters into the coordinates and we get

$$x_1 = r \cos \vartheta_1, \quad x_2 = r \sin \vartheta_1$$

and with $\varphi := \vartheta_1$ we obtain polar coordinates.

8. If γ has the components $\gamma_1, \gamma_2, \gamma_3$ we find

$$\gamma(t) = (\gamma_1(t), \gamma_2(t), \gamma_3(t)) = (\sin \vartheta(t) \cos \varphi(t), \sin \vartheta(t) \sin \varphi(t), \cos \vartheta(t)).$$

Thus we know γ with $\text{tr}(\gamma)$ when knowing the two functions $\vartheta : [0, 1] \to [0, \pi]$ and $\varphi : [0, 1] \to [0, 2\pi)$. Using Problem 6 we find further, recall $r = 1$,

$$\cos \vartheta(t) = \frac{\gamma_3(t)}{r} = \gamma_3(t)$$

674

and

$$\cot \varphi(t) = \frac{\gamma_2(t)}{\gamma_1(t)} \quad \text{or} \quad \tan \varphi(t) = \frac{\gamma_1(t)}{\gamma_2(t)}$$

depending on the zeroes of γ_1 and γ_2.

The lesson to learn is that the piece of geometric information $\mathrm{tr}(\gamma) \subset S^2$ may allow us to reduce the "complexity" of the analytic frame: from three functions $\gamma_1, \gamma_2, \gamma_3$ we can pass to two functions ϑ and φ.

9.   a) Using formula (12.20) with $u(x_1, x_2, x_3) = h(\vartheta, \varphi)$ it follows, since $\frac{\partial h}{\partial r}(\vartheta, \varphi) = 0$, for $0 \le \vartheta \le \pi$ and $0 \le \varphi \le 2\pi$ that

$$0 = \Delta_3 u(x_1, x_2, x_3) = \frac{1}{r^2} \frac{\partial^2 h(\vartheta, \varphi)}{\partial \vartheta^2} + \frac{\cos \vartheta}{r^2 \sin \vartheta} \frac{\partial h(\vartheta, \varphi)}{\partial \vartheta} + \frac{1}{r^2 \sin \vartheta} \frac{\partial^2 h(\vartheta, \varphi)}{\partial \varphi^2},$$

and multiplying with r^2 yields

$$\frac{\partial^2 h(\vartheta, \varphi)}{\partial \vartheta^2} + \frac{\cos \vartheta}{\sin \vartheta} \frac{\partial h(\vartheta, \varphi)}{\partial \vartheta} + \frac{1}{\sin \vartheta} \frac{\partial^2 h(\vartheta, \varphi)}{\partial \varphi^2} = 0$$

and for $0 < \vartheta < \pi$

$$\sin \vartheta \frac{\partial^2 h(\vartheta, \varphi)}{\partial \vartheta^2} + \sqrt{(1 - \sin^2 \vartheta)} \frac{\partial h(\vartheta, \varphi)}{\partial \vartheta} + \frac{1}{\sin \vartheta} \frac{\partial^2 h(\vartheta, \varphi)}{\partial \varphi^2} = 0.$$

   b) Using again (12.20) for $u(x_1, x_2, x_3, t) = v(r, t)$ we find

$$
\begin{aligned}
0 &= \frac{\partial^2 u(x_1, x_2, x_3, t)}{\partial t^2} - \Delta_3 u(x_1, x_2, x_3, t) \\
&= \frac{\partial^2 v(r, t)}{\partial t^2} - \frac{\partial^2 v(r, t)}{\partial r^2} - \frac{2}{r} \frac{\partial v(r, t)}{\partial r} = 0.
\end{aligned}
$$

Chapter 13

1. For $0 \le \lambda \le 1$ and $x, y \in \mathbb{R}^n$ we have

$$
\begin{aligned}
g(\lambda x + (1 - \lambda)y) &= f(\|\lambda x + (1 - \lambda)y\|) \\
&\le f(\lambda \|x\| + (1 - \lambda) \|y\|) \\
&\le \lambda f(\|x\|) + (1 - \lambda)f(\|y\|) \\
&= \lambda g(x) + (1 - \lambda)g(y),
\end{aligned}
$$

where we used in the first estimate the monotonicity while in the second the convexity of f.

2.   a) Since $\lambda_j := \frac{\kappa_j}{\kappa} \in [0, 1]$ and $\sum_{j=1}^m \lambda_j = 1$ we deduce from Jensen's inequality

$$f\left(\frac{1}{\kappa} \sum_{j=1}^m \kappa_j x_j\right) = f\left(\sum_{j=1}^m \lambda_j x_j\right)$$

$$\le \sum_{j=1}^m \lambda_j f(x_j) = \frac{1}{\kappa} \sum_{j=1}^m \kappa_j f(x_j).$$

b) We may assume $x_j > 0$ otherwise the result is trivial. Since $\exp : \mathbb{R} \to \mathbb{R}_+$ is convex we find by Jensen's inequality with $y_j := \ln x_j$ that

$$
\begin{aligned}
x_1^{\lambda_1} \cdot \ldots \cdot x_n^{\lambda_n} &= \exp(\ln(x_1^{\lambda_1} \cdot \ldots \cdot x_n^{\lambda_n})) \\
&= \exp(\lambda_1 \ln x_1 + \ldots + \lambda_n \ln x_n) \\
&\leq \lambda_1 \exp(\ln x_1) + \ldots + \lambda_n \exp(\ln x_n) \\
&= \lambda_1 x_1 + \ldots + \lambda_n x_n.
\end{aligned}
$$

The geometric-arithmetic inequality now follows with $\lambda_j = \frac{1}{n}$:

$$
(x_1 \cdot \ldots \cdot x_n)^{\frac{1}{n}} \leq \frac{x_1 + \ldots + x_n}{n}.
$$

3. Since f is convex we find for $y, z \in K_c$ and $0 \leq \lambda \leq 1$ that

$$
f(\lambda y + (1-\lambda)z) \leq \lambda f(y) + (1-\lambda)f(z) < \lambda c + (1-\lambda)c = c
$$

implying the convexity of K_c.

4. We need to prove that $(x_1, y_1), (x_2, y_2) \in \mathbb{R}^{n_1} \times \mathbb{R}^{n_2}$ such that $\left(\|x_1\|^{\alpha} + \|y_1\|^{\beta} \right)^{\frac{1}{2}} < \rho$, $\left(\|x_2\|^{\alpha} + \|y_2\|^{\beta} \right)^{\frac{1}{2}} < \rho$ implies that for $0 \leq \lambda \leq 1$ it follows that

$$
\left(\|\lambda x_1 + (1-\lambda)x_2\|^{\alpha} + \|\lambda y_1 + (1-\lambda)y_2\|^{\beta} \right)^{\frac{1}{2}} < 2^{\frac{1-(\alpha \wedge \beta)}{2}} \rho.
$$

For $0 < \alpha, \beta < 1$ we know that $\|\cdot\|^{\alpha}$ satisfies the triangle inequality, see Example 1.3.D, which implies

$$
\left(\|\lambda x_1 + (1-\lambda)x_2\|^{\alpha} + \|\lambda y_1 + (1-\lambda)y_2\|^{\beta} \right)^{\frac{1}{2}}
$$

$$
\leq \left(\lambda^{\alpha} \|x_1\|^{\alpha} + \lambda^{\beta} \|y_1\|^{\beta} + (1-\lambda)^{\alpha} \|x_2\|^{\alpha} + (1-\lambda)^{\beta} \|y_2\|^{\beta} \right)^{\frac{1}{2}}.
$$

Using that $0 \leq \lambda \leq 1$ we find for $j = 1, 2$ with $\mu_1 = \lambda$, $\mu_2 = (1-\lambda)$,

$$
\left(\mu_j^{\alpha} \|x_j\|^{\alpha} + \mu_j^{\beta} \|y_j\|^{\beta} \right) \leq \mu_j^{\alpha \wedge \beta} \left(\|x_j\|^{\alpha} + \|y_j\|^{\beta} \right)
$$

which yields

$$
\left(\|\lambda x_1 + (1-\lambda)x_2\|^{\alpha} + \|\lambda y_1 + (1-\lambda)y_2\|^{\beta} \right)^{\frac{1}{2}}
$$

$$
\leq \left(\lambda^{\alpha \wedge \beta} + (1-\lambda)^{\alpha \wedge \beta} \right)^{\frac{1}{2}} \rho.
$$

676

Now, for $p = \alpha \wedge \beta < 1$ we have $a^p + b^p \leq 2^{1-p}(a+b)^p$ for $a, b \geq 0$, implying that

$$\lambda^{\alpha \wedge \beta} + (1-\lambda)^{\alpha \wedge \beta} \leq 2^{1-(\alpha \wedge \beta)}(\lambda + (1-\lambda))^{\alpha \wedge \beta}$$

and the result follows. This example along with many other examples related to so called continuous negative definite functions is taken from S. Landwehr's Ph-D thesis, Swansea 2010.

5. By the definition of the infimum, given $x, y \in \mathbb{R}^n$, for $\epsilon > 0$ there exist points $a, b \in K$ such that $\|x - a\| < g(x) + \epsilon$ and $\|y - b\| < g(y) + \epsilon$. For $0 \leq \lambda \leq 1$ let $z = \lambda x + (1-\lambda)y$ and $w := \lambda a + (1-\lambda)b$. By the convexity of K we have $w \in K$ and further

$$\begin{aligned}
\|z - w\| &= \|\lambda(x-a) + (1-\lambda)(y-b)\| \\
&\leq \lambda \|x - a\| + (1-\lambda) \|y - b\| \\
&< \lambda g(x) + \lambda \epsilon + (1-\lambda)g(y) + (1-\lambda)\epsilon \\
&= \lambda g(x) + (1-\lambda)g(y) + \epsilon
\end{aligned}$$

implying that

$$g(z) = g(\lambda x + (1-\lambda)y) \leq \|z - w\| \leq \lambda g(x) + (1-\lambda)g(y),$$

i.e. the convexity of g.

6.   a) Already the example of two disjoint open intervals in $\mathbb{R}$ shows that the union of two convex sets need not be convex.

   b) Recall that $\lambda A = \{y \in \mathbb{R}^n | y = \lambda x, x \in A\}$. Now let $a \in A$ and $b \in B$ as well as $0 \leq \lambda \leq 1$. It follows that $\lambda a + (1-\lambda)b \in \text{conv}(A \cup B)$ since $a, b \in A \cup B$. Hence

$$\bigcup_{\lambda \in [0,1]} (\lambda A + (1-\lambda)B) \subset \text{conv}(A \cup B).$$

We denote in the following by $\overline{xy}$ the line segment connecting x and y. Now let $x, y \in \bigcup_{\lambda \in [0,1]}(\lambda A + (1-\lambda)B)$. It follows that $x \in \overline{a_1 b_1}$ and $y \in \overline{a_2 b_2}$ for some $a_1, a_2 \in A$ and $b_1, b_2 \in B$. Since A and B are convex it follows $\overline{a_1 a_2} \in A$ and $\overline{b_1 b_2} \in B$ which yields

$$\text{conv}(\overline{a_1 a_2} \cup \overline{b_1 b_2}) \subset \bigcup_{\lambda \in [0,1]} (\lambda A + (1-\lambda)B),$$

i.e.

$$\overline{xy} \subset \text{conv}(\overline{a_1 a_2} \cup \overline{b_1 b_2}) \subset \bigcup_{\lambda \in [0,1]} (\lambda A + (1-\lambda)B),$$

which means that $\bigcup_{\lambda \in [0,1]}(\lambda A + (1-\lambda)B)$ is convex and $A \cup B \subset \bigcup_{\lambda \in [0,1]}(\lambda A + (1-\lambda)B)$ from which we now deduce that $\text{conv}(A \cup B) = \bigcup_{\lambda \in [0,1]}(\lambda A + (1-\lambda)B)$.

677

7. a) The mapping $h_\lambda : K \times K \to K$, $(x, y) \mapsto \lambda x + (1 - \lambda)y$, $\lambda \in [0, 1]$, is continuous. Let $x, y \in \bar{K}$ and K convex. We find sequences $(x_\nu)_{\nu \in \mathbb{N}}$ and $(y_\nu)_{\nu \in \mathbb{N}}$, $x_\nu, y_\nu \in K$, such that $x_\nu \to x$ and $y_\nu \to y$. Now we note that $\lambda x_\nu + (1 - \lambda)y_\nu \in K$ and the continuity of h_λ yields that $\lambda x + (1 - \lambda)y \in \bar{K}$, i.e. $\bar{K}$ is convex.

    b) By part a) we know that $\overline{\text{conv}(A)}$ is convex and therefore $\bigcap \{K | A \subset K, K \subset \mathbb{R}^n$ is a closed convex set$\} \subset \overline{\text{conv}(A)}$. On the other hand $\bigcap \{K | A \subset K, K \subset \mathbb{R}^n$ is a closed convex set$\}$ is a closed convex set containing A, so

    $$\text{conv}(A) \subset \bigcap \{K | A \subset K, K \subset \mathbb{R}^n \text{ is a closed convex set}\}$$

    hence

    $$\overline{\text{conv}}(A) \subset \bigcap \{K | A \subset K, K \subset \mathbb{R}^n \text{ is a closed convex set}\}$$

    and the result is proved.

8. As a union of open balls K_r is open. It remains to prove that K_r is convex. For $x \in \mathbb{R}^n$ define as in Problem 5 $g(x) := \inf \{\|x - y\| \, | y \in K\}$. Now given $x \in K_r$ it follows the existence of $a \in K$ such that

    $$g(x) \le \|x - a\| \le r.$$

    Now take $x, y \in K_r$ and with $\lambda \in [0, 1]$ consider $z = \lambda x + (1 - \lambda)y$ to find by the convexity of g that $g(z) \le \lambda g(x) + (1 - \lambda)g(y) \le \lambda r + (1 - \lambda)r = r$, i.e. $\lambda x + (1 - \lambda)y \in K_r$.

9. a) A 0-simplex is a point, a 1-simplex is a line segment, a 2-simplex is a triangle and a 3-simplex is a tertrahedron (in $\mathbb{R}^3$).

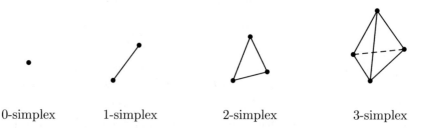

| 0-simplex | 1-simplex | 2-simplex | 3-simplex |

    b) Consider the following figure

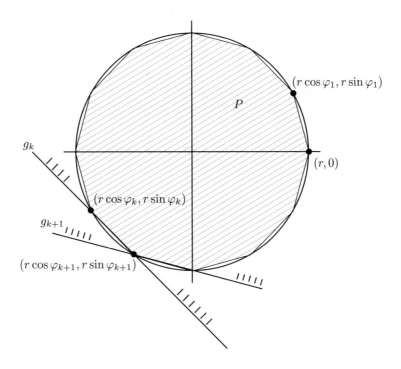

It is obvious that P belongs entirely to one of the half spaces generated by the line g_k and indeed P is the intersection of these half spaces, hence P is convex and since $P \subset \overline{B_r(0)}$ it is compact. We expect the extreme points of P to be the vertices of the polygon. If V is a vertex we claim that $P \backslash \{V\}$ is again convex. Let $x, y \in P \backslash \{V\}$ and let L be the line segment connecting x and y. If x and y are interior points of P then $V \notin L$ and the same holds if x is an interior point and y is a boundary point. For two boundary points x, y the line segment L is either an edge of ∂P, hence $V \notin L$, or $L \backslash \{x, y\} \subset \overset{\circ}{P}$, so again $V \notin L$. Hence $P \backslash \{V\}$ is convex and $V \in \text{ext}(P)$. If on the other hand $W \in \text{ext}(P)$ then W cannot be an interior point since it would be an interior point of the line segment connecting the origin $0 \in \mathbb{R}^2$ with the boundary point of P obtained as intersection of ∂P and the ray starting at $0 \in \mathbb{R}^2$ and passing through W. If $W \in \partial P$ but not a vertex it is an interior point of an edge with end points being vertices. Hence $\text{ext}(P)$ are the vertices of the polygon we started with.

Note: it is important to get used to such a "geometric" type of argument once having more routine the "algebraic" calculations can be done easily.

10. For solutions $x = (x_1, ..., x_n)$ and $y = (y_1, ..., y_n)$ we find with $0 \leq \lambda \leq 1$ that

$$a_{j1}(\lambda x_1 + (1 - \lambda)y_1) + ... + a_{jn}(\lambda x_n + (1 - \lambda)y_n)$$
$$= \lambda(a_{j1}x_1 + ... + a_{jn}x_n) + (1 - \lambda)(a_{j1}y_1 + ... + a_{jn}y_n)$$
$$\leq \lambda b_j + (1 - \lambda)b_j = b_j,$$

thus if x and y are solutions of $a_{j1}x_1 + ... + a_{jn}x_n \leq b_j$, $j \in J$, so is $\lambda x + (1 - \lambda)y$, i.e. the set of all solutions of one inequality is convex. However the set of solutions of the system is obtained as an intersection of the sets of the solutions of each inequality proving the result.

11. (This follows R. Schneider [41]) We will often use the fact that if $c \in \mathbb{R}^n$ belongs to the line segment connecting $a, b \in \mathbb{R}^n$ then $||a - b|| = ||a - c|| + ||c - b||$. Suppose that $p(\xi, K) \neq p(y, K)$. If y belongs to the line segment connecting ξ with $p(\xi, K)$ then we find

$$||\xi - p(y, K)|| \leq ||\xi - y|| + ||y - p(y, K)||$$
$$< ||\xi - y|| + ||y - p(\xi, K)|| = ||\xi - p(\xi, K)||,$$

which is a contradiction. If ξ belongs to the line segment connecting y and $p(\xi, K)$, then take q from the line segment connecting $p(\xi, K)$ and $p(y, K)$ such that the line segment connecting ξ and q is parallel to the line segment connecting y and $p(y, K)$. This implies that

$$\frac{||\xi - q||}{||\xi - p(\xi, K)||} = \frac{||y - p(y, K)||}{||y - p(\xi, K)||} < 1,$$

which is again a contradiction.

12. Let $p = \lambda p_1 + (1 - \lambda)p_2$ with $p_1, p_2 \in K$ and $0 < \lambda < 1$. Since $p \in \text{ext}(K')$ not both p_1, p_2 can belong to K'. Assume $p_1 \notin K'$ and $u(p_1) > \xi$. Since $u(p_2) \geq \xi$ it follows that

$$\xi = u(p) = \lambda u(p_1) + (1 - \lambda)u(p_2) > \xi$$

which is a contradiction.

13. We know from Problem 10 that the set of all solutions is convex, hence by assumption it is a compact set K. We set $K_0 := K$ and consider on K_0 the linear mapping $\varphi_1(x) = \sum_{l=1}^n a_{1l}x_l$. Since K_0 is compact and φ_1 is continuous there exists ξ_1 such that

$$\xi_1 := \min \{\varphi_1(x)|x \in K_0\}$$

and $\xi_1 \geq -b_1$. Now we define

$$K_1 := K_0 \cap \{x \in \mathbb{R}^n|\varphi_1(x) = \xi_1\} \neq \emptyset$$

which is compact and convex, hence we find the existence of ξ_2 such that

$$\xi_2 := \min \{\varphi_2(x)|x \in K_1\} \geq -b_2.$$

We continue this process to arrive eventually at

$$K_m := K_{m-1} \cap \{x \in \mathbb{R}^n | \varphi_m(x) = \xi_m\}$$
$$= \{x \in \mathbb{R}^n | \varphi_1(x) = \xi_1, ..., \varphi_m(x) = \xi_m\} \neq \emptyset.$$

But K_m is the solution space of a system of m linear equations, $\varphi_j(x) = \xi_j$, $j = 1, ..., m$, hence $K_m = p + W$ with some $p \in K_m$ and $W = \{x \in \mathbb{R}^n | \varphi_j(x) = 0, j = 1, ..., m\}$. Since K_m is compact it follows $W = \{0\}$ and $p \in \text{ext}(K_m)$. Now Problem 11 yields $p \in \text{ext}(K)$.

For Problems 10 and 12 we used the presentation in G. Fischer [16].

14. Let $n = 1$ and $f : K \to \mathbb{R}$ be convex. In this case K must be an interval and (13.31) reads as $f(y) - f(x) \geq f'(x)(y - x)$. The tangent to the graph $\Gamma(f)$ of f at x is given by $g(y) = f'(x)(y - x) + f(x)$, thus $f(y) - f(x) \geq f'(x)(y - x)$ or $f(y) \geq f'(x)(y - x) + f(x)$ implies $f(y) \geq g(y)$, i.e. f lies always above its tangent. When $n = 2$ we just have to replace the tangent line by the tangent plane, see Chapter 8.

Note, for $f : K \to \mathbb{R}$, $K \subset \mathbb{R}^n$, we can introduce the notion of a tangent hypersurface at $x \in K$ to $\Gamma(f)$ and (13.31) will have the same interpretation.

15. If f is strictly convex then for all $\lambda \in (0, 1)$ and $x, y \in K$, $x \neq y$ we have (at least for y sufficiently close to x)

$$(*) \qquad f(x + \lambda(y - x)) - f(x) \geq \langle \text{grad } f(x), \lambda(y - x) \rangle.$$

By the strict convexity it follows

$$f(x + \lambda(y - x)) = f(\lambda y + (1 - \lambda)x) < \lambda f(y) + (1 - \lambda)f(x)$$

hence $f(x + \lambda(y - x)) - f(x) < \lambda(f(y) - f(x))$ and with $(*)$ we arrive at

$$\lambda(f(y) - f(x)) > \langle \text{grad } f(x), \lambda(y - x) \rangle$$

or

$$f(y) - f(x) > \langle \text{grad } f(x), y - x \rangle$$

i.e. (13.31) with the strict inequality. As in the proof of Theorem 13.28 we now obtain also the strict inequality in (13.32).

16. First we note that the coercivity of f implies the existence of $R \geq 1$ such that $\|x\| > R$ implies $f(x) \geq 2|f(0)| \|x\|$ implying that $\min_{x \in \mathbb{R}^n} f(x) \geq \min_{x \in \overline{B_R(0)}} f(x)$. Hence a global minimum of f must be a local minimum too and hence if it is attained at x_0 then grad $f(x_0) = 0$. Note that the compactness of $\overline{B_R(0)}$ together with the continuity of f implies the existence of a minimum in $\overline{B_R(0)}$. Now suppose that f has two local minimum at x and y, $x \neq y$. By strict convexity we now find

$$0 = \langle \text{grad } f(x) - \text{grad } f(y), x - y \rangle > 0$$

which is a contradiction.

17. In order to prove the coercivity of g we note first

$$\frac{2}{3}(1+\|x\|^2)^{\frac{3}{4}} \geq \frac{2}{3}(\|x\|^2)^{\frac{3}{4}} = \frac{2}{3}\|x\|^{\frac{3}{2}}$$

implying that

$$\lim_{\|x\|\to\infty} \frac{\frac{2}{3}(1+\|x\|^2)^{\frac{3}{4}} - \frac{2}{3}}{\|x\|} \geq \lim_{\|x\|\to\infty} \frac{2}{3}\|x\|^{\frac{1}{2}} - \frac{2}{3}\lim_{\|x\|\to\infty} \frac{1}{\|x\|} = \infty.$$

Since g is a C^2-function we study the Hessian in order to test for convexity:

$$\frac{\partial}{\partial x_j}g(x) = x_j(1+\|x\|^2)^{-\frac{1}{4}}, j = 1, 2,$$

$$\frac{\partial^2}{\partial x_j^2}g(x) = \frac{2 + 2x_1^2 + 2x_2^2 - x_j^2}{2(1+\|x\|^2)^{\frac{5}{4}}}$$

$$\frac{\partial^2}{\partial x_1\partial x_2}g(x) = \frac{-x_1 x_2}{2(1+\|x\|^2)^{\frac{5}{4}}}$$

which gives

$$(\text{Hess } g)(x) = \frac{1}{2(1+\|x\|^2)^{\frac{5}{4}}} \begin{pmatrix} 2 + x_1^2 + 2x_2^2 & -x_1 x_2 \\ -x_1 x_2 & 2 + 2x_1^2 + x_2^2 \end{pmatrix}.$$

Since $2 + x_1^2 + 2x_2^2 > 0$ for all $x_1, x_2 \in \mathbb{R}$ and

$$\det \begin{pmatrix} 2 + x_1^2 + 2x_2^2 & -x_1 x_2 \\ -x_1 x_2 & 2 + 2x_1^2 + x_2^2 \end{pmatrix} = 4 + 6(x_1^2 + x_2^2) + 2(x_1^4 + x_2^4) + 4x_1^2 x_2^2 > 0,$$

it follows that g is convex.

In order to find the Legendre transform of g, we look at the zeroes of

$$\frac{\partial}{\partial x_j}(x \cdot \xi - g(x)) = \frac{\partial}{\partial x_j}(x \cdot \xi - \frac{2}{3}(1+\|x\|^2)^{\frac{3}{4}} + \frac{2}{3})$$

$$= \xi_j - x_j(1+\|x\|^2)^{-\frac{1}{4}}, j = 1, 2.$$

Thus we find

$$\xi_j = x_j(1+\|x\|^2)^{-\frac{1}{4}}$$

$$\|\xi\|^2 = \|x\|^2(1+\|x\|^2)^{-\frac{1}{2}}$$

$$\|x\|^2 = \frac{\|\xi\|^4 + \sqrt{\|\xi\|^8 + 4\|\xi\|^4}}{2},$$

and therefore, with these values of x_j and $\|x\|$

$$x \cdot \xi - g(x) = \|x\|^2 (1 + \|x\|^2)^{-\frac{1}{4}} - \frac{2}{3}(1 + \|x\|^2)^{\frac{3}{4}} + \frac{2}{3}$$

$$= \|x\|^2 (1 + \|x\|^2)^{-\frac{1}{2}} (1 + \|x\|^2)^{\frac{1}{4}} - \frac{2}{3}(1 + \|x\|^2)^{\frac{3}{4}} + \frac{2}{3}$$

$$= (1 + \|x\|^2)^{\frac{1}{4}} \left(\|x\|^2 (1 + \|x\|^2)^{-\frac{1}{2}} - \frac{2}{3}(1 + \|x\|^2)^{\frac{1}{2}} \right) + \frac{2}{3}$$

$$= \left(\frac{2 + \|\xi\|^4 + \sqrt{\|\xi\|^8 + 4\|\xi\|^4}}{2} \right)^{\frac{1}{4}}$$

$$\times \left(\|\xi\|^2 - \frac{2}{3} \left(\frac{2 + \|\xi\|^4 + \sqrt{\|\xi\|^8 + 4\|\xi\|^4}}{2} \right)^{\frac{1}{2}} \right) + \frac{2}{3}$$

which gives for the Legendre transform of g

$$g^*(\xi) = \left(\frac{2 + \|\xi\|^4 + \sqrt{\|\xi\|^8 + 4\|\xi\|^4}}{2} \right)^{\frac{1}{4}}$$

$$\times \left(\|\xi\|^2 - \frac{2}{3} \left(\frac{2 + \|\xi\|^4 + \sqrt{\|\xi\|^8 + 4\|\xi\|^4}}{2} \right)^{\frac{1}{2}} \right) + \frac{2}{3}.$$

Chapter 14

1. Let $x, y \in X$, $x \neq y$. Since $\mathcal{F}$ is a vector space and separates points we can find $f \in \mathcal{F}$ such that $f(x) \neq f(y)$. Now we define for $\alpha \neq \beta$ the function $g : X \to \mathbb{R}$ by

$$g(z) = \alpha \frac{f(z) - f(y)}{f(x) - f(y)} + \beta \frac{f(z) - f(x)}{f(y) - f(x)}$$

which belongs to $\mathcal{F}$. Furthermore we have $g(x) = \alpha$ and $g(y) = \beta$. Since $e \in \mathcal{F}$ it follows that $e(x)g(y) \neq e(y)g(x)$.

2. We will use Remark 14.4.B. Let $\varphi : [0,1] \to [a,b]$, $\varphi(t) = (b-a)t + a$ and note that $\varphi^{-1} : [a,b] \to [0,1]$, $\varphi^{-1}(s) = \frac{s}{b-a} = \frac{a}{b-a}$. If p is a polynomial then $p \circ \varphi^{-1}$ is a polynomial too. Furthermore we find for $u \in C([a,b])$ that $u \circ \varphi \in C([0,1])$ and by Theorem 14.3 there exists a sequence $(p_k)_{k \in \mathbb{N}}$ of polynomials p_k such that $\lim_{k \to \infty} \left\| p_k \big|_{[0,1]} - u \circ \varphi \right\|_\infty = 0$. However, for $p_k \circ \varphi^{-1}$ we have

$$\left\| u - p_k \circ \varphi^{-1} \right\|_\infty = \sup_{s \in [a,b]} |u(s) - (p_k \circ \varphi^{-1})(s)|$$

$$= \sup_{t \in [0,1]} |(u \circ \varphi)(t) - ((p_k \circ \varphi^{-1}) \circ \varphi)(t)|$$

$$= \left\| u \circ \varphi - p_k \big|_{[0,1]} \right\|_\infty$$

implying on $[a,b]$ the uniform convergence of the sequence of polynomials $p_k \circ \varphi^{-1}$ to u.

683

3. Since $[-1, 1]$ is compact we know by the Weierstrass theorem (in the form of Problem 2) that the restriction of polynomials to $[-1, 1]$ are dense in $C([-1, 1])$. Thus we can find a sequence of polynomials $(q_k)_{k \in \mathbb{N}}$ converging uniformly on $[-1, 1]$ to f. Since uniform convergence implies pointwise convergence it follows that $q_k(0)$ converges to $f(0) = 0$, and further that $(p_k)_{k \in \mathbb{N}}$, $p_k(x) = q_k(x) - q_k(0)$, converges uniformly on $[-1, 1]$ to $f - f(0)$. However $f(0) = 0$ as we find $p_k(0) = q_k(0) - q_k(0) = 0$.

4. First note that if $p(x) = \sum_{k=0}^{m} a_k x^k$ is a polynomial with real coefficients we can find sequences $(a_k^{(\nu)})_{\nu \in \mathbb{N}}$, $k = 0, ..., m$, $a_k^{(\nu)} \in \mathbb{Q}$, such that $\lim_{\nu \to \infty} a_k^{(\nu)} = a_k$. If $I \subset \mathbb{R}$ is a compact interval and $x \in I$ it follows with $p_\nu(x) := \sum_{k=0}^{m} a_k^{(\nu)} x^k$ that

$$|p_\nu(x) - p(x)| = \left| \sum_{k=0}^{m} (a_k^{(\nu)} - a_k) x^k \right| \leq \gamma \sum_{k=0}^{m} |a_k^{(\nu)} - a_k|$$

where $\gamma := \max_{0 \leq k \leq m} \max_{x \in I} |x|^k$. This implies however

$$\|p_\nu - p\|_{\infty, I} = \sup_{x \in I} |p_\nu(x) - p(x)| \leq \gamma \sum_{k=0}^{m} |a_k^{(\nu)} - a_k|.$$

Given $\epsilon > 0$ we can find $N(\epsilon) \in \mathbb{N}$ such that $\nu \geq N(\epsilon)$ implies $\gamma(m+1) \max_{0 \leq k \leq m} |a_k^{(\nu)} - a_k| < \epsilon$ implying the uniform convergence of $(p_\nu)_{\nu \in \mathbb{N}}$ to p. Now, given $f \in C(I)$, by the Weierstrass approximation theorem (in the form of Problem 2) for $\epsilon > 0$ we find a polynomial p such that $\|p_{|I} - f\|_{\infty, I} < \frac{\epsilon}{2}$ and by the previous considerations we can find a polynomial p_{ν_0} with rational coefficients such that $\|p_{|I} - p_{\nu_0|I}\| < \frac{\epsilon}{2}$ implying that $\|p_{\nu_0|I} - f\|_{\infty, I} < \epsilon$.

Note that $\mathbb{Q}$ is countable, hence the set A_l of all polynomials with rational coefficients and degree less or equal to l is countable since we can map A_l bijectively onto $\mathbb{Q}^{l+1}$. The set of all polynomials with rational coefficients is however equal to $\cup_{l \in \mathbb{N}_0} A_l$, hence countable.

5. The linearity of the integral implies that $\int_0^1 f(x) p(x) dx = 0$ for every polynomial p. By the Weierstrass approximation theorem there exists a sequence of polynomials p_k converging on $[0, 1]$ uniformly to f. Hence we find

$$\int_0^1 f^2(x) dx = \lim_{k \to \infty} \int_0^1 f(x) p_k(x) dx = 0.$$

Thus we deduce $\int_0^1 f^2(x) dx = 0$. Suppose $f(x_0) > 0$ for some $x_0 \in (0, 1)$ (the case $f(x_0) < 0$ goes similarly). Then there exists $\eta > 0$ such that $(-\eta + x_0, x_0 + \eta) \subset (0, 1)$ and $f(x) \geq \frac{f(x_0)}{2} > 0$ for all $x \in (-\eta + x_0, x_0 + \eta)$. This implies

$$0 = \int_0^1 f^2(x) dx \geq \int_{x_0 - \eta}^{x_0 + \eta} f^2(x) dx \geq \eta f^2(x_0) > 0$$

which is a contradiction. Thus f must be identically zero. Now, if $\int_0^1 g(x) x^k dx = \int_0^1 h(x) x^k dx$ for all $k \in \mathbb{N}_0$, then $\int_0^1 (g(x) - h(x)) x^k dx = 0$ for all $k \in \mathbb{N}_0$ implying that $g = h$.

684

6. a) The function defined on X by

$$h(x) = \frac{\text{dist}(x, A)}{\text{dist}(x, A) + \text{dist}(x, B)}$$

is well defined and continuous since $x \mapsto \text{dist}(x, A)$ and $x \mapsto \text{dist}(x, B)$ are continuous and $\text{dist}(x, A) + \text{dist}(x, B) > 0$ for all $x \in X$. Moreover $0 \leq h(x) \leq 1$. If $x \in A$ then $\text{dist}(x, A) = 0$, i.e. $h(x) = 0$, and if $x \in B$ then $\text{dist}(x, B) = 0$ and since $\text{dist}(x, A) > 0$ it follows that $h(x) = \frac{\text{dist}(x,A)}{\text{dist}(x,A)} = 1$.

b) It is clear that $C(X) \otimes_a C(Y)$ is a subspace of $C(X \times Y)$. Moreover let $e_X : X \to \mathbb{R}$ and $e_Y : Y \to \mathbb{R}$ be defined as $e_X(x) = 1$ and $e_Y(y) = 1$ for all $x \in X$ and $y \in Y$, respectively. It follows that $e_{X \times Y} := e_X \otimes_a e_Y$ with $(e_X \otimes_a e_Y)(x, y) = e_X(x) e_Y(y) = 1$, belongs to $C(X) \otimes_a C(Y)$. Now it remains to prove that $C(X) \otimes_a C(Y)$ is point separating on $X \times Y$, so let $(x_1, y_1), (x_2, y_2) \in X \times Y$ and assume $x_1 \neq x_2$ (the case $y_1 \neq y_2$ goes analogously). By 6a) we can find a function $g \in C(X)$ such that $g(x_1) \neq g(x_2)$ and therefore $(g \otimes_a e_Y)(x_1, y_1) = g_1(x_1) \neq g_2(x_2) = (g \otimes_a e_Y)(x_2, y_2)$ proving that $C(X) \otimes_a C(Y)$ is point separating.

7. a) By definition

$$\chi_G(x) = \begin{cases} 1, & x \in G \\ 0, & x \notin G \end{cases}$$

so $\chi_G|_G \neq 0$ and $\chi_G|_{G^c} = 0$. The support of χ_G is the closure of the set where G is not zero, hence $\text{supp}\, \chi_G = \bar{G}$.

b) Using Lemma 14.14 we see that the functions defined for $j \in \mathbb{N}$ by

$$u_j(x) = \begin{cases} \exp\left(\frac{-1}{\frac{1}{16} - |x-j|^2}\right), & |x - j| < \frac{1}{4} \\ 0, & |x - j| \geq \frac{1}{4} \end{cases}$$

are C^∞-functions with $\text{supp}\, u_j = \overline{B_{\frac{1}{4}}(j)} = [j - \frac{1}{4}, j + \frac{1}{4}]$. Now we define for $x \in \mathbb{R}$ the function $u(x) := \sum_{j=1}^\infty u_j(x)$. Since $\text{supp}\, u_j \cap \text{supp}\, u_l = \emptyset$ for $j \neq l$, for every $x \in \mathbb{R}$ the sum is well defined and has at most one non-zero term. Indeed $u|_{\overline{B_{\frac{1}{4}}(j)}} = u_j$ and $u|_{\cup_{j=1}^\infty (B_{\frac{1}{4}}(j))^c} = 0$, implying that $\text{supp}\, u = \cup_{j=1}^\infty \overline{B_{\frac{1}{4}}(j)}$ which is a non-compact but closed set.

8. We want to apply the Arzela-Ascoli theorem, Theorem 14.25, to the set K. First note that

$$K(x) := \{u(x) | u \in K\} \subset [-1, 1]$$

which yields that $K(x)$ is relative compact, note that $u \in K$ implies $|u(x)| \leq \|u\|_\infty \leq 1$. Next we prove that K is equi-continuous in $C([a, b])$. Since $u \in K$ is in $C^1([a, b])$ we know by the mean value theorem that for $x, y \in [a, b]$ we have $|u(x) - u(y)| = |u'(\vartheta)||x - y|$ for some ϑ between x and y, but for $u \in K$ and $z \in [a, b]$

685

we know that $|u'(z)| \leq \|u'\|_\infty \leq 1$, so we have for all $x, y \in [a, b]$ and all $u \in K$ that $|u(x) - u(y)| \leq |x - y|$ which implies the equi-continuity of K: for $\epsilon > 0$ we choose $\delta = \epsilon$ to find for all $y \in U_\delta(x) = (x - \delta, x + \delta)$ that $|u(x) - u(y)| \leq |x - y| < \delta = \epsilon$.

9. We want to apply the Arzela-Ascoli theorem, Theorem 14.25, to the set $H := \{F_n | n \in \mathbb{N}\} \subset C([a, b])$. Since

$$|F_n(x)| \leq \int_a^x |f_n(t)| dt \leq \int_a^b |f_n(t)| dt \leq M(b - a)$$

we now find that $\{F_n(x) | n \in \mathbb{N}\} \subset [-M(b-a), M(b-a)]$ for every $x \in [a, b]$ which implies that $\{F_n(x) | n \in \mathbb{N}\}$ is relative compact in $\mathbb{R}$. Next we observe for $x, y \in [a, b]$ that

$$|F_n(x) - F_n(y)| = \left| \int_x^y f_n(t) dt \right| \leq \int_x^y |f_n(t)| dt \leq M|x - y|$$

and for $x \in [a, b]$, $\epsilon > 0$ given it follows with $\delta = \frac{\epsilon}{M}$ for all $y \in U_\delta(x) = (x - \delta, x + \delta)$ that

$$|F_n(x) - F_n(y)| \leq M|x - y| < M\delta = \epsilon,$$

i.e. $\{F_n | n \in \mathbb{N}\}$ is equi-continuous and thus relative compact.

10. Since $\operatorname{supp} v = K_v$ is compact we find that

$$(u * v)(x) = \int_{K_v} u(x - y) v(y) dy$$

and by Theorem 14.32 we get the continuity of $u * v$ when having in mind that for $x \in \mathbb{R}$ fixed $u(x - y)$ has a compact support too. Indeed, the function $(x, y) \mapsto u(x - y)v(y)$ vanishes for $y \notin K_v$ and $x - y \notin K_u$ or $x \notin y + K_u \subset K_v + K_u$. This in turn already implies that

$$\operatorname{supp}(u * v) \subset K_u + K_v,$$

recall $K_u + K_v = \{z \in \mathbb{R} | z = x + y, x \in K_u \text{ and } y \in K_v\}$.

11. First we note that

$$(*) \qquad |\hat{u}(\xi)| = \left| \int_{\mathbb{R}} \cos(x\xi) u(x) dx \right| \leq \int_{\mathbb{R}} |u(x)| dx.$$

Next we observe that

$$|\xi| |\hat{u}(\xi)| = |\xi \hat{u}(\xi)| = \left| \int_{\mathbb{R}} \xi \cos(x\xi) u(x) dx \right| = \left| \int_{\mathbb{R}} \left(\frac{d}{dx} \sin(x\xi) \right) u(x) dx \right|.$$

We want to justify

$$\left| \int_{\mathbb{R}} \left(\frac{d}{dx} \sin(x\xi) \right) u(x) dx \right| = \left| \int_{\mathbb{R}} \sin(x\xi) \frac{du}{dx}(x) dx \right|$$

which will imply

$$|\xi||\hat{u}(\xi)| \le \int_{\mathbb{R}} |u'(x)| dx$$

and with $(*)$ we will obtain

$$(1 + |\xi|)|\hat{u}(\xi)| \le \|u\|_{L^1} + \|u'\|_{L^1} = M_u$$

or $|\hat{u}(\xi)| \le \frac{M_u}{1+|\xi|}$. Hence for $|\xi| \to \infty$ it follows that $\hat{u}(\xi) \to 0$, whereas Theorem 14.32 yields the continuity of $\hat{u}$, thus we will arrive at $u \in C_\infty(\mathbb{R})$.

Now we turn to

$$\int_{\mathbb{R}} \frac{d}{dx}(\sin(x\xi))u(x)dx,$$

and consider for $R \in \mathbb{R}$ first

$$\int_{-R}^{R} \left(\frac{d}{dx}\sin(x\xi)\right)u(x)dx = \sin(x\xi)u(x)\Big|_{-R}^{R} - \int_{-R}^{R} \sin(x\xi)\frac{du(x)}{dx}dx.$$

If we can prove that

$$(**) \qquad \lim_{R \to \infty} (\sin(R\xi)u(R) - \sin(-R\xi)u(-R)) = 0$$

we are done. Since $|\sin(R\xi)| \le 1$ it is sufficient to show that $\lim_{R \to -\infty} u(R) = \lim_{R \to \infty} u(R) = 0$. The assumption we can use is that $|u|$ and $|u'|$ are integrable over $\mathbb{R}$. For $R \in \mathbb{R}$ we have for $\rho > R$

$$u(R) = u(\rho) + \int_{\rho}^{R} u'(r)dr$$

and since $\int_{\mathbb{R}} |u'(r)| dr$ is finite we obtain as $\rho \to -\infty$ that

$$u(R) = C + \int_{-\infty}^{R} u'(r)dr.$$

This implies that $\lim_{R \to \infty} u(R)$ and $\lim_{R \to -\infty} u(R)$ exist. Moreover, for $R \to -\infty$ we find $\lim_{R \to -\infty} u(R) = C$. Suppose that $C \ne 0$. Then there exists $N \in \mathbb{N}$ such that $|u(x)| \ge \frac{1}{2}|C|$ for all $x < -N$ which contradicts the absolute integrability of u:

$$\infty > \int_{\mathbb{R}} |u(x)| dx \ge \int_{-\infty}^{-N} |u(x)| dx \ge \frac{1}{2}|C| \int_{-\infty}^{-N} 1 dx.$$

Thus $C = 0$ implying $\lim_{R \to -\infty} u(R) = 0$. By considering

$$u(R) = C - \int_{R}^{\infty} u'(r)dr$$

we find analogously that $\lim_{R \to \infty} u(R) = 0$ and $(**)$ is proved.

687

Chapter 15

1. For $0 \leq j \leq N-1$ we find

$$
\begin{aligned}
\|\gamma(t_{j+1}) - \gamma(t_j)\|^2 &= r^2(\cos t_{j+1} - \cos t_j)^2 + r^2(\sin t_{j+1} - \sin t_j)^2 \\
&\quad + h^2(t_{j+1} - t_j)^2 \\
&= r^2(\cos^2 t_{j+1} + \sin^2 t_{j+1}) + r^2(\cos^2 t_j + \sin^2 t_j) \\
&\quad - 2r^2(\cos t_{j+1} \cos t_j + \sin t_{j+1} \sin t_j) + h^2(t_{j+1} - t_j)^2 \\
&= 2r^2 - 2r^2 \cos(t_{j+1} - t_j) + h^2(t_{j+1} - t_j)^2
\end{aligned}
$$

which yields

$$
\begin{aligned}
V_{Z_N}(\gamma) &= \sum_{j=0}^{N-1} (2r^2 - 2r^2 \cos \frac{4\pi}{N} + h^2 \frac{4\pi^2}{N^2})^{\frac{1}{2}} \\
&= N(2r^2 \left(1 - \cos \frac{4\pi}{N}\right) + h^2 \frac{4\pi^2}{N^2})^{\frac{1}{2}} \\
&= (2r^2 16\pi^2 \left(\frac{1 - \cos \frac{4\pi}{N}}{\frac{16\pi^2}{N^2}}\right) + 4\pi^2 h^2)^{\frac{1}{2}} \\
&= 4\pi(2r^2 \left(\frac{1 - \cos \frac{4\pi}{N}}{\frac{16\pi^2}{N^2}}\right) + h^2)^{\frac{1}{2}}.
\end{aligned}
$$

Since $\lim_{y \to 0} \frac{1 - \cos y}{y^2} = \frac{1}{2}$ we obtain

$$
\begin{aligned}
\lim_{N \to \infty} V_{Z_N}(\gamma) &= \lim_{N \to \infty} 4\pi \left(2r^2 \left(\frac{1 - \cos \frac{4\pi}{N}}{\frac{16\pi^2}{N^2}}\right) + h^2\right)^{\frac{1}{2}} \\
&= 4\pi(r^2 + h^2)^{\frac{1}{2}}.
\end{aligned}
$$

Note that we expect this to be equal to $\int_0^{4\pi} \|\dot{\gamma}(t)\| \, dt$, and indeed $\dot{\gamma}(t) = (-r \sin t, r \cos t, h)$ implying

$$
\|\dot{\gamma}(t)\| = (r^2 \sin^2 t + r^2 \cos^2 t + h^2)^{\frac{1}{2}} = (r^2 + h^2)^{\frac{1}{2}}
$$

and therefore

$$
\int_0^{4\pi} \|\dot{\gamma}(t)\| \, dt = \int_0^{4\pi} (r^2 + h^2)^{\frac{1}{2}} \, dt = 4\pi(r^2 + h^2)^{\frac{1}{2}}.
$$

2. If γ is rectifiable and $Z(t_0, ..., t_M)$ is a partition of $[\alpha, \beta]$ we find

$$
\sum_{j=0}^{M-1} |f(t_{j+1}) - f(t_j)| \leq \sum_{j=0}^{M-1} ((t_{j+1} - t_j)^2 + (f(t_{j+1}) - f(t_j))^2)^{\frac{1}{2}}
$$

688

and it follows that f is of bounded variation. Conversely, since

$$\sum_{j=0}^{M-1} (|t_{j+1} - t_j|^2 + |f(t_{j+1}) - f(t_j)|^2)^{\frac{1}{2}} \le \sum_{j=0}^{M-1} |t_{j+1} - t_j| + |f(t_{j+1}) - f(t_j)|$$

it follows that if f is of bounded variation then γ is rectifiable.

3. Consider

$$\sum_{j=0}^{M-1} \|\gamma(t_{j+1}) - \gamma(t_j)\| = \sum_{j=0}^{M-1} \left(\sum_{k=1}^{n} |\gamma_k(t_{j+1}) - \gamma_k(t_j)|^2 \right)^{\frac{1}{2}}$$

$$\le \sum_{j=0}^{M} \sum_{k=1}^{n} |\gamma_k(t_{j+1}) - \gamma_k(t_j)|.$$

If all functions γ_k are of bounded variation, then

$$\sup_Z \left(\sum_{j=0}^{M-1} \|\gamma(t_{j+1}) - \gamma(t_j)\| \right)$$

will be finite, hence γ rectifiable. Since monotone functions are of bounded variation, under our assumption γ is rectifiable.

Now suppose that all γ_k are piecewise monotone, i.e. for $k = 1, ..., n$ there exists a partition $S_k(s_0^{(k)}, ..., s_{m_k}^{(k)})$ of $[\alpha, \beta]$ such that $\gamma_k|_{[s_j^{(k)}, s_{j+1}^{(k)}]}$ is monotone. Let $S(s_0, ..., s_N)$ be the joint partition, i.e. $S = \cup_{k=1}^{n} S_k$. Then $\gamma|_{[s_j, s_{j+1}]}$, $j = 0, ..., N$, satisfies the condition of the previous part, hence is rectifiable, but $\gamma = \gamma|_{[s_0, s_1]} \oplus \, ... \oplus \gamma_{[s_{N-1}, s_N]}$. Thus the result holds also if all γ_k are only piecewise monotone.

4. For the following also consider the figure below.

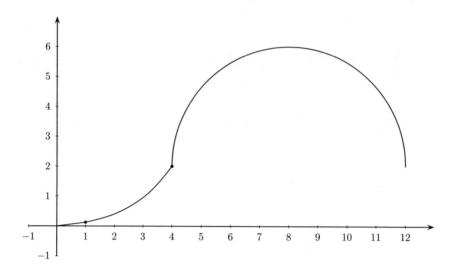

Clearly, g is continuous and $g|_{[0,1]}$, $g|_{[1,4]}$, $g|_{[4,12]}$ are continuously differentiable. Thus we have by $t \mapsto (t, g(t))$, $t \in [0, 12]$ a rectifiable curve with length

$$l_\gamma = \int_0^1 \|\dot{\gamma}(t)\| \, dt + \int_1^4 \|\dot{\gamma}(t)\| \, dt + \int_4^{12} \|\dot{\gamma}(t)\| \, dt.$$

The third integral we know: since $g|_{[4,12]}$ is a circle with radius 4, this integral is equal to 8π. Further $\dot{\gamma}|_{[0,1]} = (1, \frac{1}{8})$ and therefore $\left\|\dot{\gamma}|_{[0,1]}(t)\right\| = \sqrt{1 + \frac{1}{64}} = \frac{1}{8}\sqrt{65}$ which yields for the first integral the value $\frac{1}{8}\sqrt{65}$ too. Now, $\dot{\gamma}|_{[1,4]}(t) = (1, \frac{t}{4})$, i.e. $\left\|\dot{\gamma}|_{[1,4]}(t)\right\| = \sqrt{1 + \frac{t^2}{16}} = \frac{1}{4}\sqrt{16 + t^2}$. Therefore we find for the second integral

$$\int_1^4 \|\dot{\gamma}(t)\| \, dt = \frac{1}{4} \int_1^4 \sqrt{16 + t^2} \, dt$$

$$= \frac{1}{8}(t\sqrt{16 + t^2} + 16\ln(t + \sqrt{16 + t^2}) - \ln 4)\Big|_1^4$$

$$= \frac{1}{8}\left(4\sqrt{32} - \sqrt{17} + 16\ln\frac{4 + \sqrt{32}}{4 + \sqrt{17}}\right).$$

Eventually we find

$$l_\gamma = \frac{1}{8}\sqrt{65} + \frac{1}{8}\left(4\sqrt{32} - \sqrt{17} + 16\ln\frac{4 + \sqrt{32}}{4 + \sqrt{17}}\right) + 8\pi.$$

5. We first draft the curve γ_1:

690

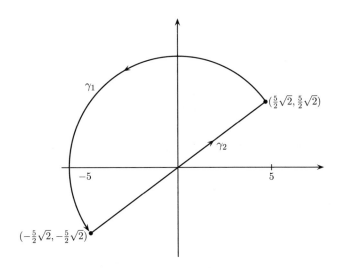

Clearly, γ_2 shall be the line segment connecting (in this order) $(-\frac{5}{2}\sqrt{2}, -\frac{5}{2}\sqrt{2})$ with $(\frac{5}{2}\sqrt{2}, \frac{5}{2}\sqrt{2})$. The equation of this line segment is $\gamma_2(t) = (t-\frac{5}{2}\sqrt{2}-\frac{5\pi}{4}, t-\frac{5}{2}\sqrt{2}-\frac{5\pi}{4})$, $t \in [\frac{5\pi}{4}, \frac{5\pi}{4} + \frac{10}{2}\sqrt{2}]$. Indeed, for $t = \frac{5\pi}{4}$ we have $\gamma_2\left(\frac{5\pi}{4}\right) = (-\frac{5}{2}\sqrt{2}, -\frac{5}{2}\sqrt{2})$ and for $t = \frac{5\pi}{4}+\frac{10}{2}\sqrt{2}$ we have $\gamma_2(\frac{5\pi}{4}+\frac{10}{2}\sqrt{2}) = (\frac{5}{2}\sqrt{2}, \frac{5}{2}\sqrt{2})$. Thus $\gamma : [\frac{5\pi}{4}, \frac{5\pi}{4}+\frac{10}{2}\sqrt{2}] \to \mathbb{R}^2$

$$\gamma(t) = \begin{cases} (5\cos t, 5\sin t), & t \in [\frac{\pi}{4}, \frac{5\pi}{4}) \\ (t - \frac{5}{2}\sqrt{2} - \frac{5\pi}{4}, t - \frac{5}{2}\sqrt{2} - \frac{5\pi}{4}), & t \in [\frac{5\pi}{4}, \frac{5\pi}{4} + \frac{10}{2}\sqrt{2}) \end{cases}$$

is a piecewise continuously differentiable curve with $\gamma = \gamma_1 \oplus \gamma_2$.

6. a) We find for $t > 0$ that $\dot{\gamma}(t) = \begin{pmatrix} 1 \\ \frac{1}{2\sqrt{t}} \end{pmatrix}$ and $\chi_{\gamma(t)} = \begin{pmatrix} e^{\gamma_1(t)} \\ e^{\gamma_2(t)} \end{pmatrix} = \begin{pmatrix} e^t \\ e^{\sqrt{t}} \end{pmatrix}$ which

yields $\langle \chi_{\gamma(t)}, \dot{\gamma}(t) \rangle = e^t + \frac{e^{\sqrt{t}}}{2\sqrt{t}}$ and therefore

$$\int_0^1 \langle \chi_{\gamma(t)}, \dot{\gamma}(t) \rangle dt = \int_0^1 \left(e^t + \frac{e^{\sqrt{t}}}{2\sqrt{t}} \right) dt = \int_0^1 e^t dt + \int_0^1 \frac{e^{\sqrt{t}}}{2\sqrt{t}} dt$$

$$= \int_0^1 e^t dt + \int_0^1 e^s ds = 2e - 2,$$

where we used the fact that $\int_0^1 \frac{e^{\sqrt{t}}}{2\sqrt{t}} dt$ is well defined as an improper integral.

b) Here is a draft of the polygon:

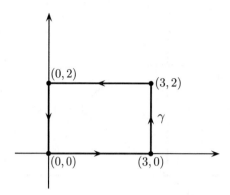

A parametrization of γ is given by $\gamma : [0, 10] \to \mathbb{R}^2$

$$\gamma(t) = \begin{cases} (t, 0), & t \in [0, 3] \\ (3, t - 3), & t \in [3, 5] \\ (-t + 8, 2), & t \in [5, 8] \\ (0, -t + 10), & t \in [8, 10] \end{cases}$$

which makes γ a piecewise continuously differentiable curve with $\dot{\gamma}\big|_{[0,3]} = \begin{pmatrix} 1 \\ 0 \end{pmatrix}$, $\dot{\gamma}\big|_{[3,5]} = \begin{pmatrix} 0 \\ 1 \end{pmatrix}$, $\dot{\gamma}\big|_{[5,8]} = \begin{pmatrix} -1 \\ 0 \end{pmatrix}$ and $\dot{\gamma}\big|_{[8,10]} = \begin{pmatrix} 0 \\ -1 \end{pmatrix}$. Further we have

$$X(\gamma(t)) = \begin{cases} \begin{pmatrix} \gamma_1(t)\gamma_2(t) \\ \gamma_2(t)e^{\gamma_1(t)} \end{pmatrix} = \begin{pmatrix} 0 \\ 0 \end{pmatrix}, & t \in [0, 3] \\[2mm] \begin{pmatrix} \gamma_1(t)\gamma_2(t) \\ \gamma_2(t)e^{\gamma_1(t)} \end{pmatrix} = \begin{pmatrix} 3(t - 3) \\ (t - 3)e^3 \end{pmatrix}, & t \in [3, 5] \\[2mm] \begin{pmatrix} \gamma_1(t)\gamma_2(t) \\ \gamma_2(t)e^{\gamma_1(t)} \end{pmatrix} = \begin{pmatrix} 2(8 - t) \\ 2e^{8-t} \end{pmatrix}, & t \in [5, 8] \\[2mm] \begin{pmatrix} \gamma_1(t)\gamma_2(t) \\ \gamma_2(t)e^{\gamma_1(t)} \end{pmatrix} = \begin{pmatrix} 0 \\ 10 - t \end{pmatrix}, & t \in [8, 10] \end{cases}$$

and therefore

$$\int_\gamma X(p)dp = \int_0^{10} \langle X(\gamma(t)), \dot{\gamma}(t) \rangle dt$$

$$= \left(\int_0^3 + \int_3^5 + \int_5^8 + \int_8^{10} \right) \langle X(\gamma(t)), \dot{\gamma}(t) \rangle dt.$$

The single integrals are:

$$\int_0^3 \langle \begin{pmatrix} 0 \\ 0 \end{pmatrix}, \begin{pmatrix} 1 \\ 0 \end{pmatrix} \rangle dt = 0,$$

$$\int_3^5 \langle \begin{pmatrix} 3(t-3) \\ e^3(t-3) \end{pmatrix}, \begin{pmatrix} 0 \\ 1 \end{pmatrix} \rangle dt = e^3 \int_3^5 (t-3) dt = 2e^3,$$

$$\int_5^8 \langle \begin{pmatrix} 2(8-t) \\ 2e^{8-t} \end{pmatrix}, \begin{pmatrix} -1 \\ 0 \end{pmatrix} \rangle dt = 2 \int_5^8 (t-8) dt = -9,$$

$$\int_8^{10} \langle \begin{pmatrix} 0 \\ 10-t \end{pmatrix}, \begin{pmatrix} 0 \\ -1 \end{pmatrix} \rangle dt = \int_8^{10} (t-10) dt = -2$$

which yields

$$\int_0^{10} \langle X(\gamma(t)), \dot\gamma(t) \rangle dt = 2e^3 - 11.$$

7. Here is first a draft of the two curves:

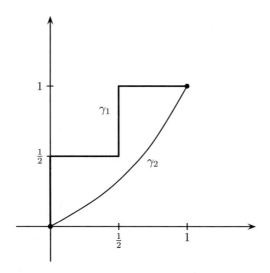

For γ_2 we already have a parametrization with $\dot\gamma_2 = \begin{pmatrix} 1 \\ 2t \end{pmatrix}$ which yields with

$$V(\gamma_1(t), \gamma_2(t)) = \begin{pmatrix} \gamma_2(t) \\ \gamma_2(t) - \gamma_1(t) \end{pmatrix} = \begin{pmatrix} t^2 \\ t^2 - t \end{pmatrix} \text{ that}$$

$$\int_{\gamma_2} V(x) dx = \int_0^1 \langle \begin{pmatrix} t^2 \\ t^2 - t \end{pmatrix}, \begin{pmatrix} 1 \\ 2t \end{pmatrix} \rangle dt$$

$$= \int_0^1 (t^2 + 2t^3 - 2t^2) dt = \frac{1}{6}.$$

Noting that $l_{\gamma_1} = 2$ we parametrize γ_1 with respect to the arc length:

$$\gamma_1(t) = \begin{cases} \begin{pmatrix} 0 \\ t \end{pmatrix}, & t \in [0, \tfrac{1}{2}] \\[2mm] \begin{pmatrix} t - \tfrac{1}{2} \\ \tfrac{1}{2} \end{pmatrix}, & t \in [\tfrac{1}{2}, 1] \\[2mm] \begin{pmatrix} \tfrac{1}{2} \\ t - \tfrac{1}{2} \end{pmatrix}, & t \in [1, \tfrac{3}{2}] \\[2mm] \begin{pmatrix} t - 1 \\ 1 \end{pmatrix}, & t \in [\tfrac{3}{2}, 2] \end{cases}$$

which gives

$$\int_{\gamma_1} V(x)dx = \int_0^{\frac{1}{2}} \langle \begin{pmatrix} t \\ t \end{pmatrix} \begin{pmatrix} 0 \\ 1 \end{pmatrix} \rangle dt + \int_{\frac{1}{2}}^1 \langle \begin{pmatrix} \tfrac{1}{2} \\ 1 - t \end{pmatrix}, \begin{pmatrix} 1 \\ 0 \end{pmatrix} \rangle dt$$

$$+ \int_1^{\frac{3}{2}} \langle \begin{pmatrix} t - \tfrac{1}{2} \\ t - 1 \end{pmatrix}, \begin{pmatrix} 0 \\ 1 \end{pmatrix} \rangle dt + \int_{\frac{3}{2}}^2 \langle \begin{pmatrix} 1 \\ 2 - t \end{pmatrix}, \begin{pmatrix} 1 \\ 0 \end{pmatrix} \rangle dt$$

$$= \int_0^{\frac{1}{2}} t\, dt + \int_{\frac{1}{2}}^1 \frac{1}{2} dt + \int_1^{\frac{3}{2}} (t - 1)dt + \int_{\frac{3}{2}}^2 1 dt = 1.$$

We deduce that $\int_{\gamma_1} V(x)dx \neq \int_{\gamma_2} V(x)dx$ and hence V cannot be a gradient field.

8. a) We connect the point $(0,0)$ to (x,y) by the line segment $\gamma(t) = \begin{pmatrix} xt \\ yt \end{pmatrix}$, $t \in$ $[0,1]$, which gives $\dot\gamma(t) = \begin{pmatrix} x \\ y \end{pmatrix}$. Now we find that

$$\int_\gamma X(p)dp = \int_0^1 \langle \begin{pmatrix} yt \\ (x - y)t \end{pmatrix}, \begin{pmatrix} x \\ y \end{pmatrix} \rangle dt$$

$$= \int_0^1 (2xyt - y^2 t)dt = xy - \frac{y^2}{2}$$

and try to see whether $\varphi(x,y) := xy - \frac{y^2}{2}$ is a potential function for X :
$(\mathrm{grad}\,\varphi)(x,y) = \begin{pmatrix} y \\ x - y \end{pmatrix} = X(x,y)$. Thus X is indeed a gradient field.

b) We again connect the point $0 \in \mathbb{R}^n$ to $x \in \mathbb{R}^n$ by the line segment $\gamma(t) = xt$, $t \in [0,1]$, and now we find

$$\int_\gamma Z(p)dp = \int_0^1 \langle f(rt)xt, x \rangle dt$$

$$= \int_0^1 f(rt)r^2 t\, dt = \int_0^r f(s)s\, ds$$

694

as a candidate for a potential function. With $\varphi(x) := \psi(r) := \int_0^r f(s)s\,ds$ we find

$$\operatorname{grad}\varphi(x) = \left(\frac{\partial}{\partial x_1} \int_0^r f(s)s\,ds, ..., \frac{\partial}{\partial x_n} \int_0^r f(s)s\,ds \right)$$

$$= \left(\frac{\partial r}{\partial x_1} f(r)r, ..., \frac{\partial r}{\partial x_n} f(r)r \right)$$

$$= \left(x_1 f(r), ..., x_n f(r) \right) = f(r)x,$$

where we used that $\frac{\partial r}{\partial x_j} = \frac{x_j}{r}$, and it follows that Z is a gradient field.

9.　a)　With $U_1(x,y) = \frac{y}{x^2+y^2}$ and $U_2(x,y) = \frac{-x}{x^2+y^2}$ we have to check whether $\frac{\partial U_1}{\partial y} = \frac{\partial U_2}{\partial x}$ holds. We find

$$\frac{\partial U_1}{\partial y} = \frac{\partial}{\partial y}\left(\frac{y}{x^2+y^2} \right) = \frac{x^2-y^2}{(x^2+y^2)^2}$$

and

$$\frac{\partial U_2}{\partial x} = \frac{\partial}{\partial x}\left(\frac{-x}{x^2+y^2} \right) = \frac{x^2-y^2}{(x^2+y^2)^2},$$

i.e. the integrability conditions are satisfied.

b)　We try to verify

$$\frac{\partial W_1}{\partial y} = \frac{\partial W_2}{\partial x}, \frac{\partial W_2}{\partial z} = \frac{\partial W_3}{\partial y}, \frac{\partial W_3}{\partial x} = \frac{\partial W_1}{\partial z}.$$

The following holds

$$\frac{\partial W_1}{\partial y} = x, \frac{\partial W_1}{\partial z} = 0, \frac{\partial W_2}{\partial x} = x, \frac{\partial W_2}{\partial z} = \frac{1}{2z}, \frac{\partial W_3}{\partial x} = 0, \frac{\partial W_3}{\partial y} = \frac{1}{2z},$$

implying

$$\frac{\partial W_1}{\partial y} = x = \frac{\partial W_2}{\partial x}, \frac{\partial W_2}{\partial z} = \frac{1}{2z} = \frac{\partial W_3}{\partial y}, \frac{\partial W_3}{\partial x} = 0 = \frac{\partial W_1}{\partial z},$$

and again the integrability conditions are satisfied.

c)　We note that F is of the type as discussed in Problem 8b) hence F is for all α a gradient field and therefore the integrability conditions are satisfied.

10.　a)　We note that $\dot{\gamma}(t) = (-5\sin t, 5\cos t, 10)$, and therefore $\|\dot{\gamma}(t)\| = (25\sin^2 t +$

$25\cos^2 t + 100)^{\frac{1}{2}} = 5\sqrt{5}$. Using formula (15.43) we find

$$\int_\gamma f(x)ds(x) = \int_0^{6\pi} (5\cos t)(5\sin t)(10t)\,\|\dot\gamma(t)\|\,dt$$

$$= 1250\sqrt{5}\int_0^{6\pi} t\cos t\sin t\,dt$$

$$= 625\sqrt{5}\int_0^{6\pi} t\sin 2t\,dt$$

$$= 625\sqrt{5}\left(\frac{1}{4}\sin 2t - \frac{t}{2}\cos 2t\right)\Big|_0^{6\pi}$$

$$= -1875\sqrt{5}\pi.$$

b) With $\|\dot\gamma(t)\|^2 = \left(\frac{2t}{1+t^2}\right)^2 + \left(\frac{2}{1+t^2} - 1\right)^2 = \frac{1+2t^2+t^4}{(1+t^2)^2} = 1$ we get

$$\int_\gamma h(x)ds(x) = \int_0^1 (2\arctan t - t + 3)e^{-\ln(1+t^2)}\,dt$$

$$= \int_0^1 \frac{2\arctan t - t + 3}{1+t^2}\,dt$$

$$= \frac{\pi^2}{16} - \frac{1}{2}\ln 2 + \frac{3\pi}{4}.$$

While the second and third integral is obvious, for the first integral the substitution $s = \arctan t$ is used.

Solutions to Problems of Part 4

Chapter 16

1.   a) The following figure shows $P(a, b, c)$

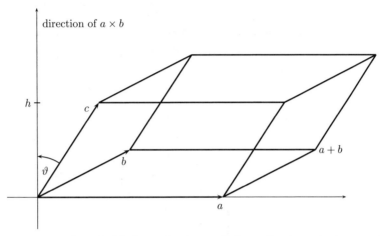

Since $\|a \times b\| = \|a\| \|b\| \sin \measuredangle(a, b)$ and $\|b\| \sin \measuredangle(a, b)$ is the length of the height of the triangle with vertices $0, a$ and b, it follows that the area A of the parallelogram with vertices $0, a, a + b, b$ is $\|a \times b\|$. If ϑ is the angle between $a \times b$ and c the length of the projection of c onto the line generated by $a \times b$ is $h = \|c\| |\cos \vartheta|$. Now, the volume of $P(a, b, c)$ is equal to the volume of any hyper-rectangle with ground surface of area A and the remaining side orthogonal to the ground surface of length h, i.e.

$$\mathrm{vol}_3(\mathrm{P(a, b, c)}) = \|\mathrm{a} \times \mathrm{b}\| \, \|\mathrm{c}\| \, |\cos \vartheta| = |\langle \mathrm{a} \times \mathrm{b}, \mathrm{c}\rangle|.$$

Now we evaluate $|\langle a \times b, c\rangle|$ for $a = (a_1, a_2, a_3), b = (b_1, b_2, b_3)$ and $c = (c_1, c_2, c_3)$:

$$a \times b = (a_2 b_3 - a_3 b_2, a_3 b_1 - a_1 b_3, a_1 b_2 - a_2 b_1)$$

$$= \left(\begin{vmatrix} a_2 & a_3 \\ b_2 & b_3 \end{vmatrix}, \begin{vmatrix} a_3 & a_1 \\ b_3 & b_1 \end{vmatrix}, \begin{vmatrix} a_1 & a_2 \\ b_1 & b_2 \end{vmatrix} \right)$$

and

$$\langle a \times b, c\rangle = c_1 \begin{vmatrix} a_2 & a_3 \\ b_2 & b_3 \end{vmatrix} - c_2 \begin{vmatrix} a_1 & a_3 \\ b_1 & b_3 \end{vmatrix} + c_3 \begin{vmatrix} a_1 & a_2 \\ b_1 & b_2 \end{vmatrix}$$

$$= \begin{vmatrix} a_1 & a_2 & a_3 \\ b_1 & b_2 & b_3 \\ c_1 & c_2 & c_3 \end{vmatrix},$$

697

so we find

$$\text{vol}_3(\text{P(a, b, c)}) = \left| \det \begin{pmatrix} a_1 \, a_2 \, a_3 \\ b_1 \, b_2 \, b_3 \\ c_1 \, c_2 \, c_3 \end{pmatrix} \right|.$$

b) In the case where $a = (a_1, a_2)$ and $b = (b_1, b_2)$ span a parallelogram P in the plane $\mathbb{R}^2$ we just consider these vectors embedded into $\mathbb{R}^3$, i.e. we consider $\tilde{a} = (a_1, a_2, 0)$ and $\tilde{b} = (b_1, b_2, 0)$. From part a) we know that the area of the parallelogram spanned by $\tilde{a}$ and $\tilde{b}$ is $\|\tilde{a} \times \tilde{b}\|$ which implies

$$\text{vol}_2(\text{P}) = \|\tilde{a} \times \tilde{b}\| = \|(0, 0, a_1 b_2 - a_2 b_1)\|$$

$$= |a_1 b_2 - a_2 b_1| = \left| \begin{pmatrix} a_1 \, a_2 \\ b_1 \, b_2 \end{pmatrix} \right|.$$

2. The parallelotop $P(a, b, c)$ is defined by

$$P(a, b, c) = \{x \in \mathbb{R}^3 \mid x = \lambda a + \mu b + \nu c, \lambda, \mu, \nu \in [0, 1]\}$$

$$= \left\{ \begin{pmatrix} \lambda a_1 + \mu b_1 + \nu c_1 \\ \lambda a_2 + \mu b_2 + \nu c_2 \\ \lambda a_3 + \mu b_3 + \nu c_3 \end{pmatrix} \bigg| \lambda, \mu, \nu \in [0, 1] \right\}$$

$$= \left\{ \begin{pmatrix} \lambda + 3\nu \\ 3\mu + \nu \\ 2\lambda + 2\mu \end{pmatrix} \bigg| \lambda, \mu, \nu \in [0, 1] \right\},$$

and therefore we find

$$T(P(a, b, c)) = \left\{ \begin{pmatrix} 1 \, 0 \, 2 \\ 0 \, 4 \, 1 \\ -3 \, 5 \, 0 \end{pmatrix} \begin{pmatrix} \lambda + 3\nu \\ 3\mu + \nu \\ 2\lambda + 2\mu \end{pmatrix} \bigg| \lambda, \mu, \nu \in [0, 1] \right\}$$

$$= \left\{ \begin{pmatrix} 5\lambda + 4\mu + 3\nu \\ 2\lambda + 14\mu + 4\nu \\ -3\lambda + 15\mu - 4\nu \end{pmatrix} \bigg| \lambda, \mu, \nu \in [0, 1] \right\}.$$

The volume of $T(P(a, b, c))$ is given by

$$\text{vol}_3(\text{T(P(a, b, c))}) = \left| \det T \right| \left| \det \begin{pmatrix} a_1 \, a_2 \, a_3 \\ b_1 \, b_2 \, b_3 \\ c_1 \, c_2 \, c_3 \end{pmatrix} \right|$$

$$= \left| \det \begin{pmatrix} 1 \, 0 \, 2 \\ 0 \, 4 \, 1 \\ -3 \, 5 \, 0 \end{pmatrix} \right| \left| \det \begin{pmatrix} 1 \, 0 \, 2 \\ 0 \, 3 \, 2 \\ 3 \, 1 \, 0 \end{pmatrix} \right|$$

$$= 19 \cdot 20 = 380$$

3. First we have to check (16.4)-(16.7) for $|\det|$.
   (16.4): This follows from Remark A.I.12.B when observing

$$|\det(a_1, \ldots, a_k + \alpha\, a_l, \ldots, a_n)| = |\det(a_1, \ldots, a_n) + \alpha \det(a_1, \ldots, a_l, \ldots, a_n)|$$

$$= |\det(a_1, \ldots, a_n)|, \qquad \uparrow k^{\text{th}} position$$

since in $\det(a_1, \ldots, a_l, \ldots, a_n)$ the vector a_l enters in the kth and l^{th} position.
(16.5): This follows immediately from Remark A.I.12.B.
(16.6): Note that $c + P(a_1, \ldots, a_n) = P(a_1 + c, \ldots, a_n + c)$ and

$$\det(a_1 + c, \ldots, a_n + c) = \det(a_1, \ldots, a_n) + \det(c, \ldots, c) = \det(a_1, \ldots, a_n).$$

(16.7): This follows again immediately from Remark A.I.12.B.
It remains to prove the uniqueness of $|\det|$ as a volume form. For this let vol_n be a function on $\mathbb{R}^n \times \cdots \times \mathbb{R}^n$ satisfying (16.4)-(16.7) and consider the function $W : \mathbb{R}^n \times \cdots \times \mathbb{R}^n \to \mathbb{R}$ defined by

$$(*) \qquad W(a_1, \ldots, a_n) = \begin{cases} \frac{\text{vol}_n(a_1, \ldots, a_n)}{|\det(a_1, \ldots, a_n)|} \det(a_1, \ldots, a_n), & \det(a_1, \ldots, a_n) \neq 0 \\ 0, & \det(a_1, \ldots a_n) = 0. \end{cases}$$

We claim that W is a determinant form in the sense of Definition A.I.11 (or Remark A.I.12) which will imply by the uniqueness of the determinant form that $\frac{\text{vol}_n(a_1, \ldots, a_n)}{|\det(a_1, \ldots, a_n)|} = 1$.

Using the characterisation of a determinant form as given in Remark A.I.40.B we note that the the right hand side in $(*)$ remains unchanged when replacing $a_k + a_l, k \neq l$, and for $\lambda \in \mathbb{R}, \lambda \neq 0$, we find

$$\begin{aligned} W(a_1, \ldots, \lambda a_k, \ldots, a_n) &= \frac{\text{vol}_n(a_1, \ldots, \lambda a_k, \ldots, a_n)}{|\det(a_1, \ldots, \lambda a_k, \ldots, a_n)|} \det(a_1, \ldots, \lambda a_k, \ldots, a_n) \\ &= \frac{|\lambda| \text{vol}_n(a_1, \ldots, \lambda a_k, \ldots, a_n)}{|\lambda| |\det(a_1, \ldots, \lambda a_k, \ldots, a_n)|} \lambda \det(a_1, \ldots, \lambda a_k, \ldots, a_n) \\ &= \lambda W(a_1, \ldots, a_k, a_n), \end{aligned}$$

the case $\lambda = 0$ is trivial. Finally (16.7) implies

$$W(e_1, \ldots, e_n) = \frac{\text{vol}_n(e_1, \ldots, e_n)}{|\det(e_1, \ldots, e_n)|} \det(e_1, \ldots, e_n) = 1.$$

Note that the extra consideration needed for the case $\det(a_1, \ldots, a_n) = 0$ is trivial.

4. Of course we cannot sketch the set A since we cannot sketch the function D. However the following figure might be helpful

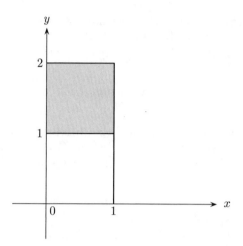

All points in the square $[0,1] \times [0,1]$ belong to A, where as only points $(x,y) \in [0,1] \times [1,2]$ with rational x belong to A. We claim

$$\partial A = [0,1] \times \{0\} \cup \{0\} \times [0,1] \cup \{1\} \times [0,1] \cup [0,1] \times [1.2].$$

Recall that $p \in \partial A$ if every open neighbourhood of p contains points from A and $A^{\complement}$. This excludes points from $(0,1) \times (0,1)$ to belong to ∂A, where points in $[0,1] \times \{0\} \cup \{0\} \times [0,1] \cup \{1\} \times [0,1]$ clearly belong to ∂A, just consider open balls with centres belong to one of these three line segments: they clearly must contain points from A and $A^{\complement}$. For points in $[0,1] \times [1,2]$ the situation is the following: every open ball with centre in $[0,1] \times [1,2]$ and positive radius must contain points with rational and irrational x-component, hence it will contain points of A and $A^{\complement}$. Hence $[0,1] \times [1,2] \subset \partial A$.

5. Consider the following figure

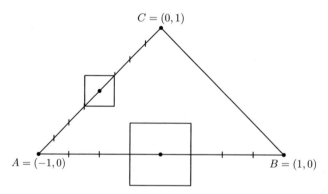

On AB we choose an equidistance partition into $2N$ intervals of length $\frac{1}{N}$. Now we make each point of the partition (including the end points) the centre of a closed square with sides parallel to the coordinate axes and side length $\frac{2}{N}$. We have $2N+1$ of such squares $Q_j^{(1)}, j = 1, \ldots, 2N + 1$, and they cover AB. The total volume of these squares is given by

$$\sum_{j=1}^{2N+1} \mathrm{vol}_2(Q_j^{(1)}) = (2N + 1)\frac{4}{N^2} = \frac{8N + 4}{N^2}.$$

Thus, given $\varepsilon > 0$, if we take $N \geq \frac{36}{\varepsilon}$ we obtain

$$\sum_{j=1}^{2N+1} \mathrm{vol}_2(Q_j^{(1)}) = \frac{8N + 4}{N^2} \leq \frac{12}{N} \leq \frac{\varepsilon}{3}.$$

Next we choose an equidistance partition of CA into M intervals of equal length. This leads to $M + 1$ points on CA each having a distance to the next point of $\frac{\sqrt{2}}{M}$. We take each of these points (including the two end points) as the centre of closed squares $Q_l^{(2)}, l = 1, \ldots, M + 1$, with sides parallel to the coordinate axes and diagonal length $\frac{2\sqrt{2}}{M}$. These squares cover CA and have side length $\frac{2}{M}$, hence area $\frac{4}{M^2}$. We have $M + 1$ of such squares and their total area adds up according to

$$\sum_{l=1}^{M+1} \mathrm{vol}_2(Q_l^{(2)}) = (M + 1)\frac{4}{M^2} = \frac{4M + 4}{M^2}.$$

For $\varepsilon > 0$ given we choose $M \geq \frac{24}{\varepsilon}$ to find

$$\sum_{l=1}^{M+1} \mathrm{vol}_2(Q_l^{(2)}) = \frac{4M + 4}{M^2} \leq \frac{8}{M} \leq \frac{\varepsilon}{3}.$$

The side BC can be treated in the same way as CA and obtain a covering with $M+1$ squares $Q_k^{(3)}, k = 1, \ldots, M+1$, with total area less than $\frac{\varepsilon}{3}$ provided $M \geq \frac{24}{\varepsilon}$. Hence we have a covering of ∂ABC by the squares $(Q_j^{(1)})_{j=1,\ldots,2N+1}, (Q_l^{(2)})_{l=1,\ldots,M+1}$, $(Q_k^{(3)})_{k=1,\ldots,M+1}$ with total area

$$\mathrm{vol}_2\left(\left(\bigcup_{j=1}^{N+1} Q_j^{(1)}\right) \cup \left(\bigcup_{l=1}^{M+1} Q_l^{(2)}\right) \cup \left(\bigcup_{k=1}^{M+1} Q_k^{(3)}\right)\right) \leq \varepsilon$$

provided that $N \geq \frac{36}{\varepsilon}$ and $M \geq \frac{24}{\varepsilon}$.

701

Chapter 17

1. The projections pr_j and $\tilde{\mathrm{pr}}_j$ are continuous and therefore they map compact sets onto compact sets. We can consider $f : K \times X \to \mathbb{R}$ as a continuous mapping $g : \mathrm{pr}_j(K) \times (\tilde{\mathrm{pr}}_j(K) \times X) \to \mathbb{R}$ and for $[a, b] \subset \mathrm{pr}_j(K)$ by Theorem 14.32 it follows that

$$G(z) = \int_a^b g(u, z) du, u \in [a, b], z \in \tilde{\mathrm{pr}}_j(K) \times X$$

is continuous since for $(y_1, \ldots, y_{j-1}, \ldots, y_{j+1}, \ldots, y_n) \in \tilde{\mathrm{pr}}_j(K), z \in X$, and

$$G(z) = F(y_1, \ldots, y_{j-1}, y_{j+1}, \ldots, y_n, z).$$

2. a) First we note that Lemma 14.33 and its proof remain unchanged if $f : I \times J \to \mathbb{R}$ is replaced by $f : I_1, \times \cdots \times I_n \times J_1 \times \cdots \times J_m \to \mathbb{R}$ and the partial derivative $\frac{\partial f}{\partial y_k}$ is investigated. However with this modification of Lemma 14.33 the proof of Theorem 14.34 works in the new situation without any changes.

    b) With f and φ as in part a) a generalisation of Corollary 14.35 would be

    $$\partial_y^\alpha \varphi(x_1, \ldots, x_{k-1}, x_{k+1}, \ldots, x_n, y_1, \ldots, y_m)$$
    $$= \int_{I_k} \partial_y^\alpha f(x_1, \ldots, x_n, y_1, \ldots, y_m) dx_k,$$

    where $\alpha \in \mathbb{N}_0^m$ and $\partial_y^\alpha = \frac{\partial^{|\alpha|}}{\partial y_1^{\alpha_1} \ldots \partial y_m^{\alpha_m}}$. The condition on f is that all partial derivations $\partial_y^\beta f, \beta \le \alpha$, exist and are continuous. Here $\beta \in \mathbb{N}_0^m$ and $\beta \le \alpha$ if and only if $\beta_j \le \alpha_j$ for $1 \le j \le m$.

3. Problem 2 allows us to extend Theorem 17.1 in the following way. Let $K \subset \mathbb{R}^m$ be a compact set and $f : [a, b] \times [c, d] \times K \to \mathbb{R}$ be a continuous function. Then for all $z \in K$ we have

$$\int_c^d \left(\int_a^b f(x, y, z) dx \right) dy = \int_a^b \left(\int_c^d f(x, y, z) dy \right) dx,$$

and the function $z \mapsto \int_a^b \left(\int_c^d f(x, y, z) dy \right) dx$ is continuous on K. Clearly in order to show (17.8) we need only to prove for $1 \le k < l \le n$ that

$$(*) \quad \int_{I_1} \cdots \left(\int_{I_k} \cdots \left(\int_{I_l} \cdots \left(\int_{I_n} f(x_1, \ldots, x_k, \ldots, x_l, \ldots, x_n) dx_n \right) \ldots dx_l \right) \ldots dx_k \right) \ldots dx_1$$

$$= \int_{I_1} \cdots \left(\int_{I_l} \cdots \left(\int_{I_k} \cdots \left(\int_{I_n} f(x_1, \ldots, x_k, \ldots, x_l, \ldots, x_n) dx_n \right) \ldots dx_k \right) \ldots dx_l \right) \ldots dx_1.$$

We prove this by induction. Assume that for $n-1$ we have

$$\int_{I_1} \cdots \left(\int_{I_k} \cdots \left(\int_{I_l} \cdots \left(\int_{I_{n-1}} g(x_1, \ldots x_{n-1}, z) dx_{n-1} \right) \ldots dx_l \right) \ldots dx_k \right) \ldots dx_1$$

$$= \int_{I_1} \cdots \left(\int_{I_k} \cdots \left(\int_{I_l} \cdots \left(\int_{I_{n-1}} g(x_1, \ldots, x_{n-1}, z) dx_{n-1} \right) \ldots dx_k \right) \ldots dx_l \right) \ldots dx_1$$

where $g : I_1 \times \ldots I_{n-1} \times K \to \mathbb{R}$ is a continuous function and $K \subset \mathbb{R}^m$ compact. Now consider the continuous function $f : I_1 \times \cdots \times I_n \times K \to \mathbb{R}$ and for $1 < k < l \le n$ the two functions

$$(x_1, z) \mapsto \int_{I_k} \cdots \left(\int_{I_k} \cdots \left(\int_{I_l} \cdots \left(\int_{I_n} f(x_1, \ldots, x_n, z) dx_n \right) \ldots dx_l \right) \ldots dx_k \right) \ldots dx_n$$

and

$$(x_1, z) \mapsto \int_{I_2} \cdots \left(\int_{I_l} \cdots \left(\int_{I_k} \cdots \left(\int_{I_n} f(x_1, \ldots, x_n, z) dx_1 \right) \ldots dx_k \right) \ldots dx_l \right) \ldots dx_n$$

Both functions are continuous on $I_1 \times K$, hence integrable on I_1 and by our induction hypothesis they coincide. Hence $(*)$ follows. (The case $k = 1$ follows analogously).

4. We evaluate $\frac{d}{dy} \int_0^{\varphi(y)} x^y \, dx$ in two different ways. First we note that

$$\int_0^{\varphi(y)} x^y dx = \frac{1}{1+y} x^{1+y} \Big|_0^{\varphi(y)} = \frac{\varphi(y)^{1+y}}{1+y}$$

and therefore

$$\frac{d}{dy} \int_0^{\varphi(y)} x^y \, dy = \frac{d}{dy} \left(\frac{\varphi(y)^{1+y}}{1+y} \right)$$

$$= \frac{d}{dy} \frac{e^{(y+1)\ln \varphi(y)}}{1+y}$$

$$= e^{(y+1)\ln \varphi(y)} \left(\frac{(1+y)\ln \varphi(y) + (1+y^2)\frac{\varphi'(y)}{\varphi(y)} - 1}{(1+y)^2} \right)$$

$$= \frac{\varphi(y)^y}{(1+y)^2} ((1+y)\varphi(y) \ln \varphi(y) + (1+y)^2 \varphi'(y) - \varphi(y)).$$

On the other hand we have using (17.11)

$$\frac{d}{dy} \int_0^{\varphi(y)} x^y dx = \int_0^{\varphi(y)} \frac{d}{dy} (e^{y \ln x}) dx + \frac{d\varphi(y)}{dy} (\varphi(y))^y$$

$$= \int_0^{\varphi(y)} x^y \ln x \, dx + \frac{d\varphi(y)}{dy} (\varphi(y))^y$$

which leads to

$$\int_0^{\varphi(y)} x^y \ln x \, dx = \frac{\varphi(y)^y}{(1+y)^2}\left((1+y)\varphi(y)\ln\varphi(y) + (1+y)^2\varphi'(y) - \varphi(y)) - \varphi'(y)\varphi(y)^y\right)$$

$$= \frac{\varphi(y)^y}{(1+y)^2}\left((1+y)\varphi(y)\ln\varphi(y) - \varphi(y)\right).$$

If we now take $\varphi(y) = 1$ and write α instead of y we find for example

$$\int_0^1 x^\alpha \ln x \, dx = \frac{1}{(1+\alpha)^2}\left((1+\alpha))1 \cdot \ln 1 - 1\right) = \frac{-1}{(1+\alpha)^2}.$$

5. Our starting point is (17.11). In the case where φ_j and f depend continuously on a further variable, say $\varphi_j : [a,b] \times K \to [c,d], K \subset \mathbb{R}^n$ compact, and $f : [a,b] \times [c,d] \times K \to \mathbb{R}$, we find without a change in the proof for $h_1(x,z)$ defined by

$$h_1(x,z) = \int_{\varphi_1(x,z)}^{\varphi_2(x,z)} f(x,y,z)dy$$

the formula

$$(*) \quad \frac{\partial}{\partial x}h_1(x,z) = \int_{\varphi_1(x,z)}^{\varphi_2(x,z)} \frac{\partial f}{\partial x}(x,y,z)dy + \frac{\partial \varphi_2}{\partial x}(x,z)f(x,\varphi_2(x,z),z)$$

$$- \frac{\partial \varphi_1}{\partial x}(x,z)f(x,\varphi_1(x,z),z),$$

provided $\frac{\partial f}{\partial x}, \frac{\partial \varphi_1}{\partial x}, \frac{\partial \varphi_2}{\partial x}$ are now continuous on $[a,b] \times K$. If $K = [\alpha,\beta], \alpha < \beta$ and $f, \varphi_1, \varphi_2, \frac{\partial \varphi_1}{\partial x}, \frac{\partial \varphi_2}{\partial x}, \frac{\partial f}{\partial x}$ have a continuous partial derivative with respect to z as has f a continuous partial derivative with respect to y we can differentiate in $(*)$ with respect to z and we find

$$\frac{\partial^2}{\partial z \partial x}h_1(x,z) = \frac{\partial}{\partial z}\int_{\varphi_1(x,z)}^{\varphi_2(x,z)} \frac{\partial f}{\partial x}(x,y,z)dy + \frac{\partial}{\partial z}\left(\frac{\partial \varphi_2}{\partial x}(x,z)f(x,\varphi_2(x,z),z)\right)$$

$$- \frac{\partial}{\partial z}\left(\frac{\partial \varphi_1}{\partial x}(x,z)f(x,\varphi_1(x,z),z)\right)$$

704

$$= \int_{\varphi_1(x,z)}^{\varphi_2(x,z)} \frac{\partial^2 f}{\partial z \partial x}(x,y,z)dy + \frac{\partial \varphi_2(x,y)}{\partial z} f(x,\varphi_2(x,z),z)$$

$$- \frac{\partial \varphi_1}{\partial z}(x,z) f(x,\varphi_1(x,z),z)$$

$$+ \frac{\partial^2 \varphi_2}{\partial z \partial x}(x,z) f(x,\varphi_2(x,z),z)$$

$$+ \frac{\partial \varphi_2}{\partial x}(x,z) \left(\frac{\partial f}{\partial y}(x,\varphi_2(x,z),z) \frac{\partial \varphi_2}{\partial z}(x,z) + \frac{\partial f}{\partial z}(x,\varphi_2(x,z),z) \right)$$

$$- \frac{\partial^2 \varphi_1}{\partial z \partial x}(x,z) f(x,\varphi_1(x,z),z)$$

$$- \frac{\partial \varphi_1}{\partial x}(x,z) \left(\frac{\partial f}{\partial y}(x,\varphi_1(x,z),z) \frac{\partial \varphi_1}{\partial z}(x,z) + \frac{\partial f}{\partial z}(x,\varphi_1(x,z),z) \right),$$

and of course there are now several ways to rearrange these terms.

Thus the conditions we need to impose on $\varphi_\nu, \nu = 1,2$ and h are for $\frac{\partial h}{\partial x_j}, \frac{\partial h}{\partial x_l}, \frac{\partial^2 h}{\partial x_j \partial x_l}$, $\frac{\partial h}{\partial y}$ to exist and to be continuous and for $\frac{\partial \varphi_\nu}{\partial x_j}, \frac{\partial \varphi_\nu}{\partial x_l}, \frac{\partial^2 \varphi_\nu}{\partial x_j \partial x_l}$ to exist and to be continuous for $\nu = 1,2$. Then we can use the formula derived above to calculate

$$\frac{\partial^2}{\partial x_j \partial x_l} \int_{\varphi_1(x_1\ldots,x_n)}^{\varphi_2(x_1,\ldots,x_n)} h(x_1,\ldots,x_n,y)dy.$$

6. By assumption we have for all $z \in \mathbb{C}$ the convergent power series

$$\exp z = \sum_{k=0}^{\infty} \frac{z^k}{k!},$$

$$\cos z = \sum_{l=0}^{\infty}(-1)^l \frac{z^{2l}}{(2l)!},$$

$$\sin z = \sum_{m=1}^{\infty}(-1)^{m-1} \frac{z^{2m-1}}{(2m-1)!}$$

For $z = i\varphi, \varphi \in \mathbb{R}, i^2 = -1$, we now find

$$\exp(i\varphi) = \sum_{k=0}^{\infty} \frac{(i\varphi)^k}{k!} = \sum_{l=0}^{\infty} \frac{(i\varphi)^{2l}}{(2l)!} + \sum_{m=1}^{\infty} \frac{(i\varphi)^{2m-1}}{(2m-1)!}$$

$$= \sum_{l=0}^{\infty} \frac{i^{2l}\varphi^{2l}}{(2l)!} + \sum_{m=1}^{\infty} \frac{i^{2m-1}\varphi^{2m-1}}{(2m-1)!}$$

$$= \sum_{l=0}^{\infty} \frac{(-1)^l \varphi^{2l}}{(2l)!} + i \sum_{m=1}^{\infty} \frac{i^{2m-2} \varphi^{2m-1}}{(2m-1)!}$$

$$= \sum_{l=0}^{\infty} \frac{(-1)^l \varphi^{2l}}{(2l)!} + i \sum_{m=1}^{\infty} \frac{(-1)^{m-1} \varphi^{2m-1}}{(2m-1)!}$$

$$= \cos \varphi + i \sin \varphi.$$

Chapter 18

1. If figure A is the figure from the problem, then Figure B shows the solution.

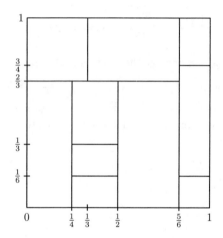 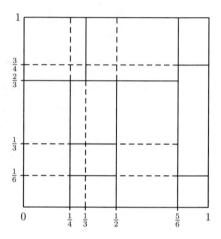

Figure A   *Figure B*

2. We have to prove that $\lambda f + \mu g \in T(K)$ and $f \cdot g \in T(K)$ for $f, g \in T(K)$ and $\lambda, \mu \in \mathbb{R}$.

First we note that if h is a step function corresponding to the partition $Z(Z_1, \ldots, Z_n)$ and the cells $K_\alpha, \alpha \in A_Z$, then h has the representation

$$h(x) = \sum_{\alpha \in A_Z} c_\alpha \chi_{\mathring{K}_\alpha}(x) + h(x) \chi_{K \setminus (\cup_{\alpha \in A_Z} \mathring{K}_\alpha)}(x)$$

and consequently, if two step functions h_1 and h_2 are given with respect to the same partition then we find for $\lambda, \mu \in \mathbb{R}$ with $h_1|_{\mathring{K}_\alpha} = c_\alpha$ and $h_2|_{\mathring{K}_\alpha} = d_\alpha$ that

$$\lambda h_1(x) + \mu h_2(x) = \lambda \sum_{\alpha \in A_Z} c_\alpha \chi_{\mathring{K}_\alpha}(x) + \lambda h_1(x) \chi_{K \setminus (\cup_{\alpha \in A_Z} \mathring{K}_\alpha)}(x)$$

$$+ \mu \sum_{\alpha \in A_Z} d_\alpha \chi_{\mathring{K}_\alpha} + \mu h_2(x) \chi_{K \setminus (\cup_{\alpha \in A_Z} \mathring{K}_\alpha)}(x)$$

$$= \sum_{\alpha \in A_Z} (\lambda c_\alpha + \mu d_\alpha) \chi_{\mathring{K}_\alpha}(x) + (\lambda h_1(x) + \mu h_2(x)) \chi_{K \setminus (\cup_{\alpha \in A_Z} \mathring{K}_\alpha)}(x)$$

and consequently $\lambda h_1 + \mu h_2$ is a step function. Analogously we find using the fact that $\chi_D(x) \cdot \chi_{D^c}(x) = 0$ for any set D that

$h_1(x)h_2(x)$

$$= \left(\sum_{\alpha \in A_Z} c_\alpha \chi_{\mathring{K}_\alpha}(x) + h_1(x)\chi_{K \setminus (\cup_{\alpha \in A_Z} \mathring{K}_\alpha)}(x) \right) \left(\sum_{\alpha \in A_Z} d_\alpha \chi_{\mathring{K}_\alpha}(x) + h_2 \chi_{K \setminus (\cup_{\alpha \in A_Z} \mathring{K}_\alpha)}(x) \right)$$

$$= \sum_{\alpha \in A_Z} c_\alpha d_\alpha \chi_{\mathring{K}_\alpha}(x) + h_1(x)h_2(x)\chi_{K \setminus (\cup_{\alpha \in A_Z} \mathring{K}_\alpha)}(x)$$

and hence $h_1 h_2 \in T(K)$. All that remains to do for proving the general case is to pass from the partitions corresponding to f and g, respectively, to their joint refinement.

3. Since $g(x) \le h(x)$ implies that $h(x) - g(x) \ge 0$ we obtain using the linearity and positivity preserving property of l that $l(h-g) = l(h) - l(g) \ge 0$ implying $l(h) \ge l(g)$.

For $f, g \in V$ we find with $\varepsilon_{\omega_0}, \omega_0$ fixed, defined by $\varepsilon_\omega(g) = g(\omega_0)$ that $\varepsilon_{\omega_0}(\lambda f + \mu g) = \lambda f(\omega_0) + \mu g(\omega_0) = \lambda \varepsilon_{\omega_0}(f) + \mu \varepsilon_{\omega_0}(g), \lambda, \mu \in \mathbb{R}$.

4. Let us first sketch the partition of K,

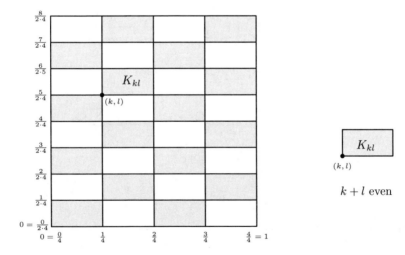

Since on half of the cells g is 1 and on the other half g is 0 we expect that $\int_K g(x)dx = \frac{1}{2}$. By definition we have

$$\int_K g(x)dx = \sum_{k=0}^{n-1} \sum_{l=0}^{2n-1} c_{kl} \left(\frac{k+1}{n} - \frac{k}{n} \right) \left(\frac{l+1}{2n} - \frac{l}{2n} \right)$$

$$= \sum_{k=0}^{n-1} \sum_{l=0}^{2n-1} c_{kl} \frac{1}{n} \frac{1}{2n} = \frac{1}{2n^2} \sum_{k=0}^{n-1} \sum_{l=0}^{2n-1} c_{kl},$$

707

where $c_{kl} = 1$ if $k + l$ is even and $c_{kl} = 0$ if $k + l$ is odd. Therefore we find

$$\sum_{k=0}^{n-1} \sum_{l=0}^{2n-1} c_{kl} = \sum_{(k,l) \in N} 1,$$

where

$$N = \{(k,l) \in \mathbb{N}_0 \times \mathbb{N}_0 \mid 0 \le k \le n-1 \; 0 \le l \le 2n-1, k+l \text{ is even}\}.$$

This set has $\frac{1}{2}n \cdot 2n = n^2$ elements and therefore we find

$$\frac{1}{2n^2} \sum_{k=0}^{n-1} \sum_{l=0}^{2n-1} c_{kl} = \frac{1}{2n^2} \cdot n^2 = \frac{1}{2}$$

as expected.

5. The midpoint of the cell $K_{k,l} = [\frac{k}{n}, \frac{k+1}{n}] \times [\frac{l}{n}, \frac{l+1}{n}], 0 \le k, l \le n-1$, is the point $\xi_{kl} = \left(\frac{2k+1}{2n}, \frac{2l+1}{2n}\right)$. The volume of such cell K_{kl} is $\text{vol}_2(K_{kl}) = \frac{1}{n^2}$ implying for z and ξ from above

$$R_n(f, z_n, \xi) = \sum_{k,l=0}^{n-1} \left(\left(\frac{2k+1}{2n}\right)^2 + \left(\frac{2l+1}{2n}\right)^2 \right) \frac{1}{n^2}$$

$$= \frac{1}{4n^4} \sum_{k,l=0}^{n-1} (2k+1)^2 + (2l+1)^2$$

$$= \frac{1}{4n^4} \sum_{k=0}^{n-1} \sum_{l=0}^{n-1} (2k+1)^2 + \frac{1}{n^4} \sum_{k=0}^{n-1} \sum_{l=0}^{k-1} (2l+1)^2$$

$$= \frac{2}{4n^3} \sum_{j=0}^{n-1} (2j+1)^2$$

$$= \frac{1}{2n^3} \left(\frac{1}{3}(4n^3 - n)\right) = \frac{2}{3} - \frac{1}{2n^2}.$$

Note that for $n \to \infty$ we find

$$\lim_{n \to \infty} R_n(f, z_n, \xi) = \frac{2}{3} = \int_0^1 \int_0^1 (x^2 + y^2) \, dx \, dy.$$

6. The purpose of this problem is to convince the student that the construction of the Riemann integral over cells in dimension N is completely analogous to that in one dimension. So we go back to the proof of Theorem I.25.13 and check the details. To prove (18.22) it is again sufficient to prove the following for every $\varepsilon > 0$

$$\int^* (f(x) + g(x)) dx \le \int^* f(x) dx + \int^* g(x) dx + \varepsilon.$$

708

We know that two step function $\varphi, \psi \in T(K)$ exist such that $\varphi \geq f, \psi \geq g$ and

$$\int \varphi(x)dx \leq \int^* f(x)dx + \frac{\varepsilon}{2} \quad \text{and} \quad \int \psi(x)dx \leq \int^* g(x)dx + \frac{\varepsilon}{2}.$$

Since $\varphi + \psi \geq f + g$ we deduce

$$\int^* (f+g)(x)dx \leq \int (\varphi + \psi)(x)dx$$

$$\leq \int \varphi(x)dx + \int \psi(x)dx$$

$$\leq \int^* f(x)dx + \int^* g(x)dx + \varepsilon.$$

7. Again we can follow closely the one-dimensional case, see Theorem I.25.19. For $\varepsilon > 0$ there exists $\varphi, \psi \in T(K)$ such that $\varphi \leq f \leq \psi$ and $\varphi^+ \leq f^+ \leq \psi^+$ and $\int_K (\psi - \varphi)(x)dx \leq \varepsilon$. This however implies that $\int_K (\psi^+ - \varphi^+)(x)dx \leq \varepsilon$ and the integrability of f^+ follows. Since $f^- = (-f)^+$ we also see that f^- is integrable implying the integrability of $|f| = f^+ + f^-$.

8.   a) We have

$$\int_0^2 \left(\int_0^3 \frac{x-y}{(x+y)^3}dy \right) dx = \int_0^2 \left(\int_0^3 \frac{2x - (x+y)}{(x+y)^3}dy \right) dx$$

$$= \int_0^2 \left(\int_0^3 \left(\frac{2x}{(x+y)^3} - \frac{1}{(x+y)^2} \right) dy \right) dx$$

$$= \int_0^2 \left(-\frac{x}{(x+3)^2} + \frac{1}{x+3} \right) dx = \int_0^2 \frac{3}{(x+3)^2}dx$$

$$= \frac{2}{5}$$

whereas

$$\int_0^3 \left(\int_0^2 \frac{x-y}{(x+y)^3}dx \right) dy = -\int_0^3 \left(\int_0^2 \frac{y-x}{(y+x)^3}dx \right) dy$$

$$= -\int_0^3 \left(\int_0^2 \left(\frac{2y}{(y+x)^3} - \frac{1}{(y+x)^2} \right) dx \right) dy$$

$$= -\int_0^3 \left(\frac{-y}{(y+2)^2} + \frac{1}{y+2} \right) dy = -\int_0^3 \frac{2}{(y+2)^2}dy$$

$$= -\frac{3}{5}$$

Note that on $(0,2] \times (0,3]$ the function $f(x,y) = \frac{x-y}{(x+y)^3}$ is not bounded.

709

b) We split the integral

$$\int_1^5 \left(\int_{-1}^1 \left(\int_2^4 ((x_1^2 + x_3^2)) \sin x_2 + e^{x_1+x_2+x_3}) dx_3 \right) dx_2 \right) dx_1$$

$$= \int_1^5 \left(\int_{-1}^1 \left(\int_2^4 (x_1^2 + x_3^2) \sin x_2 \ dx_3 \right) dx_2 \right) dx_1$$

$$+ \int_1^5 \left(\int_{-1}^1 \left(\int_2^4 (e^{x_1+x_2+x_3}) dx_3 \right) dx_2 \right) dx_1$$

and find

$$\int_1^5 \left(\int_{-1}^1 \left(\int_2^4 (x_1^2 + x_3^2) \sin x_2 \ dx_3 \right) dx_2 \right) dx_1$$

$$= \int_1^5 \left(\int_2^4 \left(\int_{-1}^1 (x_1^2 + x_3^2) \sin x_2 \ dx_2 \right) dx_3 \right) dx_1$$

but $\int_{-1}^1 (x_1^2 + x_2^2) \sin x_2 \ dx_2 = 0$. Further we have

$$\int_1^5 \left(\int_{-1}^1 \left(\int_2^4 e^{x_1+x_2+x_3} dx_3 \right) dx_2 \right) dx_1 = \int_1^5 e^{x_1} dx_1 \int_{-1}^1 e^{x_2} dx_2 \int_2^4 e^{x_3} dx_3$$

$$= (e^5 - e)(e - e^{-1})(e^4 - e^2)$$

$$= e^{10} + 2e^4 - e^2.$$

9. The conditions on f and g imply that all integrals being involved exist and can be calculated as iterated integrals in any order. Therefore we find

$$\int_K \frac{\partial f}{\partial x_{n_1}}(x)g(x)dx = \int_{K_1} \left(\int_{K_2} \left(\int_a^b \frac{\partial f}{\partial x_{n_1}}(x_1, \ldots, x_{n_1+n_2})g(x_1, \ldots, x_{n_1+n_2}) \times \right. \right.$$

$$\left. \left. \times \ dx_{n_1} \right) dx_{n_1+n_2} \ldots dx_{n_1+1} \right) dx_{dx_{n_1-1}} \ldots dx_1$$

$$= -\int_{K_1} \left(\int_{K_2} \left(\int_a^b f(x_1, \ldots, x_{n_1+n_2}) \frac{\partial g}{\partial x_{n_1}}(x_1, \ldots, x_{n_1+n_2}) \times \right. \right.$$

$$\left. \left. \times \ dx_{n_1} \right) dx_{n_1+n_2} \ldots dx_{n_1+1} \right) dx_{n_1-1} \ldots dx_1$$

$$= -\int_K f(x) \frac{\partial g}{\partial x_{n_1}}(x)dx,$$

where we used the fact that $f(x_1, \ldots, x_{n_1-1}, a, x_{n_1+1}, \ldots, x_{n_1+n_2})$ and $f(x_1, \ldots, x_{n_1-1}, b, x_{n_1+1}, \ldots, x_{n_1+n_2})$ vanish.

We now turn to (18.54). The condition imply that f and g' as well as f^2 and $(g')^2$ are integrable over K and the corresponding integrals are iterated integrals in which we may take any order of integration. Therefore we have

$$\int_K (f(x))^2 dx$$

$$= \int_{K_1} \left(\int_I \left(\int_{K_2} (f(x_1, \ldots, x_{n_1-1}, x_{n_1}, x_{n_1+n_2}))^2 \times \right. \right.$$

$$\left. \left. \times\, dx_{n_1+n_2} \ldots dx_{n_1+1} \right) dx_{n_1} \right) dx_{n_1-1} \ldots dx_1$$

$$= \int_{K_1} \left(\int_{K_2} \left(\int_I (f(x_1, \ldots, x_{n_1-1}, x_{n_1}, x_{n_1+1}, \ldots, x_{n_1+n_2}))^2 \times \right. \right.$$

$$\left. \left. \times\, dx_{n_1} \right) dx_{n_1+n_2} \ldots dx_{n_1+1} \right) dx_{n_1-1} \ldots dx_1$$

$$= \int_{K_1} \left(\int_{K_2} \left(\int_a^b \frac{\partial x_{n_1}}{\partial x_{n_1}} (f(x_1, \ldots, x_{n_1}, \ldots, x_{n_1+n_2}))^2 dx_{n_1} \right) \ldots dx_{n_1+1} \right) \ldots dx_1$$

$$= - \int_{K_1} \left(\int_{K_2} \left(\int_a^b x_{n_1} \frac{\partial}{\partial x_{n_1}} (f(x_1, \ldots, x_{n_1}, \ldots, x_{n_1+n_2}))^2 dx_{n_1} \right) \ldots dx_{n_1+1} \right) \ldots dx_1$$

where we use integration by parts for the integral with respect to x_{n_1} using the fact that the two boundary terms $f(x_1, \ldots, x_{n_1-1}, a, x_{n_1+1}, \ldots, a_{n_1+n_2})$ and $f(x_1, \ldots, x_{n_1-1}, b, x_{n_1+1}, \ldots, x_{n_1+n_2})$ vanish.
It follows from the above calculation that

$$\int_K (f(x))^2 dx \leq 2 \max(|a|, |b|) \int_{K_1} \left(\int_{K_2} \left(\int_I |f(x)| \left| \frac{\partial f}{\partial x_{n_1}}(x) \right| dx_{n_1} \right) \ldots dx_{n_1+1} \right) \ldots dx_1$$

$$= 2 \max(|a|, |b|) \int_K |f(x)| \left| \frac{\partial f}{\partial x_{n_1}} \right| dx$$

$$\leq \max(|a|, |b|) \left(\int_K |f(x)|^2 dx \right)^{1/2} \left(\int_K \left| \left(\frac{\partial f}{\partial x_{n_1}}(x) \right| ^2 \right) dx \right)^{1/2},$$

where in the last step we used the Cauchy-Schwarz inequality. Thus we arrive at

$$\int_K |f(x)|^2 dx \leq 2 \max(|a|, |b|) \left(\int_K |f(x)|^2 dx \right)^{1/2} \left(\int_K \left| \frac{\partial f}{\partial x_{n_1}}(x) \right|^2 dx \right)^{1/2}$$

or

$$\left(\int_K |f(x)|^2 dx\right)^{1/2} \leq \max(|a|, |b|) \left(\int_K \left|\frac{\partial f}{\partial x_{n_1}}(x)\right|^2 dx\right)^{1/2}.$$

(Also compare with I.26.16)

10. We note that

$$K = \{(x_1, x_2, x_3) \in \mathbb{R}^3 \mid -1 \leq x_1 \leq 1, \ 0 \leq x_2 \leq 2, \ 3 \leq x_3 \leq 4\},$$

and therefore we have

$$T(K) = \{(y_1, y_2, y_3) \in \mathbb{R}^3 \mid y_1 = 3x_1 - 1, y_2 = 4x_2 + 2, y_3 = 2x_3 + 1,$$
$$-1 \leq x_1 \leq 1, \ 0 \leq x_2 \leq 2, \ 3 \leq x_3 \leq 4\}$$
$$= \{(y_1, y_2, y_3) \in \mathbb{R}^3 \mid -4 \leq y_1 \leq 2, \ 2 \leq y_2 \leq 10, \ 7 \leq y_3 \leq 9\}.$$

Further we note that $J_T(x) = \begin{pmatrix} 3 & 0 & 0 \\ 0 & 4 & 0 \\ 0 & 0 & 2 \end{pmatrix}$, hence $|\det J_T(x)| = 24$.

Now we want to calculate the two integrals in

$$\int_{T(K)} f(x)dx = \int_K (f \circ T)(x) \det J_T(x)dx = 24 \int_K (f \circ T)(x)dx.$$

We have

$$\int_{T(K)} f(x)dx = \int_{-4}^{2} \left(\int_2^{10} \left(\int_7^9 (x_1 + x_2 + x_3)dx_3\right) dx_2\right) dx_1$$
$$= 16 \int_{-4}^{2} x_1 \, dx_1 + 12 \int_2^{10} x_2 \, dx_2 + 48 \int_7^9 x_3 \, dx_3$$
$$= -96 + 576 + 768 = 1248$$

as well as

$$24 \int_{-1}^{1} \left(\int_0^2 \left(\int_3^4 ((3x_1 - 1) + (4x_2 + 2) + (2x_3 + 1))dx_3\right) dx_2\right) dx_1$$
$$= 48 \int_{-1}^{1} (3x_1 - 1)dx_1 + 48 \int_0^2 (4x_2 + 2)dx_2 + 96 \int_3^4 (2x_3 + 1)dx_3$$
$$= -96 + 576 + 768 = 1248.$$

Chapter 19

1. a) By definition a polygon in the plane is a finite set of points connected by line segments. We know that every bounded line segment in $\mathbb{R}^2$ has Jordan content zero and the finite union of sets of Jordan content zero has Jordan content zero.

    This result implies in particular that a set G in the plane the boundary of which is a closed polygon is Jordan measurable.

b) Since $G \subset \mathbb{R}^m$ is compact we can find an open cell $K \subset \mathbb{R}^m$ such that $\overline{G} \subset K$. Further, given $\varepsilon > 0$ we can find open cells $C_1, \ldots, C_N \subset \mathbb{R}^n$ such that $M \subset \bigcup\limits_{j=1}^{N} C_j$ and $\sum\limits_{j=1}^{N} \mathrm{vol}_n(C_j) \leq \frac{\varepsilon}{\mathrm{vol}_m(K)}$. The sets $(C_j \times K)_{j=1,\ldots,N}$ are a covering of $M \times G$ with open cells in R^{n+m} and

$$\sum_{j=1}^{N} \mathrm{vol}_{n+m}(C_j \times K) = \sum_{j=1}^{N} \mathrm{vol}_n(C_j)\mathrm{vol}_m(K) < \varepsilon.$$

2. a) Note that every open cell (which in $\mathbb{R}$ is an open interval) covering a subset of R must contain also points of $R^{\complement}$. In fact any open covering of R is also an open covering of $[0,1]$. Thus, if $K_1, \ldots, K_N$ is an open covering of R it must hold that $\sum\limits_{j=1}^{N} \mathrm{vol}_1(K_j) \geq 1$. Hence R cannot have Jordan content zero.

   b) We know that $\mathbb{Q} \cap [0,1]$ is a countable set and every single point $q \in \mathbb{Q} \cap [0,1]$ has Lebesgue measure zero. Hence $\mathbb{Q} \cap [0,1] = \bigcup\limits_{q \in \mathbb{Q} \cap [0,1]} \{q\}$ is a countable union of sets of Lebesgue measure zero, hence it has Lebesgue measure zero.

3. We may consider $I = [0,1]$ and $A_j = \{j\}, j \in I$. Clearly A_j has Lebesgue measure zero. We know that $[0,1]$ does not have Lebesgue measure zero, but

$$[0,1] = \bigcup_{j \in I} \{A_j\} = \bigcup_{j \in [0,1]} \{j\},$$

hence we have a counter example.

A short argument why $[0,1]$ does not have Lebesgue measure zero is the following: $[0,1]$ is compact and if it were to have Lebesgue measure zero, by Corollary 19.9 it must have Jordan content zero. However the Jordan content of $[0,1]$ is 1.

4. First we note that for every $\rho > 0$ the set $\partial B_\rho(0)$ is of the type C^k for every $k > 0$. If $x \in \partial B_\rho(0)$ then for some $j, 1 \leq j \leq n$, it must hold that $x_j \neq 0$. This implies that in a small neighbourhood of x we find for $x_j > 0$

$$x = (x_1, \ldots, x_{j-1}, \sqrt{\rho^2 - \sum_{l \neq j} x_l^2}, x_{j+1} \ldots, x_n)$$

and for $x_j < 0$

$$x = (x_1, \ldots, x_{j-1}, -\sqrt{\rho^2 - \sum_{l \neq j} x_l^2}, x_{j+1} \ldots, x_n),$$

which gives us the representation (19.4) when we choose for $Q(x)$ a ball with centre $O \in \mathbb{R}^{n-1}$ and radius sufficiently small, i.e. less than $|x_j|$.

Now $\partial(\overline{B_R(0)} \backslash B_r(0)) = \partial B_R(0) \cup \partial B_r(0)$ and $\partial B_R(0) \cap \partial B_r(0) = \emptyset$. Thus for every $x \in \partial(\overline{B_R(0)} \backslash B_r(0))$ the construction made above applies.

5. A hyper-rectangle in a general position can be defined as

$$R(a_1, \ldots, a_{n_i} \cdot b) : \{z \in \mathbb{R}^n x = \sum_{j=1}^{n} \lambda_j a_j + b_j, \lambda_j \in [0, 1]\}$$

where $\{a_1, \ldots, a_n\} \subset \mathbb{R}^n$ is a linearly independent set, $b \in \mathbb{R}^n$ and $\langle a_j, a_l \rangle = 0$ for $j \neq l$. Thus with the notation of Chapter 16 we have

$$R(a_1, \ldots, a_n, b) = b + P(a_a, \ldots, a_n)$$

and $\mathrm{vol}_n(R(a_1, \ldots, a_n, b)) = \mathrm{vol}_n(P(a_1, \ldots, a_n))$, which is of course already the translation invariance. Further we know that

$$\mathrm{vol}_n(T(P(a_1, \ldots, a_n))) = \mathrm{vol}_n(P(a_1, \ldots, a_n))$$

for $T \in O(n)$, compare with Proposition 16.1. But $P(a_1, \ldots, a_n)$ is a non-degenerate cell. Thus the result follows almost from the definition.

However, this result leaves us with a problem which we will solve in Volume III, Part 6. A hyper-rectangle in a general position has Jordan content as defined in 19.23. In addition it has an elementary geometric volume. We need to prove that they coincide.

6.   a) The set $\left[-\frac{3}{4}, \frac{3}{4}\right] \subset (-1, 1)$ is compact and

$$f^{-1}\left(\left[-\frac{3}{4}, \frac{3}{4}\right]\right) = \left(-1, -\frac{1}{2}\sqrt{2}\right] \cup \left[\frac{1}{2}\sqrt{2}, 1\right)$$

which follows by solving the equation $y = 1 - x^4$ for $y \in \left[-\frac{3}{4}, \frac{3}{4}\right]$, compare with the figure below. Clearly the set $\left(-1, -\frac{1}{2}\sqrt{2}\right] \cup \left[\frac{1}{2}\sqrt{2}, 1\right)$ is not compact.

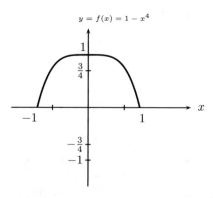

$$y = f(x) = 1 - x^4$$

b) Let X, Y, Z be Hausdorff spaces and $f : X \to Y$ and $g : Y \to Z$ proper mappings. Let $K \subset Z$ be a compact set. Then $g^{-1}(K)$ is a compact set in Y and consequently $f^{-1}(g^{-1}(K))$ is a compact set in X, hence $(g \circ f)^{-1}(K)$ is for every compact set $K \subset Z$ compact in X, which means that $g \circ f : X \to Z$ is proper.

7. Let $(U_j)_{j=I}$ be an open covering of $f^{-1}(K), K \subset Y$ compact. For $y \in K$ there exist $I_y \subset I$, a finite set, such that $f^{-1}(y) \subset \bigcup_{j \in I_y} U_j$. The set $V_y := Y \backslash f(X \backslash \bigcup_{j \in I_y} U_j)$ is open since $\bigcup_{j \in I_y} U_j$ is open and f maps closed sets onto closed sets. Further we have $y \in V_y$ and $f^{-1}(V_y) \subset \bigcup_{j \in I_y} U_j$. Since the compact set K is covered by a finite family $V_{y_1}, \ldots, V_{y_N}$ it follows that $f^{-1}(K) \subset \bigcup_{l=1}^{N} (\bigcup_{j \in I_{y_l}} U_j)$, i.e. $f^{-1}(K)$ is compact.

8. We note that for any set $K \subset \mathbb{R}$ we have
$$f^{-1}(K) = \{x \in B_R(0) \mid f(x) \in K\}$$
$$= \{x \in B_R(0) \mid g(\|x\|) \in K\}$$

We claim that $\|.\| : B_R(0) \to [0, \infty)$ is proper. Clearly $\|.\|$ is continuous and for $\rho \in [0, \infty)$ it follows that its pre-image under $\|.\|$ is $\partial B_\rho(0)$ which is compact. Now let $G \subset B_R(0)$ be a closed set and A its image under $\|.\|$, i.e. $A := \{\rho \in [0, \infty) \mid \rho = \|x\|$ for some $x \in G\}$. Let $(\rho_\nu)_{\nu \in \mathbb{N}}$ be a sequence in A converging in $[0, \infty)$ to ρ_0 We want to prove that $\rho_0 \in A$. In this case it will follow that A is closed and by Problem 7 that g is proper. For ρ_ν there exists $x_\nu \in G$ such that $\|x_\nu\| = \rho_\nu$. Since $(\rho_\nu)_{\nu \in \mathbb{N}}$ converges, the sequence $(x_\nu)_{\nu \in \mathbb{N}}$ is bounded (which follows also from the fact that $B_R(0)$ is bounded). Hence a subsequence $(x_{\nu_j})_{j \in \mathbb{N}}$ converges to some $x_{\nu_0} \in G$, recall that G is closed. It follows that $\|x_{\nu_j}\|$ converges to $\|x_{\nu_0}\| = \rho_0$. Since $x_{\nu_0} \in G$ it follows $\rho_0 \in A$. i.e. A is closed.

Thus $\|.\| : B_R(0) \to \mathbb{R}$ is proper. Further by Corollary 19.18 the mapping g is proper. Thus for $K \subset \mathbb{R}$ compact the set $L := g^{-1}(K)$ is compact in $\mathbb{R}$, implying that the set $\{x \in B_R(0) \mid \|x\| \in L\}$ is compact, which however means that $\{x \in B_R(0) \mid g(\|x\|) \in K\}$ is compact, and this gives indeed the compactness of $f^{-1}(K)$. Thus f is proper. Proposition 19.19 (or Proposition 19.21) now implies that $\partial(f^{-1}([a, b]))$ has Jordan content zero.

9. Let K_1 and K_2 be two non-degenerate compact cells with $G \subset \mathring{K}_j, j = 1, 2$, such that $\chi_G\big|_{K_j}$ is Riemann integrable. We have to prove that
$$\int_{K_1} (\chi_G\big|_{K_1})(x)dx = \int_{K_2} (\chi_G\big|_{K_2})(x)dx.$$

First we note that $G \subset K_1 \cap K_2$ and $K_1 \cap K_2$ is a non-degenerate compact cell. It follows that
$$\int_{K_j} (\chi_G\big|_{K_j})(x)dx = \int_{K_1 \cap K_2} (\chi_G\big|_{K_1 \cap K_2})(x)dx + \sum_{l=1}^{N_j} \int_{L_{l,j}} (\chi_G\big|_{L_l})(x)dx$$

where $L_{l,j}$ are non-degenerate compact cells with the property that $\overset{\circ}{L}_{l,j} \cap \overset{\circ}{L}_{l',j} = \emptyset$ for $l \neq l'$, as well as $\overset{\circ}{L}_{l,j} \cap \overset{\circ}{K}_j = \emptyset$. In addition we have

$$K_j = (K_1 \cap K_2) \cup \left(\bigcup_{l=1}^{N_j} L_{l,j} \right).$$

However, on $L_{l,j}$ it holds by construction that $\chi_G \Big|_{L_{l,j}} = 0$ and the result follows.

Chapter 20

1. This is a problem which we can solve by "reduction". If we know the triangle inequality and the Cauchy-Schwarz inequality for volume integrals over non-degenerate compact cells, we know these inequalities for Riemann integrable function on bounded Jordan measurable domains:

$$\left| \int_G f(x)dx \right| = \left| \int_K (f_G \,|_K)(x)dx \right|$$

$$\leq \int_K |\, f_G \,|_K (x)|dx$$

$$= \int_G |\, f(x) \,| \, dx,$$

and similarly

$$\int_G |f(x)g(x)|dx \leq \int_K |f_G|_K(x)g_G|_K(x)| \, dx$$

$$\leq \left(\int_K |f_G|_K(x)|^2 dx \right)^{1/2} \left(\int_K |g_G|_K(x)|^2 dx \right)^{1/2}$$

$$= \left(\int_G |f(x)|^2 dx \right)^{1/2} \left(\int_G |g(x)|^2 dx \right)^{1/2}.$$

Next we note that if we know these inequalities for the corresponding Riemann sums, then we know them for the Riemann integrals over non-degenerate compact cells. We follow the solution of Problem 8 in Chapter I.25 to show the Cauchy-Schwarz inequality. For finite sums we have, see Corollary I.14.3,

$$\sum_{k=1}^N |\, a_k b_k \,| \leq \left(\sum_{k=1}^N |a_k|^2 \right)^{1/2} \left(\sum_{k=1}^N |b_k|^2 \right)^{1/2}.$$

Now let $K \subset \mathbb{R}^n$ be a non-degenerate compact cell and f, g be two integrable functions on K. Further let $Z = Z(Z_1, \ldots, Z_n)$ be a partition of K into sub-cells $K_\alpha, \alpha \in A_Z$, compare with Definition 18.9. With $\xi_\alpha \in \overset{\circ}{K}_\alpha$ we find for the Riemann

716

sums of $f \cdot g, |f|^2$ and $|g|^2$

$$\sum_{\alpha \in A_Z} |f(\xi_\alpha)g(\xi_\alpha)| \, \text{vol}_n(K_\alpha)$$

$$\leq \left(\sum_{\alpha \in A_Z} |f(\xi_\alpha)|^2 \, \text{vol}_n(K_\alpha) \right)^{1/2} \left(\sum_{\alpha \in A_Z} |g(\xi_\alpha)|^2 \, \text{vol}_n(K_\alpha) \right)^{1/2}.$$

Passing now through a sequence of partition $(Z^{(\nu)})_{\nu \in \mathbb{N}}$ with mesh $(Z^{(\nu)}) \to 0$ we obtain

$$\int_K |f(x)g(x)| \, dx \leq \left(\int_K |f(x)|^2 \, dx \right)^{1/2} \left(\int_K |g(x)|^2 \, dx \right)^{1/2}.$$

The proof of the triangle inequality follows analogously when looking at the corresponding Riemann sums:

$$\sum_{\alpha \in A_Z} |f(\xi_\alpha) + g(\xi_\alpha)| \, \text{vol}(K_\alpha) \leq \sum_{\alpha \in A_Z} |f(\xi_\alpha)| \, \text{vol}_n(K_\alpha) + \sum_{\alpha \in A_Z} |g(\xi_\alpha)| \, \text{vol}_n(K_\alpha).$$

2. Let $x_0 \in G$. For a single point the set $\{x_0\}$ has Jordan content zero, hence Lebesgue measure zero. The function $\chi_{\{x_0\}} : G \to \mathbb{R}, \; \chi_{\{x_0\}}(x) = \begin{cases} 1, x = x_0 \\ 0, x \in G\backslash\{x_0\} \end{cases}$ is Riemann integrable with $(\int_G |\chi_{\{x_0\}}(x)|^p \, dx)^{1/p} = 0$, but $\chi_{\{x_0\}}$ is not the function being identically zero. Note that the example works for every Jordan measurable subset $F \subset G$ with $J^{(n)}(F) = 0$ and $\chi_{\{x_0\}}$ replaced by $\chi_F, \chi_F(x) = \begin{cases} 1, x \in F \\ 0, x \in G\backslash F. \end{cases}$

3. We first prove that for $\alpha, \beta \in \mathbb{N}_0^n$ a semi-norm is given on $\mathcal{S}(\mathbb{R}^n)$ by $p_{\alpha\beta}$. Clearly $p_{\alpha\beta}(u) \geq 0$ and for $\lambda \in \mathbb{R}$ and $u \in \mathcal{S}(\mathbb{R}^n)$ it follows

$$p_{\alpha\beta}(\lambda u) = \sup_{x \in \mathbb{R}^n} |x^\beta \partial^\alpha (\lambda u)(x)| = |\lambda| \sup_{x \in \mathbb{R}^n} |x^\beta \partial^\alpha u(x)| = |\lambda| p_{\alpha\beta}(u).$$

Moreover, for $u, v \in \mathcal{S}(\mathbb{R}^n)$ we find

$$\begin{aligned} p_{\alpha\beta}(u+v) &= \sup_{x \in \mathbb{R}^n} |x^\beta \partial^\alpha (u+v)(x)| \\ &= \sup_{x \in \mathbb{R}^n} |x^\beta \partial^\alpha u(x) + x^\beta \partial^\alpha v(x)| \\ &\leq \sup_{x \in \mathbb{R}^n} |x^\beta \partial^\alpha u(x)| + \sup_{x \in \mathbb{R}^n} |x^\beta \partial^\alpha v(x)| \\ &= p_{\alpha\beta}(u) + p_{\alpha\beta}(v). \end{aligned}$$

Hence, $p_{\alpha\beta}$ is indeed a semi-norm on $\mathcal{S}(\mathbb{R}^n)$. Moreover, our calculations show that for $u, v \in \mathcal{S}(\mathbb{R}^n)$ and $\lambda, \mu \in \mathbb{R}$ it follows that $\lambda u + \mu v \in C^\infty(\mathbb{R}^n)$ and

$$p_{\alpha\beta}(\lambda u + \mu v) \leq |\lambda| p_{\alpha\beta}(u) + |\mu| p_{\alpha\beta}(v),$$

which means that $u, v \in \mathcal{S}(\mathbb{R}^n)$ implies $\lambda u + \mu v \in \mathcal{S}(\mathbb{R}^n)$, i.e. $\mathcal{S}(\mathbb{R}^n)$ is a vector space over $\mathbb{R}$.

If $u \in C_0^\infty(\mathbb{R}^n)$ then for all $\alpha, \beta \in \mathbb{N}_0^n$ we find

$$p_{\alpha\beta}(u) = \sup_{x \in \mathbb{R}^n} | x^\beta \partial^\alpha u(x) | = \sup_{x \in \operatorname{supp} u} | x^\beta \partial^\alpha u(x) | < \infty$$

since $\operatorname{supp} u$ is compact and $x \mapsto x^\beta \partial^\alpha u(x)$ is continuous. Clearly $g \in C^\infty(\mathbb{R}^n)$, $g(x) = e^{-\|x\|^2}$, has no compact support so $g \notin C_0^\infty(\mathbb{R}^n)$. We want to show that $g \in \mathcal{S}(\mathbb{R}^n)$. First we note for $\alpha, \beta \in \mathbb{N}_0^n$ and $x \in \mathbb{R}^n$ that

$$x^\beta \partial^\alpha e^{-\|x\|^2} = \prod_{j=1}^n x_j^{\beta_j} \partial_{x_j}^{\alpha_j} e^{-x_j^2},$$

hence we need only to prove that $t \mapsto e^{-t^2}, t \in \mathbb{R}$, belongs to $\mathcal{S}(\mathbb{R})$ to get the result. By Problem 5 in Chapter I.9 we know that $\frac{d^k}{dt^k}(e^{-t^2}) = q_k(t)e^{-t^2}$ with suitable polynomial q_k. Thus we obtain for $1 \le j \le n$ with polynomials $q_{(\alpha_j, \beta_j)}$

$$x_j^{\beta_j} \partial_{x_j}^{\alpha_j} e^{-x_j^2} = q_{(\alpha_j, \beta_j)}(x_j) e^{-x_j^2}$$

which implies $\sup_{x_j \in \mathbb{R}} | x_j^{\beta_j} \partial_{x_j}^{\alpha_j} e^{-x_j^2} | \le C_{\alpha_j, \beta_j}$ and therefore we arrive at

$$p_{\alpha\beta}(g) \le \prod_{j=1}^n c_{\alpha_j \beta_j}.$$

4. From (20.21) we know that for some $\xi \in \overline{B_r(x_0)}$ we have

$$\frac{1}{J^{(n)}(B_r(x_0))} \int_{B_r(x_0)} f(x)dx = f(\xi).$$

Since f is by assumption continuous and $\xi = \xi(r)$ must tend to x_0 as $r \to 0$, it follows that $\lim_{r \to 0} f(\xi) = f(x_0)$. But then we must have

$$\lim_{r \to 0} \frac{1}{J^{(n)}(B_r(x_0))} \int_{B_r(x_0)} f(x)dx = f(x_0).$$

5. The proof of Lemma 20.17 extends in a straightforward way: Since $G_1, \ldots, G_N$ are mutually disjoint, $\partial(G_1 \cup \cdots \cup G_N) \subset \partial G_1 \cup \ldots \partial G_N$. By assumption $J^{(n)}(\partial G_j) = 0$ which implies $J^{(n)}(\partial G) = 0, G = G_1 \cup \cdots \cup G_N$. Further, when f is Riemann integrable over G, then $f|_{G_j}$ is Riemann integrable over G_j and for the canonical extensions we find

$$\widetilde{f_G} = \chi_G \widetilde{f_G} = \chi_{G_1} \widetilde{f_G} + \cdots + \chi_{G_N} \widetilde{f_G}$$

718

implying

$$\int_G f(x)dx = \int_G ((\chi_{G_1}\tilde{f}_G)(x) + \cdots + (\chi_{G_N}\tilde{f}_G)(x))dx$$

$$= \int_{G_1} f(x)dx + \cdots + \int_{G_N} f(x)dx.$$

6. Since each G_j is Jordan measurable, the boundary is a set of Lebesgue measure zero, and the boundness of G_j implies that ∂G_j is compact, hence ∂G_j has Jordan content zero. The set of discontinuities of f is contained in $\bigcup_{j=1}^{N} \partial G_j$ which is again a set of Jordan content zero, hence Lebesgue measure zero. Thus the function

$$h(x) = \begin{cases} \sum\limits_{j=1}^{N}(f\chi_{\mathring{G}_j})(x), x \in \bigcup\limits_{j=1}^{n} \mathring{G}_j \\ 0 \qquad\qquad , x \in G\backslash \bigcup\limits_{j=1}^{N} \mathring{G}_j \end{cases}$$

is only in a set of Lebesgue measure zero discontinuous, hence Riemann integrable over G and the following holds

$$\int_G h(x)dx = \sum_{j=1}^{N} \int_{\mathring{G}_j} f(x)dx.$$

However the set $\{x \in G \mid h(x) \neq f(x)\}$ has Jordan content zero and therefore we have $\int_G h(x)dx = \int_G f(x)dx$.

7. We use (20.28) with $f = g$ and obtain

$$\int_G \|\operatorname{grad} f\|^2 dx = \int_G \langle \operatorname{grad} f, \operatorname{grad} f\rangle dx$$

$$= -\int_G f(x)(\Delta_n f(x))dx = 0,$$

since by assumption f is harmonic in G, i.e $\Delta_n f(x) = 0$.
Thus

$$\sum_{j=1}^{n} \int_G \left(\frac{\partial f}{\partial x_j}\right)^2 (x)dx = \int_G \|\operatorname{grad} f\|^2 dx = 0,$$

and since $f \in C_0^1(G) \cap C^2(G)$, we know that $\left|\frac{\partial f}{\partial x_j}\right|^2$ is continuous and integrable

with $\int_G \left|\frac{\partial f}{\partial x_j}\right|^2 dx = 0$. This yields that for all $1 \leq j \leq n$ we have $\frac{\partial f}{\partial x_j} = 0$ in G.
Since G is convex, by Theorem 6.19 it follows that f is constant in G. But supp f is a compact subset of the open set G, so f must vanish close to ∂G implying that $f(x) = 0$ for all $x \in G$.

8. For $x = (x_1, \ldots, x_n)$ we set $\tilde{x}_j = (x_1, \ldots, x_{j-1}, x_{j+1}, \ldots, x_n)$ and with a compact Jordan measurable set $W \subset \mathbb{R}^{n-1}$ we introduce

$$G := \{(x_1, \ldots, x_n) \in \mathbb{R}^n \mid \varphi(\tilde{x}_j) \leq x_j \leq \psi(\tilde{x}_j), \tilde{x}_j \in W\}$$

where $\varphi, \psi : w \to \mathbb{R}$ and continuous functions such that $\varphi \leq \psi$. Thus G is a normal domain with respect to the e_j-axis. Now let $a \in \mathbb{R}^n$ and consider the translated set

$$G_a : a + G = \{y = (y_1, \ldots, y_n) \in \mathbb{R}^n \mid y = a + x, x \in G\}.$$

It follows that $(y_1, \ldots, y_n) = (a_1 + x_1, \ldots, a_n + x_n)$ and consequently

$$(a_1 + x_1, \ldots, a_{j-1} + x_{j-1}, a_{j+1} + x_{j+1}, \ldots, a_n + x_n) = \tilde{a}_j + \tilde{x}_j \in \tilde{a}_j + W$$

and $\tilde{a}_j + W$ is a compact Jordan measurable set. Now we define $\varphi_a : \tilde{a}_j + W \to \mathbb{R}$ by $\varphi_a(\tilde{y}_j) := \varphi(\tilde{y}_j - \tilde{a}_j) + a_j$ and $\psi_a : \tilde{a}_j + W \to \mathbb{R}$ by $\psi_a(\tilde{y}_j) := \psi(\tilde{y}_j - \tilde{a}_j) + a_j$. It follows for $y_j = a_j + x_j$ that

$$\begin{aligned}
\varphi_a(\tilde{y}_j) = \varphi(\tilde{y}_j - \tilde{a}_j) + a_j &= \varphi(\tilde{x}_j) + a_j \\
&\leq x_j + a_j = y_j \\
&\leq \psi(\tilde{x}_j) + a_j = \psi(\tilde{y}_j - \tilde{a}_j) + a_j = \psi_a(\tilde{y}_j).
\end{aligned}$$

Hence we have

$$G_a = \{(y_1, \ldots, y_n) \in \mathbb{R}^n \mid \varphi_a(\tilde{y}_j) \leq y_j \leq \psi_a(\tilde{y}_j), \, \tilde{y} + \tilde{a}_j + w\}$$

and therefore G_a is a normal domain with respect to the e_j-axis too.

9. We can represent G as

$$G = \{(x_1, x_2) \in \mathbb{R}^2 \mid x_1^2 - 1 \leq x_2 \leq \frac{1}{2} \cos(\frac{\pi}{2} x_1), \ x_1 \in ([-1, 1]\}$$

which is a normal domain with respect to x_2-axis. Now we represent G as a normal domain with respect to the x_1-axis. We choose as domain for the functions φ and ψ the interval $[-1, \frac{1}{2}]$ and we note that in this case, due to the symmetry, note that $x_1 \mapsto x_1^2 - 1$ and $x_1 \mapsto \frac{1}{2} \cos(\frac{\pi}{2} x_1)$ are even functions, we must have

$$G = \{(x_1, x_2) \in \mathbb{R}^2 \mid -\psi(x_2) \leq x_1 \leq \psi(x_2)\}.$$

By solving the equations $x_2 = x_1^2 - 1, x_2 = \frac{1}{2} \cos(\frac{\pi}{2} x_1)$ for $x_2 \in [-1, 0]$ and $x_2 \in [0, \frac{1}{2}]$, respectively we find the continuous function $\psi : [-1, \frac{1}{2}] \to \mathbb{R}$ as

$$\psi(x_2) := \begin{cases} \sqrt{x_2 + 1}, x_2 \in [-1, 0] \\ \frac{2}{\pi} \arccos(2x_2), & x_2 \in [0, \frac{1}{2}] \end{cases}.$$

10. We have

$$\int_G f(x)dx = \int_0^2 \left(\int_0^{\sqrt{x_1^2+2x_1}} (x_1 x_2^3)\ dx_2 \right) dx_1$$

$$= \int_0^2 x_1 \left(\frac{1}{4}x_2^4 \bigg|_0^{\sqrt{x_1^2+2x_1}} \right) dx_1$$

$$= \frac{1}{4} \int_0^2 x_1(x_1^2 + 2x_1)^2 dx_1$$

$$= \frac{1}{4} \int_0^2 (x_1^5 + 4x_1^4 + 4x_1^3)dx_1 = \frac{196}{15}.$$

11.   a) The domain of integration in $\int_0^1 (\int_y^1 f(x,y)dx)dy$ is the set

$$\{(x,y) \in \mathbb{R}^2 \mid y \le x \le 1, 0 \le y \le 1\} = \{(x,y) \in \mathbb{R}^2 | 0 \le y \le x,\ 0 \le x \le 1\}$$

and therefore

$$\int_0^1 (\int_y^1 f(x,y)dx)dy = \int_0^1 (\int_0^x f(x,y)\ dy)dx.$$

b) The integral $\int_1^2 (\int_1^{q^2} g(p,q)dp)dq$ is taken over the set

$$\{(p,q) \in \mathbb{R}^2 \mid 1 \le p \le q^2,\ 1 \le q \le 2\} = \{(p,q) \in \mathbb{R}^2 \mid \sqrt{p} \le q \le 2, 1 \le p \le 4\}$$

which yields

$$\int_1^2 (\int_1^{q^2} g(p,q)\ dp)dq = \int_1^4 (\int_{\sqrt{p}}^2 g(p,q)\ dq)dp$$

c) Here we integrate over

$$\{(t,r) \in \mathbb{R}^2 \mid 0 \le t \le \sin r, 0 \le r \le \frac{\pi}{2}\} = \{(t,r) \in \mathbb{R}^2 \mid \arcsin t \le r \le \frac{\pi}{2}, 0 \le t \le 1\}$$

and it follows that

$$\int_0^{\frac{\pi}{2}} \left(\int_0^{\sin r} h(t,r)\ dt \right) dr = \int_0^1 \left(\int_{\arcsin t}^{\frac{\pi}{2}} h(t,r)\ dr \right) dt.$$

12. We note for all $l \in \mathbb{N}_0$ that

$$\int_0^1 (1-x)^l dx = \int_0^1 y^l dy = \frac{1}{l+1} = \frac{0!l!}{(0+l+1)!}.$$

Now we assume the following for $k-1$ and all $l \in \mathbb{N}_0$

$$\int_0^1 x^{k-1}(1-x)^l dx = \frac{(k-1)!l!}{(k+1)!}.$$

721

Consider

$$\int_0^1 x^k(1-x)^l\,dx = \int_0^1 x^k\left(\frac{-1}{l+1}\frac{d}{dx}(1-x)^{l+1}\right)dx$$

$$= \frac{-x^k(1-x)^{l+1}}{l+1}\Big|_0^1 - \int_0^1\left(\frac{d}{dx}x^k\right)\left(\frac{-1}{l+1}(1-x)^{l+1}\right)dx$$

$$= \frac{k}{l+1}\int_0^1 x^{k-1}(1-x)^{l+1}\,dx = \frac{k}{l+1}\frac{(k-1)!(l+1)!}{(k+l+1)!}$$

$$= \frac{k!l!}{(k+l+1)!}.$$

Next we note that with $G = \{(x,y)\in\mathbb{R}^2 \mid 0\le y\le 1-x,\ 0\le x\le 1\}$ we have

$$\int_G x^n y^n\,dxdy = \int_0^1\left(\int_0^{1-x} x^n y^m\,dy\right)dx$$

$$= \int_0^1 x^n\left(\int_0^{1-x} y^m\,dy\right)dx$$

$$= \int_0^1 x^n\left(\frac{1}{m+1}y^{m+1}\Big|_0^{(1-x)}\right)dx$$

$$= \frac{1}{m+1}\int_0^1 x^n(1-x)^{m+1}\,dx$$

$$= \frac{1}{m+1}\frac{n!(m+1)!}{(n+m+2)!} = \frac{n!m!}{(n+m+2)!}$$

Chapter 21

1. a) If $\varphi = U \to V$ is a C^k-diffeomorphism for the two open sets $U,V\subset\mathbb{R}^n$, then for every point $x_0\in U$ the differential $d_{x_0}\varphi$ is invertible, Theorem 10.12 applies for all points in U, and the differential of φ at x_0 can be calculated according to (10.53) which also implies that $\varphi^{-1}:V\to U$ belongs to the class C^k.

    Conversely, Theorem 10.12 states that every point $x_0\in G$ for which $d_{x_0}g$ is invertible admits an open neighbourhood $U(x_0)\subset G$ such that $\varphi:U(x_0)\to\varphi(U(x_0))$ is a diffeomorphism (of class C^k).

    b) Since G is open, for $x_0\in G$ we can find $\varepsilon>0$ such that $B_\varepsilon(x_0)\subset G$. The open ball $B_\varepsilon(x_0)$ is an open neighbourhood of x_0 and therefore it remains to prove that every open ball $B_\varepsilon(y)\subset\mathbb{R}^n$ is diffeomorphic to $B_1(0)\subset R^n$. Every translation $\tau_a:\mathbb{R}^n\to\mathbb{R}^n, \tau_a(x)=x+a$, is a C^∞-diffeomorphism since $J_{\tau_a}=id_{\mathbb{R}^n}$. Further, the composition of two C^k-diffeomorphisms is a C^k-diffeomorphism and it remains to prove that for every $\rho>0$ the balls $B_\rho(0)$ and $B_1(0)$ are diffeomorphic. The mapping $h_\rho:B_1(0)\to B_\rho(0), h_\rho(x)=\rho x$ is a C^∞-mapping with Jacobi matrix ρid_n which is for $\rho>0$ invertible with inverse $\frac{1}{\rho}id_n$. Moreover h_ρ maps $B_1(0)$ bijectively onto $B_\rho(0)$.

2. a) We claim that $\varphi_{a,b,c} : \mathbb{R}^3 \to \mathbb{R}^3, \varphi_{a,b,c}(x,y,z) = (ax, by, cz)$, is a C^∞-diffeomorphism. First we note that $\varphi_{a,b,c}$ is a C^∞-mapping since it is linear. Hence for all $x_0 \in \mathbb{R}^3$ the differential of $\varphi_{a,b,c}$ at x_0 is $\varphi_{a,b,c}$. The matrix representation of $\varphi_{a,b,c}$ (with respect to the canonical basis) is $\begin{pmatrix} a & 0 & 0 \\ 0 & b & 0 \\ 0 & 0 & c \end{pmatrix}$ which has determinant $abc > 0$ by our assumption. Thus $\varphi_{a,b,c}$ is a C^∞-diffeomorphism from $\mathbb{R}^3$ onto itself. The image of $B_1(0)$ is calculated as follows: we know that $(x,y,z) \in B_1(0)$ if and only if $x^2 + y^2 + z^2 < 1$. Thus $(\xi, \eta, \zeta) \in \varphi_{a,b,c}(B_1(0))$, i.e. $(\xi, \eta, \zeta) = \varphi_{a,b,c}(x,y,z) = (ax, by, cz)$, if and only if

$$\frac{\xi^2}{a^2} + \frac{\eta^2}{b^2} + \frac{\zeta^2}{c^2} = \frac{(ax)^2}{a^2} + \frac{(by)^2}{b^2} + \frac{(cz)^2}{c^2} = x^2 + y^2 + z^2 < 1,$$

i.e. $(\xi, \eta, \zeta) \in \mathcal{E}_{a,b,c}$. Thus $\varphi_{a,b,c}^{-1} : \mathcal{E}_{a,b,c} \to B_1(0)$ is a C^∞-diffeomorphism and of course $\varphi^{-1}(\xi, \eta, \zeta) = \left(\frac{\xi}{a}, \frac{\eta}{b}, \frac{\zeta}{c} \right)$.

b) Note that

$$\mathbb{R}^2 \backslash \{0\} = \{ r \begin{pmatrix} \cos\varphi \\ \sin\varphi \end{pmatrix} \mid 0 < r < \infty, \varphi \in [0, 2\pi) \}$$

and

$$B_2 \backslash \overline{B_{\frac{1}{2}}(0)} = \{ \begin{pmatrix} \cos\varphi \\ \sin\varphi \end{pmatrix} \mid \frac{1}{2} < r < 2, \varphi \in [0, 2\pi) \}.$$

Therefore we only need to find a way of mapping the interval $(\frac{1}{2}, 2)$ bijectively with a C^1-mapping having a C^1 inverse onto the half line $(0, \infty)$. Clearly, there are many possibilities. One is given by $g : (\frac{1}{2}, 2) \to (0, \infty)$ where

$$g(r) := \begin{cases} \frac{1}{2-r}, 1 \le r \le 2 \\ -2r^2 + 5r - 2, \frac{1}{2} < r \le 1. \end{cases}$$

Note that if we want to construct a C^k-diffeomorpohism we must work with a polynomial of degree $k + 1$.

3. According to Theorem 13.19 a supporting hyperplane passes through every boundary point of $\overline{G}$. Since by assumption $\overline{G}$ admits a finite number of supporting hyperplanes, the boundary of $\overline{G}$ is the finite union of compact sets each belonging to a hyperplane. Hence it must have Jordan content, thus Lebesgue measure zero, implying that $\overline{G}$ is measurable.

The convex hull of a finite number of points belongs either to a hyperplane, hence has Jordan content zero, or when the interior is non-empty its boundary is the finite union of compact sets each belonging to a hyperplane.

4. With $k = 2l$ we have

$$I_k = I_{2l} = \frac{(2l)!\pi}{2^{2l}(l!)^2} \text{ and } I_{k-1} = I_{2l-1} = \frac{2^{2l-1}((l-1)!)^2}{(2l-1)!}.$$

723

Thus we have

$$I_{k-1}I_k = I_{2l-1}I_{2l} = \frac{2^{2l-1}((l-1)!)^2(2l)!\pi}{(2l-1)!2^{2l}(l!)^2}$$

$$= \frac{2^{-1}(l-1)!(l-1)!(2l-1)!2l\pi}{(2l-1)!(l-1)!(l-1)!l!^2} = \frac{\pi}{l} = \frac{2\pi}{k}.$$

The case $k-1 = 2l, k = 2l+1$ goes analogously.

5. We use spherical coordinates in $\mathbb{R}^n$ and we set for a moment

$$\Omega_n := \int_0^\pi \cdots \left(\int_0^\pi \left(\int_0^{2\pi} \left(\sin^{n-2}\vartheta_1 \sin^{n-3}\vartheta_2 \ldots \sin\vartheta_{n-2} \right) d\vartheta_{n-1} \right) d\vartheta_{n-2} \right) \ldots d\vartheta_1.$$

For f as in the problem it follows

$$\int_{B_R(0)} f^2(x)dx = \int_0^R \left(\int_0^\pi \cdots \left(\int_0^\pi \left(\int_0^{2\pi} g^2(r)r^{n-1}\sin^{n-2}\vartheta_1 \sin^{n-3}\vartheta_2 \ldots \right. \right. \right.$$

$$\left. \left. \left. \ldots \sin\vartheta_{n-2}d\vartheta_{n-1} \right) d\vartheta_{n-2} \right) \ldots d\vartheta_1 \right) dr$$

$$= \Omega_n \int_o^R g^2(r)r^{n-1}dr.$$

Consider now the remaining integral:

$$\int_0^R g^2(r)r^{n-1}dr = \int_0^R \frac{1}{n}\left(\frac{dr^n}{dr}\right)g^2(r)dr$$

$$= \frac{1}{n}r^n g^2(r)\Big|_0^R - \frac{1}{n}\int_0^R r^n \frac{d}{dr}(g^2(r))dr$$

$$= -\frac{2}{n}\int_0^R r^n g(r)g'(r)dr$$

$$= -\frac{2}{n}\int_0^R (r^{\frac{n-1}{2}}g(r))(r^{\frac{n+1}{2}}g'(r))dr$$

$$\leq \frac{2}{n}\left(\int_0^R r^{n-1}g^2(r)dr\right)^{1/2}\left(\int_0^R r^{n+1}(g'(r))^2dr\right)^{1/2}$$

$$= \frac{2}{n\Omega_n^{1/2}}\left(\int_{B_R(0)} f^2(x)dx\right)^{1/2}\left(\int_0^R r^{n+1}(g'(r))^2dr\right)^{1/2}$$

implying

$$\int_{B_R(0)} f^2(x)dx \leq \frac{2\Omega_n^{1/2}}{n}\left(\int_{B_R(0)} f^2(x)dx\right)^{1/2}\left(\int_0^R r^{n+1}(g'(r))^2dr\right)^{1/2}$$

or

$$\left(\int_{B_R(0)} f^2(x)dx \right)^{1/2} \le \frac{2\Omega_n^{1/2}}{n} R^{\frac{n+1}{2}} \left(\int_0^R (g'(r))^2 dr \right)^{1/2}.$$

6. a) The area of G is of course $\text{vol}_2(G) = \int_G 1 dx dy$. We use polar coordinates to find $G = \{(r, \varphi) \in [0, \infty) \times [0, 2\pi] \mid 0 \le r \le 3, \frac{\pi}{6} \le \varphi \le \frac{5\pi}{6}\}$, see the figure below, and it follows that

$$\text{vol}_2(G) = \int_0^3 \left(\int_{\frac{\pi}{6}}^{\frac{5\pi}{6}} r d\varphi \right) dr = \int_0^3 \frac{2\pi}{3} r dr = \frac{2\pi}{3} \frac{r^2}{2} \Big|_0^3 = 3\pi.$$

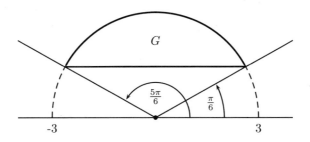

b) Using cylindrical coordinates we find

$$\text{vol}_3(H) = \int_H 1 dx dy dz = \int_0^h \left(\int_0^{2\pi} \left(\int_r^R \rho d\rho \right) d\varphi \right) dz$$

$$= 2\pi h \int_r^R \rho d\rho = \pi h (R^2 - r^2).$$

7. Again we use polar coordinates and we get

$$\text{vol}_2(G) = \int_G 1 dx dy = \int_{\varphi_1}^{\varphi_2} \int_0^{\rho(\varphi)} r dr d\varphi = \frac{1}{2} \int_{\varphi_1}^{\varphi_2} \rho(\varphi)^2 d\varphi.$$

Note that G should be seen as a normal domain given in polar coordinates.

8. The function $(x, y) \mapsto e^{\frac{y-x}{x+y}}$ is difficult to integrate either with respect to x or with respect to y. Thus writing the integral as an iterated integral will not help. However the function $(s, t) \mapsto e^{\frac{s}{t}}$ is easy to integrate with respect to s and the proposed change of coordinate will lead to this function. First we study the mapping $A : \mathbb{R}^2 \to \mathbb{R}^2, \binom{u}{v} = A\binom{x}{y}$ with $A = \begin{pmatrix} -1 & 1 \\ 1 & 1 \end{pmatrix}$ and $\det A = -2$. The line $x = 0$ is mapped onto line $s = t$, the line $y = 0$ is mapped onto the line $s = -t$, and the

725

line $x + y = 1$ is mapped onto the line $t = 1$. Hence A maps the triangle G with vertices $(0,0)$, $(1,0)$, $(0, 1)$ onto the triangle T with vertices $(0,0)$, $(1,1)$, $(-1,1)$

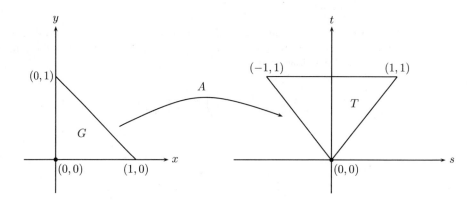

Since we want to evaluate $\int_G e^{\frac{y-x}{x+y}} dx dy$, we have to use A^{-1}:

$$\int_G e^{\frac{y-x}{x+y}} dx dy = \int_{A^{-1}T} e^{\frac{y-x}{x+y}} dx dy = \int_T e^{s/t} \,|\det A^{-1}|\, ds dt$$

$$= \frac{1}{2} \int_0^1 \left(\int_{-t}^t e^{s/t} ds \right) dt = \frac{1}{2} \int_0^1 \left(t e^{s/t} \Big|_{-t}^t \right) dt$$

$$= \frac{1}{2} \int_0^1 (e - e^{-1}) t dt = \frac{e - e^{-1}}{4} = \frac{1}{2} \sinh(1).$$

9. We use the modified spherical coordinates

$$x = 5r \sin \vartheta \cos \varphi, \, y = 4r \sin \vartheta \sin \varphi, \, z = 3r \cos \vartheta$$

which leads to the Jacobi determinant $60r^2 \sin \vartheta$ and it follows that

$$\int_{\mathcal{E} \cap \{z \geq 0\}} h(x, y, z) dx dy dz = \int_0^1 \left(\int_0^{\pi/2} \left(\int_0^{2\pi} 60(1 - r^2)^{1/2} r^2 \sin \vartheta d\varphi \right) d\vartheta \right) dr$$

$$= 120\pi \int_0^1 (1 - r^2)^{1/2} r^2 dr$$

$$= 120\pi \left(-\frac{r}{4}(1 - r^2)^{3/2} + \frac{1}{8}(r(1 - r^2)^{1/2} + \arcsin r) \right) \Big|_0^1$$

$$= 120\pi \arcsin 1 = 60\pi^2.$$

726

10. The integral we have to evaluate is

$$\int_G (x(s^2+t^2)^{1/2} + y^2(s^2+t^2)^{3/2})dxdydsdt$$

$$\int_2^7 \left(\int_3^6 \left(\int_{B_2(0)} (x(s^2+t^2)^{1/2} + y^2(s^2+t^2)^{7/2})dsdt \right) dy \right) dx.$$

We introduce polar coordinates for the inner integral to find

$$\int_2^7 \left(\int_3^6 \left(\int_{B_2(0)} (x(s^2+t^2)^{1/2} + y^2(s^2+t^2)^{3/2})dsdt \right) dy \right) dx$$

$$= \int_2^7 \left(\int_3^6 \left(\int_0^{2\pi} \left(\int_0^2 (x\rho + y^2\rho^3)\rho d\rho \right) d\varphi \right) dy \right) dx$$

$$= 2\pi \int_2^7 \left(\int_3^6 \left(x\frac{\rho^3}{3} \Big|_0^2 + y^2\frac{\rho^5}{5} \Big|_0^2 \right) dy \right) dx$$

$$= 2\pi \int_2^7 \left(\int_3^6 \frac{8}{3}x + \frac{32}{5}y^2 dy \right) dx$$

$$= 2\pi \int_2^7 8x\,dx + 2\pi \int_3^6 32y^2dy = 4392\pi.$$

11. We first prove that $N(x) = \frac{M}{\|x\|}$ and then we will justify that $\|x - \xi\|^2 = \|x\|^2 + s^2 - 2\|x\|s\cos\vartheta$. In the spherical coordinates as introduced we have with $\|\xi\| = s$

$$N(x) = \int_{B_R(0)} \frac{\rho(\|\xi\|)}{\|x-\xi\|}d\xi = \int_0^{2\pi}\int_0^\pi\int_0^R \frac{\rho(s)s^2\sin\vartheta}{\sqrt{\|x\|^2+s^2-2\|x\|s\cos\vartheta}}dsd\vartheta dy$$

$$= 2\pi \int_0^R \left(\int_0^\pi \frac{\sin\vartheta}{\sqrt{\|x\|^2+s^2-2\|x\|s\cos\vartheta}}d\vartheta \mid \rho(s)s^2ds.$$

Using the substitution $\tau = -\cos\vartheta$ we find

$$\int_0^\pi \frac{\sin\vartheta d\vartheta}{\sqrt{\|x\|^2+s^2-2\|x\|s\cos\vartheta}} = \int_{-1}^1 \frac{d\tau}{\sqrt{\|x\|^2+s^2+2\|x\|s\tau}}$$

$$= \frac{1}{\|x\|s}\sqrt{\|x\|^2+s^2+2\|x\|s\tau} \Big|_{\tau=-1}^{\tau=1}$$

$$= \frac{1}{\|x\|s}\left(\sqrt{\|x\|^2+s^2+2\|x\|s} - \sqrt{\|x\|^2+s^2-2\|x\|s} \right)$$

$$= \frac{1}{\|x\|s}\left(\|x\|+s - (\|x\|-s) \right) = \frac{2}{\|x\|},$$

implying

$$N(x) = \frac{4\pi}{\|x\|}\int_0^R \rho(s)s^2ds.$$

On the other hand we find

$$M = \int_{B_R(0)} \rho(\|\xi\|)d\xi = \int_0^R \left(\int_0^{2\pi} \left(\int_0^\pi \rho(s)s^2 \sin\vartheta d\vartheta d\varphi ds \right. \right.$$

$$= 4\pi \int_0^R \rho(s)s^2 ds$$

which eventually yields

$$N(x) = \frac{M}{\|x\|}.$$

This example has of course a physical interpretation.: $N(x)$ is the Newton potential (up to a normalisation factor) originating from a body occupying the ball $\overline{B_R(0)}$ and having density ρ experienced at a point x.

It remains to prove $\|x-\xi\|^2 = \|x\|^2 + s^2 - 2\|x\|s\cos\vartheta$. For this we use the following figure

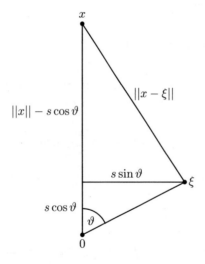

Pythagoras' theorem now yields

$$\|x-\xi\|^2 = (\|x\| - \cos\vartheta)^2 + (s\sin\vartheta)^2$$

$$= \|x\|^2 + s^2\cos^2\vartheta + s^2\sin^2\vartheta - 2\|x\|s\cos\vartheta$$

$$= \|x\|^2 + s^2 - 2\|x\|s\cos\vartheta.$$

(For this solution we use O. Forster [20]).

Chapter 22

1. a) A compact Jordan measurable exhaustion of $B_R(0) \subset \mathbb{R}^n$ is for example given by $(\overline{B_{(1-\frac{1}{k})R}(0)})_{k \in \mathbb{N}}$. Each of these closed balls is compact and Jordan measurable and included in $B_R(0)$, hence $\bigcup_{k \in \mathbb{N}} \overline{B_{(1-\frac{1}{k})R}(0)} \subset B_R(0)$. On the other hand, for $x \in B_R(0), \|x\| = \rho < R$, it follows that $R - \rho > 0$, i.e. for some $k_0 \in \mathbb{N}$ we have $R - \rho > \frac{1}{k_0}R$ or $\rho < (1 - \frac{1}{k_0})R$ implying that $x \in \overline{B_{(1-\frac{1}{k_0})R}(0)}$ which yields $B_R(0) \subset \bigcup_{k \in \mathbb{N}} \overline{B_{(1-\frac{1}{k})R}(0)}$.

    Further, for $\mathbb{R}^n$ a compact Jordan measurable exhaustion is given by $(\overline{B_k(0)})_{k \in \mathbb{N}}$. Again, each $\overline{B_k(0)}$ is compact and Jordan measurable and trivially included in $\mathbb{R}^n$, so

    $$\bigcup_{k \in \mathbb{N}} \overline{B_k(0)} \subset \mathbb{R}^n.$$

    Since $x \in \mathbb{R}^n$ is an element of $\overline{B_{[\|x\|]+1}(0)}$ the converse inclusion follows immediately.

    b) Look at the following figure:

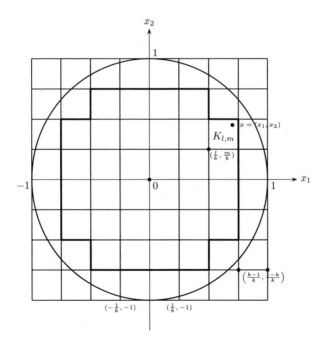

    For $k \geq 1$ we consider for the square $[-1, 1] \times [-1, 1]$ the partition induced when taking on $[-1, 1]$ the partition

729

$\{-1, -1 + \frac{1}{k}, \ldots, -\frac{1}{k}, 0, \frac{1}{k}, \ldots, 1 - \frac{1}{k}, 1\}$, see the figure above. The partition generates $(4k)^2$ compact squares $K_{l,m}^{(k)}$ with

$$K_{l,m}^{(k)} = \left[\frac{l}{k}, \frac{l+1}{k}\right] \times \left[\frac{m}{k}, \frac{m+1}{k}\right], \quad l, m = -k, \ldots, k-1.$$

We now define

$$G_k := \bigcup_{K_{l,m}^{(k)} \subset B_1(0)} K_{l,m}^{(k)} \subset B_1(0)$$

which is a compact Jordan measurable set contained in $B_1(0)$, hence $\bigcup_{k \in \mathbb{N}} G_k \subset B_1(0)$. Now let $(x_1, x_2) \in B_1(0)$. Since $x_1, x_2 \in (-1, 1)$ we can find $k \in \mathbb{N}$ such that for some $l, m \in \mathbb{Z}$, $-k \leq l < k$, $-k \leq m < k$, we have $\frac{l}{k} \leq x_1 \leq \frac{l+1}{k}$ and $\frac{m}{k} \leq x_2 \leq \frac{m+1}{k}$ implying that $(x_1, x_2) \in K_{l,m}^{(k)}$ and therefore we find $B_1(0) \subset \bigcup_{k \in \mathbb{N}} G_k$.

2. For every $k \in \mathbb{N}$ the sets $[-k, -\frac{1}{k}] \cup [\frac{1}{k}, k]$ and $[-k, -\frac{1}{k^2}] \cup [\frac{1}{k}, k]$ are compact and Jordan measurable. Further, for $x \in \mathbb{R} \setminus \{0\}$ there exists $k_1, k_2 \in \mathbb{N}$ such that $x \in G_{k_1}$ and $x \in H_{k_2}$. Thus $(G_k)_{k \in \mathbb{N}}$ and $(H_k)_{k \in \mathbb{N}}$ are compact Jordan measurable exhaustions of $\mathbb{R} \setminus \{0\}$. Further, the function $x \mapsto \frac{1}{x^3}$ is continuous on $\mathbb{R} \setminus (0)$ and for G_k we find

$$\int_{G_k} \frac{1}{x^3} dx = \int_{-k}^{-1/k} \frac{1}{x^3} dx + \int_{1/k}^{k} \frac{1}{x^3} dx$$

$$= -\frac{1}{2} \left(\frac{1}{x^2} \bigg|_{-k}^{-\frac{1}{k}} + \frac{1}{x^2} \bigg|_{1/k}^{k} \right)$$

$$= -\frac{1}{2} \left(k^2 - \frac{1}{k^2} + \frac{1}{k^2} - \frac{1}{k^2} \right) = 0$$

and therefore it follows that

$$\lim_{k \to \infty} \int_{G_k} \frac{1}{x^3} dx = 0.$$

However for H_k we have

$$\int_{H_k} \frac{1}{x^3} dx = \int_{-k}^{-1/k^2} \frac{1}{x^3} dx + \int_{1/k}^{k} \frac{1}{x^3} dx$$

$$= -\frac{1}{2} \left(\frac{1}{x^2} \bigg|_{-k}^{-1/k^2} + \frac{1}{x^2} \bigg|_{1/k}^{k} \right)$$

$$= -\frac{1}{2} \left(k^4 - \frac{1}{k^2} + \frac{1}{k^2} - k^2 \right) = \frac{k^2 - k^4}{2}.$$

Since $\lim_{k \to \infty} \frac{k^2 - k^4}{2} = -\infty$ it follows that $\lim_{k \to \infty} \int_{H_k} \frac{1}{x^3} dx$ does not exist and the function $x \mapsto \frac{1}{x^3}$ is not improper Riemann integrable over $\mathbb{R} \setminus \{0\}$.

730

3. Since $f(x) \geq 0$ for all $x \in \mathbb{R}^n$, assuming $\int_{\mathbb{R}^n} f(x)dx$ exists as a finite improper integral, it will follow that

$$\int_{\mathbb{R}^n} f(x)dx \geq \int_{\bigcup G_k} f(x)dx \geq \int_{G_k} f(x)dx \geq \int_{G_k} c \, dx = c \operatorname{vol}_n(G_k).$$

By assumption we have $\lim_{k\to\infty} \operatorname{vol}_n(G_k) = \infty$ which gives a contradiction. Hence $\int_{\mathbb{R}^n} f(x)dx$ cannot exist as finite improper Riemann integral.

4. Each integral is non-negative and we will use Proposition 22.10. The set G corresponds to $\mathbb{R}^n$ and once we have proved (22.6) to hold for all closed balls $\overline{B_R(0)}$ it follows for all Jordan measurable sets $H \subset \mathbb{R}^n$ such that $\overline{H}$ is compact. Of course we can now use the calculation of Example 21.15 and Example 21.16. Thus we find

$$\int_{\overline{B_R(0)}} e^{-x_1^2 - x_2^2}dx = \pi - \pi e^{-R^2} \leq \pi$$

and for $n + \alpha < 0$

$$\int_{\overline{B_R(0)} \backslash B_1(0)} \|x\|^\alpha dx = \frac{n\omega_n}{n+\alpha}(R^{n+\alpha} - 1) \leq -\frac{n\omega_n}{n+\alpha}.$$

Finally for $n + \alpha < 0$ we consider

$$\int_{\overline{B_R(0)}} (1 + \|x\|^2)^{\alpha/2}dx = n\omega_n \left(\int_0^1 (1+r^2)^{\alpha/2} r^{n-1}dr + \int_1^R (1+r^2)^{\alpha/2} r^{n-1}dr \right).$$

The first integral on the right hand side is estimated as follows: since $\alpha < 0$ we have $(1+r^2)^{\alpha/2} \leq 1$ and on $[0,1]$ we have also $r^{n-1} \leq 1$, hence

$$\int_0^1 (1+r^2)^{\alpha/2} r^{n-1}dr \leq \int_0^1 1 dr = 1.$$

For the second integral we find

$$\int_1^R (1+r^2)^{\alpha/2} r^{n-1}dr \leq \int_1^R r^{\alpha+n-1} dr = \frac{1}{\alpha+n} r^{\alpha+n}\Big|_1^R$$

$$= \frac{1}{\alpha+n}(R^{\alpha+n} - 1) \leq -\frac{1}{\alpha+n}$$

and eventually we have

$$\int_{\overline{B_R(0)}} (1 + \|x\|^2)^{\alpha/2}dx \leq n\omega(1 - \frac{1}{\alpha+n}).$$

731

5. Since $\|f\|_{L^1} := \int_{\mathbb{R}^n} |f(x)|dx < \infty$ we find for every Jordan measurable set $H \subset \mathbb{R}^n$ such that $\overline{H}$ is compact that

$$\int_H |\cos\langle x, \xi\rangle f(x)|dx \leq \int_H |f(x)|dx \leq \|f\|_{L^1}$$

which implies by Proposition 22.10 that for every $\xi \in \mathbb{R}^n$ the integral $\tilde{u}(\xi)$ exists and is an absolutely convergent improper integral. Hence $\tilde{u} : \mathbb{R}^n \to \mathbb{R}$ is a well defined function. Further, we can deduce from Theorem 22.14 that for every compact set $\overline{H} \subset \mathbb{R}^n$ the function $\tilde{u}_{\overline{H}} : \mathbb{R}^n \to \mathbb{R}$ defined by

$$\tilde{u}_{\overline{H}}(\xi) := \int_{\overline{H}} \cos\langle x, \xi\rangle f(x)dx$$

is continuous. Now we take the compact Jordan measurable exhaustion $(\overline{B_N(0)})_{N \in \mathbb{R}^n}$ of $\mathbb{R}^n$ and with $\tilde{u}_N := \tilde{u}_{\overline{B_N(0)}}$ we find

$$|\tilde{u}(\xi) - \tilde{u}_N(\xi)| = |\int_{\mathbb{R}^n} \cos\langle x, \xi\rangle f(x)dx - \int_{B_N(0)} \cos\langle x, \xi\rangle f(x)dx| \leq \int_{B_N^{\complement}(0)} |f(x)|dx.$$

Since

$$\int_{B_N^{\complement}(0)} |f(x)|dx = \int_{\mathbb{R}^n} |f(x)|dx - \int_{B_N(0)} |f(x)|dx,$$

but $\lim_{N \to \infty} \int_{B_N(0)} |f(x)|dx = \int_{\mathbb{R}^n} |f(x)|dx$, we find for $\varepsilon > 0$ a natural number $N = N(\varepsilon) \in \mathbb{N}$ such that $N \geq N(\varepsilon)$ implies for all $\xi \in \mathbb{R}^n$ that $|\tilde{u}(\xi) - \tilde{u}_N(\xi)| < \varepsilon$. In other words $\tilde{u}$ is the uniform limit of continuous functions, hence continuous.

6. In light of Problem 5 we need to prove that $e^{-\|\cdot\|^\alpha}$ is improper integrable over $\mathbb{R}^n$. Since $e^{-\|\cdot\|^\alpha} \geq 0$ for all $x \in \mathbb{R}^n$, if we can estimate this function against an absolutely improper Riemann integrable function, the result will follow. We know however that the function $x \mapsto (1 + \|x\|^2)^{\frac{-(n+1)}{2}}$ is absolutely improper integrable and

$$e^{-\|x\|^\alpha} = ((1 + \|x\|^2)^{\frac{n+1}{2}} e^{-\|x\|^\alpha})(1 + \|x\|^2)^{\frac{-(n+1)}{2}}.$$

Since the function $r \mapsto (1+r^2)^{\frac{n+1}{2}} e^{-r^\alpha}$ is bounded if follows that $e^{-\|\cdot\|^\alpha}$ is improper integrable over $\mathbb{R}^n$.

7. When we consider on the right hand side of (22.14) the integral $\frac{\partial f}{\partial x^{(j)}}$ as a new function, say $g, (x, y) \mapsto \frac{\partial f}{\partial x^{(j)}}(y, x)$, then g must satisfy the conditions of Theorem 22.17 for $1 \leq l \leq m$ in order that

$$\frac{\partial^2 F}{\partial x^{(l)}\partial x^{(j)}}(x) = \frac{\partial}{\partial x^{(l)}} \int_G g(y, x)dy$$

$$= \int_G \frac{\partial g}{\partial x^{(l)}}(y, x)dx = \int_G \frac{\partial^2 f}{\partial x^{(l)}\partial x^{(j)}}(y, x)dy$$

holds. This means that we require that the continuous function $f : G \times K \to \mathbb{R}$ has continuous partial derivative $\frac{\partial^2 f}{\partial x^{(l)} \partial x^{(j)}}$ for $1 \le j, l \le m$. Now if $\alpha \in \mathbb{N}_0^m$ is any multi-index, then the requirement for

$$\partial_x^\alpha F(x) = \int_G \partial_x^\alpha f(y, x) dy$$

to hold is that $\partial_x^\alpha f$ is a continuous function on $G \times K$. Note that we still assume G and K to be compact and K to be convex too.

8. Since $\operatorname{supp} u$ is a compact, say $\operatorname{supp} u = G$, we find that

$$\tilde{u}(\xi) = \int_{\mathbb{R}^n} \cos\langle x, \xi \rangle u(x) dx = \int_G \cos\langle x, \xi \rangle u(x) dx.$$

Now for $\xi \in \overline{B_R(0)} \subset \mathbb{R}^n$, a compact set, we can apply Theorem 22.17 in the modified form of Problem 7 to the C^∞-function $\partial_\xi^\alpha(\cos\langle x, \xi \rangle u(x))$. Note that $\partial_\xi^\alpha(\cos\langle x, \xi \rangle u(x)) = c_\alpha \xi^\alpha (trig_\alpha\langle x, \xi \rangle) u(x)$, where $c_\alpha \in \{-1, +1\}$ and $\operatorname{trig}_\alpha \in \{\sin, \cos\}$. Thus we find that

$$(*) \qquad \partial_\xi^\alpha \tilde{u}(\xi) = \int_G \partial_\xi^\alpha(\cos\langle x, \xi \rangle) u(x) dx, \ \xi \in \overline{B_R(0)}.$$

However, given $\xi \in \mathbb{R}^n, \xi \in \overline{B_{2\|\xi\|}(0)}$, then $(*)$ holds for all $\xi \in \mathbb{R}^n$ and $\tilde{u} \in C^\infty(\mathbb{R}^n)$.

9. First we note that the conditions on u imply that $\tilde{u}$ is well defined and a continuous function, see Problem 5. Define $\tilde{u}_N$ in a similar way as in Problem 5 by

$$\tilde{u}(\xi) = \int_{\overline{B_N(0)}} \cos\langle x, \xi \rangle u(x) dx.$$

Using Theorem 22.17 we find first for $\overline{\xi \in B_R(0)}, R > 0$, and hence for all $\xi \in \mathbb{R}^n$, that

$$\frac{\partial \tilde{u}_N}{\partial \xi_j}(\xi) = \int_{\overline{B_N(0)}} \frac{\partial}{\partial \xi_j} \cos\langle x, \xi \rangle u(x) dx = - \int_{\overline{B_N(0)}} x_j \sin\langle x, \xi \rangle u(x) dx.$$

Further we know that

$$(*) \qquad \xi \mapsto \int_{\mathbb{R}^n} (-x_j \sin\langle x, \xi \rangle) u(x) dx$$

is a continuous function on $\mathbb{R}^n$. This follows as in Problem 5 with f being replaced by $g, g(x) = -x_j u(x)$, recall that by our assumption $\|g\|_{L^1} < \infty$, when we note that in the calculations and arguments leading to the solution of Problem 5 we can replace $\cos$ by $\sin$.

Now we prove that $\frac{\partial \tilde{u}_N}{\partial \xi_j}$ converges uniformly to $(*)$. We have

$$\left| \frac{\partial \tilde{u}_N}{\partial \xi_j}(\xi) - \int\limits_{\mathbb{R}^n} (-x_j \sin\langle x, \xi\rangle u(x))dx \right|$$

$$= \left| \int\limits_{B_N^{\complement}(0)} x_j \sin\langle x, \xi\rangle u(x)dx \right|$$

$$= \int\limits_{B_N^{\complement}(0} |x_j||u(x)|dx$$

Since

$$\lim_{N\to\infty} \int\limits_{\overline{B_N(0)}} |x_j||u(x)|dx = \int\limits_{\mathbb{R}^n} |x_j||u(x)|dx$$

which follows from the solution of Problem 5 when f is replaced by $g_j(x) = x_j u(x)$, we conclude that

$$\lim_{N\to\infty} \int\limits_{B_N^{\complement}(0)} |x_j||u(x)|dx = 0,$$

thus $\left(\frac{\partial \tilde{u}_N}{\partial \xi_j}\right)_{N\in\mathbb{N}}$ converges on $\mathbb{R}^n$ uniformly to $(*)$. Eventually we want to prove that $\frac{\partial u}{\partial \xi_j}$ exists and equals $(*)$. For this we consider the corresponding (partial) difference quotient:

$$\frac{\tilde{u}(\xi + he_j) - \tilde{u}(\xi)}{h} = \frac{\tilde{u}(\xi + he_j) - \tilde{u}_N(\xi + he_j) - \tilde{u}(\xi) + \tilde{u}_N(\xi)}{h}$$

$$+ \frac{\tilde{u}_N(\xi + he_j) - \tilde{u}_N(\xi)}{h}$$

$$= \frac{1}{h} \int\limits_{B_N^{\complement}(0)} \cos\langle x, \xi + he_j\rangle u(x)dx - \frac{1}{h} \int\limits_{B_N^{\complement}(0)} \cos\langle x, \xi\rangle u(x)dx$$

$$+ \frac{\tilde{u}_N(\xi + he_j) - \tilde{u}_N(\xi)}{h}$$

$$= \frac{1}{h} \int\limits_{B_N^{\complement}(0)} \cos\langle x, \xi + he_j\rangle u(x)dx - \frac{1}{h} \int\limits_{B_N^{\complement}(0)} \cos\langle x, \xi\rangle u(x)dx$$

$$+ \frac{1}{h} \int\limits_{\xi_j}^{\xi_j+h} \frac{\partial \tilde{u}_N(\xi_1,\ldots,\xi_{j-1},\eta,\xi_{j+1},\ldots,\xi_n)}{d\eta} d\eta.$$

For fixed h we can pass to the limit as N tends to infinity and we obtain

$$\frac{\tilde{u}(\xi + he_j) - \tilde{u}(\xi)}{h} = \frac{1}{h} \int\limits_{\xi_j}^{\xi_j+h} \left(\int\limits_{\mathbb{R}^n} (-x_j \sin\langle x, (\xi_1,\ldots,\xi_{j-1},\eta,\xi_{j+1},\ldots,\xi_n)\rangle u(x)dx \right) d\eta.$$

Now we can pass to the limit as h tends to zero to find

$$\frac{\partial \widetilde{u}}{\partial \xi_j}(\xi) = \int_{\mathbb{R}^n} (-x_j \sin\langle x, \xi\rangle) u(x) dx.$$

10. First we note that

$$(*) \qquad (u * v)(x) = \int_{\text{supp } v} u(x - y) v(y) dy$$

and applying the transformation theorem with $z = x - y$ we find

$$(u * v)(x) = \int_{\text{supp } u} u(z) v(x - z) dz.$$

From Theorem 22.14 we deduce that $(u * v)$ is continuous. Next recall that

$$\text{supp } u + \text{supp } v = \{a + b \mid a \in \text{supp } u, \; b \in \text{supp } v\}.$$

Now suppose that $x \notin \text{supp } u + \text{supp } v$. If $y \in \text{supp } v$ it follows that $x - y \notin \text{supp } u$ and therefore the integral in $(*)$ vanishes whenever $x \notin \text{supp } u + \text{supp } v$. Thus $u * v$ vanishes in $(\text{supp } n + \text{supp } v)^\complement$, i.e. we have proved that $\text{supp}(u*v) \subset \text{supp } u + \text{supp } v$.

11.   a) This part is relatively straightforward: $\int j(x)dx = 1, j(x) \geq 0$ and $\text{supp } j \subset \overline{B_1(0)}$ follows from definition. To prove that j is a C^∞-function, only the points on $S^{n-1} = \partial B_1(0)$ are of interest. For $\|x\| < 1$ we find that $\partial^\alpha j(x)$ is of the type $g(x)j(x)$ where g is a rational function with singularities of inverse power-type (a pole) on S^{n-1}. But such a singularity is controlled by the behaviour of $j(x)$ as $\|x\| \to 1$.

   b) It is clear that if $j \in C^\infty(\mathbb{R}^n)$ then $j_\varepsilon \in C^\infty(\mathbb{R}^n)$ and for $\|\frac{x}{\varepsilon}\| > 1$, i.e. $\|x\| > \varepsilon$, it follows that $j_\varepsilon(x) = 0$, so $\text{supp } j_\varepsilon \subset \overline{B_\varepsilon(0)}$. Clearly $j_\varepsilon \geq 0$. Using the transformation $\frac{x}{\varepsilon} = y$ we eventually find

$$\int_{\mathbb{R}^n} j_\varepsilon(x)dx = \int_{\text{supp } j_\varepsilon} j_\varepsilon(x)dx = \int_{B_\varepsilon(0)} \varepsilon^{-n} j(\frac{x}{\varepsilon})dx$$

$$= \int_{B_1(0)} \varepsilon^{-1} j(y)\varepsilon^n dy = \int_{\overline{B_1(0)}} j(y)dy = 1.$$

   c) The relation

$$\int_{\mathbb{R}^n} j_\varepsilon(x - y)u(y)dy = \int_{\mathbb{R}^n} u(x - y)j_\varepsilon(y)dy$$

follows from Problem 10 and $J_\varepsilon(u) \in C^\infty(\mathbb{R}^n)$ follows by an application of Problem 7. Now consider $J_\varepsilon u$, using the Cauchy-Schwarz inequality we get

$$|J_\varepsilon(u)(x)|^2 = \left(\int_{\mathbb{R}^n} u(x-y) j_\varepsilon(y) dy \right)^2$$

$$\leq \left(\int_{\mathbb{R}^n} |u(x-y)|^2 j_\varepsilon(y) dy \right) \left(\int_{\mathbb{R}^n} j_\varepsilon(y) dy \right)$$

$$= \int_{\mathbb{R}^n} |u(x-y)|^2 j_\varepsilon(y) dy.$$

Now we integrate with respect to x, noting that $\operatorname{supp} J_\varepsilon(u) \subset \overline{B_\varepsilon(0)} + \operatorname{supp} u$ which is a compact set, and we find

$$\int_{\mathbb{R}^n} |J_\varepsilon(u)(x)|^2 dx \leq \int_{\mathbb{R}^n} j_\varepsilon(y) \int_{\mathbb{R}^n} |u(x-y)|^2 dx dy$$

$$= \int_{\mathbb{R}^n} j_\varepsilon(y) \left(\int_{\mathbb{R}^n} |u(z)|^2 dz \right) dy$$

$$= \int_{\mathbb{R}^n} |u(z)|^2 dz,$$

where we used the transformation theorem to get

$$\int_{\mathbb{R}^n} |u(x-y)|^2 dx = \int_{y+\operatorname{supp} u} |u(x-y)|^2 dx = \int_{\operatorname{supp} u} |u(z)|^2 dz = \int_{\mathbb{R}^n} |u(z)|^2 dz.$$

Thus we have proved

$$\int_{\mathbb{R}^n} |J_\varepsilon(u)(x)|^2 dx \leq \int_{\mathbb{R}^n} |u(x)|^2 dx.$$

d) Estimating $|J_\varepsilon(u)(x) - u(x)|^2$ we find

$$J_\varepsilon(u)(x) - u(x) = \int_{\mathbb{R}^n} j_\varepsilon(y)(u(x-y) - u(x)) dy$$

implying

$$|J_\varepsilon(u)(x) - u(x)|^2 \leq \int_{\mathbb{R}^n} j_\varepsilon(y) |u(x-y) - u(x)|^2 dy.$$

Since $\operatorname{supp} u$ is compact we can find $R > 0$ such that $u|_{B_R^c(0)} = 0$ and therefore we have for $\|y\| \leq R$

$$\int_{\|x\| \geq 2R} |u(x-y)|^2 dx \leq \int_{\|x\| \geq R} |u(x)|^2 dx = 0.$$

736

Further, since u is continuous, for $\eta > 0$ and $\varepsilon = \varepsilon(R, \eta)$ sufficiently small it follows that

$$\sup_{\|y\| \leq \varepsilon} \int_{\|x\| \leq 2R} |u(x - y) - u(x)|^2 dx < \eta.$$

For these values of ε we have

$$\int_{\mathbb{R}^n} |J_\varepsilon(u)(x) - u(x)|^2 dx$$

$$\leq \sup_{\|y\| \leq \varepsilon} \left(\int_{\|x\| \geq 2R} |u(x - y) - u(x)|^2 dx + \int_{\|x\| \leq 2R} |u(x - y) - u(x)|^2 dx \right)$$

$$= \sup_{\|y\| \leq \varepsilon} \int_{\|x\| \leq 2R} |u(x - y) - u(x)|^2 dx < \eta$$

which implies

$$\lim_{\varepsilon \to 0} \int_{\mathbb{R}^n} |J_\varepsilon(u)(x) - u(x)|^2 dx = 0.$$

Note that with small modifications this proof works also for $u : \mathbb{R}^n \to \mathbb{R}, u^2$ improper Riemann integrable. The full power of this approximation we will see after having introduced the Lebesgue integral.

Solutions to Problems of Part 5

Chapter 23

1.   a) With $\operatorname{curl}(\operatorname{grad}\varphi) = \operatorname{curl}(\frac{\partial\varphi}{\partial x_1}, \frac{\partial\varphi}{\partial x_2}, \frac{\partial\varphi}{\partial x_3})$ the following holds

$$\operatorname{curl}\left(\frac{\partial\varphi}{\partial x_1}, \frac{\partial\varphi}{\partial x_2}, \frac{\partial\varphi}{\partial x_3}\right) = \left(\frac{\partial}{\partial x_2}(\frac{\partial\varphi}{\partial x_3}) - \frac{\partial}{\partial x_3}(\frac{\partial\varphi}{\partial x_2})\right)e_1$$

$$+ \left(\frac{\partial}{\partial x_3}(\frac{\partial\varphi}{\partial x_1}) - \frac{\partial}{\partial x_1}(\frac{\partial\varphi}{\partial x_3})\right)e_2 + \left(\frac{\partial}{\partial x_1}(\frac{\partial\varphi}{\partial x_2}) - \frac{\partial}{\partial x_2}(\frac{\partial\varphi}{\partial x_1})\right)e_3$$

$$= \left(\frac{\partial^2\varphi}{\partial x_2\partial x_3} - \frac{\partial^2\varphi}{\partial x_3\partial x_2}\right)e_1 + \left(\frac{\partial^2\varphi}{\partial x_3\partial x_1} - \frac{\partial^2\varphi}{\partial x_1\partial x_3}\right)e_2 + \left(\frac{\partial^2\varphi}{\partial x_1\partial x_2} - \frac{\partial^2\varphi}{\partial x_2\partial x_1}\right)e_3$$

$$= 0$$

since for a C^2-function φ we always have $\frac{\partial^2\varphi}{\partial x_j\partial x_k} = \frac{\partial^2\varphi}{\partial x_k\partial x_j}$. However, compare with Example 5.9, if φ is not a C^2-function we can in general not expect that mixed partial derivatives are independent of the order and hence the result need not hold.

  b) We find

$$\operatorname{div}(\operatorname{curl}A) = \operatorname{div}\left(\frac{\partial A_3}{\partial x_2} - \frac{\partial A_2}{\partial x_3}, \frac{\partial A_1}{\partial x_3} - \frac{\partial A_3}{\partial x_1}, \frac{\partial A_2}{\partial x_1} - \frac{\partial A_1}{\partial x_2}\right)$$

$$= \frac{\partial}{\partial x_1}\left(\frac{\partial A_3}{\partial x_2} - \frac{\partial A_2}{\partial x_3}\right) + \frac{\partial}{\partial x_2}(\frac{\partial A_1}{\partial x_3} - \frac{\partial A_3}{\partial x_1}) + \frac{\partial}{\partial x_3}(\frac{\partial A_2}{\partial x_1} - \frac{\partial A_1}{\partial x_2}))$$

$$= \frac{\partial^2 A_3}{\partial x_1\partial x_2} - \frac{\partial^2 A_2}{\partial x_1\partial x_3} + \frac{\partial^2 A_1}{\partial x_2\partial x_3} - \frac{\partial^2 A_3}{\partial x_2\partial x_1} + \frac{\partial^3 A_2}{\partial x_3\partial x_1} - \frac{\partial^2 A_1}{\partial x_3\partial x_2} = 0$$

provided all components of A are C^2-functions and in the contrary case the remark at the end of part a) applies.

2. We start by calculating $\operatorname{grad}(\operatorname{div} A)$ and $\operatorname{curl}(\operatorname{curl} A)$:

$$\operatorname{grad}(\operatorname{div}A) = (\frac{\partial}{\partial x_1}\operatorname{div}A)e_1 + (\frac{\partial}{\partial x_2}\operatorname{div}A)e_2 + (\frac{\partial}{\partial x_3}\operatorname{div}A)e_3$$

$$= \left(\frac{\partial^2 A_1}{\partial x_1^2} + \frac{\partial^2 A_2}{\partial x_1\partial x_2} + \frac{\partial^2 A_3}{\partial x_1\partial x_3}\right)e_1$$

$$+ \left(\frac{\partial^2 A_1}{\partial x_2\partial x_1} + \frac{\partial^2 A_2}{\partial x_2^2} + \frac{\partial^2 A_3}{\partial x_2\partial x_3}\right)e_2 + \left(\frac{\partial^2 A_1}{\partial x_3\partial x_1} + \frac{\partial^2 A_2}{\partial x_3\partial x_2} + \frac{\partial^2 A_3}{\partial x_3^2}\right)e_3$$

and

$$
\begin{aligned}
\operatorname{curl}(\operatorname{curl}A) &= \operatorname{curl}\left(\frac{\partial A_3}{\partial x_2} - \frac{\partial A_2}{\partial x_3}, \frac{\partial A_1}{\partial x_3} - \frac{\partial A_3}{\partial x_1}, \frac{\partial A_2}{\partial x_1} - \frac{\partial A_1}{\partial x_2}\right) \\
&= \left(\frac{\partial}{\partial x_2}(\frac{\partial A_2}{\partial x_1} - \frac{\partial A_1}{\partial x_2}) - \frac{\partial}{\partial x_3}(\frac{\partial A_1}{\partial x_3} - \frac{\partial A_3}{\partial x_1})\right)e_1 \\
&\quad + \left(\frac{\partial}{\partial x_3}(\frac{\partial A_3}{\partial x_2} - \frac{\partial A_2}{\partial x_3}) - \frac{\partial}{\partial x_1}(\frac{\partial A_2}{\partial x_1} - \frac{\partial A_1}{\partial x_2})\right)e_2 \\
&\quad + \left(\frac{\partial}{\partial x_1}(\frac{\partial A_1}{\partial x_3} - \frac{\partial A_3}{\partial x_1}) - \frac{\partial}{\partial x_2}(\frac{\partial A_3}{\partial x_2} - \frac{\partial A_2}{\partial x_3})\right)e_3 \\
&= \left(\frac{\partial^2 A_2}{\partial x_2 \partial x_1} - \frac{\partial^2 A_1}{\partial x_2^2} - \frac{\partial^2 A_1}{\partial x_3^2} + \frac{\partial A_3}{\partial x_3 \partial x_1}\right)e_1 \\
&\quad + \left(\frac{\partial^2 A_3}{\partial x_3 \partial x_2} - \frac{\partial^2 A_2}{\partial x_3^2} - \frac{\partial^2 A_2}{\partial x_1^2} + \frac{\partial^2 A_1}{\partial x_1 \partial x_2}\right)e_2 \\
&\quad + \left(\frac{\partial A_1}{\partial x_1 \partial x_3} - \frac{\partial^2 A_3}{\partial x_1^2} - \frac{\partial^2 A_3}{\partial x_2^2} + \frac{\partial^2 A_2}{\partial x_2 \partial x_3}\right)e_3.
\end{aligned}
$$

Hence we obtain

$$
\begin{aligned}
&\operatorname{grad}(\operatorname{div}A) - \operatorname{curl}(\operatorname{curl}A) \\
&= \left(\frac{\partial^2 A_1}{\partial x_1^2} + \frac{\partial^2 A_2}{\partial x_1 \partial x_2} + \frac{\partial^2 A_3}{\partial x_1 \partial x_3} - \frac{\partial^2 A_2}{\partial x_2 \partial x_1} + \frac{\partial^2 A_1}{\partial x_2^2} + \frac{\partial^2 A_1}{\partial x_3^2} - \frac{\partial^2 A_3}{\partial x_3 \partial x_1}\right)e_1 \\
&\quad + \left(\frac{\partial^2 A_1}{\partial x_2 \partial x_1} + \frac{\partial^2 A_2}{\partial x_2^2} + \frac{\partial^2 A_3}{\partial x_2 \partial x_3} - \frac{\partial^2 A_3}{\partial x_3 \partial x_2} + \frac{\partial^2 A_2}{\partial x_3^2} + \frac{\partial^2 A_2}{\partial x_1^2} - \frac{\partial^2 A_1}{\partial x_1 \partial x_2}\right)e_2 \\
&\quad + \left(\frac{\partial^2 A_1}{\partial x_3 \partial x_1} + \frac{\partial^2 A_2}{\partial x_3 \partial x_2} + \frac{\partial^2 A_3}{\partial x_3^2} - \frac{\partial^2 A_1}{\partial x_1 \partial x_3} + \frac{\partial^2 A_3}{\partial x_1^2} + \frac{\partial^2 A_3}{\partial x_2^2} - \frac{\partial^2 A_2}{\partial x_2 \partial x_3}\right)e_3 \\
&= \sum_{j=1}^{3}(\Delta_3 A_j)e_j = \Delta_3 A.
\end{aligned}
$$

3. We insert $A + \operatorname{grad}\varphi$ into the vector-valued wave equation and find using that $\operatorname{curl}(\operatorname{grad}\varphi) = 0$ and $\operatorname{div}\operatorname{grad}\varphi = \Delta_3\varphi$

$$
\begin{aligned}
&\frac{\partial^2}{\partial t^2}(A + \operatorname{grad}\varphi) - \Delta_3(A + \operatorname{grad}\varphi) \\
&= \frac{\partial^2 A}{\partial t^2} + \frac{\partial^2}{\partial t^2}\operatorname{grad}\varphi - \operatorname{grad}(\operatorname{div}(A + \operatorname{grad}\varphi)) + \operatorname{curl}(\operatorname{curl}(A + \operatorname{grad}\varphi)) \\
&= \frac{\partial^2 A}{\partial t^2} - \operatorname{grad}(\operatorname{div}A) + \operatorname{curl}(\operatorname{curl}A) + \frac{\partial^2}{\partial t^2}\operatorname{grad}\varphi - \operatorname{grad}(\Delta_3\varphi) \\
&= \frac{\partial^2}{\partial t^2}\operatorname{grad}\varphi - \operatorname{grad}(\Delta_3\varphi),
\end{aligned}
$$

where in the last step we used that A solves the vector-valued wave equation. Now, if φ is independent of t and harmonic, i.e. $\Delta_3\varphi = 0$ the result follows. Moreover, if

740

φ depends on t and satisfies $\frac{\partial^2 \varphi}{\partial t^2} - \Delta_3 \varphi = 0$, then note that

$$\frac{\partial^2}{\partial t^2} \operatorname{grad} \varphi - \operatorname{grad}(\Delta_3 \varphi) = \operatorname{grad}\left(\frac{\partial^2 \varphi}{\partial t^2} - \Delta_3 \varphi\right) = 0.$$

Note that we gave the best known example of a "field equation" which has some "gauge invariance". It has its origin in classical electrodynamics and is the starting point of Young-Mills theory which considers non-linear field equations invariant under certain gauge transformation and eventually had deep impact in mathematical physics as well as in the theory of 4-manifolds, for example in the work of S. Donaldson.

Chapter 24

1. For a parametric surface $f : Q \to \mathbb{R}^3, Q \subset \mathbb{R}^2$, the coefficients of the first fundamental form are given by

$$E(u,v) = \langle f_u(u,v), f_u(u,v)\rangle$$
$$F(u,v) = \langle f_u(u,v), f_v(u,v)\rangle$$
$$G(u,v) = \langle f_v(u,v), f_v(u,v)\rangle.$$

Now we find

a) For $S = \Gamma(h) = \left\{ \begin{pmatrix} u \\ v \\ h(u,v) \end{pmatrix} \mid (u,v) \in Q \right\}$ we have

$$f_u(u,v) = \begin{pmatrix} 1 \\ 0 \\ h_u(u,v) \end{pmatrix}, f_v(u,v) = \begin{pmatrix} 0 \\ 1 \\ h_v(u,v) \end{pmatrix}$$

and therefore

$$E(u,v) = 1 + h_u^2(u,v),$$
$$F(u,v) = h_u(u,v)h_v(u,v),$$
$$G(u,v) = 1 + h_v^2(u,v).$$

b) For $S = \partial B_r(0) = f_r([0,\pi] \times [0,2\pi])$ where $f_r(\vartheta, \varphi)$ is given by

$$f_r(\vartheta, \varphi) = \begin{pmatrix} r \sin \vartheta \cos \varphi \\ r \sin \vartheta \sin \varphi \\ r \cos \vartheta \end{pmatrix}$$

we have

$$f_{r,\vartheta}(\vartheta, \varphi) = \begin{pmatrix} r \cos \vartheta \cos \varphi \\ r \cos \vartheta \sin \varphi \\ -r \sin \vartheta \end{pmatrix}, f_{r,\varphi}(\vartheta, \varphi) = \begin{pmatrix} -r \sin \vartheta \sin \varphi \\ r \sin \vartheta \cos \varphi \\ 0 \end{pmatrix}$$

741

and consequently

$$E(\vartheta, \varphi) = r^2 \cos^2 \vartheta \cos^2 \varphi + r^2 \cos^2 \vartheta \sin^2 \varphi + r^2 \sin^2 \vartheta = r^2,$$
$$F(\vartheta, \varphi) = -r^2 \cos \vartheta \cos \varphi \sin \vartheta \sin \varphi + r^2 \cos \vartheta \sin \varphi \sin \vartheta \cos \varphi = 0,$$
$$G(\vartheta, \varphi) = r^2 \sin^2 \vartheta \sin^2 \varphi + r^2 \sin^2 \vartheta \cos^2 \varphi = r^2 \sin^2 \vartheta.$$

c) For $S = f([a, b] \times [0, 2\pi])$, $f(u, v) = \begin{pmatrix} h(u) \cos v \\ h(u) \sin v \\ k(u) \end{pmatrix}$, it follows that

$$f_u(u, v) = \begin{pmatrix} h'(u) \cos v \\ h'(u) \sin v \\ k'(u) \end{pmatrix}, f_v(u, v) = \begin{pmatrix} -h(u) \sin v \\ h(u) \cos v \\ 0 \end{pmatrix}$$

and hence

$$E(u, v) = (h'(u))^2 \cos^2 v + (h'(u))^2 \sin^2 v + (k'(u))^2 = (h'(u))^2 + (k'(u))^2$$
$$F(u, v) = -h'(u)h(u) \cos v \sin v + h'(u)h(u) \sin v \cos v = 0$$
$$G(u, v) = h^2(u) \sin^2 v + h^2(u) \cos^2 v = (h(u))^2.$$

2. By $u \mapsto \begin{pmatrix} u \\ 0 \\ g(u) \end{pmatrix}$ a curve is given in $\mathbb{R}^3$ which we can consider as the graph of $g : \mathbb{R} \to \mathbb{R}$ embedded into $\mathbb{R}^3$. The function h has as graph the set

$$\left\{ \begin{pmatrix} u \\ v \\ h(u, v) \end{pmatrix} \mid (u, v) \in \mathbb{R}^2, h(u, v) = g(u) \right\} = \left\{ \begin{pmatrix} u \\ v \\ g(u) \end{pmatrix} \mid (u, v) \in \mathbb{R}^2 \right\}$$

which is obtained by translating the embedded graph of g. In general we know from Example 24.5 that the area of a surface S being the graph of a function $h : [a, b] \times [c, d] \to \mathbb{R}$ is given by

$$A(S) = \int_a^b \left(\int_c^d \sqrt{1 + h_u^2(u, v) + h_v^2(u, v)} dv \right) du,$$

which reduces in our case to

$$A(S) = \int_a^b \left(\int_c^d \sqrt{1 + (g'(u))^2} dv \right) du = (d - c) \int_a^b \sqrt{1 + (g'(u))^2} du.$$

Now with $\widetilde{h}(u, v) = \frac{u^2}{2}$ and $[a, b] \times [c, d] = [2, 4] \times [-1, 1]$ we find

$$A(\tilde{S}) = 2 \int_2^4 \sqrt{1 + u^2} du$$

$$= (u\sqrt{1 + u^2} + \ln(u + \sqrt{1 + u^2})) \Big|_2^4$$

$$= 4\sqrt{17} - 2\sqrt{5} + \ln \left(\frac{4 + \sqrt{17}}{2 + \sqrt{5}} \right)^2.$$

3. We have to calculate

$$\int_S \psi \cdot d\sigma = \int_2^5 \left(\int_3^4 \psi(f(u,v)) \|(f_u \times f_v)(u,v)\| dv\right) du.$$

With $f(u,v) = \begin{pmatrix} u+v \\ u-v \\ u\cdot v \end{pmatrix}$ we find $f_u(u,v) = \begin{pmatrix} 1 \\ 1 \\ v \end{pmatrix}$, $f_v(u,v) = \begin{pmatrix} 1 \\ -1 \\ u \end{pmatrix}$ and therefore

$$\|(f_u \times f_v)(u,v)\| = \left\| \begin{pmatrix} u+v \\ v-u \\ -2 \end{pmatrix} \right\| = ((u+v)^2 + (v-u)^2 + 2^2)^{1/2} = \sqrt{2u^2 + 2v^2 + 4}.$$

Moreover we have

$$\psi(f(u,v)) = \psi(u+v, u-v, uv)$$
$$= 5\sqrt{5}((u+v)^2 + (u-v)^2 + 4)^{3/2}$$
$$= 5\sqrt{5}(2u^2 + 2v^2 + 4)^{3/2}$$

which yields

$$\int_S \psi \cdot d\sigma = \int_2^5 \left(\int_3^4 5\sqrt{5}(2u^2 + 2v^2 + 4)^{3/2}(2u^2 + 2v^2 + 4)^{1/2} dv\right) du$$
$$= 5\sqrt{5} \int_2^5 \left(\int_3^4 (2u^2 + 2v^2 + 4)^2 dv\right) du$$
$$= 91996\sqrt{5}.$$

4. In general we have the formula

$$\int_S F \cdot d\sigma = \int_R \langle F(f(u,v)), (f_u \times f_v)(u,v)\rangle dvdu$$

and in our case we have $R = [a,b] \times [0, 2\pi]$,

$$(f_u \times f_v)(u,v) = \begin{pmatrix} -k'(u)h(u)\cos v \\ -k'(u)h(u)\sin v \\ h'(u)h(u) \end{pmatrix}$$

and therefore

$$\int_S F \cdot d\sigma = \int_a^b \left(\int_0^{2\pi} \langle F(f(u,v)), \begin{pmatrix} -k'(u)h(u)\cos v \\ -k'(u)h(u)\sin v \\ h'(u)h(u) \end{pmatrix}\rangle dv\right) du$$
$$= \int_a^b \left(\int_0^{2\pi} (F_3(f(u,v))h'(u)h(u) - F_2(f(u,v))k'(u)h(u)\sin v\right.$$
$$\left. - F_1(f(u,v))k'(u)h(u)\cos v)dv\right)du$$

743

In the case that $F(x, y, z) = \begin{pmatrix} x \\ y \\ z \end{pmatrix}$ we find first

$$F(f(u, v)) = F(h(u) \cos v, h(u) \sin v, k(u)),$$

hence $F_1(f(u, v)) = h(u) \cos v$, $F_2(f(u, v)) = h(u) \sin v$, $F_3(f(u, v)) = k(u)$, and consequently for this F

$$\int_S F \cdot d\sigma = \int_a^b \left(\int_0^{2\pi} k(u)h'(u)h(u) - k'(u)h^2(u) \sin^2 v - k'(u)h^2(u) \cos^2 v \right) dv du$$

$$= 2\pi \int_a^b (k(u)h'(u)h(u) - k'(u)h^2(u)) du.$$

5. In this example we have $h(u) = \sqrt{1 - u^2}$, $h'(u) = \frac{-u}{\sqrt{1-u^2}}$, and $k(u) = u$, $k'(u) = 1$. It follows that

$$\int_S F \cdot d\sigma = 2\pi \int_{-1}^1 \left(u \left(\frac{-u}{\sqrt{1-u^2}} \right) \sqrt{1 - u^2} - 1 \cdot (\sqrt{1-u^2})^2 \right) du$$

$$= 2\pi \int_{-1}^1 (-u^2 - 1(1 - u^2)) du = -4\pi.$$

Of course, S is the unit sphere S^2 since for $f(u, v) = \begin{pmatrix} \sqrt{1-u^2} \cos v \\ \sqrt{1-u^2} \sin v \\ u \end{pmatrix} = \begin{pmatrix} x \\ y \\ z \end{pmatrix}$ it

follows that $x^2 + y^2 + z^2 = (1 - u^2) \cos^2 v + (1 - u^2) \sin^2 v + u^2 = 1$. Therefore

$$F(x, y, z) = \begin{pmatrix} x \\ y \\ z \end{pmatrix}, (x, y, z) \in S = S^2,$$ is the unit normal vector field pointing

outward of S. On the other hand we have

$$(f_u \times f_v)(u, v) = \begin{pmatrix} -k'(u)h(u) \cos v \\ -k'(u)h(u) \sin v \\ h'(u)h(u) \end{pmatrix} = \begin{pmatrix} -h(u) \cos v \\ -h(u) \sin v \\ -u \end{pmatrix} = -\begin{pmatrix} x \\ y \\ z \end{pmatrix}.$$

Hence $\langle F(f(u, v)), (f_u \times f_v)(u, v) \rangle = -1$ and our calculation shows $-\int_S F \cdot d\sigma$ is the area of $S = S^2$.

6. We may calculate $\int_S F \cdot d\sigma$ by evaluating $\int_{S_1} F \cdot d\sigma$ and $\int_{S_2} F \cdot d\sigma$ separately. Now, S_1 is the unit circle in the $z = 0$ plane which we can parametrize with the help of $g : [0, 1] \times [0, 2\pi] \to \mathbb{R}^3, g(r, \varphi) = \begin{pmatrix} r \cos \varphi \\ r \sin \varphi \\ 0 \end{pmatrix}$, which gives $g_r(r, \varphi) = \begin{pmatrix} \cos \varphi \\ \sin \varphi \\ 0 \end{pmatrix}$, $g_\varphi(r, \varphi) = \begin{pmatrix} -r \sin \varphi \\ r \cos \varphi \\ 0 \end{pmatrix}$ and hence $(g_r \times g_\varphi)(r, \varphi) = \begin{pmatrix} 0 \\ 0 \\ r \end{pmatrix}$. Therefore we

744

get

$$\int_{S_1} F \cdot d\sigma = \int_0^1 (\int_0^{2\pi} \langle \begin{pmatrix} 0 \\ 0 \\ 1 \end{pmatrix}, \begin{pmatrix} 0 \\ 0 \\ r \end{pmatrix} \rangle d\varphi) dr = 2\pi \int_0^1 r dr = \pi.$$

As parametrization of S_2 we may choose $f : [0,1] \times [0, 2\pi] \to \mathbb{R}^3$, $f(r, \varphi) = \begin{pmatrix} r \cos \varphi \\ r \sin \varphi \\ \sqrt{1 - r^2} \end{pmatrix}$

which yields $f_r(r, \varphi) = \begin{pmatrix} \cos \varphi \\ \sin \varphi \\ \frac{-r}{\sqrt{1-r^2}} \end{pmatrix}$, $f_\varphi(r, \varphi) = \begin{pmatrix} -r \sin \varphi \\ r \cos \varphi \\ 0 \end{pmatrix}$ and

$(f_r \times f_\varphi)(r, \varphi) = \begin{pmatrix} \frac{r^2 \cos \varphi}{\sqrt{1-r^2}} \\ \frac{r^2 \cos \varphi}{\sqrt{1-r^2}} \\ r \end{pmatrix}$. Now it follows that

$$\int_{S_2} F \cdot d\sigma = \int_0^1 (\int_0^{2\pi} \langle \begin{pmatrix} r \cos \varphi \\ r \sin \varphi \\ \sqrt{1 - r^2} \end{pmatrix}, \begin{pmatrix} \frac{r^2 \cos \varphi}{\sqrt{1-r^2}} \\ \frac{r^2 \cos \varphi}{\sqrt{1-r^2}} \\ r \end{pmatrix} \rangle d\varphi) dr$$

$$= 2\pi \int_0^1 (\frac{r^3}{\sqrt{1 - r^2}} + r\sqrt{1 - r^2}) dr = 2\pi \int_0^1 \frac{r}{\sqrt{1 - r^2}} dr = 2\pi.$$

Hence we have $\int_S F \cdot d\sigma = 3\pi$.

7. We can parametrize S with the help of $f : [0, R] \times [0, 2\pi] \to \mathbb{R}^3$, $f(r, \varphi) = \begin{pmatrix} r \cos \varphi \\ r \sin \varphi \\ \sqrt{R^2 - r^2} \end{pmatrix}$

and as in Problem 6 we find $(f_r \times f_\varphi)(r, \varphi) = \begin{pmatrix} \frac{r^2 \cos \varphi}{\sqrt{R^2-r^2}} \\ \frac{r^2 \sin \varphi}{\sqrt{R^2-r^2}} \\ r \end{pmatrix}$. Now we find that

$$\int_S F \cdot d\sigma = \int_0^R (\int_0^{2\pi} \langle \begin{pmatrix} g_1(\rho) \\ g_2(\rho) \\ g_3(\rho) \end{pmatrix}, \begin{pmatrix} \frac{r^2 \cos \varphi}{\sqrt{R^2-r^2}} \\ \frac{r^2 \sin \varphi}{\sqrt{R^2-r^2}} \\ r \end{pmatrix} \rangle d\varphi) dr.$$

On S we have $x^2 + y^2 + z^2 = r^2 \cos^2 \varphi + r^2 \sin^2 \varphi + R^2 - r^2 = R^2$, and it follows that

$$\int_S F \cdot d\sigma = \int_0^R (\int_0^{2\pi} (g_1(R) \frac{r^2 \cos \varphi}{\sqrt{R^2 - r^2}} + g_2(R) \frac{r^2 \sin \varphi}{\sqrt{R^2 - r^2}} + g_3(R)r) d\varphi) dr = \pi g_3(R) R^2,$$

where we used that $\int_0^{2\pi} \cos \varphi \, d\varphi = \int_0^{2\pi} \sin \varphi \, d\varphi = 0$.

745

Chapter 25

1. By Gauss' theorem we have

$$\int_{\partial W} F \cdot d\sigma = \int_W \mathrm{div}\, F dxdydz$$

$$= \int_W (3z^2 + 8xy + 5y)dxdydz$$

$$= \int_{-2}^{2} (\int_{-3}^{3} (\int_{-4}^{4} (3z^2 + 8xy + 5y)dz)dy)dx$$

$$= \int_{-2}^{2} (\int_{-3}^{3} (128 + 64xy + 40y)dy)dx$$

$$= \int_{-2}^{2} 768dx = 3072.$$

2. Since $\overline{B_R(0)}$ is a Gauss domain we have

$$\int_{\overline{B_R(0)}} \mathrm{div}\, F\, dxdydz = \int_{\partial B_R(0)} \langle F, \overrightarrow{n} \rangle d\sigma.$$

Now $F\Big|_{\partial B_R(0)}$ is the outward normal vector field to $\partial B_R(0)$, note that $\begin{pmatrix} x \\ y \\ z \end{pmatrix}$ is a vector pointing outward of $\partial B_R(0)$ and for $(x, y, z) \in \partial B_R(0)$ it has length $\sqrt{x^2 + y^2 + z^2} = R$. Thus it follows that $\langle F \overrightarrow{n} \rangle = 1$ on $\partial B_R(0)$ and hence

$$\int_{\partial B_R(0)} \langle F, \overrightarrow{n} \rangle d\sigma = \int_{\partial B_R(0)} 1 d\sigma = A(\partial B_R(0)) = 4\pi R^2.$$

3. We may write

$$\int_C \left(\frac{\partial u}{\partial z} \right) vdxdydz = \int_K (\int_0^a \frac{\partial u}{\partial z}(x, y, z)w(x, y)dz)dxdy$$

$$= \int_K (u(x, y, a)w(x, y) - u(x, y, 0)w(x, y))dxdy.$$

Another way is to use Gauss' theorem for the vector field $F = \begin{pmatrix} 0 \\ 0 \\ uv \end{pmatrix}$ and to note that $\partial C = K \times \{0\} \cup K \times \{a\} \cup \partial K \times [0, a]$. For points in $\partial K \times [0, a]$ the z-component of the exterior normal vector is zero as is $\frac{\partial}{\partial x} F_1$ and $\frac{\partial}{\partial y} F_2$. The outward normal unit vector for points is $K \times \{0\}$ is $\begin{pmatrix} 0 \\ 0 \\ -1 \end{pmatrix}$ and for points in $K \times \{a\}$ it is $\begin{pmatrix} 0 \\ 0 \\ 1 \end{pmatrix}$ and $\frac{\partial}{\partial z} F_3 = \frac{\partial u}{\partial z}v$. Thus using Gauss' theorem we find

$$\int_C \left(\frac{\partial u}{\partial z} \right) vdxdydz = \int_C \mathrm{div}\, F\, dxdydz = \int_{\partial C} \langle F, \overrightarrow{n} \rangle d\sigma$$

with

$$\int_{\partial C} \langle F, \overrightarrow{n} \rangle d\sigma = \int_{\partial C} F_3 \overrightarrow{n}_3 d\sigma$$

$$= \int_K u(x, y, a) w(x, y) dx dy - \int_K u(x, y, 0) w(x, y) dx dy$$

which is a further justification of the result.

4. Since for $u, v \in \widetilde{C}_0^1(G)$ and $\lambda, \mu \in \mathbb{R}$ it follows that $(\lambda u + \mu v)\big|_{\partial G} = 0$, it is clear that $\widetilde{C}_0^1(G)$ is a vector space over $\mathbb{R}$. Following the hint we now prove that

$$\langle u, v \rangle_G := \int_G \langle \operatorname{grad} u, \operatorname{grad} v \rangle dx$$

is a scalar product on $\widetilde{C}_0^1(G)$. First of all it is clear that $\langle u, v \rangle_G$ is well defined since $x \mapsto \langle \operatorname{grad} u(x), \operatorname{grad} v(x) \rangle$ is a continuous function on $\widetilde{G}$. Furthermore we have $\langle u, v \rangle_G = \langle v, u \rangle_G$, i.e. $\langle \cdot, \cdot \rangle_G$ is symmetric. For $\lambda, \mu \in \mathbb{R}$ and $u, v, w \in \widetilde{C}_0^1(G)$ we find

$$\langle (\lambda u + \mu v), w \rangle_G = \int_G \langle \operatorname{grad}(\lambda u + \mu v), \operatorname{grad} w \rangle dx$$

$$= \lambda \int_G \langle \operatorname{grad} u, \operatorname{grad} w \rangle dx + \mu \int_G \langle \operatorname{grad} v, \operatorname{grad} w \rangle dx$$

$$= \lambda \langle u, w \rangle_G + \mu \langle v, w \rangle_G,$$

and this together with the symmetry implies that $\langle \cdot, \cdot \rangle_G$ is a symmetric bilinear form. Clearly $\langle u, u \rangle_G = \int_G \|\operatorname{grad} u\|^2 dx \geq 0$. Thus it remains to prove that if $\langle u, u \rangle_G = 0$ then $u = 0$. By the Poincaré inequality in the form of Lemma 25.12 we deduce

$$\left(\int_G |u|^2 dx \right)^{1/2} \leq 2 \max_{x \in G} |x_1| \left(\int_G \left| \frac{\partial u}{\partial x_1} \right|^2 dx \right)^{1/2}$$

$$\leq 2 \max_{x \in G} |x_1| \, \||u|\|_G = 2 \max_{x \in G} |x_1| < u, u >_G^{1/2}.$$

Thus, if $\langle u, u \rangle_G = 0$, then $\int_{\overline{G}} |u(x)|^2 dx = 0$ and since u is continuous we conclude that $u = 0$.

5. Since $\frac{\partial}{\partial x_j}\left(v \frac{\partial}{\partial x_j} u \right) = \frac{\partial v}{\partial x_j} \frac{\partial u}{\partial x_j} + v \frac{\partial^2 u}{\partial x_j^2}$ we have

$$(\Delta_3 u)v = (\operatorname{div}(\operatorname{grad} u))v = \operatorname{div}(v \operatorname{grad} u) - \langle \operatorname{grad} u, \operatorname{grad} v \rangle,$$

747

and Gauss' theorem yields

$$\int_G (\Delta_3 u)v\,dx = \int_G (\text{div}\,(v\,\text{grad}\,u) - \langle\text{grad}\,u, \text{grad}\,v\rangle)dx$$

$$= -\int_G \langle\text{grad}\,u, \text{grad}\,v\rangle dx + \int_{\partial G} \langle\overrightarrow{n}, v\,\text{grad}\,u\rangle d\sigma$$

$$= -\int_G \langle\text{grad}\,u, \text{grad}\,v\rangle dx + \int_{\partial G} v\frac{\partial u}{\partial \overrightarrow{n}}d\sigma.$$

Now if v is harmonic in G and we take $u = v$ we obtain $(\Delta_3 v)v = 0$, hence

$$\int_G \langle\text{grad}\,v, \text{grad}\,v\rangle dx = \int_{\partial G} v\frac{\partial v}{\partial \overrightarrow{n}}d\sigma$$

Since $v\frac{\partial v}{\partial \overrightarrow{n}} = \langle\overrightarrow{n}, v\,\text{grad}\,v\rangle = \frac{1}{2}\langle\overrightarrow{n}, \text{grad}\,v^2\rangle$ we eventually arrive at

$$\int_G \|\text{grad}\,v\|^2 dx = \frac{1}{2}\int_{\partial G} \frac{\partial v^2}{\partial \overrightarrow{n}}d\sigma.$$

6. When applying the first of Green's identities to u and v twice but with interchanged roles we find

$$\int_G v\Delta_3 u\,dx = -\int_G \langle\text{grad}\,u, \text{grad}\,v\rangle dx + \int_{\partial G} v\frac{\partial u}{\partial \overrightarrow{n}}d\sigma$$

and

$$\int_G u\Delta_3 v\,dx = -\int_G \langle\text{grad}\,v, \text{grad}\,u\rangle dx + \int_{\partial G} u\frac{\partial v}{\partial \overrightarrow{n}}d\sigma.$$

Subtracting these identities yields Green's second identity

$$\int_G (v\Delta_3 u - u\Delta_3 v)dx = \int_{\partial G} (v\frac{\partial u}{\partial \overrightarrow{n}} - u\frac{\partial v}{\partial \overrightarrow{n}})d\sigma.$$

For v harmonic, i.e. $\Delta_3 v = 0$ this gives of course

$$\int_G v\Delta_3 u\,dx = \int_{\partial G} (v\frac{\partial u}{\partial \overrightarrow{n}} - u\frac{\partial v}{\partial \overrightarrow{n}})d\sigma.$$

If we choose now $u = 1$, i.e. u is identically 1 for all $x \in \widetilde{G}$, then we find $\Delta_3 u = 0$ as well as $\frac{\partial u}{\partial \overrightarrow{n}} = \langle\overrightarrow{n}, \text{grad}, u\rangle = 0$ implying for harmonic v that $\int_{\partial G} \frac{\partial v}{\partial \overrightarrow{n}}d\sigma = 0$.

7. For $u(x_1, x_2, x_3) = x_1^2 + x_2^2 + x_3^2$ we find

$$\Delta_3(x_1^2 + x_2^2 + x_3^2) = \frac{\partial^2}{\partial x_1^2}x_1^2 + \frac{\partial^2}{\partial_2^2}x_2^2 + \frac{\partial^2}{\partial x_3^2}x_3^2 = 6.$$

With this u Green's second identity reads

$$6\int_{B_1(0)} v\,dx = \int_{B_1(0)} (v\Delta_3 u - u\Delta_3 v)dx = \int_{S^2} (v\frac{\partial u}{\partial \overrightarrow{n}} - u\frac{\partial v}{\partial \overrightarrow{n}})d\sigma.$$

The outward unit normal to S^2 and x is given by $\begin{pmatrix} x_1 \\ x_2 \\ x_3 \end{pmatrix}$ and therefore we find since on S^2 we have $x_1^2 + x_2^2 + x_3^3 = 1$:

$$\frac{\partial u}{\partial \overrightarrow{n}} = \langle \overrightarrow{n}, \operatorname{grad} u \rangle = \langle \begin{pmatrix} x_1 \\ x_2 \\ x_3 \end{pmatrix}, 2 \begin{pmatrix} x_1 \\ x_2 \\ x_3 \end{pmatrix} \rangle = 2$$

and

$$u \frac{\partial v}{\partial \overrightarrow{n}} = (x_1^2 + x_2^2 + x_3^2) \frac{\partial v}{\partial \overrightarrow{n}} = \frac{\partial v}{\partial \overrightarrow{n}}$$

which implies

$$6 \int_{B_1(a)} v \, dx = \int_{S^2} (v \frac{\partial u}{\partial \overrightarrow{n}} - u \frac{\partial v}{\partial \overrightarrow{n}}) d\sigma = 2 \int_{S^2} v \, d\sigma - \int_{S^2} \frac{\partial v}{\partial \overrightarrow{n}} d\sigma.$$

8. By Gauss' theorem we have

$$4\pi \int_{B_r(a)} \rho(x) \, dx = \int_{B_r(a)} \langle \overrightarrow{n}, E \rangle d\sigma = \int_{B_r(a)} \operatorname{div} E \, dx$$

or

$$0 = \int_{B_r(a)} (\operatorname{div} E(x) - 4\pi \rho(x)) dx$$

for all $a \in G$ and $r > 0$ such that $\overline{B_r(a)} \subset G$. Applying the considerations leading to (25.12) we find for all $a \in G$

$$0 = \lim_{r \to \infty} \frac{1}{\operatorname{vol}_3(\overline{B_r(a)})} \int_{B_r(a)} (\operatorname{div} E(x) - 4\pi \rho(x)) dx = \operatorname{div} E(a) - 4\pi \rho(a).$$

Chapter 26

1. The domain G is sketched in the following figure from which it is also obvious that it is a normal domain with respect to both axes.

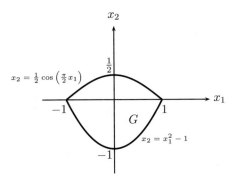

749

We can parametrize ∂G as the curve $\gamma : [-1, 3] \to \mathbb{R}^2$,

$$\gamma(t) = \begin{cases} \begin{pmatrix} -t \\ \cos \frac{\pi}{2} t \end{pmatrix}, & t \in [-1, 1] \\ \begin{pmatrix} t - 2 \\ -\sqrt{1 - (t-2)^2} \end{pmatrix}, & t \in (1, 3] \end{cases}$$

with

$$\dot\gamma(t) = \begin{cases} \begin{pmatrix} -1 \\ -\frac{\pi}{2} \sin \frac{\pi}{2} t \end{pmatrix}, & t \in [-1, 1] \\ \begin{pmatrix} 1 \\ \frac{t-2}{\sqrt{1-(t-2)^2}} \end{pmatrix}, & t \in [1, 3]. \end{cases}$$

Using Green's theorem we find for $F(x_1, x_2) = \begin{pmatrix} x_2 \\ \frac{1}{2} x_1^2 \end{pmatrix}$ that

$$\int_G (x_1 - 1) dx = \int_G \left(\frac{\partial F_2}{\partial x_1} - \frac{\partial F_1}{\partial x_2} \right) dx = \int_{-1}^3 \gamma_2(t) \dot\gamma_1(t) dt + \int_{-1}^3 \frac{1}{2} \gamma_1^2(t) \dot\gamma_2(t) dt$$

$$= \int_{-1}^1 \gamma_2(t) \dot\gamma_1(t) dt + \int_1^3 \gamma_2(t) \dot\gamma_1(t) \dot\gamma_2(t) + \int_{-1}^1 \frac{1}{2} \gamma_1^2(t) \dot\gamma_2(t) dt + \int_1^3 \frac{1}{2} \gamma_1^2(t) \dot\gamma_2(t) dt$$

$$= \int_{-1}^1 (\cos \frac{\pi}{2} t)(-1) dt + \int_1^3 (-\sqrt{1 - (t-2)^2}) 1 dt$$

$$+ \int_{-1}^1 \frac{1}{2} (-t)^2 (-\frac{\pi}{2} \sin \frac{\pi}{2} t) dt + \int_1^3 \frac{1}{2} (t-2)^2 \cdot \frac{(t-2)}{\sqrt{1-(t-2)^2}} dt$$

$$= -\int_{-1}^1 \cos \frac{\pi}{2} t\, dt - \int_1^3 \sqrt{1 - (t-2)^2} dt + \frac{1}{2} \int_1^3 \frac{(t-2)^3}{\sqrt{1-(t-2)^2}} dt$$

$$= -\int_{-1}^1 \cos \frac{\pi}{2} t\, dt - \int_{-1}^1 \sqrt{1 - s^2} ds + \frac{1}{2} \int_{-1}^1 \frac{s^3}{\sqrt{1-s^2}} ds$$

$$= -\int_{-1}^1 \cos \frac{\pi}{2} t\, dt - \int_{-1}^1 \sqrt{1 - s^2} ds$$

$$= -\frac{4}{\pi} - \arcsin 1 = -\frac{4}{\pi} - \frac{\pi}{2},$$

where we used that $\int_{-1}^1 t^2 \sin \frac{\pi}{2} t\, dt = \int_{-1}^1 \frac{s^3}{\sqrt{1-s^2}} ds = 0$ as an integral of an odd function over an interval symmetric to 0.

2. The boundary of G has two parts, the "upper part" for $x_2 \geq 0$ is the half-circle $x_1^2 + x_2^2 = 1$, and the "lower part" for $x_2 \leq 0$ is the half-ellipse $x_1^2 + \frac{x_2^2}{4} = 1$. Thus

a parametrization of ∂G is given by $\gamma : [0, 2\pi] \to \mathbb{R}^2$,

$$\gamma(t) \begin{cases} \begin{pmatrix} \cos t \\ \sin t \end{pmatrix}, & t \in [0, \pi] \\[2ex] \begin{pmatrix} \cos t \\ 2\sin t \end{pmatrix}, & t \in (\pi, 2\pi]. \end{cases}$$

With $F(x) \begin{pmatrix} -x_2 \\ x_1 \end{pmatrix}$ we find for the area of G:

$$
\begin{aligned}
A(G) &= \frac{1}{2} \int_0^{2\pi} F_1(\gamma(t))\dot{\gamma}_1(t)\,dt + \frac{1}{2} \int_0^{2\pi} F_2(\gamma(t))\dot{\gamma}_2(t)\,dt \\
&= \frac{1}{2} \int_0^{\pi} (-\sin t)(-\sin t)\,dt + \frac{1}{2} \int_\pi^{2\pi} (-2\sin t)(-\sin t)\,dt \\
&\quad + \frac{1}{2} \int_0^{\pi} \cos t \cos t\,dt + \frac{1}{2} \int_\pi^{2\pi} (\cos t)(2\cos t)\,dt \\
&= \frac{1}{2} \int_0^{\pi} \sin^2 t\,dt + \int_\pi^{2\pi} \sin^2 t\,dt \\
&\quad + \frac{1}{2} \int_0^{\pi} \cos^2 t\,dt + \int_\pi^{2\pi} \cos^2 t\,dt \\
&= \frac{1}{2} \int_0^{\pi} 1\,dt + \int_\pi^{2\pi} 1\,dt = \frac{3\pi}{2}
\end{aligned}
$$

Of course the area of G is $\frac{1}{2}$ of the area of the unit disc, which is $\frac{\pi}{2}$, plus $\frac{1}{2}$ of the area bounded by the ellipse with half axes of lengths 1 and 2 respectively, which is π.

3. The set G is given in the following figure:

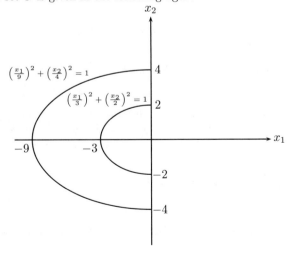

751

If we cut G by the line $x_2 = 0$ into two parts, we obtain two domains each being normal to both axes, hence we may apply Green's theorem. Now we find

$$\int_{\partial G} F_1 dx_1 + \int_{\partial G} F_2 dx_2 = \int_G (\frac{\partial F_2}{\partial x_1} - \frac{\partial F_1}{\partial x_2}) dx_1 dx_2 = \int_G (3x_2 + 2) dx_1 dx_2.$$

With $G_1 = \{(x_1, x_2) \in \mathbb{R}^2 \mid (\frac{x_1}{9})^2 + (\frac{x_2}{4})^2 \leq 1, x_1 < 0\}$ and $G_2 = \{(x_1, x_2) \in \mathbb{R}^2 \mid (\frac{x_1}{3})^2 + (\frac{x_2}{2})^2 \leq 1, x_1 < 0\}$ we find

$$\int_G (3x_2 + 2) dx_1 dx_2 = \int_{G_1} (3x_2 + 2) dx_1 dx_2 - \int_{G_2} (3x_2 + 2) dx_1 d_2.$$

For G_1 we use the elliptic polar coordinates $x_1 = 9r \cos \varphi$, $x_2 = 4r \sin \varphi$ and for G_2 we use the elliptic polar coordinates $x_1 = 3r \cos \varphi$ and $x_2 = 2r \sin \varphi$ where in each case $0 \leq r \leq 1$ and $\frac{\pi}{2} \leq \varphi \leq \frac{3\pi}{2}$ and it follows that

$$\int_{G_1} (3x_2 + 2) dx_1 dx_2 = \int_0^1 (\int_{\frac{\pi}{2}}^{\frac{3\pi}{2}} (3 \cdot 4r \sin \varphi + 2) 36 r d\varphi) dr = 36$$

and

$$\int_{G_2} (3x_2 + 2) dx_1 dx_2 = \int_0^1 (\int_{\frac{\pi}{2}}^{\frac{3\pi}{2}} (3 \cdot r \sin \varphi + 2) 6 r d\varphi) dr = 6$$

implying

$$\int_{\partial G} F_1 dx_1 + \int_{\partial G} F_2 dx_2 = 30.$$

4. Since G is a Gauss domain we can apply Gauss' theorem to find

$$\int_{\partial G} \operatorname{curl} F \cdot d\sigma = \int_G \operatorname{div}(\operatorname{curl} F) dx.$$

Since F is a C^2-vector field we have $\operatorname{div}(\operatorname{curl} F) = 0$, compare Problem 6b) in Chapter 23. Hence $\int_{\partial G} \operatorname{curl} F \cdot d\sigma = 0$. Note that ∂G has no boundary curve (it is a closed surface), or interpreted differently, the trace of the boundary curve is the empty set and hence the line integral term in Stokes' theorem formally becomes an integral over the empty set and therefore it should (formally) be equal to 0.

5. As suggested we slice S, for example along the line segment connecting $\begin{pmatrix} 1 \\ 0 \\ b \end{pmatrix}$ with $\begin{pmatrix} 1 \\ 0 \\ a \end{pmatrix}$. The boundary curve of $\tilde{S} = S \setminus \{(1, 0, z) \mid a \leq z \leq b\}$ can be parametrized

for example by $\gamma : [0, 4] \to \mathbb{R}^3$ defined as

$$\gamma(t) = \begin{cases} \begin{pmatrix} \cos 2\pi t \\ \sin 2\pi t \\ b \end{pmatrix}, & t \in [0, 1] \\ \begin{pmatrix} 1 \\ 0 \\ (a - b)t + 2b - a \end{pmatrix}, & t \in (1, 2) \\ \begin{pmatrix} \cos 2\pi t \\ -\sin 2\pi t \\ a \end{pmatrix}, & t \in [2, 3] \\ \begin{pmatrix} 1 \\ 0 \\ (b - a)t + 4a - 3b \end{pmatrix}, & t \in (3, 4]. \end{cases}$$

Since $\int_S \operatorname{curl} F \cdot d\sigma = \int_{\widetilde{S}} \operatorname{curl} F \cdot d\sigma$ we find

$$\int_S \operatorname{curl} F \cdot d\sigma = \sum_{j=1}^4 \int_{\gamma^{(j)}} F \cdot dp$$

where $\gamma^{(j)} = \gamma\big|_{[j-1,j]}$. We note that $\operatorname{tr}(\gamma^{(2)}) = \operatorname{tr}(\gamma^{(4)})$ but at $\gamma^{(2)}(t) = \gamma^{(4)}(s), t \in [1, 2], s \in [3, 4]$, we have $\dot{\gamma}^{(2)}(t) = -\dot{\gamma}^{(4)}(s)$, hence $\int_{\gamma^{(2)}} F \cdot dp = -\int_{\gamma^{(4)}} F \cdot dp$ and we arrive at

$$\int_S \operatorname{curl} F \cdot d\sigma = \int_{\gamma^{(1)}} F \cdot dp + \int_{\gamma^{(3)}} F \cdot dp.$$

However, $\gamma^{(1)}$ is an anticlockwise parametrization of the circle $\{(x, y, b) \in \mathbb{R}^3 \mid x^2 + y^2 = 1\}$ while $\gamma^{(3)}$ is a clockwise parametrization of the circle $\{(x, y, a) \in \mathbb{R}^3 \mid x^2 + y^2 - 1\}$.

Chapter 27

1. We find for Jacobi matrix of f

$$J_f(u_1, u_2, u_3) = \begin{pmatrix} 1 & 0 & 0 & \frac{\partial h}{\partial u_1} \\ 0 & 1 & 0 & \frac{\partial h}{\partial u_2} \\ 0 & 0 & 1 & \frac{\partial h}{\partial u_3} \end{pmatrix} (u_1, u_2, u_3)$$

which gives the Gram matrix.

753

$$J_f(u_1, u_2 u_3) J_f^t(u_1, u_2, u_3)$$

$$= \begin{pmatrix} 1 & 0 & 0 & \frac{\partial h}{\partial u_1} \\ 0 & 1 & 0 & \frac{\partial h}{\partial u_2} \\ 0 & 0 & 1 & \frac{\partial h}{\partial u_3} \end{pmatrix} (u_1, u_2, u_3) \begin{pmatrix} 1 & 0 & 0 \\ 0 & 1 & 0 \\ 0 & 0 & 1 \\ \frac{\partial h}{\partial u_1} & \frac{\partial h}{\partial u_2} & \frac{\partial h}{\partial u_3} \end{pmatrix} (u_1, u_2, u_3)$$

$$= \begin{pmatrix} 1 + (\frac{\partial h}{\partial u_1})^2 & \frac{\partial h}{\partial u_1} \frac{\partial h}{\partial u_2} & \frac{\partial h}{\partial u_1} \frac{\partial h}{\partial u_3} \\ \frac{\partial h}{\partial u_1} \frac{\partial h}{\partial u_2} & 1 + (\frac{\partial h}{\partial u_2})^2 & \frac{\partial h}{\partial u_2} \frac{\partial h}{\partial u_3} \\ \frac{\partial h}{\partial u_1} \frac{\partial h}{\partial u_3} & \frac{\partial h}{\partial u_2} \frac{\partial h}{\partial u_3} & 1 + (\frac{\partial h}{\partial u_3})^2 \end{pmatrix} (u_1, u_2, u_3).$$

For the Gram determinant we find now

$$g(f(u)) = \det J_f(u_1, u_2, u_3) J_f^t(u_1, u_2, u_3) = 1 + (\frac{\partial h}{\partial u_1})^2 + (\frac{\partial h}{\partial u_2})^2 + (\frac{\partial h}{\partial u_3})^2.$$

2. The unit sphere $S^3 \subset \mathbb{R}^4$ is the set $\{(x_1, x_2, x_3, x_4) \in \mathbb{R}^4 \mid x_1^2 + x_2^2 + x_3^2 + x_4^2 = 1\}$ and the upper sphere $S_+^3 = \{(x_1, x_2, x_3, x_4) \in S^3 \mid x_4 \geq 0\}$. Hence we may write $x_4 = \sqrt{1 - x_1^2 - x_2^2 - x_3^2}$ and a parametrization of S_+^3 is given by $f : \overline{B_1(0)} \to$

$\mathbb{R}^4, \overline{B_1(0)} \subset \mathbb{R}^3, f(u_1, u_2, u_3) = \begin{pmatrix} u_1 \\ u_2 \\ u_3 \\ \sqrt{1 - u_1^2 - u_2^2 - u_3^2} \end{pmatrix}$. Now we may apply the

result of Problem 1 to find for the Gram determinant

$$g(f(u)) = 1 + \sum_{j=1}^{3} \left(\frac{\partial \sqrt{1 - u_1^2 - u_2^2 - u_3^2}}{\partial u_j} \right)^2$$

$$= 1 + \sum_{j=1}^{3} \frac{u_j^2}{1 - u_1^2 - u_2^2 - u_3^2} = \frac{1}{1 - (u_1^2 + u_2^2 + u_3^2)}.$$

3. Denote by $a_{\min} := \min_{1 \leq k \leq n} a_k > 0$. Multiplying the equality $-\sum_{k=1}^{n} a_k \frac{\partial^2 u}{\partial x_k^2} = 0$ with u and integrating over G yields

$$0 = \int_G \left(-\sum_{k=1}^{n} a_k \frac{\partial^2 u}{\partial x_k^2}(x) \right) u(x) dx = -\sum_{k=1}^{n} a_k \int_G \left(\frac{\partial^2 u}{\partial x_k^2}(x) \right) u(x) dx$$

$$= \sum_{k=1}^{n} a_k \int_G \frac{\partial u}{\partial x_k} \frac{\partial u}{\partial x_k}(x) dx,$$

where we used that $u|_{\partial G} = 0$. Hence we find $0 \geq a_{\min} \int_G \| \operatorname{grad} u(x) \|^2 dx$, implying that $\int_G \| \operatorname{grad} u(x) \|^2 dx = 0$. Now we can argue as in our derivation of Corollary 27.7 to get $u = 0$ in G.

754

In this bibliography we list the works used when writing the text and we always refer to the actual copy at hand. For foreign language texts, where possible, we adopted an English translation. Very few titles below, e.g. [3], [40], [41], [43] and [48], are for further reading only.

References

[1] Abraham, R., Marsden, J. E., *Foundations of Mechanics.* 2nd ed. Benjamin/Cumming Publishing Company, Reading MA 1978.

[2] Abramowitz, M., Stegun, J. A., (eds.), *Handbook of Mathematical Functions.* 7th printing. Dover Publications, New York 1970.

[3] Appell, J., *Analysis in Beispielen und Gegenbeispielen.* Springer Verlag, Berlin · Heidelberg 2009.

[4] Armstrong, M.,A., *Basic Topology.* McGraw-Hill Book Company (U.K.), London 1979.

[5] Arnold, V. I., *Mathematical Methods of Classical Mechanics.* Springer-Verlag, New York · Heidelberg · Berlin 1978.

[6] Bauer, H., *Differential-und Integralrechung II.* Universitäts-Buchhandlung Rudolf Merkel, Erlangen 1966.

[7] Bauer, H., *Approximation and Abstract Boundaries.* Amer. Math. Monthly 85(1978), 632-647.

[8] Courant, R., *Vorlesungen über Differential-und Intergalrechmung 2. 4.Aufl.* Springer-Verlag, Berlin · Heidelberg · New York 1972.
English: *Differential and Integral Calculus II.* Interscience Publishers, New York 1936. Reprinted 1988.

[9] Crossley, M., *Essential Topology.* Springer-Verlag, London 2005.

[10] Dieudonné, J., *Grundzüge der modernen Analysis.* 2. Aufl. Vieweg Verlag, Braunschweig 1972.
English: *Foundations of Modern Analysis.* Academic Press, New York 1960.

[11] DoCarmo, M., *Differentialgeometrie von Kurven und Flächen.* Vierweg Verlag, Braunschweig · Wiesbaden 1983.
English: *Differential Geometry of Curves and Surfaces.* Prentice Hall, Englewood Cliffs NJ 1976.

[12] Endl, K., Luh, W., *Analysis II. 2. Aufl.* Akademische Verlagsgesellschaft, Wiesbaden 1974.

[13] Federer. H., *Geometric Measure Theory.* Reprint: Classics in Mathematics. Springer-Verlag, Berlin 1996.

[14] Fichtenholz, G. M., *Differential- und Integralrechnung II. 4.Aufl.* VEB Deutscher Verlag der Wissenschaften, Berlin 1972.

[15] Fichtenholz, G. M., *Differential-und Integralrechmung III. 6.Aufl.* VEB Deutscher Verlag der Wissenschaften, Berlin 1973.

[16] Fischer, G., *Analytische Geometrie* Vieweg Verlag, Braunschweig 1979.

[17] Fischer, G., *Lineare Algebra. 9. Aufl.* Vieweg Verlag, Braunschweig · Wiesbaden 1986.

[18] Fischer, H., Kaul, H., *Mathematik für Physiker 1.* Teubner Verlag, Stuttgart 1997.

[19] Forster, O., *Analysis 2.* Vieweg Verlag, Braunschweig 1979.

[20] Forster, O., *Analysis 3.* Vieweg Verlag, Braunschweig · Wiesbaden 1981.

[21] Fraenkel, L. E., *Formulae for higher derivatives of composite functions.* Math. Proc. Cambridge Phil. Soc. 83 (1978), 159-165.

[22] Giaquinta, M., Modica, G., Souček, J., *Cartesian Currents in the Calculus of Variations, I and II.* Springer-Verlag, Berlin · Heidelberg · New York 2007.

[23] Gruber, P. M., *Convex and Discrete Geometry.* Springer-Verlag, Berlin · Heidelberg · New York 1998.

[24] Heins, M., *Complex Function Theory.* Academic Press, New York · London 1968.

[25] Heuser, H., *Lehrbach der Analysis. Teil 2.* Teubner Verlag, Stuttgart 1981.

[26] Hildebrandt, S., *Analysis 2.* Springer Verlag, Berlin 2003.

[27] Hirzebruch, F., Scharlau, W., *Einführung in die Funktionalanalysis.* B. I.-Wissenschaftsverlag, Mannheim · Wien · Zürich 1971.

[28] Jackson, J. D., *Classical Electrodynamics.* John Wiley & Sons, New York · London · Sydney 1962.

[29] Kinderlehrer, D., Stampacchia, G., *An Introduction to Variational Inequalities and their Applications.* Academic Press, New York 1980.

[30] Knapp, A. W., *Basic Real Analysis.* Birkhäuser Verlag, Boston · Bassel · Berlin 2005.

[31] Lang, S., *Linear Algebra.* Addison-Wesley Publishing Company, Reading MA 1966.

[32] Lingenberg, R., *Einführung in die Lineare Algebra.* B. I.-Wissenschaftsverlag, Mannheim · Wien · Zürich, 1976.

[33] Lipschutz, M. M., *Differentialgeometrie. Theorie und Anwendurgen.* McGraw-Hill Book Company, Düsseldorf 1980.
English: *Theory and Problems of Differential Geometry,* McGraw-Hill Book Company, New York 1969.

[34] Lorenz, F., *Lineare Algebra I. Reprint.* B. I.-Wissenschaftsverlag, Mannheim · Wien · Zürich 1981, reprinted in 1985.

[35] Lorenz, F., *Lineare Algebra II. Reprint.* B. I.-Wissenschaftsverlag, Mannheim · Wien · Zürich 1982, reprinted in 1984.

[36] Morgan, F., *Geometric Measure Theory. A Beginner's Guide.* 2nd ed. Academic Press, San Diego 1995.

[37] Morse, Ph. M., Feshbach, H., *Methods of Theoretical Physics. Part 1.* McGraw-Hill Book Company, New York · Toronto · London 1953.

[38] Moskowitz, M., Paliogiannis, F., *Functions of Several Real Variables.* World Scientific, Singapore 2011.

[39] Rinow, W., *Die innere Geometrie der metrischen Räume.* Springer-Verlag, Berlin · Göttingen · Heidelberg 1961.

[40] Rudin, W., *Principles of Mathematical Analysis. International Edition.* 3rd ed. McGraw-Hill Book Company, Singapore 1976.

[41] Schneider, R., *Convex Bodies: The Brunn-Minkowski Theory.* Cambridge University Press, Cambridge 1993.

[42] Simon, B., *Convexity: An Analytic Viewpoint.* Cambridge University Press, Cambridge 2011.

[43] Spiegel, M. R., *Advanced Calculus.* McGraw-Hill Book Company, New York 1963.

[44] Spiegel, M. R., *Vector Analysis and an Introduction to Tensor Analysis. SI(metric) Edition.* McGraw-Hill Book Company, New York 1974.

[45] Stewart, J., *Calculus. Early Transcendentals. (Metric International Version.)* 6th ed. Thomson Books, Cole Belmont CA 2008.

[46] Triebel, H., *Höhere Analysis.* VEB Deutsche Verlag der Wissenschaften, Berlin 1972. English: *Higher Analysis.* J. A. Barth Verlag, Leipzig 1992.

[47] Webb, J. R. L., *Functions of Several Real Variables.* Ellis Harwood, New York 1991.

[48] Zeidler E., (ed.), *Oxford Users Guide to Mathematics.* Oxford University Press, Oxford 2004.

[49] Zorich, V. A., *Mathematical Analysis, I and II.* Springer-Verlag, Berlin 2004.

Mathematicians Contributing to Analysis (Continued)

Alembert, Jean-Baptist Le Rond d' (1717-1783).

Arnold, Vladimir Igorevich (1937-2010).

Arzela, Cesare (1847-1912).

Ascoli, Giulio (1843-1896).

Bernoulli, Daniel (1700-1782).

Cartan, Elie Joseph (1869-1951).

Cavalieri, Bonaventura (1598?-1647).

Courant, Richard (1888-1972).

Cramer, Gabriel (1704-1752).

Darboux, Jean Gaston (1842-1917).

Dini, Ulisse (1845-1918).

Fréchet, Maurice René (1878-1973).

Frenet, Jean Frédéric (1816-1900).

Fubini, Guido (1879-1943).

Galilei, Galileo (1564-1642).

Gauss, Carl Friedrich (1777-1855).

Gram, Jørgen Pedersen (1850-1916).

Grassmann, Hermann Günther (1809-1877).

Green, George (1793-1841).

Hadamard, Jacque (1865-1963).

Hamiliton, William Rowan (1805-1865).

Hausdorff, Felix (1868-1942).

Hermite, Charles (1822-1901).

Hesse, Ludwig Otto (1811-1874).

Huygens, Christiaan (1629-1695).

Jacobi, Carl Gustav Jacob (1804-1851).
Jordan, Camille (1838-1922).

Kepler, Johannes (1571-1630).
Korovkin, Pavel Petrovich (1913-1985).
Kronecker, Leopold (1823-1891).

Lamé, Gabriel (1795-1870).

Monge, Gaspard (1746-1818).

Neil, William (1637-1670).

Ostrogradskïi, Mikhail Vassilevich (1801-1862).

Poisson, Simon Denis (1781-1840).

Schwartz, Laurent (1915-2002).
Serret, Joseph Alfred (1819-1885).
Stieltjes, Thomas Jan (1856-1894).
Stokes, Georg Gabriel (1819-1903).
Stone, Marshall Harvey (1903-1989).

Young, William Henry (1863-1942).

Subject Index

世界著名数学家 Fourier 曾指出：

对自然的深刻研究是数学发现的最富饶的来源. 这种研究不仅有利地提供了一系列恰当相关的目标, 而且还有效地排除了含糊的问题和无用的计算. 它是建立分析学的方法, 也是发现至关重要的科学总要维护的概念的方法. 最基本的概念是那些代表自然事件的概念.

本书是一部英文版的分析学教程, 中文书名或可译为《分析学教程. 第 2 卷, 多元函数的微分和积分, 向量微积分》.

本书的作者有两位: 一位是尼尔斯·雅各布 (Niels Jacob), 英国数学家, 英国斯旺西大学教授; 另一位是克里斯蒂安·P. 埃文斯 (Kristian P. Evans), 也是英国数学家, 英国斯旺西大学教授.

两位作者在本书前言中介绍到:

本书是我们的《分析学教程》的第 2 卷, 它是为大学二年级以上的学生设计的. 在第 1 卷中, 特别是在第一部分中, 从高中到大学的过渡阶段决定了内容的风格和方法, 比如: 我们会慢慢地介绍抽象概念或者给出详细的 (基本的) 计算. 现在我们给出的方法是在大学研究数学的学生用的更多、更适合他们的风格的方法. 在阅读本书时, 我们的目的是引导和培养学生掌握更专业和更严谨的数学方法. 在引入新概念、通过例子探索新概念以及通过反例探索新概念的局限性时, 我们仍然会注意动机 (有些冗长). 然而, 一些常规计算被认为是理所当然的, 就像通过更抽象的概念"战斗"的意愿一样.

769

此外,我们开始改变我们使用参考文献的方式,学生应该注意这一点.一维的微积分和分析学的教学非常广泛,以至于现在在书中追溯我们如何呈现和证明结果的源头是非常困难的.关于这一点的注释我们在第 1 卷中给出了.内容越先进,就越应该更详细地指出现有文献和我们自己的资料来源.我们现在仍然处于这样的境地,即许多内容是典型的,并且已经被"标准方法"覆盖了.然而,在许多情况下,作者可能会声称他们的内容更具有原创性,学生应该了解现有文献并给予公平的评价,书的作者有义务这样做.我们希望有经验的读者会认为我们的引用是合理的,请参阅下面更多详细信息.

本卷的目的是将一个实变量实值函数的分析扩展到从 $\mathbb{R}^m$ 到 $\mathbb{R}^n$ 的映射,比如,扩展到多变量向量值映射.乍一看,我们需要提出三个更广泛的领域:收敛和连续性;线性逼近和可微性;积分.事实证明,要遵循这个程序,我们需要额外学习更多的几何学知识.一些几何学知识与拓扑概念相联系,例如,一个集合是否有很多元素呢? 一个集合有"小孔"吗? 一个集合的边界是什么? 其他的几何概念与 $\mathbb{R}^n$ 的向量空间结构相联系,比如,二次型、正交性、凸性、一些对称的类型,旋转不变量函数.但是我们还需要正确理解基本的微分几何概念,例如参数曲线和曲面(以及后来的流形和子流形).因此,从本卷开始,我们在文中包含了相当数量的几何学内容.

我们介绍积分的内容更加困难.问题是要为某些 $G \subset \mathbb{R}^n$ 中的子集定义体积或面积.一旦对相当大的子集类完成此操作,就可以沿着一维情况的线构造积分.测度与积分的 Lebesgue 理论被证明更加适合这种情况,我们将在下一卷里面研究该理论.我们遵循 Riemann 的思想(由 Darboux 转换),针对体积(面积)积分的方法只是解决积分问题的第一个不完整的尝试.然而,它基本上足以解决分析、几何以及数学物理或力学中的具体问题.

现在我们更详细地讨论本卷的内容.在前 4 章中我们讨论了收敛和连续性.虽然我们主要的兴趣是处理映射 $f: G \to \mathbb{R}^n$, $G \subset \mathbb{R}^m$,但为了为处理函数序列的收敛、线性算子的连续性等做准备,我们在介绍点集拓扑的基本概念时讨论了度量空间中的收敛性和连续性.我们也花了一些时间在赋范空间上.最后我们转向了连续映射,并在度量空间和拓扑空间的背景下研究了它们的基本性质.然后我们考虑了映射 $f: G \to \mathbb{R}^n$,并且

为此我们还研究了 $\mathbb{R}^n$ 子集的若干拓扑性质. 我们特别讨论了紧性的概念及其结果. 主要的理论概念是沿着 N. Bourbaki 的路线发展起来的, 比如, J. Dieudonné[10], 然而, 对更具体的 Euclidean 内容研究时, 我们使用了几种不同的来源, 特别是在处理连通性时, 我们更喜欢 M. Heins[24] 的方法, 一般也参考[9].

可微性是第 5 章和第 6 章的主题. 首先我们讨论了函数 $f: G \to \mathbb{R}$, $G \subset \mathbb{R}^n$ 的偏导数, 然后是映射 $f: G \to \mathbb{R}^n$, $G \subset \mathbb{R}^m$ 的可微性. 这些章节一定被视为"标准的", 并且我们的方法与已知的任何方法都没有区别. 我们一旦获得了可微性就可以转向应用方面, 在这里也需要几何知识. 第 7 章和第 8 章的合适的章名应该是 G. Monge 的经典著作《分析学到几何学的应用》. 我们处理 $\mathbb{R}^n$ 中的参数曲线, 在 $n=3$ 的情况下有更多细节, 我们首先查看 $\mathbb{R}^3$ 中的参数曲面. 除微积分的有趣且重要的应用之外, 我们还准备讨论向量微积分的积分定理, 其中我们必须将集合的边界视为参数曲线或参数曲面. 我们从 M. DoCarmo 的教科书[11]中受益匪浅.

第 9 章到第 11 章我们扩展了多元函数的微分学, 并给出了更多应用, 其中很多可以追溯到 J. d'Alembert, L. Euler 和 Bernoulli 家族的时代. 关键词是 Taylor 公式、约束条件下的局部极值(Lagrange 乘数)或包络. 但是请注意, 由于收敛域的结构, Taylor 级数或更一般的多元函数的幂级数更难以处理, 我们将这部分的讨论推迟到多元的复值函数中进行(第 3 卷). 隐函数定理及其结果、逆映射定理具有更多的理论性质, 也有很多应用. 一般来说, 在这些章节中, 我们遵循不同的来源并将它们合并在一起. 特别是由于一些经典书籍包含了很好的应用, 但理论部分现在已经过时, 我们需要付出一些努力来获得一致的观点. O. Forster[19] 的书在处理隐函数方面给了我们很大的帮助. 在第 12 章中, 我们处理了曲线坐标——一个现在经常被人们忽视的主题. 然而, 当处理包含对称性的问题时, 比如在数学物理学中, 曲线坐标是必需的. 根据我们的理解, 它也成为经典微分几何的一部分, 正如我们已经从 G. Lamé 的经典论文中了解到的知识.

几乎每本关于多元微分的书都讨论了凸性, 例如, 处理均值定理时所用的凸集, 或处理局部极值时所用的凸函数. 我们决定比其他作者更详细地处理凸集和凸函数, 包括更多关于凸集几何的内容(例如分离超平面), 其中 S. Hildebrandt 的论文[26]对我们非常有帮助. 我们通过讨论极值点(Minkowski 定理)及其在紧凸集上凸函数极值的应用来进一步做到这一点. 我们还通过变分不等式来研究了可微凸函数的表征, 就像我们讨论凸函数的 Legendre 变换(共轭函数)和凸集上的度量投影一样. 在我们课程的这个阶段, 所有这些都可以在 $\mathbb{R}^n$ 中完成, 并且对其他部分有帮助, 例如变分法、泛函分析和微分几何. 最重要的是, 这些都是很优美的结果.

在介绍了连续性和可微性(或可积性)之后, 我们可以考虑具有这些性质(和其他一些性质)的向量空间函数. 比如我们可以看一下定义在紧集 $K \subset \mathbb{R}^n$ 上的所有连续函数的空间 $C(K)$, Banach 空间具有范数 $\| g \|_\infty := \sup_{x \in K} |g(x)|$. 在经典分析中已经讨论了是否可以通过更简单的函数(如多项式或三角函数)来近似任意连续函数的问题, 还讨论了一致有界的连续函数序列是否总是具有一致收敛子序列的问题. 我们现在可以将第一个问题解释为在 Banach 空间 $C(K)$ 中找到"好的"密集子集的问题, 而第二个问题可以看作是在 $C(K)$ 中找到或表征紧集的问题. 我们特意将这些问题放到 Banach 空间的背景下, 即将这些问题视为泛函分析中的问题. 我们证明了一般的 Stone-Weierstrass 定理, 部分定理绕道而行, 首先证明了 Korovkin 近似定理, 然后证明了 Arzela-Ascoli 定理. 我们坚信在本课程的这个阶段, 学生应该开始理解将具体的经典问题重新表述为泛函分析问题的好处.

第 3 部分的最后章节讨论了线积分问题. 我们在第 3 部分中定位了线积分, 因为它最终被简化为定义在区间上而不是定义在域 $\mathbb{R}^n (n > 1)$ 上的函数的积分. 我们讨论了曲线的基本的定义、切曲线问题, 同时开始验证可积性条件.

如前所述, 定义一个界限函数 $f: G \to \mathbb{R}$, $G \subset \mathbb{R}$ 的 Riemann 积分并不像看起来那么简单. 在第 16 章中我们详细介绍了这些问题, 同时指出了克服这些困难的策略. 第一步是查看定义在超矩形(假设其轴平行)上的函数的迭代积分, 这是在参数相关积分的自然框架中完成的. 在接下来的章节中我们介绍并研究了定义在超矩形上的函数的 Riemann 积分(体积积分). 这可以按照我们在一维情况下遵循的路线来完成. 用迭代积分识别体积积分使我们能够将实际积分问题简化为一维积分.

集合 G 上的积分函数比超矩形上的积分函数会更复杂. 主要一点是我们不知道 $\mathbb{R}^n$ 中一个集合的体积是多少, 因此介绍 Riemann 积分是非常困难的,即使是阶梯函数积分的定义也会引起一些问题.事实证明,边界 ∂G 决定了我们是否可以进行定义,比如说定义一个有界连续函数 $f:G \to \mathbb{R}$ 的积分.这就导致对边界及其"内容"或"测度"的更详细的研究.基本上是拓扑概念"边界"与(隐藏的)测度理论概念"测度零集"的交织导致困难的出现.我们在第 19 章来讨论这些问题,一旦我们得到(有界)Jordan 可测集的概念,就可以为定义在有界 Jordan 可测集上的有界(连续的)函数构建积分.在这部分的介绍中,我们结合了 [20][25] 和 [26] 的部分方法.

为了求体积积分,我们需要更进一步的工具,特别是变换定理.在 Riemann 积分范围内,这个定理的证明是众所周知的困难和冗长,这主要是由于上述问题,即拓扑和测度理论概念的混合.在 Lebesgue 积分理论的背景下,变换定理有一个更易懂的证明,我们也参考了第 21 章中的注释.因此,我们在这里不再提供证明,但会清楚地陈述结果并给出许多应用.最终我们回到反常积分和参数相关积分上,但现在是在体积积分的背景下.在研究偏微分方程时,这些因素将变得非常重要.

本卷的最后一部分致力于研究 $\mathbb{R}^2$ 中的矢量积分,但大部分都是 $\mathbb{R}^3$ 中的.一个纯数学的观点是可以先介绍 E. Cartan 外演算的流形,然后介绍 n 维流形的 m 维子流形的 k 型积分,最后证明一般的 Stokes 定理.通过特殊化,我们现在可以推导出 Causs,Green 和 Stokes 的经典定理.这个过程既没有考虑历史发展,也没有考虑它是否适合二年级的学生.因此,我们决定采用更经典的方法.第 23 章在某种意义上单独介绍了第 5 部分,因此我们在这里的阐述可以更简短一些.

在第 24 章中,我们首先讨论了如何定义参数曲面面积的问题,然后讨论了标量值函数和向量场的曲面积分.通过使用线积分和曲面积分,我们可以证明 $\mathbb{R}^3$ 和 $\mathbb{R}^n$ 中的 Gauss 定理,$\mathbb{R}^3$ 中的 Stokes 定理和平面中的 Green 定理.我们研究的一部分内容致力于解决我们可以在什么类型的区域中证明这些定理,另一部分内容涉及应用.我们的目标是为对应用数学、数学物理或力学感兴趣的学生提供解决此类问题所需的工具(以及数学背景的思想).只有在第 6 卷中,我们才会对一般 Stokes 定理提供严格的证明.

在第 1 卷中我们已经提供了所有问题的解法(大约 275 个),由于依赖于线性代数的大量结果,因此在附录中收集了这些结果.由于我们的许多考虑与几何相关,因此文中包含大量数字(约 150 个). 所有这些数字都是由本书第二位作者使用 LaTex 完成的. 最后说明关于参考第 1 卷的注释,当提到第 1 卷中的定理、引理、定义等,我们会这样写,例如,定理 I. 25.9 等,当提到公式时我们这样写,例如,(I. 25.10)等.

与第 1 卷一样,用 ∗ 标记的问题更具有挑战性.

本书的目录如下:

第三部分:多元函数的微分

第四部分:多元函数的积分

本书的写作风格颇有布尔巴基的风范，抽象味道十足，有许多具体例子的部分也是一带而过，比如 p.556 中论及的 Dirichlet 问题，这是一个非常丰富的题材，许多初学者都需要一些与此相关的具体素材，比如：

1　立方体的 Dirichlet 问题

在立方体内的稳恒状态的温度分布由下列 Laplace 方程描述

$$\nabla^2 u = u_{xx} + u_{yy} + u_{zz} = 0, 0 < x < \pi, 0 < y < \pi, 0 < z < \pi$$

在立方体表面上的温度除在表面 $z = 0$ 上之外，都保持零度，即边界条件为

$$u(0,y,z) = u(\pi,y,z) = 0$$
$$u(x,0,z) = u(x,\pi,z) = 0$$
$$u(x,y,\pi) = 0$$
$$u(x,y,0) = f(x,y)$$

用分离变量法，假设解的形式为

$$u(x,y,z) = X(x)Y(y)Z(z)$$

把这样的解代入 Laplace 方程，得到

$$X''YZ + XY''Z + XYZ'' = 0$$

上式除以 XYZ，可得

$$\frac{X''}{X} + \frac{Y''}{Y} = -\frac{Z''}{Z}$$

因为这个等式的右边只依赖于 z，而其左边不依赖于 z，所以两边都必须等于常数，于是有

$$\frac{X''}{X} + \frac{Y''}{Y} = -\frac{Z''}{Z} = \lambda$$

同样，我们有

$$\frac{X''}{X} = \lambda - \frac{Y''}{Y} = \mu$$

因此，就得到下面三个常微分方程

$$X'' - \mu X = 0$$
$$Y'' - (\lambda - \mu) Y = 0$$
$$Z'' + \lambda Z = 0$$

利用边界条件，可以得到 X 的本征值问题

$$\begin{cases} X'' - \mu X = 0 \\ X(0) = X(\pi) = 0 \end{cases}$$

的本征值是

$$\mu = -m^2, m = 1, 2, 3, \cdots$$

而对应的本征函数是 $\sin mx$.

类似地，可得 Y 的本征值问题

$$\begin{cases} Y'' - (\lambda - \mu) Y = 0 \\ Y(0) = Y(\pi) = 0 \end{cases}$$

的本征值是

$$\lambda - \mu = -n^2, n = 1, 2, 3, \cdots$$

而对应的本征函数是 $\sin ny$.

因为 $\lambda = -(m^2 + n^2)$，由此可得方程 $Z'' + \lambda Z = 0$ 满足条件 $Z(\pi) = 0$ 的解是

$$Z_{mn}(z) = C_{mn} \sinh \sqrt{m^2 + n^2} (\pi - z)$$

于是 Laplace 方程满足齐次边界条件的解具有下列形式

$$u(x, y, z) = \sum_{m=1}^{\infty} \sum_{n=1}^{\infty} a_{mn} \sinh \sqrt{m^2 + n^2} (\pi - z) \sin mx \sin ny$$

利用非齐次边界条件，得到

$$f(x,y) = \sum_{m=1}^{\infty} \sum_{n=1}^{\infty} a_{mn} \sinh(\pi \sqrt{m^2 + n^2}) \sin mx \sin ny$$

于是这个二重 Fourier 级数的系数为

$$a_{mn} \sinh \pi \sqrt{m^2 + n^2} = \frac{4}{\pi^2} \int_0^{\pi} \int_0^{\pi} f(x,y) \sin mx \sin ny dx dy$$

因此,立方体的 Dirichlet 问题的形式解可写成下列形式

$$u(x,y,z) = \sum_{m=1}^{\infty} \sum_{n=1}^{\infty} b_{mn} \frac{\sinh \sqrt{m^2 + n^2}(\pi - z)}{\sinh \pi \sqrt{m^2 + n^2}} \sin mx \sin ny$$

其中

$$b_{mn} = a_{mn} \sinh \pi \sqrt{m^2 + n^2}$$

2 圆柱体的 Dirichlet 问题

例 1 考察不带电的圆柱体内的电势 u 的问题. 这时电势 u 满足 Laplace 方程

$$\nabla^2 u = u_{rr} + \frac{1}{r} u_r + \frac{1}{r^2} u_{\theta\theta} + u_{zz} = 0, 0 \leqslant r < a, 0 < z < l \quad (1)$$

设圆柱体的侧面 $r = a$ 和上底 $z = l$ 是接地的,即电势为零,而在下底 $z = 0$ 上,电势是已知的. 因而边界条件为

$$\begin{cases} u(a,\theta,z) = u(r,\theta,l) = 0 \\ u(r,\theta,0) = f(r,\theta) \end{cases} \quad (2)$$

其中 $f(a,\theta) = 0$.

设解的形式为

$$u(r,\theta,z) = R(r)\Theta(\theta)Z(z)$$

把这样的解代入 Laplace 方程,得到

$$\frac{R'' + \frac{1}{r}R'}{R} + \frac{1}{r^2} \frac{\Theta''}{\Theta} = -\frac{Z''}{Z} = \lambda$$

而且由此可得

$$\frac{r^2 R'' + rR'}{R} - r^2 \lambda = -\frac{\Theta''}{\Theta} = \mu$$

因此我们得到三个常微分方程

$$r^2 R'' + rR' - (\lambda r^2 + \mu)R = 0 \qquad (3)$$

$$\Theta'' + \mu\Theta = 0 \qquad (4)$$

$$Z'' + \lambda Z = 0 \qquad (5)$$

利用周期条件,可以得到 $\Theta(\theta)$ 的本征值问题

$$\begin{cases} \Theta'' + \mu\Theta = 0 \\ \Theta(0) = \Theta(2\pi) \\ \Theta'(0) = \Theta'(2\pi) \end{cases}$$

的本征值是

$$\mu = n^2, n = 0, 1, 2, \cdots$$

而对应的本征函数是 $\sin n\theta, \cos n\theta$. 于是

$$\Theta_n(\theta) = A_n \cos n\theta + B_n \sin n\theta \qquad (6)$$

假定 λ 是负的实数,设 $\lambda = -\beta^2$,其中 $\beta > 0$. 利用条件 $Z(l) = 0$,方程(5)的解可写成下列形式

$$Z(z) = C\sinh \beta(l - z) \qquad (7)$$

然后我们引进新自变量 $\xi = \beta r$,方程(3) 变成

$$\xi^2 \frac{d^2 R}{d\xi^2} + \xi \frac{dR}{d\xi} + (\xi^2 - n^2)R = 0$$

这个方程是 n 阶 Bessel 方程,它的通解是

$$R(\xi) = DJ_n(\xi) + EY_n(\xi)$$

其中 J_n 和 Y_n 分别是第一类和第二类 n 阶 Bessel 函数. 用原来的变量表示,我们有

$$R(r) = DJ_n(\beta r) + EY_n(\beta r)$$

因为 $Y_n(\beta r)$ 在 $r = 0$ 处是无限的,所以我们选取 $E = 0$. 另外,条件 $R(a) = 0$ 要求有

$$J_n(\beta a) = 0$$

对于每一个 $n \geqslant 0$,上式存在无穷多个正零点. 把这些正零点排列成单调增加无穷序列,我们有

$$0 < \alpha_{n1} < \alpha_{n2} < \cdots < \alpha_{nm} < \cdots$$

因此,得

$$\beta_{nm} = \frac{\alpha_{nm}}{\alpha}$$

所以

778

$$R_{nm}(r) = D_{nm}J_n(\frac{\alpha_{nm}r}{a}) \tag{8}$$

于是解最后具有下列形式

$$u(r,\theta,z) = \sum_{n=0}^{\infty}\sum_{m=1}^{\infty} J_n\left(\alpha_{nm}\frac{r}{a}\right)(a_{nm}\cos n\theta + b_{nm}\sin n\theta) \cdot \sinh\alpha_{nm}\frac{(l-z)}{a}$$

为了满足非齐次边界条件,要求有

$$f(r,\theta) = \sum_{n=0}^{\infty}\sum_{m=1}^{\infty} J_n\left(\alpha_{nm}\frac{r}{a}\right)(a_{nm}\cos n\theta + b_{nm}\sin n\theta) \cdot \sinh\alpha_{nm}\frac{l}{a}$$

因此,系数 a_{nm} 和 b_{nm} 为

$$a_{0m} = \frac{1}{\pi a^2 \sinh\left(\alpha_{0m}\frac{l}{a}\right)\left[J_1(\alpha_{0m})\right]^2}\int_0^a\int_0^{2\pi} f(r,\theta)J_0\left(\alpha_{0m}\frac{r}{a}\right)r\mathrm{d}r\mathrm{d}\theta$$

$$a_{nm} = \frac{2}{\pi a^2 \sinh\left(\alpha_{nm}\frac{l}{a}\right)\left[J_{n+1}(\alpha_{nm})\right]^2}\int_0^a\int_n^{2\pi} f(r,\theta)J_n\left(\alpha_{nm}\frac{r}{a}\right)\cos n\theta r\mathrm{d}r\mathrm{d}\theta$$

$$b_{nm} = \frac{2}{\pi a^2 \sinh\left(\alpha_{nm}\frac{l}{a}\right)\left[J_{n+1}(\alpha_{nm})\right]^2}\int_0^a\int_n^{2\pi} f(r,\theta)J_n\left(\alpha_{nm}\frac{r}{a}\right)\sin n\theta r\mathrm{d}r\mathrm{d}\theta$$

例 2 我们将讨论与上面同样的问题,但带有不同的边界条件. 考察定解问题

$$\begin{cases} \nabla^2 u = 0, 0 \leq r < a, 0 < z < \pi \\ u(r,\theta,0) = 0 \\ u(r,\theta,\pi) = 0 \\ u(a,\theta,z) = f(\theta,z) \end{cases}$$

像上面一样,用分离变量法可以得到三个常微分方程

$$r^2 R'' + rR' - (\lambda r^2 + \mu)R = 0$$

$$\Theta'' + \mu\Theta = 0$$

$$Z'' + \lambda Z = 0$$

根据周期条件,再和上面的例子一样,得 $\Theta(\theta)$ 的本征值问题的本征值是

$$\mu = n^2, n = 0, 1, 2, \cdots$$

而对应的本征函数是 $\sin n\theta, \cos n\theta$. 于是,我们有

$$\Theta_n(\theta) = A_n\cos n\theta + B_n\sin n\theta$$

现在设 $\lambda = \beta^2$,其中 $\beta > 0$,那么本征值问题

$$\begin{cases} Z'' + \beta^2 Z = 0 \\ Z(0) = 0, Z(\pi) = 0 \end{cases}$$

有解

$$Z_m(z) = C_m \sin mz$$

其中,$m = 1, 2, 3, \cdots$.

最后,我们有

$$r^2 R'' + r R' - (m^2 r^2 + n^2) R = 0$$

即

$$R'' + \frac{1}{r} R' - \left(m^2 + \frac{n^2}{r^2} \right) R = 0$$

上述方程的解是

$$R(r) = D I_n(mr) + E K_n(mr)$$

其中 I_n 和 K_n 分别是第一类和第二类 n 阶修正 Bessel 函数.

因为 R 在 $r = 0$ 处必须是有限的,我们取 $E = 0$. 因此 R 具有下列形式

$$R_{mn}(r) = D_{mn} I_n(mr)$$

利用非齐次边界条件,我们求得这个定解问题的解为

$$u(r, \theta, z) = \sum_{m=1}^{\infty} \frac{a_{m0}}{2} \frac{I_0(mr)}{I_0(ma)} \sin mz + \sum_{m=1}^{\infty} \sum_{n=1}^{\infty} (a_{mn} \cos n\theta + b_{mn} \sin n\theta) \cdot$$

$$\frac{I_n(mr)}{I_n(ma)} \sin mz$$

其中

$$a_{mn} = \frac{2}{\pi^2} \int_0^{\pi} \int_0^{2\pi} f(\theta, z) \sin mz \cos n\theta \mathrm{d}\theta \mathrm{d}z$$

$$b_{mn} = \frac{2}{\pi^2} \int_0^{\pi} \int_0^{2\pi} f(\theta, z) \sin mz \sin n\theta \mathrm{d}\theta \mathrm{d}z$$

3　球的 Dirichlet 问题

例 1　为了确定一个球内的势,我们把 Laplace 方程化为下列球坐标形式

$$\boldsymbol{\nabla}^2 u = u_{rr} + \frac{2}{r}u_r + \frac{1}{r^2}u_{\theta\theta} + \frac{\cot\theta}{r^2}u_\theta + \frac{1}{r^2\sin^2\theta}u_{\varphi\varphi} = 0 \tag{1}$$

$$0 \leqslant r < a, 0 < \theta < \pi, 0 < \varphi < 2\pi$$

设在球面上给定的势是

$$u(a,\theta,\varphi) = f(\theta,\varphi) \tag{2}$$

用分离变量法,假设解的形式为

$$u(r,\theta,\varphi) = R(r)\Theta(\theta)\Phi(\varphi)$$

把 u 代入 Laplace 方程,就得到三个常微分方程

$$r^2R'' + 2rR' - \lambda R = 0 \tag{3}$$

$$\sin^2\theta\Theta'' + \sin\theta\cos\theta\Theta' + (\lambda\sin^2\theta - \mu)\Theta = 0 \tag{4}$$

$$\Phi'' + \mu\Phi = 0 \tag{5}$$

方程(5) 的通解是

$$\Phi(\varphi) = A\cos\sqrt{\mu}\varphi + B\sin\sqrt{\mu}\varphi$$

由周期条件,得

$$\sqrt{\mu} = m, m = 0,1,2,\cdots$$

于是,我们有

$$\Phi_m(\varphi) = A_m\cos m\varphi + B_m\sin m\varphi \tag{6}$$

引进变量 $\xi = \cos\theta$ 后,方程(4) 变为

$$(1 - \xi^2)\Theta'' - 2\xi\Theta' + \left(\lambda - \frac{m^2}{1 - \xi^2}\right)\Theta = 0$$

这个方程是连带 Legendre 方程. 当 $\lambda = n(n+1), n = 0,1,2,\cdots$ 时,它的通解是

$$\Theta(\theta) = CP_n^m(\cos\theta) + DQ_n^m(\cos\theta)$$

其中 P_n^m 和 Q_n^m 分别是第一类和第二类连带 Legendre 函数.

因为 $\Theta(\theta)$ 在 $\theta = 0, \pi$ 处的有界性条件对应于 $\Theta(\xi)$ 在 $\xi = \pm 1$ 处的有界性条件,而 $Q_n^m(\xi)$ 在 $\xi = \pm 1$ 处是无限的,所以我们选取 $D = 0$. 于是方程(4) 的有界解为

$$\Theta_{nm}(\theta) = C_{nm}P_n^m(\cos\theta) \tag{7}$$

因为方程(3) 是 Euler 方程,所以解的形式为

$$R(r) = r^\beta$$

把它代入方程(3),β 就满足下列方程

$$\beta^2 + \beta - n(n+1) = 0$$

由此求得两个根 $\beta = n$ 和 $\beta = -(1+n)$. 因此方程(3) 的通解是

$$R(r) = Er^n + Fr^{-(1+n)}$$

因为 R 在 $r = 0$ 处是有限的,所以应有 $F = 0$. 于是,得到

$$R_n(r) = E_n r^n \tag{8}$$

因而,在球坐标系中的 Laplace 方程的解是

$$u(r,\theta,\varphi) = \sum_{n=0}^{\infty} \sum_{m=0}^{n} r^n P_n^m(\cos\theta)(a_{nm}\cos m\varphi + b_{nm}\sin m\varphi)$$

为了使 u 在边界上等于给定的函数 $f(\theta,\varphi)$,必须有

$$f(\theta,\varphi) = \sum_{n=0}^{\infty} \sum_{m=0}^{n} a^n P_n^m(\cos\theta)(a_{nm}\cos m\varphi + b_{nm}\sin m\varphi)$$

其中 $0 \leqslant \theta \leqslant \pi, 0 \leqslant \varphi \leqslant 2\pi$. 根据函数 $P_n^m(\cos\theta)\cos m\varphi$ 和 $P_n^m(\cos\theta)\sin m\varphi$ 的正交性,系数为

$$a_{nm} = \frac{(2n+1)(n-m)!}{2\pi a^n(n+m)!} \cdot \int_0^{2\pi}\int_0^{\pi} f(\theta,\varphi)P_n^m(\cos\theta)\cos m\varphi\sin\theta\mathrm{d}\theta\mathrm{d}\varphi$$

$$b_{nm} = \frac{(2n+1)(n-m)!}{2\pi a^n(n+m)!} \cdot \int_0^{2\pi}\int_0^{\pi} f(\theta,\varphi)P_n^m(\cos\theta)\sin m\varphi\sin\theta\mathrm{d}\theta\mathrm{d}\varphi$$

其中

$$m \leqslant n, n = 1,2,3,\cdots, m = 1,2,3,\cdots$$

$$a_{n0} = \frac{2n+1}{4\pi a^n}\int_0^{2\pi}\int_0^{\pi} f(\theta,\varphi)P_n(\cos\theta)\sin\theta\mathrm{d}\theta\mathrm{d}\varphi, n = 0,1,2,\cdots$$

例2 为了确定在一均匀电场内的接地导电球外的电势分布,要求我们解下列定解问题

$$\begin{cases} \nabla^2 u = 0, r > a, 0 < \theta < \pi, 0 < \varphi < 2\pi \\ u(a,\theta) = 0 \\ u \to -E_0 r\cos\theta, r \to \infty \end{cases}$$

设这个均匀电场是 z 方向的,从而使得电势 u 与 φ 无关,因此 Laplace 方程具有下列形式

$$u_{rr} + \frac{2}{r}u_r + \frac{1}{r^2}u_{\theta\theta} + \frac{\cot\theta}{r^2}u_\theta = 0$$

设解的形式为

$$u(r,\theta) = R(r)\Theta(\theta)$$

把这样的解代入 Laplace 方程,就得到两个常微分方程

$$r^2 R'' + 2rR' - \lambda R = 0$$

$$\sin^2\theta\Theta'' + \sin\theta\cos\theta\Theta' + \lambda\sin^2\theta\Theta = 0$$

如果我们令 $\lambda = n(n+1)$，其中 $n = 0,1,2,\cdots$，那么上面第二个常微分方程是 Legendre 方程. 这个方程的通解是

$$\Theta_n(\theta) = A_n P_n(\cos\theta) + B_n Q_n(\cos\theta)$$

其中 P_n 和 Q_n 分别是第一类和第二类 Legendre 函数. 为了使解在 $\theta = 0$ 和 $\theta = \pi$ 处是有限的，应取 $B_n = 0$（像上面的例题一样），于是，得到

$$\Theta_n(\theta) = A_n P_n(\cos\theta)$$

关于 R 的方程的解是

$$R_n(r) = C_n r^n + D_n r^{-(n+1)}$$

因而，势函数是

$$u(r,\theta) = \sum_{n=0}^{\infty}\left[a_n r^n + b_n r^{-(n+1)}\right]P_n(\cos\theta)$$

为了满足在无穷远处的条件，必须有

$$a_1 = -E_0, a_n = 0, n \geqslant 2$$

因此

$$u(r,\theta) = -E_0 r\cos\theta + \sum_{n=1}^{\infty}\frac{b_n}{r^{n+1}}P_n(\cos\theta)$$

由边界条件 $u(a,\theta) = 0$，得到

$$0 = -E_0 a\cos\theta + \sum_{n=1}^{\infty}\frac{b_n}{a^{n+1}}P_n(\cos\theta)$$

再利用 Legendre 函数的正交性，求得

$$b_n = \frac{2n+1}{2}E_0 a^{n+2}\int_{-\pi}^{\pi}\cos\theta P_n(\cos\theta)\mathrm{d}(\cos\theta)$$

$$= E_0 a^3\delta_{n1}$$

这是因为上面的积分除 $n = 1$ 外，对所有的 n 都等于零. 因此，电势分布为

$$u(r,\theta) = -E_0 r\cos\theta + E_0\frac{a^3}{r^2}\cos\theta$$

例3 一个半径为 a 的介质球放在均匀电场 E_0 内，求球内和球外的电势分布.

这个问题的定解问题是

$$\begin{cases} \nabla^2 u_1 = \nabla^2 u_2 = 0 \\ K\dfrac{\partial u_1}{\partial r} = \dfrac{\partial u_2}{\partial r},在 r = a 上 \\ u_1 = u_2,在 r = a 上 \\ u_2 \to -E_0 r\cos\theta, r \to \infty \end{cases}$$

其中 u_1 和 u_2 分别是球内和球外的电势,而 K 是介电常数.

像上面的例题一样,势函数是

$$u(r,\theta) = \sum_{n=0}^{\infty} \left[a_n r^n + b_n r^{-(n+1)} \right] P_n(\cos\theta) \tag{9}$$

因为 u_1 在原点处必须是有限的,所以取

$$u_1(r,\theta) = \sum_{n=0}^{\infty} a_n r^n P_n(\cos\theta), r \le a \tag{10}$$

对于 u_2,它必须以给定的方式趋于无穷远点,我们选取

$$u_2(r,\theta) = -E_0 r\cos\theta + \sum_{n=0}^{\infty} b_n r^{-(n+1)} P_n(\cos\theta), r \ge a \tag{11}$$

由 $r = a$ 处的两个连续性条件,得到

$$a_1 = -E_0 + \frac{b_1}{a^3}$$

$$Ka_1 = -E_0 - \frac{2b_1}{a^3}$$

$$a_n = b_n = 0, n \ne 1$$

于是求得系数 a_1 和 b_1 为

$$a_1 = -\frac{3E_0}{K+2}, b_1 = E_0 a^3 \frac{K-1}{K+2}$$

因此,当 $r \le a$ 时,电势分布为

$$u_1(r,\theta) = -\frac{3E_0}{K+2} r\cos\theta$$

而当 $r \ge a$ 时,电势分布为

$$u_2(r,\theta) = -E_0 r\cos\theta + E_0 a^3 \frac{K-1}{K+2} r^{-2}\cos\theta$$

例4 确定在两个取不同常电势的同心球之间的电势分布.

这里,需要解下列定解问题

$$\begin{cases} \nabla^2 u = 0, a < r < b \\ u = A, \text{在 } r = a \text{ 上} \\ u = B, \text{在 } r = b \text{ 上} \end{cases}$$

在这种情况下,电势只依赖于径向距离.因此,我们有

$$\frac{1}{r^2} \frac{\partial}{\partial r} \left(r^2 \frac{\partial u}{\partial r} \right) = 0$$

用初等积分法,可得

$$u(r) = C_1 + \frac{C_2}{r}$$

利用边界条件,得到

$$C_1 = \frac{Bb - Aa}{b - a}$$

$$C_2 = (A - B) \frac{ab}{b - a}$$

于是电势分布为

$$u(r) = \frac{Bb - Aa}{b - a} + \frac{(A - B)ab}{(b - a)r}$$

$$= \frac{Bb}{r} \frac{r - a}{b - a} + \frac{Aa}{r} \frac{b - r}{b - a}$$

4　球面的 Dirichlet 问题的一例[①]

例　求一个调和函数 $u(x, y, z)$,使得它在以原点为心,以 1 为半径的球 S 的内部是正则的,在 S 上等于 z^2.依次用两种方法求解.

(1)利用 Laplace 方程的线性性质,构造一个初等的二次调和多项式.

(2)利用公式

$$u(M) = \frac{1}{4\pi} \iint_S u(P) \frac{R^2 - \rho^2}{Rr^3} \mathrm{d}S$$

选取一个最方便的坐标系.

---

① 摘编自《数学习题》,J. 巴斯著,徐信之译,上海科学技术出版社,1986.

解 （1）函数 z^2 是一个二次多项式. 我们利用 Laplace 方程的线性性质来构造函数 u. 我们已经知道了如何去构造一个二次调和多项式. 立刻看出，$2z^2 - x^2 - y^2 = 3z^2 - (x^2 + y^2 + z^2)$ 就是这样的多项式. 现在，在球面 S 上，$x^2 + y^2 + z^2$ 取常数值 R^2. 因此，这个多项式在球面上取值 $3z^2 - R^2$. 这样一来，多项式

$$z^2 - \frac{x^2 + y^2 + z^2}{3}$$

是调和的，在球面上等于 $z^2 - \dfrac{R^2}{3}$.

现在，常数 $\dfrac{R^2}{3}$ 是调和的. 多项式

$$z^2 - \frac{x^2 + y^2 + z^2}{3} + \frac{R^2}{3}$$

在 S 的内部是调和且正则的，在 S 上取值 z^2. 因为 Dirichlet 问题有唯一解，所以这个多项式就是所求的解

$$u = \frac{2z^2 - x^2 - y^2 + R^2}{3}$$

（2）*球面的 Dirichlet 问题在下述意义下是可解的：存在一个显式解，求这个解归结为计算一个二重积分.*

这个公式可写为

$$u(M) = \frac{1}{4\pi}\iint\limits_S u(P)\, \frac{R^2 - \rho^2}{Rr^3}\mathrm{d}S$$

其中 ρ 是球心 O 到点 M 的距离，r 是球面上的动点 P 到 M 的距离（图 1），现在我们来运用这一公式求解. 我们有

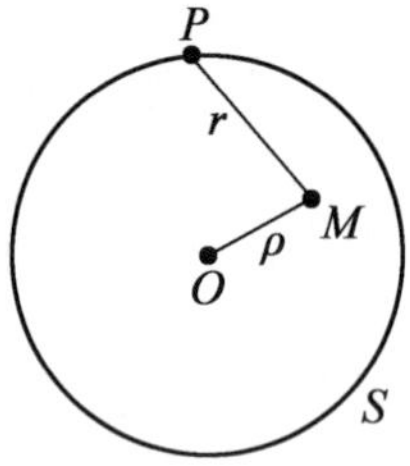

$$4\pi u = \frac{R^2 - \rho^2}{R}\iint\limits_S \frac{z^2}{r^3}\mathrm{d}S$$

我们把它化为极坐标. 如果用 x_0, y_0, z_0 表示点 M 的笛卡儿坐标，用 R, θ, φ 表示点 P 的球坐标，则有

图 1

$$z = R\cos\theta,\quad \mathrm{d}S = R^2\sin\theta\mathrm{d}\theta\mathrm{d}\varphi$$

$$r^2 = R^2 + \rho^2 - 2R(x_0\sin\theta\cos\varphi + y_0\sin\theta\sin\varphi + z_0\cos\theta)$$

由此我们得到 $4\pi u$ 的下述表达式

$$R^3(R^2 - \rho^2) \iint_S \frac{\cos^2\theta \sin\theta \mathrm{d}\theta \mathrm{d}\varphi}{\left[R^2 + \rho^2 - 2R(x_0\sin\theta\cos\varphi + y_0\sin\theta\sin\varphi + z_0\cos\theta)\right]^{3/2}}$$

这个积分看来不能化为初等表达式. 实际上, 它可以用初等方法计算, 但要做到这一点, 尚需作坐标变换.

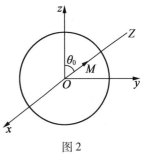

图 2

我们沿 OM 方向取 Z 轴(图 2), 使得点 M 的坐标为 $(0,0,\rho)(\rho > 0)$. 这时, z 就化为新坐标下的线性函数 $\alpha X + \beta Y + \gamma Z$, 这里 $\alpha^2 + \beta^2 + \gamma^2 = 1$. 于是

$$4\pi u = \frac{R^2 - \rho^2}{R} \iint \frac{(\alpha X + \beta Y + \gamma Z)^2}{r^3} \mathrm{d}S$$

其中

$$X = R\sin\theta\cos\varphi, Y = R\sin\theta\sin\varphi, Z = R\cos\theta$$
$$\mathrm{d}S = R^2\sin\theta\mathrm{d}\theta\mathrm{d}\varphi, r^2 = R^2 + \rho^2 - 2R\rho\cos\theta$$

这里 θ, φ 是在新坐标系下的球坐标.

这样一来, 我们有

$$(\alpha X + \beta Y + \gamma Z)^2 = \alpha^2 R^2\sin^2\theta\cos^2\varphi + \beta^2 R^2\sin^2\theta\sin^2\varphi + \gamma^2 R^2\cos^2\theta +$$
$$2\alpha\beta R^2\sin^2\theta\sin\varphi\cos\varphi + 2\alpha\gamma R^2\sin\theta\cos\theta\cos\varphi +$$
$$2\beta\gamma R^2\sin\theta\cos\theta\sin\varphi$$

这里 θ 从 0 变到 π, φ 从 0 变到 2π.

因为 r 不依赖于 φ, 所以, 我们可首先对 φ 积分

$$\iint \frac{(\alpha X + \beta Y + \gamma Z)^2}{r^3} \mathrm{d}S = R^2 \int_0^\pi \frac{\sin\theta\mathrm{d}\theta}{r^3} \int_0^{2\pi} (\alpha X + \beta Y + \gamma Z)^2 \mathrm{d}\varphi$$

现在, 关于 φ 的积分是初等的. 由于没有对应于 $\sin\varphi, \cos\varphi, \sin\varphi\cos\varphi$ 的项, 所以还剩下

$$\pi R^2(\alpha^2\sin^2\theta + \beta^2\sin^2\theta + 2\gamma^2\cos^2\theta)$$

这样一来, 我们有

$$u = \frac{R^3(R^2 - \rho^2)}{4} \int_0^\pi \frac{\left[(\alpha^2 + \beta^2)\sin^2\theta + 2\gamma^2\cos^2\theta\right]\sin\theta\mathrm{d}\theta}{(R^2 + \rho^2 - 2R\rho\cos\theta)^{3/2}}$$

我们把

$$s = R^2 + \rho^2 - 2R\rho\cos\theta$$

取作积分变量. 设

$$(\alpha^2 + \beta^2)\sin^2\theta + 2\gamma^2\cos^2\theta = \frac{As^2 + Bs + C}{4R^2\rho^2}$$

这里

$$A = 2\gamma^2 - \alpha^2 - \beta^2$$
$$B = -2(R^2 + \rho^2)(2\gamma^2 - \alpha^2 - \beta^2)$$
$$C = 2\gamma^2(R^2 + \rho^2)^2 - (\alpha^2 + \beta^2)(R^2 - \rho^2)^2$$

于是我们有

$$u = \frac{R^2 - \rho^2}{32\rho^3}\int_{(R-\rho)^2}^{(R+\rho)^2}(As^{1/2} + Bs^{-1/2} + Cs^{-3/2})\,\mathrm{d}s$$

由此我们容易求得

$$u = \frac{1}{8\rho^2}\Big[A\Big(R^2 + \frac{\rho^2}{3}\Big) + B\Big](R^2 - \rho^2) + \frac{C}{8\rho^2}$$

作某些简化后,得

$$u = \frac{(\alpha^2 + \beta^2)(R^2 - \rho^2) + (R^2 + 2\rho^2)\gamma^2}{3}$$

这个公式与(1)中所求得的公式是一样的. 这里我们有

$$u = \frac{3z^2 + R^2 - (x^2 + y^2 + z^2)}{3}$$

我们需要用 $\alpha X + \beta Y + \gamma Z$ 代替 z,用 $X^2 + Y^2 + Z^2$ 代替 $x^2 + y^2 + z^2$,然后在 $X = Y = 0, Z = \rho$ 的情况下,应用所得的公式,于是有

$$u = \frac{3\gamma^2\rho^2 + R^2 - \rho^2}{3} = \frac{3\gamma^2\rho^2 + (R^2 - \rho^2)(\alpha^2 + \beta^2 + \gamma^2)}{3}$$

$$= \frac{(\alpha^2 + \beta^2)(R^2 - \rho^2) + (R^2 + 2\rho^2)\gamma^2}{3}$$

这就是我们所要求的.

5　球面平均值与 Laplace 算子[①]

本练习是前一练习的部分补充. 在这个练习中,我们给出了一个充

---

① 摘编自《数学习题》,J. 巴斯著,徐信之译,上海科学技术出版社,1986.

分正则函数的 Laplace 算子的局部定义,并指出,在一个点的邻域内,如果一个正则函数在半径为 R 的球面(三维情况) 上的平均值不依赖于 R,那么它是局部调和的.

例 设 $U(x,y,z)$ 是一个连续函数,它在点 $M_0(x_0,y_0,z_0)$ 处有二阶连续偏导数. 我们用 $\mathfrak{M}(R)$ 表示 $U(x,y,z)$ 在以 M_0 为中心、以 R 为半径的球面 S 上的平均值,所以

$$\mathfrak{M}(R) = \frac{1}{4\pi R^2}\iint\limits_{S} U(x,y,z)\,\mathrm{d}S$$

试证明:当 $R \to 0$ 时,$\mathfrak{M}(R)$ 趋向于 $U(x_0,y_0,z_0)$,并求出无穷小

$$\mathfrak{M}(R) - U(x_0,y_0,z_0)$$

的主要部分.

解 (1) 设 $U(x,y,z)$ 是一个连续函数,在点 $M_0(x_0,y_0,z_0)$ 处有二阶连续偏导数. 设

$$\mathfrak{M}(R) = \frac{1}{4\pi R^2}\iint\limits_{S} U(x,y,z)\,\mathrm{d}S$$

是 U 在以 M_0 为中心、以 R 为半径的球面上的平均值.

当 M 趋向于 M_0 时,平均值 $\mathfrak{M}(R)$ 趋向于 U 在 M_0 处的值 U_0. 特别地,有

$$\mathfrak{M}(R) - U_0 = \frac{1}{4\pi R^2}\iint\limits_{S}\big[U(x,y,z) - U_0\big]\,\mathrm{d}S$$

从而

$$|\,\mathfrak{M}(R) - U_0\,| \leqslant \frac{1}{4\pi R^2}\iint\limits_{S}|\,U(x,y,z) - U_0\,|\,\mathrm{d}S$$

因为 $U(x,y,z)$ 是连续的,所以存在一个数 η,使得当 $R < \eta$ 时,有

$$|\,U(x,y,z) - U_0\,| < \varepsilon$$

所以

$$|\,\mathfrak{M}(R) - U_0\,| < \frac{\varepsilon}{4\pi R^2}\iint\limits_{S}\mathrm{d}S = \varepsilon$$

(2) 因为 $U(x,y,z)$ 在 M_0 处有二阶连续偏导数,所以我们可以对它用 Taylor 公式

$$U(x,y,z) = U(x_0,y_0,z_0) + (x - x_0)\frac{\partial U}{\partial x_0} + (y - y_0)\frac{\partial U}{\partial y_0} + (z - z_0)\frac{\partial U}{\partial z_0} +$$

$$\frac{1}{2}\Big[(x - x_0)^2 \frac{\partial^2 U}{\partial x'^2} + \cdots + 2(y - y_0)(z - z_0) \frac{\partial^2 U}{\partial y' \partial z'} + \cdots \Big]$$

这里括号中的表达式是 $x - x_0, y - y_0, z - z_0$ 的二次型,而二阶导数是在 M_0 与 M 之间的线段 $M_0 M$ 上的点处取得.

在 S 上对这个方程求积分

$$\mathfrak{M}(R) = U_0 + \frac{1}{4\pi R^2} \cdot \frac{\partial U}{\partial x_0} \iint\limits_S (x - x_0) \, \mathrm{d}S + \cdots$$

这些极限值是零,因为它们表示球面关于它的球心的惯性矩. 最后

$$\mathfrak{M}(R) = U_0 + \frac{R^2}{6} \Delta U_0 + R^2 \varphi(R)$$

这里当 $R \to 0$ 时,$\varphi(R)$ 趋向于 0,换言之

$$\Delta U_0 = \lim_{R \to 0} \frac{6}{R^2} [\mathfrak{M}(R) - U_0]$$

推论 如果在点 M_0 的一个邻域内,$\mathfrak{M}(R)$ 不依赖于 R,那么 $\Delta U_0 = 0$. 我们又找到了大家所熟知的函数在一点调和的性质. 在这种情况下,$\mathfrak{M}(R) = U_0$,并且 $U(x, y, z)$ 满足方程

$$U(x_0, y_0, z_0) = \frac{1}{4\pi R^2} \iint\limits_S U(x, y, z) \, \mathrm{d}S$$

若 $U(x, y, z)$ 不仅在 M_0 处满足这个方程,而且在 M_0 的整个邻域内都满足这个方程,则 U 的各阶连续导数的存在性就被证明了,而不必取为假设.

"数学分析"不仅是作为数学家的重要基础,也可以说是所有研究自然科学的必修课. 杨振宁先生晚年就曾回忆说:

当李政道和我在 1951 年研究后来被称为"单位圆定理"的时候,von Neumann 和 A. Selberg 曾建议我们去读 G. Pólya 和 Szego 的著作《分析中的问题和定理》. 1965 年 H. Whitney 曾向我和我的弟弟杨振平讲解向量场的指数的拓扑概念. 为了求解 Wiener-Hopf 积分方程,M. Kac 曾建议我们读 M. G. Krein 有关这一课题的长篇综述.

正如另一位世界著名数学家 A. Renyi 指出:

为了成功地应用数学,必须对它有深刻的理解,谁若想做出独创性的应用数学工作,他就必须是一位有创造性的数学家.反之,对应用的关心也会有助于纯数学研究.

俄罗斯的作家中有两位叫奥斯特洛夫斯基的,其中不太知名的那位曾写过一个话剧《大雷雨》,估计看过的人很少.后来根据它拍了一部影片《没有嫁妆的新娘》,其中有几句经典的对白正好可以表达我们对数学的这种情绪:

"你这么爱他,那他一定有许多优点了?"
"不,他爱我,只有这一条."
"那未免太少了."
"所以可贵."

刘培杰

2023 年 2 月 18 日

于哈工大

刘培杰数学工作室
已出版(即将出版)图书目录——原版影印

书 名	出版时间	定 价	编号
数学物理大百科全书.第1卷(英文)	2016—01	418.00	508
数学物理大百科全书.第2卷(英文)	2016—01	408.00	509
数学物理大百科全书.第3卷(英文)	2016—01	396.00	510
数学物理大百科全书.第4卷(英文)	2016—01	408.00	511
数学物理大百科全书.第5卷(英文)	2016—01	368.00	512
zeta函数,q-zeta函数,相伴级数与积分(英文)	2015—08	88.00	513
微分形式:理论与练习(英文)	2015—08	58.00	514
离散与微分包含的逼近和优化(英文)	2015—08	58.00	515
艾伦·图灵:他的工作与影响(英文)	2016—01	98.00	560
测度理论概率导论,第2版(英文)	2016—01	88.00	561
带有潜在故障恢复系统的半马尔柯夫模型控制(英文)	2016—01	98.00	562
数学分析原理(英文)	2016—01	88.00	563
随机偏微分方程的有效动力学(英文)	2016—01	88.00	564
图的谱半径(英文)	2016—01	58.00	565
量子机器学习中数据挖掘的量子计算方法(英文)	2016—01	98.00	566
量子物理的非常规方法(英文)	2016—01	118.00	567
运输过程的统一非局部理论:广义波尔兹曼物理动力学,第2版(英文)	2016—01	198.00	568
量子力学与经典力学之间的联系在原子、分子及电动力学系统建模中的应用(英文)	2016—01	58.00	569
算术域(英文)	2018—01	158.00	821
高等数学竞赛:1962—1991年的米洛克斯·史怀哲竞赛(英文)	2018—01	128.00	822
用数学奥林匹克精神解决数论问题(英文)	2018—01	108.00	823
代数几何(德文)	2018—04	68.00	824
丢番图逼近论(英文)	2018—01	78.00	825
代数几何学基础教程(英文)	2018—01	98.00	826
解析数论入门课程(英文)	2018—01	78.00	827
数论中的丢番图问题(英文)	2018—01	78.00	829
数论(梦幻之旅):第五届中日数论研讨会演讲集(英文)	2018—01	68.00	830
数论新应用(英文)	2018—01	68.00	831
数论(英文)	2018—01	78.00	832

书　名	出版时间	定　价	编号
湍流十讲(英文)	2018—04	108.00	886
无穷维李代数:第3版(英文)	2018—04	98.00	887
等值、不变量和对称性(英文)	2018—04	78.00	888
解析数论(英文)	2018—09	78.00	889
《数学原理》的演化:伯特兰·罗素撰写第二版时的手稿与笔记(英文)	2018—04	108.00	890
哈密尔顿数学论文集(第4卷):几何学、分析学、天文学、概率和有限差分等(英文)	2019—05	108.00	891
偏微分方程全局吸引子的特性(英文)	2018—09	108.00	979
整函数与下调和函数(英文)	2018—09	118.00	980
幂等分析(英文)	2018—09	118.00	981
李群、离散子群与不变量理论(英文)	2018—09	108.00	982
动力系统与统计力学(英文)	2018—09	118.00	983
表示论与动力系统(英文)	2018—09	118.00	984
分析学练习.第1部分(英文)	2021—01	88.00	1247
分析学练习.第2部分,非线性分析(英文)	2021—01	88.00	1248
初级统计学:循序渐进的方法:第10版(英文)	2019—05	68.00	1067
工程师与科学家微分方程用书:第4版(英文)	2019—07	58.00	1068
大学代数与三角学(英文)	2019—06	78.00	1069
培养数学能力的途径(英文)	2019—07	38.00	1070
工程师与科学家统计学:第4版(英文)	2019—06	58.00	1071
贸易与经济中的应用统计学:第6版(英文)	2019—06	58.00	1072
傅立叶级数和边值问题:第8版(英文)	2019—05	48.00	1073
通往天文学的途径:第5版(英文)	2019—05	58.00	1074
拉马努金笔记.第1卷(英文)	2019—06	165.00	1078
拉马努金笔记.第2卷(英文)	2019—06	165.00	1079
拉马努金笔记.第3卷(英文)	2019—06	165.00	1080
拉马努金笔记.第4卷(英文)	2019—06	165.00	1081
拉马努金笔记.第5卷(英文)	2019—06	165.00	1082
拉马努金遗失笔记.第1卷(英文)	2019—06	109.00	1083
拉马努金遗失笔记.第2卷(英文)	2019—06	109.00	1084
拉马努金遗失笔记.第3卷(英文)	2019—06	109.00	1085
拉马努金遗失笔记.第4卷(英文)	2019—06	109.00	1086
数论:1976年纽约洛克菲勒大学数论会议记录(英文)	2020—06	68.00	1145
数论:卡本代尔1979:1979年在南伊利诺伊卡本代尔大学举行的数论会议记录(英文)	2020—06	78.00	1146
数论:诺德韦克豪特1983:1983年在诺德韦克豪特举行的Journees Arithmetiques数论大会会议记录(英文)	2020—06	68.00	1147
数论:1985—1988年在纽约城市大学研究生院和大学中心举办的研讨会(英文)	2020—06	68.00	1148

刘培杰数学工作室
已出版(即将出版)图书目录——原版影印

书　名	出版时间	定　价	编号
数论:1987 年在乌尔姆举行的 Journees Arithmetiques 数论大会会议记录(英文)	2020—06	68.00	1149
数论:马德拉斯 1987:1987 年在马德拉斯安娜大学举行的国际拉马努金百年纪念大会会议记录(英文)	2020—06	68.00	1150
解析数论:1988 年在东京举行的日法研讨会会议记录(英文)	2020—06	68.00	1151
解析数论:2002 年在意大利切特拉罗举行的 C. I. M. E. 暑期班演讲集(英文)	2020—06	68.00	1152
量子世界中的蝴蝶:最迷人的量子分形故事(英文)	2020—06	118.00	1157
走进量子力学(英文)	2020—06	118.00	1158
计算物理学概论(英文)	2020—06	48.00	1159
物质,空间和时间的理论:量子理论(英文)	2020—10	48.00	1160
物质,空间和时间的理论:经典理论(英文)	2020—10	48.00	1161
量子场理论:解释世界的神秘背景(英文)	2020—07	38.00	1162
计算物理学概论(英文)	2020—06	48.00	1163
行星状星云(英文)	2020—10	38.00	1164
基本宇宙学:从亚里士多德的宇宙到大爆炸(英文)	2020—08	58.00	1165
数学磁流体力学(英文)	2020—07	58.00	1166
计算科学:第 1 卷,计算的科学(日文)	2020—07	88.00	1167
计算科学:第 2 卷,计算与宇宙(日文)	2020—07	88.00	1168
计算科学:第 3 卷,计算与物质(日文)	2020—07	88.00	1169
计算科学:第 4 卷,计算与生命(日文)	2020—07	88.00	1170
计算科学:第 5 卷,计算与地球环境(日文)	2020—07	88.00	1171
计算科学:第 6 卷,计算与社会(日文)	2020—07	88.00	1172
计算科学.别卷,超级计算机(日文)	2020—07	88.00	1173
多复变函数论(日文)	2022—06	78.00	1518
复变函数入门(日文)	2022—06	78.00	1523
代数与数论:综合方法(英文)	2020—10	78.00	1185
复分析:现代函数理论第一课(英文)	2020—07	58.00	1186
斐波那契数列和卡特兰数:导论(英文)	2020—10	68.00	1187
组合推理:计数艺术介绍(英文)	2020—07	88.00	1188
二次互反律的傅里叶分析证明(英文)	2020—07	48.00	1189
旋瓦兹分布的希尔伯特变换与应用(英文)	2020—07	58.00	1190
泛函分析:巴拿赫空间理论入门(英文)	2020—07	48.00	1191
卡塔兰数入门(英文)	2019—05	68.00	1060
测度与积分(英文)	2019—04	68.00	1059
组合学手册.第一卷(英文)	2020—06	128.00	1153
＊—代数、局部紧群和巴拿赫＊—代数丛的表示.第一卷,群和代数的基本表示理论(英文)	2020—05	148.00	1154
电磁理论(英文)	2020—08	48.00	1193
连续介质力学中的非线性问题(英文)	2020—09	78.00	1195
多变量数学入门(英文)	2021—05	68.00	1317
偏微分方程入门(英文)	2021—05	88.00	1318
若尔当典范性:理论与实践(英文)	2021—07	68.00	1366
伽罗瓦理论.第 4 版(英文)	2021—08	88.00	1408
R 统计学概论	2023—03	88.00	1614
基于不确定静态和动态问题解的仿射算术(英文)	2023—03	38.00	1618

刘培杰数学工作室
已出版(即将出版)图书目录——原版影印

书　名	出版时间	定　价	编号
典型群,错排与素数(英文)	2020—11	58.00	1204
李代数的表示:通过 gln 进行介绍(英文)	2020—10	38.00	1205
实分析演讲集(英文)	2020—10	38.00	1206
现代分析及其应用的课程(英文)	2020—10	58.00	1207
运动中的抛射物数学(英文)	2020—10	38.00	1208
2—纽结与它们的群(英文)	2020—10	38.00	1209
概率,策略和选择:博弈与选举中的数学(英文)	2020—11	58.00	1210
分析学引论(英文)	2020—11	58.00	1211
量子群:通往流代数的路径(英文)	2020—11	38.00	1212
集合论入门(英文)	2020—10	48.00	1213
酉反射群(英文)	2020—11	58.00	1214
探索数学:吸引人的证明方式(英文)	2020—11	58.00	1215
微分拓扑短期课程(英文)	2020—10	48.00	1216
抽象凸分析(英文)	2020—11	68.00	1222
费马大定理笔记(英文)	2021—03	48.00	1223
高斯与雅可比和(英文)	2021—03	78.00	1224
π 与算术几何平均:关于解析数论和计算复杂性的研究(英文)	2021—01	58.00	1225
复分析入门(英文)	2021—03	48.00	1226
爱德华·卢卡斯与素性测定(英文)	2021—03	78.00	1227
通往凸分析及其应用的简单路径(英文)	2021—01	68.00	1229
微分几何的各个方面.第一卷(英文)	2021—01	58.00	1230
微分几何的各个方面.第二卷(英文)	2020—12	58.00	1231
微分几何的各个方面.第三卷(英文)	2020—12	58.00	1232
沃克流形几何学(英文)	2020—11	58.00	1233
彷射和韦尔几何应用(英文)	2020—12	58.00	1234
双曲几何学的旋转向量空间方法(英文)	2021—02	58.00	1235
积分:分析学的关键(英文)	2020—12	48.00	1236
为有天分的新生准备的分析学基础教材(英文)	2020—11	48.00	1237
数学不等式.第一卷.对称多项式不等式(英文)	2021—03	108.00	1273
数学不等式.第二卷.对称有理不等式与对称无理不等式(英文)	2021—03	108.00	1274
数学不等式.第三卷.循环不等式与非循环不等式(英文)	2021—03	108.00	1275
数学不等式.第四卷.Jensen 不等式的扩展与加细(英文)	2021—03	108.00	1276
数学不等式.第五卷.创建不等式与解不等式的其他方法(英文)	2021—04	108.00	1277

刘培杰数学工作室
已出版(即将出版)图书目录——原版影印

书　名	出版时间	定　价	编号
冯·诺依曼代数中的谱位移函数:半有限冯·诺依曼代数中的谱位移函数与谱流(英文)	2021−06	98.00	1308
链接结构:关于嵌入完全图的直线中链接单形的组合结构(英文)	2021−05	58.00	1309
代数几何方法.第1卷(英文)	2021−06	68.00	1310
代数几何方法.第2卷(英文)	2021−06	68.00	1311
代数几何方法.第3卷(英文)	2021−06	58.00	1312

书　名	出版时间	定　价	编号
代数、生物信息和机器人技术的算法问题.第四卷,独立恒等式系统(俄文)	2020−08	118.00	1199
代数、生物信息和机器人技术的算法问题.第五卷,相对覆盖性和独立可拆分恒等式系统(俄文)	2020−08	118.00	1200
代数、生物信息和机器人技术的算法问题.第六卷,恒等式和准恒等式的相等 问题、可推导性和可实现性(俄文)	2020−08	128.00	1201
分数阶微积分的应用:非局部动态过程,分数阶导热系数(俄文)	2021−01	68.00	1241
泛函分析问题与练习:第2版(俄文)	2021−01	98.00	1242
集合论、数学逻辑和算法论问题:第5版(俄文)	2021−01	98.00	1243
微分几何和拓扑短期课程(俄文)	2021−01	98.00	1244
素数规律(俄文)	2021−01	88.00	1245
无穷边值问题解的递减:无界域中的拟线性椭圆和抛物方程(俄文)	2021−01	48.00	1246
微分几何讲义(俄文)	2020−12	98.00	1253
二次型和矩阵(俄文)	2021−01	98.00	1255
积分和级数.第2卷,特殊函数(俄文)	2021−01	168.00	1258
积分和级数.第3卷,特殊函数补充:第2版(俄文)	2021−01	178.00	1264
几何图上的微分方程(俄文)	2021−01	138.00	1259
数论教程:第2版(俄文)	2021−01	98.00	1260
非阿基米德分析及其应用(俄文)	2021−03	98.00	1261
古典群和量子群的压缩(俄文)	2021−03	98.00	1263
数学分析习题集.第3卷,多元函数:第3版(俄文)	2021−03	98.00	1266
数学习题:乌拉尔国立大学数学力学系大学生奥林匹克(俄文)	2021−03	98.00	1267
柯西定理和微分方程的特解(俄文)	2021−03	98.00	1268
组合极值问题及其应用:第3版(俄文)	2021−03	98.00	1269
数学词典(俄文)	2021−01	98.00	1271
确定性混沌分析模型(俄文)	2021−06	168.00	1307
精选初等数学习题和定理.立体几何.第3版(俄文)	2021−03	68.00	1316
微分几何习题:第3版(俄文)	2021−05	98.00	1336
精选初等数学习题和定理.平面几何.第4版(俄文)	2021−05	68.00	1335
曲面理论在欧氏空间 E_n 中的直接表示(俄文)	2022−01	68.00	1444
维纳—霍普夫离散算子和托普利兹算子:某些可数赋范空间中的诺特性和可逆性(俄文)	2022−03	108.00	1496
Maple 中的数论:数论中的计算机计算(俄文)	2022−03	88.00	1497
贝尔曼和克努特问题及其概括:加法运算的复杂性(俄文)	2022−03	138.00	1498

书　名	出版时间	定　价	编号
复分析:共形映射(俄文)	2022—07	48.00	1542
微积分代数样条和多项式及其在数值方法中的应用(俄文)	2022—08	128.00	1543
蒙特卡罗方法中的随机过程和场模型:算法和应用(俄文)	2022—08	88.00	1544
线性椭圆型方程组:论二阶椭圆型方程的迪利克雷问题(俄文)	2022—08	98.00	1561
动态系统解的增长特性:估值、稳定性、应用(俄文)	2022—08	118.00	1565
群的自由积分解:建立和应用(俄文)	2022—08	78.00	1570
混合方程和偏差自变数方程问题:解的存在和唯一性(俄文)	2023—01	78.00	1582
拟度量空间分析:存在和逼近定理(俄文)	2023—01	108.00	1583
二维和三维流形上函数的拓扑性质:函数的拓扑分类(俄文)	2023—03	68.00	1584
齐次马尔科夫过程建模的矩阵方法:此类方法能够用于不同目上的的复杂系统研究、设计和完善(俄文)	2023—03	68.00	1594
周期函数的近似方法和特性:特殊课程(俄文)	2023—04	158.00	1622
扩散方程解的矩函数:变分法(俄文)	2023—03	58.00	1623
狭义相对论与广义相对论:时空与引力导论(英文)	2021—07	88.00	1319
束流物理学和粒子加速器的实践介绍:第2版(英文)	2021—07	88.00	1320
凝聚态物理中的拓扑和微分几何简介(英文)	2021—05	88.00	1321
混沌映射:动力学、分形学和快速涨落(英文)	2021—05	128.00	1322
广义相对论:黑洞、引力波和宇宙学介绍(英文)	2021—06	68.00	1323
现代分析电磁均质化(英文)	2021—06	68.00	1324
为科学家提供的基本流体动力学(英文)	2021—06	88.00	1325
视觉天文学:理解夜空的指南(英文)	2021—06	68.00	1326
物理学中的计算方法(英文)	2021—06	68.00	1327
单星的结构与演化:导论(英文)	2021—06	108.00	1328
超越居里:1903年至1963年物理界四位女性及其著名发现(英文)	2021—06	68.00	1329
范德瓦尔斯流体热力学的进展(英文)	2021—06	68.00	1330
先进的托卡马克稳定性理论(英文)	2021—06	88.00	1331
经典场论导论:基本相互作用的过程(英文)	2021—07	88.00	1332
光致电离量子动力学方法原理(英文)	2021—07	108.00	1333
经典域论和应力:能量张量(英文)	2021—05	88.00	1334
非线性太赫兹光谱的概念与应用(英文)	2021—06	68.00	1337
电磁学中的无穷空间并矢格林函数(英文)	2021—06	88.00	1338
物理科学基础数学.第1卷,齐次边值问题、傅里叶方法和特殊函数(英文)	2021—07	108.00	1339
离散量子力学(英文)	2021—07	68.00	1340
核磁共振的物理学和数学(英文)	2021—07	108.00	1341
分子水平的静电学(英文)	2021—08	68.00	1342
非线性波:理论、计算机模拟、实验(英文)	2021—06	108.00	1343
石墨烯光学:经典问题的电解决方案(英文)	2021—06	68.00	1344
超材料多元宇宙(英文)	2021—07	68.00	1345
银河系外的天体物理学(英文)	2021—07	68.00	1346
原子物理学(英文)	2021—07	68.00	1347
将光打结:将拓扑学应用于光学(英文)	2021—07	68.00	1348
电磁学:问题与解法(英文)	2021—07	88.00	1364
海浪的原理:介绍量子力学的技巧与应用(英文)	2021—07	108.00	1365
多孔介质中的流体:输运与相变(英文)	2021—07	68.00	1372
洛伦兹群的物理学(英文)	2021—08	68.00	1373
物理导论的数学方法和解决方法手册(英文)	2021—08	68.00	1374

刘培杰数学工作室
已出版(即将出版)图书目录——原版影印

书　名	出版时间	定　价	编号
非线性波数学物理学入门(英文)	2021—08	88.00	1376
波:基本原理和动力学(英文)	2021—07	68.00	1377
光电子量子计量学.第1卷,基础(英文)	2021—07	88.00	1383
光电子量子计量学.第2卷,应用与进展(英文)	2021—07	68.00	1384
复杂流的格子玻尔兹曼建模的工程应用(英文)	2021—08	68.00	1393
电偶极矩挑战(英文)	2021—08	108.00	1394
电动力学:问题与解法(英文)	2021—09	68.00	1395
自由电子激光的经典理论(英文)	2021—08	68.00	1397
曼哈顿计划——核武器物理学简介(英文)	2021—09	68.00	1401
粒子物理学(英文)	2021—09	68.00	1402
引力场中的量子信息(英文)	2021—09	128.00	1403
器件物理学的基本经典力学(英文)	2021—09	68.00	1404
等离子体物理及其空间应用导论.第1卷,基本原理和初步过程(英文)	2021—09	68.00	1405
磁约束聚变等离子体物理:理想MHD理论(英文)	2023—03	68.00	1613
相对论量子场论.第1卷,典范形式体系(英文)	2023—03	38.00	1615
涌现的物理学(英文)	2023—05	58.00	1619
量子化旋涡:一本拓扑激发手册(英文)	2023—04	68.00	1620
非线性动力学:实践的介绍性调查(英文)	2023—05	68.00	1621
拓扑与超弦理论焦点问题(英文)	2021—07	58.00	1349
应用数学:理论、方法与实践(英文)	2021—07	78.00	1350
非线性特征值问题:牛顿型方法与非线性瑞利函数(英文)	2021—07	58.00	1351
广义膨胀和齐性:利用齐性构造齐次系统的李雅普诺夫函数和控制律(英文)	2021—06	48.00	1352
解析数论焦点问题(英文)	2021—07	58.00	1353
随机微分方程:动态系统方法(英文)	2021—07	58.00	1354
经典力学与微分几何(英文)	2021—07	58.00	1355
负定相交形式流形上的瞬子模空间几何(英文)	2021—07	68.00	1356
广义卡塔兰轨道分析:广义卡塔兰轨道计算数字的方法(英文)	2021—07	48.00	1367
洛伦兹方法的变分:二维与三维洛伦兹方法(英文)	2021—08	38.00	1378
几何、分析和数论精编(英文)	2021—08	68.00	1380
从一个新角度看数论:通过遗传方法引入现实的概念(英文)	2021—07	58.00	1387
动力系统:短期课程(英文)	2021—08	68.00	1382
几何路径:理论与实践(英文)	2021—08	48.00	1385
论天体力学中某些问题的不可积性(英文)	2021—07	88.00	1396
广义斐波那契数列及其性质(英文)	2021—08	38.00	1386
对称函数和麦克唐纳多项式:余代数结构与Kawanaka恒等式(英文)	2021—09	38.00	1400
杰弗里·英格拉姆·泰勒科学论文集:第1卷.固体力学(英文)	2021—05	78.00	1360
杰弗里·英格拉姆·泰勒科学论文集:第2卷.气象学、海洋学和湍流(英文)	2021—05	68.00	1361
杰弗里·英格拉姆·泰勒科学论文集:第3卷.空气动力学以及落弹数和爆炸的力学(英文)	2021—05	68.00	1362
杰弗里·英格拉姆·泰勒科学论文集:第4卷.有关流体力学(英文)	2021—05	58.00	1363

刘培杰数学工作室
已出版(即将出版)图书目录——原版影印

书　　名	出版时间	定　价	编号
非局域泛函演化方程:积分与分数阶(英文)	2021—08	48.00	1390
理论工作者的高等微分几何:纤维丛、射流流形和拉格朗日理论(英文)	2021—08	68.00	1391
半线性退化椭圆微分方程:局部定理与整体定理(英文)	2021—07	48.00	1392
非交换几何、规范理论和重整化:一般简介与非交换量子场论的重整化(英文)	2021—09	78.00	1406
数论论文集:拉普拉斯变换和带有数论系数的幂级数(俄文)	2021—09	48.00	1407
挠理论专题:相对极大值,单射与扩充模(英文)	2021—09	88.00	1410
强正则图与欧几里得若尔当代数:非通常关系中的启示(英文)	2021—10	48.00	1411
拉格朗日几何和哈密顿几何:力学的应用(英文)	2021—10	48.00	1412

书　　名	出版时间	定　价	编号
时滞微分方程与差分方程的振动理论:二阶与三阶(英文)	2021—10	98.00	1417
卷积结构与几何函数理论:用以研究特定几何函数理论方向的分数阶微积分算子与卷积结构(英文)	2021—10	48.00	1418
经典数学物理的历史发展(英文)	2021—10	78.00	1419
扩展线性丢番图问题(英文)	2021—10	38.00	1420
一类混沌动力系统的分歧分析与控制:分歧分析与控制(英文)	2021—11	38.00	1421
伽利略空间和伪伽利略空间中一些特殊曲线的几何性质(英文)	2022—01	68.00	1422
一阶偏微分方程:哈密尔顿—雅可比理论(英文)	2021—11	48.00	1424
各向异性黎曼多面体的反问题:分段光滑的各向异性黎曼多面体反边界谱问题:唯一性(英文)	2021—11	38.00	1425

书　　名	出版时间	定　价	编号
项目反应理论手册.第一卷,模型(英文)	2021—11	138.00	1431
项目反应理论手册.第二卷,统计工具(英文)	2021—11	118.00	1432
项目反应理论手册.第三卷,应用(英文)	2021—11	138.00	1433
二次无理数:经典数论入门(英文)	2022—05	138.00	1434
数,形与对称性:数论,几何和群论导论(英文)	2022—05	128.00	1435
有限域手册(英文)	2021—11	178.00	1436
计算数论(英文)	2021—11	148.00	1437
拟群与其表示简介(英文)	2021—11	88.00	1438
数论与密码学导论:第二版(英文)	2022—01	148.00	1423

刘培杰数学工作室
已出版(即将出版)图书目录——原版影印

书　　　名	出版时间	定　价	编号
几何分析中的柯西变换与黎兹变换:解析调和容量和李普希兹调和容量、变化和振荡以及一致可求长性(英文)	2021-12	38.00	1465
近似不动点定理及其应用(英文)	2022-05	28.00	1466
局部域的相关内容解析:对局部域的扩展及其伽罗瓦群的研究(英文)	2022-01	38.00	1467
反问题的二进制恢复方法(英文)	2022-03	28.00	1468
对几何函数中某些类的各个方面的研究:复变量理论(英文)	2022-01	38.00	1469
覆盖、对应和非交换几何(英文)	2022-01	28.00	1470
最优控制理论中的随机线性调节器问题:随机最优线性调节器问题(英文)	2022-01	38.00	1473
正交分解法:涡流流体动力学应用的正交分解法(英文)	2022-01	38.00	1475

书　　　名	出版时间	定　价	编号
芬斯勒几何的某些问题(英文)	2022-03	38.00	1476
受限三体问题(英文)	2022-05	38.00	1477
利用马利亚万微积分进行 Greeks 的计算:连续过程、跳跃过程中的马利亚万微积分和金融领域中的 Greeks(英文)	2022-05	48.00	1478
经典分析和泛函分析的应用:分析学的应用(英文)	2022-03	38.00	1479
特殊芬斯勒空间的探究(英文)	2022-03	48.00	1480
某些图形的施泰纳距离的细谷多项式:细谷多项式与图的维纳指数(英文)	2022-05	38.00	1481
图论问题的遗传算法:在新鲜与模糊的环境中(英文)	2022-05	48.00	1482
多项式映射的渐近簇(英文)	2022-05	38.00	1483

书　　　名	出版时间	定　价	编号
一维系统中的混沌:符号动力学,映射序列,一致收敛和沙可夫斯基定理(英文)	2022-05	38.00	1509
多维边界层流动与传热分析:粘性流体流动的数学建模与分析(英文)	2022-05	38.00	1510
演绎理论物理学的原理:一种基于量子力学波函数的逐次置信估计的一般理论的提议(英文)	2022-05	38.00	1511
R^2 和 R^3 中的仿射弹性曲线:概念和方法(英文)	2022-08	38.00	1512
算术数列中除数函数的分布:基本内容、调查、方法、第二矩、新结果(英文)	2022-05	28.00	1513
抛物型狄拉克算子和薛定谔方程:不定常薛定谔方程的抛物型狄拉克算子及其应用(英文)	2022-07	28.00	1514
黎曼-希尔伯特问题与量子场论:可积重正化、戴森-施温格方程(英文)	2022-08	38.00	1515
代数结构和几何结构的形变理论(英文)	2022-08	48.00	1516
概率结构和模糊结构上的不动点:概率结构和直觉模糊度量空间的不动点定理(英文)	2022-08	38.00	1517

刘培杰数学工作室

已出版(即将出版)图书目录——原版影印

书　名	出版时间	定　价	编号
反若尔当对:简单反若尔当对的自同构(英文)	2022-07	28.00	1533
对某些黎曼-芬斯勒空间变换的研究:芬斯勒几何中的某些变换(英文)	2022-07	38.00	1534
内诣零流形映射的尼尔森数的阿诺索夫关系(英文)	2023-01	38.00	1535
与广义积分变换有关的分数次演算:对分数次演算的研究(英文)	2023-01	48.00	1536
强子的芬斯勒几何和吕拉几何(宇宙学方面):强子结构的芬斯勒几何和吕拉几何(拓扑缺陷)(英文)	2022-08	38.00	1537
一种基于混沌的非线性最优化问题:作业调度问题(英文)	2023-03	38.00	1538
广义概率论发展前景:关于趣味数学与置信函数实际应用的一些原创观点(英文)	2023-03	48.00	1539
纽结与物理学:第二版(英文)	2022-09	118.00	1547
正交多项式和q-级数的前沿(英文)	2022-09	98.00	1548
算子理论问题集(英文)	2022-09	108.00	1549
抽象代数:群、环与域的应用导论:第二版(英文)	2023-01	98.00	1550
菲尔兹奖得主演讲集:第三版(英文)	2023-01	138.00	1551
多元实函数教程(英文)	2022-09	118.00	1552
球面空间形式群的几何学:第二版(英文)	2022-09	98.00	1566
对称群的表示论(英文)	2023-01	98.00	1585
纽结理论:第二版(英文)	2023-01	88.00	1586
拟群理论的基础与应用(英文)	2023-01	88.00	1587
组合学:第二版(英文)	2023-01	98.00	1588
加性组合学:研究问题手册(英文)	2023-01	68.00	1589
扭曲、平铺与镶嵌:几何折纸中的数学方法(英文)	2023-01	98.00	1590
离散与计算几何手册:第三版(英文)	2023-01	248.00	1591
离散与组合数学手册:第二版(英文)	2023-01	248.00	1592
分析学教程.第1卷,一元实变量函数的微积分分析学介绍(英文)	2023-01	118.00	1595
分析学教程.第2卷,多元函数的微分和积分,向量微积分(英文)	2023-01	118.00	1596
分析学教程.第3卷,测度与积分理论,复变量的复值函数(英文)	2023-01	118.00	1597
分析学教程.第4卷,傅里叶分析,常微分方程,变分法(英文)	2023-01	118.00	1598

联系地址:哈尔滨市南岗区复华四道街10号　哈尔滨工业大学出版社刘培杰数学工作室
网　址:http://lpj.hit.edu.cn/
邮　编:150006
联系电话:0451-86281378　　13904613167
E-mail:lpj1378@163.com